Geology and Landscape Evolution

Geology and Landscape Evolution
General Principles Applied to the United States

Second Edition

Joseph A. DiPietro
University of Southern Indiana, Evansville, IN, United States

Elsevier
Radarweg 29, PO Box 211, 1000 AE Amsterdam, Netherlands
The Boulevard, Langford Lane, Kidlington, Oxford OX5 1GB, United Kingdom
50 Hampshire Street, 5th Floor, Cambridge, MA 02139, United States

Copyright © 2018 Elsevier Inc. All rights reserved.

No part of this publication may be reproduced or transmitted in any form or by any means, electronic or mechanical, including photocopying, recording, or any information storage and retrieval system, without permission in writing from the publisher. Details on how to seek permission, further information about the Publisher's permissions policies and our arrangements with organizations such as the Copyright Clearance Center and the Copyright Licensing Agency, can be found at our website: www.elsevier.com/permissions.

This book and the individual contributions contained in it are protected under copyright by the Publisher (other than as may be noted herein).

Notices
Knowledge and best practice in this field are constantly changing. As new research and experience broaden our understanding, changes in research methods, professional practices, or medical treatment may become necessary.

Practitioners and researchers must always rely on their own experience and knowledge in evaluating and using any information, methods, compounds, or experiments described herein. In using such information or methods they should be mindful of their own safety and the safety of others, including parties for whom they have a professional responsibility.

To the fullest extent of the law, neither the Publisher nor the authors, contributors, or editors, assume any liability for any injury and/or damage to persons or property as a matter of products liability, negligence or otherwise, or from any use or operation of any methods, products, instructions, or ideas contained in the material herein.

British Library Cataloguing-in-Publication Data
A catalogue record for this book is available from the British Library

Library of Congress Cataloging-in-Publication Data
A catalog record for this book is available from the Library of Congress

ISBN: 978-0-12-811191-8

For Information on all Elsevier publications
visit our website at https://www.elsevier.com/books-and-journals

  Working together to grow libraries in developing countries

www.elsevier.com • www.bookaid.org

Publisher: Candice Janco
Acquisition Editor: Marisa LaFleur
Editorial Project Manager: Katerina Zaliva
Production Project Manager: Vijayaraj Purushothaman
Cover Designer: Christian Bilbow

Typeset by MPS Limited, Chennai, India

Contents

Preface   xi

Part I
Keys to Understanding Landscape Evolution

1. The Tortoise and the Hare

How Slow Is Slow?   3
Maps, Cross-Sections, and Scale   4
Physiographic Regions and Provinces   4
   Interior Plains and Plateaus   7
   Appalachian Mountain System   8
   Coastal Plain   8
   Cordilleran Mountain System   10
Components, Forcing Agents, Mechanisms, and Landscape Response   10
Geology, Landscape, and Tectonics   11
Geologic Time Scale   13
Questions   13

2. River Systems

Divides   15
Mississippi River System   16
Atlantic Seaboard–Gulf Coast River System   17
St. Lawrence River System   18
Rio Grande–West Texas River System   18
Colorado River System   19
Columbia River System   19
California River System   19
Great Basin River System   20
Hudson Bay River System   20
Comparison of River Systems With Physiographic Provinces   20
Questions   22

3. Component: The Rock/Sediment Type

Weathering, Erosion, and Deposition   23
The Four Rock/Sediment Types   24
   Sedimentary Rock   25
   Crystalline Rock   26
   Volcanic Rock   27
   Unconsolidated Sediment   28
The Rock Cycle   29
Rock Hardness and Differential Erosion   29
Influence of Bedrock on Landscape   30
   Landscape in Sedimentary Rocks   30
   Landscape in Crystalline Rocks   31
   Landscape in Volcanic Rocks   31
   Landscape in Unconsolidated Sediment   32
Karst Landscape   34
Distribution of Rock/Sediment Type Among the 26 Physiographic Provinces   34
Questions   39

4. Component: The Structural Form

Structural Form: The Style of Rock Deformation   41
   Folds   41
   Vertical Joint Sets   44
   Faults   44
Fault Reactivation   46
Brittle and Ductile Faults   47
Influence of Dipping Layers on Landscape   48
   Vertical to Steep-Dipping Rock Layers   48
   Horizontal to Gently Dipping Rock Layers   49
   Response of Dipping Layers to Erosional Lowering   49
   Cuestas and Hogbacks   49
Topographic Form and Structural Form   49
Recognition of Active Faults   50
Structure-Controlled and Erosion-Controlled Landscape   52
Questions   57

5. Forcing Agent: The Tectonic System

The Four Forcing Agents   59
   Tekton, the Carpenter, the Builder   59
   Climate, the Sculptor   60
   Isostasy, the Equalizer   61
   Sea Level, the Baseline   61
The Tectonic Plate   62
Plate Boundaries   62

Movement of Tectonic Plates	63
Rifting and Passive Continental Margins	64
Active Continental Margins	65
Tectonic Accretion	67
Orogeny	69
Unconformities	69
The Atlantic Passive Continental Margin	69
The Pacific Active Continental Margin	71
Thermal Plumes and Hot Spots	74
Thermal Plumes in the United States	74
Questions	76

6. Forcing Agent: The Climate System

Present-Day Climate Zones	79
Controls on Climate	81
Latitude	81
Proximity to Large Water Bodies	81
Global Wind Patterns	81
The Tilt of the Earth's Axis of Rotation	82
Mountains	84
Questions	86

7. Forcing Agent: Isostasy

Tectonic versus Isostatic Uplift	87
Elevation of Continents and Ocean Basins	89
Mountain Building and Preservation	89
Tectonic Loads	91
Thermal Isostasy	91
Glaciers	92
Deposition	92
Erosion	92
Questions	93

8. Forcing Agent: Sea Level Change

Cause of Sea Level Change	95
Measuring Sea Level and Sea Level Changes	95
Sea Level Changes over the Past 100 Million Years	98
Oxygen Isotope Record over the Past 67 Million Years	99
Influence of Earth's Orbital Parameters on Glaciation	100
Oxygen Isotope Record over the Past 1.8 Million Years	101
Sea Level over the Past 150,000 Years	101
Recent Temperature History	102
Sea Level Response to Recent Temperature History	103
The History of CO_2 in the Atmosphere	103
Questions	105

9. Mechanisms That Impart Change to Landscape

Uplift and Subsidence	107
Surface Uplift/Subsidence and Bedrock Uplift/Subsidence	107
How Does Uplift/Subsidence Occur?	108
Present-Day Uplift/Subsidence Rates	109
Measuring Ancient Uplift Rates and Elevation	110
Erosion, Deposition, and Rivers	110
Graded Rivers and Base Level	111
Base Level Changes	111
Knickpoint Migration	112
Changes in Discharge and Sediment Supply	113
The Lower Mississippi River Valley During the Most Recent Glacial Advance	113
Present-Day Erosion Rates	115
Controls on Rates of Erosion	117
Rates of Deposition	118
Exhumation	119
Erosional Exhumation	119
Calculating Rates of Erosional Exhumation	120
Tectonic Exhumation	122
Volcanism	122
Questions	122

10. Evolution of Landscape

Landscape Grows Old	125
Landscape at Topographic Steady-State	126
Steady-State as the End-Product of Growing Old	127
Rejuvenation	127
Reincarnation	129
Reincarnation While Growing Old	129
Reincarnation due to Volcanism and Tectonic Stress	131
Reincarnation due to Glaciation	131
Reincarnation due to Burial Beneath Unconsolidated Sediment	132
Summary	132
Questions	133

Part II
Structural Provinces

11. Structural Provinces, Rock Successions, and Tectonic Provinces

Structural Provinces	137
Rock Successions	141

The North American Crystalline Shield	142
Precambrian Sedimentary/Volcanic Rocks	143
The Interior Platform	143
The Miogeocline	143
Accreted Terranes	144
The Atlantic Miogeocline	144
Tectonic Provinces	**144**
Hinterland Tectonic Provinces	146
Foreland Tectonic Provinces	147
The Reactivated Western Craton and the Atlantic Marginal Basin	147
Distribution of Rock Successions and Tectonic Provinces	**148**
The Great Unconformity	**150**
Questions	**154**

12. Glacial Landscape

Effect Of Glaciation On Landscape	**157**
Landscape Development in Areas of Continental Glaciation	157
Landscape Development in Areas of Alpine Glaciation	159
A Daughter Of The Snows: Glacial Landscape In The United States	**162**
The Glacial Erosion Boundary In The United States	**164**
The Glacial Erosion Boundary Across North America	**165**
Moraines	**171**
Proglacial Lakes	**174**
Lake Agassiz	**175**
Marine Incursions	**175**
Drumlin Fields	**175**
Kame–Kettle Fields	**175**
Eskers	**176**
Sand Dune Fields	**177**
Loess Deposition	**177**
Area South Of The Glacial Limit	**178**
The Teays River	**179**
The Missouri River	**181**
Pluvial Lakes Of The Cordillera	**183**
Questions	**183**

13. Sediment and Nearly Flat-Lying Sedimentary Layers

Landscape in Nearly Flat-Lying Layers	**185**
Bench-and-Slope Landscape	185
Erosional Mountains	187
Monoclinal Slopes and Hogback Ridges	188
The Coastal Plain	**189**
Barrier Islands	190
New England	191
New Jersey to North Carolina	198
South Carolina to Florida	200
The Mississippi Embayment	203
Texas	207
Ancient Shorelines of the Coastal Plain	207
The Western Margin of Nearly Flat-Lying Sedimentary Layers	**209**
The Great Plains	**211**
The Missouri Plateau	211
The High Plains	215
The Colorado Piedmont, Pecos Valley, Plains Border, and Edwards Plateau	218
The Wyoming Basin	**219**
Uplift of the Wyoming Basin and Northern Great Plains	**221**
The Colorado Plateau	**222**
Incised Meanders	224
Bench-and-Slope Landscape	224
Mogollon Rim	227
Uplifts and Monoclines	227
Fractures and Impact Features	231
Sedimentary-Cored Anticlinal and Domal Mountains	**235**
Central Lowlands	**237**
Ozark Plateau	**241**
Salem and Springfield Plateaus	241
Boston Mountains	243
Uplift History	243
The Interior Low Plateaus	**244**
Bench-and-Slope Landscape	247
Deformed Rocks of the Shawnee Hills	248
Mammoth Cave	249
The Appalachian Plateau	**251**
Allegheny Plateau	253
Cumberland Plateau	254
Comparison of the Pottsville and Cumberland Escarpments	257
Questions	**258**

14. Crystalline-Cored Mid-Continent Anticlines and Domes

Adirondack Mountains	**261**
St. Francois Mountains	**263**
Wichita, Arbuckle, and Llano Structural Domes	**266**
Wichita Mountains	266
Arbuckle Mountains	267
Llano Uplift	268
Landscape Development	268
Western Margin of Crystalline-Cored Anticlines and Domes	**269**
Intrusive Domal Mountains	**271**
The Southern Rocky Mountains	**274**
The Front Range	276
Sawatch Mountains	279

Rio Grande Rift in Central Colorado	281
Landscape History of the Southern Rocky Mountains and Colorado Plateau	283
Cause of Accelerated Erosion in the Southern Rocky Mountains and Colorado Plateau	287
First There Is a Mountain	287
Anticlinal Mountains of the Middle Rockies	**289**
Wind River Range	290
Beartooth Mountains	293
Bighorn Mountains	295
The Black Hills	296
Water Gaps in the Rocky Mountains	**298**
Superior Upland	**298**
Geologic Overview	300
Superior Province	301
Penokean Province	302
Iron Formations	302
Sudbury Meteorite Impact Event	303
Barron and Baraboo Quartzite	303
Keweenawan Rift System	304
Questions	**306**

15. Foreland Fold and Thrust Belts

Structural Form of Foreland Thrust Faults	**309**
Comparison With the Crystalline-Cored Anticlinal Structure	**310**
Cordilleran Fold and Thrust Belt	**310**
Northern Rocky Mountains	314
The Rocky Mountain Trench	317
The Idaho-Wyoming Fold and Thrust Belt	317
Overview: Appalachian-Ouachita Fold and Thrust Belt	**318**
Valley and Ridge Fold and Thrust Belt	**319**
The Great Valley	320
Northern Appalachian Fold and Thrust Belt	321
Central Appalachian Fold and Thrust Belt	325
Southern Appalachian (Tennessee) Fold and Thrust Belt	327
Fault Zones on the Cumberland Plateau	333
Distribution of Appalachian oreland Deformation	333
Ouachita Fold and Thrust Belt	**333**
Arkansas River Valley–Northern Mountains	334
The Fourche Mountains	335
The Central Mountains	336
Athens Plateau	336
Marathon Basin Fold and Thrust Belt	**337**
Water Gaps in the Valley and Ridge and Ouachita Mountains	**337**
Questions	**340**

16. Hinterland Deformation Belts

Rocks Within Hinterland Deformation Belts	**341**
Appalachian Mountains	**342**
Physiographic Overview of the Blue Ridge	342
Geologic Overview of the Blue Ridge	342
The Blue Ridge at Roanoke	345
The Blue Ridge North of Roanoke	346
The Blue Ridge South of Roanoke	347
Level of Exhumation Across the Great Smoky Mountains	349
The Great Smoky Mountains	350
The Balsam Mountains	353
Asheville Basin	354
The Grandfather Mountain Area	355
Piedmont Plateau	358
The Blue Ridge Escarpment	359
The Fall Line	360
New England Highlands	361
Erosional History of the Appalachian Mountains	369
The Northern Rocky Mountains and North Cascades	**372**
Southern Idaho	372
Central Idaho, Montana, and Oregon	375
Northern Washington	376
The Grenville Front	**383**
Van Horn Area	384
Questions	**386**

17. Young Volcanic Rocks of the Cordillera

Magma Types and Common Volcanic Landforms	**389**
Columbia Plateau	**392**
Columbia River Flood Basalt	393
Columbia Basin	394
Blue Mountains	397
Olympic-Wallowa Lineament	400
High Lava Plains	400
Snake River Plain	**402**
Owyhee Upland	405
Yellowstone Plateau Volcanic Field	405
Origin of Volcanism on the Columbia Plateau and High Lava Plains	**406**
Cordilleran Volcanic Areas 70 to 20 Million Years Old	**407**
Northern Great Plains	407
North and South Table Mountain	407
Ignimbrite Flare-Up	409
Navaho Volcanic Field and Shiprock	418
Pinnacles, Neenach, and Nine Sisters	419
Cordilleran Volcanic Areas Younger Than 20 Million Years	**419**
Uinkaret and Markagunt Volcanic Fields	419

San Francisco Volcanic Field	420
Hopi Buttes Volcanic Field	420
Grand Mesa	421
Jemez Lineament	422
Carrizozo Lava Flow	424
Northern Nevada Rift Zone	425
The Northwest Basin and Range and Northern Sierra Nevada	425
Long Valley Caldera and the Inyo-Mono Craters	426
Sutter Buttes	427
Questions	**428**

18. Normal Fault Systems

Structural Character and Terminology of Normal Faults	**429**
Horst and Graben Structure	429
Tilted Fault Blocks, Half-Grabens, and Flexural Rebound	430
Detachments	431
Fault-Block Rotation and Rollover Anticlines	432
The Basin and Range	**432**
Physiographic Limit	432
Expansion into Surrounding Areas	438
Landscape Characteristics	438
Vertical Displacement	440
Horizontal Extension	440
Crustal Thinning and Volcanism	440
Metamorphic Core Complexes	441
Timing of Normal Faulting	442
Normal Fault Activity Verses Erosion	443
The Nevadaplano	446
Cause of Basin and Range Extension	447
Basin and Range Geology	447
The $Sr_i = 0.706$ Line	451
Rio Grande Rift	**452**
Monoclines and Normal Faults in the Big Bend Area, Texas	453
The Rio Grande Bolson Deposits	456
White Sands National Monument	458
Great Sand Dunes National Park	459
Rocky Mountain Basin and Range	**461**
The Teton Mountains	**462**
The Wasatch Mountains	**464**
Triassic Lowlands of the Appalachian Mountains	**466**
Questions	**470**

19. Cascadia Volcanic Arc System

The Juan de Fuca Plate	**473**
The Pacific Coastline	**475**
The Oregon Coast Range	**477**
Cause of Uplift Along the Oregon Coast	480
Geology of the Oregon Coast Range	481
Inland Valleys and the Forearc Basin	**482**
The Central-Southern Cascade Mountains	**483**
Geology of the Central-Southern Cascade Mountains	487
Clockwise Block Rotation	489
Normal Faults Along the Crest of the High Cascades	490
The Olympic Mountains	**491**
Geology of the Olympic Mountains	492
A Case for Topographic Steady-State	492
The Klamath Mountains	**495**
Uplift History of the Klamath Mountains	495
Geology of the Klamath Mountains	497
Questions	**498**

20. California Strike-Slip System

Landscape Associated With Strike-Slip Faults	**503**
The San Andreas Fault System	**505**
Displacement Along the San Andreas Fault	506
History of the San Andreas Fault	508
A Relict Subduction Zone Landscape	**510**
The Ancient Accretionary Prism	510
The Ancient Forearc Basin	510
The Ancient Volcanic Arc	512
The California Coast Ranges	**512**
Age of Landscape	512
Mountain Alignment Relative to the San Andreas Fault	514
Deformation History Prior to Surface Uplift	515
Mechanism and Cause of Surface Uplift	516
The Transverse Ranges and the Salton Sea	**517**
Rotation of the Transverse Block	518
Peninsular Ranges	**520**
Sierra Nevada	**521**
The Sierra Nevada Frontal Fault System	525
Sierra Nevada Uplift History	526
The Walker Lane Belt	**531**
A Tale of Three Landscapes	533
The Inyo-Mono Section	534
White Mountains	536
Inyo Mountains	540
Death Valley-Panamint Valley Region	542
Example of Active Faulting in Death Valley	**546**
Questions	**547**

21. The Grand Canyon

The Physiographic Canyon	549
Active Faults and Incision Rates	554
Hualapai Plateau	555
River Morphology	556
The Modern Colorado River	556

Argument for a 6-Million-Year-Old Canyon	558	**Finale**	563
Argument for a 70-Million-Year-Old Canyon	558	**Questions**	563
Geologic History	559	Appendix	565
Revised Arguments	560	References	583
Interpretation 1	560	Index	605
Interpretation 2	561		
Interpretation 3	562		

Preface

I had a few things in mind when I began this book. The first was that I did not want to simply tell a story. I was more interested in how the story came to be, the evidence that supports the story, and how evidence is obtained. I wanted to explain the geological logic that pertains to the story and the reasoning that allows us to make certain conclusions regarding when a mountain comes into existence and what happens to the mountain over long intervals of time. Landscape evolution implies two things: (1) that landscape undergoes change with time and (2) that landscape can completely change its look over time, relative to some previous state.

The title has changed, but this book is the Second edition to Parts I and II of my previous book entitled *Landscape Evolution in the United States: An Introduction to the Geography, Geology and Natural History*, 2013. Both Parts I and II have been completely rewritten and greatly expanded. Part III of the First edition, on mountain building, is not included here due to space constraints, but some aspects were incorporated into Part II of the revised edition. This book is written at an introductory level appropriate for first semester freshmen or for anybody with an interest in the landscape evolution, geography, and geology of the United States. However, at the same time, it is detailed enough to be useful and appropriate for upper division courses in geology, geography, and environmental science. It is also useful as a reference for teachers and professionals. This book is unique in that it provides an introduction to the general principles involved in studying landscape evolution, and then applies those principles to the varied landscape of the United States.

Part I, entitled "Keys to Understanding Landscape Evolution" examines the process of landscape evolution and how to recognize that landscape has changed from some previous state. Each chapter is independent, but readers will achieve greatest comprehension if they read Chapters 1 through 10 in sequence. Chapter 1, The Tortoise and the Hare, provides an overview of the book and introduces terminology. Chapter 2, River Systems, describes major river systems. The remaining chapters describe landscape in terms of the components that form landscape, the forcing agents that cause landscape to undergo change, and the mechanisms by which landscape undergoes change. Also discussed are the criteria used to recognize that landscape has changed from some previous state, and the paths along which landscape changes. Although the United States is used as an example, the concepts presented here can be applied to landscape anywhere on Earth. The goal of Part I is to allow you to read landscape wherever your travels take you.

Part II, entitled "Structural Provinces" applies concepts introduced in Part I to the landscape of the contiguous United States with special emphasis on the topography, rock type, rock structure, tectonic setting, climate, and recent uplift/erosion history. It is more comprehensive and with a greater detail relative to its counterpart in the first edition. The content includes detailed discussion of specific landscape areas compiled primarily from journal articles. The goal is to characterize the present-day landscape of the United States, understand its origin, how long it has been in existence, and how and why it has changed from some previous landscape. Chapter 11, Structural Provinces, Rock Successions, and Tectonic Provinces, describes the basis for dividing landscape into eight structural provinces comprising of four groups of two closely related provinces each. The eight structural provinces are discussed individually in Chapters 13 through 20. Chapter 12 discusses glacial landscape, and Chapter 21 is an updated look at the origin of the Grand Canyon. These chapters can be read in any order, but it is best to read them in sequence or at least in groups of two.

I use US Customary units of measurement (inch, foot, mile) throughout the book in order cater to a primarily US audience. I do not always show metric unit equivalents so that the instructor can quiz students on the conversion. When discussing rates, I use 100 years as the common denominator. I do this because 100 years is approximately equivalent with a human lifetime so the reader can quickly grasp the amount of change that occurs over the course of their existence.

The figures include Google Earth images, annotated landscape maps, photographs, and sketches. The figures are designed to be both simple and informative. They are an integral part of the discussion. Please take the time to examine each figure carefully. The Appendix contains uncolored full-page versions of some of the maps with the intent that they can be photocopied and hand-colored for teaching purposes. The primary landscape map used throughout this book is *Landforms of the United States*, 1957. This map, and a variety of other maps, were hand-drawn by Erwin J. Raisz from field observations, aerial photographs, and satellite imagery. Raisz was a member of the Institute of Geographical Exploration at Harvard University for nearly 20 years beginning in 1931 and is one of the founding cartographers in the United States. The first edition of his seminal map was published in 1939. The sixth and last edition was completed in 1957. It remains one of the finest landscape maps ever produced. I augment the Raisz maps with boundaries that show the distribution of physiographic provinces, rock types, structural provinces, climate zones, river systems, global wind patterns, glacial zones, and tectonic features. The complete inventory of Raisz hand-drawn landform maps is available at www.raiszmaps.com. For several figures, including Figure A.1 in the Appendix, I used a Photoshop enhanced 100-m resolution color-sliced elevation image of the United States with relief shading added to accentuate terrain features in an Albers Equal-Area Conic projection. The map was downloaded from the National Atlas of the United States of America, U.S. Geological Survey EROS Data Center, at Nationalmap.gov/small_scale/atlasftp.html.

The official vertical datum in use for the conterminous United States, and the one used whenever possible in this book, is the North American Vertical Datum of 1988 (NAVD 88). Elevations obtained from this datum are different from most USGS topographic maps, which show elevations using the National Geodetic Vertical Datum of 1929 (NGVD 29). The shift from NGVD 29 to NAVD 88 is between -2 and $+7$ ft. and, in general, the higher the peak, the greater the shift. Peaks in Colorado will gain 4 to 7 ft., while hills in Florida will lose 1 or 2 ft. For more information, visit the Peakbagger website at http://peakbagger.com/help/glossary.aspx#navd88 or the National Geodetic Survey website at https://www.ngs.noaa.gov/datums/vertical/.

I thank Anton H. (Tony) Maria for commenting on several of the chapters and Karen L. Sommer for her support. I also thank Justus C. McGill and Kevin F. Howard for reading and commenting on the chapters.

Listed below are common units of measurement, conversions, and abbreviations.

ABBREVIATIONS

millimeter (mm)	year (yr)
centimeter (cm)	million years (My)
meter (m)	million years ago (Ma)
kilometer (km)	billion years ago (Ga)
inch (in)	degrees Fahrenheit (°F)
feet (ft.)	degrees Centigrade (°C)
mile (mi)	

CONVERSIONS

1 km = 1000 m = 0.62 mi = 3280 ft.	1 mm/yr = 3.94 in/100 yr = 1 km/My
1 mi = 5280 ft. = 1.61 km = 1609.3 m	1 in/yr = 8.33 ft./100 yr = 15.78 mi/My
1 mm = 0.1 cm = 0.0394 in	1 in/yr = 2.54 m/100 yr = 25.4 km/My
1 in = 25.4 mm = 2.54 cm	°C = (°F − 32) × 0.555
1 m = 1000 mm = 3.28 ft. = 39.36 in	°F = (°C × 1.8) + 32
1 ft. = 305 mm = 0.305 m	

Part I

Keys to Understanding Landscape Evolution

Chapter 1

The Tortoise and the Hare

From California to the coast of Maine, and from Florida to the coast of Washington, the contiguous United States has some of the most spectacular scenery on Earth including the Grand Canyon, Rocky Mountains, and the majestic Appalachian Mountains. But the United States has not always looked like this. Thirty million years ago the San Andreas Fault did not exist and the state of Nevada was only about half as wide as it is today. Yellowstone National Park has literally blown up three times during the past 2.2 million years with volcanic ash spreading as far east as Iowa and as far west as the Pacific ocean. Periodically from about 2.4 million years ago to as recently as 18,000 years ago, a number of enormous ice sheets covered nearly all of Canada and a large part of the United States. During the height of these glacial advances the shoreline of the eastern United States was as much as 250 miles east of where it is today and the state of Florida was about twice its present day width. Areas of California, Oregon, and Washington have blown up repeatedly within the past few thousand years including the volcanic explosion that created Crater Lake 6870 years ago and the 1980 catastrophic explosion of Mount St. Helens. Sixteen thousand years ago giant lakes covered the desert regions of Nevada, California, and Utah, and tremendous floods poured through eastern Washington.

Without question, we can say that many of the world's landforms are only a few thousand to several tens of millions of years old. This may seem old, but it amounts to only a small fraction of the 4.55 billion year history of Earth. From this evidence alone we can surmise that landforms are ephemeral (lasting for a brief time) and are constantly in the process of change. So why is it that most people do not notice any change? The answer is that most changes occur at a rate too slow for the average person to see. For example, the United States is blessed with many great rivers and all of them carry enormous amounts of sediment from the mountains to the sea. Do we notice that the mountain mass has been reduced? The answer, of course, is no. The effect on the mountain is incremental and cumulative. If it takes 100,000 years to reshape landscape, then a 100-year-old human would have witnessed 0.1% of that change. Such a trivial amount likely would not directly impact our lives or our standard of living and therefore would not be noticed. There are, of course, catastrophic events that can shape a landform within a human lifetime. It took only a few minutes for more than 1000 feet of the Mount St. Helens volcano to blow away. Catastrophic changes are noticeable only because they occur rapidly and well within one person's lifetime. There is no doubt they contribute to the evolving landscape. But catastrophic events are periodic, and from a human perspective, do not repeat themselves for a long time.

In the legend of the tortoise and the hare, the tortoise was slow and steady; the hare was fast, but only for a short time. Spectators watching the race would have marveled at the rapid pace of the hare while perhaps not even noticing the tortoise as he passed by. But, in the end, the tortoise wins the race because of the cumulative effect of his slow and steady pace. Such are the processes of landscape evolution. Many processes, like the tortoise, produce only slow change, be it steady or not so steady. But the hare is not out of the race completely. Periodic rapid changes do occur, and these can change the look of landscape easily within a human lifetime.

HOW SLOW IS SLOW?

We must now ask ourselves: how slow is slow? In Chapter 9 we will look at actual rates of change. For now, let us first put geological time in perspective. Let's say that you live to be 100 years old. In this case, 1 million years would seem like a long time. But what if you live to be 4.55 billion years old (the age of the Earth)? From that perspective, a million years may seem trivial. If each year is counted as one second, then 4.55 billion seconds adds up to about 144 years. Using this time scale, the Earth would be 144 years old and a human would be alive on Earth for less than 2 minutes; 100,000 years would pass in about 28 hours and 1 million years in about 11.6 days. Normally, nothing physically noticeable happens to a person in 28 hours or even 11.6 days. But, relatively speaking, the Earth could change enormously. On this time scale, it might take the Earth less than a month to construct a large mountain range and only a few months to tear it down. So, in perspective, the Earth changes much faster than humans. The bottom line is that the Earth, and everything we see today, is in the process of change. Some changes are rapid enough to notice. Others are not.

MAPS, CROSS-SECTIONS, AND SCALE

To discuss landscape, it is important to understand a few terms. Landforms are described by their topography, which is the shape and form of the Earth's surface as expressed in elevation above or below sea level. Simply stated, topography is the lay of the land. Elevation refers to the height above or below sea level, whereas relief refers to the difference in elevation between any two nearby points. For example, the greatest relief in the continental United States is in eastern California along the east face of the Sierra Nevada where Mt. Whitney, at an elevation of 14,505 feet, is only 85 miles from Death Valley, at an elevation of 282 feet below sea level. Relief between these two points is 14,787 feet:

$$14{,}505 \text{ feet} - (-282 \text{ feet}) = 14{,}787 \text{ feet}.$$

It is also important to understand the difference between a map, a profile, and a cross-section. All three are illustrated in Fig. 1.1. Maps show the aerial extent of physiographic or geologic features as if you are looking at them from above; like the view from an airplane. To the uninitiated eye, a map is an underrated tool. But maps are like photographs. They hold an enormous amount of information that otherwise would be tedious and boring to convey in words. Maps give an instant visual perspective of landscape and also convey information regarding spatial relationships, size, location, topography, rock type, and rock structure. Maps have existed for hundreds of years. Before there was photography, one's vision and perspective of Earth was based largely on maps.

A profile is an outline of the shape of land as if looking from ground level. It is a mug shot that shows how topography changes along a straight line. It is like looking at the outline of a volcano. A cross-section is a profile that, in addition, shows the rock structure along a vertical slice through the Earth's interior. It is like slicing a volcano in half and looking inside.

All maps, profiles, and cross-sections have a scale that shows how distance on the map is related to distance on the ground. A fractional scale of 1:50,000 indicates that one unit on the map is equal to 50,000 units on the ground. A unit refers to any form of measurement such as an inch, foot, or centimeter. For example, at 1:50,000, one inch on a map is equal to 50,000 inches on the ground; one foot on the same map is equal to 50,000 feet on the ground. A bar scale relates distance on a map to distance on the ground. Fig. 1.2 compares the bar scale on a 1:24,000 map with the bar scale on a 1:250,000 map. One could easily see that a 1:24,000-scale map would show great detail of a small area whereas a 1:250,000-scale map would show less detail of a much larger area.

PHYSIOGRAPHIC REGIONS AND PROVINCES

A landform is an area of any size that can be visibly separated from surrounding land on the basis of its shape (e.g., its topography). A mountain peak, a valley, and even an anthill could all be considered landforms. Landscape, terrain, and physiographic province all broadly refer to the same thing: an area of land characterized by a similar set of landforms. A group of interconnected mountain peaks and valleys could be considered a

Map of a Volcano
It is as if you are looking down from an airplane.
You see the top of the volcano.

Profile of a Volcano
It is as if you are looking at the volcano from the side.
You see an outline of the volcano.

Cross-Section of a Volcano
In addition to the profile, you see the distribution of rock layers in a vertical cut through the volcano.

FIGURE 1.1 Relationship between map, profile, and cross-section.

Bar scale for 1:24,000
1 inch on the map equals 24,000 inches, 2000 feet, and 0.379 miles on the ground.
1 cm on the map is equal to 24,000 cm, 240 m, and 0.24 km on the ground.

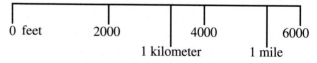

Bar scale for 1:250,000
1 inch on the map equals 250,000 inches, 20,833 feet and 3.946 miles on the ground.
1 cm on the map equals 250,000 cm, 2500 m and 2.5 km on the ground.

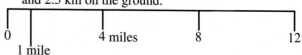

FIGURE 1.2 Comparison of map scales.

mountainous landscape. An area with many individual volcanic or glacial landforms could be referred to as a volcanic landscape or glacial terrain. Because the terms landscape and terrain have broad and varied meaning to the average person, I favor the term physiographic province to refer specifically to an area of land characterized by a similar set of landforms. We can then define a physiographic region as a larger area that groups together similar physiographic provinces. Given the definitions proposed here, each province or region must be a continuous tract of land with borders that visually separate one province or region from another. In other words, a physiographic province or region must look different from surrounding areas.

Any physiographic province, even a volcanic or glacial terrain, can broadly be classified as a plain, plateau, or mountain based on its elevation and relief in relation to surrounding land. A plain is a wide area with little relief (<500 feet) at low elevation relative to surrounding land. A plateau is a wide area at relatively high elevation bounded by steep slopes that either drop down onto plains or rise upward to a mountain range. A plateau can be flat or river dissected with considerable relief. A mountain is a landform of high relief and high elevation that rises prominently above its surroundings with relatively steep slopes and a confined summit area. The term mountain can be expanded to include a mountain range, which is a continuous line of mountain peaks, and a mountain belt (or mountain system), which is a larger landscape consisting of several semicontinuous mountain ranges separated by intermontane (between the mountain) valleys.

The primary landscape maps used throughout this book were hand-drawn between 1954 and 1957 by Erwin J. Raisz based on field observations, aerial photographs, and satellite imagery (www.raiszmaps.com). The large-format map, parts of which are shown in later chapters, remains arguably the finest US landscape map ever produced. Fig. 1.3 is a reduced copy of the Raisz map that shows the entire contiguous United States. This map can be compared directly with Fig. 1.4, which is a modern, digital, shaded-relief image that shows topography by varying brightness from an artificial sun. A quick glance at these figures suggests that, to a first approximation, we can divide the United States into four physiographic regions. Can you visualize the boundaries between these regions? There are two mountain systems, an interior plains and plateaus region, and a coastal plain. The boundaries are shown on the Raisz map in Fig. 1.5. A close look

FIGURE 1.3 The Raisz landform outline map of the United States.

FIGURE 1.4 Digital shaded-relief image of the United States from Thelin and Pike (1991).

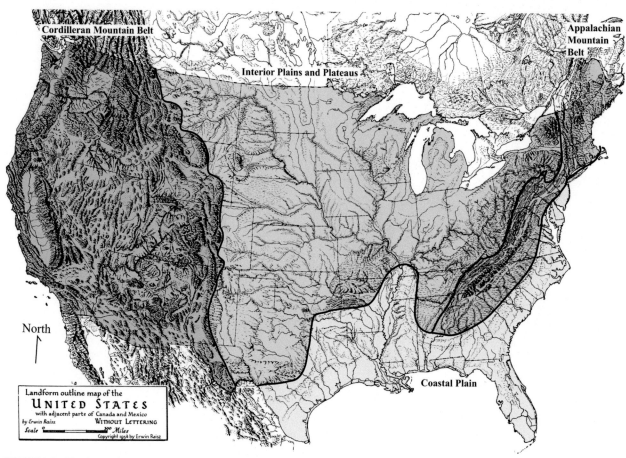

FIGURE 1.5 The four major physiographic regions of the United States.

at Fig. 1.5 suggests that there are smaller landscape provinces within the four physiographic regions. How many distinctive landscape provinces can you recognize? Obviously, the correct number is subjective depending on how specific a set of landforms one chooses to define. The US Geological Survey recognizes 25 provinces and 85 subprovinces across the contiguous United States. In this book, we recognize 26 provinces. Each is shown in Fig. 1.6 and listed in Table 1.1. They include 5 plains, 6 plateaus, and 15 mountain areas. Each is grouped into one of the four larger physiographic regions shown in Fig. 1.5. Fig. 1.7 is a simplified map of the United States in which elevation is shown with different colors. We will use Fig. 1.7 to present a brief overview of the topography in each of the four physiographic regions.

Interior Plains and Plateaus

The Interior Plains and Plateaus physiographic region encompasses the entire central part of the United States from the Rocky Mountain front in the west to the Appalachian Mountains in the east. Included within the Interior region are three plains—the Superior Upland, Central Lowlands, and Great Plains—and three plateaus, the Appalachian, Interior Low, and Ozark Plateaus. A few mountainous regions are present, such as the Wichita Mountains in Oklahoma, but only three, the Adirondack Mountains, Black Hills, and Ouachita Mountains, are deemed large enough to be shown in Fig. 1.6 as separate physiographic provinces.

As shown in Fig. 1.7, some of the lowest elevations in the Interior region, less than 1000 feet, are in the Lake Michigan-Mississippi River corridor. To the west of this corridor, the Great Plains rise gradually from about 400 feet on the Mississippi River at St. Louis to more than 5000 feet at Denver at the foot of the Rocky Mountains. Although the term plain implies low elevation, the western Great Plains forms the largest track of high ground anywhere in the Interior region. Isolated peaks in the volcanic Raton region of the Great Plains, along the Colorado-New Mexico border, rise to more than 8000 feet. The Black Hills boast several peaks above 7000 feet. Elevation also rises eastward from the Mississippi River to form hilly and mountainous terrain at the foot of the Appalachian mountain system. Much of this region is between 1000 and 2500 feet, but a few areas on the

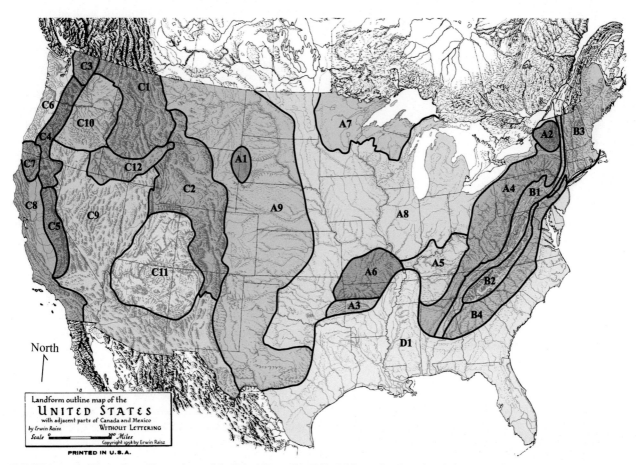

FIGURE 1.6 The 26 physiographic provinces of the United States. See Table 1.1 for explanation of symbols.

TABLE 1.1 Physiographic Provinces of the United States

A. **Interior Plains and Plateaus**
   Mountains
   1. Black Hills
   2. Adirondack Mountains
   3. Ouachita Mountains
   Plateaus
   4. Appalachian Plateau
   5. Interior Low Plateaus
   6. Ozark Plateau
   Plains
   7. Superior Upland
   8. Central Lowlands
   9. Great Plains
B. **Appalachian Mountain System**
   Mountains
   1. Valley and Ridge
   2. Blue Ridge
   3. New England Highlands
   Plateaus
   4. Piedmont Plateau
C. **Cordilleran Mountain System**
   Mountains
   1. Northern Rocky Mountains
   2. Middle-Southern Rocky Mountains
   3. North Cascade Mountains
   4. Central and Southern Cascade Mountains
   5. Sierra Nevada
   6. Washington-Oregon Coast Range and Valleys
   7. Klamath Mountains
   8. California Borderland
   9. Basin and Range
   Plateaus
   10. Columbia Plateau
   11. Colorado Plateau
   Plains
   12. Snake River Plain
D. **Coastal Plain**
   Plains
   1. Coastal Plain
   2. Continental Shelf (below sea level)

Appalachian Plateau, such as southeastern West Virginia, western Kentucky, and the Catskill Mountains of New York, top out above 4000 feet. Peaks in the Adirondack Mountains exceed 5000 feet. There are no peaks above 3000 feet in the Ouachita Mountains.

Appalachian Mountain System

The Appalachian mountain physiographic region is narrow and characterized by a strong north-northeasterly trend. It includes the Valley and Ridge, Blue Ridge, and New England Highlands mountain provinces and the Piedmont plateau. As seen in Fig. 1.7, the highest and most rugged part of the Appalachians is the southern Blue Ridge of Tennessee and North Carolina where 40 peaks rise above 6000 feet. Northward, the highest elevations are in the Valley and Ridge of Virginia where several peaks top 4000 feet. Farther north, most of eastern Pennsylvania and southern New England is below 2000 feet. Elevations rise above 4000 feet in the Green Mountains of Vermont and above 5000 feet in the White Mountains of central New Hampshire where one peak, Mount Washington, tops 6000 feet. A line of peaks above 3000 and 4000 feet extends across northern Maine from central New Hampshire to Mount Katahdin where we find the end of the Appalachian Trail and the only peak in Maine above 5000 feet. Elevations north of Mount Katahdin are mostly below 2000 feet. The Maine coastline is where the Appalachian Mountains trend directly into the ocean resulting in the highest elevations anywhere on the North Atlantic seaboard. The highest point is Cadillac Mountain in Acadia National Park at 1527 feet. Elevations across the Piedmont Plateau are below 1000 feet except along its western border in close proximity to the Blue Ridge, and in the northern Georgia-Alabama region where elevation in both the Valley and Ridge and Piedmont Plateau rarely tops 2000 feet. Cheaha Mountain, in east-central Alabama, is the southernmost peak above 2000 feet in the Appalachians Mountain system.

Coastal Plain

The Coastal Plain extends along the Atlantic seaboard from Cape Cod to the Gulf coast of Texas. There are identifiable differences between the northern and southern part of the Coastal Plain, but boundaries are subtle such that the entire region is considered to be a single physiographic province. The Coastal Plain, overall, is low-lying with beaches, swamps, and wide river valleys. Fig. 1.7 shows that nearly the entire area is below 500 feet. The surface of the Coastal Plain is inclined gently toward the shoreline and this slope continues out to sea for up to 250 miles as part of the continental shelf. Ocean depths on the shelf rarely exceed 400 feet below sea level. The continental shelf is considered to be a submerged part of the Coastal Plain. The continental slope and rise, at the outer edge of the continental shelf, mark the transition to deep ocean where depth plunges to between 13,000 and 20,000 feet below sea level.

Fig. 1.8 is a Google Earth image of the United States that shows the extent of the continental shelf. Note how far the shelf extends off the west coast of Florida and off the New England coast. Low elevation and a gentle seaward slope produce a situation where the Coastal Plain is susceptible to sea level changes. At various times in the past, the entire Coastal Plain has been submerged beneath the ocean. During other times, nearly the entire

FIGURE 1.7 Simplified map of the United States in which elevation is shown with different colors. Compiled by US Geological Survey, 1968, from the National Atlas of the United States, US Geological Survey, 1970, p. 59. Downloaded at http://www.learnnc.org/lp/multimedia/5296.

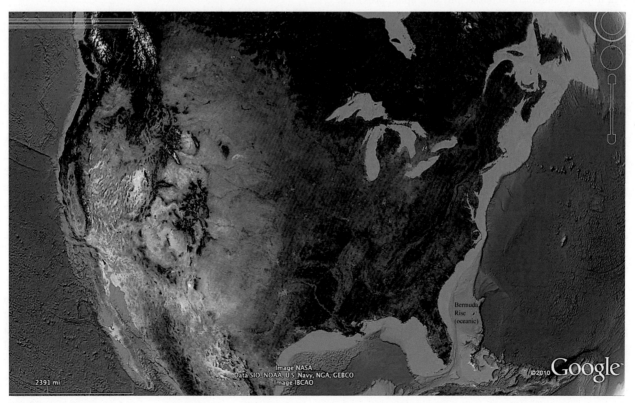

FIGURE 1.8 Google Earth image of the United States that shows the width of the continental shelf (*light blue*) along the Atlantic and Gulf coasts and its abrupt transition to the continental slope and abyssal ocean floor (*dark blue*). Note the narrow shelf along the Pacific coast.

continental shelf has been emergent. Ocean waters covered part of the Coastal Plain as recently as the last interglacial stage 128,000—118,000 years ago when sea level was 13—26 feet (4—8 m) higher than today. Sea level was about 400 feet lower than today only 18,000 years ago during the most recent glacial advance when nearly the entire continental shelf was exposed. Sea level has been rising with the melting of glaciers over the past 18,000 years resulting in progressive drowning of the Coastal Plain. Today, the entire Coastal Plain north of Cape Cod is below sea level.

Cordilleran Mountain System

The Cordillera is a complex physiographic region with one plain, two plateaus, and nine mountain provinces. There are three distinct mountain areas, the Rocky Mountains in the east, the Coast Ranges, Klamath, and Olympic Mountains along the west coast, and the Sierra Nevada and Cascade Mountains just inland from the coast. An intermontane area of plateaus, plains, and mountain blocks that includes the Columbia River and Colorado Plateaus, the Snake River Plain and the Basin and Range, is located between the Rocky Mountains and the Sierra Nevada-Cascade ranges. Landscape differences between the Northern and Middle-Southern Rocky Mountains, and between the Northern and Central-Southern Cascade Mountains, are distinct enough for each area to be considered a separate physiographic province.

The Cordilleran region is easily the highest and most rugged in the contiguous United States. Fig. 1.7 shows a vast area from eastern California to Colorado, and from Montana to New Mexico, that exceeds 5000 feet in elevation. Nearly the entire region, with the exception of the west coast, is above 2000 feet. Six of the 12 provinces, the Central-Southern Cascade Mountains, Basin and Range, Sierra Nevada, Colorado Plateau, Northern Rocky Mountains, and Middle-Southern Rocky Mountains, have peaks above 12,000 feet. The Colorado Rocky Mountains form a domal welt of high elevation with 58 recognized peaks above 14,000 feet. There are nine peaks above 14,000 feet in the Sierra Nevada including Mt. Whitney, the highest in the contiguous United States at 14,505 feet, two in the Central-Southern Cascade Mountains, Mt. Shasta and Mt. Rainer, and one, White Mountain Peak, in the Basin and Range of California.

COMPONENTS, FORCING AGENTS, MECHANISMS, AND LANDSCAPE RESPONSE

The process of landscape evolution implies that landscape undergoes change with time, and that it is possible for landscape to completely change its look relative to some previous state. One way to approach landscape evolution is to define the components that form landscape, the forcing agents that cause landscape to undergo change, the mechanisms by which landscape undergoes change, and the criteria used to recognize that landscape has changed (evolved) from some previous state.

Components are the constituent parts from which landscape is made. We will define two components: the rock/sediment type and the structural form (also known as the style of deformation or the structure of rock). The rock/sediment type is the substance that forms landscape; the structural form is the geometry of the substance. Together these components define the geology that underlies each physiographic province.

A mechanism is a process or method by which something takes place. There are five mechanisms that can exact change on landscape: uplift, subsidence, erosion, deposition, and volcanism. These mechanisms, both singularly or in combination, can, over time, destroy a preexisting landscape and create a new, entirely different-looking landscape. Uplift, subsidence, and volcanism are the mechanisms most responsible for building landscape. They tend to increase elevation and relief. Uplift pushes land to higher elevation, while subsidence lowers it. Volcanism is a relatively rapid process that can create distinctive landforms and bury preexisting landscape. Erosion and deposition are primarily responsible for the leveling of landscape. Erosion is the removal of rock and sediment from its place of origin. Deposition is the settling of eroded material in some lowland area such as a lake. The removal of rock from high elevation via erosion and the deposition of eroded material in lowland areas tend to reduce elevation and relief.

Forcing agents are the processes that cause landscape to undergo change. The forcing agents activate (or set in motion) the five mechanisms of landscape change. The two primary forcing agents are climate and the tectonic system. The tectonic system refers to the interaction of moving tectonic plates. When plates move they create internal stresses and thermal anomalies within the Earth that activate the three mechanisms primarily responsible for creating and building landscape, uplift, subsidence, and volcanism. Climate refers to the long-term condition of the atmosphere. It is the driving force most responsible for activating the two mechanisms primarily responsible for the leveling of landscape, erosion and deposition. The climatic and tectonic systems oppose each other. They interact and compete with each other to shape landscape. A significant aspect of this interaction is that their proportional effect on landscape can vary over time and from one location to another such that two areas may look different even if they have identical rock/sediment type and structural form.

The competing influences of the climatic and the tectonic systems produce secondary forcing agents that also affect landscape. The two most important secondary agents are sea level change and isostatic adjustment. Sea level is zero elevation. As sea level changes, so does the baseline to measure elevation. Areas may be drowned or become emergent. Any change in sea level, therefore, results in broad changes to worldwide elevation. Isostatic adjustment is a process related to gravity and buoyancy within the Earth that results in broad vertical uplift and subsidence of land areas, and thus, also affects elevation.

The four forcing agents are capable of influencing each other. A change in one agent can force changes to other agents, which, in turn, will cause changes in the rate at which each of the five mechanisms act upon landscape. A simple example is a change in the tectonic system that results in rapid uplift of land. The rising landmass could block wind patterns, which then alters the climate of surrounding land areas. Rates of uplift, subsidence, erosion, deposition, and possibly even volcanism, could all change, resulting in changes to the landscape. We will reintroduce and elaborate on the four forcing agents at the beginning of Chapter 5.

The final question to address is what criteria do we use to recognize that landscape has changed from a previous state. What do we see that is different about the landscape as a result of changes to components, agents, and mechanisms? Ignoring obvious and dramatic changes due to an area becoming volcanically active, the two most basic changes that are both visible and measurable are changes in elevation and relief. Two additional visible and measurable criteria are changes to both the river drainage pattern and the density of river channels. The components, agents, mechanisms, and criteria of landscape evolution are summarized in Table 1.2 and illustrated in simplified form in Fig. 1.9. With these ideas in mind we can summarize landscape development with the following statement.

The landscape that characterizes a particular area is not random, but is a direct result of the interaction of rock/sediment type and structural form with the tectonic and climatic systems, with sea level change, and with isostatic adjustment. These forcing agents interact over time to activate uplift, subsidence, erosion, deposition, and volcanism, which are the mechanisms that maintain and change landscape.

GEOLOGY, LANDSCAPE, AND TECTONICS

The overriding goal when studying geology, landscape, and tectonics is to understand something about the historical and present-day development of a small part of Earth. The subject matter of geology, landscape, and tectonics are similar, but the emphasis is on different aspects of Earth. Geology, in its purest form, is the study of the history of Earth and its life as recorded in rock and other solid matter. The goal of a geologist is to interpret

TABLE 1.2 Components, Forcing Agents, Mechanisms, and Criteria

Components That Form Landscape
  a. Rock/sediment type
  b. Structural form

Forcing Agents that Cause Landscape to Undergo Change
  Primary
  a. Tectonic activity
  b. Climate
  Secondary
  a. Sea level change
  b. Isostatic adjustment

Mechanisms by which Landscape Undergoes Change
  a. Uplift
  b. Subsidence
  c. Erosion
  d. Deposition
  e. Volcanism

Criteria to Recognize That Landscape Has Changed From its Previous State
  a. Changes in elevation
  b. Changes in relief
  c. Changes in the drainage pattern of rivers
  d. Changes in the density of river channels

geologic history based on rock type, its grain size, texture, chemistry, fossils, and mode of origin among other details. If the age of the rock can be determined, then the geologist has a clue to what the Earth looked like at the time the rock formed, which could have been hundreds of millions of years ago. The geologist gains information from a geologic map and from a detailed study of the structural, physical, and chemical properties of rocks found within the area of the map. The perfect geologic map shows the surface distribution, stacking order, and structural form (i.e., the geometry) of rock bodies without bias or interpretation. Boundaries on a geologic map follow the boundaries of distinctive rock units irrespective of landscape such that two maps of the same area, created by different authors, should look exactly the same.

The study of landscape is a branch of geology where the emphasis is on present-day Earth. Here, the geologist is most interested in how tectonics and climate interact to shape the Earth's surface. The goal is to characterize landscape, understand how it formed, why it looks the way it does, and how it evolves. The most basic form of landscape analysis does not require a detailed understanding of the chemistry, origin, or even the age of rocks. The geologist instead is interested in how surface rocks interact with climate and the prevailing tectonic regime. The study of landscape can result in the creation of a physiographic provinces map such as Fig. 1.6. Boundaries on a

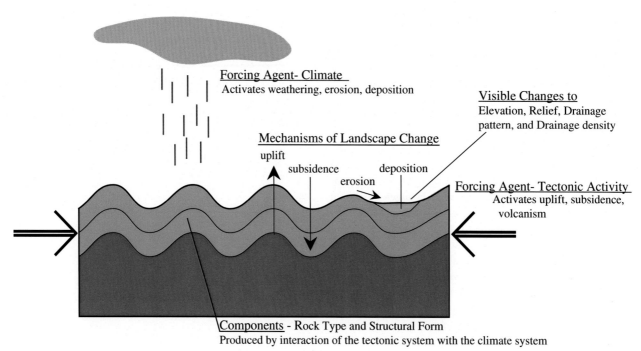

FIGURE 1.9 Sketch showing components, forcing agents, mechanisms, and visible changes to landscape (sea level change, isostatic adjustment, and volcanism are not shown).

physiographic map are based on visible changes evident in landscape. Although the underlying rock will influence landscape, boundaries are not necessarily based on the distribution or the geometry of rock. A physiographic map can cross geologic boundaries. Because physiographic boundaries are visual, they are also subjective. A map of the same area from different authors can look different dependent on the level of detail or on other aspects the mapmaker wishes to emphasize. For example, Fig. 1.6 divides the contiguous United States into 26 physiographic provinces whereas the US Geological Survey Tapestry of Time map recognizes 25 physiographic provinces and 85 subprovinces.

Tectonics is a branch of geology where the emphasis is on plate motion and the influence that plate motion has on Earth history. The goal is to develop a regional interpretation of Earth history (the term regional implies over a large area; the term local implies over a small area). This type of analysis involves geophysical and remote sensing tools that allow geologists to understand how heat and density varies within the interior Earth and how moving tectonic plates interact. The study of tectonics frequently involves the creation of a tectonic map, which is different from a geologic map. A tectonic map is a subjective interpretation of one or more geologic maps. A tectonic map unit need not be of a single rock type or structural form. The mapmaker instead will group rocks with similar geologic history or of similar origin regardless of rock type and regardless of landscape. Boundaries are often, but not necessarily, fault contacts. As such, a map of the same area from two different authors can look different dependent on the tectonic and geologic history the mapmaker wishes to emphasize.

Our goal in this book is to characterize present-day landscape, understand its origin, how long it has been in existence, and how and why it has changed from some previous landscape. At least some knowledge of geology and tectonics will be required to attain these lofty goals, but happily, the required information is contained in this book.

The connection between landscape, geology, and tectonics is different in each of the four physiographic regions. Old rocks and a tectonic regime that has been inactive for a long time characterize the Appalachian Mountains and most of the Interior Plains and Plateaus. Tectonics, because it is no longer active, takes on a diminished role in landscape evolution such that the primary forcing agent is climate. In such areas there may be a disconnect between the geology-tectonics of the region and the present-day landscape. In other words, the landscape that formed when the tectonic regime was active is not the landscape we see today. Such a disconnect may not exist in the Coastal Plain and especially in the Cordillera. The tectonic regime and the rocks in these regions are significantly younger, and in some cases the age of rock and sediment progresses from millions of years ago right up to the present day. We, therefore, cannot study landscape evolution without paying close attention to the recent geology and tectonics of these areas.

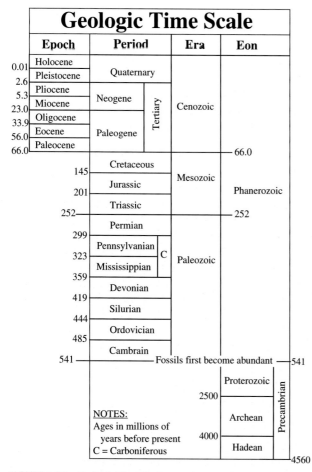

FIGURE 1.10 Geologic Time Scale.

Beginning with Chapter 11, we will make use of a combination geologic/physiographic map that I refer to as a structural provinces map. It is less subjective than a physiographic map because rather than subdividing areas based on visual similarities, landscape is divided based on similar rock/sediment type and structural (geometric) form. We will discover that rock type and structural form have a strong influence on landscape. The premise, therefore, is that areas underlain with similar rock type and structural form will have at least some physiographic similarity, and that differences will be due largely to climatic and tectonic factors, which we can gauge. Such a classification implies that landscape may vary within a single structural province and that the province may not be continuous from one location to the next.

GEOLOGIC TIME SCALE

The Geologic Time scale is a division of time based primarily on the evolution of fossils and supplemented with radiometric dates. As shown in Fig. 1.10 it is divided into eons, eras, periods, and epochs. Specific intervals of time are indicated with names rather than numbers. For example, the Mesozoic era is defined based on a specific set of fossils for which radiometric dating has shown to be between 252 and 66 million years old. The Cretaceous period represents the interval of time within the Mesozoic era from 145 to 66 million years ago when a specific subset of those fossils was alive. The Geologic Time scale is introduced because it will be useful when describing the age of rock units that form landscape.

The remaining chapters in Part I of this book describe in more detail the components, agents, and mechanisms of landscape change. Following this discussion, Chapter 10 provides some insight as to how components and forcing agents interact to create paths by which landscape evolves. However, before we embark down this path, Chapter 2 provides a short discourse on river systems of the United States and their influence on landscape. Beginning with Chapter 11, Part II of this book introduces the structural provinces of the United States and provides a detailed discussion of landscape in each province.

QUESTIONS

1. Review the following conversions: 1 inch = ___ cm, 1 cm = ___ inch, 1 mile = ___ feet, 1 mile = ___ km, 1 km = ___ mile, 1 foot = ___ m, 1 m = ___ feet.
2. Landscape is ephemeral. What does this mean?
3. What does a scale of 1:40,000 mean?
4. What is the fractional scale of a map where 1 inch on the map equals 82 miles on the ground?
5. Define the following: Topography, Elevation, Relief, Landform, Physiographic Region, Physiographic Province, Plain, Plateau, Mountain, Mountain Range, Mountain belt (Mountain system).
6. Name the four major Physiographic regions in the contiguous United States.
7. What is the relief between Mt Whitney in the Sierra Nevada and Owens Lake (elevation 3563 feet) 17.2 miles to the southeast in Owens Valley?
8. Name three plateau provinces that are entirely east of the Mississippi River.
9. In what part of the country is the Coastal Plain currently entirely below sea level? Your choices are Tennessee; North Carolina, Florida, Georgia, Maine, Indiana, Texas, Louisiana, Mississippi.
10. Name the physiographic province composed of small isolated mountain blocks and intervening valleys that occupies a vast area of Nevada.
11. Make a copy of Appendix Fig. A.3A or A.3B. Use Fig. 1.6 and Table 1.1 as a guide and label each individual physiographic province using their complete names. Draw the boundaries of the four physiographic regions with a thick heavy line.

12. In which physiographic province do you live?
13. The process of landscape evolution implies two things. What are they?
14. Name the two components of landscape evolution that define the geology that underlies each physiographic province.
15. Name the four forcing agents of landscape evolution.
16. Name the five mechanisms of landscape change.
17. Name four criteria that we can use to recognize landscape change?
18. Tectonic activity is the driving force most responsible for activating which of the five mechanisms of landscape change?
19. Climate is the driving force most responsible for activating which of the five mechanisms of landscape change?
20. How are boundaries on a physiographic province map defined?
21. Why should several geologic maps of the same area, but authored by different geologists, look the same?
22. How is a tectonic map different from a geologic map?
23. Use Fig. 1.7 as a guide and write a paragraph or two that discusses areas in the Cordillera where elevation is less than 500 feet. Begin in the south and work your way northward. Name specific physiographic provinces and specific rivers and valleys.
24. Use Fig. 1.7 as a guide and write a paragraph or two that discusses significant areas in the Cordillera where elevation is more than 9000 feet. Begin in the south and work your way northward. Name specific physiographic provinces and specific rivers and valleys.
25. Using Fig. 1.7 as a guide, what is the elevation surrounding the Great Lakes? Where is it lowest, and where is it highest?
26. Using Fig. 1.7 as a guide, name the three states east of the Mississippi River that have the most land area above 2000 feet in elevation?
27. Use Fig. 1.7 to describe the distribution of highest elevation in the eastern United States.
28. In Google Earth, use the ruler to measure the width of the continental shelf at various locations along the east, Gulf, and west coasts of the United States. Where is it widest? Speculate as to why it is so much narrower on the west coast.
29. Why is landscape different in each of the 26 physiographic provinces?

Chapter 2

River Systems

A physiographic subdivision of the United States is based on the topographic expression of land in which similar landforms are grouped to form a province. An alternative subdivision is based on the distribution of major river systems with boundaries that correspond with drainage divides. A river system is a network of stream channels that either converge into a major river to form a drainage basin (a watershed) or that enter the same major body of water. River systems are separated by drainage divides, which are continuous ridges of high ground where water on either side is drained into a different river system. Fig. 2.1 is a schematic drawing that shows several river systems separated by divides. In addition to the major divides shown in this figure, there must be smaller drainage divides between each and every stream valley regardless of its size. The head (or headwaters) of the river (H in Fig. 2.1) is close to a divide where the stream channel begins to take form. The mouth is where the river ends.

In this chapter we separate the continental United States into nine major river systems. Each river system is shown with a different color in Fig. 2.2 and separated by drainage divides shown with thick black lines. The small sections of the Mississippi River and Colorado River systems that extend into Canada and Mexico respectively are colored. Parts of the California, Rio Grande, Columbia, St. Lawrence, Hudson, and Atlantic Seaboard River systems that extend into Mexico and Canada are not colored but the divides are shown. Uncolored areas in the United States represent areas of internal drainage, meaning they have no outlet to an ocean. Any water entering the area, such as during a rainstorm, is trapped in low areas where it forms a permanent lake or a playa (a temporary lake that eventually dries up). Table 2.1 lists each river system along with major rivers and a few other features. Below is a brief description of each river system. For reference, we define discharge as the volume of water that passes a certain point per second. The measurements quoted in this chapter are at the mouth of the river except where noted.

DIVIDES

The best-known drainage divide in North America is the Continental Divide (or Great Divide), which trends through the Rocky Mountains. This divide separates water that will eventually reach the Pacific Ocean from water that will reach either the Atlantic or Arctic Oceans. In Fig. 2.2, the Continental Divide is located along the western boundaries of the Mississippi and Rio Grande—West Texas river systems. A small area of internal drainage, known as the Great Divide Basin, straddles the Continental Divide in the Red Desert area of Wyoming. A second area forms part of the San Luis Valley in Colorado. Four of the nine river systems shown in Fig. 2.2 drain into the Atlantic Ocean: the Mississippi, Atlantic Seaboard-Gulf Coast, St. Lawrence, and Rio Grande—West Texas. Three river systems drain into the Pacific Ocean: the Colorado, Columbia, and California. One river system west of the Continental Divide, known as the Great Basin, is an area of internal drainage.

The Northern (or Laurentian) Divide separates water that will eventually reach the Arctic Ocean (or Hudson Bay) from water that will reach the Atlantic Ocean. This divide forms the northern boundaries of the Mississippi

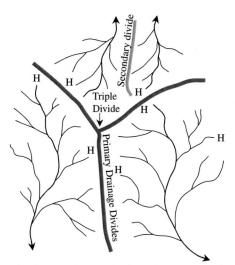

FIGURE 2.1 Three river systems separated by drainage divides. Smaller divides separate each stream. The stream channel begins to take form at its head (H).

FIGURE 2.2 Major river systems of the United States. Areas of the Pacific Northwest not included in the Columbia River system drain directly into the Pacific Ocean. Areas in southern Arizona not part of the Colorado River system drain toward Mexico to the Gulf of California.

and St. Lawrence River systems. In Fig. 2.2, only the Hudson Bay river system reaches the Arctic Ocean.

A glance at Fig. 2.2 shows that the Northern Divide meets the Continental Divide in northern Montana at a location known as a triple divide because it separates water flowing in three directions. Several triple divides are shown in Fig. 2.2, but the triple divide in Montana is unique because it separates water that will eventually reach three oceans, the Atlantic, Arctic, and Pacific. This divide is located at the appropriately named Triple Divide Peak (8020 feet) in Glacier National Park.

The Eastern Continental Divide separates water draining into the Atlantic Ocean from water draining into the Gulf of Mexico. In Fig. 2.2, this divide can be traced along the eastern boundary of the Mississippi River system from Pennsylvania to eastern Georgia. The divide continues through central Georgia and through central Florida, but is not shown in Fig. 2.2 because rivers in Georgia, Alabama, Florida, and Mississippi that empty directly into the Gulf of Mexico are shown as part of the Atlantic Seaboard−Gulf Coast river system.

MISSISSIPPI RIVER SYSTEM

Extending across nearly the entire Interior Plains and Plateaus region, the Mississippi River system is the largest in the United States, draining 1150 square miles with an average discharge of 593,000 ft^3/s (16,792 m^3/s). The northern boundary of the Mississippi River system extends across the Canadian border in Montana before dropping south of the Red River, which forms the border between North Dakota and Minnesota. It then wraps around the southern margin of the Great Lakes coming to within only a few miles of the Lake Michigan shoreline. Included in the Mississippi River system as shown in Fig. 2.2 are the Atchafalaya River and a few smaller rivers that do not flow directly into the Mississippi but help to build the Mississippi River Delta region.

The Mississippi River begins in and around Lake Itasca and flows north then east, before turning southward, eventually flowing to the Atlantic Ocean via the Gulf of Mexico. Along the way, some of the country's largest rivers drain directly into the Mississippi. The Missouri River, the longest in the United States at

TABLE 2.1 Major River Systems of the United States

1. Mississippi River System

Drains to Gulf of Mexico	
Arkansas River	Red River (Oklahoma)
Mississippi River	South Platte/Platte River
Missouri River	Tennessee River
Ohio River	

2. Atlantic Seaboard–Gulf Coast River System

Many separate river basins, drain to Atlantic Ocean and Gulf of Mexico	
Alabama/Mobile River	James River
Broad/Santee River	Pearl River
Cape Fear River	Potomac River
Connecticut River	Roanoke River
Chattahoochee/Apalachicola River	Savannah River
Delaware River	Susquehanna River
Hudson River	Suwannee River

3. St. Lawrence River System

Drains to Gulf of St. Lawrence	
Drains the Great Lakes	
St. Lawrence River	Lake Champlain
Finger Lakes	

4. Rio Grande–West Texas River System

Drains to Gulf of Mexico	
Encloses an area of internal drainage	
Brazos River	Rio Grande
Colorado River (Texas)	Sabine River
Nueces River	Trinity River
Pecos River	

5. Colorado River System

Drains to Gulf of California	
Colorado River	Little Colorado River
Gila River	San Juan River
Green River	White River
Gunnison River	

6. Columbia River System

Drains to Pacific Ocean	
Crooked/Deschutes River	Snake River
Columbia River	Willamette River
Owyhee River	

(Continued)

TABLE 2.1 (Continued)

7. California River System

Drains to Pacific Ocean	
Kern River	Sacramento River
Klamath River	San Joaquin River
Rogue River	

8. Great Basin River System

Internal drainage (no outlet)	
Bear River	Lake Tahoe
Humboldt River	Pyramid Lake
Owens River	Utah Lake
Sevier River	Walker Lake
Great Salt Lake	

9. Hudson Bay River System

Drains to Hudson Bay	
Rainy River	Red River

2540 miles, flows out of the Northern Rockies of Montana and enters the Mississippi at St. Louis, contributing 76,200 ft^3/s (2158 m^3/s) of average discharge. The Ohio River, the second largest in the United States in terms of average discharge, flows out of Pennsylvania and enters the Mississippi at Cairo, Illinois where it contributes 281,000 ft^3/s (7957 m^3/s) of discharge, almost half of all water in the Mississippi. The Arkansas River, flowing eastward out of the Colorado Rockies, enters the Mississippi River near Pine Bluff, Arkansas contributing 41,000 ft^3/s (1161 m^3/s) of discharge.

The most common drainage pattern along the Mississippi and its tributaries is the classic dendritic pattern, which looks similar to the branching pattern of a tree. A dendritic pattern is visible along the Mississippi River in Fig. 2.2 and is also shown in Fig. 2.1. This type of pattern forms in areas where exposed rock is relatively homogeneous and where rock structure has little or no influence on the location or arrangement of streams. In this case, the underlying rock consists of fairly uniform, nearly flat-lying, sedimentary layers overlain with unconsolidated sediment.

ATLANTIC SEABOARD–GULF COAST RIVER SYSTEM

As shown in Fig. 2.2, the Atlantic Seaboard–Gulf Coast river system includes all rivers east of the Mississippi River system that flow directly into the Atlantic Ocean. The divide that separates this system from the Mississippi does not follow along the crest of the Appalachian Mountains. Instead, it cuts diagonally across the Appalachians. The divide begins in the Green Mountains of Vermont, extends westward across the Adirondack Mountains to the Appalachian Plateau and then eastward eventually to the eastern margin of the Blue Ridge in North Carolina. All 40 of the recognized 6000-foot peaks in the southern Blue Ridge are located west of the divide within the Mississippi river system. South of the Blue Ridge, the divide extends across the southern Appalachians, the Appalachian Plateau, and the Coastal Plain to an area just east of the Mississippi River (Fig. 2.2).

The Atlantic Seaboard—Gulf Coast river system does not possess a trunk river into which other rivers drain. Instead, it is composed of many smaller, nearly parallel rivers each of which drain across the Appalachian Mountains and Coastal Plain directly into the Atlantic Ocean. Some of the largest rivers, from north to south, include the Connecticut, Hudson, Delaware, Susquehanna, Potomac, James, Cape Fear, Savannah, Suwannee, Apalachicola, Alabama, and Pearl Rivers. The Susquehanna River, in Pennsylvania and Maryland, is the largest with an average discharge of 38,200 ft^3/s (1081.7 m^3/s). Of the others mentioned, only the Hudson (21,900 ft^3/s (620 m^3/s)), Connecticut (17,070 ft^3/s (483 m^3/s)), and Apalachicola rivers (16,600 ft^3/s (470 m^3/s)) have average discharges greater than 13,100 ft^3/s (371 m^3/s).

ST. LAWRENCE RIVER SYSTEM

The natural flow of the Great Lakes is northeastward through southeastern Canada via the St. Lawrence River system to the Gulf of St. Lawrence and the Atlantic Ocean. In addition to the Great Lakes, the St. Lawrence River system drains southeastern Canada, part of the Lake Region of northeastern Minnesota, the Finger Lakes region of New York, Lake Champlain along the Vermont-New York border, and part of the Adirondack Mountains. The Great Lakes form the largest body of fresh water in the world. As such, average discharge through the St. Lawrence River is an enormous 348,000 ft^3/s (9854 m^3/s), second only to the Mississippi River in North America and almost as much as the Ohio and Missouri Rivers combined. When measured from its farthest headwaters located west of Lake Superior, it is 1900 miles long, the fourth longest within (or partly within) the United States behind the Missouri and Mississippi Rivers and only 80 miles less than the Yukon River.

The divide that separates Lake Superior, Lake Michigan, and Lake Erie from the Mississippi River system consists of a series of low-lying hills, some of which are glacial moraines. The separation is so slight that several man-made diversions into and out of the St. Lawrence drainage basin have been created in the past 200 years. One of the largest is a canal that was completed in 1900 by the city of Chicago that linked Lake Michigan with the Mississippi River system. The canal had the effect of reversing the flow of the Chicago River away from Lake Michigan (into which it previously emptied) and into the Mississippi River system. The canal was built to protect Lake Michigan, a source of drinking water, from the city's sewage, and to open a shipping lane between the two waterways. Today, this connection is a gateway for the dreaded Asian carp to enter the Great Lakes. Asian carp have already invaded the Mississippi River system. Protections, such as an electric current, are in place across the canal, but it seems a few of the fish have managed to pass through. If these fish invade the Great Lakes, they could potentially overcrowd and destroy native species that includes the lake sturgeon.

RIO GRANDE—WEST TEXAS RIVER SYSTEM

The only other US river system shown in Fig. 2.2 that empties into the Atlantic Ocean is the Rio Grande system of Colorado, New Mexico, and southwest Texas. Included in the system are rivers in southern Texas west of the Mississippi River such as the Nueces, Trinity, and Sabine Rivers that drain directly into the Gulf of Mexico.

The Rio Grande begins along the Continental Divide in the San Juan Mountains of southwest Colorado, flows southward through central New Mexico, and then along the Texas-Mexico international border to the Atlantic Ocean via the Gulf of Mexico. At 1900 miles, it is approximately the same length as the St. Lawrence River, but has nowhere near the same discharge. The Rio Grande flows through arid and semi-arid regions where so much water is diverted for agricultural irrigation that discharge sometimes goes to zero. Maximum discharge at Rio Grande City, Texas, about 120 miles from its mouth, is 3500 ft^3/s (99 m^3/s). Average discharge at Brownsville, 30 miles from its mouth, is less than 1000 ft^3/s (28 m^3/s).

The most important tributary is the Pecos River, which flows parallel with the Rio Grande through New Mexico and southern Texas. As shown in Fig. 2.2, the Pecos is separated from the Rio Grande by a large area of internal drainage that extends from just east of Albuquerque to the area south of Van Horn, Texas. This area of internal drainage includes part of Guadalupe Mountains National Park, the Tularosa Basin including White Sands National Monument, and the Salt Basin. A second area of internal drainage is located in the southwest corner of New Mexico continuing southward into the Chihuahuan Desert of Mexico. The Rio Grande in westernmost Texas forms a narrow corridor between the two areas of internal drainage (Fig. 2.2).

COLORADO RIVER SYSTEM

The Colorado River has its headwaters in the Colorado Rocky Mountain Front Range along the continental divide surrounding Grand Lake. From its headwaters, the river flows westward through Grand Junction and southwestward through Utah before turning west through the Grand Canyon in Arizona, and then along the Arizona-Nevada and Arizona-California borders to the Pacific Ocean via the Gulf of California. The Colorado River drains the entire area shown in Fig. 2.2. Major tributaries include the Gunnison River in Colorado, the Green and San Juan Rivers in Utah, the Little Colorado and Gila Rivers in Arizona, and the White River in Nevada, which forms the long, narrow, northward indentation into the Great Basin. Small areas of the United States along the Arizona-Mexico border are shown with a different color because they drain directly to the Gulf of California.

At 1450 miles, the Colorado is the seventh longest river in the United States, yet its average discharge is both small and highly variable depending on the size of the winter snowpack, the degree of drought that the area is experiencing, and the amount of water drawn for human consumption. Average discharge in recent years is less than 15,000 ft^3/s (425 m^3/s) at Davis Dam, Arizona, 80 miles south of Las Vegas and more than 250 miles from the Gulf of California.

An unfortunate feature of the Colorado River is that the volume of water decreases in a downstream direction from Davis Dam such that during at least part of the year it dries-up near its mouth in Mexico before reaching the Pacific Ocean. By contrast, the Mississippi river becomes larger in a downstream direction due to rainfall and the addition of water from downstream tributary rivers. There are three reasons why the Colorado system is different. The first is the size of the Colorado. Even where it contains its greatest volume of water, the Colorado is small in terms of total discharge. The second is that the Colorado River loses water through evaporation and ground infiltration as it flows across the desert region of Utah and Arizona. The third is that seven water-starved states (Wyoming, Colorado, Utah, Arizona, New Mexico, Nevada, and California) all draw water from the river and its tributaries. The Colorado River is the life-blood of these states, but there just isn't enough water for each of the seven states. A similar situation affects the Rio Grande, which also becomes smaller and dries-up near its mouth.

COLUMBIA RIVER SYSTEM

The Columbia River drainage basin covers the highlighted area shown in Fig. 2.2. The drainage basin extends northward well into Canada, and eastward to the continental divide incorporating most of the Northern Rocky Mountains, the Columbia Plateau, Snake River Plain, and the Central Cascade Mountains. Areas of the Pacific Northwest shown with a different color drain directly into the Pacific Ocean. These areas include part of the Oregon Coast Range, the Olympic Mountains, and part of the North Cascade Mountains as far east as Ross Lake and as far south as Mt. Rainier. The Columbia River begins in Canada and flows northward before turning sharply south, entering the United States near Northport, Washington. It continues south through mountains before turning west along the northern edge of the Columbia Plateau and then south-southeast along the western edge of the plateau. It then turns westward to form the Washington-Oregon border on its way to the Pacific Ocean. It is the fourth largest river in the United States in terms of average discharge at 265,000 ft^3/s (7504 m^3/s). Its tributaries are some of the largest rivers in the west including the Snake, Kootenay, Willamette, Spokane, Okanogan, John Day, and Deschutes Rivers. The Snake River has its headwaters in the southern part of Yellowstone National Park. It flows through Jackson Lake at the eastern base of the Teton Mountains and then westward along the Snake River Plain through Idaho Falls and Twin Falls. In western Idaho it turns northward to form the Idaho-Oregon border through Hell's Canyon before turning westward and joining the Columbia River at Kennewick, Washington. It is the twelfth largest river in terms of average discharge at 56,900 ft^3/s (1611 m^3/s). The Eastern Snake River Plain Aquifer, one of the largest in the world, provides about 7800 ft^3/s (221 m^3/s) of water to the Snake River through springs particularly in the vicinity of Thousand Springs and American Falls. The Kootenay River flows southward from Canada. It enters the United States in western Montana where it forms the 90-mile long Lake Koocanusa. It then makes a U-turn through the northeast corner of Idaho at Bonners Ferry and flows back into Canada where it enters the Columbia River. The Spokane and Okanogan Rivers enter the Columbia River in Washington along the northern edge of the Columbia Plateau. The Willamette River flows northward from the vicinity of Eugene, Oregon through Willamette Valley entering the Columbia River north of Portland. Its average discharge is 37,400 ft^3/s (1059 m^3/s). The John Day and Deschutes Rivers flow across central Oregon.

CALIFORNIA RIVER SYSTEM

The California River system encompasses the Klamath Mountains, the California Borderland, Sierra Nevada, and the Southern Cascade Mountains. The divide that separates the Sierra Nevada from the Great Basin passes along the crest of the Sierra Nevada for about 400 miles. Rivers in this part of the west are smaller than those in the Columbia River system. The Klamath River flows southwest from the vicinity of Klamath Falls, Oregon across the northern Klamath Mountains to the Pacific Ocean near Klamath, California. Its average discharge is 17,080 ft^3/s (484 m^3/s). Two other rivers, the Sacramento and the San Joaquin, drain the northern and southern part of California's Central Valley respectively. Neither is exceedingly large. The Sacramento has an average discharge of 23,500 ft^3/s (665 m^3/s) and the San Joaquin only 5100 ft^3/s (144 m^3/s).

The southern Sierra Nevada west of Mount Whitney, and the southernmost Central Valley surrounding Bakersfield, together forms an area of internal drainage shown in Fig. 2.2. The area is drained by the Kern River, the only south-flowing river in the Sierra Nevada. Normally, the Kern River flows into ephemeral lakes in the Central Valley south of Bakersfield; lakes that in the past would occasionally overflow into the San Joaquin drainage system. Today, nearly all of the water is diverted for irrigation and drinking.

GREAT BASIN RIVER SYSTEM

A rather unique drainage basin occupies the desert regions of California, Nevada, Utah, Oregon, and a small part of Wyoming and Idaho. It is known as the Great Basin; and it truly is a great basin because any water that enters the area is trapped with no outlet to the sea. It is a huge area of internal drainage, second in the world to the much larger Tarim Basin in China (among internal basins not occupied by a large lake such as the Caspian Sea or Lake Chad). Although predominantly dry, a few lakes exist close to mountain fronts where they receive water from melting snow. The largest include Lake Tahoe, Pyramid Lake, Walker Lake, Utah Lake, and the Great Salt Lake. These lakes are remnants of much larger and more numerous lakes that existed during and following the last major glacial advance 18,000 years ago (Fig. 12.27). There are no large rivers. The Humboldt River, with an average discharge of only 390 ft^3/s (11 m^3/s), parallels Interstate 80 in northern Nevada between Elko and Lovelock before emptying into the intermittently dry Humboldt Sink (near Carson Sink). The Bear River flows northward along the Utah-Wyoming border to Idaho south of Pocatello, and then southward to the Great Salt Lake. With an average discharge of 2400 ft^3/s (68 m^3/s), it is the largest tributary of the Great Salt Lake and the longest river in North America that does not reach an ocean. The Sevier River receives water from the Colorado Plateau in Utah, flows northward and then southward to Sevier Lake south of Provo. The lake is dry most of the year due to irrigation. Owens River flows southward along the base of the eastern slope of the Sierra Nevada through Bishop and Lone Pine to Owens Lake. It has an average discharge of 390 ft^3/s (11 m^3/s), but most of the water is diverted for agriculture and the Los Angeles Aqueduct such that the lake is predominantly dry.

HUDSON BAY RIVER SYSTEM

North of the Mississippi River System, the northern part of North Dakota and Minnesota drains northward through Canada to the Hudson Bay. The largest river, with an average discharge of approximately 20,200 ft^3/s (572 m^3/s), is the Rainy River, which flows westward along the border with Canada from International Falls to Lake of the Woods.

Much of the Hudson Bay River system in Canada and the United States was flattened beneath continental glaciers and then occupied by a large lake known as Lake Agassiz. The result is a very flat landscape with large areas of internal drainage particularly in Canada north of Montana. One such area in the United States is shown in Fig. 2.2. The diminutive Red River, with an average discharge of 8600 ft^3/s (244 m^3/s), flows north on the floor of ancient Lake Agassiz along the Minnesota-North Dakota border to form a deep incursion of the Hudson Bay River system into the United States.

COMPARISON OF RIVER SYSTEMS WITH PHYSIOGRAPHIC PROVINCES

Fig. 2.3 shows the boundaries of the 26 physiographic provinces in heavy black lines superimposed on the same colored map of river systems shown in Fig. 2.2, except without the underlying Raisz landform map. It is not hard to notice the absence of correlation between river systems and physiographic provinces. There are several reasons for the lack of correlation. Water will take any downhill path to reach an ocean even if it means crossing geological and landscape boundaries. Additionally, and perhaps more importantly, drainage divides between river systems frequently change their location via headward erosion and stream piracy. Headward erosion is the ability of a river to lengthen its channel at its head. Recall from Fig. 2.1 that the head of a river is located near its drainage divide where the channel first begins to take form. Headward erosion causes the channel to lengthen in a direction farther and farther uphill closer to the drainage divide. During this process, it is possible for headward erosion to erode across a drainage divide and steal water from

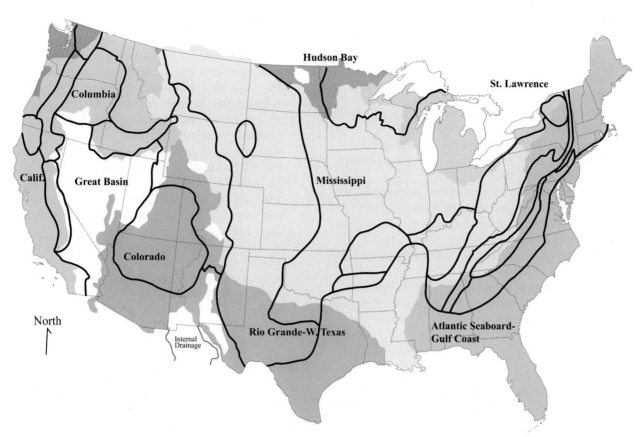

FIGURE 2.3 The boundaries of the 26 physiographic provinces in heavy black lines superimposed on a colored map of the major river systems of the United States.

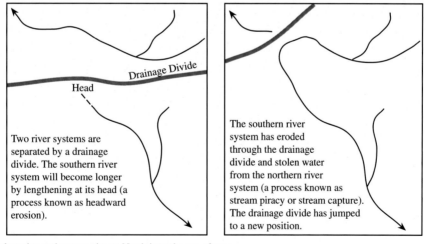

FIGURE 2.4 Headward erosion and stream piracy. North is to the top of page.

another river, a process known as stream piracy (or stream capture). The act of stream piracy causes the divide between river channels to jump to a new location. The process is shown in Fig. 2.4. Stream piracy is a common phenomenon; meaning drainage divides are not stable features relative to physiographic boundaries. Because river systems cannot be described as physiographic or geologic regions, we will end our discussion of river systems and return to our discussion of physiographic provinces.

QUESTIONS

1. What is a river system?
2. What is a drainage divide? What does the continental (Great) divide separate?
3. There are at least nine major river systems in the contiguous United States. Name the river systems that empty into the Atlantic Ocean.
4. Draw an example of a dendritic drainage pattern.
5. In what types of rock or rock structure do dendritic drainage patterns form?
6. Name the only major river system in the contiguous United States that does not empty into an ocean.
7. Name the river systems that drain Glacier National Park in Montana.
8. What is an aquifer? Research the Eastern Snake River Plain Aquifer.
9. Make a copy of Appendix Fig. A.4. Pick one or more river systems from Table 2.1 and trace the path of each river directly on the map. You may want to enlarge that part of the map. Use a different colored pencil for each river system. Label the major rivers with a letter designation and create a key for your lettering system that shows the name of each river. You may have to consult additional references to complete this question, but most of the major rivers are labeled on the detailed Raisz landform maps provided throughout the book.
10. Make a copy of Appendix Fig. A.2. Use Fig. 2.2 and Table 2.1 as a guide and draw the boundaries of all the major US river systems. Be careful not to draw a divide across any river or stream. Label all the river systems. Label the continental divide. Label the five Great Lakes.
11. In what mountain range is the continental divide located?
12. What is special about the triple divide in Montana?
13. What is special about the Great Basin drainage system?

Chapter 3

Component: The Rock/Sediment Type

Three rock types are important in landscape evolution: sedimentary, crystalline, and volcanic. Examples of each type are listed in Table 3.1. Each has a strong influence on landscape wherever and whenever they are exposed at the Earth's surface. Rocks not exposed at the surface are, instead, buried beneath an apron of unconsolidated (loose) sediment and soil. If the apron of unconsolidated sediment is thick enough to completely cover any trace of the underlying rock type, then sediment will impart its own unique look upon the land, thus creating a fourth landscape-forming rock/sediment type. This chapter provides an overview of the four rock/sediment types and their distribution across the United States. The critical aspect of the rock/sediment type with respect to landscape is its resistance to weathering and erosion relative to surrounding rock.

WEATHERING, EROSION, AND DEPOSITION

The weathering process includes all changes that result from exposure of rock material to the atmosphere. If you leave your bicycle outside for a year or two, it will rust, which is a form of weathering. The weathering process includes physical changes that break the rock into smaller pieces, and chemical changes by which rock reacts with water, air, and organic acids, and partly or wholly dissolves. Physical weathering is equivalent to hitting a rock with a hammer. Chemical weathering is equivalent with pouring acid on a rock. The residual product of weathering is unconsolidated sediment.

An important form of physical weathering is frost cracking (also known as ice wedging). Frost cracking occurs where water collects in a small crack in a rock and then freezes. Upon freezing, water expands by about 9%. The pressure generated during expansion causes the crack to propagate (become wider and longer) such that the rock eventually breaks in half. Frost cracking is particularly active along mountaintops in the west where temperature is above freezing during the day and below freezing at night for much of the year. Piles of broken rock litter summit areas above tree line. An example is Mount Audubon in the Front Range of Colorado as shown in Fig. 3.1.

Chemical weathering is the partial or complete dissolution of rock. Water is naturally slightly acidic and becomes more acidic when in contact with dead and decaying plant matter. Many minerals only partially dissolve, and in doing so, leave behind clay residue. An example is the partial dissolution of potassium feldspar, which is a major mineral in granitic rock and which is also common in sandstone and some types of metamorphic rock. The following chemical equation describes the process:

$$2KAlSi_3O_8 + (2H^+ + 9H_2O) = Al_2Si_2O_5(OH)_4 + (2K^+ + 4H_4SiO_4)$$

Potassium feldspar + acidic water = clay + components in solution in water

TABLE 3.1 Landscape-Forming Rock/Sediment Types

1. Sedimentary (sandstone, limestone, and shale)
2. Crystalline (plutonic granite and gabbro; metamorphic slate, schist, gneiss, and marble)
3. Volcanic (nonexplosive basalt; explosive silicic andesite-rhyolite-tuff)
4. Unconsolidated Sediment (alluvium, glacial drift, coastal deposits, eolian deposits, and soil)

FIGURE 3.1 Photograph looking along the ridge of Mount Audubon, Front Range, Colorado. The large rock pile that litters the ridge formed via frost cracking.

The reaction removes potassium (K) and some of the silica (SiO_2) from feldspar. Both go into solution in water. The potassium is now available for plants to absorb. The dissolved silica (silicic acid; H_4SiO_4) may enter the groundwater system and precipitate around sand grains to form the cement that binds sandstone together.

Quartz, which is crystalline silica, is the only common mineral that is not strongly affected by chemical weathering. It does not dissolve in water. All other common minerals dissolve completely or are partially dissolved and reduced to clay minerals. For this reason quartz and clay are the two most abundant minerals in sedimentary rock. Chemical weathering is most effective when water is present. Therefore, the amount of chemical weathering is controlled largely by the amount of available water.

The weathering process disaggregates rock but it does not remove rock material from its original location. Erosion is the physical removal of rock and sediment from its original location by an agent such as water, ice, air, gravity, or animal/human interference. The breakdown of rock through the weathering process greatly facilitates erosion. If erosion does not occur, the weathered remnants will remain in place, mix with organic material, and become soil.

Deposition is the accumulation of eroded sediment such as in a lake or subsiding basin. Deposition can occur in one of three environments, marine, nonmarine, and transitional. The term marine refers to something found in, or produced by, the sea. Marine environments include shallow marine, deep marine, and reef. Nonmarine environments are those that form on land such as a desert, lake, river, and glacier. Transitional environments form along coastlines where there is a component of both land and ocean. Such environments include beach, delta, and estuary.

THE FOUR ROCK/SEDIMENT TYPES

The rock/sediment type forms one of two components of landscape evolution. Landscape will evolve differently depending on whether it is underlain with sedimentary, crystalline, or volcanic rock, or if it is underlain with a thick layer of unconsolidated sediment. This section discusses the origin of each of the four rock/sediment types and introduces some of the common rocks that make up each type.

FIGURE 3.2 Photograph of a thick sandstone layer above thinner beds of sandstone and shale, Sunset Bay State Park, Oregon. The beds are tilted slightly to the left. The shale layer is about one foot thick.

Sedimentary Rock

Sedimentary rock consists of unconsolidated sediment and organic material originally deposited at the Earth's surface and subsequently buried, compressed, and cemented into rock. Sedimentary rocks are distinctive because they are almost always stratified, meaning they are layered. An individual layer is referred to as a bed, and each bed has a composition, color, or texture different from underlying and overlying beds. Each bed was originally deposited as a flat-lying (horizontal) or nearly flat-lying layer. A succession of beds stacked on top of one another is referred to as a stratigraphic sequence or simply as stratigraphy. Each bed within a stratigraphic sequence can vary in thickness from less than one inch to significantly more than two hundred feet. Large areas of the United States consist of interlayered beds of the three most common types of sedimentary rock, sandstone, shale, and limestone. These three rocks are described next.

Sandstone is composed of sand-sized rock particles and minerals (mostly quartz), mixed with variable amounts of silt and clay, all cemented together to form rock. The sand on the beach that you might visit this summer will eventually become sandstone. If you walk from the beach into the water, the sandy beach quickly turns to mud. Shale, the most abundant of all sedimentary rocks, is compressed and hardened mud. Fig. 3.2 shows a tilted sequence of sandstone and shale. The rocks were tilted after they were deposited. Limestone is composed mostly of the mineral calcite ($CaCO_3$). It forms at the bottom of a warm, clear, shallow ocean, primarily from the accumulation of shell, skeletal, and fecal debris. Limestone tends to be highly fossiliferous. Coral reefs, including the Florida Keys, are made of limestone.

In addition to sandstone, shale, and limestone, other less common sedimentary rocks include conglomerate, breccia, dolostone, and coal. A conglomerate contains rounded pebbles and cobbles in addition to sand-sized material and is often associated with sandstone. Breccia is similar to conglomerate except that the pebbles and cobbles are angular rather than rounded. Conglomerate, breccia, sandstone, and shale are known as clastic rocks because they are composed of broken fragments (or clasts) of preexisting rock. The term clastic is from the Greek *klastos* which means broken. The material that forms a clastic rock, the cobbles, the pebbles, the sand,

FIGURE 3.3 Photograph of crystalline rock. Metamorphic gneiss (dark rock) intruded by granitic igneous rock (white), Sawatch Range, Colorado. Together, these rocks form the crystalline rock type.

silt and clay, are known as detritus, which in Latin means to wear away (detrital is the adjective). These rocks form in a great variety of environments that include river, glacial, desert, shallow marine, and beach.

Limestone, dolostone, and coal are not clastic rocks. Rather than form from broken pieces of preexisting rock, they form through chemical and biochemical precipitation, and from the accumulation of organic material. Dolostone is closely associated with limestone but with a slightly different composition. It is composed almost entirely of the mineral dolomite ($CaMg(CO_3)_2$). As noted earlier, limestone forms primarily in shallow marine water through the action of sea animals. Dolostone forms in brine waters (evaporating sea water) where calcium in limestone is replaced with magnesium. Together, limestone and dolostone are referred to as carbonate rocks because their chemical composition includes CO_3. Coal consists almost entirely of the decayed remains of trees and other vegetation. It forms in swampy environments and is often associated with shale.

Crystalline Rock

The term crystalline rock, as used in this book, refers to rocks that have crystallized at depth below the Earth's surface. Igneous plutonic rocks and metamorphic rocks both crystallize at depth. These two rocks are combined because they often occur together and because they have similar landscape significance. The two rocks are shown together in Fig. 3.3.

Plutonic rocks are one of two distinctive types of igneous rock, the other being volcanic rocks, which form a separate landscape rock type discussed below. There is no chemical or mineralogical difference between plutonic and volcanic rocks. Both crystallize from magma (liquid rock). The primary difference is textural. A plutonic rock is relatively coarse-grained whereas a volcanic rock is fine-grained. A plutonic rock crystallizes from molten magma below the Earth's surface, usually at depths between 1 and 40 miles (1–65 km), and for this reason they are often referred to as igneous intrusive. The Earth's interior is warm. Under these conditions, magma cools slowly enough so that all minerals in the rock grow large enough to be visible to the naked eye. A volcanic rock crystallizes from molten magma (lava) at the Earth's surface and therefore is often referred to as igneous extrusive. In this situation magma cools quickly so that minerals remain small or invisible to the naked eye. The only large minerals in a volcanic rock are those that crystallized at depth before the magma reached the surface. The rock surrounding these crystals is fine-grained. Now that we have established the difference between plutonic and volcanic igneous rocks, let us discuss some common plutonic (intrusive) rocks.

A pluton is a generic term for any single intrusive rock body. A batholith is a large body composed of many intrusions that can cover nearly an entire mountain range. The most common and most familiar plutonic and batholithic rock is granite, which is light-colored and composed of glassy quartz, white plagioclase, pink K-feldspar, and black specks of biotite. Individual minerals are large enough to be seen with the naked eye. True granite contains roughly equal amounts of quartz, plagioclase, and K-feldspar. Other granite-like plutons, such as granodiorite, syenite, and diorite, contain variable amounts of these minerals or, in some cases may lack one or two of the minerals, or may contain other minerals such as hornblende. We do not need to make sharp a distinction between these various rock types, as all are relatively light-colored, have a similar plutonic mode of origin, and a similar effect on landscape. We will refer to them collectively as granitic or granitoid.

A second, far less common type of intrusive rock is gabbro. In contrast to granitic rock, gabbro is dark-colored, without K-feldspar or biotite, and with very little, if any, quartz. It is composed primarily of the minerals plagioclase and pyroxene with or without olivine.

Metamorphic rocks form at depth below the Earth's surface, and for this reason they are often associated with plutonic rocks. The major difference is that metamorphic rocks do not crystallize from magma. Instead they form by solid-state recrystallization of preexisting rock in response to changes in temperature and pressure during burial. If a sedimentary or volcanic rock is buried, or if an intrusive rock is moved to a new depth (and therefore to new pressure-temperature conditions), the minerals in the rock may become unstable and recrystallize thereby forming a new texture and probably a new set of minerals.

The process of solid-state recrystallization is similar to baking bread. Before it is baked, dough consists of flour, yeast, sugar, salt, and water. Let's pretend that this mixture is a sedimentary rock. Nothing happens to this rock during the time it sits on the countertop. However, as soon as the rock is placed in a hot oven, the minerals begin to react to form a "metamorphic rock" that we refer to as bread. Obviously, bread has a very different texture and taste then the original dough. However, if we were to analyze both the dough and the bread, we would find that both have the same chemical composition. The only difference is the loss of water during the heating process. The original ingredients never melted; they simply reacted with each other to form bread. The reaction occurred in the solid state and resulted in a change of both texture and mineralogy. The reaction occurred because bread (not dough) is the stable "rock" at the elevated temperature in the oven.

Real rocks behave in a similar way. The original chemical composition of rock does not change during metamorphism except for the loss of water and other fluids. It is the minerals in the rock that change. The original minerals react in the solid state to produce a new set of minerals and a new texture that is stable at the new pressure and temperature.

A simple example is the reaction of calcite and quartz at high temperature to form the mineral wollastonite. Wollastonite looks nothing like calcite or quartz, but it has the same chemical composition as the two minerals combined minus the loss of CO_2. The reaction is as follows.

$$\text{Calcite} + \text{Quartz} = \text{Wollastonite} + \text{fluid}$$
$$CaCO_3 + SiO_2 = CaSiO_3 + CO_2$$

There are many types of metamorphic rocks. Slate is derived from the metamorphism of shale and looks very much like shale except it is harder. The best pool tables and the best roofing tiles are made of slate. Schist is a shiny, biotite-muscovite-rich rock derived from the metamorphism of shale or sandstone at higher temperature-pressure conditions than those of slate. Gneiss is an even higher temperature-pressure rock with well-developed light and dark bands. Gneiss can be derived from the metamorphism of shale, sandstone, or granite. Quartzite is composed primarily of crystalline quartz and is derived from the metamorphism of quartz-rich sandstone. Marble is derived from the metamorphism of limestone or dolostone. Finally, greenstone, blueschist, and amphibolite are derived from the metamorphism of the volcanic rock, basalt. As the names imply, greenstone tends to be green, and blueschist tends to be metallic blue. Amphibolite is dark green or black. Again, we need not make sharp distinctions between all the various metamorphic rocks. They can be grouped with granitoid rocks and referred to collectively as crystalline for the purpose of landscape analysis.

Volcanic Rock

Volcanic (or extrusive igneous) rocks are the final major rock type in landscape development. These are igneous rocks that crystallize from lava at the Earth's surface rather than below the surface like plutonic rocks. They are fine-grained or glassy relative to plutonic rocks. We will divide volcanic rocks into two groups, basalt (or basaltic) and silicic (or andesite-rhyolite-tuff).

Basalt is by far the most common volcanic rock. It is a dark, drab rock composed of the minerals plagioclase and pyroxene with or without green olivine. Notice that the mineralogy of basalt is the same as the plutonic rock, gabbro. If basaltic magma does not reach the Earth's

FIGURE 3.4 Photograph of a basalt flow at Lava Butte, Oregon. The basalt is approximately 7000 years old.

surface it will crystallize at depth to form gabbro. In the case of basalt, the mineralogy hardly matters because individual minerals typically are too small to see with the naked eye. Basalt forms the world's oceanic crust, and for this reason, it is the most abundant rock of any kind close to the Earth's surface. It is the rock that you would see if you visit the volcanoes of Hawaii. A small basalt flow is shown in Fig. 3.4.

When basalt erupts to the surface, it tends to be non-explosive and fluid. It is capable of flowing for distances of more than 100 miles and can form layers similar to sedimentary rock. Most eruptions, including those in Hawaii, are safe enough to observe from a fairly close distance.

Another important volcanic rock is the andesite-rhyolite-tuff family, which collectively will be referred to as silicic rock. These rocks are the volcanic equivalent of granitic rock. They have the same composition as granitic rock; the only difference is that they crystallize at the Earth's surface rather than at depth below the surface. Rapid crystallization at the Earth's surface results in a glassy texture in which individual minerals are so small they are difficult to see.

Unlike basalt, silicic volcanic rocks tend to be explosive. You do not want to be too close to these when they erupt. Both andesite and rhyolite are lighter in color, more quartz-rich, and do not flow as easily as basalt. Tuff is a name used for explosive volcanic ash that has solidified into rock. It is these rocks, and not the more fluid basaltic rocks, that form the classic, pyramid-shaped volcanic cones known as strato or composite volcanoes. Examples include Mount St. Helens and Mt. Rainier in the Central and Southern Cascade Mountains.

Unconsolidated Sediment

Unconsolidated sediment, also known as surficial material or simply as sediment, is not rock. It is the weathered, eroded remnants of preexisting rock. Much of the sediment in the United States consists of alluvium (sediment reworked, moved, and deposited by rivers), glacial drift (sediment moved and deposited by glaciers), coastal deposits (sediment moved by ocean currents and waves), eolian deposits (sediment moved by wind such as across a desert or along a shoreline), and soil (sediment weathered in place and mixed with organic material).

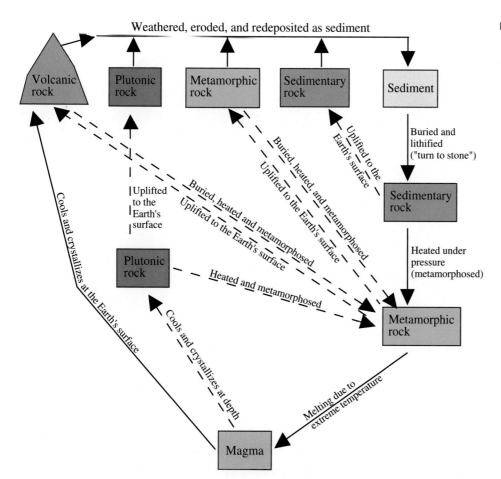

FIGURE 3.5 The rock cycle.

Unconsolidated sediment, regardless of its origin, represents sedimentary rock in the process of formation.

THE ROCK CYCLE

The three rock types, along with unconsolidated sediment, are related to each other by the rock cycle shown in Fig. 3.5. All rocks ultimately begin as magma (liquid rock) at some depth within the Earth that can vary from a few miles to 1800 miles. Magma is less dense than solid rock; therefore, once formed, it begins to rise. As it rises, it either cools at depth as a plutonic (intrusive) rock such as granite, or it reaches the Earth's surface as volcanic rock. If it crystallizes at depth, the plutonic rock could eventually reach the Earth's surface through erosion of overlying rock. Once rock is at the surface, it will be weathered and eroded, and the remains will be deposited as sediment. As more and more sediment is deposited, the bottom of the pile becomes compacted and cemented to form sedimentary rock. Sedimentary rock can then be driven deep into the Earth by tectonic processes where it is metamorphosed. As temperature increases, the rock can melt, become magma, and begin the rock cycle anew.

There are shortcuts in the rock cycle. Rather than continuing deep into the Earth and melting, sedimentary and metamorphic rocks could instead be exhumed to the Earth's surface where they would be weathered and eroded to form a second-generation sedimentary rock. A volcanic rock could be buried and metamorphosed rather than weathered and eroded. A plutonic rock could be metamorphosed rather than brought to the surface.

ROCK HARDNESS AND DIFFERENTIAL EROSION

Bedrock is a solid mass of rock that is physically connected to the interior Earth. When bedrock is exposed at the Earth's surface it is known as outcrop. Of primary importance in landscape development is the way that an outcrop reacts to weathering and erosion. Rocks that resist weathering and erosion are said to be hard, strong, or resistant. We could say that the rock has low erodability meaning that it is not easily weathered and eroded. Rock that is easily weathered and eroded is said to be soft, weak, or nonresistant. It has high erodability.

The absolute hardness or softness of a rock is important in landscape evolution, but more important is the relative hardness of two rocks that lie adjacent to each other. The interlayering of rocks with different degrees of erodability creates a landscape of differential erosion, which can be defined as a location where adjacent rocks weather and erode differently such that the more resistant rock protrudes above the less resistant rock.

The resistance of rock to weathering and erosion is largely a function of rock type and texture. Certain rocks, because of the way they are held together, and because of the minerals they possess, resist weathering and erosion better than others. One can argue that the most important characteristic of rock type in terms of landscape evolution is its absolute and relative resistance to weathering and erosion. In the next section we examine how rock hardness and texture play a role in landscape evolution.

INFLUENCE OF BEDROCK ON LANDSCAPE

It is important to emphasize that the rock type and the structural form of rock both have a strong influence on landscape. These are the components from which landscape is made. Here we discuss the influence of rock type. Structural form is discussed in Chapter 4.

If we consider each rock type as a whole, we would discover that crystalline rocks, in general, are most resistant to weathering and erosion. Volcanic rocks are less resistant, and sedimentary rocks are least resistant. We can consider this hierarchy to be a generally accurate, but keep in mind that there are many exceptions. As an example, a common situation across the United States is where crystalline rocks underlie sedimentary rocks. If part of this rock sequence were elevated, we might expect the less resistant sedimentary rocks to erode quickly off the top of the uplift, thus exposing crystalline rocks underneath. Once crystalline rocks are exposed at the surface, we might expect differential erosion to continue to wear down the relatively soft sedimentary rocks while leaving the resistant crystalline rocks as highlands. The resulting landscape is depicted in Fig. 3.6A.

In addition to considering each of the three rock types as a whole, we must also consider how individual rocks respond to weathering and erosion. The most common sedimentary rocks are limestone, sandstone, and shale. The most common crystalline rocks are granite and certain metamorphic rocks that include slate, schist, gneiss, and marble. The most common volcanic rocks are basalt lava flows, andesite-rhyolite (silicic) lava flows, and layers of silicic tuff (volcanic ash). Each is different with respect to composition and texture, and therefore different with respect to their erodability. In order to better understand landscape evolution, we must consider how weathering and erosion affects each individual rock.

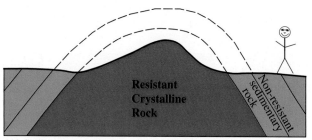

(A) Sedimentary rocks have been eroded off the top of the fold leaving resistant crystalline rock to form highlands

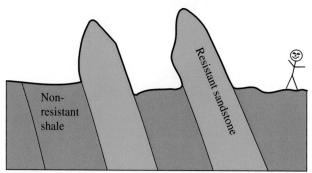

(B) Resistant sandstone protrudes above non-resistant shale

FIGURE 3.6 Differential erosion. (A) Resistant crystalline rock protruding above less resistant sedimentary layers. (B) Two resistant layers of sandstone protruding above nonresistant layers of shale.

Landscape in Sedimentary Rocks

Sedimentary rocks have two important characteristics. They are stratified (layered), and some layers are more resistant to weathering and erosion than others. Of the three most common sedimentary rocks, shale is least resistant. Limestone is resistant in arid climates but less resistant in humid climates due to the development of karst features as explained below. Sandstone is likely to be the most resistant sedimentary rock, especially in humid climates. Layer thickness is also a consideration. Thick layers of sandstone are likely to be more resistant than thin layers, especially where some of the layers are shale. When these three rocks are interlayered, differential erosion creates a landscape where the more resistant rock layers protrude above weaker layers. An example is shown in Fig. 3.6B.

The second important characteristic of sedimentary rocks is that they are capable of imparting a strong pattern to the landscape. When a layered sequence of sedimentary rocks is tilted, the eroded edge of resistant layers will form ridges. Conversely, the eroded edge of nonresistant layers will likely form valleys. Alternating ridges and valleys creates a pattern to the landscape and, as a result, the landscape appears ordered, predictable, or at least not random. There are many examples in this book, such as Figs. 13.34 and 15.10, where differential erosion of folded and tilted sedimentary layers has created a rather ordered

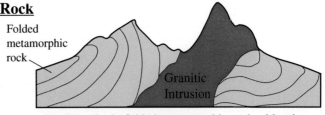

Two General Categories of Crystalline Rock

(A) Batholith composed of granitic rock without internal structure and minor metamorphic rock.

(B) Complexly folded metamorphic rock with minor intrusion of granitic rock.

FIGURE 3.7 The two general categories of crystalline rock. (A) Granitic rock. (B) Meta-morphic rock.

pattern to the landscape. Sedimentary rocks make up as much as 70% of the surface of the United States. Differential erosion between shale, limestone, and sandstone, therefore, is a major contributor to landscape.

Landscape in Crystalline Rocks

Crystalline rocks generally fit into one of two categories; batholiths, composed dominantly of granitic rock with little in the way of metamorphic rock as shown in Fig. 3.7A, and areas of metamorphic rock with only a few granitic intrusions as shown in Fig 3.7B. In both cases, the rocks are capable of producing a monotonous, somewhat random, less predictable landscape relative to what might be seen in sedimentary rocks. An explanation and examples of both categories follows.

As noted earlier, a batholith is a large area composed of many granitic intrusions. There are several batholiths in the United States that cover, or nearly cover, an entire mountain range. Examples include the Sierra Nevada Batholith, which forms most of the Sierra Nevada in California, the Idaho Batholith, which forms part of the Northern Rocky Mountains in Idaho, and the White Mountain Batholith, which forms part of the White Mountains of New Hampshire. Many batholiths originate as underground feeder chambers below a chain of active volcanoes. They are brought to the surface following extinction and erosion of the overlying volcanic chain. Metamorphic rocks are present in batholiths but are of secondary importance. Granitic batholiths tend to be homogeneous in the sense that there is no layering, no obvious folds, and few faults. The rocks are virtually structureless. Fractures (joint sets) may be the only conspicuous structure. The absence of layering is most responsible for producing random, unpredictable landscape without a consistent structural trend. Fig. 3.8 is a photograph of the Sierra Nevada Batholith near Whitney Portal, California. Note that the granitic rock is homogeneous and nonlayered. The only conspicuous trends present in the rock are vertical fractures.

In contrast to a batholith, thick sequences of metamorphic rocks are typically layered, especially if the rocks were of sedimentary origin prior to metamorphism. Layered metamorphic rocks are shown in Fig. 3.7B. If the metamorphic sequence is layered, it is capable of imparting a pattern to the landscape similar to what is seen in sedimentary rocks. Metamorphic rocks, however, have at least three characteristics that make such a pattern less prominent. The first is that the rocks may contain a complex fold pattern such that the layers do not have a consistent trend. The second is that the layering may be disrupted by granitic intrusions and, thus, not as continuous as in sedimentary rocks. The third has to do with the metamorphic process, which tends to lower contrasts in erodability between adjacent layers such that the effect of differential erosion is diminished. Shale, for example, becomes harder during the metamorphic process. Adjacent layers of similar erodability (similar hardness) will behave, from a landscape point of view, as if they are a single homogeneous layer. All three of these characteristics are depicted schematically in Fig. 3.7B. The rugged crystalline landscape of the North Cascade Mountains, shown in Fig. 3.9, is an example of a metamorphic area that is without a consistent pattern to the landscape.

The combination of resistance to erosion and absence of differential erosion within crystalline rocks results in some of the most rugged highland landscapes on Earth. A majority of the mountainous regions of the United States are underlain with crystalline rock including the North Cascades region of Washington, the Sierra Nevada of eastern California, most of the Colorado Rocky Mountains, the Teton Mountains and Wind River Range of Wyoming, the Blue Ridge Mountains in the Appalachians, the Adirondack Mountains, and most of the New England Highlands.

Landscape in Volcanic Rocks

Volcanic rocks are rather unique because they form from magma within the Earth like plutonic rocks, and are deposited at the Earth's surface like sedimentary rocks. However, unlike layered sedimentary rocks, volcanic rocks are known for their specific landforms such as individual volcanic cones and lava flows. In order for

FIGURE 3.8 Photograph of Mt. Muir (14,015 feet) on the east-face of the Sierra Nevada batholith from Whitney Portal area, California. Mt. Muir is located about one mile south of Mt. Whitney. The granitic rock is massive, homogeneous, nonlayered, and fractured. It produces random, unpredictable landscape without consistent structural trend relative to well-layered sedimentary rock.

volcanic rocks to show layering similar to what is seen in sedimentary rocks, the lava flows must be large enough and fluid enough to completely bury (flood) landscape. Such a possibility generally does not occur with the sticky, more explosive silicic (andesite/rhyolite) flows. However, outpourings of basalt are fluid enough and voluminous enough to create a layered sequence. In Chapter 17 we will discover that basalt layers several thousand feet thick cover nearly all of the Columbia Plateau.

A major difference between a layered sequence of sedimentary rocks and one of basalt is that all of the basalt layers are more-or-less the same composition. Each layer has the same relative resistance to erosion, and as a result, they do not create a landscape of strong differential erosion such as what is seen in sedimentary rocks. There are weak zones between basalt flows due to fracturing during flow and soil development between flows, but these create only a limited amount of differential erosion.

Lava flows, particularly basaltic lava flows less than 20 million years old, are often interlayered with young, poorly consolidated and, therefore, weak sedimentary rock. Because it is more resistant than sedimentary rock, a large basalt flow, or series of basalt flows, will resist erosion relative to surrounding, sedimentary layers. As surrounding sedimentary layers erode down, the resistant layer of basalt will evolve into a wide, flat, topographically high bench (a tableland or mesa) across which the basalt at the surface protects underlying sedimentary layers from erosion. An example is Grand Mesa, near Grand Junction, Colorado, reportedly the largest mesa in the world. The circa-10 million-year-old basalt that caps Grand Mesa is visible in Fig. 3.10. The basalt flowed initially into a lowland valley. Subsequent erosion of the surrounding sedimentary rocks has created the highland mesa. Young, landscape-forming volcanic rocks less than 20 million years old are common only in the Cordillera. Basaltic mesas and the origin of many other volcanic landforms in the Cordillera are discussed in Chapter 17.

Landscape in Unconsolidated Sediment

Unconsolidated sediment is widespread in all 26 physiographic provinces. It forms a thin mantle in arid climates, but can be more than 300 feet thick in humid climates. In most areas, unconsolidated sediment is draped over

FIGURE 3.9 Google Earth image looking west across the North Cascade Mountains. The complex structure in metamorphic and plutonic rock produces random, unpredictable landscape relative to well-layered sedimentary rock. Mt. Baker (a Cascade volcano) is the high peak at distant center. Mount Shuksan, a crystalline mountain, is the high peak in front and slightly to the right of Mt. Baker.

FIGURE 3.10 Google Earth image looking southeast across part of Grand Mesa near Grand Junction, Colorado. The thin, resistant layer that caps the mesa is c.10 million-year-old basalt. There is an abundance of landslide deposits located just below the mesa. The Powderhorn Mountain Resort ski area is seen at left-center.

bedrock such that the underlying bedrock still controls the shape of the land even though it is not visible at the surface. In this case, it is like throwing a sheet over a table. The table controls the shape even though the table cannot be seen. If, on the other hand, we were to completely bury the table in sand, then the sand, and not the table, would control the shape. Sediment-controlled landscape forms only in areas where sediment is thick enough to completely bury the underlying rock such that the rock exerts no influence on landscape. This concept is illustrated in Fig. 3.11. Examples include thick glacial deposits in the northern Central Lowlands as shown in Fig. 3.12, alluvial (river) deposits in the lower Mississippi River valley as shown in Fig. 3.13, coastal deposits along the eastern seaboard and Gulf coast, and thick soil on the Piedmont Plateau and Coastal Plain. None of these areas constitute an entire physiographic province.

(A) Thin layer of unconsolidated sediment (uc) draped over bedrock. Bedrock controls landscape.

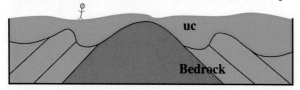

(B) Thick layer of unconsolidated sediment (uc) covers bedrock. Unconsolidated sediment controls landscape.

FIGURE 3.11 Sketch showing control of unconsolidated sediment. (A) A thin layer of unconsolidated sediment covers bedrock but does not control landscape. (B) A thick layer of unconsolidated sediment completely buries bedrock and controls landscape.

KARST LANDSCAPE

Limestone in arid environments tends to be as resistant to weathering and erosion as sandstone, or more so. Limestone, however, has another property unlike sandstone or shale. Limestone is composed mostly of calcite (calcium carbonate) and will dissolve in acidic water. Dolostone (calcium-magnesium carbonate) has similar properties, but it does not dissolve quite as readily. Rainwater is naturally slightly acidic. If limestone is present at or close to the Earth's surface, water will seep into cracks and bedding planes and dissolve the rock to produce voids, which we know as caves. As the voids get larger, part of a cave might collapse to produce a sinkhole, which is either an open hole or a pond at the Earth's surface. In this type of landscape it is common for streams to disappear into an open sinkhole, flow underground through a cave system, and return to the surface as a spring. Landscape with caves, sinkholes, disappearing streams and springs is referred to as karst landscape.

Karst landscape develops where limestone is exposed at or close to the Earth's surface. In this case, sinkholes dominate due to constant collapse of near-surface caves. The Lime Sink region of Florida between Tallahassee and Orlando is a classic example. Limestone in this region is famous for its natural springs and for development of sinkholes that have swallowed roads, houses, and anything else that just happens to be built above them. Fig. 3.14 is a Google Earth image that shows the Lime Sink area just north of Tampa. The many ponds shown in the image are sinkholes.

Sinkholes are far less common in areas where limestone is covered with an overlying surface layer of insoluble rock such sandstone or shale. The caves that form in the underlying limestone are, in this situation, protected from collapse by the overlying insoluble surface layer. Very large caves can form under these conditions. Mammoth Cave, in Kentucky, is an example. The difference can be striking between landscape with limestone at the surface and one where limestone underlies a surface layer of insoluble rock. Fig. 3.15 is a Google Earth image of an area west of Bedford, Indiana that shows an abrupt change from a sinkhole-filled terrain underlain with limestone on the right, to landscape consisting of dissected hills underlain primarily with sandstone and shale on the left.

Some of the best-developed karst landscape in the world occurs in the humid climates of the Valley and Ridge, Ozark Plateau, Interior Low Plateaus, and Florida Coastal Plain. Karst landscape is also well developed in many areas of the west including southwest Texas, southeast New Mexico, Idaho, Montana, and the Black Hills. In the majority of these areas, limestone is oriented in nearly horizontal (flat-lying) layers. One exception is the Valley and Ridge where layers in some areas are inclined at steep to vertical angles.

Limestone is the most common rock to form karst landscape, but it is not the only rock. Evaporite rocks, composed of the minerals halite, anhydrite, and gypsum, will also dissolve in water to form caves and other karst features. Evaporite caves are present in New Mexico, Texas, and the Black Hills, as well as other areas in the Cordillera. There is one area in Minnesota, southwest of Duluth, near the town of Sandstone, where karst features, such as sinkholes and springs, are developed in quartz sandstone. In this case, the quartz sandstone does not dissolve, but instead is strongly fractured.

DISTRIBUTION OF ROCK/SEDIMENT TYPE AMONG THE 26 PHYSIOGRAPHIC PROVINCES

Fig. 3.16 shows the general distribution of the three major rock types. We can see that sedimentary rocks are widespread especially in the central and southeastern part of the United States. Crystalline rocks dominate the Appalachian Mountains, are rare in the central part of the United States, and are distributed across the Cordillera. Young volcanic rocks less than 70 million years old are restricted to the Cordillera and western Great Plains. Volcanic landscape-forming rocks are located in Fig. 3.16 and discussed in Chapter 17. Older volcanic rocks in the Cordillera and in areas east of the Great Plains are

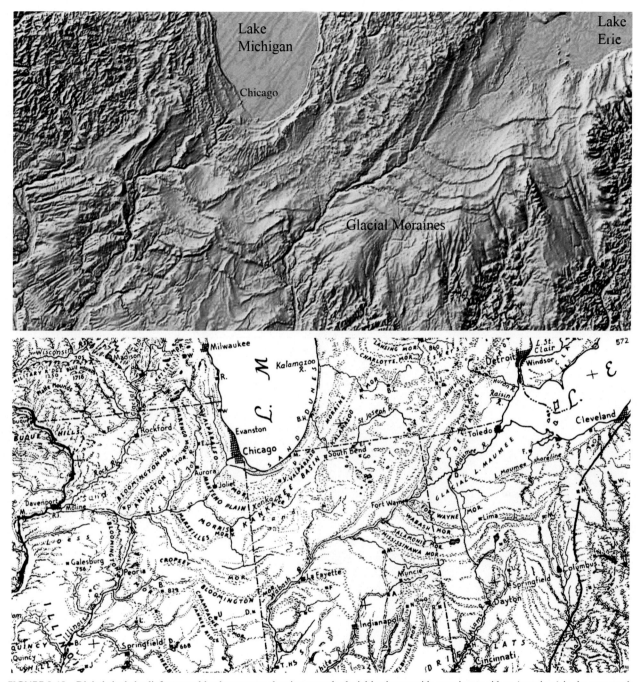

FIGURE 3.12 Digital shaded-relief map and landscape map showing smooth glacial landscape with conspicuous ridges (moraines) in the area south of Lake Michigan and Lake Erie. Bedrock topography is completely buried across most of this area. The highly dissected area at upper left is the Driftless area, which was not glaciated. Indiana state boundaries parallel north direction. The shaded-relief image is from Thelin and Pike (1991).

intermingled with sedimentary and plutonic rocks, variably deformed, or are metamorphosed. Only three of these areas are shown in Fig. 3.16. All three are east of the Cordillera and are discussed in the context of their structural setting. Volcanic rocks in the Superior Upland and Ozark Plateau are discussed in Chapter 14. Those in the Piedmont Plateau and New England Highlands are discussed in Chapter 18.

In Fig. 3.17, the boundaries of the 26 physiographic provinces, shown as thick black lines, are superimposed onto Fig. 3.16 with the underlying Raisz landform map removed. Question 24, at the end of the chapter, will ask you to study

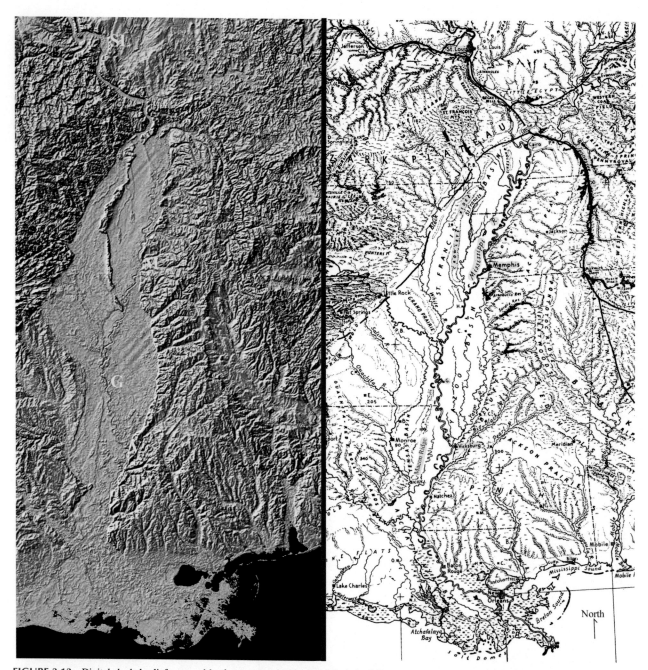

FIGURE 3.13 Digital shaded-relief map and landscape map showing the Mississippi flood plain in the Mississippi Embayment. With the conspicuous exception of Crowley's Ridge, the flood plain completely buries bedrock. St. Louis (SL) marks the confluence of the Missouri and Mississippi Rivers. The Mississippi River flows southward to Cairo (C) where it enters the Embayment, joined by the Ohio River. The Mississippi then flows along bluffs along the eastern edge of the floodplain to Memphis (M) where it turns westward forming an arc into the floodplain at Greenville (G) that loops back to the bluffs at Vicksburg (V). Southward, the river maintains its course along the eastern edge of the floodplain until it reaches the Mississippi Delta and turns eastward. Note also the large sediment accumulation along the shoreline west of the Mississippi Delta with sparse accumulation to the east. The shaded-relief image is from Thelin and Pike (1991).

Fig. 3.17 and make a list of physiographic provinces based on dominant rock type. It then asks you to use this information to color the physiographic map shown in Fig. A.3A. The exercise is meant to better facilitate a correlation between rock type and physiographic province.

Notice in Fig. 3.17 that there is a strong correlation between rock type and physiographic province in the eastern United States but not in the west. Throughout this chapter we have emphasized that rock hardness (erodability) is a primary influence on landscape development and that hardness

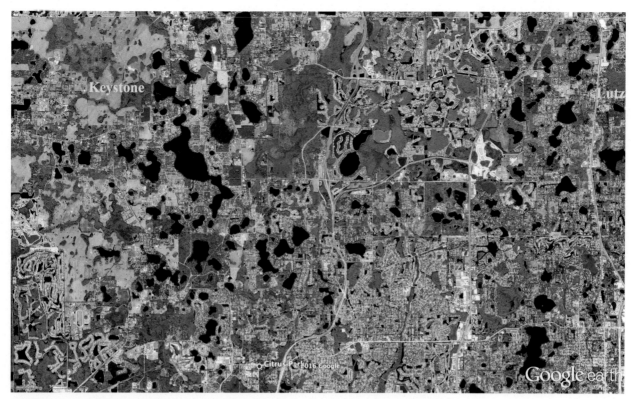

FIGURE 3.14 Google Earth image looking north at sinkholes in the northern outskirts of Tampa, Florida.

FIGURE 3.15 Google Earth image looking north at two contrasting terrains, a sinkhole-filled karst terrain underlain with limestone at right, and a hilly, dissected upland terrain underlain with sandstone and shale at left. The location is a few miles west of Bedford, Indiana.

is a function of rock type and rock texture. The correlation between rock type and physiographic province in the eastern United States implies that rock hardness and differential erosion, to a large extent; control the shape of the land. The weaker correlation in the western United States suggests that, in addition to rock hardness, the other component of landscape, the structural form, has a prominent role in shaping the land. Chapter 4 discusses structural form in more detail.

FIGURE 3.16 Landscape map showing distribution of rock types.

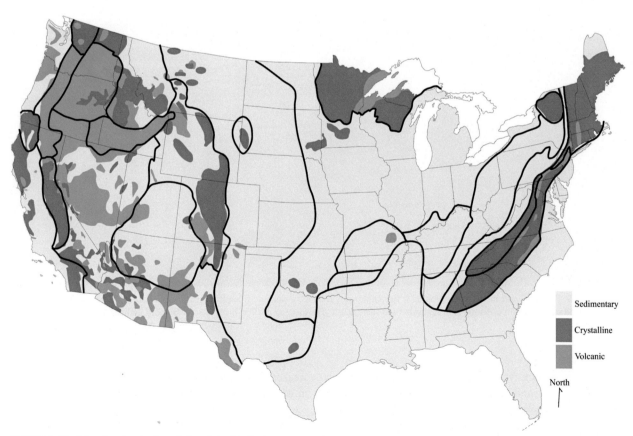

FIGURE 3.17 Map showing the boundaries of the 26 physiographic provinces (*heavy black lines*) superimposed on a colored map (Fig. 3.16) that shows the distribution of rock types.

QUESTIONS

1. What two rock types form crystalline rocks? What is their erodability relative to other rock types? In what part of the Earth do crystalline rocks form? Name a common crystalline rock.
2. What is the erodability of sedimentary rocks relative to other rock types? What does it mean that the rocks are stratified? Name the three most common sedimentary rocks. In what part of the Earth do sedimentary rocks form?
3. Of the three most common sedimentary rocks, which is least resistant? Which is composed dominantly of calcite? Which forms karst topography? Which two are resistant in arid climates? Which is likely to be most resistant in humid climates?
4. Do volcanic rocks form at the Earth's surface or at depth below the surface? Name a common nonexplosive volcanic rock. Name a common explosive volcanic rock. Which volcanic rock is fluid enough to flow great distances? What is the most common volcanic rock?
5. Of the three rock types (crystalline, sedimentary, volcanic), which is most resistant to erosion? Which forms at depth below the surface? Which typically is least resistant to erosion? Which is deposited from water or wind at the Earth's surface? Which flows as lava on the Earth's surface?
6. Name some areas of the United States where landscape is controlled largely by unconsolidated sediment.
7. Explain the following terms: Weathering, Erosion, Bedrock, Outcrop, Resistant, Nonresistant, Differential Erosion, Alluvium, Glacial Drift, Soil, marine, carbonate, clastic, and stratigraphy.
8. What is karst topography? How does it form?
9. When discussing landscape evolution, why is it reasonable to group igneous intrusive rocks and metamorphic rocks into a single category of crystalline rock?
10. An individual layer of sedimentary rock is referred to as a _____.
11. Explain how caves form in limestone.
12. When crystalline and sedimentary rocks sit adjacent to each other, why do crystalline rocks typically form highlands and sedimentary rocks form lowlands?
13. What is a plutonic rock and where does it crystallize? Name a common intrusive (plutonic) igneous rock.
14. What is a batholith, where do batholiths form, and what type of rock characterizes batholiths?
15. How does a metamorphic rock form?
16. What is the textural difference between volcanic and plutonic rocks? Are they different chemically?
17. Name three rock types that form silicic volcanic rock.
18. In which physiographic province are silicic volcanic rocks abundant?
19. Of the three major rock groups (excluding unconsolidated sediment) which is most widespread in the United States?
20. Under what conditions will the type of rock (sedimentary, crystalline, volcanic) have no influence on the shape of landscape?
21. Describe the weathering process. If weathered material remains in place it will form what?
22. Name a physiographic province where glacial landforms are well developed.
23. Name a physiographic province where alluvial landforms are well developed.
24. Use Fig. 3.17 as a guide to subdivide the 26 physiographic provinces into one of the following dominant categories: (1) Sedimentary, (2) Crystalline, (3) Volcanic, (4) Sedimentary and Crystalline, (5) Sedimentary and Volcanic, or (6) Crystalline and Volcanic. Use a copy of Fig. A.3A to color each physiographic province based on its rock category and provide a key to your map. Discuss the distribution of rocks relative to physiographic province. Which provinces contain the least variety of rock type; which are most variable? Note that the groupings are subjective. For example, the Piedmont Plateau and New England Highlands can be listed as crystalline even though a small amount of volcanic and sedimentary rocks are present. Similarly, the Colorado Plateau could be listed as sedimentary, and the Superior Upland as the only crystalline-volcanic province.
25. The three rock groups, along with unconsolidated sediment, are related to each other by the _____.
26. Regarding Fig. 3.13, explain why it is advantageous for Memphis and Vicksburg to be located where they are.
27. Name some typical marine, nonmarine, and transitional depositional environments.
28. Fig. A.5B is an uncolored version of Fig. 3.16. Make a photocopy of this figure and color each of the three rock types. Write a paragraph or two on the distribution of each rock type.
29. There are two basic volcanic rock types in landscape evolution. Name and describe some of the characteristics of both.
30. Define unconsolidated sediment

31. What word is used to describe sediment deposited from the following source: river, glacial, weathered rock, wind.
32. Using Google Earth, measure the distance from Milwaukee to Indianapolis in Fig. 3.12 in miles and kilometers, and determine the scale of the map. One inch on the map equals _____ miles. One cm on the map equals _____ km.
33. Using Google Earth, measure the distance from St. Louis to New Orleans in Fig. 3.13 in miles and kilometers, and determine the scale of the map. One inch on the map equals _____ miles. One cm on the map equals _____ km.
34. In what physiographic province is Fig. 3.15 located?
35. Research Topic: Describe the types of unconsolidated sediments that might be produced in the following environments: river, lake, delta, beach, glacial, shallow marine, tidal flat, swamp, deep marine, volcanic, reef, desert.

Chapter 4

Component: The Structural Form

Deformation is defined as any change in the orientation, position, or shape of a rock body that has resulted from stress. Changes in orientation, position, and shape are expressed respectively as rotation, translation, and strain, each of which is shown schematically in Fig. 4.1A. Stress is the intensity of force acting on rock. It is defined as force per unit area. Fig. 4.1B shows that stress can be in the form of compression, where rock is squeezed; tension, where rock is pulled apart; or shear, where one rock slides horizontally past another. In geology, we do not see stress; we see only the resulting deformation. The type of stress (compressional, tensional, or shear) is inferred based on the type of deformation. Stress creates deformation only if the magnitude of stress is greater than the strength of rock. For example, you can squeeze a rock in your hand so that it is under compressional stress, but if the rock does not break or change shape, then there is no deformation. Stresses strong enough to deform rock are imposed primarily through plate tectonic movement. Plates move horizontally across the Earth; therefore all three types of stress (compressional, tensional, and shear) typically occur in the horizontal plane as shown in Fig. 4.1B, although stresses in the vertical plane, such as above a rising granite intrusion, are also important.

The structural form (or style of rock deformation) is the geometry that the rock body assumes following deformation. In this chapter, we are interested in the limited number of structural forms that are large enough and widespread enough to impart a strong influence on landscape. The most common structural forms are folds, joints, and faults. There are three fold types, one joint type, and four fault types that strongly influence landscape. Each is listed in Table 4.1 and illustrated in Fig. 4.2. We begin this chapter with a description of each structural form.

STRUCTURAL FORM: THE STYLE OF ROCK DEFORMATION

Folds and faults are best displayed in sedimentary rock because the layering accentuates the structure.

Sedimentary rocks are originally deposited in planar layers that are close to horizontal. When deformed, these layers become inclined (no longer horizontal). The inclination is described by the strike and dip of the layer. If you were to hold a pencil perfectly horizontal on the surface of an inclined plane, the compass direction of a vertical plane that passes through the pencil would define the strike of the rock. The dip is the maximum slope of the layer. It is the angle of inclination when measured perpendicular to the strike of the plane. The dip is the direction that water flows down the inclined layer. A dip of 0 degree is horizontal. A gentle dip is between 0 and 30 degrees. A moderate dip lies between 30 and 60 degrees. A steep dip is between 60 and 90 degrees. A dip of 90 degrees is vertical. An example of the strike and dip of inclined layers is shown in Fig. 4.3.

Folds

A fold is a bend in an originally planar layer of rock. Most folds form via horizontal compressional stress, but vertical (up/down) stresses are also important. Folds large enough to form landscape are of three types, anticlinal, monoclines, and nearly flat-lying.

For simplicity, we use the category name, anticlinal, to refer to the collection of folds that includes anticlines synclines, domes and structural basins. In this category, the dip of beds on the limbs of each of these folds is sufficiently steep (usually >20 degrees) that one would readily notice that the beds are inclined. An anticline is formed when layers are bent into an upward arch such that layers dip away from a long center hinge line. Grab your shirt with two hands, one below your neck, the other near your belly button, and pull up to form an anticline. A syncline is a downward bend such that layers dip toward a long center hinge line. Lay down across a soft mattress to form a syncline. Anticlines and synclines often occur in sequence as shown in Fig. 4.2A (left side) such that they resemble the crinkling of a thin stack of paper. These folds typically form through horizontal compression. A single fold can form an elongate mountain or

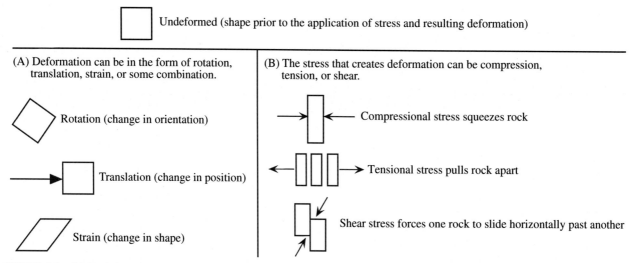

FIGURE 4.1 (A) Depiction of the three components of deformation. (B) Depiction of horizontally oriented compressional, tensional, and shear stress, and resulting deformation.

TABLE 4.1 Structural Form—Landscape-Forming Structures

1. Folds
    a. Anticlinal (includes anticlines, synclines, domes, and basins)
    b. Monoclines
    c. Nearly Flat-Lying (includes flat-lying and gently tilted rock, and broad anticlines, synclines, domes, and basins)
2. Vertical joint sets
3. Faults
    a. Normal faults (also known as block faults, gravity faults, or extensional faults)
    b. Reverse faults (these are high-angle compressional faults)
    c. Thrust faults (these are low-angle compressional faults)
    d. Strike-slip faults (also known as shear faults)

basin more than 100 miles long and 50 miles wide. It is not uncommon for normal or reverse faults to be present on the flanks of these folds. Many of the mountain ranges in the Middle-Southern Rocky Mountains have an anticlinal structural form. A dome is an upward bulge in which layers are bent into the shape of a dome. Grab your shirt with one hand and pull the shirt away from your body; the shirt will form a dome. A structural (or geologic) basin is a downward bowl-shaped indentation (or sag) such that sedimentary layers dip toward a low point near the center of the fold. A person sitting on a mattress would form a structural basin. Domes and basins can form in sequence via compressional stress similar to anticlines (Fig. 4.2A, left side), but more commonly form via vertical stress or in association with an intrusion such that beds surrounding the fold remain horizontal as shown in Fig. 4.2A (right side). Associated mountains and basins are roundish in shape and can vary in size from more than 100 miles to less than 50 miles in diameter. Examples include small intrusive domes on the Colorado Plateau such as the La Sal Mountains.

A monocline is a single, sharp, moderate to steep dip in otherwise nearly flat-lying layers. These folds are less common than the anticlinal folds, but are well developed on the Colorado Plateau. It is common for a monocline to develop above a blind (hidden) fault as depicted in Fig. 4.2B.

Nearly flat-lying layers are transitional to the previously described steeper-dipping anticlinal fold structures. In this category, layers remain at such gentle dip (usually <10 degrees) that they appear nearly flat-lying. The gently dipping layers can be in the form of a very broad anticline, syncline, dome, or structural basin. The key word here is broad. These folds can span a distance of more than 200 miles (322 km) as shown in Fig. 4.2C. An anticline can be so broad that it is referred to as an arch. Nearly flat-lying structures are widespread across the Interior Plains and

Folds (shown prior to erosion)

(A) Anticlinal folds

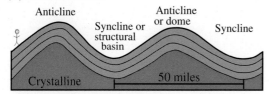

(A) Anticlinal folds

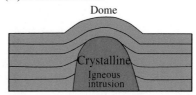

b) Monocline (underlain by a reverse fault)

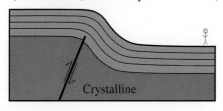

(C) Nearly flat-lying layers (deformed into broad folds)

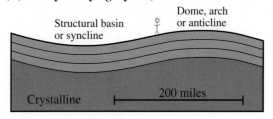

Joints and Faults (shown prior to erosion)

(D) Joint set
(evenly-spaced vertical fractures)

(E) Normal (gravity, block) faults
(steep dip near the surface, horizontal dip at depth)

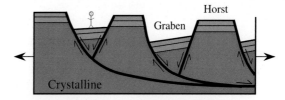

(F) Reverse faults

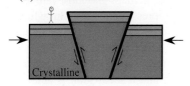

(G) Thrust fault

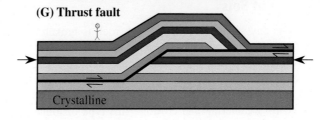

(H) Right-Lateral strike-slip fault

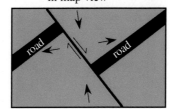

(I) Right-lateral strike-slip fault in map view

(J) Left-lateral strike-slip fault in map view

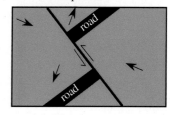

FIGURE 4.2 Structural forms. Panels A through J show each of the structural forms listed in Table 4.1. All structural forms are shown in cross-section except I and J.

Plateaus region. For example, rocks across the entire state of Michigan are down-folded into a large structural basin known as the Michigan Basin. This basin is partly responsible for the semicircular shape of surrounding Lake Michigan and Lake Huron (Fig. 3.16). In contrast to the broad folds displayed across the Interior Plains and Plateaus, nearly flat-lying sedimentary layers on the Coastal Plain tilt consistently toward the ocean. The nearly

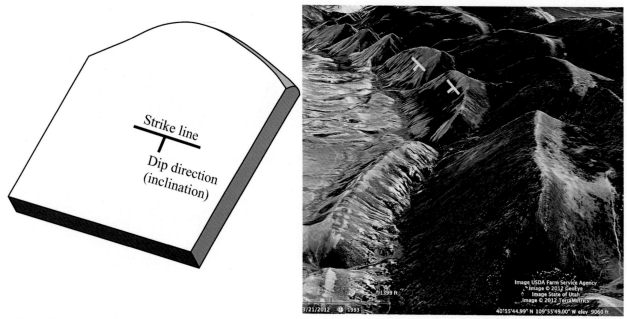

FIGURE 4.3 Strike and dip of layers. The Google Earth image is of inclined layers along the north flank of the Uinta Mountains in northeastern Utah. The symbols show the approximate strike and dip of layers.

flat-lying category represents either the original structural form of sedimentary layers without any deformation, or mild deformation typically resulting from vertical stress.

Vertical Joint Sets

A fracture is any separation along which rock is broken. The type of joint discussed here is an extensional fracture with movement perpendicular to the fracture surface such that rock on either side separates as shown in Fig. 4.2D. Separation is on the order of an inch or two, but the space is widened by weathering and erosion. This type of joint typically occurs in sets of parallel, evenly spaced, vertical, or near-vertical clusters as shown in Fig. 4.2D. They are particularly well developed in granitic rock as shown in Fig. 4.4, and in thick layers of nearly flat-lying sandstone as shown in Fig. 4.5. Extensional joint sets can form as a result of pure tension where rocks are literally pulled apart as in Fig. 4.6A, or as a result of bending or buckling due to vertical or horizontal stress as shown in Fig. 4.6B and C where the rock layer undergoes extension along the crest of a fold.

Although joint sets influence and modify many areas, they are not the primary landscape-controlling structure within any individual physiographic province. Joint sets, however, are responsible for landscape in several of our most famous national parks, including Bryce Canyon and Arches National Park, as discussed in Chapter 13.

There are other types of joints. An exfoliation joint affects primarily granitic rock. When exposed at the Earth's surface, large bodies of granite tend to peel almost like an onion. This type of joint is oriented parallel with the ground surface such that the rock breaks in sheets from a few inches to a few feet thick (they are not vertical joints). The result is a granite outcrop in the shape of a dome. The granite domes of Yosemite National Park, shown in Fig. 4.7, are perhaps the most famous. A shear joint forms via compressional stress in which vertical fractures typically form in two directions oriented about 60 degrees to each other. In this case there is a small amount of displacement parallel with the fracture surfaces, which implies that a shear joint is actually a small fault (see the next section).

Faults

A fault is a fracture along which there has been movement (displacement) of rock parallel with the fracture surface. As indicated in Table 4.1 and Fig. 4.2, there are four different types of faults that influence landscape. They are normal, reverse, thrust, and strike-slip faults.

A normal fault forms where tensional stress pulls rock apart. It is different from a joint in the sense that there is enough extension for the rock to slide down the dip of the fracture surface under the force of gravity. Thus, a normal fault is "normal" in the sense that the major displacement force is gravity. Although steeply dipping (>60 degrees) near the Earth's surface, many normal faults flatten to near horizontal dip at depth. Normal faults occur along the margins of folds, below monoclines, and wherever rocks are pulled apart. This type of faulting is also referred to as block faulting because when one side drops

FIGURE 4.4 Photograph of a closely spaced joint set in granitic rock at City of Rocks National Reserve, Idaho.

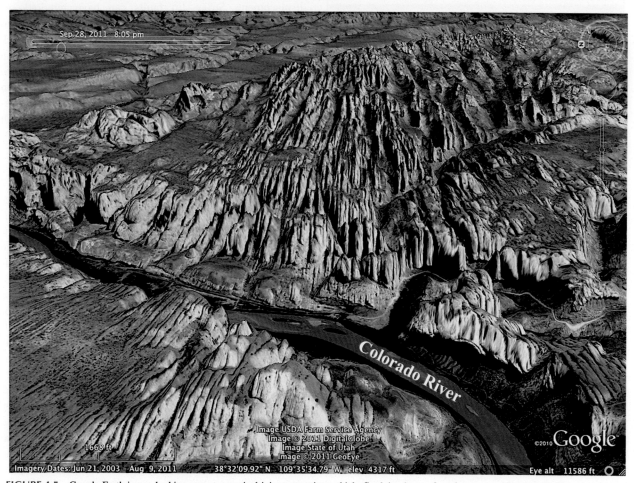

FIGURE 4.5 Google Earth image looking east at a vertical joint set cutting a thick, flat-lying layer of sandstone near Moab, Utah. Note that the vertical joint set extends across the Colorado River without deflection.

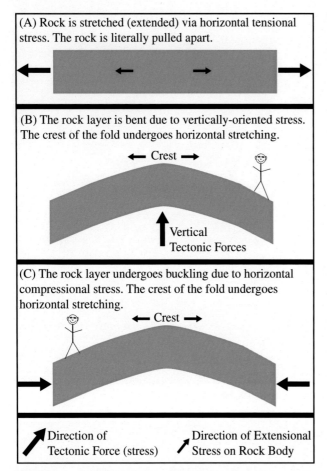

FIGURE 4.6 Vertical joint sets can form as a result of (A) pure extension, (B) extension in the crest of a bent rock layer, or (C) extension in the crest of a buckled rock layer.

down under the influence of gravity, the rock on the other side is left standing as a high isolated mountain block. Normal faults commonly occur in semi-parallel sets that form alternating valleys (basins) and mountain blocks. The down-thrown block is known geologically as a graben; the up-thrown block is a horst. Typical horst and graben structure is illustrated in Fig. 4.2E. Normal faults have a major effect on landscape and are the primary structure in the Basin and Range province where total vertical displacement on a normal fault can be more than 30,000 feet. Normal faults in The Grabens area of Canyonlands National Park on the Colorado Plateau, shown in Fig. 4.8, create a mini-version of the Basin and Range province. The process of normal faulting is described more fully in Chapter 18.

Reverse faults are the opposite (the reverse) of normal faults in the sense that they work against gravity as shown in Fig. 4.2F. This type of fault forms where rocks undergo compression. In this case, the rocks are pushed up the dip of the fracture plane against the force of gravity. Reverse faults are defined by their steep dip (usually >60 degrees). They occur along the margins of folds, below monoclines, and wherever rocks are compressed. Reverse faults rarely create enough displacement to have a major effect on landscape, but they are significant in a few areas that include the California Coast Ranges and the Middle-Southern Rocky Mountains.

Thrust faults form where rocks undergo compressional stress. They are a special type of reverse fault in which the dip of the fault alternates from near horizontal, to a dip between 30 and 45 degrees, and then back to near horizontal; a fault geometry referred to as flat-ramp-flat. The flat-ramp-flat geometry, as shown in Fig. 4.2G, results in development of anticlines above ramp areas and the stacking of older rock layers above younger rock layers. Displacement on thrust faults can be on the order of 50 miles or more implying that this type of fault is an important contributor to landscape. It is the major landscape-forming structure in the Valley and Ridge and in the eastern part of the Northern Rocky Mountains.

Shear stress produces strike-slip faults. In this case, the rock breaks along a vertical or near-vertical fracture surface and the two sides are displaced horizontally past each other as shown in Fig. 4.2H. Displacement occurs parallel to the strike of the fault typically with little associated upward (reverse) or downward (normal) dip-slip displacement. Strike-slip faults can have hundreds of miles of displacement. They are major contributors to landscape. A right-lateral strike-slip fault displaces rock to your right as you look across the fault as depicted in Fig. 4.2I. A left-lateral strike-slip fault displaces rock to your left as depicted in Fig. 4.2J. The famous San Andreas Fault is a landscape-forming right-lateral strike-slip fault. Right-lateral offset along the San Andreas Fault at Wallace Creek is clearly visible in Fig. 4.9.

FAULT REACTIVATION

It is not uncommon for high-angle normal, reverse, or strike-slip faults to undergo several periods of reactivation; meaning an active fault becomes inactive for tens or even hundreds of millions of years, and then becomes active again under a different tectonic stress regime. Faults are susceptible to reactivation because the process of faulting creates planar zones of weakness in otherwise strong rock. These planar zones are the first to fail upon the application of stress. An example of fault reactivation is shown in Fig. 4.10 with respect to the development of a monocline. In this example, a series of high-angle normal faults form under tensional stress in crystalline rocks. These faults become inactive, and following erosion to a flat plain, the faults and the crystalline rock are covered with sedimentary rock layers. One of the faults is then reactivated, this time as a reverse fault under compressional stress. Had the new stress regime been extensional or shear, the fault could have reactivated as

FIGURE 4.7 Google Earth image looking northeastward at granite domes in Yosemite National Park, California. The two prominent domes at the left are North Dome and Basket Dome. Half Dome is at right.

FIGURE 4.8 Google Earth image looking NNE at The Grabens, Canyonlands National Park, Utah. The valleys form down-dropped grabens bordered on one or both sides by normal faults.

a normal fault or a strike-slip fault. In this example, the reactivated fault does not cut all the way through the sedimentary layers to reach the surface. Instead it cuts only partway through and remains at depth to form what is known as a blind fault. Sedimentary layers not cut by the fault are draped (folded) over the blind fault producing a monocline at the surface. In other examples, reactivated faults can cut all the way to the surface. Reactivated high-angle faults are present in the California Borderland, the Middle-Southern and Northern Rocky Mountains, the Colorado Plateau, and the Interior Low Plateaus among many other areas. In some cases, the faults have undergone multiple periods of reactivation. Monoclines on the Colorado Plateau are, in many cases, underlain with reactivated blind faults. Fault reactivation can occur along low-angle faults, but is more common along high-angle faults because the application of horizontal stress, in virtually any direction, will more easily cause displacement on high-angle faults.

BRITTLE AND DUCTILE FAULTS

Temperature within the Earth increases with depth. The rate of increase with depth (known as the geothermal gradient) varies with location. Rocks heat up quickly at

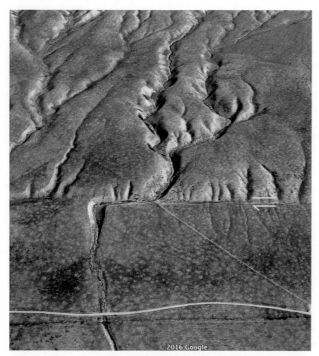

FIGURE 4.9 Google Earth image looking NNE at right-lateral offset of Wallace Creek and truncation of small streams along the San Andreas Fault at Carrizo Plain National Monument, Temblor Range, California. The fault forms a lineament (a straight, steep escarpment) that extends from left to right across the middle of the image. Relative displacement along the fault is shown with *half-arrows*.

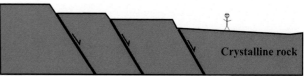

Crystalline rock undergoes a period of active normal faulting.

The faults became inactive. The crystalline rock is eroded to a flat plain and younger sedimentary rocks cover the faults. The normal faults create zones of weakness in the crystalline rock.

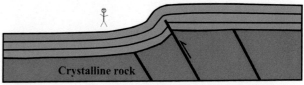

Following deposition of sedimentary layers, one of the older faults is reactivated. The fault cuts upward into overlying sedimentary layers, but does not reach the surface. A fault that does not reach the surface is known as a blind fault. Sedimentary layers are draped over the blind fault to create a monocline.

FIGURE 4.10 Series of cross-sections that show development of a reactivated fault. Sedimentary layers form a monocline above the reactivated fault in the bottom panel.

Yellowstone National Park due to the presence of magma at depth, and very slowly in central Illinois where no magma body exists. In the absence of an underlying magma body, rocks within about 10 to 15 miles of the Earth's surface are cold and brittle. When brittle rocks break during fault displacement they separate along a discrete planar fracture or zone of fractures. These rocks are visibly broken to create a brittle fault zone. Most of the faults discussed in this book are brittle faults. We will refer to these simply as faults. Rocks below about 10 to 15 miles depth are warm enough to recrystallize or partially recrystallize during deformation. The rocks deform in a ductile manner. They bend and flow such that they are not visibly fractured or broken. Instead, they appear streaked, lineated, and fine-grained relative to surrounding rocks. Such rocks are referred to as mylonite (or mylonitic). We will refer to this type of fault as a ductile fault zone or as a shear zone.

INFLUENCE OF DIPPING LAYERS ON LANDSCAPE

Any type of planar structure in any type of rock, be it a dipping rock layer, a joint or a fault, is capable of imparting a strong imprint on the land. In order to better understand how planar structures interact with landscape, we will look at geometric differences between steep to vertical-dipping planar structures, and nearly flat-lying to gently dipping planar structures.

Vertical to Steep-Dipping Rock Layers

Geometry predicts that if a vertical plane intersects the Earth's surface, it will create an absolute straight line (a lineament) across all land areas regardless of the magnitude of topographic relief. This requirement is illustrated in Fig. 4.11A, which shows a vertical plane (thick line) crossing a stream, and in Fig. 4.5, which shows vertical joints forming straight lines where they cross the Colorado River. As described by geologists, a lineament is any visible scar, any alignment of features, or any straight line that can be followed across the Earth's surface. A lineament could represent an alignment of folds, faults, fractures, volcanic features, springs, or bedding planes. A lineament could have no known explanation or could conceivably be a figment (an optical illusion) of the camera angle.

Joint sets and strike-slip faults characteristically dip at vertical to near-vertical angles. Both create narrow planar zones of fractured rock near the Earth's surface that can easily erode into straight, narrow, lowland valleys

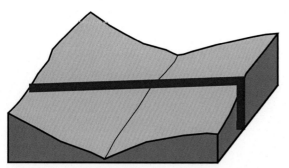

(A) A vertical layer cuts straight across the river valley.

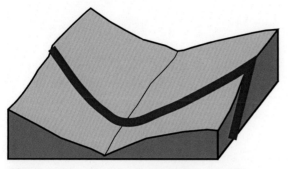

(B) A gently dipping layer deflects in the direction of dip as it crosses the river valley.

FIGURE 4.11 Dipping layers and topography. The vertical layer in (A) is not deflected. The dipping layer in (B) is deflected in the direction of dip.

(lineaments). The narrow, straight valley shown in Fig. 4.9, for example, was created by the vertical-dipping San Andreas Fault. Tomales Bay, shown in Fig. 4.12, is another example of a narrow, straight valley created by the San Andreas Fault.

Normal and reverse faults also characteristically dip steeply near the Earth's surface (usually > 60 degrees). They too are capable of forming lineaments. An example, shown in Fig. 4.8, is the series of lineaments created by normal faults in The Grabens area. Faults and fractures are not the only structures capable of producing a strong lineament. Steeply dipping rock layers characteristically produce a linear landscape that follows the strike of the inclined beds. Resistant rock layers produce distinct ridges. Weak rock layers create narrow, linear valleys. Fig. 4.13 shows a steep, straight ridge along the Colorado Front Range south of Denver created by resistant, nearly vertical layers of sedimentary rock.

Horizontal to Gently Dipping Rock Layers

In strong contrast to the outcrop pattern of near-vertical rock layers, geometry predicts that rock layers oriented horizontal or near horizontal will create a highly sinuous (curvy) path that follows topography across the Earth's surface. This requirement is illustrated in Fig. 4.11B, which shows a gently dipping plane (thick line) crossing a stream. The plane is deflected in the direction of dip where the rock layer crosses the stream. As a rule of thumb, moderately dipping layers are deflected in the direction of dip (downstream in the case of Fig. 4.11B) as they cross low areas such as river valleys.

Fig. 4.14 shows a thick, nearly flat-lying, dark layer cropping out between two thin white layers along the side of a river valley on the Colorado Plateau near Kanab, Utah. The rock units create a sinuous, zigzag pattern on the Earth's surface as they cross several small side valleys. The rock layers are deflected upstream as they cross each valley suggesting that they dip very gently toward the top of the image. In addition to rock layers, thrust faults dip at horizontal to moderate angles and therefore also form a sinuous pattern across the Earth's surface.

Response of Dipping Layers to Erosional Lowering

Geometry also predicts how a dipping rock layer will respond to erosional lowering. A vertical rock layer maintains its surface position during erosional lowering, whereas a gently dipping rock layer will migrate (retreat) in the direction of dip. This relationship is illustrated in Fig. 4.15.

Cuestas and Hogbacks

The terms cuesta and hogback are used repeatedly throughout this book so you need to understand exactly what they refer to. Both are landforms, not geologic structures. Cuesta is Spanish for slope. The term is used most commonly in the United States for a long gentle slope that ends on one side along a steep slope. In this book, we will describe a cuesta as an escarpment, steep slope, or cliff that separates two nearly flat or gently inclined areas at different elevation. Cuestas typically form along the eroded edges of nearly flat-lying sedimentary rock layers. The edge of Black Mesa, shown in Fig. 4.16, is an excellent example.

A hogback is a steep, narrow ridge that forms where steeply dipping resistant rock layers protrude above the surrounding land. The ridge in Fig. 4.13 is a hogback. Two hogbacks are shown in Fig. 3.6B.

TOPOGRAPHIC FORM AND STRUCTURAL FORM

It is important to understand the difference between the topographic form of landscape (the shape of the land) and the structural form of the underlying bedrock (the shape of

FIGURE 4.12 Google Earth image looking SSE at Tomales Bay, California. The straight, narrow valley marks the location of the vertical San Andreas Fault. San Francisco is at distant center. Point Reyes National Seashore is at right.

rock). These are two very different, independent, entities. Landscape, for example, can be described as a mountain, plain, or topographic basin regardless of the structure of rock. Structure can be described as an anticline, syncline, or structural basin regardless of the shape of landscape. Fig. 4.17 shows how any of four structural forms (flat-lying layers, anticline, syncline, structureless crystalline) can produce any of three landscape forms (mountain, valley, plain).

We must also understand the difference between a topographic high and a structural high. A topographic high refers specifically to the highest elevation of land such as the top of a mountain or hill. A structural high refers to the highest part of a particular rock layer within the structure even if the highest part of the rock layer has been eroded. The crest of an anticline, for example, is a structural high regardless of the level of erosion.

Because they refer to different aspects of the Earth's surface, landscape and structural form can be described in combination. An anticlinal mountain is one in which topographic and structural highs coincide. An anticlinal valley is one in which a topographic low is paired with a structural high. Both are shown in Fig. 4.17. Split Mountain, near the Colorado-Utah border, is an anticlinal mountain in one area, and an anticlinal valley in another, as shown in Fig. 4.18.

RECOGNITION OF ACTIVE FAULTS

Active faults are present across nearly the entire Cordillera, so we need to understand how they are recognized at the Earth's surface. A primary criterion used to locate active faults is displaced markers such as the offset stream valley in Fig. 4.9. In the absence of displaced markers, one of the best and most widespread criteria is displaced unconsolidated sediment. Most unconsolidated sediment is young, less than 2 million years old. If it is displaced along a fault, then that displacement must have occurred sometime after the sediment was deposited. An obvious goal would be to date the sediment to constrain more accurately the age of last movement along the fault.

Active faults at the Earth's surface typically form steep planar zones of broken, crushed rock. As such, they

FIGURE 4.13 Google Earth image looking north at a linear ridge of steeply dipping, resistant layers along the Colorado Front Range between Castle Rock and Littleton, Colorado. The ridge forms a hogback.

are capable of producing lineaments of various types that include linear valleys, linear scarps of steep relief, linear mountain fronts, faceted spurs, and areas of aligned features such as springs, landslides, or volcanic cones. For example, strike-slip faults, because they tend to be vertical, produce straight, narrow valleys such as the valley that marks the San Andreas Fault in Figs. 4.9 and 4.12. However, when dealing with lineaments, one must be careful because not all lineaments are necessarily related to faulting. The origin of a lineament as an active fault has to be confirmed by additional criteria such as offset unconsolidated sediment.

Let us look at some examples of lineaments associated with active faults. A linear scarp can be defined as a narrow line of abnormally steep slope. When high-angle faults break the surface they can displace topography both vertically and horizontally. In most cases, the dip of the fault is steeper than the slope of the landscape. Thus, they create a narrow line of abnormally steep slope (a linear scarp). An example from Death Valley is located at the white arrow in Fig. 4.19. Repeated vertical displacement along active normal faults (and to a lesser extent, reverse faults) will create a linear scarp that can elevate with repeated displacement to form a steep, straight mountain front. An example of a straight mountain front along an active normal fault is shown with the orange arrow in Fig. 4.19. Active faulting along the mountain front ends at the yellow arrow, and at the same time, the mountain front becomes sinuous (not straight).

Fig. 4.20 shows how faceted spurs and a straight mountain front might form along an active normal fault. Fig 4.20A shows a sinuous, low-lying mountain front prior to active faulting. In Fig. 4.20B, the mountain front is steepened and straightened due to displacement along the planar fault surface. Streams attempting to keep pace with displacement are forced to cut downward into the fault scarp, thus creating a narrow v-shaped gap through the base of the scarp as shown in Fig. 4.20B. The part of the scarp face located between stream valleys will develop into a characteristic triangular shape known as a faceted spur (f) or triangular facet. With erosion, the faceted spur retreats but maintains its triangular shape as shown in Fig. 4.20C.

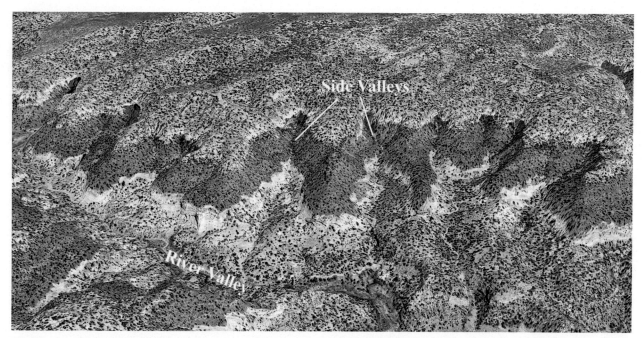

FIGURE 4.14 Google Earth image that shows the relationship between topography and gently dipping (near horizontal) beds. The image is located on the Colorado Plateau near Kanab, Utah. A dark-colored rock unit crops out along side valleys. The outcrop pattern of the dark unit follows topography creating a sinuous (curved) pattern that v's upstream as it crosses numerous tributary valleys. The rock pattern suggests that the beds dip very gently toward the top of the image.

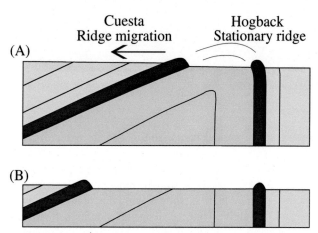

FIGURE 4.15 (A) Cross-section of a thin resistant layer (blue) between thick nonresistant layers. The rocks are deformed into an anticline with a moderate dip on the left side and a vertical dip on the right side. The resistant rock layer forms a cuesta on the left side and a hogback on the right. (B) The same cross-section following erosional lowering. The cuesta retreats in the direction of dip. The hogback remains stationary.

Fig. 4.20D is a Google Earth image of a faceted spur along the western flank of the Oquirrh Mountains, Utah.

STRUCTURE-CONTROLLED AND EROSION-CONTROLLED LANDSCAPE

In Chapter 3, we concluded that rock hardness is a primary influence on landscape and that hardness is a function of rock type and rock texture. In this chapter we have seen that structural form is also a controlling factor on landscape and that, together, these two properties form the components of landscape evolution. Given the prominence that these two components have on controlling the shape of land areas, we can informally define two special styles of landscape: structure-controlled and erosion-controlled.

A landform or landscape can be referred to as structure-controlled if the shape of the land mimics the shape of the underlying structural form. Under these conditions, topographic highs correspond with structural highs and topographic lows with structural lows. The most obvious circumstance under which structure-controlled landscape might exist is where there has been a minimal amount of erosion on newly developed or active structural forms. Fig. 4.21A shows ideal structure-controlled landscape composed of folds (left) and normal faults (right) without any erosion. The landscape exactly replicates the underlying rock structure.

A second situation that could create structure-controlled landscape is where erosion strips off layers of folded rock in sequence so that landscape continues to mimic the structure of rock even though erosion has occurred. The anticlinal mountain in Fig. 4.18 is one such example. The surface of the mountain appears to be underlain by a single folded rock layer, but note the presence of eroded rock layers surrounding the perimeter of the mountain. Such a pattern can form if the folded

FIGURE 4.16 Google Earth image looking southeast at Black Mesa, Arizona near the 4-Corners region of the Colorado Plateau. Black Mesa, on the right side, slopes gently to the right. The edge of the mesa forms an escarpment (a cuesta) that separates the mesa from a lower plateau level on the left.

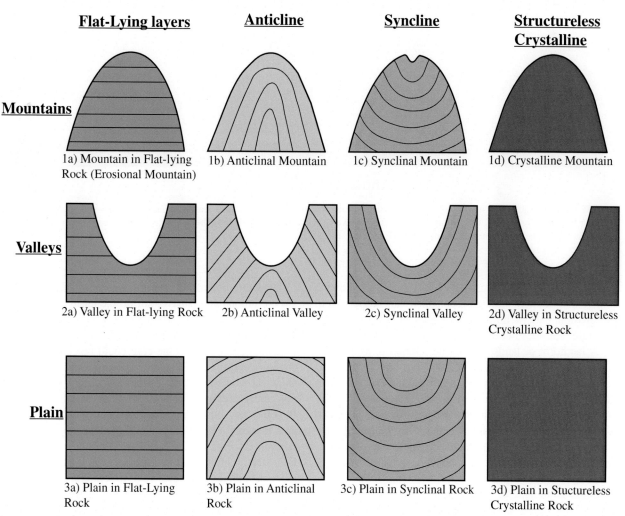

FIGURE 4.17 Series of cross-sections that show how any of four structural types (flat-lying layers, anticline, syncline, structureless crystalline) can produce any of three landscape forms (mountain, valley, plain).

FIGURE 4.18 Google Earth image looking west at the Split Mountain anticline and Split Mountain Canyon located at Split Mountain, near Dinosaur National Monument, Utah. An anticlinal mountain is developed at the top of the image. The Green River cuts through the mountain forming an anticlinal valley near the center of the image.

surface layer strongly resists erosion. Fig. 4.21B shows a folded and faulted structure-controlled landscape where the erosion pattern mimics the shape of the underlying structure, but does not exactly replicate the structure.

A landform or landscape can be referred to as erosion-controlled (or hardness-controlled) if erosion has reduced areas of soft rock, leaving only relatively hard rock to form highlands. Figs. 4.21C and 3.6B are examples. Erosion-controlled landscape is favored where differential erosion has had ample time to wear down and modify exposed rock, perhaps in the absence of tectonic activity. Under these conditions, a topographic high may not correspond with a structural high. For example, the left side of Fig. 4.21C shows synclinal mountains capped with resistant rock. The right side of the figure shows a topographic highland on the downthrown side of the fault due to the presence of a thick resistant rock layer. The resistant rock layer has already been eroded from the up-thrown side of the fault.

Structure-controlled and erosion-controlled landscapes are not mutually exclusive, nor do the terms fit every possible landscape. A flat plain is not necessarily structure-controlled or erosion-controlled. An anticlinal mountain with crystalline rock in the core could be considered both erosion-controlled, and structure-controlled. Fig. 4.22 shows how erosion of an anticlinal mountain might progress under conditions of no tectonic stress. Fig. 4.22A and B could be considered structure-controlled. Fig. 4.22D, and possibly Fig. 4.22E, could be considered erosion-controlled. Fig. 4.22F could be considered both, and Fig. 4.22C and G could be considered neither.

It is logical to suggest that structure-controlled landscape would dominate in tectonically active areas where the structural form is young and not yet strongly eroded, and that erosion-controlled landscape would dominate in tectonically inactive areas where erosion and deposition have had ample time to modify the landscape. But such is

FIGURE 4.19 Google Earth image looking north at the Grapevine Mountains (Amargosa Range), Death Valley region, California. The Northern Death Valley right-lateral strike-slip fault forms the linear scarp that cuts the alluvial fan (*white arrow*). The Grapevine normal fault is present along the linear mountain front (*orange arrow*). The fault ends where the mountain front ends (*yellow arrow*).

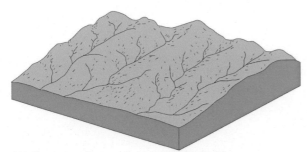

(A) Stream valleys prior to normal faulting.

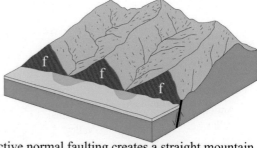

(B) Active normal faulting creates a straight mountain front. Streams cut through the mountain front leaving remnants of the scarp face to form faceted spurs (f).

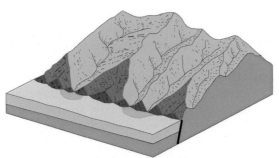

(C) Repeated fault displacement maintains a straight mountain front. Older faceted spurs are worn down and recede.

(D) Faceted spurs along the Oquirrh normal fault along the western flank of the Oquirrh Mountains, Utah.

FIGURE 4.20 Sketches that show formation of a straight mountain front and a faceted spur (f). (A) Prior to faulting. (B) and (C) During active faulting. (D) Example of a faceted spur. Based on Hamblin and Christensen (2001).

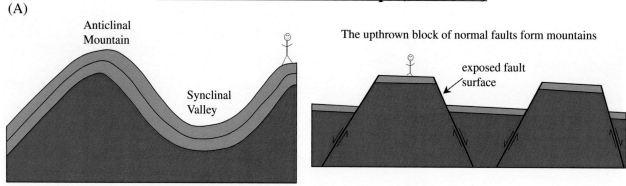

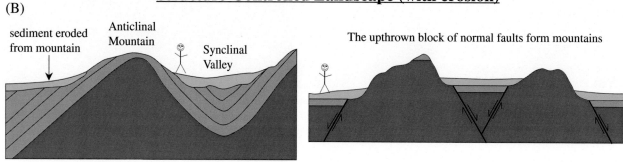

FIGURE 4.21 Structure-controlled landscape (A) with no erosion and (B) with some erosion. (C) Erosion-controlled landscape.

not necessarily the case. For example, one part of Split Mountain in Fig. 4.18 appears to be structure-controlled (the anticlinal mountain) and the other part does not. We need to emphasize that the terms structure-controlled and erosion-controlled are used strictly to describe the interaction of the two rock components with the shape of the land. They are purely descriptive terms whose sole purpose is to present the reader with a description of how landscape interacts with rock hardness and structural form. They have no other stated or implied meaning. They imply nothing about the age of landscape and nothing regarding active versus inactive tectonic stress. Such conclusions must be determined independently.

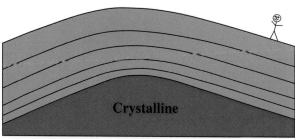

(A) Anticlinal mountain of sedimentary rock with no erosion.

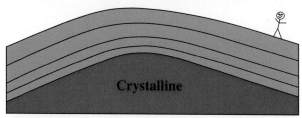

(B) Anticlinal mountain of sedimentary rock in which the top sedimentary layer has been eroded.

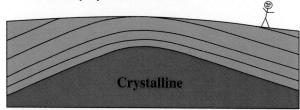

(C) Erosion to a nearly flat plain in folded sedimentary layers.

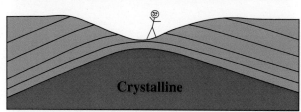

(D) Erosion to form an anticlinal valley in sedimentary rock.

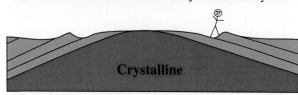

(E) Erosion to a near flat plain across an anticline cored with crystalline rock and with cuestas.

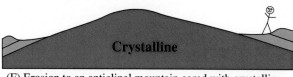

(F) Erosion to an anticlinal mountain cored with crystalline rock.

(G) Erosion to a nearly flat plain in crystalline rock.

FIGURE 4.22 Panels (A) through (G) show how erosion of an anticlinal mountain might progress under conditions of no tectonic stress.

QUESTIONS

1. Describe the three components of deformation.
2. Describe the three types of stress: compression, tension, and shear, and the different types of faults that result from each.
3. What is the strike and dip of a rock layer?
4. Why is a normal fault sometimes referred to as a gravity fault?
5. Sketch a cross-section of an anticlinal mountain, an anticlinal valley, and a mountain composed of flat-lying sedimentary rock.
6. Draw the following structures: syncline, anticline, monocline, normal fault (label a horst and graben), reverse fault, thrust fault.
7. Explain with the aid of sketches how a monocline can form from either compressional or tensional stress.
8. This type of fault typically has vertical dip.
9. What type of joint characterizes the granite domes at Yosemite National Park
10. What is a lineament? Name various geological features that can create a lineament. Why do strike-slip faults typically form lineaments?
11. How does one distinguish a brittle fault from a ductile fault?
12. Describe various criteria to recognize an active fault.
13. Name the type of fault where one side has moved to the right relative to the other side.
14. What is the name for a fracture in which displacement is perpendicular to the fracture surface?
15. What is a cuesta? What is a hogback?
16. Use a protractor to measure the angle of dip of the hogback in Fig. 4.13.
17. What is the strike of the hogback in Fig. 4.13? Use the north arrow at upper right of image.
18. What is the strike of the vertical joints in Fig. 4.5? Use the north arrow at upper right of image.
19. What is the difference between a topographic high and a structural high?
20. Describe how landscape composed of tilted sedimentary layers might be different from landscape composed of massive granitic rock.
21. Describe with a sketch how a ridge composed of steeply dipping beds might erode differently from a ridge composed of gently tilted beds.
22. Describe how landscape with a vertical fault might be different from landscape with a gently tilted fault or bedding plane (compare Figs. 4.13 and 4.16).
23. In Google Earth, fly to the following areas and interpret the nearby features: Leadville, Colorado, Woodland Park, Colorado, Big Pine, California, Mount Tobin, Nevada, Dixie Valley, Nevada. Use the U.S. Geological Survey, 2006, Quaternary fault and fold database as a guide (https://earthquake.usgs.gov/hazards/qfaults/).
24. Explain how the mountain shown in Fig. 4.22 could erode from panel A to B, to D, and then to F.

Chapter 5

Forcing Agent: The Tectonic System

The Earth's landscape is in a constant state of change due to the interaction of internal forces as exemplified by the tectonic system and external forces as exemplified by the climate system. The tectonic and climatic systems are the primary agents that force landscape to change. Together they create the components that form landscape and they control the rate, intensity, and longevity of the mechanisms that exact change on landscape. Without these forces of nature, the Earth would be an unchanging, stagnant planet. Sea level change and isostatic adjustment, the other two forcing agents, are a consequence of the tectonic-climate interplay.

This chapter provides a short introduction to the fundamental characteristics of the tectonic system and an overview of the present-day tectonic setting of the United States. We noted in Chapter 4 that the movement of tectonic plates is the primary cause of stress in Earth. A critical aspect of the tectonic system with respect to landscape in the United States is the distinction between the Cordillera where active tectonic stresses within the past 17 million years have been strong enough to change the structural form of rock, cause volcanism, and cause uplift/subsidence of land areas, versus areas outside the Cordillera where tectonic stresses are weak such that there has been very little volcanism, mountain uplift, or changes in structural form for hundreds of millions of years. However, before we discuss tectonics, now is a good time to define and summarize interrelationships among the four forcing agents.

THE FOUR FORCING AGENTS

In Chapter 1 we outlined the components, forcing agents, and mechanisms of landscape evolution, and noted that any change in landscape can be recognized by changes in elevation, relief, drainage pattern and drainage density. In Chapters 3 and 4 we introduced the two components of landscape, the rock/sediment type and the structural form. In this chapter, and in the following three chapters, we introduce the two primary forcing agents of landscape evolution, the tectonic and climate systems, and the two secondary agents, isostatic adjustment and sea level change. However, before we discuss each forcing agent individually, we need to point out interrelationships among the four agents and to list the mechanisms that are activated by each agent. We do this at the outset because this information otherwise would be lost or difficult to find if presented individually. Fig. 5.1 is a more detailed version of Fig. 1.9 that summarizes the information discussed in this section. For clarity, only first-order processes in landscape evolution are shown in Fig. 5.1. There are many feedback processes that are not shown. For example, a change in sea level brought on by climate change could alter ocean currents, which in turn affect climate. For now, we will ignore such feedback processes because they are ultimately put into motion by one of the two primary forcing agents, climate and tectonics.

Tekton, the Carpenter, the Builder

The tectonic system refers to the movement of tectonic plates. The energy to drive the tectonic system emanates from the Earth's interior. The surface expression of the tectonic system is uplift, subsidence, and volcanism of land areas, and as such, it is the carpenter, the builder of Earth. The interaction of two adjacent tectonic plates moving in different directions creates stresses that are strong enough to fold and fault large masses of rock and thus change the structural form of rock. These are the only stresses on Earth that can be applied rapidly enough and sustained long enough to create a mountain range.

Plate interaction also creates thermal anomalies. The interior Earth, to a depth of 1800 miles (2897 km), is a nicely layered sequence of primarily solid rock. If left without a tectonic system, there would be no impetus for the transfer of heat or the melting of rock. It is the tectonic system, the interaction of moving tectonic plates, that perturbs the interior Earth. The subduction of one plate beneath another pushes cold rock deep into the Earth; the breaking apart (rifting) of tectonic plates causes warm rock to rise toward the surface. Both disturb the perfect layering of Earth, creating volcanic, plutonic, and

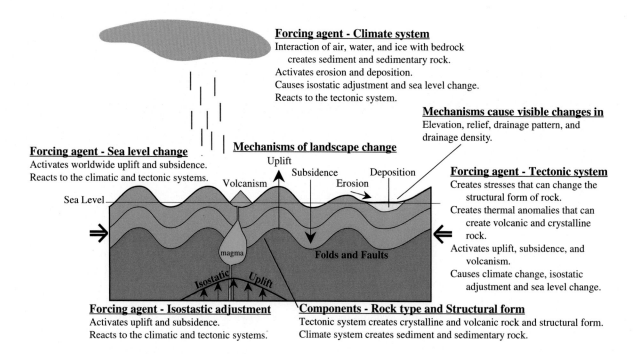

FIGURE 5.1 Sketch showing components, forcing agents, mechanisms, and visible changes to landscape.

metamorphic rocks, and ultimately resulting in uplift, subsidence, and volcanism of land areas.

The tectonic system is the strongest and the most independent of forcing agents. Given a lowland area of low tectonic stress such as the Central Lowlands, it is unlikely that any change in climate, sea level, or isostasy, no matter how drastic, can increase tectonic stress enough to create mountains. An active tectonic stress, on the other hand, can cause significant isostatic, climatic, and sea level changes particularly at time scales of hundreds of thousands to millions of years. Tectonic stress, for example, can displace large masses of rock and cause temperature changes in the Earth's crust and upper mantle, both of which result in isostatic changes. The uplift of mountains and the movement of tectonic plates can block air and ocean currents, which in turn, will alter climate. The subduction and rifting of tectonic plates, and the initiation of oceanic volcanism, can change the size and depth of ocean basins. As the size of an ocean basin changes, water is displaced onto or off continents, which causes sea level to change.

As summarized in Fig. 5.1 the tectonic system activates three of the five mechanisms of landscape formation and change: uplift, subsidence, and volcanism. It is also responsible for the creation of plutonic and metamorphic rocks, and for changes in the structural form of rock. The tectonic system is independent and largely unaffected by changes in climate, isostasy, or sea level, but it can cause significant change to each of these forcing agents. It is the tectonic system that is most responsible for building elevation and relief.

Climate, the Sculptor

Climate is the long-term condition of the atmosphere. It affects only the Earth's surface. The interaction of air, water, and ice with bedrock results in the weathering and erosion of bedrock and subsequently the deposition of weathered rock as sediment and sedimentary rock. The combination of weathering, erosion, and deposition reduces elevation and relief by lowering highlands and raising lowlands. Climate modifies and shapes landscape, and in doing so, it is the sculptor of Earth.

The tectonic system, in a sense, feeds the climate system. Without an active tectonic system to vertically and horizontally displace rock, the climate system would have no competition and, through erosion and deposition, would have long ago worn landscape down to a flat surface. There would be no mountains and few if any hills. Much of the land area would be at sea level or below a shallow ocean perhaps only a few feet deep. There would be no crystalline rock exposed at the surface, no volcanic rock, and very little if any sedimentary rock—only unconsolidated sediment. The Earth would literally be a rock in space covered in dirt and water. The climate system would have nothing to do on a flat Earth except to storm around frustrated at the absence of exposed bedrock to erode. Whereas the tectonic system is largely responsible for building landscape, the climatic system is most responsible for modifying and flattening landscape. The tectonic and climatic systems compete with each other, and in doing so they drive the rock cycle (Fig. 3.5).

Fig. 5.1 summarizes the effect of climate at the Earth's surface. We have already noted that mountain building and the movement of tectonic plates can result in climate change. Conversely, the climate system, on its own (without a preexisting active tectonic system), has little if any first-order effect on the tectonic system. This is not to say that climate has no effect on tectonics or the rate at which tectonic activity occurs. It is possible, for example, for climate-driven erosion-deposition processes to affect rates of tectonic deformation. But these are second-order effects that occur only in the presence of an active, preexisting tectonic regime. The point here is that climate and climate change alone cannot activate (or initiate) the tectonic system. It cannot cause the movement of tectonic plates.

In terms of other forcing agents, climate can cause both isostatic and sea level change. Erosion and deposition of material redistributes weight, thereby causing (initiating) isostatic changes. Climate affects ocean temperature, and because of the huge volume of ocean water, even small changes in temperature can cause enough thermal expansion or contraction to change sea level. Changes in long-term ocean deposition rates could conceivably also affect sea level by changing ocean basin volume. Climate change in the form of glaciation has a profound and lasting effect on landscape. The accumulation and melting of glaciers can cause hundreds of feet of both sea level change and isostatic uplift/subsidence over short time periods of hundreds to thousands of years. In terms of mechanisms of landscape change, climate activates erosion and deposition, which flattens landscape and results in the creation of sediment and sedimentary rock.

Although tectonic and climate processes compete with each other to change elevation and relief, their effect over time is often different. Tectonic processes such as earthquakes, volcanism, and uplift/subsidence of land areas are periodic in nature and can be absent for long periods of time. Climate-driven processes, such as erosion and deposition, on the other hand, are unceasing and are always active to some degree regardless of the tectonic condition. The comparison reminds us again of the tortoise and the hare. When tectonic processes are active, they are capable of creating landscape at a rate faster than climatic processes can tear them down. In this case, the hare outpaces the tortoise and mountains are born. But climate-driven processes are unceasing even in areas where tectonic processes lose strength or lie dormant. As the hare sleeps, the tortoise slowly gains ground. The mountain loses elevation through erosion and is eventually flattened. The tortoise wins the race. But should the hare awaken, the mountain will rise again.

Isostasy, the Equalizer

Isostasy is perhaps the least appreciated of the forcing agents, but it is extremely important in landscape development. We will explain isostasy in Chapter 7. For now, we can say that isostasy, like tectonic activity, is a process that emanates from the Earth's interior, and as such, is capable of activating uplift and subsidence.

Isostasy is the great equalizer. If weight is added to the Earth's crust, the crust sinks. If weight is removed, the crust rises. Tectonic stress and climate are both capable of redistributing weight and, therefore, both cause isostatic changes. Tectonic stress can displace rock and change the temperature of rock (and therefore its density). Climate activates erosion and deposition and can cause glaciation. Given its dependence on climate and tectonic stress, we will consider isostasy to be a secondary forcing agent. Sea level change can also redistribute weight and thus also cause isostatic changes. Conversely, it is possible for isostatic changes to affect sea level. We will ignore these possibilities in Fig. 5.1 because such changes ultimately result from climate and tectonic processes.

As summarized in Fig. 5.1, isostasy reacts to tectonic and climatic processes by activating uplift and subsidence. Although it is possible for isostatic changes to occur in the absence of a climate system (through tectonic processes), isostatic changes cannot activate erosion and deposition without an active climate. Isostasy, in the absence of climate, has no influence on erosion and deposition.

Isostasy affects landscape in much the same way as tectonic stress. That is, through uplift and subsidence. Because of this, it is sometimes difficult to separate uplift/subsidence brought on by isostatic changes from those brought on directly by tectonic activity. This problem is addressed in Chapter 7.

Sea Level, the Baseline

Sea level is the Earth's readily available baseline. It is an easily located surface from which other measurements, such as elevation, can be made. We noted earlier that sea level could change dramatically in response to changing climate (glaciation), and more slowly in response to the tectonic system. For this reason, we will consider sea level to be a secondary forcing agent dependent on the climate and tectonic systems. Sea level, like climate, affects only the Earth's surface. If the baseline for elevation is sea level, than rising sea level represents a lowering of worldwide average elevation, which in turn could cause transgression of seawater onto a continent. Conversely, the lowering of sea level exposes land, thus increasing worldwide average elevation. Thus, as summarized in Fig. 5.1, sea level reacts to tectonic and climatic processes, and can activate two of the five mechanisms of landscape change, worldwide uplift and subsidence. It is possible for sea level change to affect rates of erosion and deposition, but these are second-order effects that occur ultimately in response to changes in the tectonic or climatic systems. Let us now turn to a more detailed discussion of the tectonic system.

THE TECTONIC PLATE

Plate tectonic theory states that the outer layer of Earth, known as the lithosphere, is broken into brittle tectonic plates that move slowly as a semirigid unit over a weak, partially melted layer of the mantle known as the asthenosphere. Energy for the tectonic system is provided by the outward flow of heat from the hot center of Earth and by heat-producing radioactive decay of elements within Earth. The theory evolved over several decades and was accepted by most of the geological community in the late 1960s on the basis of newly discovered data on the structure and composition of ocean basins. Today, with the advent of satellites and global positioning systems, present-day rates of tectonic plate movement are easily measured. The interaction of these plates as they move toward, away, or laterally past each other produces most of the world's earthquakes and volcanoes and is ultimately responsible for the rock type and the structural form that underlies individual landscape provinces. The tectonic stresses and thermal anomalies generated by displacement of tectonic plates play a key role in the development of present-day landscape.

The tectonic system results from the interaction of the top three layers of Earth: the crust, the solid upper mantle, and the asthenosphere. The uppermost layer, which includes the land surface, is the Earth's crust. The crust that forms continental areas is different from the crust that forms ocean basins. In continental areas the crust averages between 18 and 30 miles thick (30–48 km) and is granitic in composition. Oceanic crust averages between 2 and 4.5 miles thick (3–7 km) and is basaltic in composition. Basaltic (oceanic) crust is dense and heavy when compared to an equal volume of granitic (continental) crust.

The solid upper mantle underlies both continental and oceanic crust. It is composed of peridotite, which is a heavy, magnesium-rich, silica-poor rock that is rare at the Earth's surface. Like the Earth's crust, the solid upper mantle is thicker below continents than below ocean basins. Continental crust and oceanic crust are both embedded into the solid upper mantle and together these two layers form the rigid, brittle lithosphere. In plate tectonic theory, the lithosphere is the tectonic plate. Or, to put it more accurately, the lithosphere is broken into sections, each of which represents a tectonic plate.

Continental lithosphere is, on average, between 90 and 100 miles thick (145–161 km), but can be thinner or can be as much as 175 miles thick. Oceanic lithosphere typically is less than 50 miles thick. Continental and oceanic lithospheres are both underlain by the asthenosphere, which is a partially melted part of the mantle (1%–2% partial melt) that acts like a lubricated surface upon which the lithosphere (the tectonic plate) can slide.

A single tectonic plate can consist of both oceanic and continental lithosphere. The eastern United States and the western Atlantic Ocean are both part of the same North American tectonic plate. Fig. 5.2 is a cross-section that shows the transition from continent to ocean within the North American plate. When we introduced the Coastal Plain in Chapter 1 we noted that part of the Coastal Plain was currently below sea level as the continental shelf. In Fig. 5.2, the continental shelf is the shallow portion of the ocean underlain by continental crust. The edge of the continental shelf is marked by the continental slope and rise, which is the offshore region where the depth of the ocean basin increases relatively abruptly from a few hundred feet to several thousand feet. The transition from continental to oceanic crust within a single tectonic plate coincides topographically with the slope and rise. Note the change in thickness between continental and oceanic crust (and lithosphere) at this location even though they are part of the same North American plate. In this figure, the plate boundary with the Eurasian plate is located in the middle of the Atlantic Ocean at the Mid-Atlantic ridge (right side of figure). Notice also that oceanic lithosphere becomes progressively thinner with proximity to the Mid-Atlantic ridge plate boundary.

PLATE BOUNDARIES

Worldwide, there are seven large tectonic plates and many additional smaller ones. As plates move, they interact with adjacent plates such that two adjacent plates will diverge (move away from each other), converge (move toward each other), or slide horizontally past each other (known as shear or transform). Because each plate is rigid and moves as a single unit, most of the world's earthquakes, rock deformation, plutonic activity, metamorphism, and volcanism occur at or near plate boundaries. Plate collision is a major driving force in mountain building.

Fig. 5.3 is a map (top diagram) and cross-section (bottom diagram) that shows all three plate boundaries. When two plates diverge, they create a void that is filled with rising magma. As the magma cools, it attaches itself to the two diverging plates thereby creating oceanic crust that enlarges the size of both plates. In Fig. 5.3, plates B

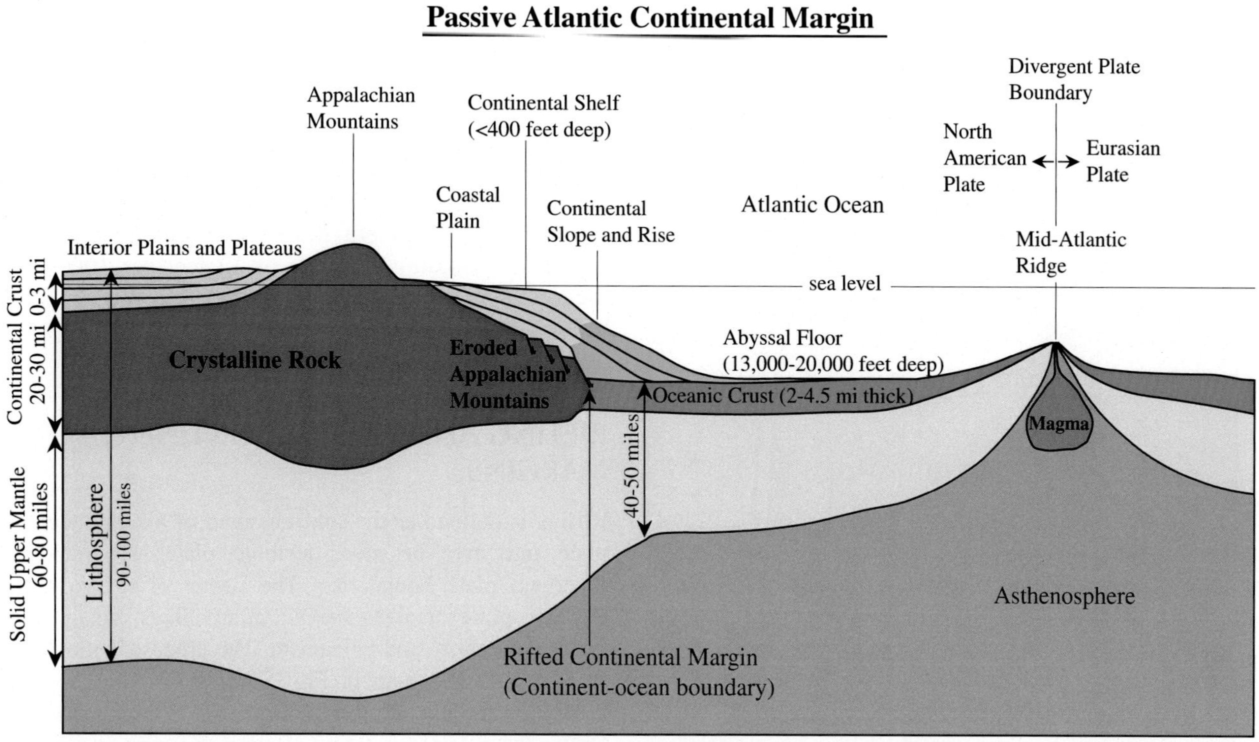

Passive Atlantic Continental Margin

FIGURE 5.2 Cross-section of the United States passive Atlantic continental margin. The North American continental crust and the western Atlantic oceanic crust reside on the same (North American) tectonic plate.

and C are diverging from each other. The Mid-Atlantic ridge, as shown in Fig. 5.2, is a divergent plate boundary that separates the North American plate from the Eurasian plate.

Convergence of two plates often results in the sinking (subduction) of one plate beneath the other. The subducting plate is consumed and, therefore, becomes smaller. Interestingly, this process can enlarge the overriding plate if material is scraped off the subducting plate and attached (accreted) to the overriding plate. The concept of tectonic accretion is discussed in more detail below. In Fig. 5.3, plate B is converging with and subducting below plate A. Gravity predicts that the heavier (denser) plate is subducted.

A transform fault is a strike-slip fault that also forms a plate boundary. In its purest form, a transform boundary conserves mass because nothing is added or removed as one plate slides horizontally past another. However, in many cases, there is a certain amount of divergence or convergence between plates. Divergence along a transform could potentially create subsidence, normal faulting, and volcanism. Where volcanism occurs, the transform is referred to as a leaky transform. Convergence along a transform creates folds, thrust faults, tectonic uplift, and mountains. In Fig. 5.3, transform faults are shown in map view (top diagram) between plates B and C, and between plates B and A.

MOVEMENT OF TECTONIC PLATES

When tectonic plates move, they do not move across the Earth in a line as if pushing a book across a table. The Earth is spherical, which requires that plates rotate about a pole of rotation that emanates upward from the center of Earth to intersect the Earth's surface. The location of a pole at the Earth's surface is referred to as an Euler pole (pronounced *oiler* pole). The Euler pole for plate B is shown in Fig. 5.3. An Euler pole can be used to determine both the direction and velocity of movement of one plate relative to another. Because the plate is rotating about an axis, the relative velocity, and therefore the rate of divergence and convergence, increases with distance from the Euler pole. This phenomenon is illustrated in Fig. 5.3 with dashed lines emanating from the Euler pole. Notice that the lines become farther apart with distance from the pole.

Because a transform plate boundary is, by definition, a location where two plates are sliding past each other, the orientation of a transform is approximately (or ideally) parallel with the relative movement direction of the two plates at that location. Arrows and half-arrows in Fig. 5.3 show the direction of plate movement. Note that these arrows are approximately parallel with the orientation of transforms. In an ideal situation, a transform forms a broad, curved arc consistent with rotation about the Euler

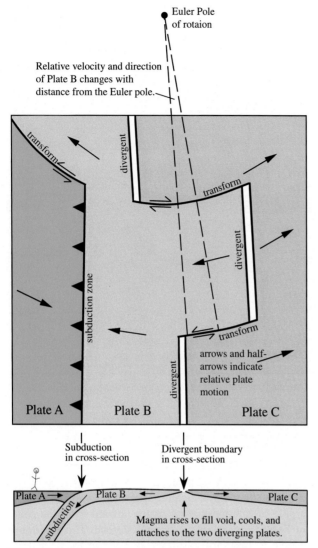

FIGURE 5.3 Map (*top diagram*) and cross-section (*lower diagram*) that shows interaction of the three plate tectonic boundaries.

million years, a tectonic plate can move into an entirely different climate regime. Plate movement can block wind and ocean currents potentially resulting in global climate changes. We will discover in Chapter 8 that the closing of the Panama Seaway between the Caribbean Sea and the Pacific Ocean resulted in a change in ocean currents to the point that conditions became favorable for our recent northern hemisphere glaciation. If two plates are colliding at a rate of 33 feet per 100 years, then 62.5 miles of Earth must be destroyed every 1 million years.

RIFTING AND PASSIVE CONTINENTAL MARGINS

Rifting is defined as the splitting apart of a single tectonic plate into two or more tectonic plates separated by divergent plate boundaries. The rifting of a continental tectonic plate creates normal fault valleys, small tilted block mountains, and volcanism. The process is illustrated in Fig. 5.4. The map in Fig. 5.4A shows the continent breaking initially in three directions, forming what is known as a triple junction. Below the map is a cross-section that shows the lithosphere stretching, upwarping, breaking along normal faults, and becoming thinner. In many cases, only two of the three arms of the triple junction develop into a divergent plate boundary. The third arm that does not go to completion is known as a failed rift or aulacogen (Fig. 5.4B).

The map in Fig. 5.4B shows that the once single tectonic plate has now split into two tectonic plates separated by a divergent plate boundary and a narrow ocean basin. As the two plates diverge, the warm asthenosphere rises and partially melts creating basaltic magma that attaches to the two plates and enlarges the ocean basin. The ocean basin will continue to widen as the two plates to drift apart. With distance from the divergent plate boundary, the continental margin will become cooler, collect sediment, and isostatically sink to create a passive continental margin.

A passive continental margin is one in which continent and adjacent ocean basin resides on the same tectonic plate. Passive continental margins are developed on both sides of a divergent plate boundary as a result of rifting as indicated in Fig. 5.4B. The absence of a plate boundary at the continental margin implies that the continent moves in tandem with the adjacent ocean basin such that tectonic stresses generated between the continent and ocean basin are minimal. The stresses are not of sufficient magnitude to modify or change the structural form of rock or to elevate land areas into a mountain range. There are few earthquakes and no active volcanoes. Slow subsidence

pole. If one were to draw a series of lines emanating perpendicular from a transform fault, those lines would intersect at a point, which is the Euler pole. For example, the two dashed lines drawn in Fig. 5.3 intersect at the Euler pole. Note also in Fig. 5.3 that divergence and convergence of two plates is not necessarily head-on, but more normally includes a component of shear (transform) displacement.

Tectonic plates move at rates that vary from less than 6 feet per 100 years to 66 feet per 100 years (1.83–20.1 m/100 years); and these rates may have been faster in the ancient past. At an average rate of 33 feet per 100 years (about 10 cm/year), a tectonic plate can move 62.5 miles (about 100 km) in 1 million years. Such rates seem slow, but over the course of several

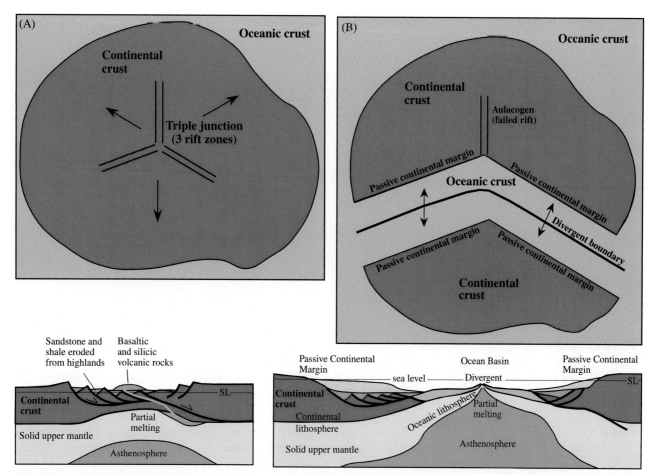

FIGURE 5.4 Rifting in map view and in cross-section. (A) The map shows initial development of a triple junction. The cross-section below the map shows bulging and stretching of the crust and development of normal faults and volcanism. (B) The map shows development of a divergent plate boundary between two continental fragments as well as a failed rift. The cross-section shows development of a passive continental margin on both sides of the divergent plate boundary. Based on Van der Pluijm and Marshak (2004, p. 398).

and wave erosion over many millions of years will flatten and bevel the continental margin, especially in the face of rising sea level. The result is a broad coastal lowland and a wide continental shelf underlain with sedimentary layers that tilt gently toward the ocean. The US Atlantic seaboard, as seen in Fig. 5.5, is a passive continental margin. Note the wide continental shelf and the location of the Mid-Atlantic Ridge (the divergent plate boundary, MAR) in the middle of the Atlantic Ocean. We discuss development of the Atlantic passive continental margin later in this chapter.

ACTIVE CONTINENTAL MARGINS

An active (as opposed to passive) continental margin is one in which the transition from continent to ocean basin coincides with a plate boundary. In this case, one plate is either subducting beneath another, or the two plates are sliding past each other along a transform fault. The interaction of two or more plates moving in different directions creates stresses and thermal anomalies in the Earth strong enough to produce volcanism, metamorphism, earthquakes, intense deformation, and areas of rapid uplift and subsidence. Tectonic stresses generated at plate boundaries can modify or change the structural form of rock and create mountains.

The subduction of one tectonic plate beneath another produces several geologic landforms and phenomena that include an oceanic trench, an accretionary prism, a forearc basin, earthquakes, and a mountain range composed of explosive volcanic cones. Fig. 5.6A is a cross-section that shows the connection between tectonics and landscape. A narrow oceanic trench, usually less than 50 miles wide and up to 36,000 feet deep (more than twice the normal depth of an ocean basin), marks the location where one plate subducts beneath another. Friction between the two plates at the trench causes part of the subducting oceanic lithosphere to be scraped off and added to the underside of the overriding plate, a process known as underplating. The underplated rocks are strongly deformed and, in some cases,

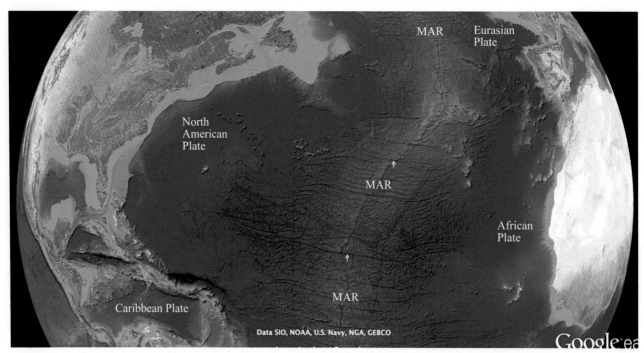

FIGURE 5.5 Google Earth image looking north that shows the location of the Mid-Atlantic ridge (MAR) relative to the eastern North American seaboard. Notice that the Mid-Atlantic ridge is offset along transform faults (*yellow arrows*) which are oriented approximately perpendicular with the ridge system. The *light* color surrounding the continents is the continental shelf.

metamorphosed, producing a jumbled, chaotic, thrust faulted mélange of accreted oceanic rock known as an accretionary prism (or accretionary wedge). Most of the accretionary prism material remains below sea level; however, the underplating process produces a wedging affect that causes uplift and tilting that can elevate part of the accretionary prism above sea level. The process of underplating is shown in Fig. 5.7.

As the subducting oceanic lithosphere sinks into the mantle, it begins to break apart producing earthquakes that can extend as deep as 435 miles (700 km). Subducting lithosphere is the only tectonic phenomenon that produces earthquakes deeper than about 44 miles (70 km). Earthquakes occur at various depths within the subducting plate creating a string of earthquakes known as the Benioff zone, which collectively define the angle at which the plate is subducting (Fig. 5.6A). As the subducting plate heats up, it releases fluids that cause part of the mantle above the subducting slab to melt. The rising magma eventually reaches the surface to produce an arc of explosive silicic volcanic cones. Underplating and uplift at the accretionary prism, coupled with inland development of a volcanic arc, produces a lowland area between the two topographic highs. This lowland is referred to as the forearc basin (meaning basin in front of the volcanic arc).

Transform plate boundaries displace lithosphere horizontally such that ideally there is no divergence, convergence, or vertical displacement and therefore no subduction, mountains, or volcanism. However, because rocks are broken and crushed during displacement, there is the potential for strong, shallow (<44 miles deep) earthquakes. Transform faults typically are oriented vertical or near vertical close to the surface such that the primary landscape feature is a strong lineament usually in the form of a narrow valley. The Mid-Atlantic Ridge is offset along transform faults that create lineaments at the bottom of the Atlantic Ocean. Two lineaments are located with yellow arrows in Fig. 5.5. As noted earlier, slight divergence or convergence along a transform will cause vertical displacement in the form of normal fault valleys or thrust faulted and folded mountains.

The US Pacific seaboard is an active continental margin marked by the San Andreas Fault in California and the Cascadia trench in Washington, Oregon, and northern California. The San Andreas Fault is a right-lateral transform fault that separates the Pacific Plate from the North American plate. The Cascadia trench marks the location where the small oceanic Juan de Fuca plate is subducting below the North American plate. Characteristic features of an active continental margin are mountains near the coastline and a narrow continental shelf. Both features result from uplift associated with plate interaction and both are characteristic features of the US Pacific coastline. We discuss development of the Pacific active continental margin later in this chapter.

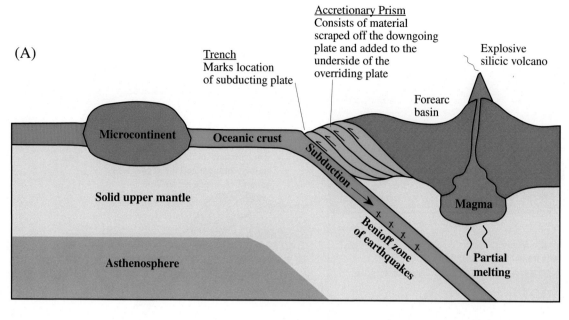

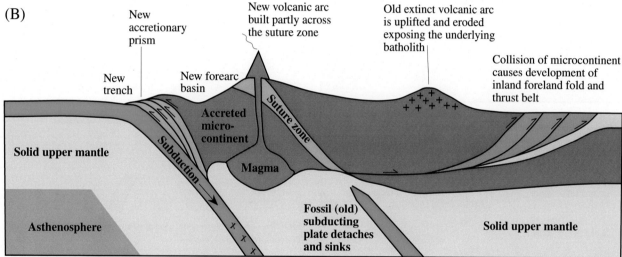

FIGURE 5.6 Cross-sections that show subduction characteristics and accretion. (A) Characteristics prior to accretion of a microcontinent. (B) Characteristics following accretion of a microcontinent.

TECTONIC ACCRETION

The term accretion (or tectonic accretion) is used for any process (including underplating) that transfers material from the subducting plate to the overriding plate. The process of accretion, in effect, enlarges the overriding plate. In addition to small fragments of oceanic lithosphere, it is possible for a large terrane to be accreted to the edge of the overriding plate. The large accreted terrane may be in the form of a volcanic island, a volcanic seamount chain, an oceanic plateau, a divergent spreading center, or a small continental mass known as a microcontinent. These large, coherent blocks accrete to the edge of the overriding plate because they are too buoyant to subduct.

Accretion of a large lithospheric mass to an existing continent is a major driving force in mountain building. The ultimate accretion event occurs when two large continental plates collide. In this case, nether plate will easily subduct. The result is continental collision and development of a high mountain chain. The Himalayan Mountains are a present-day example of continental collision between India and Asia beginning about 54 million years ago. The Appalachian Mountains also culminated with continent—continent collision as discussed below.

Accretion stuffs the subduction zone forcing active subduction to either jump to the oceanward side of the accreted terrane as shown in Fig. 5.6B, or to flip direction as shown schematically in Fig. 5.8. In either case, the

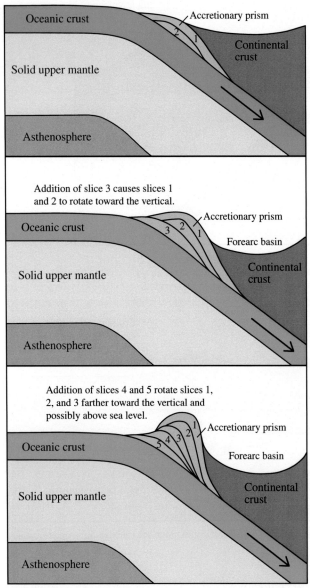

FIGURE 5.7 Series of cross-sections that show the process of underplating.

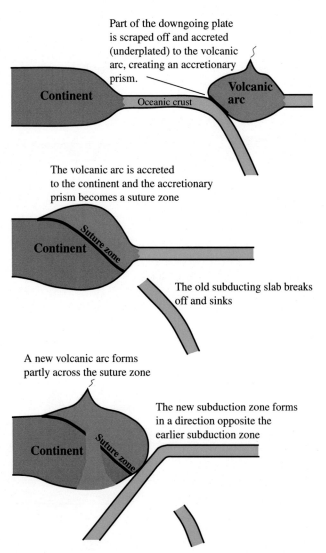

FIGURE 5.8 Cross-section sketch showing a switch in subduction direction following accretion.

accreted terrane is welded onto an enlarged overriding plate and the old intervening accretionary prism becomes a suture zone. A suture zone, therefore, separates what were once two independent tectonic plates. It can be defined as an ancient, inactive accretionary prism (an inactive subduction zone) that marks the location where a preexisting oceanic basin was completely subducted and destroyed. Suture zones are recognized in the field as a jumbled, chaotic zone composed of sheared blocks of oceanic rock surrounded by fine-grained oceanic sediment. Both the blocks and the fine-grained oceanic sediment could be metamorphosed to various degrees. The entire rock sequence (the ancient accretionary prism) is referred to as a mélange zone. The Cordilleran and Appalachian Mountain systems were built in part by collision of accreted terranes, each originally separated by a suture (mélange) zone. In the Cordillera, most of Washington, Oregon, and California, and part of Nevada and Idaho consist of accreted terranes. In the Appalachians, most of the Piedmont Plateau and New England Highlands, and part of the southern Blue Ridge consist of accreted terranes.

Note the spelling of the word terrane, which is different from terrain. Terrane is used in geology to signify an area of similar geologic history distinct from the geologic history of surrounding areas. A tectonic terrane is separated from surrounding areas by one or more suture zones. The word terrain is used to signify a physical stretch of land. It has geographic meaning but no unique geologic meaning.

OROGENY

Orogeny (or orogenesis) derives from the Greek *oros*, which means mountain, and *genesis*, which means origin or mode of formation. The term mountain building implies that the rate of surface uplift is greater than the rate of erosion such that, over time, a lowland area evolves into a mountain system. Orogeny refers specifically to deformation imposed during mountain building. Although mountains form in a variety of ways, most geologists associate orogeny with continental-size mountain systems that develop along an entire continental margin as a result of the convergence and accretion of two or more tectonic plates. Such a compressional mountain system is known geologically as an orogenic belt (or orogenic system). An orogenic belt is a deformational belt. The energy for orogeny is derived from horizontal compression, gravity, heat, and climate, particularly climate-driven erosion. Together, these energy sources disturb the thermal structure of the lithosphere and create stresses strong enough to produce folds, thrust faults, strike-slip faults, normal faults, metamorphism, granitic intrusion, isostatic adjustment, mountain uplift, and if one plate is pushed (subducted) beneath the other, explosive silicic volcanism.

Deformation associated with an orogenic event does not necessarily occur across an entire mountain range, but rather may affect only one part of a mountain belt or may propagate from one area to the next to affect different parts of a mountain at different times. Deformation in any one area and during a single orogeny can occur as a series of short pulses of stress separated by periods of quiescence that can last several million years during which time the rocks could undergo metamorphism, isostatic adjustment, volcanism, intrusion, and erosion. The rocks can be folded or deformed multiple times during a single orogeny. Orogeny is exacerbated through the addition of accreted terranes. Rather than subduction alone, it is the sheer mass of an accreted terrane that creates mountains when they enter a subduction zone and resist subduction. The Appalachian and Cordilleran Mountain belts have both undergone several orogenic events during the Phanerozoic eon (meaning they occurred less than 541 million years ago, Fig. 1.10), each lasting 20–30 million years more or less.

Orogeny ends usually with erosion of the orogenic mountain and deposition of undeformed sediment above the eroded remnants of deformed rock. A depositional contact that separates deformed rocks below from less deformed rocks above is known as an unconformity (described in the following section). Mountain systems can undergo multiple orogenic events that are distinguished most easily where they are separated by unconformities. Each orogenic event is given a proper name so that it can be distinguished and discussed.

UNCONFORMITIES

An area that is below sea level typically undergoes sediment deposition. If the area is elevated above sea level and becomes dry land, then deposition will cease and erosion is likely to ensue. Following erosion, land may again sink below sea level and sediment deposition will again ensue, thus covering and preserving the erosion surface. An unconformity is a depositional contact between two rocks of different age across which there is gap in the history of deposition. In most cases, as in the example just described, an unconformity is an erosion surface. It represents what was once the surface of Earth.

Unconformities can be used to date orogenic events. Fig. 5.9 shows a possible sequence by which an unconformity can form. The numbers indicate the age of rock layers (in millions of years ago). The first panel shows deposition of older rocks. The second panel implies that the older rocks underwent a period of nondeposition during which time the rocks were faulted, folded, exposed as land, and eroded. The 240 million-year-old rock layer was eroded and removed from the rock record. Given the degree of deformation, we can suggest that deformation was associated with orogeny. The third panel shows subsequent deposition of younger sedimentary rocks, thus creating an unconformity. In this example, the period of time not represented by rock can be bracketed between 270 million years ago (the age of the youngest remaining rock below the unconformity) and 110 million years ago (the age of the oldest rock above the unconformity). Folding, faulting, and erosion must have occurred between 270 and 110 million years ago. Unconformities do not directly date the time of folding and faulting (orogeny) because part of the time gap records erosion following orogeny. We can narrow the time during which orogeny occurred by dating sedimentary rocks derived from erosion of the rising orogenic mountains (and deposited elsewhere), and by obtaining radiometric dates on the age of metamorphism and intrusion (if any).

THE ATLANTIC PASSIVE CONTINENTAL MARGIN

The US Atlantic seaboard is a passive continental margin because continent and ocean are both part of the North American plate. As expected, it is tectonically quiet. There are few earthquakes, no volcanoes, and no active mountain building. Currently, the North American plate is

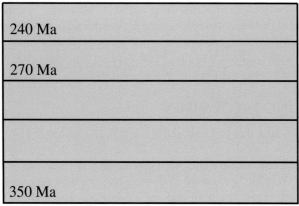

Deposition from 350 to 240 million years ago beneath a shallow ocean.

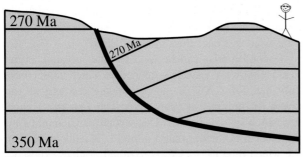

Rocks are faulted, folded, uplifted, and exposed as land area. The youngest (240 Ma) rocks are removed (eroded) from the rock record.

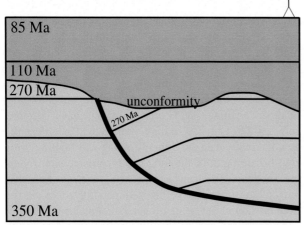

Deposition of overlying rocks at 110 Ma creates an unconformity. The 240 Ma rocks have been lost from the rock record. Existing rocks above and below the unconformity bracket the time of faulting, folding, uplift, and erosion to be after 270 Ma and before 110 Ma.

FIGURE 5.9 Sequence that shows the formation of an unconformity. Numbers represent the age of rock units in millions of years. Notice in the third panel that the underlying land surface was not completely eroded to flatland prior to deposition of younger rock units.

moving southwestward at an approximate rate of 8.5 feet per 100 years. Slow subsidence, wave erosion, and deposition are primarily responsible for development of the Coastal Plain and the wide continental shelf.

The plate tectonic events that created the passive continental margin include orogeny, accretion, uplift of the Appalachian and Ouachita Mountains, and subsequent rifting and development of a divergent plate boundary at the Mid-Atlantic Ridge. Orogeny and terrane accretion began about 480 million years ago and culminated between 335 and 265 million years ago with continental collision. There have been no fewer than five orogenic events in the Appalachian Mountains (note that numeric ages are approximate). They are the Middle to Late Ordovician Taconic orogeny (480–446 Ma), the Late Ordovician to Middle Silurian Salinic orogeny (446–423 Ma), the Late Silurian to Middle Devonian Acadian orogeny (421–387 Ma), the Late Devonian to Early Mississippian Neoacadian orogeny (385–345 Ma), and the Middle Mississippian to Middle Permian Alleghany orogeny (335–265 Ma). We can add the Ouachita orogeny to this list, which affected the Gulf coast region, and which occurred at about the same time as the Alleghany orogeny.

Given the timing for each of the five orogenic events, it would appear that orogeny was more-or-less continuous from about 480–265 million years ago. Such a broad statement, however, is not exactly true because each orogeny affected different areas with different levels of intensity. For example, the Alleghany orogeny was not felt, or was lightly felt across western New England. The Salinic and Acadian orogenies were not widely felt in the southern Appalachians. Thus, while some areas underwent orogeny, other areas experienced quiet deposition or erosion.

The Alleghany and Ouachita orogenies culminated with continental collision of North America, Europe, Africa, and South America to form a single worldwide supercontinent known as Pangea (meaning all Earth or all lands). Fig. 5.10 shows the configuration of Pangea during final collision. One can see that the US Appalachian Mountains were formed primarily by collision with Africa, that South America was involved in the Gulf of Mexico, and that Europe and Greenland were involved in Canada. North America straddled the equator and was rotated clockwise from its present orientation. The Atlantic Ocean did not exist. The collision created the Appalachian Mountains as one of the largest and highest mountain ranges to have ever existed on Earth. The mountain belt stretched from Greenland to South America covering all four continents. In the United States, the combined Appalachian-Ouachita Mountain system covered all of the present-day Appalachian-Ouachita Mountain region, the Coastal Plain, and the Atlantic continental shelf. In discussions that follow, we will refer to the collection of orogenic events that

FIGURE 5.10 Sketch that shows the supercontinent Pangea c.265 million years ago. Note that the southern part of the United States was on the equator and rotated clockwise at this time. Based on Chernicoff and Whitney (2007, p. 29).

created the Appalachian-Ouachita Mountains as the Appalachian-Ouachita orogeny.

The existence of the supercontinent Pangea was short-lived. Rifting of Pangea beginning about 237 million years ago resulted in the separation of North America from other continents, the opening of the Atlantic Ocean, and the development of the Atlantic passive continental margin. Actual separation of North America from Africa, South America, and Europe occurred below what is now the Coastal Plain and continental shelf. About half of what were once the Appalachian Mountains, and most of what were once the Ouachita Mountains, underwent erosion and subsidence as a result of rifting. This part of the mountain belt is now buried beneath rocks of the Coastal Plain and continental shelf. The exposed remnants of the orogenic belt are the Appalachian Mountains, the Ouachita Mountains, and a small area in southwest Texas known as the Marathon region. Collectively, these remnant regions form the Appalachian-Ouachita orogenic system.

It is important to realize that the Appalachian-Ouachita orogenic system has remained largely tectonically unaltered since 265 million years ago. The modern-day structural form present in the rocks had already developed prior to 265 million years ago. The only major tectonic episodes since 265 million years ago have been small granitic intrusions in the White Mountains of New England, and development of normal fault-sedimentary/volcanic valleys in New England and the Piedmont Plateau known as the Triassic Rift Valleys or the Triassic Lowlands. But even these tectonic episodes are ancient. The intrusions are more than 100 million years old, and the Triassic Rift Valleys developed between 237 and 174 million years ago during rifting of Pangea and initial opening of the Atlantic Ocean. Most of the rocks on the Coastal Plain and continental shelf were deposited following rifting during slow subsidence associated with the drifting of North America away from the Mid-Atlantic Ridge. Although older rocks are present at depth, most of the exposed sedimentary rocks on the Coastal Plain are less than 130 million years old. These rocks have remained largely undeformed since deposition.

As noted in Fig. 5.4, the rifting of a continent can result in development of a failed rift, an aulacogen. Aulacogens create weak zones in the middle of a continent that often subside and develop into a major river valley. North America prior to Appalachian-Ouachita orogeny and the formation of Pangea underwent a failed rifting event about 525 million years ago. Subsidence associated with this failed rift has created the Mississippi Embayment where the Coastal Plain swings northward almost to St. Louis (Figs. 1.5 and 3.13). The Mississippi River follows the failed rift zone through the Mississippi Embayment. Buried faults that were active during this ancient rifting event will periodically reactivate. The most significant reactivation produced the highly destructive New Madrid earthquakes of 1811 and 1812.

The Interior Plains and Plateaus are composed of nearly flat-lying sedimentary layers, most of which have never experienced Appalachian or Cordilleran orogenic mountain building and deformation. Many of these rocks were present during mountain building, but were far enough inland from the collision zone to have remained undeformed or to have undergone only mild deformation throughout the mountain building process.

THE PACIFIC ACTIVE CONTINENTAL MARGIN

The western seaboard of the United States, as shown in Fig. 5.11, is an active continental margin that involves three tectonic plates: the North American plate, the Pacific plate, and the Juan de Fuca plate. The boundary between the Pacific and North American plates in California is a transform plate boundary marked by the San Andreas strike-slip fault. Along this fault, the Pacific plate is moving northwestward relative to the North American plate at an approximate rate of 16.4 feet per 100 years (5 cm/year). A small piece of California, including Los Angeles, is west of the San Andreas Fault. San Francisco is east of the fault. If present-day plate motions continue, Los Angeles will slide

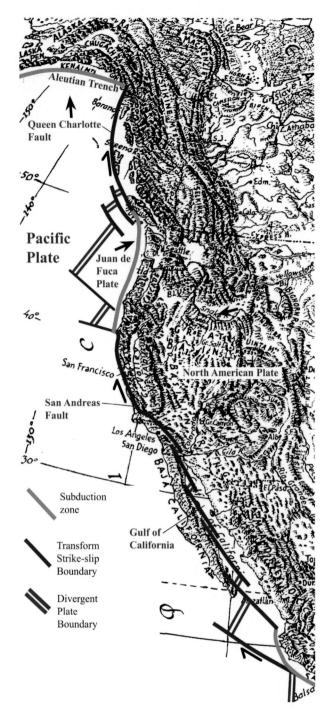

FIGURE 5.11 Map of western North America showing the plate tectonic configuration.

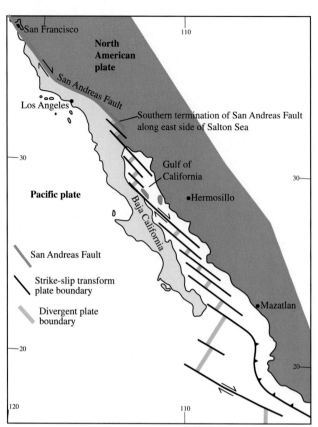

FIGURE 5.12 Map showing the tectonic setting of the Gulf of California. Based on Wallace (1990, p. 76).

northward and will reach the city of San Francisco in about 12 million years.

The San Andreas Fault extends from the vicinity of Cape Mendocino southward to the Salton Sea. Fig. 5.12 shows the southern termination. The San Andreas transform plate boundary continues southward as a series of transform fault segments within the Gulf of California that connect small divergent plate segments. The transforms have collectively forced the rifting (the separation) of Baja California from Mexico resulting in the opening of the Gulf of California. The Gulf began to open only 5–6 million years ago and is the most recent rifting event to affect the United States. Prior to 6 million years ago, Baja California was part of Mexico and the Gulf of California did not exist. Over time, the Gulf will continue to open and Baja California will slide northward with Los Angeles, eventually reaching San Francisco and beyond.

A small tectonic plate, referred to as the Juan de Fuca plate, is present north of the San Andreas Fault off the northern California-Oregon-Washington coastline. This plate is moving northeastward relative to North America along a convergent plate boundary such that the Juan de Fuca plate is subducting beneath the North American plate at the Cascadia trench at an approximate rate between 9.8 and 13.8 feet per 100 years. Subduction of the Juan de Fuca plate causes melting and magma generation in the mantle, which rises to the surface to create the Cascade volcanoes.

As shown in Fig. 5.11, Juan de Fuca plate ends in southern Canada and the Cascadia trench is replaced by the Queen Charlotte transform fault. Much like the San Andreas Fault, the Queen Charlotte is a right-lateral transform that separates the North American and Pacific plates. However, unlike the San Andreas, it is located mostly offshore. It extends to Alaska where the Pacific-North American plate boundary makes a 90-degree turn allowing the Pacific plate to subduct beneath the North American plate at the Aleutian trench.

Notice that landscape along the Oregon-Washington coast shown in Fig. 5.11 (or Fig. 1.3) mimics the landscape shown in cross-section in Fig. 5.6B. Both consist of coastal mountains, an inland valley, and an inland volcanic mountain range. This landscape is a direct consequence of subduction of the Juan de Fuca plate. California has a similar landscape except that the inland mountain range (the Sierra Nevada) is not volcanic and the plate boundary is the San Andreas transform fault rather than a subduction zone. In Chapter 20 we will discover that the Juan de Fuca plate was once part of a larger plate known as the Farallon plate. For more than one hundred million years prior to initiation of the San Andreas Fault (at 29 million years ago), the Farallon plate underwent subduction beneath California. The California landscape is, in part, a relict of this ancient subduction zone setting.

Features that characterize an active continental margin include frequent earthquakes, recent volcanism, areas where folds and faults are actively forming, and rapid uplift/subsidence. Such features are expected to be present within about 200 miles inland of an active continental coastline. However, in the case of the United States, we find evidence to classify the entire Cordillera as an active tectonic landscape, a distance inland of more than 1000 miles. Every state in the Cordillera has experienced active faults, earthquakes, volcanism, and tectonic uplift/subsidence within the past 1 million years. There are several probable causes for active deformation so far inland. Prior to initiation of the San Andreas Fault, the Farallon plate was undergoing subduction beneath the North American plate along the entire US Pacific coastline. Beginning about 80 million years ago, rather than subducting directly into the mantle at a (normally) steep angle as shown in Fig. 5.6B, there is evidence that the Farallon plate subducted at such a shallow angle that it underplated the Cordilleran crust as depicted in Fig. 5.13. The underplated slab remained coherent all the way to Wyoming and Colorado. The forcing of this slab across the Cordillera was the likely impetus that elevated the Middle-Southern Rocky Mountains from sea level beginning 75 million years ago as discussed in Chapter 14. Additionally, the eventual sinking of this slab into the mantle was the likely

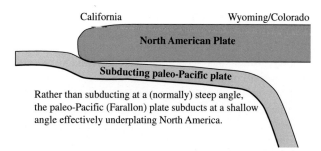

FIGURE 5.13 Schematic cross-section showing shallow subduction (underplating) of the Paleo-Pacific (Farallon) plate beneath the North American plate.

impetus that created volcanism across the entire Cordillera as discussed in Chapter 17.

There are additional reasons for an active tectonic landscape across the Cordillera. We will discover in Chapter 18 that active normal faults are present, not only in the Basin and Range, but in all 12 of the Cordilleran provinces. One probable reason for normal fault activity is the present-day interaction of the Pacific and North American plates. Earlier, we noted that the North American plate is moving southwestward relative to the Pacific plate at an approximate rate of 8.5 feet per 100 years, and that the Pacific plate is moving northwestward relative to the North American plate at an approximate rate of 16.4 feet per 100 years. The result of this interaction is that the fast moving Pacific plate is diverging slightly from the North American plate. As the Pacific and North American plates diverge slightly, the remnants of the underplated Paleo-Pacific (Farallon) plate may be acting as a mechanical couple that helps drag the North American plate westward with the Pacific plate. This westward pull stretches the North American plate, thus creating normal faults. A contributing factor is volcanism associated with the sinking of the Farallon plate and with the Yellowstone hotspot. The heat from volcanic activity softens and weakens the interior Cordilleran lithosphere allowing it to deform more readily.

Even though it has been under active tectonic development for hundreds of millions of years, we will discover in Part II of this book that much of the present-day Cordilleran structural form and landscape has developed during only the past 40 million years, and in many places, within the past 17 million years. The modern Cascade volcanic arc, for example, began to develop only 48 million years ago, the San Andreas Fault initiated 29 million years ago, and widespread normal faulting in the Basin and Range, volcanism in the Columbia Plateau and Snake River Plain, and strong uplift of the Colorado Plateau and Middle-Southern Rocky Mountains all occurred within the past 17 million years. Tectonic change has been so complete that landscape that existed prior to 17 million years ago has mostly vanished from the Cordillera. Such recent tectonic change is in contrast to the rest of the country, which has seen little tectonic change over the past several hundred million years.

Prior to its present-day tectonic setup, the Cordillera experienced at least five orogenic events, and similar to the Appalachians, none of the events affected the entire Cordilleran Mountain system. The five orogenic events are the Late Devonian to Early Mississippian Antler orogeny (360–347 Ma), the Late Permian to Early Triassic Sonoma orogeny (260–247 Ma), the Late Jurassic Nevadan orogeny (164–145 Ma), the Late Cretaceous to Early Eocene Sevier orogeny (115–52 Ma), and the Late Cretaceous to Middle Eocene Laramide orogeny (75–45 Ma). To these events we can add Late Pennsylvanian-Early Permian (c.300 Ma) uplift of a series of mountains from Colorado to Texas known as the Ancestral Rocky Mountains. Let us also not forget the modern-day orogeny occurring right now along the west coast resulting from interaction of the North American, Pacific, and Juan de Fuca plates. The Cordillera, during this time, was involved in numerous terrane accretion events, but was never involved in major continent−continent collision of the magnitude seen in the Appalachians. A subduction zone and an ocean basin have existed continuously off the west coast since at least the Antler orogeny. The most recent large-scale terrane accretion in the United States ended along the Washington-Oregon coast about 49 million years ago as discussed in Chapter 19. Accreted terranes in the Cordillera (as well as in the Appalachians) were later dismembered and shuffled along strike-slip faults, thus complicating the collision history. In discussions that follow, we refer to the collection of orogenic events that created the Cordilleran Mountains as the Cordilleran orogeny.

THERMAL PLUMES AND HOT SPOTS

A thermal plume is a column of hot (but not liquid) rock that rises vertically from the base of the mantle. As the plume encounters the rigid lithosphere, upward motion is temporarily halted, causing the hot rock to flatten against the base of the lithosphere. The rising plume takes the shape of a mushroom with a large head and a long tail. As more and more hot rock accumulates, it begins to melt, which stretches and weakens the overlying lithosphere eventually allowing magma to reach the surface. Because of the mushroom shape of the plume, the initial outpouring of lava is tremendous, producing what are known as flood basalts. The mushroom head and flood basalt are depicted in Fig. 5.14A. The mile plus thickness of basalt that forms the Columbia Plateau is considered by many to be an example of hot spot flood basalt. Once the plume head is depleted, the tail of the plume will continue to create volcanism at the surface over the course of millions of years, but the total amount of extruded volcanic material is far less voluminous than the initial outpouring. Constant thermal plume-fed volcanism at the Earth's surface creates what is known as a hot spot.

A thermal plume, a hot spot, is traditionally thought to be stationary or nearly stationary although there is evidence that at least some plumes move. In any case, active volcanism is restricted only to that part of the tectonic plate that is directly above the hot spot. If we assume that the thermal plume remains stationary, then as the tectonic plate moves, the area of active volcanism on the tectonic plate will move off the hot spot and become inactive. If the hot spot is below oceanic lithosphere, the inactive volcanic island will erode, cool, and isostatically sink to become a seamount (a generic term for an undersea mountain). At the same time, the adjacent part of the tectonic plate, which is now above the hot spot, will become the active volcanic site. In this manner, as shown in Fig. 5.14B, a hot spot is capable of creating a chain of volcanoes that are progressively older with distance from the active hot spot. It is like moving a piece of paper across a lit match.

Hot spot volcanism can occur below oceanic and continental lithosphere. Volcanism below oceanic lithosphere tends to be basaltic and nonexplosive because basaltic magma generated in the upper mantle can easily rise through the dense, thin oceanic crust. Hot spot volcanism beneath continental lithosphere tends to be bimodal (two distinct compositions) consisting both of nonexplosive basaltic lava and explosive rhyolite volcanism. Basaltic magma has a harder time rising through the thick, less dense continental crust. Consequently, some of the magma reaches the surface, but large quantities may crystallize at depth at the base of the continental crust. The heat generated from crystallization is enough to partially melt continental crust, thus creating rhyolite magma.

Thermal Plumes in the United States

The best-known hot spot is the active Hawaiian hot spot, which produced the Hawaiian Island−Emperor Seamount Chain shown in Fig. 5.15. The volcanic chain stretches from the big island of Hawaii directly above the hot spot, to a seamount known as Meiji at the western end of the Aleutian Islands. Active volcanism is restricted to the big island and to an undersea volcano known as loihi, which is destined to be the next Hawaiian Island in the chain. Meiji was directly above the hot spot about 85 million years ago and was one of the first volcanic islands in the chain. Note that the seamount chain changes orientation in the vicinity of Yuryaku. This bend has been taken to indicate that Pacific plate motion abruptly changed direction about 43

(A) A plume of hot rock rises and accumulates at the base of the lithosphere creating a large body of magma. Heat from the magma stretches and weakens the lithosphere eventually allowing the magma to reach the surface. Initial outpouring is in the form of flood basalt as the entire plume head is extruded.

(B) Subsequent volcanism is fed by the tail of the plume. A volcano is extinguished when it moves off the hot spot creating a chain of extinct volcanoes that are progressively younger toward the active volcano above the hot spot.

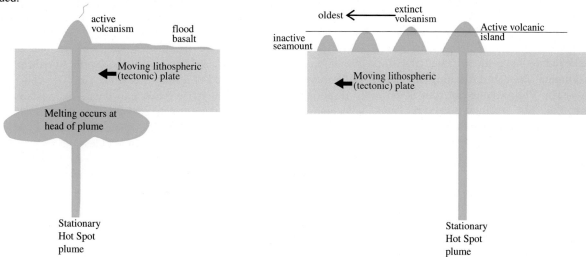

FIGURE 5.14 Origin of hot spot volcanism. (A) Initial outpouring of magma in the form of flood basalt. (B) Subsequent volcanism fed by the tail of the plume.

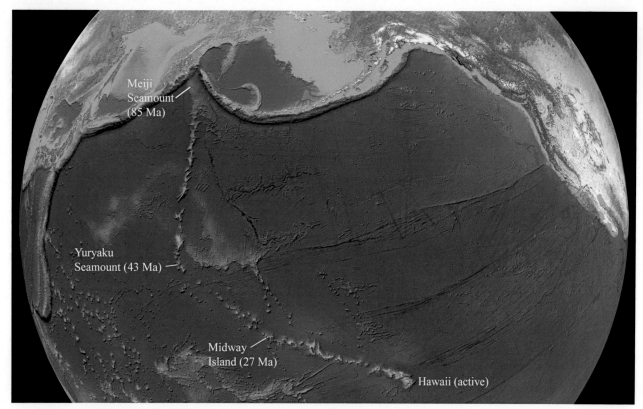

FIGURE 5.15 Google Earth image looking slightly west of north at the Hawaiian Island–Emperor Seamount chain. The long scar that extends across the center of the ocean toward Yuryaku Seamount is the Mendocino Fracture zone.

million years ago. However, there is also evidence that the hot spot plume itself migrated between 81 and 47 million years ago and that the change in Pacific plate motion may have been more gradual than previously supposed.

Several hot spots, in addition to the Hawaiian hot spot, have influenced or created landscape in the United States. The active Yellowstone hot spot, discussed in Chapter 17, is responsible for the creation of volcanic rocks on the Snake River Plain and is believed to have also created the flood basalt on the Columbia Plateau beginning about 17 million years ago. The Yellowstone hot spot is currently responsible for the hot springs, geysers, and volcanism in Yellowstone National Park.

The Great Meteor (or New England) hot spot is responsible for the New England seamount chain, the longest volcanic chain in the North Atlantic Ocean. The seamount chain extends from Georges Bank off the coast of Massachusetts all the way to the Great Meteor Seamount in the eastern Atlantic Ocean off the northwest coast of Africa. Part of the seamount chain is visible in Fig. 5.5 located just above the words North American plate. This part of the chain began to form about 100 million years ago. Prior to creating volcanism in the Atlantic Ocean, the New England hot spot began life on the North American continent in Canada possibly as early as 200 million years ago. It is believed to be responsible for the younger (130–100 million years) phase of magmatism in the White Mountain batholith of central New Hampshire and possibly for uplift of the Adirondack Mountains as discussed in Chapter 14.

Some workers suggest that a now extinct thermal plume, known as the Bermuda hot spot, may be responsible for part of the landscape in the lower (southern) Mississippi River valley as discussed in Chapter 13. Other workers question the existence of this plume.

QUESTIONS

1. Make an enlarged copy of the west side of Fig. 1.3 (Appendix) with plenty of open space on the left side. Draw the following and provide a short explanation of each feature: the Juan de Fuca plate boundaries including the Cascadia (Juan de Fuca) subduction zone, the San Andreas fault, and the rifted plate boundary in the Gulf of California.
2. Speculate as to why the Hawaiian Island–Emperor Seamount chain bends at Yuryaku. Could it be it due to movement of the hot spot? All of the islands and seamounts are part of the Pacific plate.
3. What is a triple junction? What is an aulacogen?
4. Use Google Earth to find the last island in the Hawaiian Island–Emperor Seamount chain. What is the name of the island? Estimate its age. The big island of Hawaii, although still active, is currently moving off the Hawaiian hot spot. If we presume that volcanism on the island will end within 1 million years, how many years can we expect Hawaii to exist as an island once it has moved off the hot spot?
5. Using Google Earth, locate the New England hot spot seamount chain off the coast of Massachusetts. Measure the visible length, from one end to the other. Assuming that the North American plate has moved at a rate of 8.5 feet per 100 years, how long would it have taken to create the seamount chain? Assuming a stationary hot spot, what direction would the North American plate have moved? Which side of the seamount chain is oldest?
6. Describe the difference between a passive continental margin and an active continental margin. Which is the most appropriate label for the Texas Gulf coast and why?
7. What is the tectonic system? Where does it obtain its energy? How does movement of tectonic plates build landscape?
8. What is the climatic system? Where does it obtain its energy? How does climate shape landscape?
9. Describe the competition between the tectonic system and the climate system.
10. How does tectonic plate movement affect climate? Isostasy?
11. What are typical rates of plate movement in inches per year, inches per 100 years, millimeters per year, centimeters per year, centimeters per 100 years, miles per million years, and kilometers per million years? What is the correlation between millimeters per year and kilometers per million years?
12. Describe, with a drawing, the processes of underplating and tectonic accretion.
13. What is an accretionary prism? What is a suture zone?
14. Describe the three major types of plate boundaries. Which are responsible for major volcanism? Which are responsible for major, destructive earthquakes?
15. What is a thermal plume?
16. Where are some of the major active hot spots in the United States?
17. What is flood basalt and under what tectonic environment do they form? Where can we find flood basalt in the United States?
18. Explain why an ideal transform fault will form a broad curved arc and not a straight line across the Earth's surface.
19. Suggest several explanations as to why the strike of the San Andreas Fault does not form a broad curved arc across the Earth's surface.

20. How does one date orogeny?
21. List the five orogenic events that created the Appalachian Mountains. Include their age according to the Geologic Time Scale and in millions of years before present. Do the same for the Cordillera. When did the Ouachita orogeny occur and what event in the Cordillera occurred at about the same time? What was the final outcome of the Appalachian orogenies?
22. What tectonic event(s) occurred along the Gulf coast to create the present-day landscape? Approximately when did they occur and what was its final outcome.
23. Explain why the Mississippi River is located where it is.
24. Speculate as to why active Cordilleran deformation extends so far inland from the Pacific coast.
25. What major structures form the Basin and Range and what factors contributed to its development?

Chapter 6

Forcing Agent: The Climate System

Climate is the long-term condition of the atmosphere. It is weather averaged over a long period of time, usually 30 years. Two of the most important measures of climate are temperature and seasonal precipitation, but other measures, such as humidity, wind velocity, fog, frost, and sunshine, are also considered.

Climate is a surface agent driven by the heat of the sun. Disproportionate heating of one area relative to another, along with the Earth's rotation, drives atmospheric and oceanic circulation, which along with gravity, activates two of the five mechanisms of landscape formation and change: erosion and deposition. In terms of landscape modification, climate creates conditions where air, water, and ice can disintegrate bedrock, and transport and deposit the detritus. Climate controls the type and rate of weathering, the depth of soil development, the type of vegetation, and rates of erosion and deposition. It also influences other Earth surface phenomena such as mass wasting processes, the development of river, groundwater, karst, glacial, and desert systems, short-term sea level changes, and isostasy.

Climate is the sculptor of landscape. Climate-driven erosion and deposition removes material from highlands and deposits material in lowlands ultimately reducing topographic relief to a nearly flat plain close to sea level. In the process, the climate system creates unconsolidated sediment, which upon burial, forms nearly flat-lying sedimentary rock.

As discussed at the beginning of Chapter 5, climate competes with tectonic activity to shape landscape. In steady climate, and in the absence of strong tectonic stress, landscape tends toward equilibrium with climate in which rates of erosion and deposition remain relatively constant. Any change in landscape is slow and predictable. In Chapter 10 we refer to slow, predictable change as landscape that slowly grows old.

A critical aspect of climate with respect to landscape is climate change. Climate change disrupts landscape equilibrium causing rates of weathering, erosion, and deposition to change. But, barring a catastrophic event such as meteorite impact, rate changes alone do not necessarily cause profound change in landscape. Instead, they may simply cause landscape to evolve along a similar path either faster or slower than its previous evolution depending on whether rates of weathering, erosion, and deposition increase or decrease. There is perhaps only one form of climate change that has a profound and lasting effect on landscape, and that is glaciation. The advance of glaciers across New England, the north-central United States, and mountainous areas of the Cordillera within the past 2.4 million years has caused significant changes in elevation, relief, drainage pattern, and drainage density, not only in areas that have been glaciated, but also in nonglaciated areas due to changes in sea level, isostasy, river volume, and river sediment load. The effect of glaciation on landscape is discussed in Chapter 12. In this chapter, we restrict our discussion to present-day climate. In Chapter 8 we discuss paleoclimate, the origin of glaciation, and its effect on sea level.

PRESENT-DAY CLIMATE ZONES

Fig. 6.1 shows present-day climate zones in the United States, each of which is listed in Table 6.1. The United States is approximately 3000 miles east to west and 2000 miles north to south. Climate varies dramatically across this large area. Climate is warm and humid in the southeast (wet-dry savanna, humid subtropical) and becomes cooler and somewhat less humid in the northeast (humid continental). Both areas are wet enough to support thick soil and a lush hardwood deciduous forest of maple, oak, ash, beech, and hickory. Much of the farmland region of Ohio, Indiana, Illinois, Iowa, and Missouri was once covered with forest. Farther north, the deciduous forest mixes and finally gives way, in eastern Canada, to a coniferous forest of spruce, fur, pine, hemlock, and cedar. This region is cooler and drier than the northeastern United States producing subpolar climate.

FIGURE 6.1 Landscape map showing present-day climate regions of the United States.

TABLE 6.1 Climate Zones in the United States

1. Wet-dry savanna
2. Humid subtropical
3. Humid continental
4. Steppe
5. Desert
6. Mediterranean
7. Moist coastal
8. Mountain
9. Subpolar (Northeastern Canada)

Climate becomes markedly drier as we head west to the Great Plains where the lush forests of the east give way to open grasslands and thinner soil. Arid and semiarid conditions prevail in lowland areas across nearly the entire Great Plains and Cordillera producing steppe and desert climates. Adjacent mountain areas produce their own (mountain) climate, which tends to be cooler and wetter than lowland areas. Moist air coming off the Pacific Ocean keeps the coastal areas of Washington, Oregon, and northern California very wet (moist coastal), and southern California seasonably wet (Mediterranean).

An area subject to humid climate will evolve differently than an area subject to arid climate because each will experience different rates of weathering, erosion, and deposition. Weathering (but not necessarily erosion) tends to be most intense (and therefore most rapid) in areas across the eastern half of the United States and along the Pacific coast where climate is wet for most of the year. A wet climate promotes lush vegetation, which in turn adds acids to the water that promote chemical weathering. Roots hold weathered material in place, inhibiting erosion and promoting development of a deep, rich soil. Rocks are poorly exposed.

Weathering tends to be less intense across much of the Great Plains and Cordillera where the climate is dry and temperature alternates between hot and cold. In this situation, frost cracking dominates, vegetation is sparse, chemical weathering is curtailed, and only a thin soil develops such that underlying rocks are widely exposed. Erosion, particularly in mountains, can be far more intense than in the eastern United States due to high relief and glaciation at high elevation.

CONTROLS ON CLIMATE

The climate zones outlined in this brief synopsis of the United States are controlled largely by latitude, proximity to large water bodies (including the ocean and ocean currents), global wind patterns, the tilt of the Earth's axis of rotation, and mountains.

Latitude

The effect of latitude is obvious. Areas close to the equator receive more direct rays of the sun and therefore, are warmer. The sun is lower on the horizon with distance from the equator, and therefore the northern part of the United States tends to be cooler than the south.

Proximity to Large Water Bodies

Proximity to a large water body has several effects, one of which is lake-effect snow for which Buffalo, New York, is famous. More importantly, large bodies of water have the effect of moderating air temperature. If two cities are located the same distance from the equator, a coastal city such as Seattle is cooler in summer and warmer in winter relative to an inland city such as Minneapolis. Seattle is located at 47°36′ north latitude. July average high temperature is 74.5°F (23.6°C). January average high temperature is 47°F (8.3°C). Minneapolis is located at 44°59′ north latitude, about 181 miles (291 km) south of Seattle. Minneapolis has a July average high temperature of 83°F (28.3°C) and a January average high temperature of 22°F (−5.5°C).

The reason for the greater extremes in Minneapolis is its location far from a large body of water. Relative to land, water takes longer to heat and to cool. This is why, when the temperature outside is 95°F, the water in your pool remains relatively cool. If hot air moves across a body of cool water, the air will become cooler. Seattle benefits from its proximity to the cool Pacific Ocean during summer months. Minneapolis receives only hot inland continental air. Conversely, in winter, the Pacific Ocean warms frigid air blowing toward Seattle whereas Minneapolis receives no such moderating effect.

The moderating influence of the ocean is felt most strongly on the west coast because air masses in the United States typically move across the Pacific Ocean onto land. The effect in southern California in particular is to produce relatively moderate and pleasant year-round temperatures. Temperatures on the eastern seaboard are also moderated, but the effects are less pronounced because air typically moves from land areas toward the Atlantic Ocean. In New England, several inches of snow could fall in Vermont, while at the same time, coastal areas receive only rain. Similarly, coastal areas remain cool relative to inland areas during summer months. Boston lies at 42°22′ north latitude, 180 miles (290 km) south of Minneapolis, yet its July average high temperature is 82°F (27.8°C) and its January average high temperature is 36.5°F (2.5°C). The effect of summer cooling is not obvious along the Gulf coast because perennially warm water in the Gulf of Mexico does little to moderate the hot, humid, summer air.

Ocean currents also influence coastal areas. The eastern seaboard is influenced by the Gulf Stream, which carries warm equatorial waters northward. The Gulf Stream adds warmth and moisture to the air increasing both temperature and humidity relative to what conditions along the eastern seaboard would be like if the Gulf Stream did not exist. Occasionally, as we all know, the warm, moist air can manifest itself into a tropical storm or hurricane, especially along the southern coastline.

The western seaboard of the United States is moderated by the California Current, which carries cold water southward along the US Pacific coastline. It is one reason why ocean water is so much colder along the west coast relative to the east. The California Current is associated with upwelling of nutrient-rich cold water, which creates a biologically diverse marine ecosystem. Upwelling also cools the already cold water creating coastal fog.

Global Wind Patterns

Global wind patterns have a huge influence on climate. But, before we discuss wind patterns, we need to know a few basics. Warm air can hold more water vapor (dissolve more water) than cold air. Warm air is also less dense than cold air; therefore, it rises. As warm air rises, it replaces cold, dense air at altitude, causing low-pressure atmospheric conditions. At the same time, the rising warm moist air will cool as it rises causing humidity in the air to increase, possibly leading to the formation of clouds and precipitation. This explains why you can expect precipitation when the weatherman indicates low-pressure atmospheric conditions.

Cool air has the opposite effect. Cool air is dense; therefore, it sinks. As it sinks it replaces warm, light air causing high-pressure conditions. At the same time, the cool, dense air will warm as it sinks and thus, become drier. Such conditions favor brilliant, cloudless, high-pressure conditions, perfect for outdoor activity.

Uneven heating of the Earth's atmosphere by the sun causes air pressure in one area to be higher than in another area. Wind is the movement of air from high-pressure areas to low-pressure areas. Global wind patterns have their origin at the equator where the sun is constantly warming the air. As the air warms, it rises and cools, causing low-pressure conditions and rain. The absence of wind and the predictable occurrence of afternoon showers near the equator is the reason why this

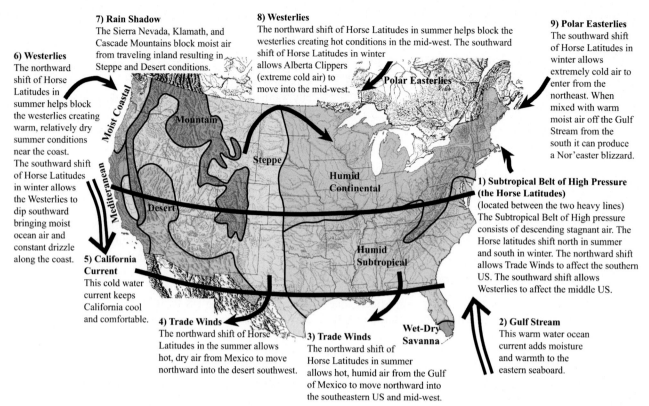

FIGURE 6.2 Landscape map showing present-day wind and ocean current patterns in the United States. The figure should be read clockwise beginning with #1, the Subtropical Belt of High Pressure.

region is known as the Doldrums. As the rising air reaches high atmosphere, it begins to move laterally toward the north and south poles. This lateral movement away from the equator causes the air to cool even more, and eventually to sink. The sinking air creates a subtropical belt of high pressure in an area known as the horse latitudes. The horse latitudes are located approximately between 30 and 38 degrees north and south of the equator. In the northern hemisphere, the horse latitudes cover the southern part of the United States.

As air in the horse latitudes descends, it warms and becomes dry. Many of the world's deserts, including the Mojave Desert, lie within the dry, high pressure descending air of the horse latitudes. Because the movement of air is vertical (downward), there is little in the way of wind associated with the horse latitudes except along its margins where air moves horizontally northward and southward toward regions of low pressure. The north and south horizontal movement of air creates global wind patterns. In the northern hemisphere, the Westerlies lie to the north of the horse latitudes, and the Trade Winds to the south. All northern hemisphere winds (and ocean currents) rotate clockwise due to the counterclockwise (eastward) rotation of Earth. The rotation deflects the wind such that the Trade Winds blow mainly out of the east, and the Westerlies out of the west. The Westerlies produce the prevailing west to east movement of weather systems in the northern part of the United States and are partly responsible for jet streams with wind speeds up to 250 mph. The Trade Winds affect the very southern part of the United States with a steady, comfortable breeze.

A third prevailing wind pattern, the Polar Easterlies, emanates from high-pressure areas near the Arctic Circle and blows mainly out of the east. The Easterlies primarily remain north of the United States, affecting Canada, but on occasion they dip southward into New England and into the north-central part of the country.

Of particular importance is the seasonal shift of all wind currents. All northern hemisphere wind currents shift northward 5 to 10 degrees during summer months, and shift southward 5 to 10 degrees during winter months as explained below. A synopsis of global wind patterns and their effect on the United States is presented in Fig. 6.2. Please take a moment to read though this figure beginning with #1, the subtropical belt of high pressure.

The Tilt of the Earth's Axis of Rotation

The summer warmth and winter cold that we all experience is due to the fact that, over the course of a year, the direct rays of the sun will migrate from north of the equator where the sun is high in the sky creating warm summer conditions, to south of the equator where the sun is low in the sky creating cool winter conditions. The reason the direct rays of the sun migrate north and south of the equator is that the Earth's axis of rotation is tilted 23.5 degrees to the plane of its orbit around the sun. As shown in Fig. 6.3, the orientation of the axis does not change position as the Earth rotates around the sun. The north axis will point directly away from the sun one day of the year, and point directly toward the sun one day of the year. On these days, the sun's direct rays will shine respectively on the Tropic of Capricorn 23.5 degrees below (south of) the equator, and on the Tropic of Cancer 23.5 degrees above (north of) the equator. In the northern hemisphere, we refer to these times as the winter and summer solstice. They are the shortest and longest days of the year. As the sun's direct rays migrate from the northern to the southern hemisphere and back again, they pass the equator twice. These days represent the equinox when the sun's direct rays fall exactly on the equator and the day is approximately 12 hours long everywhere.

In a matter of speaking, all global wind patterns follow the sun. As the vertical rays of the sun shift northward during the northern hemisphere summer, the global wind patterns follow, and vice versa in winter. The shift, however, is delayed from that of the sun and is generally only 5 to 10 degrees rather than 23.5 degrees.

The shift in global wind systems produces major climatic effects as described in Fig. 6.2. The summer northward shift of dense, descending air associated with the subtropical belt of high pressure curtails the Westerlies in the northern United States creating warm, relatively dry summer conditions along the Washington-Oregon

The Seasons and The Shift in Global Wind Patterns

The Earth's N-S axis of rotation is tilted 23.5 degrees to the plane of its orbit around the Sun.
The axis does not change position with respect to the Sun such that the north axis points directly
away, and directly toward the Sun once per year during the Winter and Summer solstice respectively.
The north-south axis is oriented exactly perpendicular to the Sun's rays twice per year during the Spring
and Autumn equinox.

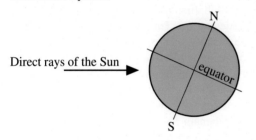

Winter arrives in the Northern Hemisphere when the Earth's north axis points away from the Sun and direct rays of the Sun are below the equator. The Winter solstice occurs on or near December 22 when direct rays of the Sun are at their southern-most latitude, the Tropic of Capricorn. The Horse Latitudes, Trade Winds, Westerlies, and Polar Easterlies all follow the Sun and shift southward 5-10 degrees. The southward shift allows the Westerlies, and to a lesser extent the Polar Easterlies, to affect weather patterns in the US.

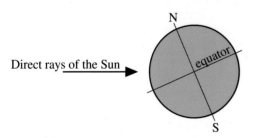

Summer arrives in the Northern Hemisphere when the Earth's north axis points toward the Sun and direct rays of the Sun are above the equator. The Summer solstice occurs on or near June 22 when direct rays of the Sun are at their northern-most latitude, the Tropic of Cancer. The Horse Latitudes, Trade Winds, Westerlies, and Polar Easterlies all follow the Sun and shift northward 5-10 degrees. The northward shift allows the Trade Winds to affect weather patterns in the US.

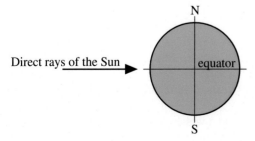

Direct rays of the Sun fall squarely on the equator during Spring and Autumn equinox on or near March 21 and September 22 respectively when the Earth's north-south axis is oriented perpendicular to the Suns rays.

FIGURE 6.3 The seasons and the shift in global wind patterns.

FIGURE 6.4 Looking southward at mountains in the vicinity of La Junta Peak near Telluride, Colorado. Note the change in vegetation from deciduous trees near the base of the mountain to coniferous trees near the top.

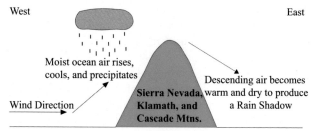

FIGURE 6.5 Development of a rain shadow east of the Cascade-Klamath-Sierra Nevada mountain ranges.

coast, and hot conditions in the continental interior (#6 and 8, Fig. 6.2). At the same time, the warm Trade Winds rotate northward. In the southeastern United States, the Trade Winds move over the Gulf of Mexico where they pick-up moisture producing hot, humid conditions (#3, Fig. 6.2). In the western United States, these same winds move over the dry desert region of Mexico, resulting in hot, dry air over Arizona and New Mexico (#4, Fig. 6.2).

The winter southward shift of the horse latitudes allows the Westerlies to rotate into the heartland of the United States. On the west coast they bring moist air from the Pacific Ocean resulting in cool (but not freezing) days of constant drizzle, especially in the Pacific northwest (#6, Fig. 6.2). In the central United States, Westerlies bring frigid Alberta Clippers of extreme cold, windy air (#8, Fig. 6.2). The Polar Easterlies also shift southward during winter months, carrying with them frigid temperatures. In New England, Polar Easterlies can, on occasion, mix with moist coastal air emanating from the Gulf Stream, creating major blizzards and cold snaps known as Nor'easters (#9, Fig. 6.2).

Mountains

Mountains affect climate in at least two ways. Firstly, temperature decreases by about 3.57°F for every 1000 feet of elevation gain (6.5°C/1000 m). Vegetation is temperature and precipitation dependent and, as a result, vegetation changes dramatically in a vertical direction up a mountain. Fig. 6.4 is a photograph from near Telluride, Colorado that shows the vertical progression from deciduous forest near the base of the mountain, to mixed forest, to coniferous forest, to barren land above tree line at the top of the mountain. Secondly, mountains block air masses from moving across a region. Mountains force air masses to rise in order to pass over them. As air rises, it cools to its saturation point, which first produces large cumulus clouds and then thunderstorms. This is why clouds tend to be present in the vicinity of a mountain and why a passing afternoon mountain thunderstorm is commonplace.

The forcing of precipitation as air moves over a mountain range produces a climatic effect known as a rain shadow. The effect is shown schematically in Fig. 6.5. The Cascade-Klamath-Sierra Nevada ranges, which extend nearly the length of the US Cordillera, produce a strong rain shadow that partly explains the arid and semi-arid climate throughout the west. As air moves inland from the Pacific Ocean, it rises, cools, and precipitates. The air is dry by the time it reaches the eastern side of the mountain range. To see the effect of a rain shadow, one could contrast the thick, lush, forest region west of the Cascade Mountains in Washington and Oregon shown in Fig. 6.6A, with the arid grasslands, wheat fields, and vineyards east of the mountain range shown in Fig. 6.6B. The Appalachian Mountains, by contrast, do not produce a pronounced rain shadow because prevailing Westerly winds blow from west to east across dry land.

The situation just described represents only the present-day climate, which has been fairly stable for the past 7000–10,000 years. Before that, for the past 2.4 million years, we have experienced a series of glacial advances and retreats, during which time our climate has sometimes been warmer but, more commonly, cooler than today. At certain times during the past, and most recently only

FIGURE 6.6 The rain shadow effect in Oregon. (A) Deep forest in the Coast Range west of Corvallis, Oregon. (B) Sparse vegetation east of the Cascade Mountains near Bend Oregon.

18,000 years ago, nearly all of Canada and a large part of the northern United States was covered in ice (glaciers) up to 2 miles thick. The glaciations have had a dramatic and lasting effect on our landscape as discussed in Chapter 12.

QUESTIONS

1. How is climate different from weather?
2. Describe the global wind patterns and the resulting climate in the area where you live.
3. At what latitude are the Tropics of Cancer and Capricorn located and what is their significance with respect to climate?
4. At what latitude are the Arctic and Antarctic Circles located and what is their significance with respect to climate?
5. Explain why rising warm air results in low-pressure conditions and rain. Globally, where does rising warm air occur?
6. If the weatherman says that an area is under a high-pressure system, what does he mean and what type of weather should we expect?
7. What is wind? Why would wind be stronger in one area relative to another?
8. Describe how the climate where you live would be different if, (a) the Earth's axis was oriented exactly perpendicular to the sun's rays during the entire year, or (b) if the Cascade-Klamath-Sierra Nevada ranges did not exist, but a large mountain range did exist extending from Minnesota to Louisiana.
9. Research topic. Nine climate zones were mentioned briefly at the beginning of this chapter. Research two of these and determine on what basis they are defined and how one is different than the other.
10. Research topic. Find the average temperature and precipitation for mid-August and mid-January in (or near) the town you live in and compare them with those in Seattle, Minneapolis, and Boston. Speculate as to why they are similar or different.
11. Fig. A.6 in the Appendix is an uncolored version of Fig. 6.1. Color this map, research the climate zones, and provide a description of each.
12. Why is climate different at the same latitude on the west coast versus the east coast?
13. Speculate as to why Seattle receives less than an inch of precipitation in July and more than 5 inches in January.

Chapter 7

Forcing Agent: Isostasy

As introduced in Chapter 5, isostasy is the equalizer. It is from the Greek *isos*, which means equal, and *stasis*, which means standing. It is the rising or settling of a portion of the Earth's lithosphere that occurs when weight is removed or added in order to maintain equilibrium between buoyancy forces that push the lithosphere upward and gravity forces that pull the lithosphere downward. When these two forces balance, the lithosphere is said to be in isostatic equilibrium, which implies that it is in gravitational equilibrium such that the lithosphere literally floats on the underlying weak, malleable asthenosphere.

The concept is similar to how a boat behaves in water. As shown in Fig. 7.1, an empty boat will weigh less and be less dense than the same boat loaded with cargo. It will float high on the water. However, as the boat is loaded with cargo, it will become dense and heavy. It sinks slightly into the water to compensate. Unload the boat and it rises again. Isostatic adjustment occurs instantaneously and the boat maintains isostatic equilibrium with the water. In the same manner, most of the Earth's lithosphere is at or close to isostatic equilibrium with the underlying asthenosphere.

Complete isostatic adjustment of the Earth's surface does not occur instantaneously as it does with a boat in water, but it does occur rather quickly. If a load were suddenly placed on the Earth's surface, isostatic adjustment would begin almost immediately. For example, measurable subsidence occurred around Lake Mead within a few years following the building of Hoover Dam. But, depending on the magnitude of the imbalance, it could take hundreds or thousands of years before isostatic equilibrium is achieved.

Isostasy is a secondary forcing agent that reacts to both climate and tectonics. Any form of weight redistribution within the lithosphere is capable of triggering isostatic adjustment. Climate-driven processes redistribute weight through erosion, deposition, and glaciation. Plate tectonic activity redistributes weight through the faulting and folding of rock, the extrusion of thick piles of volcanic rock, and by altering the thermal character of the lithosphere. In this chapter, we begin with a discussion that outlines significant differences between tectonic and isostatic uplift. We then survey tectonic processes that cause isostatic adjustment including a brief discussion of thermal isostasy. We end the chapter with a survey of climatic processes that cause isostatic adjustment.

TECTONIC VERSUS ISOSTATIC UPLIFT

Tectonic stress and isostatic compensation can both activate uplift (and subsidence). We can refer to one as tectonic uplift and the other as isostatic uplift. When dealing with an area undergoing uplift, it is difficult to separate the tectonic component from the isostatic component. There are, however, significant differences in their mode and rate of uplift (and subsidence), their sustainability, and their final outcome.

Tectonic uplift is activated by tectonic stresses in the Earth created by the interaction of moving tectonic plates. Tectonic uplift can be regional across an entire mountain range or plateau, or focused on a single fault zone or fold crest. It is associated with strong deformation of rock, including the development of folds and faults that change the structural form of rock. Uplift generated by tectonic stress can be an ongoing continuous process, but more typically is episodic in the sense that periods of no uplift alternate with periods of rapid uplift generated during intermittent earthquakes. Tectonic uplift through intermittent earthquakes can be sustained over the course of hundreds of thousands to tens of millions of years at rates that average between 4 and 10 inches per 100 years. Tectonic uplift will continue for as long as the plate interaction that created the stress persists. The potential final outcome is a high mountain range.

Isostatic adjustment is density (or weight) driven. Isostatic uplift can be focused along fault zones, but is more typically associated with broad warping of rock across an entire mountain range or plateau. The end result is broad regional uplift (or subsidence) of land areas in which rock layers remain nearly flat-lying. Isostatic adjustment is not capable of strong deformation or the production of a high mountain range. In contrast to periodic earthquake-generated pulses of tectonic uplift, isostatic adjustment is more typically an ongoing continuous process. Isostatic adjustment often occurs over a period of hundreds to thousands of years, but can occur indefinitely because once a density imbalance presents itself, isostatic adjustment will begin and will continue for as long as the imbalance is present. Continuous erosion of a mountain range will drive continuous isostatic uplift. The rate of isostatic uplift can vary from less than an inch to several tens of feet per 100 years depending on the magnitude of the imbalance. The rate of uplift gradually decreases as isostatic equilibrium is approached and will end once isostatic equilibrium is achieved.

The response of land areas during isostatic adjustment is a function of the strength of the underlying lithosphere. Fig. 7.2 shows how the lithosphere might respond to the addition of weight. Given similar rock type, the strength of lithosphere is a function of thickness and temperature. A thick or cold lithosphere is strong. It would resist bending resulting in a broad wavelength syncline with a possible flexural bulge as depicted in Fig. 7.2A. In Fig. 7.2B, the lithosphere shown is thin, warm, and weak, such that the lithosphere bends beneath the weight. A flexural bulge is shown in the figure, but may not necessarily be present. Rocks in both cases maintain a nearly flat-lying structural form. In the third, less common scenario shown in Fig. 7.2C, a rigid block bound along vertical faults is depressed beneath a heavy weight. The presence of faults allows the lithosphere on opposite sides of the fault to respond differently. Lithosphere below the weight is depressed. Lithosphere on the outside of either fault undergoes flexural uplift. In this case, the response to

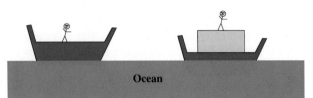

A boat with no cargo floats high on the water. The same boat loaded with cargo is heavy and sinks deeper in the water. Both boats are in isostatic equilibrium with the water.

FIGURE 7.1 Example of isostatic equilibrium.

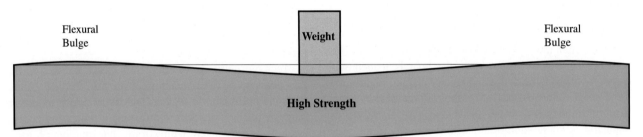

(A) Response of a cold, thick (high strength) lithosphere to the addition of weight.

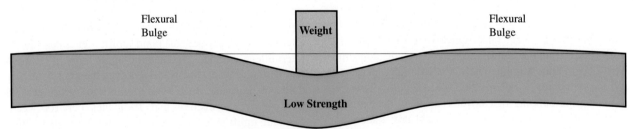

(B) Response of a warm, thin (low strength) lithosphere to the addition of weight.

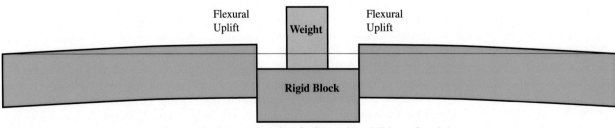

(C) Response of lithosphere with vertical through going faults to the addition of weight.

FIGURE 7.2 Response of lithosphere to the addition of weight. The thin line is a reference horizontal line.

isostatic loading is similar to tectonic displacement along a normal fault.

ELEVATION OF CONTINENTS AND OCEAN BASINS

If we were to remove all ocean water from the surface of Earth, we would see an obvious difference in height between continental and oceanic crust. Isostasy explains this topographic difference. Continental crust is granitic in composition. It is less dense than oceanic crust, which is basaltic in composition. The relatively light continental crust is thick relative to oceanic crust; therefore, it replaces heavier rock at depth (Fig. 5.2). Similar to a boat in water, the lighter, thicker continental crust reaches isostatic equilibrium higher on the asthenosphere than thin, dense oceanic crust.

A graph that shows the distribution of elevation on the Earth's surface is presented in Fig. 7.3. The graph clearly shows two distinct elevations. The elevation of continents is mostly between sea level and one km above sea level (0–3280 feet). Ocean basins are mostly between 4 and 5 km below sea level (13,120 and 16,400 feet). This bimodal distribution of Earth's surface elevation is a direct result of the density difference between oceanic and continental crust. Oceanic lithosphere can become so heavy that it literally sinks (subducts) into the asthenosphere, producing a subduction zone. Sinking is initiated and facilitated particularly where two tectonic plates collide. In this case, buoyancy contrasts dictate that the dense heavy lithosphere must subduct below the light lithosphere.

MOUNTAIN BUILDING AND PRESERVATION

The existence of high mountains is broadly attributed to the convergence of tectonic plates, a process that drives tectonic uplift. Isostasy also causes uplift and is responsible, in many respects, for the preservation of high mountains long after tectonic uplift has ceased. The normal thickness of continental crust is about 25 miles (40 km). During tectonic plate convergence, the continental crust is thickened to between 30 and 55 miles (48–89 km). Crustal thickening results from a number of processes that include folding, thrust faulting, metamorphism, intrusion, and tectonic accretion. Crustal thickening below the mountain occurs in the form of a root of relatively light (less dense) continental crust that sticks downward into the heavier (more dense) mantle much like the root of an iceberg and similar to what is shown in Fig. 7.4. The light crustal root replaces surrounding heavier mantle and, as a result, the land rises in order to maintain isostatic equilibrium. It is this isostatic rise that, in part, creates the mountain in the first place, even during tectonic collision.

Theoretically, we would expect a mountain range to gain or maintain elevation for as long as tectonic collision continues and the rate tectonic uplift is greater than or equal to the rate of erosion. Once the collision process ends, erosion should continuously lower elevation until the mountain no longer exists as a major topographic feature.

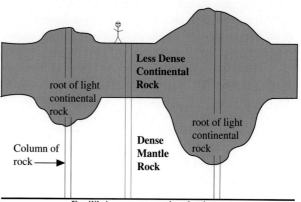

A deep root of continental crust is more buoyant because it displaces a greater amount of dense mantle rock. As a result, the continental crust floats higher on the asthenosphere to form mountains. The deeper the crustal root, the higher the mountain. As the mountain is eroded, the crustal root isostatically uplifts to compensate for the loss of weight. For every 5 feet of erosion, the crustal root (and the mountain) will isostatically uplift between 4.0 and 4.25 feet. Isostatic uplift continues until the crustal root is gone and the mountain is reduced to a nearly level plain at an elevation similar to the surrounding area. Isostatic equilibrium is maintained throughout the erosional process. All three columns of rock weigh the same down to the equilibrium compensation depth.

FIGURE 7.4 Example of a low-density crustal root below a mountain.

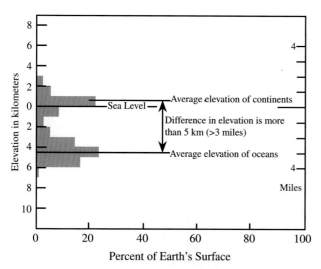

FIGURE 7.3 Graph that shows the distribution of elevation at the Earth's surface. Based on Moores and Twiss (1995, p. 8).

Isostasy does not allow erosional lowering to happen; or to put it more accurately; isostasy allows erosional lowering to happen, but at a much slower rate than it would without isostasy.

Under conditions of ideal isostatic equilibrium, a column of rock that extends down to a certain datum line (usually the asthenosphere) is said to be in floating equilibrium such that the weight of that column at a given latitude is the same everywhere. Fig. 7.4 shows three columns of rock. Theoretically, each column weighs the same. The added weight of the mountaintop on the right side of the figure is compensated at depth by the presence of a light crustal root of continental (granitic) rock that has replaced the heavier rock of the surrounding mantle.

Erosion lowers the elevation of the mountain, but at the same time it removes weight. The removal of weight from the top of the mountain is compensated for by isostatic uplift of the crustal root, which causes uplift of the mountain. Isostatic equilibrium is maintained and the weight of the column remains constant because the weight that is lost from the top of the mountain due to erosion is gained at the base of the mountain by uplift of the light crustal root and the replacement of light, crustal root rock with heavier mantle rock. The mountain shown on the left side of Fig. 7.4 is lower in elevation, but the column of rock weighs the same as the column on the right side of the figure.

As erosion continues, isostatic uplift will continue until the crustal root is reduced to normal crustal thickness, at which time the mountain would likely be reduced to a nearly flat plain at an elevation similar to surrounding land (as represented by the middle column in Fig. 7.4). An iceberg floating in water undergoes a similar process. Most of the iceberg forms a root below surface water. As ice (weight) is removed from the iceberg through melting, the root below the water surface will rise to compensate. The root will continue to rise until the iceberg melts completely.

The amount of isostatic rise is a function of the ratio of the density of the material being removed from the top of the mountain (from the top part of the column) and the density of material added at the base of the lithosphere (at the base of the column). The density of crustal granitic rock is roughly 2.7 g/cm^3. The density of mantle material added to the base of the column is roughly 3.3 g/cm^3. The ratio is $2.7/3.3 = 0.82$. Thus, given these densities, for every five feet of elevation removed from the top of the mountain by erosion, the mountain will isostatically uplift 4.1 feet ($5 \times 0.82 = 4.1$) such that mountain elevation is decreased by only slightly less than 11 inches. As a rule of thumb, we can say that for every 5 feet of elevation removed by erosion, a mountain range will isostatically uplift by approximately 4.0–4.25 feet. Thus, the overall height of the mountain is lowered, but the lowering is at a much slower rate than if erosion alone were the only factor.

The process by which the lithospheric root of a mountain is removed via simultaneous erosion and isostatic uplift is shown schematically in Fig. 7.5. The entire process is possible because isostatic adjustment is rapid and

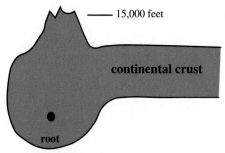

Mountain range immediately after the end of continental collision and tectonic uplift. The continental crust has been thickened to more than twice its normal thickness. The black dot represents a metamorphic rock. The mountains are more than 15,000 feet high.

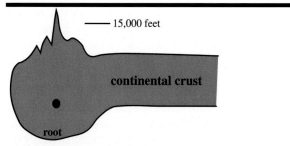

Erosion without tectonic uplift reduces the weight of continental crust in the mountain range which results in isostatic uplift. For every five feet of erosion, the mountain isostatically uplifts about four feet. If erosion occurs primarily in valleys so that isolated peaks are not eroded, it is possible for a peak to gain elevation even though tectonic uplift has ceased. This is a short-term phenomenon that lasts only until the mountaintop is eroded. The metamorphic rock (black dot) has exhumed toward the surface as a result of isostatic uplift and erosion.

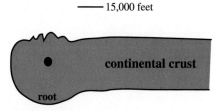

The mountain root has almost disappeared and mountain elevations have been reduced to low-lying hills. The metamorphic rock has risen closer to the surface.

The mountain root is completely removed and the mountain has been reduced to a flat plain at about the same elevation as adjacent land. The rate of erosion becomes very slow as does any isostatic adjustment. The metamorphic rock has been exhumed to the surface.

FIGURE 7.5 Isostatic compensation following a mountain building event.

responsive enough to keep pace with erosion. The top panel in Fig. 7.5 shows a mountain range above 15,000 feet immediately following the end of tectonic collision and uplift. The lower three panels show progressive erosion and isostatic compensation in the absence of tectonic uplift. Under these conditions, average elevation of the mountain range as a whole is progressively lowered, but notice in the second panel that an isolated peak has risen to higher elevation, well above 15,000 feet, and that topographic relief has increased. This scenario is possible if the mountain peak initially escapes erosion and erosion is concentrated in river valleys. The focused removal of rock from river valleys will lessen the overall weight of the land, which in turn triggers broad isostatic uplift of the entire mountain range, including the uneroded peak. It is a short-lived phenomenon that will last only until the mountain peak is removed via erosion. Isostatic uplift cannot create a mountain range, but it can maintain a mountain range for longer than if the mountain was subject to erosion alone.

An important outcome of this process is that it results in the surface exposure of once deeply buried crystalline rock. Note how the metamorphic rock in Fig. 7.5 (the black dot) is brought to the surface. Uplift coupled with simultaneous erosional removal of overlying rock explains why crystalline rock is often found on mountaintops. Isostatic compensation of a thickened crust is partly responsible for the persistence of the Appalachian Mountains 265 million years after tectonic collision had ended, and why once deeply buried crystalline rock now makes up most of the mountain belt. Isostatic adjustment slows the process of mountain decay, but the end result is the same. Given the continued absence of tectonic uplift, the combination of erosion and isostatic uplift will eventually erode a mountain, such as the Appalachians, to a flat plain just as if erosion alone were acting on the mountain.

Given the height of a mountain and knowledge of the density of rock, one might expect that it would be straightforward to calculate the thickness of a mountain root. But such calculations often diverge from actual geophysical measurements for potentially several reasons. One possibility is that some of the input parameters are incorrect. Perhaps there is a particularly dense rock layer that isn't accounted for in the calculations. The beauty of science is that input parameters can be changed to fit observations and measurements. Such back and forth between observation, measurement, and modeling leads (in this case) to a better understanding of Earth's interior. It would, for example, lead to the discovery of that dense rock layer.

A second possibility is that the mountain is held up partially by active tectonic stress. The idea is that tectonic stress acts like a vise that compresses the mountain and holds it higher than what would be expected by isostasy

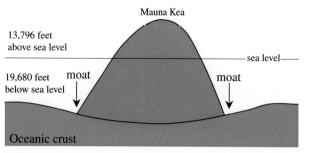

FIGURE 7.6 Isostatic depression of oceanic crust surrounding Hawaii.

alone. If the tectonic stress is reduced, the mountain will sink to isostatic equilibrium. Such a process might result in the production of normal faults and graben structures in an otherwise compressional mountain belt.

TECTONIC LOADS

The final isostatic response brought on by the tectonic system discussed here is the accumulation of tectonic loads. The piling of rock due to tectonic activity will result in a strong isostatic response. One example is the accumulation of thick piles of volcanic rock. The Columbia Plateau and Snake River Plain have isostatically subsided due to the accumulated weight of a mile or more of basalt. The massive volcanic pile that forms the island of Hawaii is another example. Fig. 7.6 shows how the weight of the pile has isostatically depressed the sea floor below it, creating a moat around the island. When measured from the bottom of the ocean to the summit of Hawaii's highest point (Mauna Kea), the Hawaiian island is about 33,476 feet high. On this basis, Mauna Kea is the tallest structure on Earth.

In Chapter 15, we will discuss foreland fold and thrust belts where thick piles of rock are thrust one upon another in a manner depicted in Fig. 4.2G, thus creating a mountain range. Similar to the moat that surrounds the big island of Hawaii, the weight of the mountain range will create a depression known as a foredeep basin located adjacent to the mountain as shown in Fig. 7.7 where detritus from erosion of the rising mountain accumulates. Areas on the Appalachian Plateau and Great Plains have acted as foredeep basins to the rising Appalachian and Rocky Mountains respectively.

THERMAL ISOSTASY

The subduction of oceanic lithosphere, the development of hot spots, and the divergence of tectonic plates all contribute to changes in the thermal condition of the lithosphere. Elevated temperature causes thermal expansion, which lowers the density of rock. The lithosphere responds isostatically by uplifting. Conversely, cooling of the

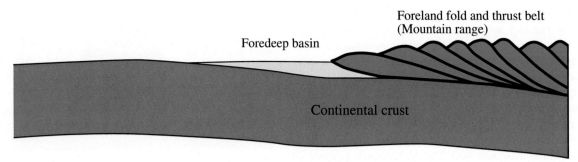

FIGURE 7.7 Example of a foredeep depositional basin in front of a rising mountain belt.

lithosphere will result in thermal contraction, higher density, and subsidence. We can refer to temperature-controlled density differences as thermal isostasy or thermal isostatic uplift (subsidence).

In the case of thermal isostatic compensation, it is possible to create (by added heat) and maintain (by sustained heat) areas of high elevation for as long as the lithosphere remains abnormally warm, even if the crust is relatively thin. Thermal isostatic uplift explains the abnormally high average elevation of the Basin and Range as seen in central and eastern Nevada even though the region is underlain with thin crust (Fig. 1.7). Thermal isostatic buoyancy helps explain the height of the Colorado Rocky Mountains where crustal thicknesses of 25–28 miles (40–45 km) are not enough to explain the high elevation. Thermal isostasy via warm, rising magma, explains the relatively high elevation surrounding Yellowstone National Park. Conversely, slow cooling of the lithosphere helps explain subsidence along the Atlantic and Gulf coast passive margins. In Chapters 14 and 16, we will see that thermal isostasy has played a key role in the evolution of both the Adirondack and Blue Ridge Mountains.

GLACIERS

Perhaps the most obvious isostatic response to climate-driven processes is the addition or subtraction of weight due to the accumulation or melting of continental glaciers. Glaciers can accumulate rapidly. The preservation of just 12 inches of snow per year will create a snowpack almost a mile thick in 5000 years. The weight of a glacier sitting on a continent is enough to cause several hundred to several thousand feet of isostatic subsidence over the course of a few thousand years. The Antarctic ice sheet has depressed part of the crust an estimated 3000 feet. The lithosphere below the Hudson Bay region in central Canada is understood to have been depressed more than 1300 feet during the height of the last major glacial advance 22,000–18,000 years ago. The removal of glaciers since about 8,000 years ago has resulted in rebound (isostatic uplift), which, in turn, has caused successive shorelines to emerge from the sea. These elevated shoreline terraces can be dated to determine the rate of uplift.

The analysis suggests that at least 935 feet of isostatic uplift has occurred along the Hudson Bay shoreline over the past 8,000 years. This is an amazingly high overall average uplift rate of almost 12 feet per 100 years. Even more amazing is that the rate must have been much higher when isostatic uplift first began because rates will slow as land areas approach equilibrium. It is estimated that uplift rates were initially as high as 39 feet per 100 years.

Today, the Hudson Bay shoreline continues to rise at rates of up to 4.3 feet per 100 years. In this case, we see that several thousand years are required for the lithosphere to completely rebound to its original equilibrium elevation. Uplift rates are highest in the area of Hudson Bay because this is where the continental ice sheet was thickest and subsequently where the magnitude of isostatic subsidence was greatest. The rate of isostatic uplift decreases southward where the thickness of the ice sheet was less.

DEPOSITION

A large delta such as the Mississippi river delta in Louisiana can deposit such an enormous amount of sediment that the weight causes the continental margin to slowly subside. Isostatic subsidence, in this case, is rapid enough to keep pace with deposition. The delta sinks as more sediment is deposited. The possible result is the accumulation of many thousands of feet of sediment, all deposited in shallow water. Such is the current situation along the Atlantic and Gulf coast shorelines where shallow-water deposition and slow cooling of the lithosphere has driven isostatic subsidence since the opening of the Atlantic Ocean about 170 million years ago.

EROSION

Erosion removes weight from a location, thus causing isostatic uplift. The focused removal of material from one area but not another could, therefore, cause focused isostatic uplift. For example, if mountain elevation results in a rain

FIGURE 7.8 Google Earth image looking NNE along the Colorado River just north of the confluence with the Green River on the Colorado Plateau, Utah. The sedimentary layers are flexed upward slightly where they cross the river as part of the Meander anticline. The La Sal Mountains form the skyline at upper right.

shadow, then the wet, windward, side of the mountain could conceivably undergo erosion at a faster rate than the dry side. This could cause the wet side to isostatically uplift at a higher rate, thus altering the shape of the mountain.

Another example has to do with river incision and the focused removal of rock from a river canyon. Such a situation could potentially cause the lithosphere directly below the canyon to isostatically rise to a greater extent than areas surrounding the canyon where bedrock has not been removed. The result could be the formation of a broad anticline whose crest follows the center of the river canyon and where bedrock layers in the canyon walls dip gently away from the center of the canyon. In this case, the anticline is a product of focused erosion in which formation of the river canyon creates the anticline. Here, we have a situation in which focused climate-driven erosion causes both isostatic uplift and broad folding.

A superb example of an anticline that follows the path of a river canyon is the Meander Anticline, which precisely follows meanders of the Colorado River for about 25 miles in the area north and south of its confluence with the Green River in Canyonlands National Park, Utah. In this case, the river flows across an area where low-density evaporite (salt) beds are present at depth. The unloading (erosion) of overlying rock has caused the lighter evaporite rock to rise toward the surface and, in doing so, has arched overlying beds at the rim of the canyon into an anticline. Part of the Meander Anticline is shown in Fig. 7.8. Note that the layers are flexed upward on either side of the river.

QUESTIONS

1. Under what circumstances does isostatic uplift (or subsidence) occur?
2. What are the significant differences between tectonic uplift and isostatic uplift in terms of their mode and rate of uplift/subsidence, their sustainability, and their final outcome?

3. How is isostatic uplift different from tectonic uplift in terms of its longevity?
4. Explain why ocean basins are, on average, more than 13,000 feet below sea level, and continents are above sea level.
5. If two oceanic lithospheres were to collide, what would you have to know in order to predict which lithosphere would subduct beneath the other?
6. How soon after applying weight to the Earth's surface would you expect isostasy to respond? For how long will isostasy respond?
7. What is thermal isostasy? Why might the introduction of heat into the lithosphere cause uplift?
8. Explain how an individual mountain peak can increase its elevation in a mountain range with no tectonic uplift?
9. Research the Meander Anticline in Utah. What is it? Where is it exactly? How did it form?
10. Explain why isostasy allows mountains to maintain elevation over a much longer time period than might otherwise be expected.
11. Provide some examples of isostatic compensation.
12. How would isostasy respond to each of the following: the accumulation of ice, the cooling of the lithosphere, river incision, sediment deposition, the replacement of mantle rock with continental crustal rock, melting of a glacier, the accumulation of volcanic rock, the removal of rock by erosion, and lithospheric heating,
13. Can isostatic uplift create a mountain? Explain why or why not.
14. The Basin and Range is an area of thin crust. Why are many basins above 3000 feet in elevation?
15. Assume that 1 cm in Figure 7.4 is equivalent to 25 km depth. Also assume that the width of each column of rock in Figure 7.4 is vanishingly small (the column has no width). Assume a constant density for crustal rock of 2700 kg/m^3, and a constant density of 3300 kg/m^3 for mantle rock. Calculate the weight of the middle column by multiplying length times density for both the crust and mantle and adding the two products together. Do the same for the other two columns. If we assume that the middle column is in perfect isostatic equilibrium, can we also state that the other two columns are in perfect isostatic equilibrium? If not, is there excess mass, or is there mass deficiency? If we assume no change in the thickness of mantle rock, calculate the correct thickness of crustal rock for both outside columns assuming they are in equilibrium with the middle column. Show your math.
16. It was stated that isostatic uplift can sustain a mountain range, but is not capable of creating a high mountain range. Can you think of a scenario where isostatic uplift of a plateau can result in mountainous topography? Name the physiographic province where these types of mountains are present.

Chapter 8

Forcing Agent: Sea Level Change

Elevation is measured relative to sea level; therefore, any change in the absolute worldwide level of the sea will result in worldwide changes in elevation. A rise in sea level is equivalent with worldwide subsidence of land and could result in river and shoreline flooding. A drop in sea level is equivalent with worldwide uplift of land and could result in river incision and expansion of land areas across the continental shelf. Sea level change is a secondary forcing agent that occurs in response to tectonic plate movement, and more importantly, to climate change in the form of glaciation and deglaciation. This chapter discusses sea level change within the past 100 million years with respect to paleoclimate, atmospheric CO_2, and the recent glaciation.

CAUSE OF SEA LEVEL CHANGE

Global sea level is dependent on the total volume (area times depth) of the Earth's ocean basins, the amount of water available on earth, and the temperature of the ocean. All have varied throughout geologic time. Sea level, therefore, has also varied. Ocean basins open via rifting, and close via subduction. Old ocean basins are deeper than young ones because the rock that underlies old ocean basins is relatively cold and dense. The density (weight) of the rock causes it to isostatically sink. The age of ocean basins is dependent on the velocity of plate movement and on the total length and location of subduction zones relative to divergent plate boundaries. If tectonic movement results in ocean basins becoming younger, smaller, and more shallow over time, seawater will be displaced onto continents and global sea level will rise. The opposite occurs if ocean basins become, older, larger, and deeper. Tectonic plates move at rates that vary between about 11 and 125 miles per million years (6–66 feet per 100 years). The effect on sea level is estimated to be on the order of 0.039 inches (1.0 mm) per 100 years. Such a slow rate of change has had minimal effect on recent sea level change, but over the course of tens or hundreds of millions of years, such rates can potentially create as much as 1000 feet of sea level change. At a rate of 0.039 inches per 100 years, sea level can change by 325 feet (99 m) over a 10 million year period.

At the global scale, climate change refers primarily to temperature changes that affect the ocean and atmosphere. The most profound climate change with respect to landscape and sea level change is glaciation. Glaciation affects both the amount of available water on Earth and the temperature of the ocean. During a cycle of glacial advance and retreat, sea level can change at rates that exceed 6 feet per 100 years. Part of this change (as much as 50%) can be due to thermal contraction and expansion of seawater as it cools and warms. Glaciation, coupled with thermal expansion and contraction of seawater, and with variations in groundwater and lake storage, can potentially change sea level by as much as 700 feet over the course of thousands to tens of thousands of years. This rate of sea level change is orders of magnitude faster than rates described above for plate tectonic movement.

Although plate movement cannot very quickly alter the size of an ocean basin, plate movement can alter climate and possibly induce glaciation over relatively short time intervals of thousands to millions of years. Landmasses, for example, can be displaced to high latitude where they are more susceptible to glaciation. Displaced landmasses can block or deflect both surface and deep ocean currents, thereby altering the climate of continents, again, possibly resulting in glaciation. Such a scenario may have caused the recent Northern Hemisphere glaciation as discussed below.

MEASURING SEA LEVEL AND SEA LEVEL CHANGES

Throughout the 20th century, tide gauges positioned along shorelines have been used to record mean sea level. In the United States, mean sea level is determined by recording the height of the surface of the sea at many locations at hourly intervals over a 19-year period. The results are then averaged. The frequency of measurement is required in order to average out the tidal highs and lows caused by gravitational forces from the moon and sun. More

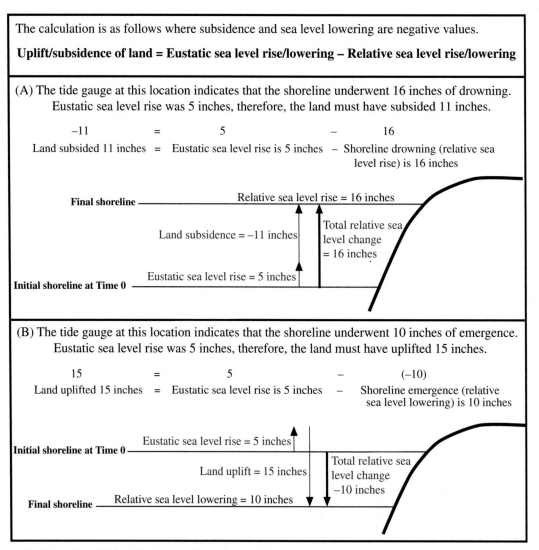

FIGURE 8.1 Calculation of vertical land displacement if eustatic and relative sea level change are known.

recently, satellite altimetry measurements have supplemented tide gauge measurements. Sea level change is indicated if mean sea level changes over an interval of time.

Measuring changes in mean sea level at any one location using tide gauges is not straightforward because, in addition to actual changes, one must also account for vertical displacement of land (uplift/subsidence), which will affect sea level at that location. From a global perspective we want to measure eustatic (global or real) sea level changes. These are actual worldwide changes in the volume of available water regardless of whether land at any one location is rising or sinking.

From a landscape perspective, we need to make a distinction at any one location between sea level change due to eustatic changes in the worldwide volume of water and sea level change due to vertical displacement of land. Tide gauge stations measure relative sea level change, which can be defined as the sum effect of eustatic sea level change and vertical displacement (uplift/subsidence) of land at that location. Thus, we are dealing with three variables at any one location, eustatic sea level change, relative sea level change, and vertical displacement of land. If two of the three variables are known, the third can be calculated as follows where subsidence and sea level lowering are negative values (examples are given in Fig. 8.1).

Relative sea level rise/lowering = Eustatic sea level rise/lowering − Uplift/subsidence of land
  or
Uplift/subsidence of land = Eustatic sea level rise/lowering − Relative sea level rise/lowering

When measuring ancient eustatic sea level changes, one way to mitigate the effect of vertical land displacement is to map the worldwide distribution and age of marine sedimentary rocks (rocks deposited from ocean water). By studying and correlating the distribution of marine sedimentary sequences worldwide, it is possible, to a first approximation, to remove the effects of vertical land displacement and study long-term eustatic sea level changes. Such measurements extend back hundreds of millions of years. Given the numbers presented at the beginning of this chapter, it is possible for sea level to change by as much as 1700 feet if we assume that tectonic plate motion and climate simultaneously create conditions for maximum change. Early studies of sea level change implied this much variation or more. More recent studies suggest that sea level has varied by less than 1000 feet over the past 541 million years and probably by less than 700 feet, especially over the past 100 million years.

To understand the effect of sea level change on landscape, we need more precise and accurate estimates of sea level, particularly over the past several hundred thousands to several tens of millions of years. The most direct method is to locate and date strandlines and marine terraces that formed at a time when sea level was higher than today. A strandline is a shoreline high water mark, now elevated above sea level. A marine terrace is a flat or gently inclined surface that ends abruptly against higher ground and which represents a wave-cut abrasion platform cut at sea level but now elevated above sea level. A marine terrace can be thought of as a raised beach. Dating is best accomplished along low-lying coastlines where vertical tectonic/isostatic displacement can be calibrated. Unfortunately (with some exceptions), this technique is available with high confidence only as far back as the previous sea level high stand, about 125,000 years ago.

Fortunately, there now exists a variety of dating techniques that can be used to estimate sea level change of as little as a few feet. One technique employs radioactive uranium. Uranium is dissolved in seawater allowing marine animals, particularly coral, to extract it and incorporate it into their shell. By analyzing the parent/daughter ratios of certain decay products of uranium, it is possible to date marine sediments as far back as 500,000 years. Coral reefs are particularly helpful because they grow to the height of sea level. One must be careful with this technique because reefs can be eroded as sea level decreases, and the rate of reef growth cannot always keep pace with sea level rise.

The most widely used method to track changing ocean and air temperature employs the stable oxygen isotopes ^{18}O and ^{16}O. The ratio of ^{18}O to ^{16}O is used as a proxy (a substitute) for direct measurement of temperature because it can be reliably measured in the rock record and because the ratio changes in a predictable way with changes in temperature. These changes, when combined with field data, can be correlated with the growth and melting history of continental glaciers and with changes in sea level.

Both isotopes are present in water. An important consideration is how the isotopes behave during evaporation and precipitation and how this behavior changes with temperature. The isotopes do not weigh the same, and as a result, they fractionate (they behave differently) during evaporation and precipitation. The ^{16}O isotope is preferentially evaporated from ocean water because it is lighter than ^{18}O. The opposite effect occurs during precipitation. The heavier ^{18}O isotope is preferentially precipitated. As an air mass moves from a warm region to a cold region, early precipitation will have a high concentration of ^{18}O such that later precipitation from the same air mass will be progressively depleted in ^{18}O. By the time the air mass reaches the poles where glaciers form, much of the ^{18}O in the air mass will have already precipitated in the form of rain that will return to the ocean. The ^{16}O isotope, on the other hand, is sequestered in glacial ice and not returned to the ocean. Fractionation is measured by recording the $^{18}O/^{16}O$ ratio relative the $^{18}O/^{16}O$ ratio of standard seawater. If the ^{16}O isotope is sequestered in glacial ice, the $^{18}O/^{16}O$ ratio will increase in seawater and decrease in glacial ice, and these ratios continue to change as glaciers increase in volume.

Oxygen isotope ratios can be extracted both from ocean-dwelling fossils and from ice cores. Ratios extracted from fossils provide a record of temperature change in the ocean. The primary fossils used for oxygen isotope analysis are the shells of deep water and sediment-dwelling (benthic) foraminifera. The foraminifera are extracted from drill cores and are dated either directly or through magnetic stratigraphy principles. The process of calculating temperature is not straightforward because the $^{18}O/^{16}O$ ratio incorporated by foraminifera is not an exact match to the ratio in ocean water, and because the incorporation ratio in foraminifera changes with water temperature. Fortunately, scientists are able to correct for these discrepancies. With proper procedure, this technique can be used as far back in time as unaltered foraminifera are found, which can be more than one hundred million years. Glacial advances and retreats inferred from these data are correlated and calibrated with field data to show that this method is both accurate and robust. Hydrogen also has a heavy isotope, known as deuterium, which allows hydrogen isotopes to be used in a manner similar to oxygen isotopes.

Ice cores are a gold mine of information because the oxygen isotope ratio they provide is the same oxygen isotope ratio that was in the atmosphere at the time the snow fell. We noted earlier that there is a preference for the heavier ^{18}O isotope to be concentrated in early precipitation (rain) within an air mass. An important factor is that the amount of ^{18}O concentrated in early precipitation will increase as temperature decreases. What this implies is that, as worldwide climate cools, less of the ^{18}O isotope will reach the continental glacier. In other words, the $^{18}O/^{16}O$ ratio in glacial ice decreases as climate becomes

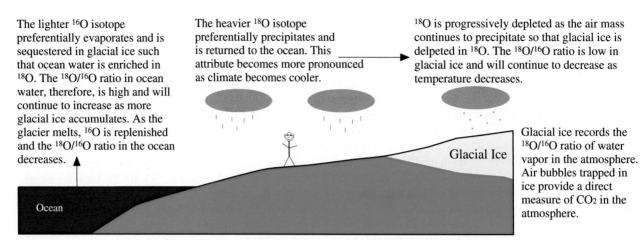

FIGURE 8.2 Diagram that shows how evaporation, precipitation, and temperature affect oxygen isotopes.

cooler. Ratios found in ice cores thus provide a continuous record of temperature change in the atmosphere. The relationships are summarized in Fig. 8.2.

In addition to direct measurement of $^{18}O/^{16}O$ ratios, ice cores also provide direct measurement of atmospheric dust and CO_2. Air bubbles trapped in glacial ice record the amount of CO_2 in the atmosphere at the time the bubble was trapped. Air temperature, the amount of CO_2, and the thickness of accumulated dust and snow will vary from one season to the next during the course of a year. This variation is recorded in the ice core allowing scientists to see how climate changes on a seasonal and yearly basis. The seasonal variation also provides a method to date an ice core similar to counting tree rings. Radiometric dating of volcanic ash layers within the ice core can also determine the age of ice. Drilling in Antarctica has provided data as far back as 850,000 years. The Greenland ice core record extends back to about 100,000 years.

SEA LEVEL CHANGES OVER THE PAST 100 MILLION YEARS

Based on the location, age, and continuity of marine sedimentary rocks, we can say with a high degree of certainty that sea level has been higher than present-day levels throughout nearly all of the past 541 million years. The implication is that global temperature has also been higher throughout this time span. Let us summarize what we know about the past 100 million years.

On the basis of oxygen isotope analysis, the period between 100 and 34 million years ago was warmer than today with continuous high sea level and few glaciers anywhere on Earth. Estimates of sea level vary widely. Some workers have suggested maximum estimates in the range of 820 to 1050 feet above present day, but these estimates might be too high. Studies using more recent cumulative data suggest that sea level was between 165 and 500 feet above present day throughout the 100 million- to 34 million-year period with a maximum sea level high-stand occurring between 50 and 55 million years ago. There were no permanent ice caps on Greenland or Antarctica during the 100 to 34 million-year time interval, and although the entire Earth was ice-free for most of that time, evidence of sea level oscillations on the order of 50 to 100 feet (15–30 m) imply periodic growth and decay of small inland glaciers on Antarctica.

The amount of continental drowning during a rise in sea level is dependent on the mean elevation of continents, the area of low-lying coastal areas, and the magnitude of uplift or subsidence along the coast. Fig. 8.3 shows areas in the United States that are inferred to have been below sea level during three snapshots of time 80, 50, and 15 million years ago. The figure is based on the distribution of marine sedimentary rocks. Do not infer from this sequence that sea level has steadily lowered over this time span, as this is not what happened. Note the presence 80 million years ago of a large inland sea that covered most of the Great Plains, the Colorado Plateau, and the Middle and Southern Rocky Mountains. Obviously, these high elevation landscapes were not in existence 80 million years ago. This inland waterway is known as the Western Interior Seaway. It was the last inland sea to have inundated the United States. The Coastal Plain, on the other hand, has been lowland throughout its existence and has frequently been inundated during high sea level stands. Much of Florida was below sea level just 3 million years ago.

Sea level changes on the west coast are not shown in Fig. 8.3 because of ongoing mountain building and tectonic modification. The west coast consists of accreted terranes and some areas of California, Oregon, and Washington did not exist as a part of the North American continent prior to 50 million years ago.

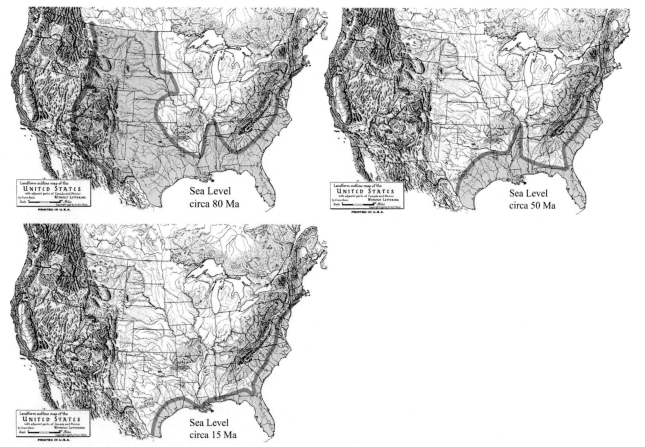

FIGURE 8.3 Sea level 80, 50, and 15 million years ago.

OXYGEN ISOTOPE RECORD OVER THE PAST 67 MILLION YEARS

A graph of deep-sea oxygen isotope ratios over the past 67 million years is shown in Fig. 8.4. The graph shows only the very broadest, general trends at the one million-year scale or greater. It does not show the many oscillations and perturbations that occurred on the 100,000-year scale. Nevertheless, it shows an overall warming trend from 67 million years ago to 51 million years ago during what is known as the Early Eocene climatic optimum. Following the Early Eocene peak, climate began to cool slowly, leading to the appearance of small glaciers on Antarctica by about 37 million years ago. A rapid cool-down at 34 million years ago led to a full-scale ice cap on Antarctica that lasted until about 26 million years ago when a nearly equally rapid warm-up caused Antarctic glaciers to shrink and possibly to disappear completely for short time intervals. The warm period lasted until 15 million years ago when a strong cool-down established the permanent glaciers on Antarctica that we see today. This general cooling trend has continued to the present day (perhaps excluding the last several decades). Continental glaciers appeared in the northern hemisphere possibly as early as 8.0 million years ago and became permanent by 3.2 million years ago. As discussed in Chapter 12, glaciers covered large areas of the United States beginning about 2.4 million years ago.

The rapid cool-down at 34 million years ago and the ensuing glaciation of Antarctica resulted from a combination of factors. A decrease in CO_2 in the atmosphere may have played a role, but perhaps the most important factor was plate tectonic movement and its effect on ocean currents. Prior to 34 million years ago, Antarctica was close enough to the southern tip of South America that cold-water ocean currents circling Antarctica were forced northward up the west coast of South America toward the equator where they would warm. Between 34 and 29 million years ago, South America moved northward thereby opening Drakes Passage between the two continents. This opening allowed cold-water to continuously and completely encircle Antarctica, forming the Antarctic Circumpolar Current and sending Antarctica into a deep freeze. At the same time, CO_2 levels may have decreased

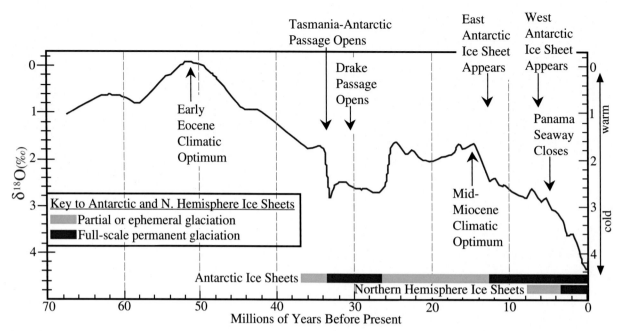

FIGURE 8.4 Global deep-sea oxygen isotope records based on data compiled from more than 40 Deep Sea Drilling Project and Ocean Drilling Project sites. *Redrawn from Figure 2 of Zachos et al. (2001).*

due to increased rates of weathering following the rise of the Himalayan Mountains (weathering of silicate-rich rocks extracts CO_2 from the atmosphere) and to the proliferation of our old friend, the foraminifera, whose calcium carbonate shells extract CO_2 from sea water. Sea level varied by 100–200 feet (30–60 m) between 34 and 2.7 million years ago coincident primarily with the growth and decay of Antarctic ice sheets.

Some of the largest oscillations in sea level of more than 400 feet have occurred in the past 3.2 million years in conjunction with glacial advances and retreats not only on Antarctica, but also on northern hemisphere landmasses. As with the glaciation of Antarctica, the reason for the deeper plunge into the ice age has its roots in plate tectonics and ocean currents. By 3.2 million years ago plate tectonics had already placed large landmasses at northern latitudes. Between 6 and 4 million years ago, tectonic plate movement closed the Panama Seaway that once connected the Caribbean Sea with the Pacific Ocean. The closing of the seaway altered ocean currents and created conditions conducive for northern hemisphere glaciation.

INFLUENCE OF EARTH'S ORBITAL PARAMETERS ON GLACIATION

Once global conditions became conducive to glaciation, the pace of glaciation, and the concomitant fall and rise of sea level, became strongly influenced by three of Earth's orbital parameters: eccentricity, obliquity, and precession, which collectively make up the Milankovitch climate cycle. Together, these parameters affect the distribution and amount of solar radiation reaching the Earth. Eccentricity refers to the shape of Earth's orbit around the sun. Eccentricity varies from a nearly circular orbit to an elliptical one over a period of 100,000 years. Obliquity refers to the tilt of Earth's axis relative to the vertical rays of the sun. In Chapter 6 we saw that present-day obliquity is 23.5° thereby defining the Tropics of Cancer and Capricorn. Obliquity varies from 22.1° to 24.5° over a period of 41,000 years. Generally, the greater the angle of obliquity, the greater the seasonal contrast. Winters become colder while summers become warmer as obliquity increases. Precession refers to the wobble of the Earth's axis. In Chapter 6 we described the Earth's axis as fixed in space relative to Earth's orbit around the sun thereby creating the solstice and the equinox twice per year. Rather than being strictly fixed, the Earth's axis wobbles in a circular pattern not unlike a slowly spinning top. Precession varies with eccentricity on roughly 22,000-year periods. If summer solstice occurs when Earth is relatively far away from the sun (in an elliptical orbit), then summer will be cool. If summer solstice occurs when Earth is relatively close to the sun (in a near circular orbit), then summer will be hot. These three parameters can combine to create conditions of consecutive cool northern hemisphere summers, which results in less glacial melting over the course of a year, greater ice accumulation, and a concomitant fall in sea level.

OXYGEN ISOTOPE RECORD OVER THE PAST 1.8 MILLION YEARS

The Milankovitch parameters, in and of themselves, are not strong enough to produce glaciation, but they do produce a periodicity to the glaciation, and this periodicity is reflected in oxygen isotope ratios when viewed at a fine scale. Fig. 8.5 shows oxygen isotope ratios for the past 1.8 million years. Notice the strong oscillations in the data. These oscillations trace out glacial cycles. Prior to one million years ago, a complete glacial cycle of advance and retreat occurred roughly every 41,000 years in sync with the obliquity of Earth.

Between 1 million and 800,000 years ago, the graph shows that the periodicity of oxygen isotopes changed from 41,000-year cycles to 100,000-year cycles. It is not certain why this happened, but one theory has to do with the thickness of glaciers and the land surface over which the glaciers moved. For millions of years prior to the appearance of continental glaciation, the land area in northern Canada, where the ice sheet originated, underwent weathering and presumably the formation of a thick, well-developed soil. The earliest glacial advances removed this soil by scraping the land surface down to bare bedrock. Subsequent glaciers would have to develop directly on bedrock. Glaciers move both along the base of the ice sheet and within the ice body itself. Early glacial advances would have found it easy to move along their base because the underlying soil would be weak and easily deformed. Ice would have moved rapidly southward without much resistance. The result was thin, relatively fast-moving glaciers that carried great piles of sediment southward. Once bedrock became exposed, the base of the ice sheet could become frozen on bedrock and basal movement would be curtailed. If the base of an ice sheet were frozen, ice within the body of the ice sheet would become the primary form of movement. Such conditions would require ice to accumulate to a much greater thickness prior to the ice sheet advancing southward. The theory suggests that a thick ice sheet would survive warm intervals and remain thick enough to dampen sea level change (oxygen isotope ratios) on the 41,000-year cycle, but not on the stronger 100,000-year cycle.

If we use the oxygen isotope variations shown in Fig. 8.5 as a proxy for glacial advance and sea level change, then over the past 500,000 years we can delineate glacial maximums and low sea level stands at approximately 430, 340, 250, 140, and 20 thousand years ago, and high sea level stands during the present-day and at 490, 405, 325, 240, and 120 thousand years ago. Notice that how quickly sea level rises following a glacial maximum. For each of the five major glacial cycles depicted on the graph, sea level rose to its maximum level within 25,000 years following the glacial maximum. Also notice the many minor peaks and valleys depicted on the graph. From this, we can conclude that there were numerous minor advances and retreats superimposed on the major cycles. One of the largest of these secondary glacial advances occurred 220,000 years ago.

SEA LEVEL OVER THE PAST 150,000 YEARS

Fig. 8.6 shows a detailed sea level curve for the past 150,000 years based on oxygen isotope proxies, strandlines, and marine terraces. Here, we get a more detailed record of sea level rise and fall. With respect to this figure, sea level rose 361 feet (110 m) in 5000 years at the end of the previous glaciation between 140,000 and 135,000 years ago at an average rate of more than 7 feet per 100 years. A rapid drop in sea level 131,000 years ago was followed immediately by a rise that exceeded present-day levels by 13 to 26 feet (4–8 m). This high sea level stand lasted from about 128,000–118,000 years ago. Oxygen isotope and sediment core studies indicate

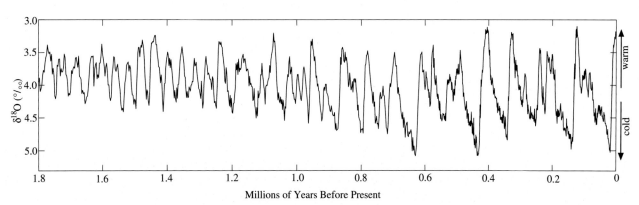

FIGURE 8.5 Stable oxygen isotope ratios for the past 1.8 million years (the LR04 benthic $\delta^{18}O$ stack) constructed by the graphic correlation of 57 globally distributed benthic $\delta^{18}O$ records. *Redrawn from Figure 4, Lisiecki and Raymo (2005).*

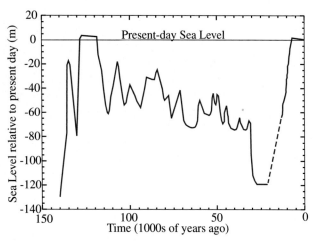

FIGURE 8.6 Relative sea level curve for the past 150,000 years based on the Huon Peninsula, Papua New Guinea, and Bonaparte Basin, Australia. *Redrawn from Figure 2 of Lambeck et al. (2002).*

that mean global surface temperature at this time was at least 2°C warmer than present. The result was substantial drowning of the Coastal Plain including all of southern Florida south of Palm Beach.

Sea level fell between 118,000 and 30,000 years ago but with many oscillations. Fig. 8.5 shows 11 high stands during this time interval with oscillations between 30 and 116 feet (10–35 m) at rates that could have exceeded 6.5 feet per 100 years for short periods of time. A sharp drop of 100–130 feet (30–40 m) within a period of 1000–2000 years beginning 30,000 years ago signaled the beginning of the final glacial advance that lasted in the United States until about 18,000 years ago. During that time (and during the previous glaciation that ended 135,000 years ago) glaciers covered three times as much land area as they do today and sea level was about 400 feet (122 m) lower than it is today, exposing nearly all of the Atlantic continental shelf (e.g., Fig. 1.7). It was at this time that Florida was about twice the size it is today and New York City was more than 100 miles from the coastline. The Pacific continental shelf was deeply incised by rivers during the glacial advance but, because the shelf is so narrow (Fig. 1.8), the location of the coastline did not change appreciably.

RECENT TEMPERATURE HISTORY

Greenland is fertile ground (or should I say fertile ice?) for the study of temperature history over the past 100,000 years, particularly with respect to northern hemisphere climate. It is clear from deep-sea sediment cores and ice cores that Greenland has seen stronger and more rapid temperature swings relative to Antarctica. The major events are given names. There were as many as 25 Dansgaard-Oeschger events recorded in Greenland between 100,000 and 10,000 years ago where air temperature increased by 5 to 10°C (9°–18°F) in only 10 to 50 years, and then remained warm for several centuries. These events appear to have been cyclic, occurring either every 1470 years, or in multiples of 1470 years (i.e., every 2940 or 4410 years). Additional warm periods in the northern hemisphere occurred from about 8000 to 6700 years ago, from 1100 to 600 years ago, and from 150 years ago to the present. There have also been at least six Heinrich events between 65,000 and 14,500 years ago in which atmospheric temperatures cooled rapidly. The most recent cold snaps were the Younger Dryas between 12,800 and 11,500 years ago and the Little Ice Age between 700 and 150 years ago.

Although these events were felt globally, their effect was different or diminished beyond the North Atlantic Ocean. For example, ice cores from Antarctica, in some cases, indicate a reverse reaction to that of Greenland. Antarctica underwent warming or was hardly affected when Greenland slipped into a cold phase, or underwent cooling as Greenland returned to a warm phase. The limited extent and different outcomes of these rapid temperature swings suggest that they were not brought on by rapid changes in CO_2, which would have a global effect felt in Antarctica as well as Greenland. Instead, the evidence suggests that these events have their origin in the disruption and shifting of warm- and cold-water ocean currents brought on by the release of large volumes of ice in the form of icebergs, or from floodwaters released suddenly from failed glacier-damned lakes. Physical evidence for this interpretation includes the presence of coarse-grained layers of glacial sediment within deep-sea sediment cores located far from any landmass. These sediments must have been transported across the open ocean by icebergs. Additionally, there is evidence for catastrophic flooding on land that can be tied to some of the events. Variations in solar radiation output, volcanism, and CO_2 levels may have also contributed. The Little Ice Age is thought to have been a consequence of unusually high volcanic activity.

A compilation of globally distributed temperature records for the past 11,300 years is shown in Fig. 8.7. The figure shows the change in temperature relative to the mean annual temperature recorded between 1961 and 1990 (set at zero). The data suggest that global temperature increased by about 0.6°C from 11,300 years ago to 9600 years ago and then remained on a warm plateau until 5300 years ago with a slight peak just prior to 7000 years ago. Global temperature, at times during this interval, may have been higher than present-day global temperature. This warm interval was followed by long-term global cooling of 0.43°C between 5300 and 1000 years ago followed by additional cooling of 0.28°C at the peak of the Little Ice Age.

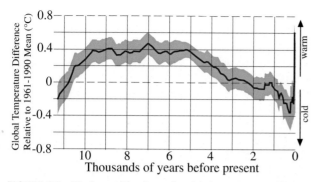

FIGURE 8.7 The heavy line shows the global temperature difference for the past 11,300 years relative to the 1961–1990 mean temperature, based on 73 globally distributed records. The shaded area represents 1σ uncertainty. *Redrawn from Figure 1B of Marcott et al. (2013).*

Within the past 150 years, and particularly during the past three decades, the data indicate a steep upward trend in temperature (as seen at the right edge of the graph) in which each of the last three decades have been hotter than the previous, and the hottest on record. From 1880 to 2013, average global temperature has risen 0.8°C (1.4°F). To put this in perspective, since 1880, average global temperature has increased from nearly the coldest level recorded in the past 11,300 years to nearly the warmest. Even conservative estimates indicate that by 2100, global temperature will have exceeded anything seen in the past 11,300 years. Such rapid global warming is not seen in any record of the past 541 million years of Earth history.

SEA LEVEL RESPONSE TO RECENT TEMPERATURE HISTORY

Although not steady, sea level has been on the rise for the past 18,000 years due, not only to glacial melting, but also to thermal expansion of ocean water as ocean temperature becomes warmer. A total sea level rise of 400 feet (122 m) over the past 18,000 years implies an average rise of more than 26 inches per 100 years. The highest rates of more than 3 feet per 100 years occurred between about 9600 and 5500 years ago when global temperatures periodically rose above present-day temperature and when there was considerably more glacial ice on continents to melt than there is today. Rates may have approached 15 feet (4.5 m) per 100 years for a short time around 7600 years ago when northern hemisphere ice sheets underwent rapid decay. According to Fig. 8.6, sea level rose slightly above present-day levels at about that time. It is lucky for us that sea level is not rising at such a high rate today as it would create havoc for the many cities and the billions of people living along coastlines.

Nearly the entire North American continental ice sheet had melted by 5,500 years ago, and since that time, sea level has increased an average of only 2 to 4 inches per 100 years (0.5–1.0 mm/year). However, the rise in sea level has accelerated over the past 150 years. Sea level is estimated to have risen 7 to 8 inches during the 20th century and is expected to rise a total of 15 to 20 inches during the 21st century. But even this estimate might be conservative. Some estimates suggest that sea level rise could reach 3 to 6 feet by the end of the 21st century, especially if the Antarctic ice sheet were to destabilize at an accelerated rate. Sea level is currently rising at a rate of nearly 13 inches per 100 years based on satellite altimetry with thermal expansion of the ocean accounting for about 44% of that rise. According to the US Geological Survey, if the Greenland ice sheet were to melt completely, sea level would rise by 21 feet (6.5 m). This is more than enough to drown all the world's coastal cities and cause severe river flooding. About 90% of the world's ice is contained in the Antarctic Ice Sheet. Sea level would rise another 240 feet (73.3 m) or so if this ice sheet were to melt. Such a catastrophic rise would drown nearly all of the Coastal Plain.

It should be noted that global sea level change does not affect all locations evenly. Gravitational forces due to latitude and the presence of large landmasses will pull the ocean higher at one location relative to another. Thus, a 10-inch eustatic (worldwide) sea level rise can result in a 12-inch rise at one location and an 8-inch rise at another. As noted earlier, the relative sea level rise (or fall) at any one location is dependent, not only on eustatic sea level rise at that location, but also on the amount of vertical land displacement.

Eustatic (worldwide) sea level change affects not only shoreline regions but also river systems particularly with respect to whether a river cuts downward into bedrock or deposits sediment in a river channel. We discuss the effect of sea level on rivers in Chapter 9.

THE HISTORY OF CO_2 IN THE ATMOSPHERE

From the above discussion, we can infer that there are many factors that control climate, glaciation, and sea level change. Changing ocean currents appear to have a rapid and profound effect on climate that is most severe along landmasses that border the altered current. The Earth's orbital parameters create a distinct periodicity to climate that is clearly reflected in the oxygen isotope record. We have seen, however, that the magnitude of this periodicity is not enough to force glaciation unless additional influences are at work such as the positioning of continents at high latitudes and the altering of ocean currents.

So far, we have not discussed in detail the role that CO_2 and other greenhouse gases play in our climate system. A greenhouse gas is defined as any gaseous compound that is capable of absorbing infrared radiation thereby trapping and holding heat in the atmosphere. The most potent greenhouse gas is water vapor. However, water vapor is held in the atmosphere for only a short period of time and the amount of water vapor held in the atmosphere is limited by saturation and precipitation. The most dangerous greenhouse gas at the moment is CO_2 because it is abundant relative to other greenhouse gases (excluding H_2O); because the amount of CO_2 that can be held in the atmosphere is virtually unlimited; because CO_2 in the atmosphere holds heat, which creates conditions whereby more and more water vapor can be held in the atmosphere, thereby exacerbating global warming; and because CO_2, when dissolved in the ocean at an increasing rate, creates conditions that acidify the ocean. The acid dissolves calcium carbonate, which in turn causes calcium carbonate-shelled animals such as coral, most shellfish, and foraminifera to die, thereby destroying the ocean ecosystem.

The carbon cycle is similar to the water cycle. Water is cycled from the ocean to the atmosphere and back to the ocean. Water can be stored for a period of time in such intermediaries as living organisms, glaciers, lakes, rivers, and groundwater. CO_2 also cycles between the ocean and the atmosphere. It can be stored for short periods of time in living organisms, or for potentially very long periods of time in the form of limestone, coal, and petroleum.

The primary natural mechanisms by which CO_2 is released into the atmosphere include respiration by plants and animals, soil respiration, and ocean-atmosphere exchange. Soil respiration is any respiration that occurs below ground by organisms including those involved in the decomposition of organic matter. Ocean-atmosphere exchange is an attempt to equalize CO_2 concentration between surface waters of the ocean and the lower part of the atmosphere. It is a reciprocal process. Atmospheric CO_2 is dissolved into ocean surface waters, while at the same time, dissolved CO_2 is released by the ocean back into the atmosphere. Additional natural mechanisms by which CO_2 is released to the atmosphere include the weathering and dissolution of carbonate rocks, volcanic eruptions and the cooling of magma, the release of CO_2 along faults and fractures in sedimentary basins, and the metamorphism of limestone.

Primary mechanisms by which CO_2 is removed from the atmosphere include ocean-atmosphere exchange, plant photosynthesis, the production of calcium carbonate shells, the biochemical precipitation of limestone by algae, the burial of organic matter to form coal and petroleum, and the chemical weathering of silicate rocks.

In the natural world (excluding mans contribution to CO_2 in the atmosphere), the carbon cycle appears to have reached a near steady-state over the past one million years and for most of the past 20 million years such that the amount of CO_2 added to the atmosphere by all sources has equaled the amount of CO_2 removed from the atmosphere by all sources. The evidence for this is the fact that CO_2 in the atmosphere has remained fairly constant between 180 and 300 ppm for all of the past 850,000 years and for most of the past 20 million years. We will discuss this evidence in the next few paragraphs. Of course, we are all aware that man's recent CO_2 contribution has upset this balance.

Before we look at the most recent history of CO_2 in the atmosphere, let us look at how CO_2 levels have changed over the past 450 million years. The amount of CO_2 in the atmosphere can be determined directly via ice core analysis for only the past 850,000 years. Beyond that time, proxies are used. Some CO_2 proxies include changes in ^{13}C within certain minerals in soils, in certain types of algae, and in liverworts (similar to moss). Other proxies include changes in the stomatal (pore) density of fossil leaves, and changes in ^{11}B and B/Ca in foraminifera.

It is important to understand the limitations of CO_2 proxies. The data with few exceptions become increasingly sparse with distance into the past. As a result, the resolution of data from several hundred million years ago is such that any change in CO_2 levels that occurred over time spans of less than one million years is difficult or impossible to detect. Precise temperature data also are not available for the distant past; therefore we use sea level curves and the geological record of glaciation to correlate changes in CO_2 with changes in temperature. The data are much better for the past 80 million years; even better for the past 20 million years; and is excellent over the past one million years. With increasingly better and more abundant data, the resolution will also improve.

It is also important to understand that one cannot compare the effect of atmospheric CO_2 concentrations from the ancient past with present-day concentrations. The distribution of continents, the albedo (the fraction of solar radiation reflected from Earth back into space), vegetative cover, the continent to ocean ratio, the amount of volcanism, the composition of the oceans and atmosphere, and the solar luminosity have all varied throughout geologic history.

Solar luminosity alone is very important. Solar luminosity is the total amount of energy produced by the sun and radiated into space in the form of electromagnetic radiation. It is well known that solar luminosity has steadily increased with time. The sun was cooler 4.55 billion years ago and has grown hotter. Increasing solar luminosity means that the amount of solar radiation reaching the Earth over time has steadily increased. If present-day Earth had existed 450 or 350 million years ago in the exact form it exists today with 400 parts per million (ppm) CO_2 in the atmosphere, the Earth would be an icebox with global surface temperature well below what we are experiencing today. For example, the estimated threshold for glaciers to form in the northern hemisphere today is about 500 ppm CO_2 in the atmosphere. Glaciers are unlikely to form above that threshold. Four hundred fifty million years ago this same threshold is estimated to have been about 3000 ppm CO_2 and 350 million years ago the threshold was about 2000 ppm CO_2.

Overall, as might be expected from its physical properties, a correlation exists over the entire 450 million-year time period between high atmospheric CO_2 concentrations and high globally averaged surface temperature. The spread in CO_2 concentration is large. The data suggest that CO_2 concentrations were in the neighborhood of 2000 ppm to more than 5000 ppm between 450 and 326 million years ago dipping to between 2000 and 500 ppm during cold spells and glaciation. By 326 million years ago, at the beginning of a major glaciation, CO_2 levels had dropped below 500 ppm. CO_2 concentrations remained low for most of the following 59 million years, rising again at the end of the glaciation beginning 267 million years ago to variable levels between 500 and well above 3000 ppm. Levels remained high for most of the next 167 million years before consistently dropping to between 100 and 1500 ppm beginning 100 million years ago where it remained until the most recent glacial stage beginning 34 million years ago. The data suggest that CO_2 levels have remained below 500 ppm for most of the glacial stage and have not been above 500 ppm in at least the past 20 million years. Present-day values of 400 ppm were exceeded only between 14 and 16 million years ago when average global temperature was 3 to 6°C warmer than present, and possibly between about 5 and 3 million years ago (the warmest period was 4.4 to 4.0 million years ago), just prior to the final glacial advance, when estimated average global temperature was 2 to 3°C warmer than present (this warm period shows only as a small bump in Fig. 8.4).

Trapped air bubbles within the Antarctic ice core provide a direct measurement of CO_2 over the past 850,000 years. Throughout that time, CO_2 levels have varied between a low of 180 ppm and a high of 300 ppm. CO_2 levels have not even approached present-day levels of more than 400 ppm in at least the past 850,000 years. Some of the most rapid swings in CO_2 that are recorded in ice cores prior to 1866 have been on the order of 10 ppm per 100 years. These numbers are in stark contrast to present-day rates of CO_2 accumulation. In the 50-year period from 1866 to 1916, the amount of CO_2 in the atmosphere as recorded from ice cores increased 15 ppm from 287 to 302 ppm. In the following 50 years, from 1916 to 1966, CO_2 in the atmosphere as recorded from ice cores, and beginning in 1958 as recorded directly at the Mauna Loa observatory in Hawaii, increased 20 ppm to 322 ppm. From 1966 to 2016, CO_2 in the atmosphere as recorded directly at Mauna Loa, increased 81 ppm to 403 ppm. The rise from January 2015 to January 2016 alone was 2.5 ppm.

It is interesting that, in the past 450 million years, the only time CO_2 levels were consistently below 500 ppm was during the two longest-lived glaciations; the first occurring 326–267 million years ago and the second beginning 34 million years ago and continuing to the present. A huge volume of the world's coal reserves formed during the first glaciation, including nearly all of the eastern US coal fields. Coal is fossilized carbon. It is composed of the remains of plants that did not decay but instead were buried. Plants consume CO_2 through photosynthesis, and release CO_2 upon death and decay. In this case, the carbon cycle was broken. Rather than being released back into the atmosphere, CO_2 was sequestered in the ground causing or at least contributing to low CO_2 levels and glaciation. We were gradually lifted out of this glacial period by a combination of factors that likely had to do with shifting continents and volcanism. The human species evolved during the second major glaciation when global temperatures were some of the coolest in history and when there existed abundant habitable landmass. As noted earlier, low CO_2 levels during this glacial episode may be related to increased weathering associated with the rise of highland regions such as the Himalayas, and with the proliferation of plants and animals. In only the past 50 years we have seen an unprecedented increase of CO_2 in the atmosphere, most of which can be directly linked to the activities of nearly 8 billion people. In this case we seem to be artificially and very abruptly lifting ourselves out of the ice age by releasing quantities of CO_2 into the atmosphere that had been sequestered in the ground for tens and hundreds of millions of years.

QUESTIONS

1. How is sea level determined?
2. What are eustatic sea level changes?
3. What are strandlines and marine terraces? What can they be used to date?
4. What methods are used to track present-day eustatic sea level change?
5. What variables affect the amount of continental drowning during a rise in sea level?
6. Describe how the melting of the Greenland glacier might affect worldwide sea level. How might this melting affect isostatic equilibrium on Greenland?
7. Based on what you have read in this chapter, how can we be reasonably sure that the Colorado Rockies did not exist 80 million years ago?

8. How has sea level changed over the past 120,000 years?
9. What are some of the highest inferred rates of sea level rise over the past 18,000 years?
10. Given the fact that sea level has risen about 400 feet in the past 18,000 years, how much higher could it rise if all of the world's glaciers were to melt?
11. Go to http://geology.com/sea-level-rise/ and describe how a sea level rise of between 3.3 and 98.4 feet (1–30 m) might affect Florida and the Chesapeake Bay area. Do the same for other areas of your choice.
12. Go to http://tidesandcurrents.noaa.gov/sltrends/sltrends.html and pick five tide gauge locations around the United States. Assuming that eustatic sea level rise over the past 100 years has been 8 inches; determine the amount of land uplift/subsidence at each of those stations.
13. Give two reasons why glaciation causes such extreme changes in sea level.
14. What caused the glaciation of Antarctica?
15. What caused glaciation in the northern hemisphere?
16. Greenland has experienced rapid shifts in climate. Explain why these shifts are not related to CO_2 in the atmosphere.
17. Explain the Milankovitch climate cycle. What does it tell us regarding glaciation?
18. What do relatively high ratios of $^{18}O/^{16}O$ in seawater imply?
19. Does a relatively high $^{18}O/^{16}O$ ratio in ocean-dwelling foraminifera imply low eustatic sea level, or does it imply high eustatic sea level.
20. How much has sea level changed in the past 18,000 years? What has caused this change?
21. How do we know the amount of CO_2 in the atmosphere 50 million years ago, 500,000 years ago, and today?
22. Why is it incorrect to compare CO_2 levels from hundreds of millions of years ago to today's CO_2 levels?
23. How do we determine sea level 50 million years ago and 100,000 years ago?
24. Explain what may have caused the shift in glaciation beginning about one million years ago from 41,000-year cycles to 100,000-year cycles.
25. Why is the current estimate of a 12- to 13-inch rise in sea level from 2000 to 2100 probably an underestimate?
26. Research several predictions regarding estimates of sea level rise this century. Why are some estimates so much higher than others?
27. What is significant regarding temperature swings over the past 10,000 years verses the past 140 years.
28. If we assume that sea level has risen at a rate of 0.267 inches per year for the past 18,000 years, how many feet has sea level risen in the past 18,000 years?

Chapter 9

Mechanisms That Impart Change to Landscape

The concept of landscape evolution implies that landscape undergoes change with time through one or more of five mechanisms; uplift, subsidence, erosion, deposition, or volcanism. It is possible, therefore, to measure landscape evolution by determining the rate, calculated over specific time intervals of the past, at which each of these mechanisms occur. Volcanism is a unique and rapid form of landscape evolution. Excluding volcanism, the two most important mechanisms are uplift and erosion. Unfortunately, it is difficult to measure uplift rates from the geologic past. We can, however, measure rates of erosion as explained later in this chapter, and we can extrapolate these rates to infer the timing of mountain uplift. The basic premise in this extrapolation is that rates of erosion increase as relief and elevation increase, particularly if higher elevation results in glaciation. In other words, a discrete period of rapid erosion can be correlated with a discrete period of mountain uplift. We will discuss the reasoning behind this premise later in the chapter. As an example, if measured rates of erosion were 2 inches per 100 years (0.51 mm/year) from 20 to 10 million years ago, and 10 inches per 100 years (2.54 mm/year) from 10 to 5 million years ago, we can presume that mountain uplift was strongest from 10 to 5 million years ago. In this example, we were able to deduce when rapid uplift may have occurred, however, we must not assume without additional evidence, that the rate of uplift was equal to the rate of erosion. In fact, the inference that the mountain has gained elevation implies that the rate of uplift was greater than the rate of erosion for at least part of its history.

In the case of a lowland area uplifting to form a mountain, the most obvious visible modifications to landscape are changes in elevation and relief. Additional visible modifications could include changes in river drainage pattern and river drainage density. For example, mountain uplift could potentially block or divert rivers, thus altering the river drainage pattern. Continental glaciation is capable of destroying entire river systems through erosion or through the burial of river systems via deposition. An entirely different drainage pattern will develop once the glacier retreats. The presence of abandoned river channels filled with sediment hint at a previous landscape that may no longer exist. Thus, we can gauge the extent to which landscape evolves by the extent to which elevation, relief, drainage density, and drainage pattern are changing or have changed over time.

This chapter explains the effect that each of the five mechanisms has on landscape, and provides some insight as to how each is measured. Special emphasis is given to uplift and erosion, the two most important mechanisms with respect to landscape evolution. In Chapter 10 we will look specifically at how these mechanisms combine to create one or more paths by which landscape evolves.

UPLIFT AND SUBSIDENCE

Surface uplift and surface subsidence refer to the raising and lowering of part of the Earth's surface relative to mean sea level. For convenience, we will refer to surface uplift/subsidence simply as uplift and subsidence. Uplift and subsidence can be regional in the sense that they encompass an entire mountain range, plateau, or valley, or they can refer to a small area or point that has been displaced such as along a fault where one side drops down relative to the other (and relative to mean sea level).

Surface Uplift/Subsidence and Bedrock Uplift/Subsidence

We can consider uplift and subsidence to be equal but opposite mechanisms of vertical land displacement. Uplift results in the creation of land areas, therefore, we will concentrate on uplift in this chapter, but keep in mind that the conclusions can also be applied to subsidence.

It is important to understand the difference between surface uplift (or subsidence) and bedrock uplift (or subsidence). Surface uplift refers to a topographic increase in surface elevation. Land surfaces can rise in elevation due to tectonic stress, isostatic compensation, and deposition. Land surfaces can lower in elevation due to tectonic stress, isostatic compensation, compaction, and erosion. Bedrock uplift/subsidence refers to the raising or lowering of bedrock relative to a medium such as mean sea level. It is the vertical displacement of rock with or without a corresponding change in surface elevation. Bedrock can rise due to tectonic stress and isostatic compensation, and can subside due to tectonic stress and isostatic compensation. The vertical position of bedrock relative to mean sea level is not affected by deposition, compaction, or erosion. We will use the term bedrock uplift/subsidence when referring specifically to the vertical displacement of bedrock. The following equation describes surface uplift:

Surface uplift = bedrock uplift + deposition − compaction − erosion

In the following discussion, we will ignore the effects of deposition and compaction, as neither are highly significant in a mountain environment undergoing uplift. By ignoring deposition and compaction, the equation for surface uplift is as follows:

Surface uplift = bedrock uplift − erosion

Thus, bedrock uplift creates surface uplift, some or all of which is reduced by erosion. Fig. 9.1 illustrates this relationship. The situation prior to uplift is shown in Fig. 9.1A. Bedrock (rock 1) is 150 feet below sea level (−150 ft). The Earth's surface is at sea level (0 ft). The situation following uplift is shown in Fig. 9.1B. Bedrock uplift is 200 feet (rock 1 was uplifted from −150 ft to 50 ft in elevation). The original land surface was also uplifted 200 feet (from 0 to 200 ft in elevation), but 50 feet of rock was eroded so that total surface uplift is 150 feet (from 0 to 150 ft in elevation).

The figure also illustrates how erosion has the effect of bringing deeply buried rocks to the surface. Prior to uplift and erosion, rock 1 was 150 feet below the surface. Following uplift and erosion, rock 1 is 100 feet below the surface. The process by which deeply buried rocks are brought to the surface is known as exhumation. Notice also that surface uplift in the absence of erosion is equal to bedrock uplift such that the buried rock remains just as deeply buried as it was prior to uplift.

How Does Uplift/Subsidence Occur?

Uplift (subsidence) results from tectonic stress and isostatic adjustment. In both cases, uplift (subsidence) can occur through earthquakes (seismic) or without earthquakes (aseismic; also known as creep). Uplift can be continuous or discontinuous, and may or may not occur in cycles or pulses.

Tectonic uplift (subsidence) is typically discontinuous and seismic although continuous aseismic creep can also occur. For example, a land area can uplift or subside by more than 20 feet in less than a minute during an earthquake and then remain relatively stable for hundreds or thousands of years until the next earthquake. Tectonic uplift (subsidence) may occur in cycles or pulses. For example, an area can be subject to a cluster of earthquakes or to rapid aseismic creep events over a period of tens of thousands of years, become dormant for an equal or unequal amount of time, and then awaken again.

Isostatic uplift/subsidence is more typically aseismic, although it can be associated with faults and earthquakes. It is a continuous process but not at a constant rate. An

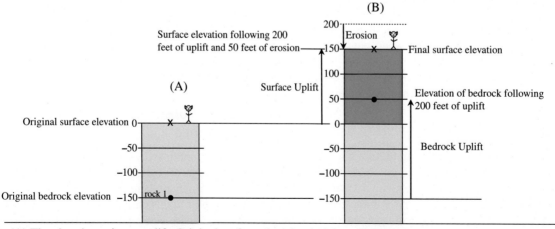

(A) The situation prior to uplift. Original surface elevation is 0 feet. Original bedrock elevation (rock 1) is 150 feet below the surface (-150 feet). (B) The situation following uplift. Bedrock is uplifted 200 feet to an elevation of 50 feet. The surface is uplifted 200 feet but 50 feet is removed via erosion. Bedrock uplift is 200 feet. Surface uplift is 150 feet. Bedrock has moved 50 feet closer to the surface.

FIGURE 9.1 Relationship between bedrock uplift and surface uplift.

area will undergo continuous isostatic adjustment, but the rate of adjustment will likely decrease as isostatic equilibrium is approached. For example, isostatic uplift of the Hudson Bay area has been continuous following the retreat of glaciers but at progressively slower rates.

Present-Day Uplift/Subsidence Rates

Present-day rates of uplift and subsidence are easily measured using space-based satellites and ground-based leveling techniques. Such measurements by their nature are of short-term duration, typically from 1 year to about 100 years. Because of the discontinuous nature of particularly tectonic uplift, these measurements are not necessarily indicative of long-term rates measured over thousands or millions of years. One must be careful not to extrapolate measured present-day vertical displacement rates deep into the geological past unless these rates can be shown to have been constant over the long-term.

Fig. 9.2 illustrates average rates of present-day regional uplift and subsidence across the United States based on pre-1972 leveling techniques. It shows that much of the Cordillera, the Great Lakes region, and part of the southeastern United States are currently experiencing broad uplift at rates between 4 and 59 inches per 100 years. Part of the Central Valley of California and most of the Atlantic and Gulf coast areas are subsiding at rates between 4 and 39 inches per 100 years. Not shown on this diagram are small areas of local uplift and subsidence such as along part of the California coastline. This is likely due to the regional scale of the data in which uplift of a small ridge is canceled by subsidence of an adjacent valley. These areas, taken together, create a larger area with zero net elevation change.

The highest uplift rates in the United States are in the Lake Superior region. Much of the uplift is due to isostatic adjustment following the retreat of glaciers beginning 18,000 years ago. Ice thickness was greater in the northern Lake Superior region; therefore, this area was depressed to a greater degree and is now rebounding (uplifting) at a higher rate relative to surrounding areas to the south. These relatively high uplift rates extend into Canada such that the area north of the Great Lakes is uplifting faster than the area to the south. The result is an overall southward tilt of the Great Lakes region. If this continues, Lake Michigan may eventually naturally drain into the Mississippi River Valley.

Isostatic adjustment of the Great Lakes area is an example of rapid present-day uplift rates that cannot be sustained over the long-term. The land will likely rise perhaps a few 100 feet in the next 1000 or several 1000 years but uplift will slow and eventually cease as isostatic equilibrium is restored. This type of uplift is not an indication of mountain building, which instead, is driven by tectonic stress that can be sustained for millions of years.

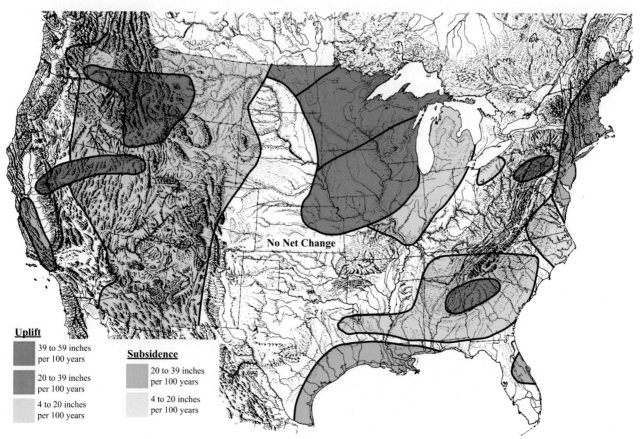

FIGURE 9.2 Areas of active uplift and subsidence. *Modified from Keller and Pinter (2002, p. 93).*

The point here is that measured present-day rates could merely be the result of short-term or transient causes. They are not necessarily equal to long-term rates measured over thousands or millions of years.

Measuring Ancient Uplift Rates and Elevation

Although present-day uplift rates can be measured, there are no unequivocal methods by which to measure ancient uplift rates or to measure maximum elevation attained by ancient mountain belts such as the Appalachian Mountains. One can use indirect methods, but the problem is that a perceived indicator of uplift, such as a change in fossil population from warm temperature species to cold temperature species, could instead result from a change in climate without any change in elevation. We must also keep in mind that sea level can change markedly (by at least 600 ft) over relatively short periods of time (20,000–40,000 years) due to expansion and retreat of glaciers. If we measure surface uplift based on mean sea level, then a relative drop in sea level is equivalent to worldwide uplift.

We have seen in Chapter 8 that oxygen isotopes are a powerful tool when applied to glaciation and sea level change. During the past two decades, some progress has been made using stable oxygen and hydrogen isotopes to infer ancient uplift rates and elevation. It can be shown that there is a predictable relationship between the isotopic composition of water and elevation. In Chapter 8 we noted that the lighter oxygen isotope (^{16}O) is preferentially evaporated. Conversely, the heavy oxygen isotope (^{18}O), and the heavy stable hydrogen isotope, deuterium (D), are preferentially concentrated in the earliest precipitation and are progressively depleted with additional precipitation. This property implies that if an air mass moves from ocean to continent, the air mass will progressively become depleted in heavy isotopes as more and more precipitation occurs. High mountains amplify the progressive depletion effect because they produce a rain shadow in which air is forced to rise and precipitate on one side of the mountain, leaving only deuterium-depleted precipitation on the rain shadow side. By studying the effects of this depletion across a present-day mountainous region, one can show that the magnitude of depletion is related to the height of the mountain. In other words, there is a predictable, quantitative relationship between isotopic depletion and elevation change.

In these types of studies, oxygen and hydrogen isotopes are obtained from authigenic minerals (and volcanic glass) of different ages located in different places across an area. The term authigenic implies that the mineral forms in place in equilibrium with the atmosphere. These minerals preserve the isotopic composition of the surface water they absorb. Authigenic minerals of the same age, but from different locations, can show how deuterium depletion, and therefore elevation, changes across a region. Authigenic minerals from the same location, but of different age, can show how deuterium depletion (elevation) changes with time. If, for example, the magnitude of deuterium depletion increases in progressively younger authigenic minerals from a single location, then uplift of a mountain can be inferred and elevation can be estimated.

Unfortunately, there are many variables that must be considered when using oxygen and hydrogen isotopes for ancient uplift rates and elevation. Such variables include differences or changes in prevailing temperature and humidity over time, sea level changes, and changes in the flow direction of air masses. As a result, these studies can have large error bars attached to them. Rather than absolute values, they can more accurately indicate relative gains or losses of elevation. It is therefore possible to estimate ancient elevation if, for example, a starting elevation (such as present-day elevation) is known. This method has been used to infer absolute elevation and elevation changes in the Cordillera particularly over the past 80 million years. But, the ability to quantitatively suggest maximum elevation for an ancient mountain system such as the Appalachian Mountains remains elusive.

Although values for ancient elevations are difficult to obtain, we can at least determine average values for the rate at which mountains come into existence. On the basis of present-day uplift rates, and on geological evidence pertaining to ancient uplift and exhumation rates, we can suggest that long-term tectonic uplift of a mountain system has occurred at rates of less than 25 inches per 100 years with typical rates on the order of 4 to 10 inches per 100 years (1.0–2.54 mm/year). However, it should be noted that maximum uplift rates experienced within an actively deforming and uplifting mountain range can exceed 100 inches per 100 years especially for short time periods of several thousand or several tens of thousands of years. Uplift can be seismic, aseismic, or both. With limited erosion, and an uplift rate of 10 inches per 100 years, a 5000-foot mountain range can emerge from sea level in 600,000 years. We could expect similar rates for subsidence.

EROSION, DEPOSITION, AND RIVERS

As noted in Chapter 3, erosion is the removal of rock and sediment from its original location on the Earth's surface by an agent such as water, ice, air, gravity, or animal/human interference. Deposition is the accumulation of eroded sediment such as in a lake or subsiding basin. River action is one of the most widespread forms of both erosion and deposition. Rivers respond quickly to tectonic- and climate-induced forcing agents and therefore, are capable of rather quickly changing the look of landscape. Two of the most visible and measurable criteria for landscape change are changes in river drainage pattern and changes in the density of river channels. Rivers also affect elevation and relief. In this section, we look at some of the most basic circumstances that would cause a river to erode bedrock or to deposit sediment.

Graded Rivers and Base Level

An important characteristic of a river is its ability to cut downward into bedrock and form a narrow V-shaped valley, a process known as downcutting or incision. As downcutting occurs, the river channel progressively reaches lower elevation. Downcutting is a form of erosion. The lowest elevation that a river channel can erode is referred to as ultimate base level. For rivers that empty into an ocean, ultimate base level is equal to sea level. At this elevation, the gradient (slope) of the river is equal to zero.

For most rivers, including those that empty into an ocean, there are local base levels along the river where the gradient (slope) decreases to zero or nearly so. Local base level represents the lowest elevation that the river upstream from that point can erode. Situations that produce local base level include an upstream lake or dam, a landslide, a location where the river flows over especially hard flat rock, uplift along part of a river channel where it crosses an active fault, or any place where the steepness of the river gradient becomes markedly lower. Local base levels are temporary and, once removed by erosion, a stream can cut downward as far as the next local base level downstream.

Fig. 9.3 is a longitudinal profile of a river with examples of both local and ultimate base level. A longitudinal profile shows the elevation of a river as if you are walking along the river bottom beginning at its head (where the river first forms) and ending at its mouth (where the river empties into an ocean).

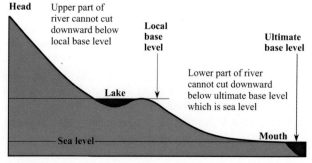

FIGURE 9.3 Longitudinal profile of a river from head to mouth showing local and ultimate base level.

In theory, a river can develop a smooth concave-upward longitudinal profile that is steep near its head and flat near its mouth so that it merges gently with the ocean. Under such conditions, it is possible for the river to reach a graded (equilibrium) condition where neither erosion nor deposition is occurring. A graded river has no local base levels per se. Instead, the longitudinal profile can be regarded as an infinite succession of local base levels. The river cannot erode or deposit material unless conditions change. The longitudinal river profile is said to be at its base level of erosion.

Whereas incision lowers the elevation of a river channel, deposition raises its elevation. A graded river will begin to erode its channel or deposit sediment in its channel dependent on changes primarily to three factors: base level, the amount of water passing a certain point in the river (e.g., its discharge), and the amount of sediment supplied to the river. Below, we discuss in simplified form how a graded river might react to each of these changes.

Base Level Changes

Base level changes affect stream velocity, which has a direct effect on erosive power. A relative drop in base level can be accomplished by uplift of land areas upstream from base level, the destruction of local base level, or by the lowering of sea level. Such action has the potential to increase stream gradient, which increases stream velocity, which favors incision as the river attempts to reach the new, lower base level of erosion, and return to a graded condition. Conversely, subsidence of a river channel below that of a graded condition, a rise in sea level, or formation of local base level would likely lower stream velocity. Deposition is favored under these conditions in order to raise the channel back to its base level of erosion. In any given scenario, a river will be affected initially at the location where the change in base level is most pronounced. The effect would then propagate upstream over time. If, for example, the lake in Fig. 9.3 were to suddenly empty (local base level is destroyed), incision would be concentrated initially at the steep area immediately downstream from the lake, and over time would propagate up river in an attempt to reach the new base level of erosion consistent with the lowered base level.

Because ultimate base level coincides with sea level, any change in sea level will result in a change in the dynamics of all river systems that empty directly into an ocean. Thus, a change in sea level not only profoundly affects coastal landscape, it can also affect inland landscape. Fig. 9.4A shows a smooth, concave-upward, longitudinal river profile. Because zero elevation begins at mean sea level, a rapid drop in sea level is equivalent (from the perspective of the river) with a global rise of land area relative to base level. Such a possibility would

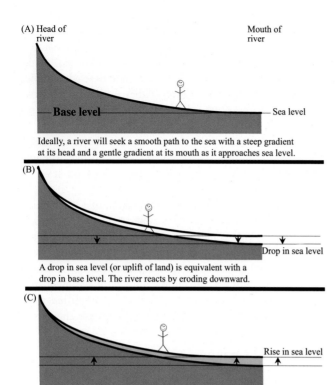

FIGURE 9.4 Longitudinal profiles that show the response of a graded river to sea level change. (A) Graded River profile. (B) Incision in response to a drop in sea level. (C) Deposition in response to a rise in sea level.

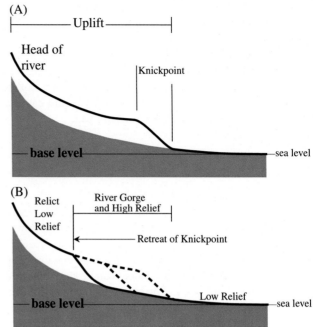

FIGURE 9.5 (A) Uplift along a longitudinal stream profile results in a knickpoint (a steep area) along an otherwise smooth gradient. (B) The knickpoint migrates upstream creating a river gorge and high relief in its wake. Based on Burbank and Anderson, 2012, Figure 8.9.

perturb the graded profile of the river by increasing its gradient (water velocity) resulting in incision in order to reach the lower sea level. Incision would begin at the mouth of the river where the change in gradient is most pronounced until the river could once again enter the ocean smoothly at the lower elevation. Incision would then propagate upstream in an attempt to reach the base level of erosion and graded conditions. This situation is shown in Fig. 9.4B.

A rapid rise in sea level is equivalent with global subsidence of land areas relative to base level. Such a scenario would likely result in a lowered river gradient. The velocity of the river would decrease causing deposition. As the river channel fills with sediment, the channel itself becomes smaller and the amount of water the channel can hold becomes less. The displaced river water could cause larger and more frequent inland floods. Such a situation could materialize in the United States if worldwide glaciers continue to melt. This situation is shown in Fig. 9.4C.

Knickpoint Migration

Let's look in more detail at the response of an initially graded river to a phase of surface uplift occurring mid-stream, and the propagation of that response upstream. Fig. 9.5A shows a river profile that was disturbed near its midpoint by a phase of uplift due perhaps to displacement along an active fault that cuts across the river. In this case, incision would begin at the edge of the uplifted area where the stream gradient (slope) is steepest and, therefore, where stream velocity and cutting power is greatest (other factors being equal). Geologists use the term knickpoint to refer to any location along a stream channel where the gradient is steep relative to the gradient above and below. For example, a waterfall along an otherwise smooth stream gradient is considered to be a knickpoint. In Fig. 9.5A, the river gradient below the knickpoint has not changed. The river gradient directly above the knickpoint has decreased slightly, creating a local base level. The river gradient farther above the knickpoint has not changed appreciably.

The presence of a knickpoint along a river profile has no genetic meaning. Its cause must be investigated and determined. Situations that can produce a knickpoint include vertical displacement of the ground surface along part of a stream channel such as Fig. 9.5, a change in climate, isostatic or sea level changes, and a change in rock type over which the river flows. A knickpoint can form, for example, where the river flows from very resistant rock onto less resistant rock. Although it is possible under certain conditions for a knickpoint to remain stationary over time, most migrate upstream due to erosion and slope retreat as shown in Fig. 9.5B. Once a knickpoint reaches a tributary stream, it will migrate up both the tributary stream and the main stream.

The effect of a migrating knickpoint is to incise its channel, thereby creating a deep river valley, gorge, or canyon in the area where the knickpoint has passed, thus increasing relief between river valley and surrounding landscape. Let us suggest a scenario in which a river basin in a low-relief terrain undergoes uplift. If a knickpoint develops along a stream channel and retreats upstream as shown in Fig. 9.5B, it will separate a downstream high relief landscape with gorges, rapids, and waterfalls, from the older, preexisting, low-relief landscape above the knickpoint. Under these conditions it may be possible to date the initiation of uplift by dating how long ago the knickpoint began to form.

Changes in Discharge and Sediment Supply

Discharge is the amount of water passing a certain point along the river measured over a specified interval of time. Discharge can be estimated by multiplying the depth of the river channel times the width times stream velocity. If the depth and width of a river channel are held constant, then discharge is a function of stream velocity. Under these conditions, an increase in the amount of water entering a graded river channel creates greater discharge in the form of increased water velocity. Greater water velocity results in greater erosional power, which favors erosion. Conversely, a decrease in discharge would favor lower water velocity and deposition.

Changes in sediment supply have the opposite effect. An increase in sediment supply, perhaps by removal of hillside vegetation, favors deposition because there is not enough water in the river channel to carry additional sediment. Conversely, a decrease in sediment supply will allow the excess water to erode its banks or incise its river channel.

Keep in mind that changes in base level, discharge, and sediment supply become complicated, and in some cases counterintuitive, especially if all three change at the same time. For example, the melting of a glacier could potentially increase sediment load in a river channel thereby favoring deposition. But, at the same time, the melting glacier could also increase discharge, thereby favoring erosion. Other factors to consider are changes to the depth and width of the river channel and changes to the smoothness of the river bottom. Next, as an example, we will look at how the lower Mississippi River responded to the most recent glacial advances and retreats.

The Lower Mississippi River Valley During the Most Recent Glacial Advance

The lower Mississippi River Valley is an excellent location to see how changes in base level, discharge, and sediment supply affect deposition and erosion. The lower Mississippi Valley lies south of St. Louis in the Mississippi Embayment area of the Coastal Plain. The region is shown in Fig. 3.13. The Mississippi River follows a meandering pattern typical of rivers that flow across areas of low relief. A meandering river is one with a series of bends such that the river forms a sinuous path as shown in Fig. 9.6A. A meandering pattern is somewhat self-sustaining because, once a river establishes a meander, the displacement of water around the bend favors erosion along the outside of the bend and deposition along the inside. The outer bank is known as the cut bank. The area of deposition on the inner bank is known as a point bar. Simultaneous cut bank erosion and point bar deposition results in migration of the river channel in the direction of the cut bank. On large meanders, it is possible for two cut banks to erode toward each other, thereby cutting off and abandoning a meander bend and forming an oxbow lake. The eventual abandonment of a meander bend is depicted in Fig. 9.6A. Following abandonment, the path of the river is temporarily straightened until a new meander begins to form, possibly in a different direction. The process of meandering levels the ground creating a

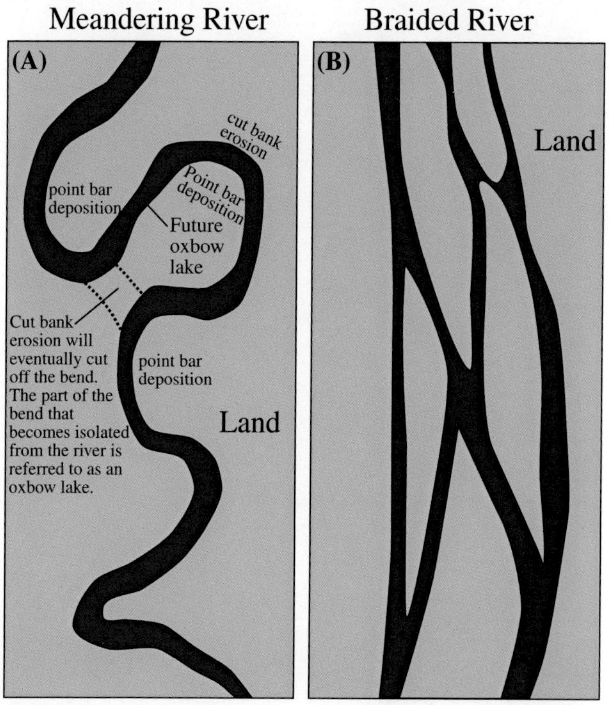

FIGURE 9.6 Examples of (A) a meandering river, and (B) a braided river.

floodplain, which can be defined as low-lying, relatively flat ground lying immediately above normal river level, and subject to flooding. The lower Mississippi River floodplain is obvious in the digital shaded-relief image shown on the left side of Fig. 3.13. The floodplain is an area where unconsolidated sediment (not bedrock) controls landscape.

In situations where sediment load overwhelms a river channel or where discharge decreases substantially, a meandering river pattern can switch to a braided pattern. A braided river is one in which a network of many river channels are separated by small islands usually composed of sand and gravel. A braided pattern is shown in Fig. 9.6B. Notice that the braided pattern is dominated by deposition and that the meandering pattern favors simultaneous lateral erosion and deposition. Neither implies incision.

Numerous meander bends, abandoned river segments, and oxbow lakes are visible in Fig. 9.7, which looks due north at the meander belt just south of Greenville, Mississippi. The bright white areas in the image are sand

FIGURE 9.7 Google Earth image looking north along the Mississippi River in the area south of Greenville, Mississippi showing numerous meander bends. Lake Village is next to an oxbow lake. Additional oxbows are visible surrounding Glen Allan. The bright white areas are sand bars, mostly point bars. Macon Ridge extends from Eudora, south-southwestward to the edge of the image.

bars. Most form point bars on the inside of a meander bend. The image encompasses parts of three states. Mississippi lies east of the Mississippi River, Arkansas lies to the west, and Louisiana covers the southwestern corner beginning about 7 mi south of Eudora. Meander bends in this area migrate rapidly. The border between the states was placed originally in the 1800s along the Mississippi River pathway. However, since that time, the river has changed course several times. Boundaries between states, however, have not changed, and as a result, there are numerous areas west of the present-day river course that belong to the state of Mississippi, and areas east of the river course that belong to Arkansas and Louisiana.

Studies have revealed a strong correspondence between depositional history on the lower Mississippi River and glacial and sea level history. River sediment less than 100,000 years old has been dated using a variety of methods that include relative dating, radiocarbon dating, and most recently, optically stimulated luminescence, which provides an age estimate of the time since sediment was deposited and shielded from sunlight. Recall that Fig. 8.6 shows the variation of sea level over the past 140,000 years. Based on this figure, we can say that between 100,000 and 30,000 years ago, sea level oscillated between -20 and -75 meters (-65 and -246 ft) below present-day sea level with changes brought on by the advance and retreat of glaciers. Between about 30,000 and 18,000 years ago, the accumulation of ice on land resulted in a sea level drop of at least 120 meters (394 ft) below present-day levels. Sea level has risen continuously since 18,000 years ago.

Based on our simple summary of base level change, we would expect periods of incision and deposition during oscillating sea level change from 100,000 to 30,000 years ago, followed by incision from 30,000 to 18,000 years ago as sea level fell, followed by deposition from 18,000 years ago to the present-day during sea level rise. However, the actual history appears to have been influenced more by sediment supply than by sea level (base level) history because there is little evidence of incision during the entire 100,000-year period. Recall that neither a meandering pattern nor a braided pattern implies incision.

Floodplain deposits dated 84,000 to 77,000 years ago indicate a meandering river pattern much like the modern Mississippi except that the river was 8 to 21 m (26–69 ft) below present-day elevation. In this case, a lower river channel would be expected because, according to Fig. 8.6, sea level dropped from -24 to -65 m (-79 to -213 ft) below present-day levels during that time span.

Beginning 77,000 years ago, a large amount of glacier-derived sediment began reaching the river channel. The result was deposition. Between 77,000 and 64,000 years ago, while sea level oscillated between -42 and -74 m (-138 to -243 ft), the Mississippi River switched from its characteristic meandering pattern to a braided pattern. Apparently, there was so much available sediment that deposition occurred in spite of a presumed increase in glacial meltwater discharge. The river maintained a braided pattern well past the maximum drop in sea level 18,000 years ago, returning to its characteristic meandering pattern only about 11,000 years ago in response to decreasing sediment load.

Today, much of the lower Mississippi River Valley is filled with braided stream deposits, some of which stand as terraces above the present-day floodplain. An example of a braid belt is Macon Ridge, a 135-mile-long ridge up to 25 miles wide located west of Vicksburg, Mississippi and extending southward to Sicily Island. Macon Ridge is labeled in Fig. 3.13. Part of the ridge is visible in Fig. 9.7.

Although much of the lower Mississippi Valley was dominated by braided river deposition between 77,000 and 11,000 years ago, there is at least a suggestion that the southernmost part of the river did respond to sea level (base level) drop by incising its channel. As noted earlier, we would expect incision to begin at the mouth of the river (where a drop in elevation is most noticeable) and then propagate upstream. Detailed fieldwork indicates that exposed braided river belts in the vicinity of the Mississippi–Louisiana state line dip steeply southward and disappear completely near Baton Rouge below modern meandering river deposits. Such a steep dip for the braided river deposits indicates that the river was seeking a lower base level (a lower sea level) than present-day. One possible interpretation is that incision was occurring, or had occurred, in the southernmost part of the valley at times between 77,000 and 11,000 years ago while at the same time, braided river deposition was occurring in the northern part of the valley. Beginning about 11,000 years ago, the modern day rise in sea level caused the southern part of the valley to switch to meandering deposition, thus burying evidence of both incision and braided river deposition south of Baton Rouge. Let us now return to a more general discussion concerning rates and controls on erosion and deposition.

PRESENT-DAY EROSION RATES

It is possible to measure present-day erosion and deposition rates; however, as with uplift and subsidence, these rates may not accurately reflect long-term rates that have occurred over the past 100,000 or million years. However, unlike uplift/subsidence rates, it is possible to estimate ancient erosion/deposition rates with reasonable accuracy. We will explore how to do this later in the chapter under the heading exhumation. For now, we will look at ways to measure present-day rates of erosion and deposition and how to extrapolate these measurements into the recent past.

From a purely qualitative perspective, the existence of high mountains suggests that rates of erosion are slower than rates of uplift. If the opposite were true, then mountains would not exist and most of the Earth's land surface would lie close to sea level. We will discover, however, that there is a limit to the height of a mountain.

One way to measure present-day erosion rates is to measure the volume of sediment carried away by rivers within a specified drainage basin. This amount can be divided into the surface area of the drainage basin in order to arrive at an average denudation rate, which is the rate of overall vertical (erosional) lowering of land. The calculation is equivalent with taking all the sediment removed over the course of a year and spreading it out evenly over the entire drainage basin. The amount of erosional lowering (the denudation rate per year) would be equivalent to the thickness of the sediment layer. The procedure is illustrated in Fig. 9.8. In practice, the material in the river that can be measured is the suspended load (sediment in the water column of the river). The dissolved load and the bed load (the material rolling and bouncing along the bed of the river) are more difficult to measure. This limitation is not a major concern in many rivers where most of the sediment is moved by suspension. Additionally, much of the erosion on hill slopes may not reach the stream and is instead deposited at the base of the hill or stored as alluvium on floodplains, and thus does not leave the drainage basin. Such complications underestimate total erosion rates.

The present-day average denudation rate across the United States is about 0.24 inches per 100 years. What this means is that, on average, a 0.24-inch-thick layer of rock and sediment has been removed from all land areas in the United States over the past 100 years. If we assume zero uplift, it means that the average elevation of the United States has been lowered by 0.24 inches. I stress that this is an average rate. Some areas undoubtedly were eroded at much higher rates while other areas were not eroded at all or experienced a gain in elevation through deposition. It has been suggested that this rate is twice as high as it was only a few 100 years ago. Apparently, the disruption of the land surface by agricultural practices has substantially increased the rate of denudation.

A different method by which to measure erosion is the bedrock incision rate, which is the rate at which a river erodes downward (incises) into bedrock. It is a measure of erosion on a smaller scale than regional denudation, on the scale of an individual river or stream. As rivers cut downward, they meander slightly, thereby cutting a terrace, known as a strath terrace, directly into bedrock. These surfaces are left elevated above the present-day river channel as the river continues to incise its channel. If the radiometric age of the strath terrace can be

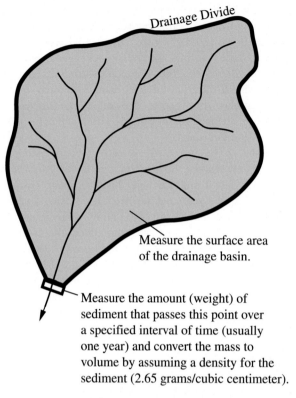

Calculation of Denudation Rate

Denudation rate is the rate of overall vertical (erosional) lowering of land.

$$\text{Volume} = \frac{\text{weight}}{\text{gravity} \times \text{density}}$$

$$\text{Denudation rate} = \frac{\text{volume of sediment}}{\text{area of drainage basin}}$$

FIGURE 9.8 Calculation of denudation rate.

determined along with the elevation of the present-day channel, then the average rate of river incision can be measured as shown in Fig. 9.9. Incision rates can be extrapolated into the recent past if the radiometric age of two abandoned strath terraces at different elevations can be measured. Using this method, we can extrapolate as far back as the age of the oldest strath terrace, which can be several million years or more. Unfortunately, terraces are often destroyed via erosion and landslides, thus limiting how far back in time they can be preserved.

Strath terraces are not easy to date. If the strath is cut into easily dated volcanic rock, then incision must have occurred after deposition of the volcanic rock. Conversely, if a lava flow partially covers a strath terrace, then the terrace must have formed prior to the lava flow. Alternatively, fossils, if present, can provide constraints on the age of a strath terrace. None of these methods

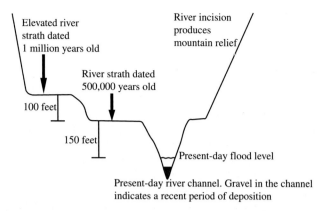

FIGURE 9.9 Calculation of bedrock river incision rates.

provide a precise date as to when the river abandoned the strath terrace. Instead, we can use a complicated technique known as cosmogenic radionuclide dating. Cosmic rays emanating from the center of the Milky Way galaxy are used to date how long a surface has been free of water and subject to long-term cosmic ray bombardment. An age is obtained by measuring daughter products produced by the collision of cosmic rays with the dry rock surface. The method dates how long ago the surface last emerged from below water (river) level. Typical nuclides used in this dating technique include ^{3}He, ^{10}Be, and ^{26}Al.

River incision can occur due to any number of factors already discussed in this chapter. One possible factor is the lowering of base level either through uplift of the surrounding land or a drop in sea level. River incision occurs because the river channel becomes elevated above base level, which has the potential to increase stream velocity and, therefore, cutting power (Fig. 9.4B). If uplift occurs, the river could maintain its pre-uplift elevation by downcutting into the rising land area at the same rate as uplift, thus creating a canyon or gorge. River incision, however, does not imply uplift. Incision can occur if the surrounding land is stationary or even subsiding if, for example, base level drops faster than the subsiding land. Under these conditions, the river will be forced to cut downward in order to create a smooth, graded path.

Similar with uplift and subsidence, the rate of incision is not necessarily steady. Instead, there may be periods of rapid incision, periods of little or no incision, and periods of deposition during the measured time interval. In Fig. 9.9, for example, it is possible that nearly all of the incision from 500,000 years ago to the present occurred only in the past 100,000 years, or that incision was periodically interrupted by deposition. A good indication that a river is presently incising and eroding its channel is where it flows over bare bedrock. In this case, the river must be removing material rather than depositing. If sediment is present at the bottom of a river channel, then it indicates either that the river is currently depositing material or that it has recently begun to erode its channel but has not yet removed all previously deposited material.

Typical incision rates vary from less than 1 inch to several inches per 100 years and thus, can be much faster than regional denudation rates to the point where they approach or exceed uplift rates. As with uplift, maximum incision in mountain belts can be 10s of inches or more per 100 years, especially at high elevation where powerful glacial incision occurs, or where rocks in the river channel are strongly weathered (recall from Chapter 3 that weathering facilitates erosion). Because river incision is restricted to the river channel itself, it results in a carving effect that shapes the mountain producing steeper slopes and greater relief. The steeper slopes, in turn, promote gravity-driven landslides, slope retreat, and potentially increased denudation rates.

Periods of rapid river incision, when extrapolated into the past, can be used to indirectly determine periods of rapid uplift and therefore periods of mountain building. For example, if a period of rapid river incision can be isolated between two periods of relatively slow river incision, then we can suggest that rapid incision is due to uplift and mountain building. This procedure must account for climate and sea level change, and changes in rock type, because such changes could potentially trigger the same effect without uplift.

CONTROLS ON RATES OF EROSION

We have discussed two very different methods of measuring erosion rates, the denudation rate across an entire drainage basin, and river incision rate, on the scale of an individual river. We have also seen that rates can vary from very low to those exceeding rates of uplift. Now we can ask ourselves, what causes rates of erosion to be slow in some instances and rapid in others? Based on observation and logic, we can conclude that rates of erosion are dependent on rock hardness, vegetation, climate, elevation, and relief. Obviously, soft rock, or strongly weathered rock, will erode at a higher rate than hard or unweathered rock. An area without vegetation can potentially erode faster than one with vegetation because roots hold soil in place, thus curtailing erosion. Water on the other hand, especially running water, will increase erosion rates simply because it is capable of moving loose soil. Thus, other factors being equal, an ideal situation that allows for rapid erosion is a semihumid climate where there is running water but limited vegetation. As a rule of thumb, we can say that, at any given velocity, the greater the volume of water, the more capable the ability to erode. Conversely, we can say that for any given volume of water, the greater the velocity, the more capable the ability to erode.

In any climate, the rate of erosion increases with elevation and particularly with relief. High elevation forces air to rise, which potentially creates precipitation. Cold air at high elevation limits the amount of vegetation and creates the potential for glaciation. Glaciers are powerful erosional agents. On balance, they are significantly more effective than running water. High elevation alone, however, does not necessarily cause an increase in erosion rates, especially without glaciers. A flat plateau at high elevation could conceivably erode at about the same rate as a flat plain at low elevation simply because there is little relief and little, if any, moving water. It is relief (the change in elevation between two points) that allows gravity to work in favor of erosion. High relief implies steep slopes, which allow for less vegetation, faster moving streams, and greater potential for landslides. The highest rates of erosion are typically found in mountain belts because we have the combination of elevation and relief, precipitation with limited vegetation, fast-moving water, and the potential for landslides and glaciers. Still, the fact that we have high mountains with glaciers implies that rates of erosion are, to a first approximation, slower than rates of uplift.

Rates of erosion can also be highly variable across an individual mountain range, especially if that range creates a major rain shadow such as the Cascade Mountains in Washington, or if the range is oriented in an east-west direction, such as the Uinta Mountains in Utah and the Ouachita Mountains in Arkansas-Oklahoma. In both situations, different sides of the mountain receive different amounts of precipitation and sunlight, which affects the types of plants that can grow, the rates of weathering and erosion, and ultimately the rate of isostatic compensation. Rates of erosion would be expected to be higher on the windward (wet) side of a rain shadow and on the north (wet) side of an east-west-oriented mountain range. Higher rates of erosion on one side of a mountain result in higher rates of isostatic uplift. The wet side of the mountain rises at a slightly faster rate. This asymmetric uplift causes headward erosion and the pirating of river channels on the wet side of the mountain such that the crest of the range migrates toward the dry side. The result is an asymmetric mountain range in which the dry side is steeper and not as wide as the wet side. Such migration has been shown to occur in the North Cascades. Both the Uinta and Ouachita Mountains exhibit climate and vegetation differences on their north verses their south flanks.

The preceding paragraphs beg the question, is there an erosional limit to how high a mountain can grow? The answer is yes. As the mountain grows in elevation and relief, the rate of erosion will increase to where it is equal to the rate of uplift. The mountain continues to undergo bedrock uplift, but any additional surface uplift is nullified by an equal amount of erosion creating a seemingly unchanging landscape. Such landscape is said to be in topographic steady state. This type of landscape evolution is discussed in Chapter 10.

There is another limit to the height of a mountain and it has to do with gravity and temperature. Rocks are only so strong. As a mountain grows higher, the deeply buried rocks at the root of the mountain are under increasingly greater weight (a mountain root is shown in Fig. 7.4). The rocks are warm at depth, which weakens them. At some point, the weight of the mountain can become greater than the strength of the deeply buried rock allowing the warm ductile rock to flow from beneath the mountain. In other words, the buried root could potentially expand horizontally. The mountain reacts by collapsing (losing elevation) possibly through normal faulting at the surface.

RATES OF DEPOSITION

Deposition is the opposite of erosion in the sense that sediment accumulation raises land areas relative to surrounding land. Average rates of deposition for both the present-day and the geological past are estimated in a fairly simple manner by measuring the thickness of a sequence of rock (or sediment) and dividing by the time interval over which the rocks were deposited. In this calculation, the amount of compaction that occurs when sediment is turned to rock must be taken into account. When this is done, it is found that rates of deposition are roughly equal to rates of erosion. This is not surprising given that any eroded material must be deposited somewhere. However, it should be realized that a single mountain range, or perhaps even a single mountain peak, could erode into several separate depositional basins. Conversely, a single depositional basin can receive sediment from multiple sources.

As land area subsides, it will form a basin that will fill with water and receive sediment from erosion of surrounding highlands. Any change in surface elevation is dependent on the rate of subsidence verses the rate of sediment deposition. If the amount of sediment deposition equals the amount of surface subsidence, then elevation will not change. Surface subsidence in this situation is equal to zero and the basin could be said to be in topographic steady state. If the rate of deposition is faster than subsidence, or occurs without subsidence, then the basin will fill with sediment and any body of water, such as a lake, will disappear to form dry land. Deposition increases weight, which causes isostatic subsidence. Deposition and simultaneous slow isostatic subsidence over the course of millions of years will produce huge thicknesses of sediment and, eventually, sedimentary rock. An example of deposition during slow subsidence is along the eastern North American coastline where tens of thousands of feet of sedimentary rocks have accumulated, all in shallow-water. Measuring the time interval over which these rocks were deposited gives an indication of long-term subsidence rates.

Subsidence without deposition will result in loss of elevation. Such a situation can occur in desert areas where rates of erosion (and, therefore, rates of deposition) are slow. In this case, it is possible for land areas to sink below sea level. An example is Death Valley.

EXHUMATION

Exhumation is the process by which once deeply buried rocks are brought to the surface. There are two primary mechanisms by which exhumation can occur. The first is the erosional removal of overlying rock—a process referred to as erosional exhumation. Erosion is particularly effective when coupled with simultaneous uplift. The process of erosional exhumation is probably valid for many mountain belts including the Appalachians. It is the most important and most common process by which exhumation occurs. Faulting, particularly normal faulting, can also result in the exhumation of rock—a process referred to as tectonic exhumation.

Erosional Exhumation

We have seen in Fig. 9.1 that as bedrock uplift occurs, the surface of a mountain is uplifted and eroded thereby exposing previously buried rock underneath. Buried bedrock is uncovered (exhumed) at the same rate that overlying bedrock is removed (eroded). In other words, the rate of exhumation is equal to the rate of erosion. The relationship between uplift, erosion, and exhumation is illustrated in more detail in Fig. 9.10. The top panel shows three rocks prior to uplift and erosion with rock 1 (R1) at the surface. The second panel from the top shows 400 feet of bedrock uplift and 400 feet of surface uplift. There is no erosion. The three rocks (R1, R2, R3) have been uplifted to higher elevation, but remain as deeply buried

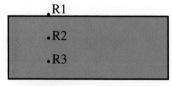

Three rocks before uplift and erosion. Rock 1 is at the surface,
Rock 2 is 200 feet below the surface, and Rock 3 is 400 feet below
    the surface.
Rocks 1 and 2 are sedimentary. Rock 3 is crystalline.

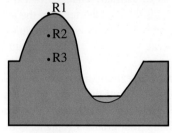

The rocks undergo 400 feet of both surface uplift and bedrock uplift
    with no erosion.
Rock 1 is still at the surface. Rock 2 remains 200 feet below the
    surface, and Rock 3 remains 400 feet below the surface.
Bedrock uplift is 400 feet. Surface Uplift is 400 feet.
Erosion = Exhumation = 0 feet.

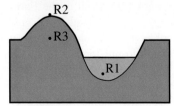

The mountain is lowered 200 feet by erosion. Rock 1 is eroded
    and deposited in a basin. Rock 2 is at the surface and Rock 3 is
    200 feet below the surface.
Bedrock uplift is 0 feet (there has been no vertical movement of bedrock).
Surface uplift is minus 200 feet (surface is lowered 200 feet via erosion).
Erosion = Exhumation = 200 feet (200 feet was removed from the moutain top).

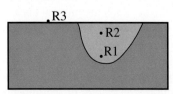

The mountain is lowered another 200 feet by erosion. Rock 2 is
    eroded and deposited in a basin which is now filled with sediment.
Rock 3, the crystalline rock, has been exhumed to the surface.
Bedrock uplift is 0 feet (there has been no vertical movement of bedrock).
Surface uplift is minus 200 feet (surface is lowered another 200 feet via erosion).
Erosion = Exhumation = 200 feet (another 200 feet was removed from the moutain top).

The mountain underwent a total of 400 feet of bedrock uplift and 400 feet of erosion for a total surface uplift of 0 feet.

FIGURE 9.10 Example of erosional exhumation.

as they were prior to uplift. In the third panel, rock 2 (R2) has been exhumed to the surface following 200 feet of erosion without any additional bedrock uplift. Rock 1 (R1) has been eroded and deposited in a nearby basin. Rock 3 (R3) is now 200 feet closer to the surface. Overall, the mountain underwent 400 feet of bedrock uplift and 200 feet of erosion for a total surface uplift of 200 feet. All buried bedrock in the mountain, including rocks R2 and R3, underwent 200 feet of exhumation at the same rate as erosion. The bottom panel shows an additional 200 feet of erosion without bedrock uplift thereby eroding rock 2 (R2) and exhuming rock 3 (R3) to the surface. Overall, the mountain underwent 400 feet of bedrock uplift and 400 feet of erosion for a total surface uplift of 0 feet. The mountain underwent 400 feet of exhumation at the same rate and magnitude as erosion.

Uplift and erosion are strongest in mountain belts. Thus, it is common to find crystalline rock (rock that originally formed deep within the earth) exposed at the top of a mountain. Some lowland areas, such as the Superior Upland province and the Piedmont Plateau province, are composed of crystalline rock. The presence of these rocks at the Earth's surface suggests that, at one time long ago, a mountain range was present.

In Figs. 9.1 and 9.10 bedrock is exhumed at the same rate as erosion. We shall see in the discussion below that it is possible to calculate ancient rates of exhumation. These values can then be used to infer ancient rates of erosion. We have also seen that the rate of erosion increases with elevation and especially with relief. Using this information, it is possible to infer how long ago and at what rate a mountain was uplifted based on changes in the rate of bedrock exhumation. We can suggest that periods of rapid exhumation (that is, rapid erosion) correspond with periods of rapid mountain uplift, and that periods of slow exhumation correspond with little or no mountain uplift.

Calculating Rates of Erosional Exhumation

The rate of exhumation is the rate at which buried bedrock is brought to the surface. This rate, in the absence of tectonic exhumation, is equal to the rate of erosion. For many rocks, it is possible to accurately estimate the pressure at which the rock formed as well as its age. Pressure is determined by obtaining the exact composition of coexisting minerals in a rock and then applying thermodynamic principles to determine the pressure at which those coexisting mineral compositions are at equilibrium. Age is determined through radiometric (isotopic) dating. It is important to understand that nearly all radiometric dating techniques use specific minerals in a rock to determine age. They do not use the entire rock. Pressure, once it is calculated, is a function of rock density, depth, and the force of gravity. Rock density and the force of gravity are well known variables that do not change very much, at least within the top 40 miles of Earth. They can be treated as constants. If pressure is also known, then depth (d) of burial is easily calculated as pressure (P) divided by the product of rock density (r) and gravity (g).

$$\frac{P}{r \times g} = d$$

For example, if $P = 2.47$ kbar (or 247,000,000 Pa), $r = 2750$ kg/m^3 and $g = 9.8$ m/s^2, then depth is equal to 9165 m (30,061 ft). Because the rock is presently at the Earth's surface, the average rate of exhumation is simply depth divided by age. Although crude, this simple relationship allows geologists to estimate average exhumation rates deep into the geological past. For example, if a rock presently at the Earth's surface crystallized 56 million years ago at a depth of 30,061 feet, the average rate of exhumation is 30,061 feet per 56 million years (or 0.64 in. per 100 years). This, of course, is exhumation averaged over the entire 56 million-year interval. It assumes that exhumation began soon after the rock formed, which is not necessarily true. Exhumation could have started at any time after the rock formed. If, for example, exhumation occurred entirely during the last 14 million years, then the actual average rate of exhumation over the shorter time period would be four times the previously calculated rate. Fortunately, sophisticated dating techniques allow geologists to estimate changes in exhumation rates over time periods that are shorter than the age of the rock. Thus, it is possible to determine when, during the 56 million-year interval, most of the exhumation occurred.

In order to estimate rates of exhumation over specific time intervals, geologists make use of the geothermal gradient and the closure temperature of certain individual minerals in a rock. The geothermal gradient represents the gradual increase in temperature with depth in the Earth's interior. The assumption regarding the geothermal gradient is that the rock will cool as it moves closer to the surface during exhumation. The closure temperature is the time and temperature at which the radiometric clock for that mineral begins. The closure temperature is a most important factor to consider when dating minerals. The closure temperature of a mineral does not necessarily correspond with the temperature at which the mineral crystallized. Different minerals have different closure temperatures, and an individual mineral's closure temperature will vary depending on the dating method used. A mineral with a high closure temperature, one that is close to or above the crystallization temperature of the mineral, can be used to determine the age of crystallization. For exhumation, we are interested in minerals and dating techniques that use a low closure temperature. In this case, the radiometric clock begins long after crystallization, at a time when the mineral is at low temperature close to the surface. The radiometric clock for that mineral begins at the depth that corresponds with the closure temperature. If geologists date several minerals from the same rock, but with different closure temperatures, they can determine how long ago the rock was at various depths on its way to the surface. From this information, they can calculate the rate at which the rock made its way to the surface at various times in the past. We elaborate on this procedure below.

Common dating methods include argon 40-argon 39 ($^{40}Ar/^{39}Ar$), uranium−thorium/helium (U−Th/He), and fission track. Argon is a product of the radioactive decay of potassium, which is a common element in a variety of minerals. Because it is an inert (nonreactive) noble gas, all argon will diffuse (move) out of a mineral when that mineral is at high temperature. The closure temperature of a mineral is the temperature at which argon is retained in the mineral. It is at this time and temperature that the radiometric clock begins. From this time forward, the mineral will retain ever-increasing amounts of argon as more and more potassium undergoes decay. In the simplest case, an argon age is the amount of time it would take for the amount of argon in the mineral to accumulate. This age corresponds with the amount of time that has passed since the sample first cooled below the closure temperature (and first started to accumulate argon) during erosional exhumation to surface. The mineral hornblende has a relatively high closure temperature in the argon-argon system of around $500 \pm 25°C$; muscovite and biotite have intermediate closure temperatures of about $425 \pm 25°C$ and $330 \pm 25°C$, respectively; and K-feldspar has a low but variable closure temperature between 150 and 300°C.

The same principle is employed with the (U−Th)/He method except that the mineral apatite is used and the inert gas is helium rather than argon. Helium is a product of the radioactive decay of uranium and thorium. It is expelled from apatite at temperatures above 70°C and is completely retained within the mineral at temperatures below 30°C. Helium is partially retained between these temperatures.

Fission tracks are lines (tracks) of near constant length that represent damage within the minerals apatite and zircon caused by radioactive decay of small amounts of enclosed uranium. The tracks are completely erased in apatite at temperatures above about 110°C, partially retained between 110 and 50°C (track lengths are shortened), and fully retained below 50°C. Fission tracks recorded in zircon are erased above about 310°C, retained below about 210°C, and shortened between these two end-member temperatures. The temperature range across which track lengths are shortened is referred to as the partial annealing zone.

Although it is possible in many instances to calculate the pressure (depth) at which a rock forms, it is not possible to calculate the pressure at the time the rock passes through these rather low closure temperatures. The exhumation rate, therefore, is determined by inferring a geothermal gradient. Once established, a geothermal gradient is used to infer depth of burial at the time the closure temperature was reached. Normal geothermal gradients are between 20 and 30°C/km of depth. Thus, if several dating methods are employed within an area, average exhumation rates can be determined at various stages of exhumation.

For example, if we infer a geothermal gradient of 25°C/km and determine an $^{40}Ar/^{39}Ar$ date on biotite with a closure temperature of 330°C, a fission track date on zircon with a closure temperature of 210°C, and a (U−Th)/He date on apatite with a closure temperature of 30°C, we can determine the rate at which the rock was exhumed from a depth of 13.2 km (the depth at which the rock is at 330°C) to 8.4 km (the depth at which the rock is at 210°C) to 1.2 km (the depth at which the rock is at 30°C). If the biotite age is 17 million years, the zircon age is 14 million years, and the apatite age is 8 million years, the exhumation rate from 13.2 to 8.4 km depth is 4.8 km per 3 million years (or 6.3 in. per 100 years). From 4.8 to 1.2 km depth, the exhumation rate is 3.6 km/6 My (or 2.36 in. per 100 years). If we can show that the rock has only recently reached the Earth's surface, we can also calculate the rate of exhumation from 1.2 km depth to the surface as 1.2 km/8 My (or 0.59 in. per 100 years). If we relate increased rates of erosional exhumation with increased rates of uplift, we can infer that the mountain came into existence through rapid uplift between 17 and 14 million years ago (resulting in a 6.3 in. per 100 years exhumation rate). Uplift slowed over the next 6 million years and became very slow (or nonexistent) over the final 8 million years.

There are inherent problems and assumptions when using these techniques. The closure temperature can vary by 25°C or more depending on the size of the mineral grain, the speed at which ions move through the mineral (a process known as diffusion), the mineral purity, and the rate of cooling. Furthermore, it is possible for some of the inert gas to be retained prior to reaching the closure temperature or for inert gas to be expelled for a short time after reaching the closure temperature. Additionally, there is no straightforward correlation between the temperature of rock and the depth at which rock is buried. Thus, barring independent pressure data, the geothermal gradient (and, therefore, the depth) at the time of closure must be estimated. It is also possible for the geothermal gradient to change during the process of exhumation. These problems create potentially large uncertainty in the age of the sample and the depth at which the sample reached its closing temperature. Such problems, therefore, create uncertainty in the absolute rate of erosional exhumation. Fortunately, geologists have developed techniques and models to account for these and other uncertainties to the point where they allow for tight constraints on what the isotopic age means in terms of an exhumation rate. Regardless of the uncertainties, the relative rates between the different minerals, using the same assumptions, are, in most cases, precise enough to show trends.

It is important to realize that what is being measured is the rate of erosional exhumation averaged over a certain time period. The rate of bedrock uplift, the rate of surface uplift, and the elevation achieved by the mountain are not calculated. They can only be estimated. The underlying assumption is that the cause of rapid erosional exhumation is rapid bedrock and surface uplift (mountain building). This assumption is valid for narrow mountain ranges where uplift is expressed by an increase in surface relief and slope such that rates of erosion also increase, ideally to the point where the mountain reaches topographic steady state such that rates of uplift and erosion are equal. This assumption is not necessarily valid for plateau uplift of a wide area where erosion is curtailed and surface relief and slope are not substantially increased.

Tectonic Exhumation

Tectonic exhumation is the process by which rocks are brought closer to the Earth's surface without erosion. Rather than erosion, the overlying rocks are removed by some other method, most typically by normal faulting as shown in Fig. 9.11. Tectonic exhumation via normal faulting can be very rapid, resulting in the generation of topographic relief and in the rapid cooling of rocks as they are quickly brought to the surface. Such a relationship allows us to use low temperature cooling data such as apatite and zircon fission track and (U—Th)/He ages as a proxy to date when normal faulting was active and when landscape was undergoing topographic change. Clearly, when calculating exhumation rates using low-temperature geochronology, it is important to distinguish between erosional and tectonic exhumation.

VOLCANISM

Volcanism is the process by which magma and associated gases rise to the surface to form lava flows, volcanoes, and other forms of magma eruption. This is a special type of landform formation and modification that can shape or reshape a landscape potentially in a matter of days, months, or years, rather than the thousands to hundreds of thousands of years required for uplift, subsidence, erosion, and deposition. Volcanic landscapes and the consequences of volcanism are discussed in Chapter 17.

QUESTIONS

1. Define the following: uplift, subsidence, erosion, deposition, denudation, exhumation.
2. Explain differences between surface uplift, bedrock uplift, and exhumation.
3. Are present-day uplift rates indicative of long-term uplift rates? Explain.
4. What is meant by the statement "mountain elevation is at topographic steady state."
5. If rocks that originally formed deep in the earth are exposed on a flat plain close to sea level, what type of landscape likely was present at that location a long time ago (hint: in what type of landscape are deeply buried rocks brought to the surface).
6. What are inferred typical rates of tectonic uplift for the geological past? State your answer in inches per 100 years, cm per 100 years, mm/year, and km/My.
7. Is it possible to determine ancient rates of uplift/subsidence? Explain.
8. It is not possible to determine the maximum elevation attained by the Appalachian Mountains. State and explain some criteria we can use to estimate its maximum elevation.
9. Which is typically faster at low elevation, rates of uplift or rates of erosion? Provide evidence for your answer.
10. Explain how the rate of erosion varies with the following: climate, hardness of rock, intensity of weathering, vegetation, elevation, and relief.
11. If relief between river bottom and a river terrace is 120 feet, and the age of the terrace is 420,000 years, what is the average rate of river incision in inches per 100 years and millimeters per year?
12. Does river incision necessarily imply uplift of land? Explain.

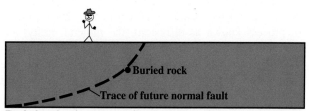
Buried rock prior to tectonic exhumation along a normal fault.

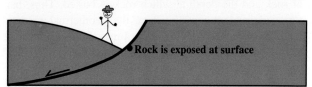
The rock is exhumed to the surface without erosion.

FIGURE 9.11 Example of tectonic exhumation.

13. What is a knickpoint? Suggest several mechanisms by which a knickpoint might develop?
14. What is the exhumation rate for a rock buried 4 mi deep, 9 million years ago (in inches per 100 years; cm per 100 years, and mm/year)?
15. What is the rate of deposition for a 5-mi-thick sequence of rocks deposited over a time interval of 270 million years (in inches per 100 years; cm per 100 years; mm/years and km/My)? What factor must you account for in order to make an accurate calculation?
16. True or False. The only way to expose deeply buried rocks is through erosion. Explain your answer.
17. How can a strath terrace be used to estimate the average incision rate of a river?
18. The chapter states that typical river incision rates are much faster than regional denudation rates. Explain why this would logically make sense.
19. How might a major river near where you live be affected by rising sea level? How might the effect be different if the river emptied directly into the ocean verses directly into another river?
20. What can be inferred regarding the history of a mountain if exhumation rates are found to be (a) very slow over a period of time and (b) very fast over a period of time?
21. Given that ancient rates of uplift are difficult if not impossible to measure, what criteria is used to gauge how long ago and at what rate a mountain undergoes uplift?
22. In Google Earth, fly to Niagara Falls, New York. Explain the river setting in terms of erosion, deposition, and the development of the Niagara River gorge.
23. What are some of the inherent problems when calculating exhumation rates.
24. What is erosional exhumation?
25. Can you think of any form of tectonic exhumation other than through normal faulting? For example, would a diapiric rise of warm (but solid) rock constitute a form of tectonic exhumation?
26. Why is it important to set aside preconceived notions and expectations when approaching a geologic problem? How might prefabricated opinions affect observation and understanding?
27. How are present-day rates of uplift and subsidence measured?
28. What type of field evidence might you look for to distinguish erosional exhumation from tectonic exhumation?
29. What specific regions of the United States are experiencing high rates of uplift and why?
30. Discuss elevation and relief in relation to landscape change. How are they measured? How are rates obtained?

Chapter 10

Evolution of Landscape

Based on previous chapters, we can conclude that landscape evolves primarily through interaction between the tectonic and climate systems. Together, these forcing agents activate the five mechanisms of landscape change. We can also conclude that it is possible for landscape over time too completely change its look from some previous landscape. Such conclusions beg two questions, what criteria can we use to determine when or how long ago landscape first comes into existence, and what are the paths by which landscape can evolve?

Barring volcanic catastrophe, the time it would take for landscape to completely change its look can be on the order of millions to tens of millions of years. Therefore, we cannot precisely determine how long ago landscape first comes into existence. We can, however, use the fact that landscape is shaped in large part by its components, its rock type, and structural form. On this basis, we can be reasonably certain that the landscape we see today is no older than the age of the near-surface bedrock that underlies the landscape, and no older than the age of the youngest structural form that has shaped the near-surface bedrock.

These criteria provide only a maximum age for when landscape first began to develop. They do not necessarily provide an accurate age. If rocks and structural forms are both ancient, as in the Appalachian Mountains, then landscape could potentially be far younger than the indicated age of either component. On the other hand, if either the bedrock or the most recent structural form is young, then a more accurate age for landscape is indicated. For example, the most recent structural form in the Basin and Range are normal faults. These faults overprint older thrust faults, but because they represent the most recent structural form, it is the normal faults, and not the older thrust faults, that control the shape of landscape. A date on the initiation of normal faults, therefore, provides a maximum age for the associated landscape.

Regarding the paths by which landscape may evolve; landscape can change its shape in response to any change in the near-surface rock/sediment type or structural form, to changes in climate, to changes in the rate or direction of tectonic stress, to changes in isostasy, or to changes in sea level. Each is capable of altering the rate at which one or more of the five mechanisms act on landscape. Landscape can also evolve in response to steady climatic conditions in which steady rates of erosion and deposition progressively lowers elevation and relief. In the case of the Basin and Range, we can say that a change in tectonic stress from compression (thrust faults) to tension (normal faults) created conditions that allowed landscape to evolve from some previous form dominated by thrust faults to a recognizably different form controlled by normal faults.

In this chapter, we characterize four basic paths by which landscape can evolve, each dependent on the degree and rate to which the components and forcing agents change. The four types are (1) growing old, (2) topographic steady-state, (3) rejuvenation, and (4) reincarnation. We also recognize relict landscape as vestiges of preexisting landscape. Keep in mind that the five mechanisms of landscape change are activated by changes to components and forcing agents, and that criteria used to recognize landscape evolution are visible changes in elevation, relief, river drainage pattern, and river drainage density.

LANDSCAPE GROWS OLD

In this book, we use the term growing old to denote a process that could also be referred to as erosional lowering or erosional decay. Landscape grows old by slowly losing elevation and relief through erosion and deposition such that it approaches the shape of flatland, or evolves to a low relief terrain composed of alternating hard and soft rock. Growing old implies that change occurs slowly and in a predictable way so as to remain recognizable compared to its earlier state.

Under ideal circumstances, landscape will grow old in areas where bedrock and structural form do not change appreciably, and where there are no forcing agents acting on landscape other than steady climate. There is no tectonic activity or glaciation, sea level is reasonably steady or a non-factor, and isostatic changes are in response to erosion and deposition only. Under these conditions, rates of erosion

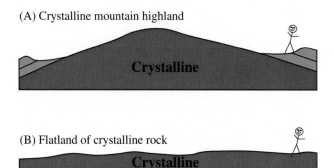

FIGURE 10.1 An example of growing old in which a crystalline mountain highland slowly erodes to a nearly flat plain.

and deposition are faster than rates of uplift and subsidence such that land areas slowly and predictably lose elevation and relief. Any change in drainage pattern or drainage density will also be slow and predictable.

The ideal evolution is like a person growing old. If you knew a person when he was young, chances are you would recognize him 10 or 20 years later because you know the changes that are expected. We also know that some people appear to grow old faster than others. The same is true of landscape. Some areas are worn down faster than other areas. The rate of erosion and deposition depends on a number of factors discussed in Chapter 9 that include climate, vegetation, rock hardness, elevation, and relief. Landscape composed of soft rock in a humid climate would likely wear down faster than one of hard rock in a dry climate. Under most conditions, we would expect the process of growing old to occur slowly and predictably over periods millions to hundreds of millions of years.

Fig. 10.1 shows an example of growing old in which a crystalline mountain highland evolves to a nearly flat plain. Bedrock is relatively homogeneous in this example. In areas where there are strong contrasts in erodibilty, differential erosion may not allow erosion to a flat plain. In this case, hard rocks will form highlands, but overall elevation is progressively reduced and relief remains gentle and low.

Another aspect to keep in mind regarding Fig. 10.1 is that rates of erosion and deposition decrease as elevation and relief decrease. In other words, the rate of erosional lowering decreases as landscape approaches a nearly flat plain. Later in this chapter, we discuss the possibility that the rate of erosional lowering can decrease to the point where landscape appears to be unchanging. In this sense, landscape has evolved to topographic steady-state.

LANDSCAPE AT TOPOGRAPHIC STEADY-STATE

Steady-state refers to topographic equilibrium in which landscape does not appear to change over a long period of time. Topography is neither gaining nor losing elevation, and other visible signs of change such as changes in relief, drainage pattern, and drainage density are not occurring at any measurable rate. The basic shape of landscape remains constant. Topographic steady-state is achieved and maintained when rates of uplift are balanced by equal rates of erosion. The dependence of steady-state on equal rates of uplift and erosion implies that rocks are under tectonic stress and are undergoing tectonic uplift.

Steady-state does not imply perfect balance between uplift and erosion at all times. For example, one might expect greater rates of erosion during a rainy season than during a dry season. Or, perhaps a large landslide alters the balance. Tectonic uplift often occurs in spurts, usually during a series of earthquakes rather than continuously. The temporary nature of these imbalances suggests that steady-state is not achieved over short time intervals, but can be achieved over periods of thousands of years or longer.

We can determine if a mountain is at steady-state if we can measure rates of uplift and rates of erosion over various time scales from thousands to millions of years. If we can show that these rates are of similar magnitude over various time scales, we can say that the rate of erosion is equal to the rate of uplift and that the mountain has achieved steady-state over the measured time interval. As noted in Chapter 9, it is possible to measure present-day uplift rates, but direct measurement of ancient uplift is difficult at best. In practice, one could measure exhumation rates over various time scales and compare them to present-day erosion rates. If all of the measured values are similar, then one could suggest that steady-state has been achieved. In Chapter 19, we will describe evidence to suggest that the Olympic Mountains have achieved topographic steady-state over the past 14 million years at an elevation of 8000 feet.

A mountain under ideal steady-state conditions does not vary in elevation. It is possible, however, for a mountain to rapidly change elevation and then return to steady-state. One scenario where this can occur is where the rate of erosion is controlled not by rock hardness, but by the rate of uplift. In other words, the erosive limit of bedrock has not been reached and erosion could occur at a faster rate if there were greater erosive power. Uplift potentially increases elevation and relief, which in turn increases erosive power. If long-term uplift rates begin to increase, a mountain previously at steady-state will experience a net gain in elevation and presumably also relief. As elevation and relief increase, the rate of erosion will also increase such that the mountain will reach and maintain steady-state at a higher elevation and at higher rates of uplift and erosion. Conversely, if long-term rates of uplift decrease, erosion will, at first, outpace uplift and mountain elevation will decrease, which in turn will lower the rate of erosion until steady-state is again achieved at the lower elevation. A similar reaction could occur if the rate of erosion slows or increases due to a change in the prevailing climate.

Steady-state may also be affected if erosion uncovers a different rock type or structural form. Such a scenario could either increase rates of erosion or slow rates of erosion, which would cause the mountain to lose or gain elevation prior to regaining steady-state conditions at the new elevation. Thus, it is possible for a steady-state mountain to exist over a range of elevations whose height is dependent on the rate of uplift relative to its erosive power.

Steady-State as the End-Product of Growing Old

Growing old implies progressive loss of elevation and relief such that landscape evolves to a nearly flat plain or one of low relief. During the final stages of growing old, it is possible for landscape to approach quasitopographic steady-state. As landscape is worn to low relief, the rate of erosion slows and therefore the rate of landscape modification also slows. Under these conditions, it is possible for landscape to become accustomed to the prevailing climatic rates of erosion and deposition such that, within certain limits, elevation, relief, drainage pattern, and drainage density remain constant or seemingly remain constant over long periods of time measured in millions to hundreds of millions of years. Any loss of elevation is slow and slight and is mostly compensated for by isostatic adjustment. Landscape is at equilibrium with climate-driven erosion and deposition and does not appear to change. Such a condition might be favored on the stable continental interior where relief and rates of erosion are low.

An example might be the Interior Low Plateaus (Chapter 13) where landscape is one of low elevation and relief, which implies low rates of erosion and isostatic adjustment. This area has evolved for hundreds of millions of years without a strong tectonic influence under climatic conditions that probably have not been more extreme than what it has experienced during the glacial/interglacial period of the past 2.4 million years. Even if we take into account sea level and isostatic changes, we can suggest that over a period of 100 to 200 million years, landscape of the Interior Low Plateaus has not evolved very much in terms of elevation, relief, and drainage characteristics. Different rocks have been uncovered during the long period of erosional lowering, but all of the rocks are sedimentary and most are nearly flat-lying such that they have similarly influenced landscape. We can suggest that such landscape has reached a quasitopographic steady-state due to very slow rates of erosion and isostatic response without the addition of tectonic stress.

The Piedmont Plateau has decayed (grown old) for so long that elevation and relief are low. One could suspect that this province also has reached or approached quasitopographic steady-state between climate-driven erosion and isostatic adjustment.

REJUVENATION

Rejuvenation refers specifically to incision into a preexisting river system and the creation of deep valleys, gorges, and canyons where none existed before. Rejuvenation can occur in response to isostatic uplift, broad tectonic uplift, or the lowering of sea level. All three processes lowers base level, which causes rivers to incise their channel, but does not substantially change the structural form of bedrock. Under these conditions, landscape in the process of growing old could undergo rejuvenation in the sense that landscape reverts to an earlier higher-elevation—higher-relief-time-period. In other words, lost elevation and relief are restored. Landscape, in large part, remains recognizable from its previous state because erosion is concentrated in river valleys. Existing stream channels are incised (deepened) rather than substantially changed or abandoned. The drainage pattern and drainage density remain recognizable from its previous state although some changes, such as stream piracy might be expected (Fig. 2.4). In the absence of strong tectonic stress, rejuvenation adds new life to the process of growing old. It prolongs the process. Rejuvenation occurs ideally in areas where incision does not uncover a different rock type or structural form, and where strong tectonic stress, glaciation, and volcanism are absent.

The Colorado plateau is an example of landscape rejuvenation. In only the past 6 million years, the Colorado plateau has experienced thermal isostatic and broad tectonic uplift on the order of several thousand feet without a substantial change in structural form. Fig. 10.2 compares lazy river meanders on the Mississippi River in Missouri with deeply entrenched meanders on the Green River in Utah. It is likely that meanders on the Green River established themselves at a time of low elevation and relief on the Colorado plateau when surrounding landscape resembled landscape surrounding the Mississippi River. Uplift of the plateau has resulted in rejuvenation in the form of river incision, entrenchment of meanders, and development of deep canyons. If left solely to erosion and deposition, we would expect this area to grow old by eventually evolving back to a Mississippi-style meandering river.

A rise in sea level will have an effect on landscape similar to isostatic or broad tectonic subsidence. It would result in loss of elevation and a rise in base level, which

FIGURE 10.2 Comparison between river meanders on the Coastal Plain with those on the Colorado Plateau. (A) Google Earth image looking east at meander bends on the Mississippi River at New Madrid, Missouri. River elevation is between 265 and 290 feet, which is slightly lower than the surrounding farmland at 275–300 feet. Reelfoot lake was formed or enlarged during a series of three major earthquakes and several strong aftershocks that occurred between December 16, 1811 and February 7, 1812. The earthquakes were responsible for as much as 20 feet (6 m) of vertical displacement, although no faults are known to have broken the surface. (B) Google Earth image looking westward at entrenched canyon meanders on the Green River just north of Canyonlands National Park. The river is at an elevation of about 3958 feet. The abandoned meander at center of photo is between 4050 and 4200 feet. The canyon rim is between 4750 and 5000 feet. Spring Canyon is at lower right, Hell Roaring Canyon is at lower left, and Horseshoe Canyon is at upper left-center where Barrier Creek occupies part the abandoned meander.

could trigger an increase in rates of deposition and a loss of relief. This type of change, when applied to landscape in the process of growing old could cause essentially the opposite reaction to rejuvenation. It could cause landscape to appear to grow old at a faster rate.

The southern Blue Ridge is an example of landscape that, during the process of growing old, has undergone several periods of base level rise and base level lowering. This region has grown old over a period of 265 million years, evolving from mountains as much as 20,000 feet high to their present-day appearance. The overall change has been one of decreasing elevation and relief via erosion without a strong tectonic influence. But the evolution has not been linear. The Blue Ridge has undergone periods of isostatic and sea level change as well as climate change that have periodically caused both lower elevation and relief, and higher elevation and relief. In Chapter 16 we will suggest that 50 million years ago the Blue Ridge was lower in elevation and relief than it is today and that it has only recently gained its present-day elevation and relief through isostatic uplift and landscape rejuvenation.

REINCARNATION

Reincarnation implies such severe change in landscape that the previous landscape is no longer recognizable except possibly as isolated relicts. Reincarnation refers specifically to the development of mountains and valleys with a different style or different orientation relative to preexisting landscape. There are recognizable differences in elevation, relief, drainage pattern, and drainage density. Reincarnation can occur if erosion uncovers a different rock type or structural form, if tectonic stress is strong enough to change the structural form of rock, if there is volcanism, or if glaciation occurs. Reincarnation is favored in areas where tectonic uplift is faster than erosion, but can occur under other conditions as well. We will first discuss reincarnation due to a change in components while growing old, followed by discussions of reincarnation due to volcanism, tectonic stress, glaciation, and burial beneath unconsolidated sediment. The process of reincarnation is not instantaneous under any circumstance. One must be aware; therefore, that relict landscape might persist long after the younger landscape begins to take shape.

Reincarnation While Growing Old

We have stated repeatedly that two of the strongest controls on landscape are rock type and structural form. During the process of growing old, a change in either component will result in different patterns of differential erosion and deposition, which could potentially change (reincarnate) landscape. Under these conditions, without strong tectonic stress, volcanism, or glaciation, the process of reincarnation will be exceedingly slow, on the order of tens to hundreds of millions of years. The process will be so slow as to be predictable.

Fig. 10.3 shows the possible landscape evolution of an anticline composed of sedimentary rock layers underlain with crystalline rock. Fig. 10.3B shows little change from Fig. 10.3A except for the erosional stripping of the top sedimentary layer. Stripping occurred because the top layer was nonresistant relative to the more strongly resistant layer underneath. The top layer was removed prior to significant erosion of the underlying resistant layer. Fig. 10.3C shows how landscape can evolve from an anticlinal mountain to an anticlinal valley through differential erosion of nonresistant sedimentary rock layers at the crest of the anticline. Split Mountain, as seen in Fig. 4.18, shows erosional stripping of an anticlinal layer as depicted in Fig. 10.3B as well as the development of an anticlinal valley as depicted in Fig. 10.3C. In Fig. 10.3D, landscape has evolved to one of low relief with cuestas and with crystalline rock at or just below the surface. Through continued erosion, Fig. 10.3E shows how landscape has evolved to a crystalline-cored anticlinal mountain. In this example, the weaker sedimentary layers have been stripped off the much harder crystalline rock without significant erosion of the crystalline rock, similar to Fig. 10.3B. Such a situation is especially possible if the contact between the two rocks is a relatively smooth unconformity. Fig. 10.3F shows the final stage of erosional lowering in which landscape evolves to a nearly flat plain. Each stage of landscape development shown in Fig. 10.3 evolved slowly enough (through erosion only) to be predicted. Most of the stages result in landscape that is significantly different from preexisting landscape. In Chapters 13 and 14, we will explore several anticlinal areas that have evolved, primarily through erosional lowering, to one of the stages shown in Fig. 10.3.

In Fig. 10.3, we saw how reincarnation could occur if erosion uncovers a different rock type. Reincarnation will also occur if erosional lowering uncovers rock units with a markedly different structural form. The Athens Plateau is an example. The Athens Plateau is located at the southern margin of the Ouachita Mountains bordering the Coastal Plain. Fig. 10.4 is a Google Earth image looking eastward at De Queen Lake on the Athens Plateau west of Hot Springs, Arkansas. The structural form surrounding De Queen Lake is one of thrust faults, folds, and tilted sedimentary rock layers deformed between about 320 and 265 million years ago. Differential erosion of these rocks likely created a landscape of long, narrow, mountain ridges composed of resistant rock with intervening valleys composed of nonresistant rock. However, by 80 million years ago, erosional lowering had reduced the landscape

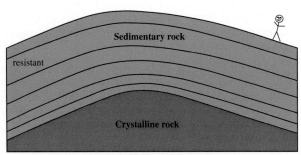

(A) Anticlinal mountain of sedimentary rock.

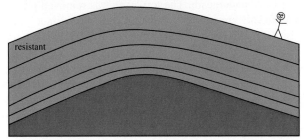

(B) Anticlinal mountain of sedimentary rock.

(C) Anticlinal valley in sedimentary rock.

(D) Near flatland across an anticline with cuestas and cored with crystalline rock.

(E) Anticlinal mountain cored with crystalline rock.

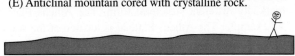

(F) Flatland of crystalline rock.

FIGURE 10.3 Panels (A) through (F) are a series of cross-sections that show the evolution of landscape during the process of growing old.

FIGURE 10.4 Google Earth image looking eastward at ridges and valleys that cross De Queen Lake, Arkansas, on the Athens Plateau, Ouachita Mountains. The landscape results from the erosional uncovering of folded, thrust faulted, and tilted layers. The landscape changes near the southern (right) margin of the image due to the presence at the surface of nearly flat-lying sedimentary rocks of the Coastal Plain.

to lowland that was subsequently inundated by a shallow ocean and covered with nearly flat-lying sedimentary rock layers. A second round of erosional lowering has since removed the overlying canopy of nearly flat-lying rock layers, thus reexposing the folded, tilted, and thrust faulted layers underneath. As a result, the landscape is now transforming into one of resistant linear ridges and less-resistant valleys as expressed by the many arms of the lake. De Queen Lake is one of several lakes on the plateau that accentuates this emerging landscape. Note how the linear landscape is lost at the extreme south (right) of the image. Here, nearly flat-lying sedimentary rocks remain at the surface just above the deformed sedimentary rocks. The change in structural form creates the physiographic boundary between landscape of the Ouachita Mountains and that of the Coastal Plain.

Although somewhat controversial, the linear Appalachian Valley and Ridge landscape may have undergone a history similar to that of the Athens Plateau. The Valley and Ridge consists of thrust faulted, folded, and tilted sedimentary rock layers that may have, at one time, been covered either with layers of nearly flat-lying sedimentary rock associated with the Appalachian Plateau, or with folded layers of homogeneously nonresistant shale. In either case, landscape associated with these overlying rocks may have more closely resembled the flat landscape of the Coastal Plain. Similar with the history surrounding De Queen Lake, the overlying flat-lying rocks may have been stripped during erosional lowering, thus exposing tilted resistant and nonresistant layers underneath to create the present-day valley and ridge landscape. In this scenario, erosion of tilted hard and soft layers has created relief, perhaps accentuated due to periods of broad isostatic uplift.

The type of reincarnation discussed above occurs entirely via erosion. It delays the process of growing old by creating intermediate landscape forms, but it does not change the end result. Without tectonic activity, volcanism, or glaciation, the end result likely will still be erosional lowering to a nearly flat plain or one of low relief with protrusions of hard rock as shown in Fig. 10.3F.

Reincarnation due to Volcanism and Tectonic Stress

We have seen that reincarnation can occur under conditions of growing old if during the process, a different landscape component is uncovered. The process of reincarnation under these conditions is very slow, on the order of tens to hundreds of millions of years, and is predictable. The swiftest, most drastic, and least predictable changes in landscape occur under conditions of tectonic stress, volcanism, and glaciation. Reincarnation under these conditions can be so thorough as to leave no trace or clue to the shape of the previous landscape.

The most obvious and profound form of reincarnation is volcanism. With a time scale that can be as short as years or thousands of years, volcanism is one of the fastest and most thorough mechanisms by which landscape can reincarnate. Volcanism can cover large areas, disrupt river patterns, and potentially create volcanic mountains at rates that could exceed 25 feet per 100 years. For example, prior to 17 million years ago, the Columbia Plateau likely consisted of rugged highlands similar in appearance to the surrounding Northern Rocky Mountains. This landscape, within a period of about 1.2 million years, became buried beneath a flood of basalt more than a mile thick in some areas. The outpouring of basalt changed both the near-surface bedrock and the structural form resulting in a new, completely different landscape. The central-southern Cascade landscape has been in existence for as much as 48 million years, however, all of the major present-day volcanoes are younger than 1.6 million years old. Older volcanic cones have blown up, gone extinct, and eroded. Volcanism is an obvious landscape changer.

Perhaps the most widespread mechanism by which landscape reincarnation has occurred is through changes in the near-surface structural form of rock due to active tectonic stress. The process of folding and faulting activates discrete areas of uplift and subsidence at rates that initially far outpace rates of erosion and deposition. The result is the appearance of mountains and valleys with a style or orientation that in most instances contrasts sharply with preexisting landscape. Normal faults, for example, create individual mountain blocks separated by wide intervening basins. Strike-slip faults displace landforms laterally. Thrust faults create high mountain ridges and narrow intervening valleys. Anticlines and synclines create alternating mountains and valleys. All of these result in drastic changes in elevation, relief, drainage pattern, and drainage density. Landscape reincarnation can be in the form of a shift from an inactive to an active tectonic system, or from a change in stress direction within an already active tectonic system.

The time scale for tectonic stresses to create a structural form capable of reincarnating landscape is on the order of hundreds of thousands to tens of millions of years. Many areas in the Cordillera have undergone recent topographic reincarnation due to changes in structural form including the California Borderland where the San Andreas Fault and associated strike-slip faults have been active for 29 million years. The structural form, and therefore the landscape of the Basin and Range physiographic province, the largest province in the Cordillera, has developed largely in the past 17 million years.

The Cordilleran landscape is the only region in the United States currently under active tectonic stresses strong enough to have recently changed the structural form of near-surface bedrock, and therefore, the only region to have undergone recent tectonic reincarnation. With few exceptions, tectonic reincarnation of the presently exposed Appalachian-Ouachita Mountain landscape ended approximately 265 million years ago. The only major exception has been the development of narrow, isolated normal fault valleys known as the Triassic Lowlands or Triassic Rift Valleys, but even these are more than 170 million years old. As can be inferred from Fig. 5.2, there is a part of the Appalachian-Ouachita belt that has experienced complete tectonic reincarnation, but that part presently lies below the Coastal Plain, and thus, can no longer be considered part of the mountain belt. The landscape of the Coastal Plain itself, as well as that of the Interior Plains and Plateaus, with few exceptions, has not experienced tectonic reincarnation for at least the past 541 million years. The rocks have remained nearly flat-lying since that time, implying that tectonic reincarnation has not been a major factor in the landscape development. One exception is the Great Plains where young volcanism and tectonic stresses related to Cordilleran tectonics have crept into the region.

Reincarnation due to Glaciation

Glaciation, like volcanism, can cause rapid changes in landscape in a time frame of hundreds to thousands of years. Glaciation is an extreme form of climate change. Here, we are referring to areas that were glaciated. Climate change has certainly affected nonglaciated areas, but in the absence of tectonic stress, volcanism, glaciation, or any change in rock type or structural form, it is unlikely that climate change alone is strong enough to reincarnate landscape. Nonglacial climate change will more likely speed-up, slow-down, or rejuvenate the process of growing old. Examples of landscape reincarnation due to glaciation are present across the north-central United States where continental glaciers, discussed in

Chapter 12, have flattened and buried preexisting landscape, thus completely altering the preexisting drainage pattern. Glacial erosion and deposition have modified mountainous areas in the Cordillera as well as mountainous areas in New England and the Adirondacks.

Reincarnation due to Burial Beneath Unconsolidated Sediment

The accumulation of unconsolidated sediment thick enough to completely bury bedrock-controlled landscape can also result in reincarnation. The most widespread examples are the continental glacial landscapes of the north-central United States. Another example is the southern Mississippi River floodplain shown in Fig. 3.13.

One might consider shorelines as a possible location where nonglacial climate processes can result in reincarnation. Waves and tides combine to produce one of the most powerful and relentless natural forces on Earth. The constant pounding of wave action can reshape a shoreline in one person's lifetime particularly in areas where there is an abundance of unconsolidated material or soft sedimentary rock. Hurricanes can reshape a shoreline in a matter of hours. Reshaping, however, does not necessarily mean reincarnation such that the shoreline takes on an entirely different look. We will learn in Chapter 13 that the morphology and character of the Atlantic shoreline may not have undergone significant change over the past 100,000-plus years despite sea level change during that time of more than 400 feet. Instead, it appears that the beaches, estuaries, and barrier islands present along the shoreline today have remained present simply by retreating as the shoreline retreated and by advancing as the shoreline advanced. In other words, the location of the shoreline has changed, but the morphology of the shoreline may have remained relatively constant.

SUMMARY

What constraints can we use to determine when a landscape comes into existence? If landscape has been carved and shaped from bedrock that possesses a certain structural form, it stands to reason that landscape can be no older than the near-surface bedrock that forms the landscape, and no older than the structural form that shapes the bedrock. Once formed, we can recognize four basic paths by which landscape can evolve, growing old, topographic steady-state, rejuvenation, and reincarnation.

Fig. 10.5 shows how landscape could evolve over time. An initially low elevation, low relief landscape rapidly gains elevation and is reincarnated into a mountain range due to the imposition of strong tectonic stress. Elevation gain (surface uplift) slows as the rate of erosion approaches the rate of bedrock uplift. The mountain achieves steady-state at time **T1**. Steady-state is interrupted at time **T2** by an increase in the rate of long-term bedrock uplift. The mountain gains elevation until time **T3** when the rate of uplift is balanced by the rate of erosion and the mountain once again achieves steady-state. Tectonic stress ends at time **T4** and the mountain begins to grow old via erosional lowering. The process is interrupted by a period of rejuvenation at time **T5**, but eventually reaches quasi-steady-state. The process can be interrupted, delayed, or halted at any time by reincarnation initiated by volcanism, the initiation of a new or different tectonic stress, glaciation, or the erosional uncovering of a different rock type or structural form. Glaciation between times **T3** and **T4** can disrupt steady-state and result in glacial modification or reincarnation of landscape, but these changes will be lost during the process of growing old because all mountain evidence of glaciation will be removed.

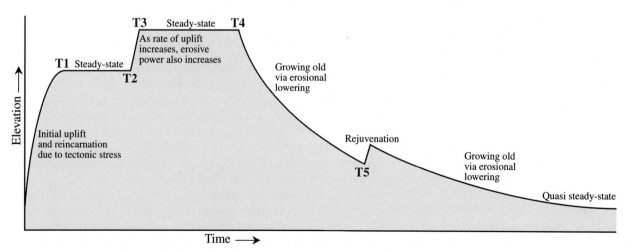

FIGURE 10.5 Elevation-time diagram showing reincarnation, steady-state, growing old, and rejuvenation.

QUESTIONS

1. Name the criteria used to determine the maximum age of landscape.
2. Describe the general age of rock and the age of the structural form in the Cordillera, the Interior Plains and Plateaus, Coastal Plain, and Appalachians.
3. Is it possible for the rock and the structure of the rock to be the same age? Explain.
4. Define each of the following: erosional lowering, rejuvenation, reincarnation, topographic steady-state.
5. What criteria can be used to determine whether or not a mountain is at topographic steady-state?
6. Name the criteria required for reincarnation during erosional lowering. By what other processes could reincarnation occur?
7. Explain why ideal topographic steady-state landscape requires tectonic uplift as opposed to isostatic uplift.
8. When can landscape undergoing only isostatic changes reach, or nearly reach topographic steady-state?
9. Can topographic steady-state be achieved or approached without tectonic stress?
10. What are the general rates of landscape reincarnation for the following processes, erosional lowering, tectonic stress, glaciation, and volcanism?
11. Explain why steady-state topography is not truly steady-state during short time intervals.
12. How is it possible for a steady-state mountain to exist at different elevations at different times?
13. In Fig. 10.3B, why is it more likely for the resistant layer to be breached at the crest of the anticline rather than anywhere else?
14. What is the only region in the United States to have undergone recent tectonic reincarnation?
15. Has tectonic reincarnation occurred in the Appalachian Mountains since the time mountain building ended?

Part II

Structural Provinces

Chapter 11

Structural Provinces, Rock Successions, and Tectonic Provinces

There are a variety of ways to categorize and divide land areas of the United States. In Part I of this book we used physiographic provinces, rivers, rock type, and climate. In the first chapter of Part II, we introduce three additional classifications, structural provinces, rock successions, and tectonic provinces. Structural provinces are an alternative form of landscape classification based on similar rock/sediment type and structural form. This classification is used throughout the remainder of the book. There are eight structural provinces each of which is discussed in separate chapters that follow. Rock successions and tectonic provinces are geologic classifications based respectively on how the rocks and the structural form are interpreted. These classifications are independent of present-day landscape and instead provide information on Earth history. They are introduced here so that structural provinces can be placed in their proper geological context. In this chapter, we discuss the three classifications and then describe how structural provinces relate to rock successions and tectonic provinces. We end the chapter by discussing the concept of an unconformity, including the Great Unconformity, which is present across the entire central part of the United States.

STRUCTURAL PROVINCES

A physiographic subdivision is one that separates land areas based on their topographic expression. In essence, a physiographic province is one that can be defined because it looks different from surrounding land. It is a descriptive and somewhat subjective approach to the classification of landscape that allows us to locate and study individual provinces. The problem is that it does not lend itself very easily to a general understanding of landscape because it does not predict similarities between provinces or why one is different from another.

What if we used a different set of criteria to classify landscape? We have seen that landscape is strongly influenced by its two components, the rock/sediment type and structural form. We can use these components to define a new set of provinces, which I refer to as structural provinces. We will define structural provinces based on having the same rock/sediment type and structural form without regard for topography or location. Under this classification, we eliminate the two components as unknown entities and allow widely separated areas to be grouped into a single structural province. Although not defined by topography, we will discover that areas with similar rock/sediment type and structural form possess similar landscape traits irrespective of the forcing agents and mechanisms of change that are acting on the landscape. A structural province classification will allow us to gauge the extent to which forcing agents and mechanisms of change influence landscape and how landscape might evolve when subject to different combinations of forcing agents and mechanisms.

We will divide the contiguous United States into eight structural provinces shown in Fig. 11.1 based on similar rock type and structural form. In areas where several structural generations have developed, we use the structural form that exerts greatest control on landscape, usually the youngest. Unconsolidated sediment is thick enough in a few areas to completely control landscape. As such, it can logically be considered a ninth structural province. Such areas are not shown in Fig. 11.1, but they include landscape in the Superior Upland and Central Lowlands composed of glacial sediment, landscape in the Piedmont Plateau and Coastal Plain composed of alluvial (river), beach, and soil sediment, and landscape in northern Nebraska, eastern Colorado and elsewhere composed of sand dunes. Glacial landscape, in particular, is widespread and creates both depositional and erosional landforms. Because it is such an important form of landscape modification and reincarnation, it will be discussed separately in Chapter 12. Other areas of unconsolidated sediment will be discussed in the context of their underlying rocks, primarily in Chapter 13.

In order to facilitate our understanding of structural provinces, Fig. 11.2 shows the general characteristics of each province including unconsolidated sediment. Each panel in

138 PART | II Structural Provinces

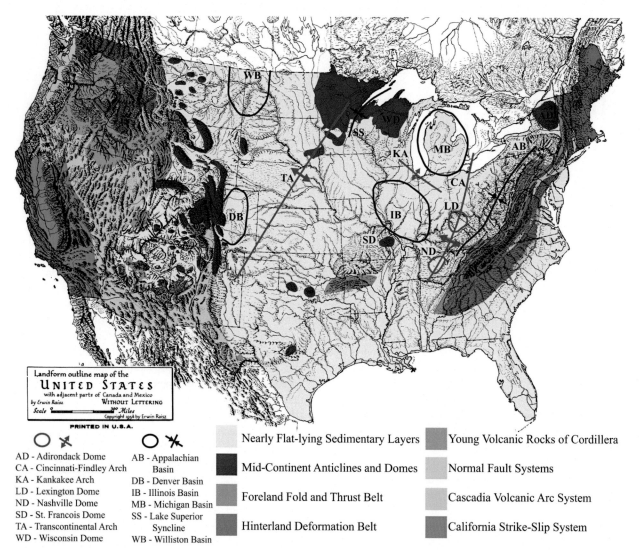

FIGURE 11.1 Landscape map showing structural provinces. The distribution of young volcanic rocks in the California strike-slip system and normal fault systems provinces are not shown.

the figure includes a cross-section that depicts the characteristic rock type, structural form, and resulting landscape traits. The four lines below the cross-section provide (in words) the rock type, structural form, landscape traits, and representative physiographic provinces, respectively.

Fig. 11.3 shows the boundaries of the 26 physiographic provinces superimposed on a colored map of structural provinces without the underlying Raisz landform map. This map allows us to predict the dominant rock type (or types) and structural form within each physiographic province. In some instances, the boundaries of structural provinces mimic those of physiographic provinces. Such a correlation highlights the strong influence that rock type and structural form have on topography and landscape evolution.

Table 11.1 subdivides the eight structural provinces into groups of two based on similar or transitional characteristics. The Interior United States and Coastal Plain group is comprised of the nearly flat-lying sedimentary layers and the crystalline-cored, mid-continent anticlines and domes structural provinces. These provinces form most of the interior United States including the Middle-Southern Rockies and Colorado Plateau. The cross-section in Fig. 11.2 implies that the nearly flat-lying sedimentary province consists of broad domes and basins across which sedimentary rocks remain nearly flat-lying. Several domes, basins, arches, and synclines are labeled in Fig. 11.1. These folds transition into crystalline-cored anticlines and domes in areas where folds are tight enough and erosion is deep enough to expose underlying crystalline rocks. Sedimentary rocks of the Coastal Plain are also part of the nearly flat-lying structural province. These rocks form a modern-day passive continental margin that developed subsequent to the rifting of Pangea

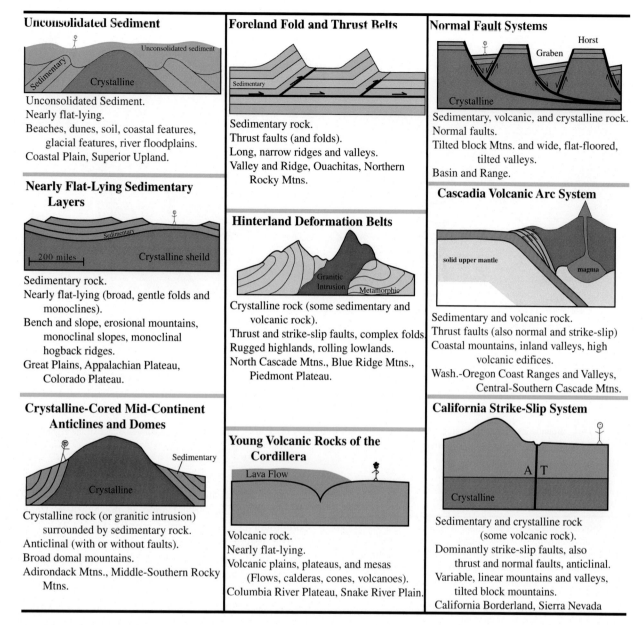

FIGURE 11.2 Schematic depiction of the eight structural provinces and unconsolidated sediment. Each panel shows a cross-section that depicts the characteristic rock type, structural form, and resulting landscape. The four lines below the cross-section provide (in words) the rock type, structural form, landscape, and representative physiographic provinces, respectively.

during the opening of the Atlantic Ocean. The Middle-Southern Rockies and Colorado Plateau were affected by the Laramide orogeny ending about 45 million years ago, which elevated and exposed crystalline-cored anticlines and domes. Elsewhere, the rocks in this group of provinces have not undergone strong, orogenic deformation.

The Appalachian–Ouachita and Cordilleran Mountain belts group of provinces represent areas of compressional mountain building whose orogenic structure and resulting landscape have not been greatly modified by postorogenic deformation, volcanism, or deposition. This group comprises the foreland fold and thrust belt and the hinterland deformation belt structural provinces. In the Appalachians, the foreland fold and thrust belt corresponds almost exactly with the Valley and Ridge and Ouachita Mountains physiographic provinces as well as with a small area in the Marathon region of southwest Texas. The Blue Ridge, Piedmont Plateau, and New England Highlands correspond with the hinterland deformation belt structural province. The rocks and structures that form these areas have remained largely unaltered since their formation more than 265 million years ago. Much of the landscape change since that time has been

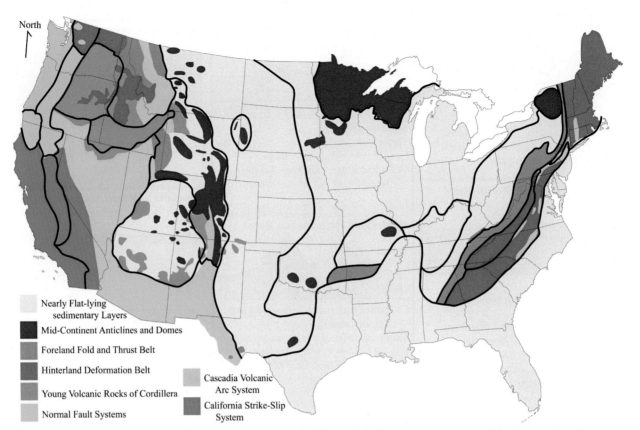

FIGURE 11.3 Boundaries of the 26 physiographic provinces, shown with heavy black lines, superimposed on a colored image of the eight structural provinces. The underlying landscape map is removed.

TABLE 11.1 Structural Provinces

The Interior United States and Coastal Plain
  Nearly flat-lying sedimentary layers
  Crystalline-cored, mid-continent anticlines, and domes
Appalachian–Ouachita and Cordilleran Mountain Belts
  Foreland fold and thrust belts
  Hinterland deformation belts
Post-Orogenic Tectonic Reincarnation
  Young volcanic rocks of the Cordillera
  Normal fault systems
Modern-Day Orogenic Mountain Belt
  Cascadia volcanic arc system
  California strike-slip system

through erosion, isostasy, and sea level change. The Cordilleran fold and thrust belt and hinterland deformation belt developed throughout the Cordilleran orogeny, ending with the Sevier orogeny about 52 million years ago. Since that time, the Cordillera has undergone extensive postorogenic reincarnation via normal faulting and volcanism such that nearly all of the landscape characteristics that developed in association with compressional, orogenic mountain building have since been destroyed. The landscape that remains includes foreland areas in the Northern and Middle Rocky Mountains, and hinterland areas in the Northern Rockies, Columbia Plateau, and North Cascades (Fig. 11.3).

The young volcanic rocks of the Cordillera and the normal fault systems structural provinces represent areas within the Appalachian and Cordilleran orogenic belts where tectonic and volcanic reincarnation has destroyed the orogenic landscape and created a completely different landscape in its place. These two provinces together form the post-orogenic tectonic reincarnation group of structural provinces as outlined in Table 11.1. As shown in Fig. 11.3, normal faults are the dominant landscape-forming structure in the Basin and Range and in part of the Northern and Southern Rocky Mountains. Young volcanic rocks cover nearly all of the Columbia Plateau and Snake River Plain as well as areas across the Colorado Plateau and Rocky Mountains. Both structural provinces developed largely but not entirely in the past 17 million years. The only postorogenic reincarnation in the exposed part of the Appalachians are the small Triassic lowland rift valleys in New England and the Piedmont Plateau, shown as part of the normal fault systems province in Fig. 11.3. These areas developed between 237 and 174 million years ago during the rifting of Pangea.

The Cascadia volcanic arc system and the California strike-slip system represent a modern-day orogenic mountain belt along the Pacific continental margin as defined by the interaction of two or more tectonic plates. The Cascadia volcanic arc system developed largely in the past 48 million years. The California strike-slip system developed only in the past 29 million years. Both structural provinces are associated with rock/sediment types and structural forms that create unique landscape traits as outlined in Fig. 11.2. This modern orogenic belt interacts with active normal faults to reshape landscape along the Pacific coastline.

ROCK SUCCESSIONS

Thus far, we have looked at rocks only with respect to how they shape landscape. But rocks have another very important function, and that is the ability to reveal Earth history. Revelations of Earth history from the study of rocks requires an increasingly complex degree of detail in order to test theory and interpretation. We will not delve deeply into this subject in this book, but it is informative to have at least a first order understanding of the origin and general framework of the rocks that make up the United States. To this end, we will define a rock succession as a stack (a succession) of rocks, be they sedimentary, crystalline, volcanic, or some combination, that are related to each other by their age, thickness, and geologic interpretation. They are similar to the previously described landscape-forming rock/sediment types except that we will group different rock types together and apply a genetic interpretation to the rock succession.

The most general classification subdivides the United States into as few as six rock successions listed in Table 11.2. They are the North American crystalline shield, Precambrian sedimentary/volcanic rocks, interior platform, miogeocline, accreted terranes, and the Atlantic miogeocline. The first four developed on the North American continent mostly prior to and during Appalachian and Cordilleran orogeny. They can be referred to as Native (or North American) rock successions. The accreted terrane rock succession developed initially offshore either on oceanic crust or as a foreign continent. This rock succession is exotic to North America and, as the name implies, was added (accreted) to North America during Appalachian and Cordilleran orogeny. The Atlantic miogeocline represents rocks of the Coastal Plain, all of which were deposited along the passive Atlantic continental margin following Appalachian orogeny and associated with the opening of the Atlantic Ocean.

Fig. 11.4 is a schematic west-to-east cross-section of the United States that shows the original location and stacking order of rock successions prior to Appalachian and Cordilleran deformation. The Atlantic miogeocline is not represented in this figure because it postdates Appalachian orogeny. Note that the two sides are mirror images. The North American crystalline shield (or more succinctly, the crystalline shield) forms the substratum of the entire continent. Shield rocks in the continental interior are overlain primarily by the interior platform rock succession. Shield rocks at the continental margins are overlain by the Precambrian sedimentary/volcanic and miogeoclinal rock successions. Accreted terranes are separated from native North American rocks by an ocean. Continental slope-rise rocks are transitional between native and accreted terrane rock successions. The following discussion defines each rock succession separately.

TABLE 11.2 Rock Successions

North American Crystalline Shield
  Igneous and metamorphic rock >1000 Ma
Precambrian Sedimentary/Volcanic Rocks
  Sedimentary/volcanic rock 1800 to 541 Ma
Interior Platform
  Sedimentary (continental interior) rock <541 Ma
Miogeocline
  Sedimentary (continental shelf) rock <541 Ma
Accreted Terranes
  Volcanic arc terranes, ocean basin rocks, and microcontinents
Atlantic Miogeocline
  Sedimentary (modern continental shelf) rock <130 Ma

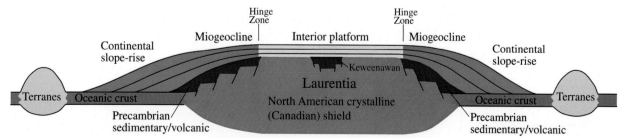

FIGURE 11.4 Conceptual west-to-east cross-section across the United States with Appalachian and Cordilleran deformation removed. The cross-section shows the original distribution and stacking of rock successions.

The North American Crystalline Shield

The North American crystalline shield represents the hardened crystalline core, the nucleus of North America. It is the original (or native) North American continent that existed prior to tectonic accretion associated with Appalachian and Cordilleran orogeny. Today, it is a slab generally 12–35 miles thick (19-56 km) composed almost entirely of crystalline rocks that vary in age from 3.6 to 1.0 billion years old. It is sometimes referred to as crystalline basement. The crystalline shield rock succession developed as a result of several orogenic events, the strongest of which ended roughly 2.5, 1.6, and 1.0 billion years ago. Each of these orogenic events culminated in the formation of a giant landmass (a supercontinent) that included all of the Earth's landmasses that were in existence at the time. Rifting and separation of continents followed each orogenic event. The last major event, known as the Grenville orogeny, affected the eastern part of North America between 1.3 and 1.0 billion years ago culminating with formation of the supercontinent Rodinia. Because Rodinia formed previous to Pangea, it is also known as proto-Pangea. Beginning less than 800 million years ago, Rodinia began to break apart, and the part that would become the North American crystalline shield became isolated from other continental fragments. This fragment of shield rock is known as Laurentia.

The known extent of the crystalline shield in the United States is shown with blue highlight in Fig. 11.5. Most of the shield is hidden below younger rock layers. The red areas show where shield rock is exposed at the surface. Crystalline shield rocks are more widely exposed in central Canada where they are known as the Canadian Shield. As implied in Fig. 11.5, the Superior Upland and Adirondack Mountains form a direct southern continuation of the Canadian Shield. For most of our discussion, we will treat the crystalline shield as single entity, however, keep in mind that it is composed of a great variety of rocks and structures that cover two-thirds of Earth history involving numerous collisional, intrusive, and rifting events.

The western boundary of buried shield rocks in the US Cordillera has been determined based on geochemical evidence discussed in Chapter 18. As shown in Fig. 11.5, the boundary lies close to the western Idaho border extending southward through Nevada and across southern California. It is unlikely that buried shield rocks extend far beyond this boundary.

FIGURE 11.5 Landscape map that shows the known extent of the North American crystalline shield in the United States in blue highlight. Red highlight shows areas where the crystalline shield is exposed at the surface. *Based on Whitmeyer and Karlstrom (2007).*

In the southern and eastern United States, the shield is known to extend as far south as the Ouachita Mountains and as far east as the Valley and Ridge. In contrast to the Cordillera, it is not possible to geochemically determine the eastern and southern boundaries of buried shield rock; therefore, the full extent of shield rock in the southern and eastern United States is poorly known. Surface exposures, shown in red, are present along the entire western Blue Ridge and in the Berkshire/Green Mountains of western Massachusetts and Vermont. These rocks were carried westward along thrust faults, which implies that buried shield rocks are present east of these exposures below the Piedmont Plateau and possibly below the Coastal Plain.

Precambrian Sedimentary/Volcanic Rocks

The Precambrian sedimentary/volcanic rock succession was deposited in basins primarily along the margins of the crystalline shield at various times, between 1.8 billion and 541 million years ago. These are the oldest unmetamorphosed sedimentary and volcanic rocks in the United States. The oldest part of this rock succession was deposited in basins prior to the rifting of Rodinia. The younger rocks form a rift-related clastic sequence on both sides of the continent associated with breakup of Rodinia. The rocks are largely absent in the central part of the continent except for the 1.1-billion-year-old Keweenawan rock sequence in the Lake Superior region shown in Fig. 11.4, and in a few additional areas primarily on the Colorado Plateau. The Keweenawan rock sequence represents a failed attempt at continental rifting apparently just prior to the climax of Grenville orogeny! The Precambrian sedimentary/volcanic rock succession is present in the Cordillera, primarily in the foreland fold and thrust belt. Variably metamorphosed equivalent rocks are present in the western part of the Appalachian hinterland deformation belt and are scattered across the Cordillera.

The Interior Platform

By the beginning of the Phanerozoic, 541 million years ago, all mountains in the United States had eroded to a relatively flat surface (a platform) bordered on all sides by a passive continental margin (Chapter 5). The only rock successions in existence at this time were shield rocks exposed across the continental interior, and Precambrian sedimentary/volcanic rocks exposed at the surface or hidden beneath a shallow ocean along the continental margin. Throughout the Paleozoic and Mesozoic eras, the continental interior would periodically subside below sea level and be covered by a shallow ocean that deposited layer after layer of sedimentary rock directly on top of the eroded stump of the crystalline shield. These rocks form the interior platform rock succession as shown in Fig. 11.4. They represent the progression of landforms and inland seas that developed on top of the shield prior to, during, and following Appalachian and Cordilleran orogenic events. The environments recorded by platform rocks include shallow inland seas, rivers, deltas, deserts, swamps, and lakes. Some of the sediment for these depositional environments was derived from erosion of adjacent Appalachian and Cordilleran orogenic highlands. The total thickness of the sedimentary rock succession in most areas is between 5,000 and 15,000 feet. Areas that subsided continually throughout most of the Paleozoic, such as the Michigan and Illinois Basins, received a thick succession of platform rock. Areas that underwent periods of uplift, such as the Transcontinental and Cincinnati Arches, received a much thinner rock succession (Fig. 11.1). The defining characteristics of the interior platform rock succession are that they are sedimentary, Phanerozoic in age (<541 Ma), and typically less than 15,000 feet thick. With the exception of the eastern Cordillera, these rocks were not widely involved in Appalachian or Cordilleran orogenesis. Along with the underlying crystalline shield, they form much of the nearly flat-lying sedimentary and crystalline-cored midcontinent anticlines and domes structural provinces.

The Miogeocline

Rocks of the miogeocline represent a continuation of interior platform deposition along a passive continental margin where the crystalline shield begins to thin and merge with oceanic lithosphere. The rocks are the same age as those of the interior platform, but because they were deposited along the continental margin, there are some differences. A major difference is the thickness of the miogeoclinal succession, which can exceed 50,000 feet. The great thickness results from slow isostatic sinking due to the cooling of the lithosphere following rifting, and to the weight of the accumulating sediment. Another difference is that miogeoclinal rocks are almost everywhere underlain by the Precambrian sedimentary/volcanic rock succession rather than directly by shield rocks.

Miogeoclinal rocks, in spite of their great thickness, were deposited in shallow water. The rocks represent sediment shed from the continental interior and deposited along the continental shelf, much like the present-day Atlantic continental shelf. Depositional environments include shallow marine shelf, reef, beach, barrier island, estuary, tidal zone, and deltas. Constant deposition in a slowly subsiding basin produced a rapid transition between the thin interior platform rock succession and the much thicker miogeoclinal succession. This original transition is referred to as a hinge line (or hinge zone; Fig. 11.4). The hinge line forms an abrupt transition because it also marks the approximate craton-ward limit of rifting and deposition of the sedimentary/volcanic rock succession associated with the breakup of Rodinia. Major thrust faults in both the Appalachian and Cordilleran orogenic belts would later take advantage of this hinge line by pushing the miogeoclinal rock succession over the interior platform succession. The sedimentary miogeocline forms the primary rock succession within both the Cordilleran and Appalachian fold and thrust belt structural province.

Accreted Terranes

The accreted terrane rock succession is a complex group of rocks associated with such diverse settings as island volcanic arcs, ocean basins, volcanic seamounts, and oceanic plateaus. Nearly all of the rock is oceanic in origin, but continental masses (microcontinents) that include transported plutonic rocks are also present. Dozens of distinct terranes were accreted at different times during Appalachian and Cordilleran orogeny. All of these rocks became part of North America via subduction, collision, and tectonic accretion, and in doing so, have undergone variable amounts of deformation, metamorphism, and intrusion. These rocks are surrounded by suture zones and have a stratigraphy and geologic history that is different from surrounding rocks. They are terranes in the truest sense. Accreted terranes in some areas are covered or intruded by younger rocks that formed on the North American continent following tectonic accretion. Nearly all of the sedimentary and volcanic accreted terrane rock in the Appalachians has been metamorphosed. Unmetamorphosed sedimentary and volcanic accreted terrane rocks are widespread across the western part of the Cordillera with metamorphosed rocks present primarily in the northern California Coast Ranges, the northern Sierra Nevada, the Blue Mountain region of Oregon, the Klamath Mountains, and the North Cascades. Both the Appalachian and Cordilleran accreted terrane belts contain abundant intrusive rock.

The Atlantic Miogeocline

The entire Coastal Plain and adjacent Atlantic continental shelf is an actively subsiding passive continental margin that has undergone continuous deposition since initial rifting of Pangea beginning in the Triassic some 237 million years ago. It is a modern-day miogeocline, referred to here as the Atlantic miogeocline. The rocks bury and hide the eroded stump of the Appalachian–Ouachita Mountain belt, which originally extended to the edge of the continental shelf. The weight of accumulating sediment coupled with slow cooling has caused isostatic subsidence, which has provided room for increasingly greater depositional thicknesses toward the ocean. The resulting geometry is that of a wedge that thickens from zero at the boundary with the Piedmont Plateau to more than 40,000 feet in some areas below the continental shelf. Periodic incursion of the ocean onto the Coastal Plain has created a diverse set of marine, transitional, and continental depositional environments that includes those previously mentioned for the interior platform and miogeocline. The sedimentary rocks of the Coastal Plain, because they are universally young, tend to be softer than those of the miogeocline and interior platform. The oldest exposed rocks are no more than about 130 million years (Cretaceous).

TECTONIC PROVINCES

Before we discuss the distribution of rock successions across the United States and their relationship to structural provinces, we need to have a general understanding of the structural/tectonic setup of the United States. The United States consists of a stable interior region bound on its eastern side by the ancient, inactive Appalachian–Ouachita orogenic belt and Coastal Plain, and on its western side by the wide, actively developing Cordilleran orogenic belt. Recall from Chapter 5 that orogeny refers to a period of time during which compressional rock deformation and mountain building takes place. The ultimate driving force for orogeny is the convergence of two or more tectonic plates. Convergence creates stresses strong enough to cause rocks to fold, fracture, and fault. The resulting impact on landscape is uplift, subsidence, and in some areas, volcanism. In most cases, as in the Appalachians and Cordillera, plate convergence occurs along an entire continental margin resulting in a continental-scale compressional mountain belt that is longer than it is wide.

Although different in detail, all convergent mountain belts, including the Cordillera and Appalachians, show characteristic changes across their width from the continental interior toward the ocean. These changes can be followed for hundreds of miles along the length of the mountain belt creating what we refer to here as tectonic provinces. Tectonic provinces are defined primarily based on a certain structural style, but each province is also composed of one or more characteristic rock successions. In this book, we define six tectonic provinces, four of which, the Craton, the Foreland Fold and Thrust Belt, the Native North American belt, and the Accreted Terrane Belt, are present in both the Appalachian and Cordilleran Mountain belts, and are present in other mountain belts of the world. These provinces collectively form the characteristic features of an idealized orogenic belt. Fig. 11.6 is a cross-section that shows the structural attributes and characteristic rock successions of each of the four tectonic provinces. Missing from this cross-section is the reactivated western craton, which is specific to the Cordillera,

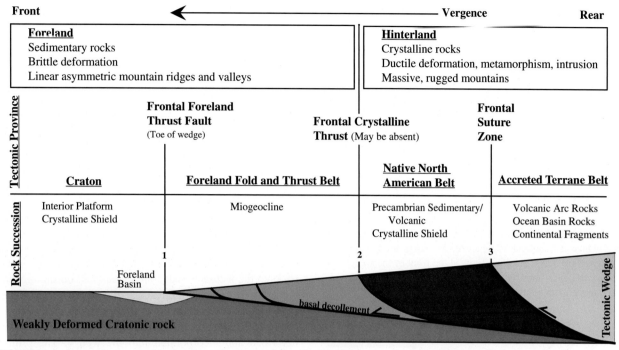

FIGURE 11.6 Schematic cross-section that shows the characteristics of an idealized orogenic tectonic wedge and a correlation of rock successions and tectonic provinces.

TABLE 11.3 Tectonic Provinces with Associated Rock Successions and Structural Attributes

FORELAND
  Craton
    Crystalline Shield, Interior Platform (+Foreland Basin)
    Weak deformation since 541 Ma
  Reactivated Western Craton
    Crystalline Shield, Interior Platform (+Foreland Basin)
    Anticlinal (± thrust faults)
  Foreland Fold and Thrust Belt
    Miogeocline (+Foreland Basin)
    Thrust faults and folds
HINTERLAND
  Native North American Belt
    Crystalline Shield, Precambrian Sedimentary/Volcanic
    Thrust faults, metamorphism, intrusion
  Accreted Terrane Belt
    Accreted Terranes
    Thrust and strike-slip faults, metamorphism, intrusion
  Atlantic Marginal Basin
    Atlantic Miogeocline (Post-Orogenic)
    Weak deformation since 130 Ma

Fig. 11.6 implies that the idealized orogenic mountain belt can be separated into a foreland region and a hinterland region and that the overall shape is that of a wedge, thin in the foreland, thick in the hinterland, with a well-defined thrust fault, known as the basal décollement, that separates rocks within the tectonic wedge from weakly deformed or undeformed rocks underneath. Normal thickness of continental crust is about 26 miles Crustal thickness at the heel of a tectonic wedge can be more than two times this amount. Parts of both the Appalachian and Cordilleran systems presently reach crustal thicknesses of about 34 miles.

The tectonic wedge is one of the most important concepts in the origin of compressional mountain belts. Development of the tectonic wedge is similar to pushing a fresh layer of snow to the edge of your driveway. The process is illustrated in Fig. 11.7. Keep the angle of the shovel constant as you push across the driveway. Snow initially builds against the shovel. As the wedge builds, it will reach critical taper, which refers to the shape of the wedge when tectonic pushing forces are equal to tectonic resisting forces. The wedge at this point is in a state of equilibrium such that it will slide without internal deformation. As the wedge slides forward, snow no longer builds against the shovel. Instead, the entire wedge slides as a rigid coherent mass, and fresh snow is broken at the front (the toe) of the wedge. Applying this analogy to a mountain belt, the hinterland represents the rear (or heel) of the mountain belt

and the Atlantic Marginal Basin, which postdates (i.e., occurred after) Appalachian orogeny. The six tectonic provinces, their associated rocks successions and their structural attributes, are listed in Table 11.3.

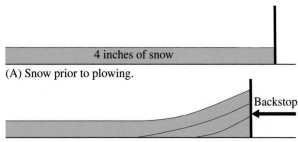

(A) Snow prior to plowing.

(B) Snow piles up at the back of the wedge (the hinterland) until the wedge achieves critical taper.

(C) Once critical taper is achieved, the wedge moves as a single rigid mass, breaking snow at the front of the wedge (the foreland). Snow at the front of the wedge breaks in sequence from oldest to youngest as shown.

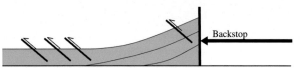

(D) Faulting in the foreland causes the wedge to change shape. If the wedge loses critical taper, then faulting may occur at the rear of the wedge in order to reestablish critical taper.

FIGURE 11.7 Formation of the tectonic wedge using snow as an example.

where the tectonic wedge thickens. The foreland represents the front of the mountain belt that forms after the wedge reaches critical taper and is pushed as a rigid mass into the continent. The foreland is where previously undeformed rock is broken at the toe of the wedge.

The direction that rocks are transported within the tectonic wedge is known in geology as vergence. Most of the thrust faults in the Appalachian–Ouachita belt dip eastward or southward and verge westward or northward. Those in the Cordillera dip west and verge east. It can be inferred from this geometry that the tectonic wedge in both mountain belts has been pushed toward the continental interior (toward the foreland). Tectonic wedge theory therefore predicts that orogenic mountain building begins in the hinterland and migrates to the foreland after the tectonic wedge reaches critical taper. The hinterland thickens through volcanism, terrane accretion, metamorphism, igneous intrusion, folding, and faulting. It could take several orogenic cycles for the hinterland to build to critical taper.

Given the above description, we can define the hinterland as shown in Fig. 11.6 as the rear or interior part of a mountain belt, close to the collision zone where crystalline rocks are brought to the surface and where volcanic and sedimentary rocks may also be present. The primary mode of deformation is thrust faulting; however, strike-slip faulting, multiple fold generations, metamorphism, and intrusion are significant. The hinterland is where rocks initially thicken into the shape of a wedge, where crystalline rocks are exposed, where rugged mountains develop, and potentially where mountains reach their highest elevation.

The term foreland technically refers to undeformed rocks in the continental interior region directly in front of the mountain system. It is also the location where debris, shed from erosion of rising hinterland mountains, is deposited. Such depositional basins are variably referred to as foreland basins, a foreland clastic wedge, a molasse basin, or a foredeep. All of these terms imply more-or-less the same thing, a depositional basin at the front of a mountain belt. The foreland is the primary direction toward which rocks in the tectonic wedge are transported during orogeny. Once the tectonic wedge reaches critical taper, it will push into the foreland such that previously undeformed sedimentary rocks, along with foreland basin deposits, will be deformed and incorporated into the wedge. The primary mode of deformation is thrust faults. We can therefore define the foreland as shown in Fig. 11.6 as the front of a compressional mountain belt where sedimentary rocks are either undeformed or have been incorporated into the toe of the tectonic wedge primarily along thrust faults. Typically, there is no metamorphism or igneous intrusion.

Hinterland Tectonic Provinces

The idealized hinterland shown in Fig. 11.6 consists of two distinct tectonic provinces, the native North American belt toward the front, and the belt of accreted terranes at the rear. Rocks that form the native North American belt consist mostly of the crystalline shield and the Precambrian sedimentary/volcanic rock succession, but miogeoclinal rocks may also be also present. The belt of accreted terranes at the rear of the tectonic wedge consists of oceanic volcanic arcs, ocean basin rocks, and small continental masses separated from North American rock successions and from each other by suture zones. Rocks in both tectonic provinces of the Appalachians were variably folded, faulted, metamorphosed, and intruded during Appalachian orogeny and, therefore, are dominantly crystalline. Those in the Cordillera were deformed, but not everywhere metamorphosed or intruded, and so are variably sedimentary and crystalline. Mountains in the hinterland are high, massive, and rugged, primarily because of compression, isostatic uplift of thickened crust, and the presence of poorly layered, resistant crystalline rock. This type of landscape forms the hinterland deformation belt structural province as shown in Fig. 11.2 and described in Chapter 16.

Foreland Tectonic Provinces

The idealized foreland shown in Fig. 11.6 consists of two tectonic provinces, the foreland fold and thrust belt and the craton. The foreland fold and thrust belt forms the front of the tectonic wedge. In an idealized orogenic system, the fold and thrust belt does not develop until the tectonic wedge reaches critical taper and begins to push into previously undeformed sedimentary rocks of the foreland. Hinterland mountains, by this time, may have already risen to great heights. The fold and thrust belt develops primarily in the well-layered miogeocline rock succession because the nearly horizontal bedding planes allow for easy slip, and because the beds thicken toward the hinterland. Precambrian sedimentary/volcanic rocks, and foreland basin rocks derived from erosion of hinterland mountains, may also be present.

The fold and thrust belt develops initially at the rear of the miogeocline and propagates (moves) across the miogeocline toward the craton as the tectonic wedge grows and is pushed inland. Early foreland basin rocks are deformed and incorporated into the fold and thrust belt and younger foreland basins develop closer to the craton. The fold and thrust belt continues to expand toward the craton until the frontal thrust fault, at the toe of the wedge, reaches the hinge zone (Fig. 11.4). The shallow depth of crystalline shield rock at the hinge zone causes the frontal thrust to deflect to the surface such that the interior platform rock succession is not greatly involved in compressional mountain building.

As rocks in the miogeocline are progressively deformed, a sedimentary mountain range develops parallel to the preexisting hinterland mountain range. Thrust faults impart an asymmetry to the mountain range because they transport rock layers mostly in one direction: toward the craton. The resulting foreland fold and thrust belt landscape is one of asymmetrical, linear mountain ridges that are steep on the cratonic side and tilted with a more gentle dip on the oceanward side. The mountain ridges alternate with long, narrow valleys. This type of landscape forms the foreland fold and thrust belt structural province as shown in Fig. 11.2 and described in Chapter 15. Thus, the end result of orogeny and the building of the tectonic wedge are two parallel mountain belts, a crystalline hinterland mountain belt and a younger sedimentary foreland fold and thrust mountain belt.

The craton is the stable interior part of the continent that was not directly involved in Appalachian or Cordilleran orogeny. Characteristic rock successions on the craton are the crystalline shield and the interior platform, but Precambrian sedimentary/volcanic rocks are also present, most notably the Keweenawan rock sequence in the Lake Superior region.

A foreland basin, as shown in Fig. 11.6, can develop on the craton directly adjacent to the mountain belt where it receives detritus (erosional debris) from eroding highlands. This basin represents the final foreland basin to develop, as earlier-formed basins would have been incorporated into the fold and thrust belt. The basin forms at the foot of the mountain due to isostatic principles. Imagine placing a stack of heavy books in the middle of a long shelf. The shelf will sag under the weight. Similarly, the weight of the mountain causes the Earth's crust to sag, thus creating the foreland basin.

Foreland basins are important for at least two reasons. The first is that the age of sediment in the basin dates the existence of the mountain belt. The second is that the type of sediment provides information on the timing and progression of unroofing (exhumation). For example, if sediment in the lower part of the foreland basin consists entirely of sedimentary rock debris, and sediment in the upper part consists of crystalline rock debris, it may be possible to date the time when crystalline rock was first exposed, eroded, and deposited in the basin.

The Reactivated Western Craton and the Atlantic Marginal Basin

The formation of a tectonic wedge and the building of a compressional mountain belt as described above do not include two of the six tectonic provinces, the reactivated western craton, and the Atlantic marginal basin. The reactivated western craton is unique to the Cordillera. It is an aberration to the idealized compressional orogenic belt described above. From a rock succession point of view, the reactivated western craton is part of the craton. It is located east of the Cordilleran fold and thrust belt, and it consists primarily of interior platform sedimentary rocks underlain with rocks of the crystalline shield. The unique aspect of the reactivated western craton is that it became involved in Cordilleran mountain building following development of the Cordilleran fold and thrust belt. Deformation in the reactivated western craton is associated with the circa 75- to 45-million-year-old Laramide orogeny, the youngest orogeny to affect the interior Cordillera. The Cordilleran fold and thrust belt developed primarily during the Sevier orogeny between 115 and 52 million years ago.

The Atlantic marginal basin coincides with the Coastal Plain physiographic province. We noted earlier that the Appalachian—Ouachita Mountains at one time extended across the Coastal Plain all the way to the edge of the continental shelf. The Atlantic marginal basin represents a return of part of the Appalachian hinterland to a passive continental margin following rifting, erosion, and subsidence. It is a modern-day miogeocline composed entirely of the Atlantic miogeocline rock succession. The rocks unconformably overlie the eroded stump of the Appalachian—Ouachita hinterland mountains.

DISTRIBUTION OF ROCK SUCCESSIONS AND TECTONIC PROVINCES

It is important to understand the difference between structural provinces and rock successions/tectonic provinces. Structural provinces provide information on landscape. Rock successions and tectonic provinces are defined based on an interpretation of the rocks and their structure. They are purely geological. They provide information on geologic history, but no direct information on present-day landscape.

It is also important to understand the close association between tectonic provinces and rock successions. We will discover that boundaries separating tectonic provinces coincide, in some cases exactly, with boundaries that separate rock successions. The reason for this correlation has to do with the original location of rock successions as outlined in Fig. 11.4, and the end result of mountain building as outlined in Fig. 11.6. Collisional mountain building progresses from ocean to continent. Accreted terranes collide at the continental margin, thereby adding mass at the heel of the tectonic wedge. The accreted terranes are then pushed into the Precambrian sedimentary/volcanic and crystalline shield rock successions at the margin of the continent, piling these rocks on top of each other and subsequently pushing them onto the miogeocline as the tectonic wedge grows to critical taper. The miogeocline is then pushed over the interior platform rock succession. Thus, we see that orogeny stacks the rock successions one above the other as shown in Fig. 11.6 but leaves their original continent to ocean distribution largely intact.

With these ideas in mind, we can now look at the distribution of rock successions and tectonic provinces across the United States. The two maps are shown respectively in Figs. 11.8 and 11.9. In the construction of these maps, we have ignored all reincarnation that followed orogenic mountain building except for the Atlantic miogeocline rock succession and the Atlantic marginal basin tectonic province. The two maps show the full extent of the Appalachian–Ouachita and Cordilleran rock successions and tectonic provinces as defined by the rocks, without the complicating effects of later reincarnation.

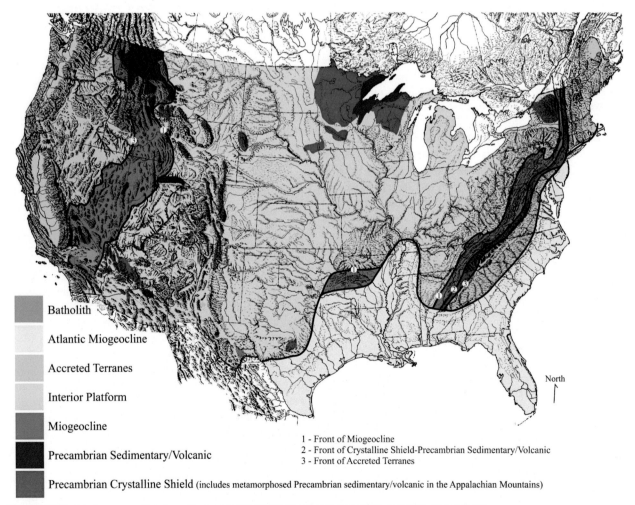

1 - Front of Miogeocline
2 - Front of Crystalline Shield-Precambrian Sedimentary/Volcanic
3 - Front of Accreted Terranes

Precambrian Crystalline Shield (includes metamorphosed Precambrian sedimentary/volcanic in the Appalachian Mountains)

FIGURE 11.8 Landscape map that shows the present-day distribution of rock successions across the United States.

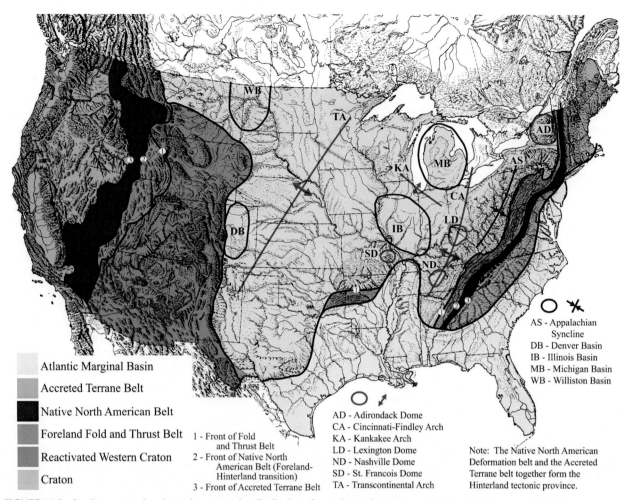

FIGURE 11.9 Landscape map that shows the present-day distribution of tectonic provinces across the United States.

The location of lines numbered 1, 2, and 3 in the Appalachian–Ouachita system, and 1 and 3 in the Cordillera, are identical on the two maps. These lines represent major rock succession and tectonic boundaries, and are also shown in Fig. 11.6. Boundary line 1 on the tectonic provinces map (Fig. 11.9) forms the front of the foreland fold and thrust belt. It separates the foreland fold and thrust belt from the relatively undeformed craton in the Appalachian–Ouachita belt, and from primarily the reactivated western craton in the Cordillera. In most areas, this boundary is represented by a thrust fault (the frontal foreland thrust). One exception is Pennsylvania where deformation in the foreland fold and thrust belt gradually diminishes toward the craton without a major bounding fault. The identically located boundary line 1 on the rock successions map (Fig. 11.8) separates the miogeocline from the interior platform rock succession. This boundary, in Fig. 11.4, is located at the hinge zone that originally separated the miogeocline from the interior platform.

Boundary line 2 on the tectonic provinces map (Fig. 11.9) forms the front of the Native North American belt. It marks the first appearance of crystalline rock in the orogenic belt, and as such, it represents the foreland-hinterland boundary. This line is variably a thrust fault (the frontal crystalline thrust), a depositional contact between crystalline rock and overlying sedimentary rock, or an intrusive contact. In the Appalachians, an identically located boundary line 2 on the rock successions map (Fig. 11.8) marks the frontal boundary of a semicontinuous belt of crystalline shield and metamorphosed Precambrian sedimentary/volcanic rock that extends along the western Blue Ridge northward to the Green Mountains of western Vermont. The change in rock type across this boundary line creates a topographic step that helps separate the physiographic Valley and Ridge from the Blue Ridge. There is no semicontinuous belt of crystalline rock west of the miogeocline in the Cordillera. Instead, we define boundary line 2 on the tectonic provinces map (Fig. 11.9) as marking the first appearance of crystalline rock west of the fold and thrust belt. The crystalline rocks are represented by isolated occurrences of shield rock, intrusive rock, and rock metamorphosed during Cordilleran orogeny. Because the crystalline rocks do not form a continuous or semicontinuous belt, but instead are surrounded by sedimentary rocks of the miogeocline, a Cordilleran boundary line 2 cannot be defined on the rock successions map (Fig. 11.8).

Boundary line 3 on the tectonic provinces map (Fig. 11.9) marks the front of the accreted terrane belt. An identically located line on the rock successions map (Fig. 11.8) marks the frontal boundary of rocks that form accreted terranes. This boundary line forms the frontal suture zone that separates North American rock successions from accreted terranes. It marks the location of an ancient subduction zone where landmasses on opposite sides were once separated by an ocean. In its unaltered state, it marks the location of an ancient accretionary prism, described in Chapter 5 as a mélange zone. The mélange zone, however, is often covered, offset, reactivated, or otherwise obscured by later faulting, intrusion, and deposition. In the Appalachians, boundary line 3 corresponds with the Taconic suture zone, a major tectonic boundary that formed during Taconic orogeny. Accreted terranes directly east of this suture zone are some of the earliest to have been accreted and incorporated into the Appalachian orogenic belt. Most are metamorphosed and intruded. These rocks, together with the native North American belt, form one of the largest continuous areas of crystalline rock in the world, encompassing most of the Blue Ridge, Piedmont Plateau, and New England Highlands. Boundary line 3 in the Cordillera corresponds in part with a major thrust fault known as the Roberts Mountains thrust. Both sedimentary and crystalline rocks form the Cordilleran accreted terrane belt. Fig. 11.10 is a schematic present-day cross-section across the northern United States with subsequent reincarnation ignored. This figure should be studied carefully because it shows the correlation between physiographic regions, tectonic provinces, structural provinces, rock successions, and physiographic provinces.

THE GREAT UNCONFORMITY

In Chapter 5, we defined an unconformity as a depositional contact between two rocks of different age across which there is gap in the history of deposition. We implied that in most cases, an unconformity marks a period of erosion and therefore represents what was once the surface of Earth. There are many unconformities in the rock record. An obvious example is the contact that separates deformed Appalachian rock successions from the undeformed Atlantic miogeocline (Fig. 11.8). This unconformity is obvious in Alabama because the trend of Appalachian rock successions is abruptly terminated (buried) beneath younger rock of the Atlantic miogeocline. We can say with certainty that deformation in the Appalachian Mountains occurred prior to deposition of the oldest rock in the Atlantic miogeocline.

Let us now turn our attention to the contact that separates the crystalline shield from the interior platform. We noted previously that shield rocks are more than one billion years old and that the overlying interior platform rock succession is less than 541 million years old. At a minimum, there is nearly half a billion years of time that cannot be accounted for in the rock record. This particular unconformity not only marks a huge gap in the geologic record, it is also widespread across the entire central United States. Fig. 11.11 is a reduced copy of the structural provinces map without the underlying Raisz landform map and with two cross-section lines shown as thick, heavy lines. The cross-section in Fig. 11.11B extends from the Colorado Plateau eastward to Lake Superior. The cross-section in Fig. 11.11C extends from Lake Superior southeastward to the Valley and Ridge. The unconformity between the crystalline shield and the interior platform is present across the length of both cross-sections.

John Wesley Powell coined the term, Great Unconformity, for spectacular exposures of this contact in the Grand Canyon. Powell led the first river expedition through the Canyon in 1869. Fig. 11.12 shows the Great Unconformity as seen at Powell Point in Grand Canyon National Park. The arrows in the photograph point to the contact between 1.84 and 1.66 billion-year-old shield rocks that include the Vishnu Schist and Zoroaster Granite, and the overlying Tapeats Sandstone at the base of the interior platform rock succession. The Tapeats is between 541 and 485 million years old. Here, the unconformity represents a gap in the rock record of more than one billion years. Fig. 11.13 is a simplified cross-section of the Grand Canyon that shows this unconformity on the lower left-hand side.

In modern usage, different workers use the term Great Unconformity somewhat differently. Some restrict the term to represent the gap in the rock record that separates Cambrian and younger sedimentary rocks (<541 Ma) of the interior platform rock succession from Precambrian crystalline shield rocks that are typically more than 1 billion years old. Others use the term to represent a gap in the rock record that separates Cambrian sedimentary rocks (<541 Ma) from Precambrian rocks of any type including sedimentary. Still others use the term to represent the largest gap in time below Cambrian sedimentary rocks. In the eastern part of the Grand Canyon, the

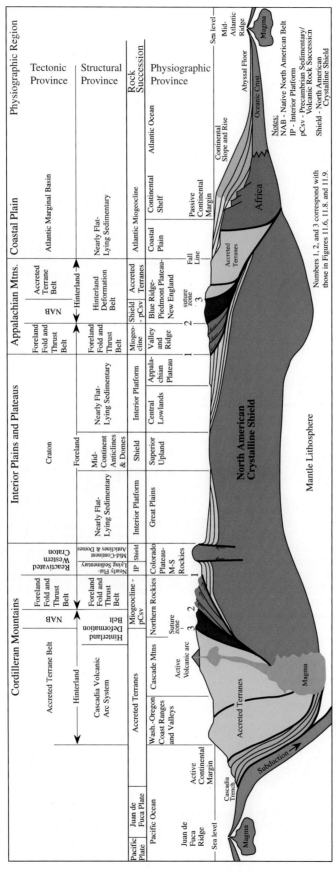

FIGURE 11.10 Schematic cross-section that shows the present-day distribution and correlation of physiographic regions, tectonic provinces, structural provinces, rock successions, and physiographic provinces across the northern United States.

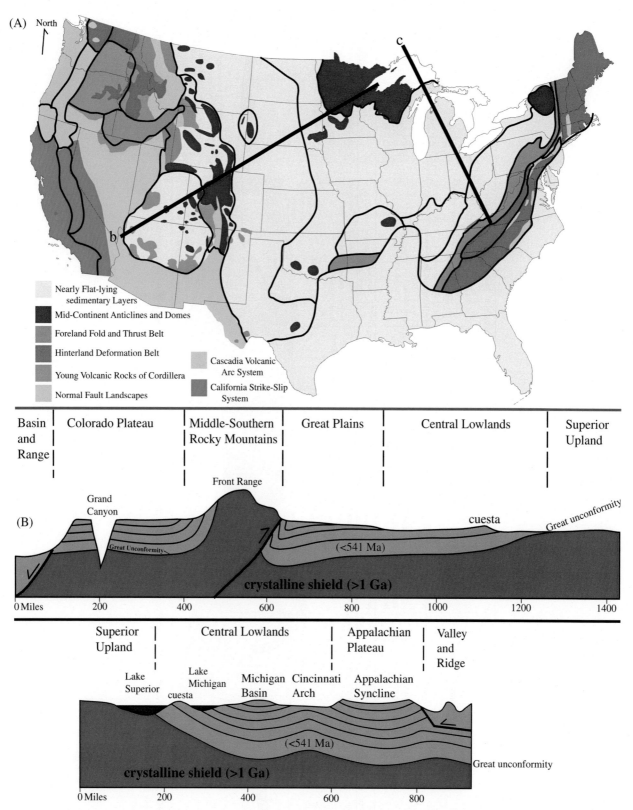

FIGURE 11.11 A) Structural province map showing the location of two cross-sections. (B) Cross-section from the eastern edge of the Basin and Range to the Superior Upland. (C) Cross-section from the Superior Upland to the western margin of the Valley and Ridge.

FIGURE 11.12 Photograph looking northeastward across the Grand Canyon at the Great Unconformity between Vishnu Schist (crystalline shield) and nearly flat-lying Tapeats Sandstone. White arrows point to the contact. The Colorado River flows from center right to lower left. Bright Angel Creek has cut vertically downward along a high-angle fault to form the prominent, straight Bright Angel Canyon visible near the center of the photograph.

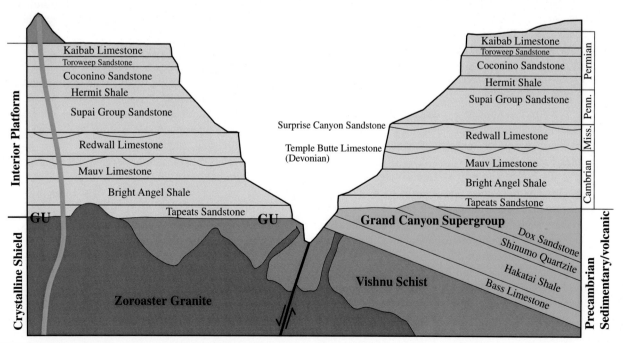

FIGURE 11.13 Simplified cross-section of the Grand Canyon that shows rock successions. The Great Unconformity (*GU*) forms the upper surface of the crystalline shield. *Based on O'Dunn and Sill (1988, p. 175)*

Precambrian sedimentary/volcanic rock succession (known here as the Grand Canyon Supergroup, 1235–735 Ma) partially fills the time gap between the crystalline shield and interior platform rock successions as seen in the lower right of Fig. 11.13. In this case, there are unconformities at both the upper and lower contacts of the Grand Canyon Supergroup. The lower contact with crystalline shield rocks represents the longest gap in time and therefore could be considered the Great Unconformity. Others might apply the term to the upper contact directly below the Cambrian Tapeats Sandstone. We will restrict usage of the term to the contact that separates interior platform rocks from those of the crystalline shield. Across the United States, the gap in time represented by this contact can vary from more than three billion years to less than a half-billion years. Although the Great Unconformity typically forms a flat surface, it does possess topographic relief in some areas that include the Grand Canyon as depicted in Fig. 11.13.

QUESTIONS

1. Explain how a physiographic province is different from a structural province.
2. On what basis is a structural province defined?
3. How is a rock succession defined?
4. Explain how a tectonic province is different from a structural province?
5. How is a rock succession different from the rock/sediment type as used in this book?
6. What type of fault is shown in Fig. 11.13?
7. What was Rodinea? When did it exist, How was it different from Pangea? What happened to it?
8. Fig. A.7C in the Appendix is an uncolored version of Fig. 11.1. Photocopy and color the map. Describe each structural province and its distribution across the United States.
9. Fig. A.8B in the Appendix is an uncolored version of Fig. 11.8. Photocopy and color the map. Describe each rock succession and its distribution across the United States.
10. Fig. A.9B in the Appendix is an uncolored version of Fig. 11.9. Color this map and describe each of the tectonic provinces. For example, what type of rocks, structures, and tectonic setting are expected in each area?
11. What are the defining characteristics of the interior platform rock succession?
12. Using Fig. 11.3, describe how the basin and range physiographic province is different from the Normal Fault systems structural province? How is the Middle-Southern Rocky Mountain province portrayed on the structural provinces map? Which physiographic provinces contain crystalline-cored mid-continent anticlines and domes? How are the Appalachian physiographic provinces portrayed on the structural provinces map? How is the California borderlands-Sierra Nevada physiographic region different from the California strike-slip structural province? Which structural provinces form the Northern Rocky Mountain physiographic province? In terms of their structural provinces, how is the Northern Rocky Mountains-North Cascades region similar with the Valley and Ridge-Blue Ridge/Piedmont Plateau region? How are these areas different? How is the Central-Southern Cascade physiographic province different from the Cascadia Volcanic arc structural province? What does the Great Plains province have in common with the Appalachian Plateau?
13. The Grand Canyon was cited as one area where the Precambrian sedimentary/volcanic rock succession partially fills the depositional gap between the crystalline shield and interior platform rock successions. Name one other location in the interior United States where this occurs.
14. What does the term critical taper imply?
15. The foreland basin is often in the shape of a wedge that thickens toward the mountain. Why? What controls its shape?
16. Using the Google Earth ruler, measure the width of the Appalachian Mountain belt at Boston, New York, Washington, DC, Charlotte, and Atlanta. How did you define the eastern and western limit of the origin? Why do these values not reflect the true width of the orogenic belt?
17. Describe each of the six rock successions.
18. What is the Cordilleran reactivated western craton? Speculate on its origin. Why is there no counterpart in the Appalachians?
19. Describe differences between the foreland and hinterland of a mountain belt in terms of rock type, style of deformation, and metamorphism.
20. How is the Appalachian hinterland different from the Cordilleran hinterland?
21. With respect to isostasy and the strength of the Earth's crust and lithosphere, what factors might contribute to the development of a deep foreland basin.
22. Define the terms clastic and detritus.
23. Explain how detritus in a foreland basin can be used to better understand the exhumation history of a now vanished mountain belt.
24. What are the parts of an idealized orogenic belt?
25. Describe the Great Unconformity.

26. Speculate as to why the Appalachians have such a wide belt of crystalline rock, and the Cordillera does not.
27. Compare Figs. 11.9 and 1.6. Name the physiographic provinces that form part of the reactivated western craton, the Cordilleran fold and thrust belt, native North American belt, and accreted terrane belt. Do the same for the Appalachians—Ouachita belt
28. On what basis can we consider the reactivated western craton to be part of the craton?
29. Why is the Cordilleran foreland fold and thrust belt in Fig. 11.9 (tectonic provinces map) different from the foreland fold and thrust belt in Fig. 11.1 (structural provinces map), and why are they the same in the Appalachian belt?
30. Why are there large areas of crystalline shield rock (shown in Fig. 11.8) exposed across the reactivated western craton (shown in Fig. 11.9)?
31. Why is there no present-day miogeocline (similar to the Atlantic miogeocline) on the west coast?
32. Compare Figs. 11.8 and 1.6, and name the physiographic provinces that include Precambrian sedimentary/volcanic rock. Do the same for rocks of the crystalline shield.

Chapter 12

Glacial Landscape

Fifty million years ago North America and the rest of the world was much warmer than today. At that time alligators and turtles could have sunned themselves close to the Arctic Circle. The Earth has cooled since then, reaching a low point during the Ice Age beginning about 3.2 million years ago and continuing until very recently.

The glaciers that affected North America were of two types, continental glaciers, also known as ice sheets, and alpine (or valley) glaciers. Continental glaciers, as the name suggests, cover large regions of continents. Alpine glaciers are restricted to mountaintops and mountain valleys, although several glaciers can coalesce to form a larger glacier. The continental glaciers that affected the US formed initially in the Hudson Bay area and in the Canadian Cordillera. They are known as the Laurentide and Cordilleran Ice Sheets respectively. The Laurentide Ice Sheet was by far the largest and had the greatest impact on the US, extending deep into the central US and covering all of New York and New England. The Cordilleran Ice sheet covered mountains and valleys of the Canadian Cordillera. This ice sheet periodically moved southward from Canada, but reached only the northern fringe of the US Cordillera. Alpine glaciers were widespread across high mountain regions of the US Cordillera south of the Cordilleran Ice Sheet and were present briefly in the northeast following retreat of the Laurentide Ice Sheet. Alpine glaciers still exist in the Cordillera.

EFFECT OF GLACIATION ON LANDSCAPE

Glaciers are one of the most powerful erosional and depositional forces of nature. The mass and volume of a glacier is sufficient to flatten landscape, carve mountains, and move anything from a grain of sand to a rock larger than a building. Glaciers disrupt river drainage patterns and can bury landscape in sediment so thick that bedrock no longer influences the shape of land. Glaciers are a far more powerful erosional agent than rivers. If one had a five-pound brick of butter, a river would be like slicing into the butter with a hot knife; a glacier would be like gouging the brick with a hot spoon and ripping out the center. The erosional power of alpine glaciers has been likened to a buzz saw that limits the elevation of uplifting mountains. Glaciers are capable of modifying and completely changing (reincarnating) landscape. In this section we describe, in general terms, the erosional and depositional character of continental and alpine glaciation.

Landscape Development in Areas of Continental Glaciation

Continental glaciers are thick ice masses that originate on land and show evidence of flow. They originate in the zone of accumulation, which is the area where, over the course of a year, the amount of snow accumulation exceeds the amount of melting. If snow accumulates at an average rate of only six inches per year, a snow pack one mile thick will have formed in less than 11,000 years. As snow accumulates and turns to ice, the weight of the snow causes ice at the bottom of the pile to flow outward from under the snow pack into areas that otherwise would not be covered with ice. Once a snowpack of this type begins to flow outward in this manner, a continental glacier is born. As long as snow continues to accumulate at the top of the pile, ice will continue to flow outward from below the pile and move into warmer areas until it eventually melts. Normally, the movement of ice is in the form of lobes that follow lowland regions (valleys). The warmer region into which a glacier flows and melts is known as the zone of ablation (ablation means to remove by melting, evaporation, and vaporization). Year-round ice would not exist in the zone of ablation without glaciation.

The zone of accumulation is associated with erosion. As glaciers accumulate to thicknesses between 1 and 3 miles (5000-16,000 feet), the outward flow of ice from the bottom of the pile will scour the land surface. Loose rock and soil are incorporated into the glacier greatly increasing its mass and erosive power. High spots are streamlined and flattened, and weak rock is gouged to create depressions. Eroded material is then transported into the area of ablation by the moving ice and deposited once the ice melts. What is left in the area of accumulation after glacial retreat is a nearly flat, shapeless landscape of bare often polished and striated bedrock. The river network that existed prior to glaciation is destroyed in favor of a deranged network of slow-moving streams that connect thousands of lakes and ponds that fill depressions where weak rock was gouged. The shear abundance of small lakes, the deranged drainage, and the widespread exposure of bedrock are characteristic features of continental glacial erosion.

Fig. 12.1 shows before and after cross-sections and an example of glacial erosion in the zone of accumulation. The image looks north across northern Minnesota at a flat glacial-scoured landscape composed of crystalline, sedimentary, and volcanic rocks of the Canadian shield and the Keweenawan rift system. Gouging of weak rock has created an abundance of linear, aligned lakes that follow the trend of rock units thereby revealing part of the structure of the area.

The removal of soil and the plucking and scraping of bedrock in the zone of accumulation produces a huge volume of loose material that is incorporated into the glacier and transported southward to the zone of ablation. As glaciers move into the zone of ablation, they flatten and scour high spots, but rather than expose bedrock, upon melting they cover the entire region with glacial sediment to thicknesses of more than 100 feet. The result is a low-relief landscape of thick, richly fertile glacial soil that covers nearly all bedrock. Fig. 12.2 shows before and after cross-sections and an example of glacial deposition in the zone of ablation. The image looks northwest across a drumlin field east of Rochester, New York. Drumlins are streamlined, asymmetric hills composed of glacial sediment oriented in the direction of glacier flow. Major differences between this area and the area of erosion are fewer lakes and no exposed bedrock. Fig. 12.3 summarizes processes affecting the zones of accumulation and ablation.

All glacial depositional landforms are composed of drift. The term drift is used for any sediment of glacial origin. There are two types. Unstratified drift (or till) is dropped directly as the glacier melts. Stratified drift is till that has been reworked by running water and layered. There are a number of depositional landforms in addition to drumlins that are associated with continental glaciation.

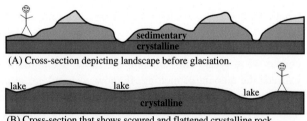

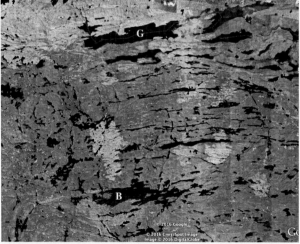

FIGURE 12.1 Depiction of landscape (A) before glaciation, and (B, C) following glacial erosion. The image looks north across northern Minnesota at a flat glacial-scoured landscape composed of crystalline, sedimentary, and volcanic rocks of the Canadian shield and the Keweenawan rift system.

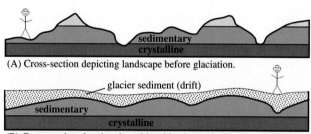

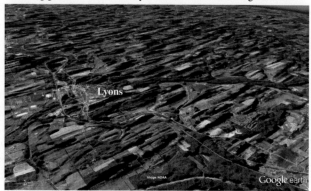

FIGURE 12.2 Depiction of landscape (A) before glaciation, and (B, C) following glacial deposition. The image looks northwest across a drumlin field east of Rochester, New York.

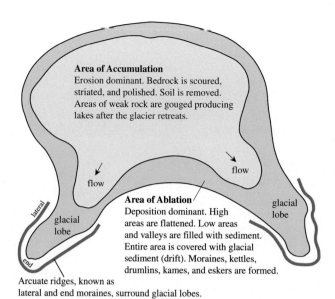

FIGURE 12.3 Representation of continental glaciation.

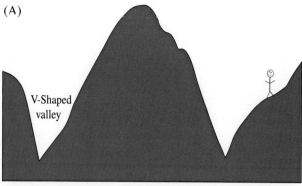

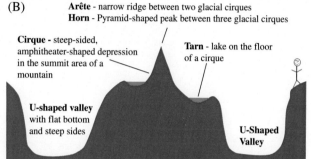

FIGURE 12.4 Sketch showing landforms (A) before alpine glaciation, and (B) following alpine glaciation.

All of them form primarily when the glacier is stagnant or in retreat because an advancing glacier will bulldoze and destroy any previously deposited glacial landform. Deposition during retreat implies that the oldest landforms occur farthest down valley.

Moraines are one of the most common glacial depositional landforms. There are several types of moraines, but the most conspicuous form high, steep, arcuate ridges, and hummocky, uneven ground along the sides and front of a retreating glacial lobe. They are known as lateral and end moraines respectively and are shown schematically in Fig. 12.3. Moraines, because of their curved shape, are capable of trapping water to form a lake. Moraines and drumlins are composed of unstratified drift. Other glacial depositional landforms, including kettles (depressions, often filled with water), kames (flat-topped, steep-sided hills), and eskers (long sinuous ridges) are composed of stratified drift. Examples and further explanations of each of these are given later in the chapter.

Landscape Development in Areas of Alpine Glaciation

Alpine (or valley) glaciers form on mountaintops in a manner similar to continental glaciers. As snow accumulates, a glacial lobe will begin to flow downward into a valley under the force of gravity where it eventually melts. This type of glacier is confined to individual mountain valleys although in some instances, several alpine glaciers will coalesce on mountaintops, high plateaus, or across several valleys to produce a small ice cap.

The erosive and depositional power of alpine glaciers is equal to that of continental glaciers, but their extent is limited. Rather than the flattening effect of continental glacial erosion, alpine glaciers have a carving effect that shapes the mountain. Fig. 12.4 is a schematic illustration that shows the difference between a non-glaciated mountain and one that was glaciated. Prior to glaciation, many of the mountains in the western US were rounded in a manner similar to the present-day southern Appalachians. They were, to put it bluntly, far less spectacular in their appearance. Alpine glaciers create erosional landforms that include u-shaped valleys, cirques (steep-sided, amphitheater-shaped depressions in the summit area of a mountain), tarns (lakes on the floor of a cirque), arêtes (narrow ridges between glacial cirques), and horns (pyramid-shaped mountain peaks carved on all sides by glaciers). Fig. 12.5 shows a deeply gouged, snowless cliff face (a cirque) carved into the side of Mt. Rainier by the active Carbon Glacier (C). The glacier itself can be seen extending down the mountain. Notice in this figure the clear distinction between fresh snow in the zone of accumulation, and the dark, dirty snow in the zone of ablation. Notice also, the wide u-shaped valley occupied by the glacier in the zone of ablation. Glaciers have carved all of the high mountains in the west providing naturalists, photographers, hikers, and alpine climbers with some of the most spectacular mountain scenery on earth.

FIGURE 12.5 Google Earth image looking south at Mt. Rainier. Carbon glacier (C) is on the right side, Winthrop glacier (W) is near center, and Emmons Glacier (E) is at the left. Goat Rocks (G) is an extinct volcano.

Moraines are the most common alpine glacial depositional landform. Other depositional landform types, such as drumlins and kettles, are more common in continental glacial settings. Fig. 12.6 looks northward at lateral and end moraines surrounding a u-shaped valley partly occupied by Willow Lake on the west flank of the Wind River Range. Note the sharp crest of the lateral moraines and the hummocky ground that forms the end moraine. Continental glacial moraines are of similar shape, but typically are broader, not as sharp-crested, and with less relief. Let us now look in more detail at how, and to what extent, alpine glacial erosion shapes a mountain.

Alpine glacial erosion features are especially well displayed in mountains where rocks are without layering or strong contrasts in erodability. Granitic rock, in particular, is non-layered and massive in the sense that it does not necessarily weather and erode in any particular direction related to rock structure.

A granitic intrusion will punch its way through sedimentary layers and, following erosion, will create a crystalline-cored domal mountain with a broad summit area. As ice accumulates on the broad summit, the weight of the ice will cause the ice to flow outward from the base of the pile toward a river valley. The outward movement of ice at the bottom of the pile will cause ice at the head of the valley to rotate backwards into the mountain creating a spooning effect that cuts a steep, amphitheater-shaped headwall cirque into the top of the mountain, leaving a small basin on the floor of the cirque that later will be occupied by a lake (tarn). The process is illustrated in Fig. 12.7. Over time, the headwalls of several cirques will retreat (erode back) leaving a narrow ridge (arête) between them along the top of the mountain. A similar shape could be achieved if one were to continuously carve into a mound of ice cream with a spoon.

An excellent example of glacial erosion in a massive domal exposure of crystalline rock is the Capitol Peak area of the Elk Mountains east of Aspen, Colorado. A Google Earth image of the area looking southwest is shown in Fig. 12.8. A young granitic intrusion underlies the entire area from Mount Daly (D, at the left edge of the figure) to Snowmass Peak (SP). Included are two 14,000-foot peaks, Capitol Peak (CP) and Snowmass Mountain (S).

FIGURE 12.6 Google Earth image looking just east of north at lateral and end moraines surrounding a u-shaped valley partly occupied by Willow Lake on the west flank of the Wind River Range, Wyoming.

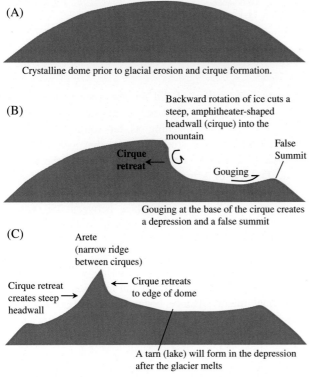

FIGURE 12.7 Schematic diagrams that show how alpine glaciers gouge a rounded mountaintop to create cirques, tarns, and arêtes.

Note the continuous narrow ridge that connects Mount Daly (D) with Snowmass Peak (SP). Note how elevation drops precipitously on both sides of the ridge. This sharp ridge is an arête and its presence results in Capitol Peak (CP) being one of the most difficult mountains in Colorado to climb. If you look carefully you will see that the ridge is composed of a series of curved, segments. There are at least six curved segments between Capitol Peak (CP) and Snowmass Peak (SP). Each curved, amphitheater-shaped segment is the headwall of a cirque and the birthplace of a glacier. Additional cirques are readily visible on either side of Clark Peak (C). Several glaciers coalesced below Capitol Peak to form a small ice cap now occupied by the Pierre Lakes (P). The basin represents the floor of a composite cirque and the lakes are tarns. The elevation of the basin is above 12,000 feet, which is more than 2500 feet above the valley shown in the upper left of the image. The glaciers have literally gouged-out and removed the top section of the dome structure as illustrated in Fig. 12.7. The glaciers are now mostly gone, leaving a large snowfield in their wake. If you look closely at the lower right of the image, you can see two tongue-like features labeled rock glacier. A rock glacier can be thought of as a pile of rocks internally cemented in a matrix of ice. It is likely that these were once ice glaciers in which much of the ice has since melted. Whereas ice glaciers flow, rock glaciers move via creep, which in this case is a slow downward movement of rock particles brought on by gravity and facilitated by internal ice deformation.

Notice in Fig. 12.8 that the northern (left) side of the mountain is occupied by glacial cirques and that the south (right) side is a single steep slope. This type of geometry is not unusual. The north side of a mountain will get less sun and therefore will be wetter and cooler year-round.

C-Clark Peak, CL-Capitol Lake, CP-Capitol Peak, D-Mount Daly, H-Hagerman Peak, P-Pierre Lake, S-Snowmass Mountain, SL-Snowmass Lake, SP-Snowmass Peak.

FIGURE 12.8 Google Earth image looking southwest at the Capitol Peak area, Colorado.

Glaciers are more likely to survive and grow relative to the sunny south side. As north-side glaciers grow larger, constant erosion in the cirque area will wear the headwall back to the edge of the mountain as illustrated in Fig. 12.7. In the case of the Capitol Peak dome, ice has also cut into the sunny side of the mountain creating very steep headwall cirques and at least one tarn (Capitol Lake, CL). The present-day shape of the mountain is therefore due to a combination of headwall retreat by glaciers on the north side, and strong scouring on the south side.

A DAUGHTER OF THE SNOWS: GLACIAL LANDSCAPE IN THE UNITED STATES

Fig. 12.9 and Table 12.1 subdivides glacial landscape in the US into several regions. The Laurentide Ice Sheet is responsible for the area of erosion and for two areas of glacial deposition, old drift and young drift. The Ice Age did not consist of a single glaciation, but rather there were multiple glacial advances and retreats, and during some of the glacial retreats the Earth was warmer than today. The Laurentide Ice Sheet refers specifically to the most recent glacial advance from the Hudson Bay area, but there were earlier iterations. ^{26}Al-^{10}Be cosmogenic radionuclide dating of till in the central part of the US suggests that the first major advance took place 2.4 million years ago, reaching as far south as St. Louis, Missouri. This same study suggests that a second major advance south of 45 degrees north latitude (the approximate latitude of Minneapolis, Minnesota) did not occur until about 1.3 million years ago and that at least three more major advances took place between 750,000 and 128,000 years ago. The last prolonged interglacial warm stage occurred between 128,000 and 118,000 years ago. This warm period was followed by the last glacial stage beginning 118,000 years ago during which time the Laurentide Ice Sheet advanced and retreated several times. The final major advance began about 30,000 years ago, reaching maximum advance less than 22,000 years ago. At that time, glaciers covered the area marked in Fig. 12.9 as young drift. This final glacial advance did not push as far south as some of the earlier advances, which are shown in the figure as old drift. Final retreat began 18,000 years ago and by 10,000 years ago the Laurentide Ice Sheet had retreated well into Canada. By 6,000 years ago large areas of glacial ice had disappeared from Canada.

The area of erosion corresponds to the cumulative effect of all advances and retreats of the Laurentide Ice Sheet. As shown in Fig. 12.9, the area covers Canada

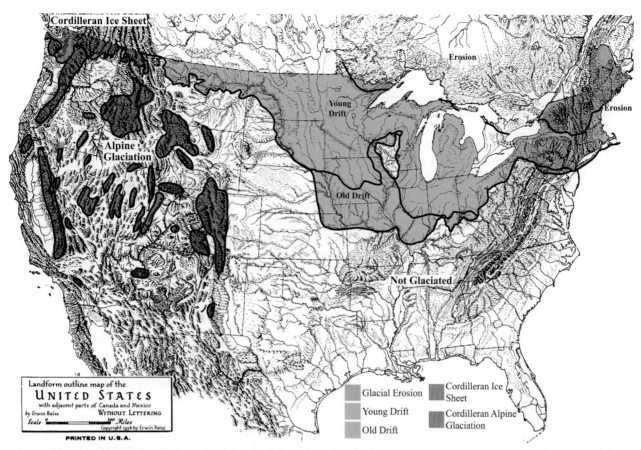

FIGURE 12.9 Landscape map showing areas in the United States that were glaciated in the past 2.4 million years. These areas are listed in Table 12.1. Non-glaciated areas in the US are uncolored.

TABLE 12.1 Glacial Zones

1. Areas of Continental Glacial Erosion
2. Areas of Continental Glacial Deposition
   a. Young Glacial Drift (<118,000 years old)
   b. Old Glacial Drift (>118,000 years old)
3. Driftless Area
4. Not glaciated
5. Cordilleran Ice Sheet
6. Cordilleran Alpine Glaciation

where it coincides roughly with what was once the zone of accumulation. In the US, the area of erosion includes only the northern fringe of Minnesota and northern New York/New England.

As noted previously, the area of young glacial drift outlines the extent of the last major advance of the Laurentide Ice Sheet. The sediment that forms young drift was deposited primarily during final retreat between 22,000 and 10,000 years ago. Because of their young age, moraines and other glacial landforms associated with young drift are well preserved. We will look at examples later in the chapter.

The area of old glacial drift shown in Fig. 12.9 corresponds with glaciations that occurred between 2.4 million and 118,000 years ago. In the central part of the US, some of these glaciers advanced farther south than glaciers associated with young drift. These older glacial advances left many depositional landforms including moraines, kettles, kames, eskers, and drumlins. However, those in the northern US were bulldozed and destroyed beneath younger glaciers, and those in the southern US were eroded, leaving mostly flat plains composed of glacial drift. Thus, the primary difference between old and young drift is the preservation of glacial landforms in the area of young drift, and the near complete erosion and destruction of glacial landforms in the area of old drift.

The Cordilleran Ice Sheet was smaller than the Laurentide and affected only the northern fringe of the US Cordillera leaving behind glacial drift but little in the way of glacial depositional landforms. Fig. 12.9 shows its maximum southward extent in the US Cordillera. Similar with the Laurentide Ice Sheet, there were several glacial advances. The final advance is known as the Fraser glaciation. It began in Canada 28,000 years ago flowing westward onto the continental shelf (sea level was lower), eastward where it merged with the Laurentide Ice Sheet, and southward to the northern margin of the US covering mountains and valleys. The ice sheet, on the basis of radiocarbon dating, reached the Puget Sound lowland and lowland valleys of the Northern Rocky Mountains north of the Columbia Plateau by 20,300 years ago. It reached its greatest southward extent in the area of the Puget Sound lowland, advancing beyond the latitude of Seattle by 17,600 years ago, and advancing just south of Olympia by 16,950 years ago. It then retreated rather quickly. In Canada, the ice sheet had largely disappeared from lowland areas by 12,500 years ago. Mountainous areas affected by the Cordilleran Ice Sheet were later subject to alpine glaciation.

Alpine glaciers were widespread in the US Cordillera, particularly in the Cascade Mountains, Sierra Nevada, and Rocky Mountains, including glaciers as far south as New Mexico. Glaciers are still present in these areas although they are far less extensive and no longer include New Mexico. Glaciers also, at one time, covered the Yellowstone area, the high plateaus and mountain areas of the Colorado Plateau, the high peaks of the Basin and Range, Olympic Mountains, and Klamath Mountains, and even modified San Gorgonio Mountain, an 11,503-foot peak in the San Bernardino Mountains of Southern California. However, with the exception of the Olympic Mountains, few, if any, of these glaciers are present today.

Alpine glaciers also briefly occupied the White Mountains of New Hampshire, the Mt. Katahdin area of Maine, and probably the Mt. Marcy region of the Adirondack Mountains. The best-known alpine glacial area in the east is Tuckerman Ravine, an east-facing cirque on the flank of Mount Washington in the White Mountains where spring skiers can climb to the top and ski down through deep snow long after commercial ski slopes have shut down.

Non-glaciated areas shown in Fig. 12.9 include areas south of continental glaciation, areas below alpine glaciation, and the Driftless area located along the Mississippi River in southwestern Wisconsin. The Driftless area is unusual because it represents a non-glaciated island surrounded by areas of glaciation. There is no doubt the area was covered in snow during much of the Ice Age; however, in order to be a glacier, the snowpack must show evidence of flow. Such evidence, in the form of glacial deposits (drift), is absent in the Driftless area. Glacial lobes during the most recent glaciation advanced southward though the Lake Superior trough into Iowa, and southward through the Lake Michigan trough into southern Indiana. The Driftless highland between the two lobes was bypassed. Earlier glacial advances likely took the same route. Landscape in the Driftless area and in areas directly south of old drift are relatively hilly and river-dissected. These areas offer a hint at what the central US may have looked like prior to glaciation.

For the remainder of this chapter we will look in some detail at the glacial landscape of the US. We will concentrate our effort on the effects of the Laurentide Ice Sheet where erosional and depositional landforms are preserved. River systems were destroyed during glacial advances, but reestablished themselves with a different pattern on the newly created landscape. There is evidence that the Mississippi and Ohio River systems did not extend as far north prior to glaciation and that river systems in the northern part of the US, including the Missouri, upper Mississippi and upper Ohio Rivers, flowed northward rather than their present-day southward direction. The Great Lakes were north-flowing rivers before being gouged into lakes. Parts of several rivers, including the Kansas, Missouri, and Ohio Rivers, formed or reformed partly at the southern margin of the glacial advance. We will look in particular at the history of two rivers, the Teays River, which no longer exists, and the Missouri River.

In order to facilitate our discussion, we will use six overlapping Raisz landform maps that cover the entire glaciated region east of the Cordillera. The maps are placed together, in sequence from west to east for easy reference. Figs. 12.10 and 12.11 cover the northwestern and northeastern Great Plains respectively. Figs. 12.12 and 12.13 cover the Lake Superior region southward. Fig. 12.14 covers the Lake Michigan region and overlaps slightly with Fig. 12.15, which covers the northeastern US. Glacial moraines on these maps are shown with a stippled dot pattern. Many are named and abbreviated, Mor. Many of the glacial features discussed below are highlighted in yellow on these maps. The explanation to all of the maps is given at the bottom of Fig. 12.10.

THE GLACIAL EROSION BOUNDARY IN THE UNITED STATES

The area of continental glacial erosion in central North America barely reaches the United States. It is shown in Fig. 12.12 along the northeastern edge of Minnesota from International Falls to Isle Royal. A close-up view of the area around Gunflint Lake is shown in Fig. 12.1A. Characteristic features are streamlined topography, an abundance of exposed bedrock, and many small lakes that mimic the structural form of rock. This area is home to Voyageurs National Park and to the Boundary Waters Canoe Area Wilderness.

Contrasts in landscape between areas of glacial erosion and glacial deposition are well displayed in Figs. 12.12 and 12.14. The density of lakes decreases abruptly south of the boundary where glacier deposition features such are moraines, drumlins, and till plains dominate, and exposed bedrock is largely absent. Fig. 12.16 is a Google Earth image of northeastern Minnesota looking north that shows an abrupt change in landscape across the glacial erosion-deposition boundary

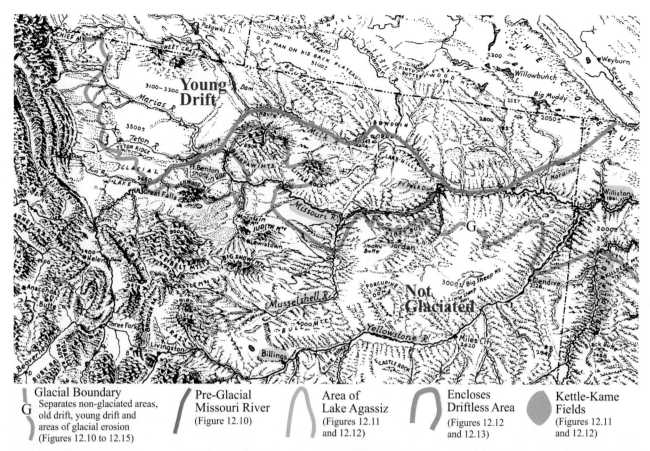

| Glacial Boundary G — Separates non-glaciated areas, old drift, young drift and areas of glacial erosion (Figures 12.10 to 12.15) | Pre-Glacial Missouri River (Figure 12.10) | Area of Lake Agassiz (Figures 12.11 and 12.12) | Encloses Driftless Area (Figures 12.12 and 12.13) | Kettle-Kame Fields (Figures 12.11 and 12.12) |

FIGURE 12.10 Raisz landform map showing glacial features on the northwestern Great Plains. The explanation covers Figs. 12.10 through 12.15.

(thick white line). The faint white line is the international border. Access to the area can be made via the Gunflint Trail (route 12), which begins at Grand Marais (GM) and ends at the Canadian border on Saganaga Lake (SL).

A large area of glacial erosion is shown in the New York-New England region of Fig. 12.15. Continental glaciation was different in this area due to the presence of hilly and mountainous terrain. Glaciers were unable to thoroughly flatten the rough topography. Although subtle, one can still notice a change in topography from relatively rugged landscape in the area of glacial erosion to more subdued landscape in the area of young drift. Perhaps the most conspicuous landscape change occurs along the coastline north and south of Portland, Maine. The New England coast north of Portland, in the area of erosion, is rocky with many inlets. The coastline to the south, in the area of deposition, is markedly smoother with long sandy beaches. Much of the beach sediment was deposited initially by retreating glaciers and subsequently reworked by ocean currents. Within the area of glacial erosion, the western Adirondack Mountains were heavily scraped and gouged creating a landscape with hundreds of lakes similar to what is seen in northern Minnesota except with greater relief. In western New York, glaciers gouged river valleys into steep-sided, u-shaped troughs to create the eleven Finger Lakes. The two largest, Lake Seneca and Lake Cayuga, are highlighted in Fig. 12.15. Glacial erosion in northern New England removed soil and exposed more bedrock than what is seen in the non-glaciated southern Appalachians.

The hilly landscape in the north caused ice in topographically low areas to stagnate and melt in place resulting in deposition within the area of erosion. Conversely, some areas in the south were scraped clean, creating pockets of erosion within the area of young drift. For example, Fig. 12.17 looks northwest across the Reading Prong in the area of young drift at the New Jersey—New York state line. The Reading Prong is a highland composed of crystalline and sedimentary rock elevated 300 to 900 feet above the surrounding landscape. The abundance of lakes suggests that this area was not deeply covered in drift, and instead shows the effects of glacial erosion.

THE GLACIAL EROSION BOUNDARY ACROSS NORTH AMERICA

The glacial erosion boundary in the Superior Upland and in the northeastern US crosses different rock types and several structural forms implying an absence of bedrock

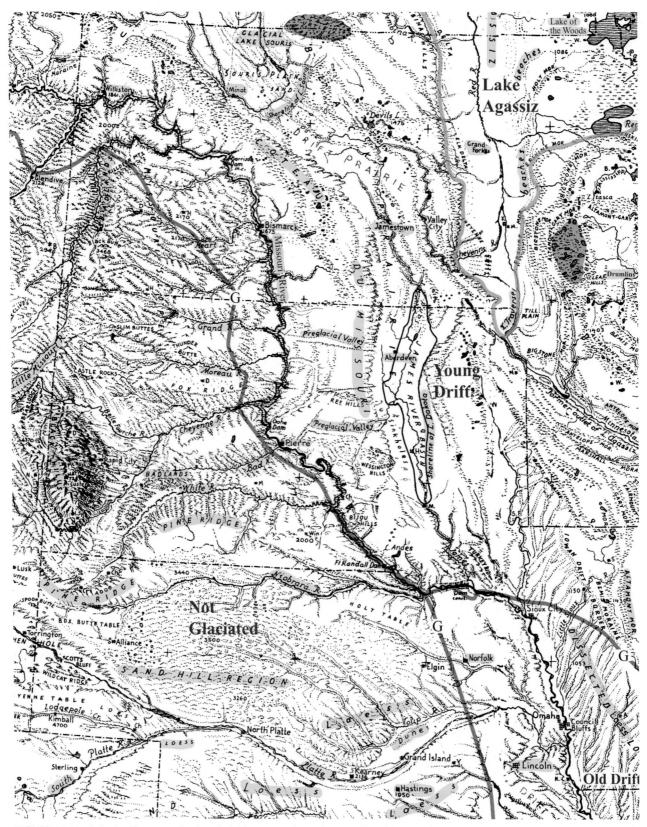

FIGURE 12.11 Raisz landform map showing glacial features on the northeastern Great Plains.

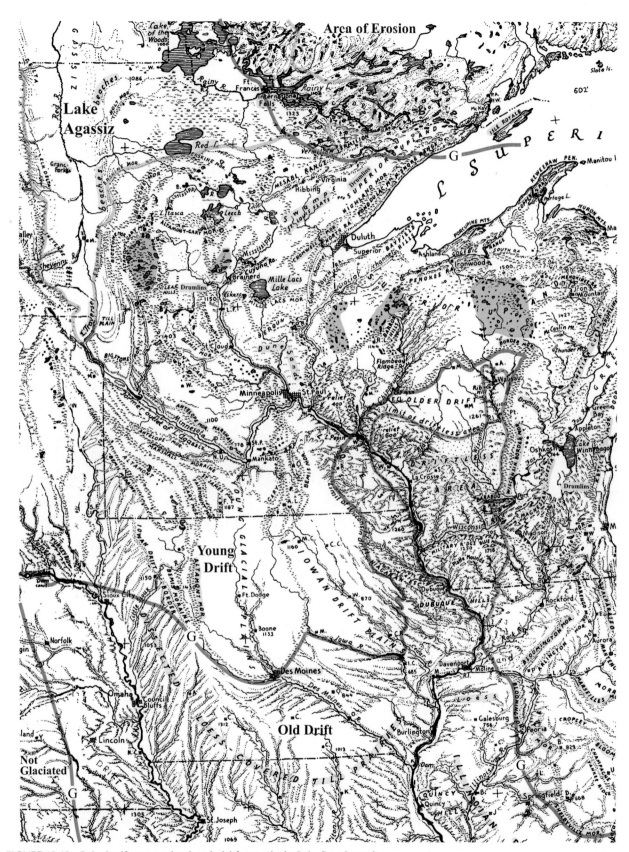

FIGURE 12.12 Raisz landform map showing glacial features in the Lake Superior region.

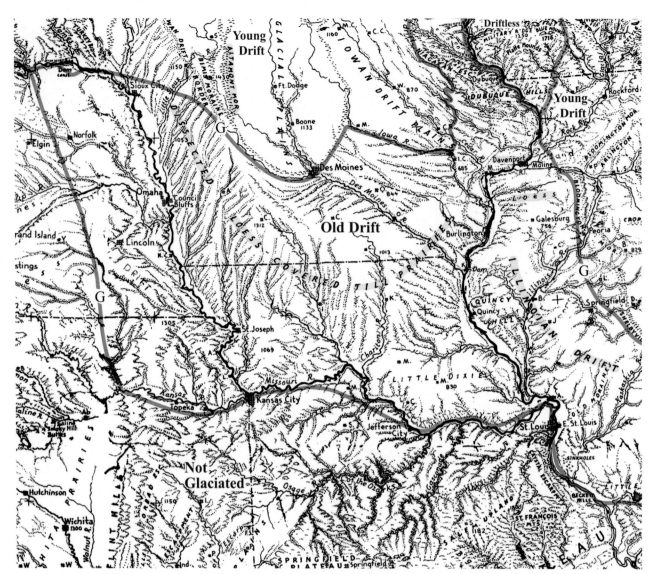

FIGURE 12.13 Raisz landform map showing glacial features in the southwestern Central Lowlands.

control on the glacial boundary. There does, however, appear to be at least some bedrock control at the scale of the North American continent.

Fig. 12.18 is a Google Earth image of Canada and the United States. The heavy yellow lines in the figure separate ancient crystalline rocks of the Canadian shield from nearly flat-lying sedimentary rocks of the Interior platform. The thick red-orange line encloses the outcrop extent of the Keweenawan Rift system within the Canadian shield. The white line forms the glacial erosion boundary. The location of this boundary is based on the density of small lakes, which is far greater in the area of erosion relative to the area of deposition.

Notice that the western limit of glacial erosion in Canada mimics the western limit of crystalline rocks. The reason for this is not abundantly clear. One possibility has to do with the fact that sedimentary rocks are soft, and as such, they would be expected to weather and disintegrate (crumble) into sediment more readily than crystalline rocks. This sediment would then be incorporated into the glacier and available for deposition as the glacier melts. Crystalline rocks would produce less sediment.

Notice also that the glacial erosion boundary closely coincides with a series of large lakes that extend in a line from the Great Lakes northwestward into Canada. Included are Lake Winnipeg (w), Lake Athabasca (a), the Great Slave Lake (s), and the Great Bear Lake (b). These lakes are a product of glacial erosion. As the glacier moved outward over crystalline rock, it encountered thin, overlying layers of soft sedimentary rock and promptly gouged a basin that was later filled with water

FIGURE 12.14 Raisz landform map showing glacial features in the Lake Michigan region.

FIGURE 12.15   Raisz landform map showing glacial boundaries across New York-New England.

to create the lakes. Notice, for example, the sharp increase in size of the Great Slave Lake (s) where sedimentary rocks are encountered. It is probable that many of the large lakes (including the Great Lakes) were river valleys prior to glaciation. The size and shape of Lake Superior (S), for example, is clearly controlled by the outcrop pattern of relatively weak rocks of the Keweenawan Rift system. The depth of Lake Superior is an example of the predaceous power of glaciers. Lake Superior lies at an elevation of 602 feet above sea level but was gouged to depths greater than 700 feet below sea level. Other Great Lakes, such as Lake Michigan (M) and Lake Huron (H) were also gauged from relatively soft rocks.

Lake Athabasca (a) and the eastern part of the Great Slave (s) and Great Bear lakes (b) are underlain with crystalline rock. As just mentioned, it is possible that these areas were deep river valleys gouged and widened during glaciation. But there are other explanations as well. For example, crystalline rock on these areas may be weaker due to the presence of numerous fracture and fault zones, or these areas may have originally been covered with a thin layer of sedimentary rock that was gouged and removed.

In the Superior Upland province, the line of glacial erosion lies well to the north of the southernmost extent of crystalline rocks. Here we can define a clear distinction between areas of glacial erosion and glacial deposition without the added influence brought on by a change in rock type.

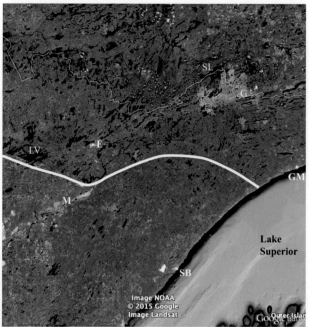

E-Ely, G-Gunflint Lake, GM-Grand Marais, LV-Lake Vermilion, M-Mesabi Iron Range, SB-Silver Bay, SL-Saganaga Lake.

FIGURE 12.16 Google Earth image looking north across Minnesota and Canada at the boundary (thick white line) separating areas of glacial erosion from areas of glacial deposition. The thin white line is the international border.

MORAINES

Glacial moraines in the central US are shown with stippled pattern in Figs. 12.11, 12.12, and 12.14, many of which are named and abbreviated Mor. They were

FIGURE 12.17 Google Earth image that looks northwest across an area of young drift. The many lakes on the Reading Prong (Hudson Highlands) suggest this area underwent glacial erosion. The black lines separate young drift, old drift, and not glaciated. The thin white line is the New Jersey—New York border.

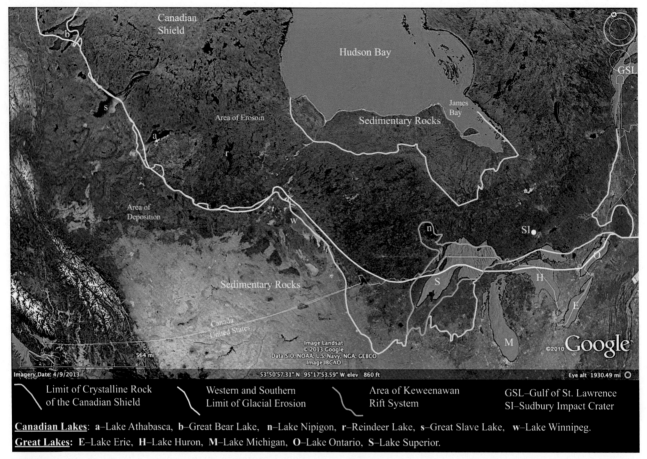

FIGURE 12.18 Google Earth image of Canada and the northern US showing the extent of the Canadian Shield, the western and southern limit of Laurentide glacial erosion, and the extent of the Keweenawan Rift system. Limit of glacial erosion is based on Sugden (1978).

deposited during glacial retreat, less than 22,000 years ago, implying they are restricted to areas of young drift and that the oldest moraines are those farthest south where they mark the boundary with old drift. Most of the moraines are curved to reflect deposition around glacial lobes. Their curvature outlines the major lobes that existed during the most recent glaciation.

Moraines in the central US typically show less than a few hundred feet of relief and are located in plowed farmland. They are not obvious on the ground or in Google Earth images. We can, however, use the patterns shown on the Raisz maps to locate major lobes. With respect to Fig. 12.14, a glacial lobe occupied present-day Lake Erie as indicated by the Fort Wayne, Wabash, Salamonie, and Mississinawa moraines. Another lobe existed in present-day Saginaw Bay as indicated by the Lansing, Charlotte, West Branch, and Port Huron moraines. A large glacial lobe occupied present-day Lake Michigan as indicated by the Valparaiso, Marseilles, Bloomington, and Shelbyville moraines. On the basis of the moraine pattern, it appears that this lobe split into three sub-lobes. One lobe advanced southwestward toward Peoria as indicated by the Bloomington Moraine (shown in Fig. 12.12), one advanced southward toward Springfield as indicated by the Shelbyville Moraine, and one advanced southeastward toward Indianapolis as indicated also by the Shelbyville Moraine. In Fig. 12.12, a glacial lobe occupying Lake Superior appears to have combined with one from the Minnesota-North Dakota area to extend all the way to Des Moines, Iowa as suggested by the Cromwell, Antelope, Marshall, and Altamont Moraines among others.

Moraines, because they form curved ridges, can trap water behind them to create lakes. Several large lakes in Minnesota, shown in Fig. 12.12, have formed behind glacial moraine ridges including Red Lake, Leech Lake, and Mille Lacs Lake. The Wisconsin Border Moraine extends southward from Wausau to form the northeastern border of the Driftless Area, and in doing so cuts directly across the Baraboo Range where it traps Devils Lake with no river inlet or outlet (Devils Lake is too small be seen in Fig. 12.12, but is shown in Fig. 14.43).

The largest moraine on the central US is the Coteau Du Missouri, shown in Fig. 12.11. Coteau means hill or hillside in French. The Coteau Du Missouri is a long, continuous glacial moraine that formed when part of a glacial lobe stagnated and melted in place. Its eastern side forms a gently sloping escarpment 300 to 600 feet high known as the Missouri Escarpment. The base of this escarpment forms the boundary between the Great Plains and Central Lowlands. The Coteau marks the approximate western limit of abundant glacial depositional landscape features in this part of the central US.

As seen in Figs. 12.19, the surface of the Coteau Du Missouri forms an irregular hummocky landscape with thousands of small lakes and ponds, most of which are kettles, a glacial depositional landform described below. The lakes are situated along the crest of the Coteau Du Missouri at an elevation between 2000 and 2200 feet. The sharp break in the number of lakes coincides with the upper edge of the sloping Missouri Escarpment, which leads down to Minot at an elevation between 1550 and 1800 feet. Apparently, kettles were unable to form on the sloping land surface.

Moraines are present in the area of Fig. 12.15, but they are not labeled as such except on Long Island where end moraines mark the southern extent of glaciation on land. Additional end moraines, crossing southern Connecticut, Rhode Island, Martha's Vineyard, Nantucket and Cape Cod, are traced in the figure. These moraines are important because they help build the islands above sea level and because coarse glacial sediment allows rainwater to percolate downward to be trapped as a source of fresh drinking water.

Three end moraines are present on Long Island and are clearly visible in a digital elevation model (DEM) shown in Fig. 12.20. Notice in the DEM that the glacial moraines form all of the highland ridges on the island creating as much as 250 feet of relief. Orient Point and Montauk Point both exist due to the presence of the Roanoke Point and Ronkonkoma moraines respectively. Cross-cutting relationships can be employed to determine the relative ages of the three moraines. Keep in mind that end moraines form during glacier retreat and are destroyed (bulldozed) when the glacier advances. The Ronkonkoma is the oldest moraine because it is the farthest south. The Roanoke Point moraine is cut-off by the Harbor Hill moraine, implying that Harbor Hill is the youngest. Following deposition of the Ronkonkoma and Roanoke Point moraines, the relationships suggest that the Harbor Hill glacier advanced southward in western

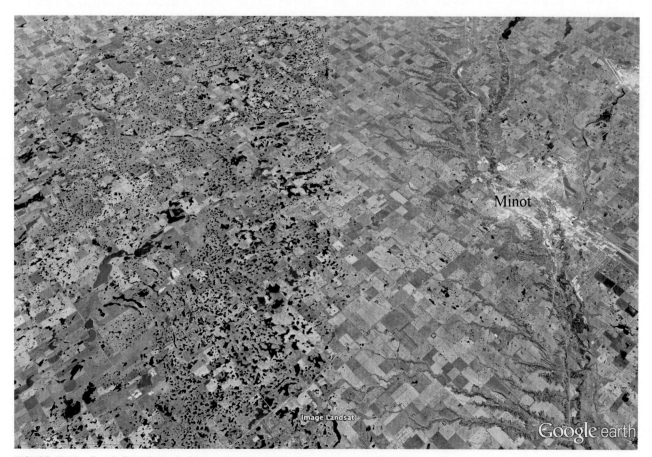

FIGURE 12.19 Google Earth image looking northwest at kettle lakes on the Coteau Du Missouri. The lakes end along a sharp boundary that coincides with the top of a gently sloping 300 to 600-foot escarpment that leads down to Minot.

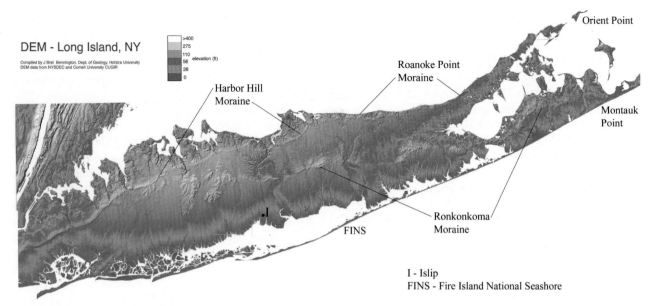

FIGURE 12.20 Digital elevation model (DEM) of Long Island. Compiled by J. Bret Bennington and downloaded at http://people.hofstra.edu/J_B_Bennington/research/long_island/li.html.

Long Island bulldozing and destroying the western part of the other two moraines. In eastern Long Island, the Harbor Hill glacier did not extend far enough south to reach the Roanoke Point and Ronkonkoma moraines thus preserving both moraines in that area. These moraines do not mark the southernmost advance of glaciation. Sea level was down during glacial advances, implying that part of the New England and Long Island continental shelf was exposed and glaciated.

PROGLACIAL LAKES

Lakes that form at the margin of retreating glaciers are known as proglacial lakes. They form for two reasons. The first has to do with isostatic sinking under the weight of the glacier, which causes marginal land areas to slope toward the glacier trapping water as ice melts. Secondly, lateral and end moraines form ridges around retreating glacial lobes that help to contain the accumulating water. Lakes also form where rivers, particularly north-flowing rivers, are blocked and damned by advancing glaciers.

There is no doubt proglacial lakes were present throughout the 2.4 million year glacial history, but those that developed during final retreat less than 22,000 years ago are best preserved. Lakes were widespread in the central US especially from 17,000 to 10,000 years ago. Large lakes would follow the retreating glacier, growing and shrinking in size, until the glacier would retreat far enough for the lake to drain into a river system. Other lakes would dry-up possibly leaving behind one or more smaller lakes, or fill with sediment. Vanished lakes leave behind remarkably flat landscape composed of silt and clay that represents what was once the floor of the lake. These areas now form farmland and swampy lowland.

It is likely that the glacial troughs now occupied by the Great Lakes were dug-out initially from river valleys during the first glacial advance 2.4 million years ago and subsequently enlarged during later glaciations. The present-day Great Lakes formed during glacial retreat beginning about 17,000 years ago. They reached their maximum size roughly 12,500 years ago when lake level was more than 400 feet higher than today. The lakes, at that time, spilled across lowlands interconnecting with each other. Fig. 12.14 shows that Lake Erie once extended southwest into Indiana as glacial Lake Maumee. Clay plains between Lake Erie and Lake Huron imply these lakes were once connected. Saginaw Bay extended farther to the southwest as indicated by glacial lake plains in that area. Lakes covered part of Michigan's Upper Peninsula as suggested by the presence of swamps. Lake Ontario spilled across lowland regions of northern New York (Fig. 12.15).

The Raisz maps show several additional areas once occupied by lakes. The shoreline of glacial Lake Dakota, now occupied by the James River Basin, is highlighted in Fig. 12.11. In the same figure, glacial Lake Souris once occupied a large area along the US-Canada border near Minot, North Dakota, an area presently occupied by the Souris Plain and Souris River. In Fig. 12.12, the area west of Duluth, labeled swamps and St. Louis Flats, was the site of glacial lakes Upham and Aikin. South of Green Bay, Lake Winnebago was once part of the much larger glacial Lake Oshkosh. Glacial Lake Wisconsin once occupied the northeastern part of the driftless area along the present-day Wisconsin River. Lakes, at one time also covered much of the Illinoian Drift Plains in southern Illinois and southwest Indiana shown in Figs. 12.14 and 12.13. In Fig. 12.15, large lakes formed along the Hudson River lowland and in the Connecticut River valley of Massachusetts.

Lakes formed across the Montana Great Plains surrounding Great Falls where the Missouri River was blocked, near Jordan where the Musselshell and smaller tributary rivers were blocked, and at Glendive where the Yellowstone River was blocked (Fig. 12.10). Additional lakes existed in the Northern Rocky Mountains and the Pacific Northwest where rivers were dammed by the Cordilleran Ice Sheet. Periodic failure of these dams released great volumes of water that poured through the northern part of the Columbia Plateau sculpting the landscape and creating what are known as the Channeled Scablands. These great floods and their effect on topography are discussed in Chapter 17.

LAKE AGASSIZ

One of the largest lakes to have existed in North America was glacial Lake Agassiz. The lake formed initially about 14,500 years ago and expanded as glaciers retreated into Canada. Throughout its life the lake would vary in size depending on the movement of glaciers and the availability of drainage outlets. At its maximum size between 11,000 and 10,000 years ago, the lake was 700 miles long and 250 miles wide extending from South Dakota well into Canada. Most of the glacial ice had melted by 8,000 years ago and the giant lake drained northward into the newly opened Hudson Bay.

Figs. 12.12 and 12.11 show the extent of Lake Agassiz in the US with a thick tan line. The most prominent remnant is along the North Dakota-Minnesota border where an arm of the lake extended southward to the Mississippi River system divide along the south shore of Lake Traverse near the border with South Dakota. A flat plain now occupies the North Dakota-Minnesota border area along with the undersized, north-flowing, Red River, which meanders incessantly through the middle of the valley. As shown in Fig. 12.11, ancient deltas are present along west side of the valley and beaches along the east side. Fig. 12.21 looks north along the meandering Red River, which is highlighted. Note the very flat terrain and the long sandy beach ridges along the east side. Fig. 12.12 shows the eastward extent of Lake Agassiz into an area of Minnesota now occupied by swamps and many hundreds of lakes including Lake of the Woods, Red Lake, and Rainy Lake. Just across the border in Canada, Lake Winnipeg, Lake Winnipegosis, and Lake Manitoba are all remnants of glacial Lake Agassiz.

MARINE INCURSIONS

The worldwide melting of ice between 17,000 and 11,000 years ago caused sea level along coastal New Hampshire and Maine to rise faster than isostatic uplift resulting in shoreline drowning. The Maine interior was drowned during the height of oceanic incursion between 16,000 and 15,000 years ago, including areas well to the north of Lewiston and Augusta along the Androscoggin and Kennebec Rivers, and well north of Bangor along the Penobscot River as far as Millinocket (M, Fig. 12.15). By 13,800 years ago, the rate of isostatic uplift had caught up with sea level rise along the Maine coast and the ocean began retreating, reaching a low stand of 170 feet below present-day sea level about 12,800 years ago. Sea level has risen ever since, reaching near present-day levels by about 3000 years ago. A similar scenario resulted in ocean water invading the St. Lawrence River and inundating Lake Champlain between 13,000 to 9000 years ago.

DRUMLIN FIELDS

In addition to moraines and ancient lake deposits, other glacial deposition features are displayed on the Raisz maps including several large drumlin fields. A drumlin is an elongated, streamlined hill composed of till that tapers in the direction of ice movement. They have the shape of an upturned boat. A drumlin field near Rochester, New York is shown in Figs. 12.2C and highlighted in Fig. 12.15. Well-preserved drumlin fields are highlighted in Fig. 12.12 north of Duluth, south and east of Green Bay, and west of Mille Lacs Lake where it is known as the Wadena-Brainerd-Pierz field. Drumlin fields are present in Fig. 12.14 along the northeastern shore of Lake Michigan near Traverse City. Bunker Hill and the Dorchester Heights near Boston, Massachusetts, and the Boston Harbor Islands, are drumlins (not labeled in Fig. 12.15). Plymouth Rock, also near Boston, is a glacial erratic, a large rock transported many miles by glaciers and set down when the ice melted. Drumlins are also present north of the Little Rocky Mountains in Fig. 12.10, but are not labeled.

KAME–KETTLE FIELDS

The area of glacial deposition is not without numerous lakes and ponds. There are several areas in Minnesota and Wisconsin shown in Fig. 12.12 that boast a large density of lakes, five of which are highlighted with bluish tint. Previously, we noted the large density of lakes on the Coteau Du Missouri in Fig. 12.19. Another area forms the Turtle Mountains, located along the North Dakota-Canada border in Fig. 12.11. Most of the lakes embedded in these glacial sediments are kettles. A kettle

176 PART | II Structural Provinces

FIGURE 12.21 Google Earth image looking north at the Red River Valley south of Grand Forks, North Dakota. Note the flat landscape, the highly meandering Red River (highlighted in white), and the long beach ridges on the eastern side. The Red River forms the North Dakota-Minnesota border.

is (was) a buried block of stagnant ice that, upon melting, creates a steep-sided hole that often fills with water. The fact that the kettles form a line in Fig. 12.12 suggests that pockets of stagnant ice were left behind as the main glacier retreated to Canada. Fig. 12.22 looks north at the kettle field shown in Fig. 12.12 located southeast of Ironwood, Michigan. Because they form randomly wherever ice is buried, the kettles produce a random pattern when compared to the bedrock-controlled lakes in the area of glacial erosion as seen in Fig. 12.1C.

Also present in kettle fields, but not visible in Fig. 12.22, are steep-sided, flat-topped isolated hills composed of stratified drift, known as kames. The melting of stagnant ice creates steep-sided holes in the ice that fill with sediment. The mound of sediment is left behind as a kame after the surrounding ice melts.

ESKERS

An esker is a long, sinuous, snake-like ridge composed of stratified drift that forms beneath a stagnant glacier. As a stream flows below a stagnant ice pack, sediment falls to the bottom and builds into a long sinuous pile. Rather than moving off the pile as would be expected, the stream continues to flow over the pile because it is locked in position by the walls of ice. Once the ice melts, the pile of sediment is left as an esker. Eskers are common glacial features and are often found in kame—kettle fields. Parnell Esker in Kettle Moraine State Forest north of Milwaukee, Wisconsin, is a well-known example. Another well-developed esker more than two miles long and 70 feet high forms part of Great Esker Park along Weymouth Back River, south of Boston.

Eskers can form stream networks similar in geometry to surface stream systems. An esker system in the Adirondack Mountains trends northeast-southwest through the center of the region west of Mt. Marcy. It is about 85 miles (137 km) long and has the shape of a single trunk stream with several tributaries. These eskers are notable because many traverse lakes accessible by canoe or kayak. Fig. 12.23 shows an esker located in Sabattis, New York along Lows Lake and Hitchins Pond.

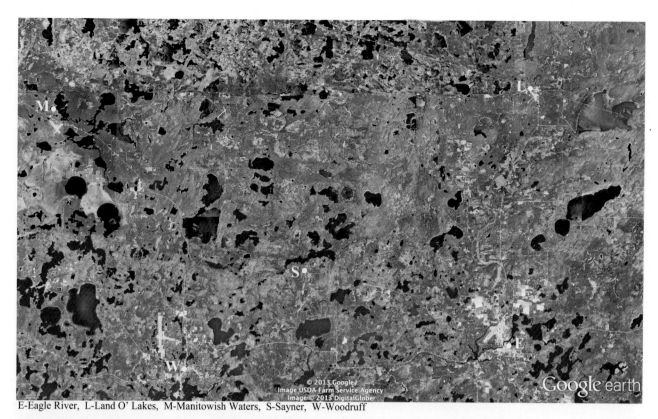

E-Eagle River, L-Land O' Lakes, M-Manitowish Waters, S-Sayner, W-Woodruff

FIGURE 12.22 Google Earth image looking north at kettles in Wisconsin southeast of Ironwood, Michigan. Land O' Lakes (L) is on the Michigan border, which is shown with a faint white line. Compare with lakes north of the glacial erosion boundary in Fig. 12.1C.

The esker is 7 miles long and 40-45 feet above the lake. In Maine, a large number of eskers are present from Holton near the Canadian border (H, Fig. 12.15), through Bangor to south of Portland. These eskers are noticeable because roads and trails are often built atop them. Route 16 northward out of Stillwater (near Orono) is an example. In both Maine and the Adirondacks, eskers are developed in topographically low areas where glaciers stagnated. They are examples of glacial deposition in areas that are otherwise characterized by erosion. In Fig. 12.14, a dense network of eskers stretches from south of Lansing, Michigan northeastward beyond Flint (F).

SAND DUNE FIELDS

Glaciers are capable of carrying copious amounts of sediment. Under certain conditions, this sediment, once deposited, can be reshaped into sand dunes. The Anoka Sand Plain located along the Mississippi River between Saint Cloud and Minneapolis is one example highlighted in yellow as Dune Sand in Fig. 12.12. The sand was dropped by retreating glaciers and then reworked and abandoned by the Mississippi River. Wind subsequently pushed some of the sand into dunes creating Sand Dunes State Forest among other features.

Sand dunes, highlighted in Fig. 12.14, cover a large area of the southern and eastern Lake Michigan shoreline east of Chicago including Indiana Dunes National Lakeshore and Warren Dunes State Park in Michigan. Sand initially collected about 6500 years ago at Indiana Dunes National Lakeshore as high lake levels allowed waves to push sand ashore. Dunes began to form in earnest beginning 4500 years ago as lake levels dropped, allowing large quantities of exposed sand to be blown inland by strong winds from the north and west.

LOESS DEPOSITION

Most of the area of old drift is covered in drift plains where landforms such as moraines either were not developed or have since been destroyed. The area is dissected in some areas with rivers, but otherwise is quite flat. Large areas of loess overlie parts of these old drift plains. Loess is a deposit of fine-grained wind-blown sediment. The rubbing of rocks at the bottom a glacier produces a large amount of fine-grained material that is left behind when the glacier retreats. Over time, loess dries-out and is carried downwind to be deposited elsewhere, commonly south of the area of glaciation. The loess shown in the Raisz maps was deposited above

FIGURE 12.23 Google Earth image looking eastward at the snake-like esker in Lows Lake and Hitchins Pond, Sabattis, New York. The esker can be followed in the distance along the south (right) side of the pond.

old drift following the most recent glacial advance less than 22,000 years ago. Examples include the large dissected loess-covered till prairies south of Des Moines and its continuation eastward south of Davenport in Figs. 12.12 and 12.13, and the large loess fields west of Indianapolis in Fig. 12.14.

AREA SOUTH OF THE GLACIAL LIMIT

Areas immediately south of glacial deposition in the central US include the Interior Low Plateaus and Ozark Plateau. Much of this area is hilly and heavily dissected by rivers. The change in landscape from relatively flat drift plains of the Central Lowlands to the hilly Ozark Plateau and Interior Low Plateaus region is best seen in Figs. 12.13 and 12.14 between St. Louis and Louisville. A Google Earth image of southwestern Indiana in Fig. 12.24 looks north across this abrupt boundary. The area of old drift, at upper left in the figure, consists mostly of flat farmland. The non-glaciated, river-dissected area to the south is less conducive to farming due to the presence of hills, which form forested regions. It is amazing to think that old drift in this area is probably younger than 310,000 years, whereas landscape to the south has not changed appreciably in more than 200 million years.

For reasons unknown, there appears to be no change between glaciated and non-glaciated landscape in the area between Louisville and Cincinnati shown in Fig. 12.14. There is evidence in the form of glacial erratics (transported glacial rocks) that glaciers reached south of the Ohio River, as far south as Carlisle and Frankfort, Kentucky, but there are no obvious topographic glacial features. One possibility is that thin glaciers advanced southward into the Bluegrass Region of Kentucky creating a diffuse boundary and leaving little evidence of their existence. Similarly, there are few obvious differences between glaciated and non-glaciated areas west of Kansas City as seen in Fig. 12.13. Perhaps here too the glacier was thin and left little or no evidence of its existence. Note that the boundary roughly follows the Kansas and Missouri Rivers. The boundary shown in Figs. 12.13

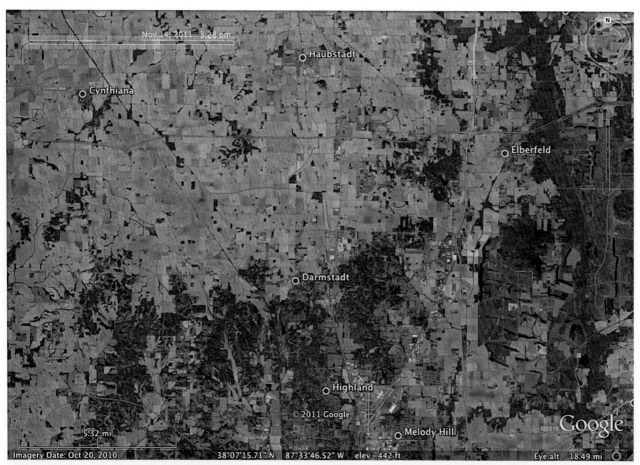

FIGURE 12.24 Google Earth image looking north at the transition between hilly, forested, unglaciated Interior Low Plateaus in the south, and flat, glaciated farm fields of the Illinoian Drift Plains in the north (old drift). Location is just north of Evansville, Indiana. Much of the lowland area was periodically occupied by lakes.

and 12.14 is the generally accepted southern limit of glaciation.

THE TEAYS RIVER

One of the largest river systems to have existed in the central US prior to glaciation was the Teays River. This river system was destroyed during the glaciation. The favored interpretation is that the ancient river system originated south of the area of continental glaciation in the Blue Ridge Mountains of North Carolina and flowed northward and westward through Virginia, West Virginia, Ohio, and Indiana, reaching as far north as the present-day Wabash River south of Fort Wayne before turning southward and entering the ancient Mississippi River just north of St. Louis. The ancient Teays River is known as the Mahomet River in Illinois. The upstream portion of the river in North Carolina, Virginia, and West Virginia still exists as the New and the Kanawha Rivers. (It is important in our discussion to realize that the New and the Kanawha Rivers are one and the same. The name changes from New River to Kanawha River at Gauley Bridge, West Virginia where the Gauley River enters the New/Kanawha River 23 miles north (downstream) of New River Gorge. The downstream portion of the Kanawha River was buried in glacial drift and destroyed during an early glacial advance, possibly as early as 2.4 million years ago. The buried river system can be followed below young glacial drift where geophysical and drilling evidence indicate the existence of a network of buried river valleys cut into bedrock. The largest valley is more than a mile wide and in some areas forms a gorge several hundred feet deep.

The location where the ancient Teays river channel becomes buried beneath glacial drift is spectacularly displayed in the Google Earth image shown in Fig. 12.25 surrounding Chillicothe, Ohio. The image looks north across the Ohio-West Virginia-Kentucky border. Areas of old drift, young drift, and non-glaciated areas are separated with thick white lines. Note the marked change in

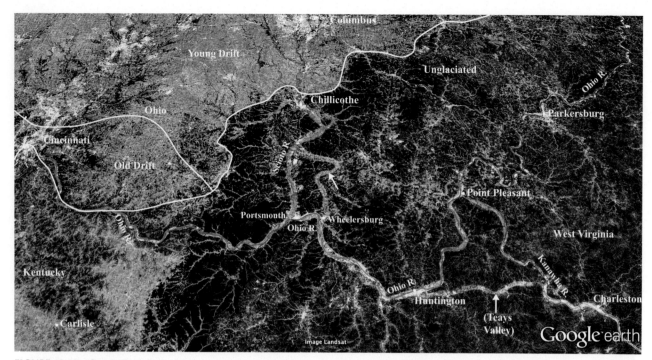

FIGURE 12.25 Google Earth image looking north across the Ohio-Kentucky-West Virginia border. White lines are glacial boundaries. The southern area, including Parkersburg and Carlisle, is unglaciated. The ancient Teays River valley disappears northward beneath glacial drift at Chillocothe. Prior to glaciation the Teays River flowed from Charleston through two now abandoned river valleys (shown with arrows) to Chillocothe and beyond.

landscape at the glacial boundary in the vicinity of Chillicothe. The glaciated area consists of relatively flat farmland covered in drift. The non-glaciated area is hilly, more thoroughly dissected by streams, and forested. The glacial boundary is far less distinct along the west (left) side of the image between Ohio and Kentucky where, as previously mentioned, there is evidence to suggest that thin glacial sheets may have extended farther south than shown.

Three major present-day river valleys and two now abandoned river valleys are shown in Fig. 12.25. The Ohio River flows through the towns of Parkersburg, Point Pleasant, Huntington, Wheelersburg, Portsmouth and Cincinnati. The Kanawha River flows from Charleston to Point Pleasant where it enters the Ohio River. The Scioto River flows southward from Columbus through Chillicothe to Portsmouth where it joins the Ohio River. The two abandoned river valleys are located with arrows. The ancient Teays River followed the present-day Kanawha River through Charleston, turning west into the now abandoned Teays Valley near Huntington (arrow), and then followed the present-day Ohio River to Wheelersburg where it turned northward and followed the abandoned river valley northward (arrow). It then continued north along the present-day Scioto River valley to Chillicothe where the channel becomes buried beneath glacial drift. Note that the Teays River flowed northward, opposite the present-day flow direction of the Scioto River.

The Teays River, prior to glaciation, appears to have been the major river to drain the area west of the Appalachian Mountains. Today, that river is the Ohio River with its headwaters in Pennsylvania. The Ohio River was in existence prior to glaciation, but did not extend nearly as far north. Its headwaters at that time were probably in the vicinity of Cincinnati and Carlisle. It was smaller than the ancient Teays River. The Ohio River established itself as the major river in the region following destruction of the Teays. This interpretation implies that the northern part of the Ohio River, north of Huntington, came into existence sometime after the beginning of glaciation less than 2.4 million years ago, but that the southern part of the river is much older. Part of the older river channel west of Louisville may have been pushed southward by advancing glaciers, and thus may also be young.

The New (and Kanawha) river channel, from its origin in North Carolina to where it empties into the Ohio River at Point Pleasant, is considered by some to be one of the oldest rivers still in existence in the US, possibly dating back more than 265 million years. Because it is located south of the glacial limit, glacial advances never modified this part of the river channel. If the New/Kanawha River now occupies the upper reaches of what was once the Teays River Valley, then the Teays River must have been hundreds of millions of years old prior to destruction of its lower reaches during glaciation.

FIGURE 12.26 Google Earth image looking west along the Wabash River. The town of Wabash, Indiana is just below (east of) the bottom of the image. The Wabash River valley widens downstream from 0.33 miles at the lower pair of white arrows to 2.5 miles at the upper pair where it enters a sediment-filled remnant of the Teays River valley.

The ancient Teays River valley is partially exhumed in the area just west of Wabash, Indiana. Fig. 12.26 is a Google Earth image looking west along the Wabash River. The town of Wabash is located east of (just below) the image. The white arrows point at the banks of the river valley. As seen in the eastern (lower) part of the figure, the Wabash River flows in a narrow valley that is 0.33 miles (0.54 km) wide at the white arrows. A short distance farther west (downstream), the valley abruptly widens to 2.5 miles (4.0 km) between the arrows. Here, the Wabash River flows into and partially excavates a sediment-filled remnant of the much larger Teays Valley.

THE MISSOURI RIVER

The history of the Missouri River, like the Teays River, is revealed through the presence of buried, partially buried, and abandoned pre-glacial river valleys. Two partially buried and abandoned pre-glacial river valleys are highlighted in Fig. 12.11 east of the Missouri River in South Dakota. Others are known to be present. The configuration of the valleys indicates that the Missouri River did not exist in its present form prior to glaciation. The present-day river instead appears to be a composite of three separate pre-glacial river systems. The pre-glacial northern and middle river systems flowed northward toward the Hudson Bay. The southern system flowed eastward and southward into a shortened Mississippi River system. The present-day river cuts a new path that, in part, occupies and exhumes sections of each of the three pre-glacial river courses.

The inferred pre-glacial northern Missouri River system is shown between Great Falls and the Canadian border with a thick light green line in Fig. 12.10. From Great Falls, the pre-glacial river flowed northward along the west side of the Bear Paw Mountains to Havre. It then followed the present-day Milk River valley eastward to the Fort Peck Dam. From Fort Peck Dam it followed its present-day course to Poplar (P, Fig. 12.10) where it turned and flowed northeastward toward Canada.

Evidence of northward flow in the northern segment is in the form of the Little Missouri River. Best seen in Fig. 12.11, the Little Missouri River flows northward in North Dakota just east of the border with Montana, and then abruptly turns eastward close to the line of glaciation. Buried valleys indicate the river originally continued northward at this location, presumably to join the north-flowing pre-glacial Missouri River, but was deflected eastward by glacial advances.

With the exception of a few moraines, there are no obvious glacial landforms in Montana and no significant landscape differences north and south of the line of glaciation as depicted in Fig. 12.10. The present-day Missouri River flows across the glacial boundary at Great Falls and remains north of it for most of its path to Williston. Given the fact that the Missouri River flows through an area of young drift, one could argue that this part of the river valley must be less than 22,000 years old. However, the degree of surrounding river dissection suggests instead that this river valley is older than 22,000 years. One possibility is that the Missouri River cut and occupied this section of river valley during an early glaciation when the northern (Milk River) valley was blocked. This section of river valley was subsequently buried during final glacial advance, but then exhumed and reoccupied by the Missouri River sometime thereafter.

The pre-glacial middle river system extended in Fig. 12.11 approximately from east of Williston to just south of Pierre, South Dakota. Buried river valleys in this area suggest that the Heart, Grand, Moreau, Cheyenne, and Bad Rivers flowed eastward across the present-day Missouri River and then northward toward Canada. For example, the pre-glacial valley west of Aberdeen appears to be a continuation of the Grand River. Similarly, the pre-glacial valley near Pierre is in line with the Bad River. Evidence for northward flow is the abrupt angle at which several of the rivers enter the Missouri River. The Moreau, Cheyenne, and Bad Rivers all flow northeastward, requiring a change in direction to enter the present-day south-flowing Missouri River. The implication is that the Missouri River was not present in this

FIGURE 12.27 Landscape map showing glacial lakes in the Great Basin, Nevada-Utah.

area prior to glaciation consistent with evidence outlined above that the river originally flowed northward from Montana into Canada.

Also evident in Fig. 12.11 is the strong contrast between areas east of the Missouri River where glacial deposition landforms are common, and areas west of the river where the degree of river dissection is more extensive and there are no obvious glacial landforms. It appears that the many rivers coming from the west were able to exhume and reoccupy their original pre-glacial valleys once they crossed the glacial boundary. Thus, there are no buried pre-glacial valleys west of the Missouri River between Williston and Pierre. These same pre-glacial valleys have remained buried east of the present-day Missouri River due to the absence of tributary rivers entering from the east.

Fig. 12.11 shows the White, Niobrara, and Platte Rivers all entering the Missouri River in a southward-flowing direction suggesting these rivers flowed south prior to glaciation. A south-flowing southern river system implies that there must have been a river divide located just south of Pierre that separated pre-glacial east- and north-flowing rivers such as the Bad River, from pre-glacial east- and south-flowing rivers, such as the White River.

Taken together, the evidence suggests that the Missouri River, as we see it today, is young and was not established until glaciation forced integration of three separate river systems into a single south-flowing system. Prior to glaciation, the northern and middle segments, as far south as Pierre, flowed northward as separate river systems. These segments were forced to integrate and to flow southward due to the presence of a large glacial lobe in Canada that extended southward to form the Coteau du Missouri.

The present-day northern segment of the Missouri River, from Great Falls to Williston, follows a path that reoccupies part of its older pre-glacial valley, as well as a valley that was established during glaciation. Between Williston and Pierre, the redirected present-day, south-flowing Missouri River established a new glacial or post-glacial path along the western margin of the Coteau du Missouri. South of Pierre, the Missouri River makes sharp turns at its confluence with the White, Niobrara, and Kansas Rivers as seen in Figs. 12.11 and 12.13. Rather than establishing a new route, it is conceivable that the Missouri River reclaimed and followed part of the pre-glacial, south- and east-flowing channels of all three rivers.

PLUVIAL LAKES OF THE CORDILLERA

We end our discussion of glaciation with pluvial lakes of the Cordillera. *Pluvial* is Latin for rain. A pluvial lake is one that comes into existence when the climate is wet, but shrinks or disappears as the climate becomes dry. The western US was wetter during the last glaciation than it is today. One area that hosted a large number of lakes was the Great Basin, centered in Nevada and Utah, where perhaps more than one thousand lakes existed between 25,000 and 10,000 years ago. A map that shows the location of some of the larger lakes is presented in Fig. 12.27. Most of these lakes have simply evaporated in today's arid climate. Particularly large pluvial lakes in western Utah and western Nevada were Lake Booneville and Lake Lahontan respectively. Lake Bonneville began to form as early as 32,000 years ago, and by 20,000 years ago was more than 1000 feet deep and covered more than 20,000 square miles extending northward to Idaho, westward to Nevada, and southward almost to Arizona (ten times the size of the Great Salt Lake). The lake reached its maximum size about 14,500 years ago when it flooded northward through Red Rock Pass, Idaho into the Snake and Columbia River drainages. Along with the Lake Missoula floods, discussed in Chapter 17, this is one of the largest freshwater floods known in the geologic record. The lake shrank drastically following the flood, and by 10,000 years ago Utah Lake, Sevier Lake, and the Great Salt Lake were its only remnants. Today, much of the dried remnants form the Great Salt Lake Desert, host to the famous Bonneville Salt Flats. Evidence of the size and depth of Lake Bonneville is seen in the many terraces and wave-cut beaches along hillsides that surround Salt Lake City. Lake Lahontan, in northwestern Nevada, reached its peak between 12,000 and 13,000 years ago when it was as much as 900 feet deep. Remnants of Lake Lahontan include Walker Lake, Carson Sink, Pyramid Lake, and Honey Lake. A large lake also covered the southern part of California's Central Valley.

QUESTIONS

1. How long ago did the most recent glacial advance on the US begin and when did it end? How are you defining the beginning and end of the glacial advance?
2. Name the two major ice sheets that formed during the most recent glacial advance and describe their extent across the United States.
3. Notice in Fig. 12.11 that deltas are present along the west side of ancient Lake Agassiz in the Red Valley whereas beaches characterize the eastern side. Suggest several scenarios that could explain this relationship.
4. Notice in Fig. 12.11 that beaches are present in the middle of Lake Agassiz in the area southwest of the Lake of the Woods. Suggest several scenarios that could explain this relationship.
5. Notice the area of glacial erosion in the northeast corner of Fig. 12.12. Describe how this area is different from areas of glacial deposition to the immediate south.
6. What evidence is there in Fig. 12.12 that a large glacier occupied Lake Superior. How far south did this glacier advance?
7. Study the patterns of lakes in Fig. 12.1C. Describe what you see. Are there dominant and secondary

trends? What can you say about the rocks based on these patterns?

8. Do you live in an area that was glaciated? Describe the landscape and how it may have been modified during the glaciation. Are there glacial deposits or landforms near to where you live? Describe them.
9. Can you find any glacial landforms in the area of old drift in Figs. 12.13 and 12.14?
10. Can you name the three volcanoes seen in the distance in Fig. 12.5?
11. Which of the following were glaciated? New England Appalachians, Southern Appalachians, Ozark Plateau, Klamath Mountains, Sierra Nevada, Colorado Rockies.
12. In Google Earth, fly to the following locations in New York and see if you can locate eskers. The locations are Five Ponds, Massawepie, Spectacle Ponds, Onchiota, and Mountain Pond. How far can you reasonably follow eskers in each area?
13. What types of landforms are associated with glacial stagnation and retreat?
14. What is the primary physiographic difference between old drift and young drift?
15. Write a short paragraph that explains the history of the Teays River.
16. What is the driftless area? Why is it unusual?
17. Research Topic. What would you expect unstratified drift to look like? How would it be different from stratified drift.
18. Research Topic. Describe how moraines, kettles, kames, eskers, and drumlins form.
19. Research Topic: Describe how cirques, tarns, arêtes, and horns form.
20. Research Topic: Research the history of Lake Lahontan or Lake Bonneville. When did they exist? What happened to them? Did they cause floods?
21. Research Topic: A large lake once existed in Death Valley. What was the name of this lake? When did it exist? What happened to it? Did it cause floods?
22. Fig. A.10 in the Appendix is an uncolored version of Fig. 12.9. Color each glacial region and provide a short paragraph on each.
23. Regarding Fig. 12.14, what evidence is there that Lake Erie was once larger than it is today?
24. In Google Earth, fly to Fort Benton, Montana. Follow the Missouri River northeastward until it begins to turn south. According to Fig. 12.10, this is the location where the pre-glacial Missouri continued north toward Big Sandy and Havre to the present-day Milk River. The present-day south-flowing channel of the Missouri, therefore, is relatively young. Compare the older northeast flowing Missouri River channel at Fort Benton with the younger south-flowing channel. Can you see any differences in their morphology? If so, what are they? Can you trace the ancestral Missouri channel through Big Sandy to Harve?
25. In Google Earth, fly to Pinedale, Wyoming. North and east of Pinedale are a series of four lakes at the foot of the Wind River Range. Examine these lakes and explain their origin.
26. What is a drumlin? How do they form and what are they composed of?
27. In Google Earth, fly to Doubtful Lake, Washington (in North Cascades National Park). Name the various glacial landforms in the area.

Chapter 13

Sediment and Nearly Flat-Lying Sedimentary Layers

The nearly flat-lying sedimentary layers structural province characterizes a vast area of the United States including most of the Interior Plains and Plateaus region, part of the eastern Cordillera, and the entire Coastal Plain. With the exception of the Laramide orogeny in the eastern Cordillera, the rocks largely escaped deformation associated with Appalachian and Cordilleran orogeny. That is not to say the rocks are entirely undeformed. The term nearly flat-lying implies that rocks are tilted at a gentle angle, typically less than 10°. Within the Interior Plains and Plateaus region and eastern Cordillera, this gentle tilt creates broad anticlines, domes, synclines, and structural basins, some of which are shown in Fig. 11.1. Monoclines are also present. In this case, the rocks are nearly flat-lying except in the hinge region of the fold where the dip is steeper than 10°. Monoclines are particularly well developed on the Colorado Plateau.

The Interior Plains and Plateaus region, together with the eastern Cordillera, form the craton and reactivated western craton tectonic provinces (Fig. 11.9). The sedimentary rocks are part of the thin interior platform rock succession and are almost everywhere underlain either with rocks of the crystalline shield or with younger intrusive rocks (Fig. 11.8). In areas where folds are tight and erosion is deep, the underlying crystalline shield (or underlying intrusive rocks) crop out in the core of domal structures surrounded by sedimentary rocks (Fig. 11.1). These areas form the mid-continent, crystalline-cored anticlines and domes structural province discussed in Chapter 14. The two structural provinces are transitional, and together they form the Interior US and Coastal Plain group of structural provinces (Table 11.1).

The Coastal Plain is not part of the craton, nor is it underlain with rocks of the interior platform rock succession. Instead, it forms the Atlantic marginal basin tectonic province and the Atlantic miogeocline rock succession (Figs. 11.9 and 11.8). Rather than forming basins and domes, the rocks more typically tilt gently and consistently toward the ocean. The rock succession thickens toward the ocean such that nowhere are underlying crystalline rocks exposed.

We begin our discussion with a description of the four landscape types that typify the nearly flat-lying sedimentary structural form. They are bench-and-slope, erosional mountains, monoclinal slopes, and monoclinal hogback ridges. We then discuss the Coastal Plain in New England and proceed southward along the Atlantic and Gulf coasts to Texas. Our discussion then moves to the interior part of the United States beginning in the Great Plains, Wyoming Basin, and Colorado Plateau, and continues eastward across the Central Lowlands, Ozark Plateau, Interior Low Plateaus, and Appalachian Plateau.

Unconsolidated sediment covers nearly all of the United States but creates its own landscape only in areas where it is thick enough to completely bury underlying bedrock (Fig. 3.11). There are several areas in the United States where unconsolidated sediment completely buries nearly flat-lying sedimentary rock to form its own landscape. We will describe beaches along the Atlantic and Gulf coast shorelines, river deposition in the Mississippi Embayment area, and sand dunes on the Great Plains and elsewhere. Glacial landscape was discussed in Chapter 12.

LANDSCAPE IN NEARLY FLAT-LYING LAYERS

Nearly flat-lying layers erode to form one of four prominent landform types: bench-and-slope, erosional mountains, monoclinal slopes, and monoclinal hogback ridges. Fig. 13.1 shows how these landform types can evolve from erosion of broad structural basins and domes (left side), and from a monocline (right side). In this section, we briefly introduce each of these landform types.

Bench-and-Slope Landscape

Bench-and-slope landscape consists of a series of flat or nearly flat bedrock benches that step down across a steep

Structural Basins and Domes

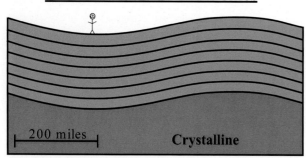

(A) Broad basins and domes within nearly flat-lying sedimentary layers prior to erosion.

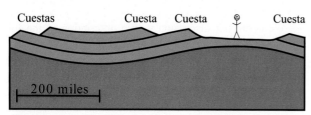

(B) Bench-and-slope landscape formed from erosion of basins and domes shown in (A).

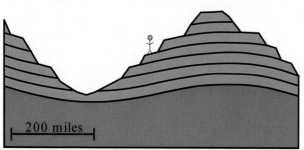

(C) Erosional mountains formed from erosion of basins and domes shown in (A).

Monoclines

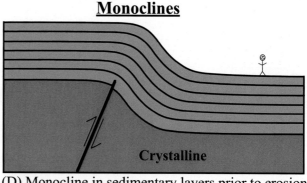

(D) Monocline in sedimentary layers prior to erosion.

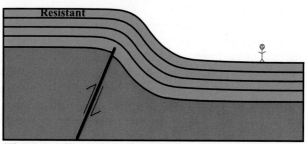

(E) Monoclinal slope formed from erosional stripping of the top two layers shown in (D).

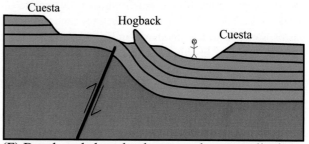

(F) Bench and slope landscape and a monoclinal hogback ridge formed from erosion of the monocline shown in (D).

FIGURE 13.1 Panel (A) shows basins and domes in nearly flat-lying layers. Panels (B) and (C) show erosion of those layers. Panel (D) shows a monocline. Panels (E) and (F) show erosion of the monocline.

escarpment (a cuesta) to a lower bedrock bench. This type of landscape forms where sideways erosion across bedrock is more prominent than downcutting (incision) into bedrock. Sideways erosion strips rocks layer by layer such that an original stack of sedimentary layers as shown in Fig. 13.1A is reduced to the stack of layers shown in Fig. 13.1B.

Bench-and-slope landscape can form where resistant layers such as sandstone and limestone alternate with nonresistant layers such as shale. The process is illustrated in Fig. 13.2. The initial situation depicted in Fig. 13.2A shows shale at the surface prior to erosion. Because shale is weak, rivers and streams meander freely across the surface progressively removing the layer. Where present at the surface, shale produces a landscape of gentle, low-lying hills covered with sediment and meandering streams such as what is shown in Fig. 13.2B.

As more shale is removed, an underlying resistant layer of sandstone or limestone is uncovered. The hard underlying layer will initially resist incision, forcing streams to meander sideways and sweep away any remaining shale at the surface. The result is a nearly flat plateau or plain underlain with a surface layer of resistant rock as shown in Fig. 13.2C. The hard, resistant surface layer is referred to as caprock. The caprock layer forms the bench in bench-and-slope landscape.

In order to maintain slope, the stream will eventually have to cut into the resistant caprock layer where it will encounter strong walls that impede meandering. The stream is forced to cut downward producing a deep

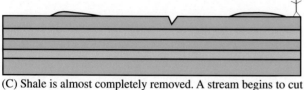

(A) Alternating layers of weak shale and resistant sandstone with shale at the surface prior to erosion.

(B) Shale is eroded to low-lying hills covered in sediment.

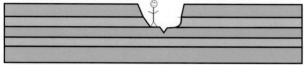

(C) Shale is almost completely removed. A stream begins to cut vertically into the sandstone possibly along a vertical joint.

(D) The stream cuts downward through resistant sandstone and through weak underlying shale to the next resistant layer of sandstone creating a bench. The weak walls of the shale allow undercutting of sandstone causing the bench to retreat and the valley to widen.

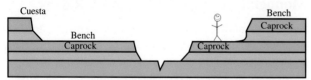

(E) A series of benches is produced, each composed of resistant (caprock) sandstone and underlying weak shale. The high bench erodes, retreats, and is eventually removed (stripped) such that landscape is lowered to the next caprock bench.

FIGURE 13.2 Panels (A) through (E) show the progressive formation of bench-and-slope landscape.

valley. Many of the famous gorges in the United States, including Zion Canyon in Utah and the Grand Canyon, result from downcutting through resistant, nearly horizontal, rock layers. In some cases, downcutting is facilitated by the presence of vertical joints or faults that form narrow weak zones. Bright Angel Canyon, shown in Fig. 11.12, is an example of downcutting along a fault.

As the stream cuts downward through the caprock layer, it may encounter an underlying layer of shale. In this situation, the weak walls of the shale allow the stream to once again meander. This action undercuts the sandstone, producing cliffs that break and retreat as depicted in Fig. 13.2D. The landscape is sequentially lowered as higher benches retreat and are eventually removed. We can refer to the process of bench retreat and removal as horizontal stripping. In this way, a plateau can evolve to a plain. A mature bench-and-slope landscape is shown in Fig. 13.2E. It is characterized by wide benches underlain with resistant caprock layers that end at an escarpment (cuesta) that steps down to another bench. The geology is predictable in the sense that if you step down across an escarpment (cuesta) from one caprock surface to another, you are also stepping down onto a lower (older) rock layer within the sedimentary sequence.

Erosional Mountains

Erosional mountains in nearly flat-lying sedimentary layers represent an extreme form of bench-and-slope landscape where erosion cuts consistently downward rather than sideways, leaving scattered isolated areas of the original elevated land surface to form mountaintops as depicted in Fig. 13.1C. These mountains are erosional in the sense that uplift is not necessary for their formation. The mountains are what remain of an original plateau surface following erosional downcutting. The formation of erosional mountains is favored over bench-and-slope landscape in areas of regional uplift, falling sea level, or along the edge of a high plateau surface crossed by rivers where there is a steep drop to low elevation.

One could argue that any highland area is subject to erosion; therefore, any mountain could be considered an erosional mountain, especially those where tectonic processes are dormant. In this book we restrict the term to an originally flat plateau area so thoroughly dissected by stream erosion that only isolated parts of the original surface are left to form highlands. Dissection occurs in stages as shown in Fig. 13.3 in which streams continually invade flat areas until the entire surface is dissected. Erosional mountains have characteristics that distinguish them from tectonically uplifted mountains. Rather than sharp, steep summit areas, erosional mountains characteristically develop broad, flat summit areas composed of remnants of the original flat plateau surface. River valleys between summit areas are often sharp and steep. Roads and houses, therefore, are built on the wide summit area rather than in valleys, thus creating a second distinction with tectonically uplifted mountains.

Erosional mountains are a characteristic feature of plateau areas underlain with sedimentary rock; however, they can form in any rock type. The Salmon and Clearwater Mountains in the Idaho Batholith are an example of erosional mountains in granitic rock (Chapter 16). The Klamath Mountains are a second example developed

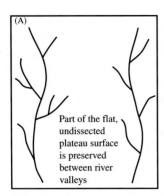

FIGURE 13.3 Sketches showing (A) intermediate stream dissection. Part of the undissected plateau surface is preserved between river valleys. (B) Mature stage of stream dissection. The original plateau surface is preserved only as isolated, flat-topped erosional mountains.

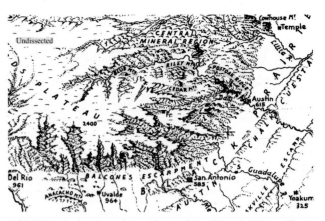

FIGURE 13.4 Landscape map of part of the Edwards Plateau, Texas, showing the progressive dissection of a flat plateau to form erosional mountains.

in strongly deformed sedimentary and crystalline rock (Chapter 19). With respect to nearly flat-lying sedimentary rock, the Appalachian Plateau is the quintessential location in the United States for erosional mountains. Classic examples include the Cumberland, Allegheny, Pocono, and Catskill Mountains, all of which are part of the Appalachian Plateau. Because they are so thoroughly river dissected, these areas form some of the roughest landscape in the eastern US.

Earlier, we suggested that bench-and-slope landscape represents a process by which a plateau can evolve into a plain. Erosional Mountains can be thought of in the same way. If the process of river dissection were to go to completion without a major change in forcing agents, then the mountains themselves will eventually be eroded such that the entire plateau region is lowered to a nearly flat-lying plain. The amount of time it takes to produce erosional mountains is dependent on climate, the number of through-going rivers and streams, the hardness of rock, the amount of relief between the edge of the plateau and adjacent plain, and how uplift/subsidence and sea level change affects base level. Given these variables, it is easy to see how one area could evolve faster than another.

Let us trace the evolution of erosional mountains in nearly flat-lying sedimentary rock progressing from an originally flat plateau surface, through the intermediate stage shown in Fig. 13.3A, and finally to the stage of dissected mountains as depicted in Fig. 13.3B. The edge of a plateau, where it drops to a plain, is a prime location for development of erosional mountains because the steep slope provides velocity and increased cutting power to streams. Under these conditions, streams cut progressively into the plateau surface such that it is possible to see the entire transition from an undissected flat plateau to dissected erosional mountains at one location.

The Edwards Plateau in south-central Texas is an example. The beginning stage in the formation of an erosional mountain is illustrated by the flat, undissected landscape in the northwest corner of Fig. 13.4. The intermediate stage is shown in the area of the Blue Mountains where streams have cut into the plateau leaving isolated flat areas between valleys. Stream dissection along the eastern edge of the plateau is nearly complete with the formation of the Cedar and Riley Mountains, both of which can be considered erosional mountains.

Fig. 13.5 shows the area west of Austin, Texas. Note the dissected erosional mountains at the front of the image and the undissected plateau at the rear. Also note that most of the homes and roads are built across flat summit areas. Erosion and dissection will, over time, migrate into the flat undissected part of the plateau.

Monoclinal Slopes and Hogback Ridges

Fig. 13.1D shows a monocline prior to erosion. In some instances, it is possible for landscape to mimic the shape of the monocline. In other words, the landscape looks just as it does in Fig. 13.1D. Such a situation can arise if erosion has not progressed very far, or if erosion has managed to strip nonresistant folded layers off the top of a particularly resistant layer such as what is shown in Fig. 13.1E. We can refer to this type of landform as a monoclinal slope.

If erosion breaches the monocline, then a hogback ridge composed of resistant rock, can be produced in the area of steep dip while, at the same time, bench-and-slope landscape is produced in the flat-lying parts of the fold as shown in Fig. 13.1F. We can refer to this type of hogback ridge as a monoclinal hogback ridge.

Other landscape forms are possible. If the erodability of rocks is similar across the monocline (there is no differential erosion), then the monocline could erode to form a flat plain. Now that we understand the general landscape types, we can proceed with a discussion of the Coastal Plain.

FIGURE 13.5 Google Earth image looking northward at erosional mountains along the edge of Edwards Plateau in the area west of Austin, Texas. Undissected Edwards Plateau is seen at upper right.

THE COASTAL PLAIN

The Coastal Plain extends along the Atlantic coastline from Cape Cod to Texas. It is a low-lying nearly flat plain less than 500 ft above sea level that slopes gently toward the ocean. Relief is rarely more than a few tens of feet. The Coastal Plain extends offshore below sea level as the continental shelf for distances that vary from a few miles to over 200 mi (Fig. 1.8). Much of the shelf region is less than 400 ft deep, which is important because the shallow water, combined with the warm Gulf Stream, provides some of the best fishing grounds in the world. The inland boundary of the Coastal Plain is marked by the Fall Line, which extends continuously from Cape Cod to at least as far south as Montgomery, Alabama. The Fall Line is an escarpment 100–300 ft high that marks the boundary between resistant crystalline rock of the Piedmont Plateau and soft sedimentary rock of the Coastal Plain. It is an important boundary that will be discussed in more detail in Chapter 16.

Sedimentary rocks on the Coastal Plain are younger and weaker than those in the Interior Plains and Plateaus region. Some are so poorly consolidated that they can be confused with recently deposited unconsolidated sediment. Nearly all of the rocks are less than 130 million years old. The rock layers tilt gently toward the ocean at a slightly higher angle than the slope of the land such that the oldest rocks are close to the Fall Line and the youngest are close to the ocean. The entire rock sequence thickens toward the ocean from zero thickness at the Fall Line to more than 13,000 ft below the Atlantic continental shelf and to more than 40,000 ft off the Gulf coast producing an overall wedge-shaped geometry. Where the tilted rock layers are of sufficient hardness, they form small, subtle, inland-facing cuestas, typically with only a few tens of feet of relief. Two locations where inland facing bedrock cuestas are particularly well displayed are the Black Belt in the Mississippi Embayment region and the Texas Coastal Plain. Both are discussed below. The rocks form the Atlantic miogeocline rock succession and the Atlantic marginal basin tectonic province.

Extensive drilling and seismic profiling indicate that there are sedimentary and crystalline rocks buried deep below the Coastal Plain and continental shelf that are older than those exposed at the surface. The oldest sedimentary rocks were deposited beginning about 237 million years ago and are associated with initial rifting and the opening of the Atlantic Ocean. Below these rocks are deformed sedimentary and crystalline rocks that represent the eroded remnants of the Appalachian/Ouachita orogenic belt. The presence of deformed Appalachian/Ouachita rocks at depth indicate that some of the orogenic mountains had already eroded and subsided to sea level by 237 million years ago. Fig. 5.2 is a cross-section of the Atlantic coast that shows buried Appalachian rocks. It is possible that rocks of the North American crystalline shield are present below the Appalachian rocks across at least part of the Coastal Plain.

Grain size, composition, sorting, and fossils in exposed sedimentary rock layers suggest that the Coastal Plain has been a low relief coastal landscape throughout its history (more than 130 million years), whose size has varied with sea level changes. Three major depositional environments are preserved, and their distribution is an indicator of sea level oscillation. The depositional environments are shoreline, inland continental, and shallow marine. Shoreline rocks contain sediment deposited in estuaries, lagoons, tidal marshes, beaches, and barrier islands. Inland continental rocks were deposited in river, delta, and lake environments with sediment derived from erosion of the Appalachian Mountains. Shallow marine environments include rocks such as limestone and dolostone that were deposited as reefs when a shallow ocean covered the Coastal Plain. The age and distribution of these rocks indicate that sea level was quite high between 100 and 50 million years ago. A few of the sea level high stands are shown in Fig. 8.3.

Weak sedimentary rocks in a humid climate have produced a thick cover of unconsolidated sediment that buries bedrock to the point of controlling most, but not all of the bedrock landscape features. Unconsolidated sand in the form of sand bars, barrier islands, and beaches cover large areas of the Coastal Plain, both along the present-day coastline and inland where it was deposited during high sea level stands. Sediment left by inland continental and marine environments is also present as are thick soil horizons.

We begin our discussion by introducing one of the most widespread landforms on the Coastal Plain, the barrier island. We then discuss landscape features beginning in New England. We have already noted that sea level variations play a key role in Coastal Plain landscape. Presently, the entire Atlantic-Gulf coast is losing ground to rising sea level. Table 13.1 shows tide-gauge readings, in millimeters per year and in inches per 100 years, at various locations beginning in New England and ending in Texas as recorded by the National Oceanic and Atmospheric Administration (NOAA). The tide gauge readings record relative sea level rise as defined in Chapter 8. Actual (eustatic) sea level rose 7 to 8 inches worldwide over a 100-year period ending in 2000. Sea level is currently rising at a rate of nearly 13 inches per 100 years based on satellite altimetry. Tide gauge readings greater than about 8.2 inches per 100 years suggest land subsidence in addition to sea level rise.

Barrier Islands

The barrier island is the most consistently present landform on the Atlantic-Gulf coastline. A barrier island is a narrow strip of sandy beach, with or without vegetation, that forms an offshore island less than 20 ft above sea level separated from the main shoreline by a narrow lagoon or marsh area that lies roughly at sea level, followed by the mainland, which can rise fairly quickly to between 10 and 50 ft above sea level. Barrier islands are present from Long Island, New York to Cape Fear, North Carolina. They form nearly the entire eastern coastline of Florida, part of the western coastline between Cape Romano and St. Petersburg, and most of the Florida panhandle westward to the Mississippi Delta. They also form nearly the entire Texas coastline from Galveston southward. Daytona Beach, shown in Fig. 13.6, is a fabulous example.

Barrier islands are not stable landforms. Instead, they migrate and change shape due to longshore drift and changes in sea level. Longshore drift is a process whereby wave action moves sand laterally along the shoreline. The process is shown in Fig. 13.7. Longshore drift can move sand along the shoreline at rates that exceed 3000 ft/day. Longshore drift along the entire Atlantic seaboard is consistently toward the south.

Sea level changes affect the position of barrier islands. Recall from Chapter 8 that 18,000 years ago sea level was about 400 ft lower than it is today and that the outer edge of the Atlantic continental shelf was nearly fully exposed. Given this circumstance, one might think that the coastline looked very different 18,000 years ago. But this may not have been the case, at least as it concerns the presence of barrier islands. Rather than forming within the past few hundred or few thousand years, it is possible that barrier islands have been in continuous existence over the entire 18,000-year rise of sea level. Two characteristics may have combined to progressively push the barrier islands inland, an abundance of sand and a gently sloping coastal shelf. Given that there is an abundance of older inland barrier island sand deposits on the Coastal Plain, we can surmise that barrier islands have been a part of the Atlantic coastline for much longer than 18,000 years; perhaps for hundreds of thousands or millions of years, and that they have simply migrated with sifting sea level.

TABLE 13.1 Relative Sea Level Change Based On Tide-Gauge Measurements

Tide Gauge	mm/yr	in/100yrs.	Measurement	Interval
New England				
Eastport, ME	2.12	8.3	1929–2016	Maine/Canada Border
Portland, ME	1.87	7.4	1912–2017	
Boston, MA	2.82	11.1	1921–2017	
The Battery, NY	2.84	11.2	1856–2017	New York City
New Jersey to North Carolina				
Atlantic City, NJ	4.08	16.1	1911–2017	
Ocean City Inlet, MD	5.59	22.0	1975–2017	
Washington DC	3.24	12.8	1924–2017	
Chesapeake Bay Bridge	5.92	23.3	1975–2017	
Oregon Inlet Marina, NC	4.36	17.2	1977–2017	
Wilmington NC	2.30	9.1	1935–2017	Near Cape Fear
South Carolina to Florida				
Charleston, SC	3.25	12.8	1921–2017	
Mayport, FL	2.59	10.2	1928–2017	Jacksonville, FL
Lake Worth Pier, FL	3.70	14.6	1970–2017	West Palm Beach
Key West, FL	2.42	9.5	1913–2017	
Fort Meyers, FL	3.10	12.2	1965–2017	
St. Petersburg, FL	2.75	10.8	1947–2017	Tampa, FL
Apalachicola, FL	2.23	8.8	1967–2017	
Mississippi Embayment				
Grand Isle, LA	9.08	35.7	1947–2017	Miss. Delta, LA
Texas				
Galveston Pleasure Pier, TX	6.62	26.1	1957–2011	
Rockport, TX	5.58	22.0	1937–2017	
Port Isabel, TX	3.98	15.7	1944–2017	Brownsville, TX

http://tidesandcurrents.noaa.gov/sltrends/sltrends.html.
Source: Based on NOAA Tides&Currents.

Thus, if given enough sand and a gentle coastal slope, it is possible for barrier islands and other shoreline landforms to migrate inland with rising sea level, and to reform ocean ward as sea level drops.

In order to facilitate discussion of the Coastal Plain, we will use a series of five Raisz landform maps labeled Fig. 13.8–13.12, on which certain features and names are highlighted. The maps are grouped together for easy reference.

New England

Most of the Atlantic and Gulf coast shoreline is currently losing ground to the ocean through subsidence, sea level rise, or a combination of both (Table 13.1). Such coastlines tend to be highly irregular because the rising ocean claims coastal low areas particularly river valleys, producing deep, inland bays, coves, marshes, and near-shore islands that were once areas of high ground connected to the mainland.

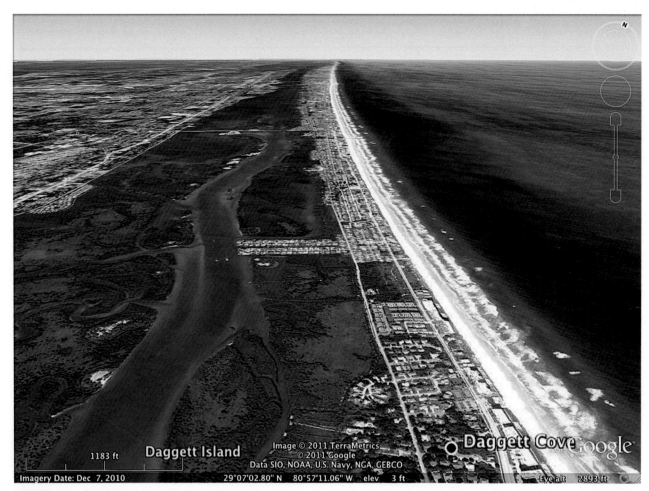

FIGURE 13.6 Google Earth image looking northward along Daytona Beach, Florida. The barrier island protects the inland tidal marsh from wave action.

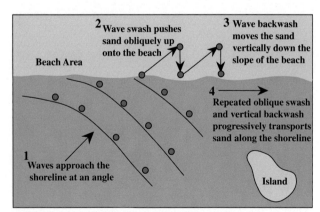

FIGURE 13.7 Sketch map of ocean water breaking on a beach creating longshore drift. The sequence is numbered.

Evidence of recent drowning is best seen along the north Atlantic coast. The entire New England Coastal Plain north of Cape Cod is below sea level. Recall from Chapter 12 that the Maine coastline experienced initial drowning with sea level rise due to glacial melt, followed by ocean retreat due to isostatic uplift, and then drowning beginning 12,800 years ago due to glacial melt and thermal expansion of warming ocean water.

Today, ocean waves in Maine lap directly against crystalline rock of the Appalachian Mountains producing a highly irregular rocky coastline visible in Fig. 13.8. Long inlets represent drowned glacial/river valleys. Small coastal mountains and islands represent highland areas not yet covered with water. In Fig. 13.13 we see a strong Appalachian structural fabric that trends at an angle to the Maine coastline. Cadillac Mountain, at 1528 ft, forms the high point on Mt. Desert Island in Acadia National Park, and is the highest point along the Atlantic seaboard. This area, as they say in New England, is where the mountains meet the sea.

The Coastal Plain emerges from below sea level at Cape Cod and the coastal landscape changes to one dominated by unconsolidated sand and silt brought to the coastline by glaciers and rivers. Once deposited, the sediment is shaped, transported, and reshaped by waves and currents

Sediment and Nearly Flat-Lying Sedimentary Layers **Chapter | 13** 193

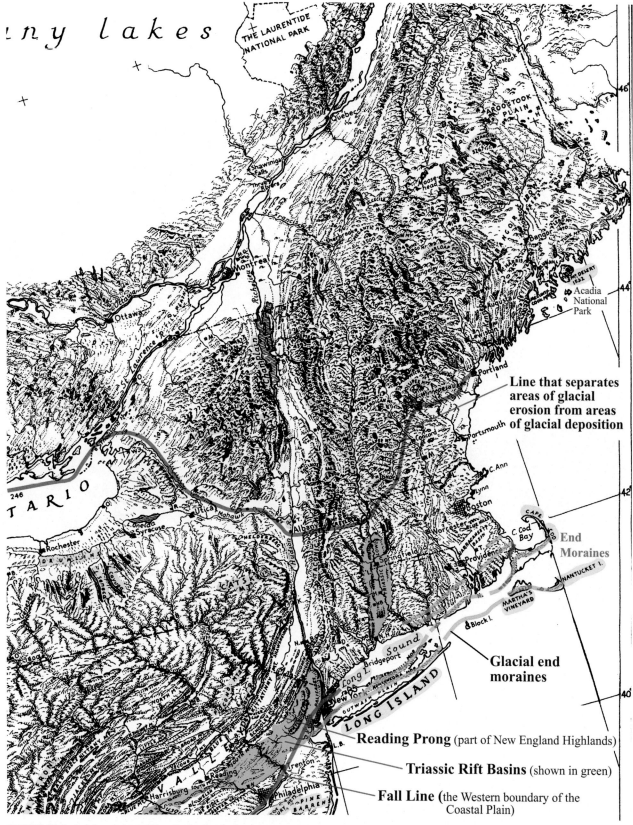

FIGURE 13.8 Landscape map of New England and adjacent areas.

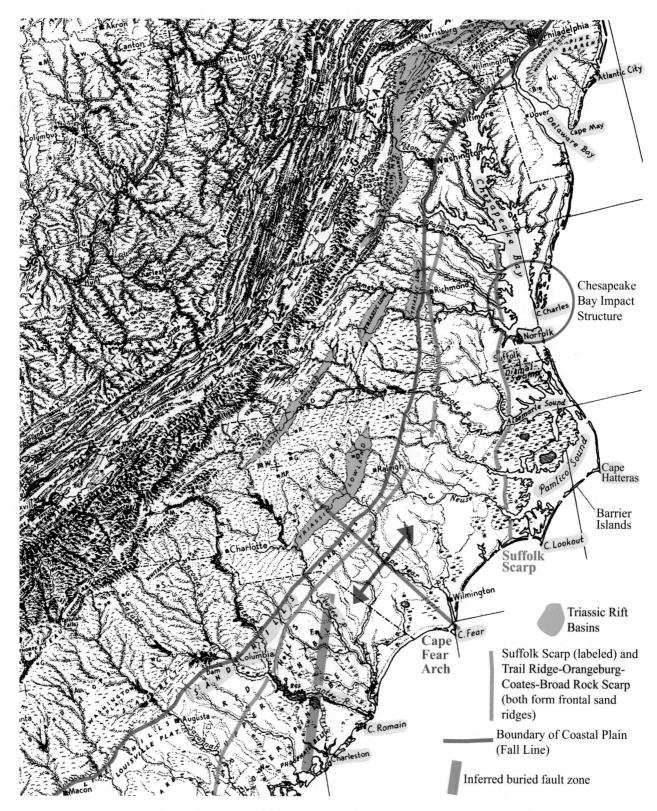

FIGURE 13.9 Landscape map of the Mid-Atlantic Coastal region from New Jersey to Georgia.

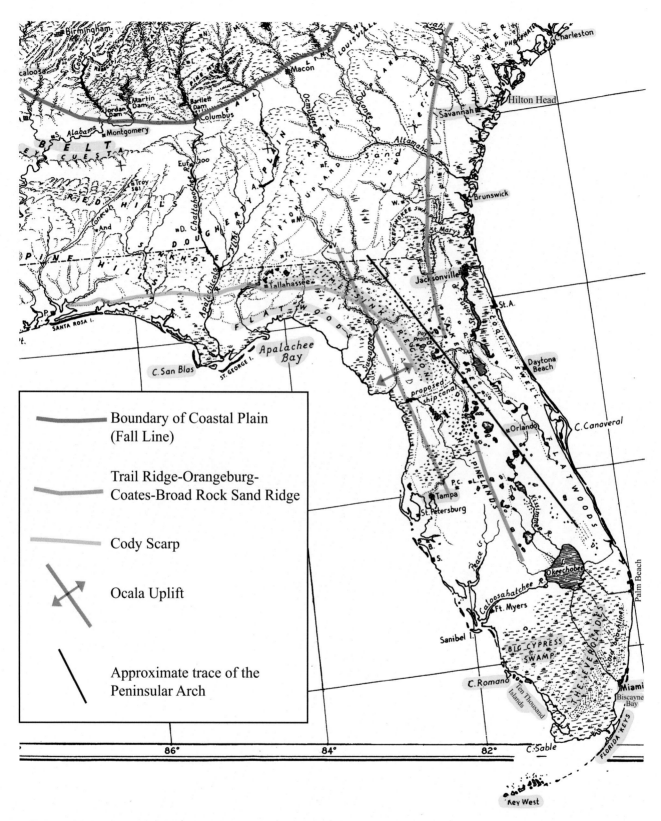

FIGURE 13.10 Landscape map of the South Carolina–Florida coastal area.

FIGURE 13.11 Landscape map of the Mississippi Embayment area. Heavy line outlines the boundary of the Coastal Plain.

FIGURE 13.12 Landscape Map of the Texas coast. Heavy lines outline the boundary of the Coastal Plain and separate the Great Plains from the Central Lowlands.

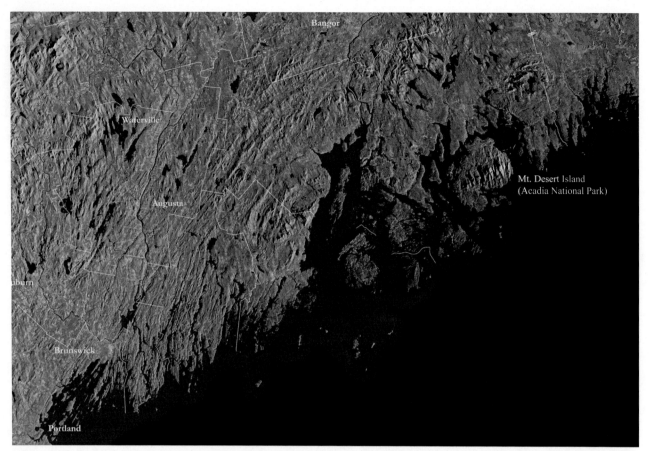

FIGURE 13.13 Image of the drowned Maine coastline from Portland to Acadia National Park. *Image from USGS, EROS image gallery, Landsat State Mosaics.* .

via longshore drift to form sandy beaches and barrier islands. As noted in Chapter 12, the southern New England—Long Island area marks the southernmost extent of continental glaciation on the east coast. The glaciation produced thick mounds of sediment in the form of end moraines that have helped shape and build Long Island, Martha's Vineyard, Nantucket, and Cape Cod. The end result of this deposition is that much of the Coastal Plain landscape in southern New England and Long Island is controlled by unconsolidated shoreline and glacial sediment rather than by the underlying sedimentary and crystalline rock.

The Fall Line extends through Long Island Sound where it marks the boundary between crystalline rock of the New England Highlands and sedimentary rock of the Long Island Coastal Plain. Long Island Sound itself is a product of glaciation. Weak sedimentary rock at the sedimentary-crystalline rock contact was gouged-out to create the ocean inlet and, at the same time, left a rare sedimentary bedrock feature on the north side of Long Island; an inward-facing cuesta of sedimentary rock topped with glacial drift. This feature is shown in Fig. 13.14. Other aspects of the glaciation in the northeast were discussed Chapter 12.

New Jersey to North Carolina

The Atlantic shoreline from Cape May, New Jersey, to Cape Hatteras, North Carolina, shown in Fig. 13.9, shows the characteristics of a drowned coastline. The Delaware Bay, Chesapeake Bay, Albemarle Sound, and Pamlico Sound are all drowned river valleys protected in part by barrier islands. According to Table 13.1, this area appears to be losing ground to sea level faster than both the New York—Boston area to the north and the Cape Fear area to the south.

The Pine Barrens in southern New Jersey form a large part of the Coastal Plain north of Delaware Bay. Here, one can drive for miles through a dense forest of Pitch Pine. The area is underlain principally by the Cohansey Sand, a formation composed largely of unconsolidated medium to coarse-grained quartz-rich sand with quartz pebble

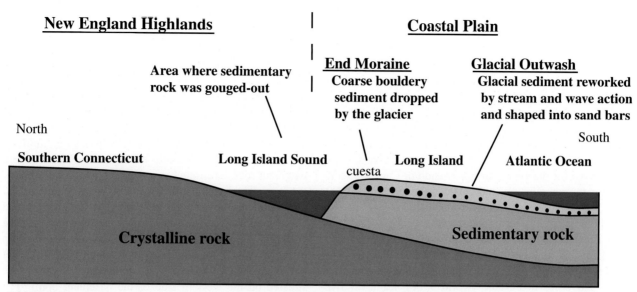

FIGURE 13.14 North to south cross-section from Connecticut to Long Island. The Fall Line trends through Long Island Sound separating crystalline rock from tilted sedimentary rocks that form a north-facing cuesta. An end moraine is built on the cuesta.

conglomerate lenses. The age of the Cohansey is debatable. It likely formed between 11 and 14 million years ago, but some of it could be as young as 5 million years. It appears to be a river deposit mixed with beach sands and other marine deposits. Maximum thickness is 351 ft (107 m), but most of it is less than 170 ft thick. The sandy composition creates a permeable, acidic soil with few organic nutrients. Water drains through the sand too quickly for most deciduous trees to take root; thus, the prevalence of pine trees. Early settlers coined the term, barrens, because they found the soil unsuitable for cultivation. Sandy regions are common elsewhere on the Coastal Plain, and these areas also typically support pine forests.

The western side of the Pine Barrens is marked by a small, inland-facing bedrock cuesta shown in Fig. 13.9 as Woodbury Heights. The cuesta approximately marks the boundary between the Cohansey Sand and older rock units to the west. As best seen in the upper right of Fig. 13.9 (and lower right of Fig. 13.8), the Delaware River at Trenton makes a 90° turn from a southeasterly direction to a southwestward direction parallel with the Fall Line through Philadelphia to the Delaware Bay. The Woodbury Heights cuesta may have formed a barrier that played a role in this deflection.

The Coastal Plain sedimentary sequence was disrupted 35 million years ago when a meteorite, estimated to be between 1.8 and 3.1 miles in diameter, crashed offshore just north of Norfolk, Virginia near what is today the mouth of Chesapeake Bay. This area is known as the Chesapeake Bay impact site. Its location is shown in Fig. 13.9. Sea level was higher at the time of impact such that the shoreline was in the vicinity of the Fall Line. The impact produced a crater more than 53 miles (85 km) in diameter and about 5000 ft deep, penetrating through sedimentary rock into crystalline rock of the eroded Appalachian Mountains. It is the largest impact structure in the United States and the sixth largest in the world. It likely vaporized both itself and the surrounding ocean water on impact. Broken rock material, thrown up as a result of impact, came down in a pile that filled the crater. At the same time, ocean water swept into the pile of broken rock creating a water-saturated slurry of sand and rubble known today as the Exmore Breccia.

The crater is now buried and out of sight under several thousand feet of sediment that accumulated following impact. Most of what we know of the crater was gathered from drill core and seismic reflection records. Shocked quartz and melted fragments from drill cores confirm that it is indeed an impact structure. The 35-million-year age of impact is based on radiometric dating of melted fragments. The impact permanently disrupted groundwater flow by truncating fresh water aquifers. Water within the Exmore Breccia is 1.5 times saltier than normal seawater making it useless for drinking and for industry. Today, as shown in Table 13.1, there is significantly greater subsidence over the impact structure than in surrounding areas.

South of Pamlico Sound, the smooth scalloped shape of the coastline between Cape Lookout and Cape Romain shown in Fig. 13.9 is rather unique for the Atlantic seaboard. The scallops are undergoing erosion and debris is being transported to the capes. Erosion coupled with the smoothness of the scallops is more indicative of uplift than of subsidence. According to Fig. 9.2, this area is transitional between areas of active uplift and subsidence. However, the area is located directly on the Cape Fear

Arch (shown in Fig. 13.9 with a thick line and double arrows), which displays both long-lived and active deformation. Rock units across the Cape Fear Arch form a broad anticlinal flexure. The size of the arch is indicated by the presence, in its center, of rock layers elevated 2500 ft above the same rock layers in surrounding areas. The high elevation of these rock layers, coupled with the absence of high topographic elevation, suggests slow, long-lived, sustained uplift that accumulated over tens of millions of years. Uplift associated with this arch may be responsible for the oceanward protrusion of Cape Hatteras and for the warping of sand ridges as young as 110,000 years old as discussed later in this chapter.

The area in and surrounding the Cape Fear Arch not only shows evidence of long-lived deformation, but also of recent deformation. According the Table 13.1, relative sea level rise at Cape Fear is much less than surrounding areas and close to eustatic (world wide) sea level rise implying the possibility of active, though slight, uplift. The Cape Fear area is also one of recent earthquake activity that includes the largest historic earthquake on the eastern seaboard, the September 1, 1886 Charleston, South Carolina earthquake with an estimated magnitude of 7.3. One suggestion is that the Charleston earthquake, along with numerous smaller more recent earthquakes, may be associated with a buried fault zone that extends in a north-northeast direction from just east of Charleston to the Pee Dee River as shown with the thick shaded line in Fig. 13.9. Rivers that cross this zone, including the Santee and Pee Dee, show anomalous characteristics that suggest it is a zone of upwarping. Rivers are incised as they cross the inferred fault zone, and several rivers including the Santee and Pee Dee, are deflected northward to the northern edge of their floodplain where they cross the zone.

Given the above evidence, we can suggest that the scalloped shoreline between Cape Lookout and Cape Romain is the result of long-lived and active uplift associated with earthquakes and development of the Cape Fear Arch. Although a buried fault zone may be present, there is no evidence of active faults breaking the surface.

South Carolina to Florida

The South Carolina-Georgia coastline near Savannah shows evidence of recent drowning although not to the same extent as Chesapeake Bay. A glance at Fig. 13.10 may help you understand why this part of the coastline is known as the Sea Island section. Drowned river valleys have left a series islands just offshore that include Sullivan's Island near Charleston, St. Simons Island near Brunswick, and Hilton Head near Savannah. A Google Earth image of Hilton Head is shown in Fig. 13.15. Note the narrow waterways and widespread marsh areas (gray) that create the island.

The northern part of Florida is situated on a broad anticlinal structure known as the Peninsular Arch that has been present for more than 100 million years. The center of the arch extends nearly the length of Florida from the Georgia border southward through the vicinity of Payne Prairie and Orlando in Fig. 13.10 to the area east of Lake Okeechobee. It is defined by uplift of crystalline basement rocks in part composed of rocks that correlate with the Appalachian orogenic belt. The basement uplift is broad, extending the width of northern Florida where basement rocks are found at a depth of 1 to 2 miles below sea level. The basement drops quickly toward the east and more gradually toward the west where it supports a broad continental shelf as seen in Fig. 1.8 that is wider than the emergent state of Florida itself. The average regional dip of rocks away from the center of the arch is about 5 feet per mile. The Ocala Uplift is a satellite anticlinal structure situated on the western side of the basement uplift. The axis of the uplift is approximately located in Fig. 13.10. The Ocala Uplift is much younger than the Peninsular Arch. The age of deformed beds suggests that uplift may have begun prior to 40 million years ago and continued until about 5 million years ago.

Nearly all of Florida is underlain with limestone, which dissolves in the humid climate to produce karst topography. Karst is present wherever limestone is at or close to the surface. In these areas, the scourge of Florida, sinkholes, are present nearly everywhere except in the mangrove swamps and everglades of southernmost Florida. They are particularly abundant in the Tampa area stretching eastward to Lakeland and Orlando, and northward to the Payne Prairie area (Fig. 13.10). They are also common along the Cody Scarp as far west as Tallahassee as described below. Fig. 3.14 looks northward across a myriad of sinkholes in the northern outskirts of Tampa. Areas where sinkholes are less common are covered in clay, mud, and sand. These sediments protect the underlying limestone from dissolution. Associated with karst is one of the largest aquifers in the country, the Floridan aquifer, which underlies nearly all of Florida and extends into neighboring states. An aquifer is a body of permeable rock or sediment that contains groundwater. The Floridan aquifer is the primary source of fresh water, and is responsible for the great number springs of which Florida is famous.

Inland areas of south Florida, west of Miami, rarely rise more than 25 ft above sea level. Here, we enter the freshwater wetlands of the Florida Everglades and the Big Cypress Swamp. This region forms Everglades National Park, Everglades Wildlife Management area, and Big Cypress National Preserve. The wetlands are situated on a platform, much of which is underlain by the Miami Limestone, a young rock unit deposited during the previous high sea level stand 125,000 years ago. A drop in sea

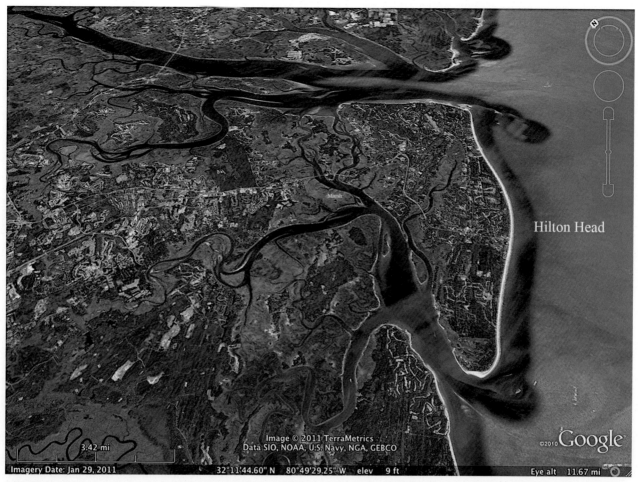

FIGURE 13.15 Google Earth image looking NNE at Hilton Head Island, South Carolina. Port Royal Sound is at top of image. Note the sandy beach along the shoreline.

level during the final glacial advance beginning 118,000 years ago resulted in subaerial exposure and cementation of the limestone, creating a reddish, impermeable crust. The flat topography close to sea level, the slow rise in sea level over the past 7000 years, the presence of surrounding higher elevation, the abundant rainfall, and the impermeable underlying limestone, have all combined to create conditions conducive to formation of wetlands. The impermeable limestone substrate allows an overlying sheet of fresh water to move slowly southward through the wetlands supporting primarily saw grass. All of southern Florida is within a few tens of feet of sea level, implying that it is susceptible to sea level rise. According to Table 13.1, sea level at Lake Worth Pier, Key West, and Fort Meyers is rising at rates of 9.4 to 14.5 inches per 100 years. The highest elevation on Key West is 18 ft, but most of the town sits at less than 5 ft.

The eastern (Atlantic) coastline of Florida is remarkably straight and marked with barrier islands along nearly its entire length all the way to Miami Beach (Fig. 13.10). The abundant quartz-rich sand likely arrived via longshore drift and possibly river transport. South of Miami, the embayed shape of Biscayne Bay protects it from strong wave action or longshore drift. Rather than barrier islands, Biscayne Bay hosts the northern end of the Florida Keys as well as the Great Florida Reef, the most extensive living barrier reef system in North American waters and the third largest in the world. A key (also spelled cay) is a small, flat island close to the mainland. The entire Florida Key system extends as an arc from Virginia Key, just off the coast of Miami, to Loggerhead Key in Dry Tortugas National Park 72 mi (116 km) west of Key West. The key system is shown in Fig. 13.16. Similar to the Everglades and Big Cypress Swamp, much of the area of the Keys is federal and state-protected.

The Keys are separated approximately at the Seven Mile Bridge into the Upper Keys to the east, and the Lower Keys to the west. In Fig. 13.16 one can instantly see the morphologic difference between these two sets of keys. Neither is currently part of the living reef system.

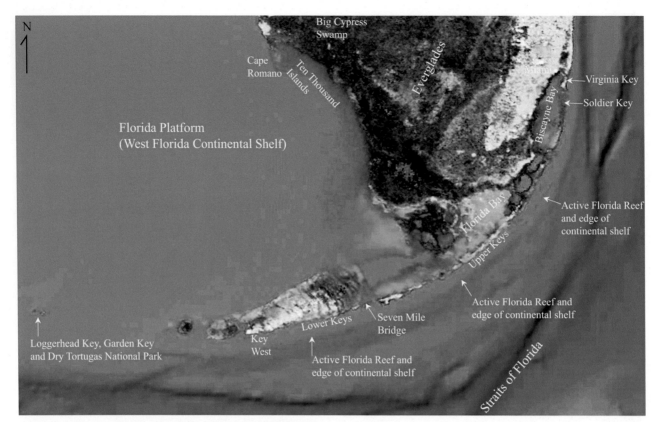

FIGURE 13.16 Google Earth image of the Florida Keys oriented north.

The Lower Keys, which includes Key West, are composed of the same Miami Limestone that underlies the Everglades. The Miami Limestone consists of small, rounded limestone granules, known as oolites, along with varying amounts of skeletal material. The limestone was deposited, c. 125,000 years ago, not as a living reef, but on a limestone shoal (sandbar) probably shoreward of a living reef. These rocks also underlie the city of Miami, Virginia Key, and Key Biscayne. The Upper Keys are built on the Key Largo Limestone, which is, in part, the same age as the Miami Limestone. The Key Largo Limestone is a fossil reef that became emergent within the past 100,000 years. The limestone consists of coral, limestone sand, and quartz sand. The Key Largo Limestone, and the many coral species that form the ancient reef, are beautifully displayed at Windley Key Fossil Reef Geological State Park.

Living reefs require warm, shallow water without an abundance of silt. The active Great Florida Reef (or simply Florida Reef) forms an arc a few miles seaward and parallel with the Florida Keys from Soldier Key, 11 mi (23 km) south of Virginia Key in Biscayne Bay National Park, to Garden Key, 68 mi (110 km) west of Key West in Dry Tortugas National Park. The reefs are young, generally less than 7000 years old, and best developed oceanward of the largest keys. They came into existence during the warming phase following the last major glaciation.

Two areas along the west (Gulf) coast of Florida are dominated by marshes and mangrove swamps. One area is south of Cape Romano near the Big Cypress Swamp. The other extends from Tampa northward to Apalachee Bay. As shown in Fig. 13.10, both areas are located along embayed parts of the Florida coast such that they are protected from strong wave action and longshore drift. Large accumulations of sand are absent.

The coastline south of Cape Romano is a labyrinth of mangrove islands and waterways known as the Ten Thousand Islands. Limestone in this area is covered with 10 to 15 feet of mangrove plant matter. A mangrove is a type of plant that can grow in warm, muddy, salty water. In contrast to other parts of Florida that are losing ground to rising sea level, the proliferating mangrove forest is building seaward. The plants have an interlocking root system that traps river mud before it can reach the open ocean. The trapped mud eventually builds above sea level thus adding to the coastline.

The marshland at Apalachee Bay sits on a platform of karsified limestone less than 50 ft above sea level that tilts gently westward due to its location on the west side of the Peninsular arch (Fig. 13.10). Sinkholes are clearly evident in the Google Earth image of the area shown in Fig. 13.17, as are tree-filled islands known as marsh hammocks that form on isolated limestone knobs high enough above the

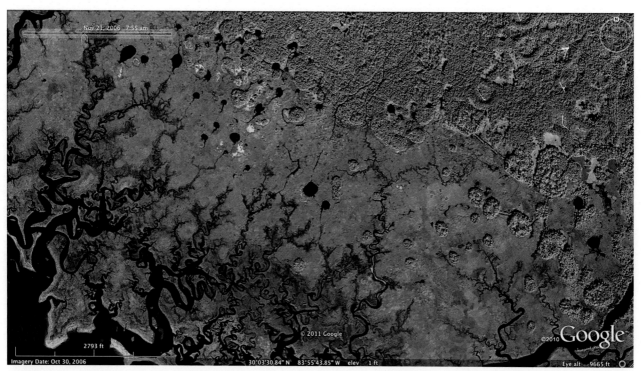

FIGURE 13.17 Google Earth image oriented north and looking vertically downward at marshlands in Apalachee Bay southeast of Tallahassee, Florida. Sinkholes on the limestone platform are evident at upper left-center. Marsh hammocks are at lower right and at upper left-center intermingled with sinkholes. The Econfina River is at lower right. The shoreline is at lower left. The Econfina River is not labeled in Fig. 13.10, but is drawn where it crosses the first e in Tallahassee.

salt water to support trees. The limestone in this area is Eocene in age (56.0–33.9 Ma) and is some of the oldest exposed rock in Florida. It is elevated and exposed due to its location along the crest of the Ocala Uplift. The limestone platform extends as far inland as the Cody Scarp, which forms the most dramatic break in topography anywhere in Florida. The Cody scarp is an eroded cuesta that steps up to a higher bench area upon which Tallahassee and Gainesville are built. The scarp in a few areas has as much as 100 feet of relief. Both cities are more than 150 ft above sea level. The Cody Scarp separates a decidedly hilly, river-dissected upland composed primarily of younger sand, clay, and mud sediment, and poorly consolidated rocks, from the older, lower elevation karsified limestone to the west. The layer of sand/clay/mud rock protects the underlying limestone so that there are fewer sinkholes.

The Cody Scarp formed originally at the edge of the wave-cut limestone platform at a time when sea level was higher. Waves rolled across the platform undercutting and eroding the overlying sand/clay/mud rocks and exposing the underlying limestone. Subsequent river erosion along the scarp face and in highlands east of the scarp has continued to remove the sand/clay/mud rocks thereby exposing limestone and causing the scarp face to retreat further. Sinkholes develop in exposed limestone along the retreating scarp, and with the exception of the Suwannee River, streams disappear into the sinkholes as they cross the scarp only to reappear downstream at a spring.

The Mississippi Embayment

The Mississippi Embayment is the region of the Coastal Plain that swings northward into the mid-section of the United States. It is a superb example of an unconsolidated alluvial (river) landscape shaped by the meandering Mississippi River. Fig. 13.11 shows a wide, flat, swampy, floodplain covered with abandoned stream channels and oxbow lakes. Included in the floodplain are the Atchafalaya, Tensas, Yazoo, and St. Francis Basins. The floodplain is more plainly visible in Fig. 3.13.

Crowley's Ridge is a remnant highland located west of the river within the floodplain that has not yet been worn down. It is underlain with young (<50 million years old) sedimentary rocks and partly covered with glacial-derived, wind-blown silt (loess). Crowley's Ridge is 150 mi long, up to 12 mi wide, and 250 to 550 feet above the floodplain. Other, less obvious ridges on the floodplain consist of old river deposits including braided stream deposits such as Macon Ridge discussed in Chapter 9 and visible in Figs. 3.13, 9.7, and 13.11.

Escarpments define both the eastern and western margins of the Mississippi River floodplain. The eastern escarpment is a bluff that extends almost continuously for 330 miles from Natchez, Mississippi to Gilt Edge, Tennessee. In places, the bluff rises more than 200 ft above the valley floor. Baton Rouge, Natchez, Vicksburg, and Memphis are all located where the river swings eastward against the bluff offering access to the river as well as relative safety above the floodplain. The bluff consists of young sedimentary rocks (<56 Ma) and is known as Bluff Hills at Vicksburg where it too is capped with loess. The well-developed bluff is shown in the vicinity of Vicksburg in Fig. 13.18.

The escarpment at the western edge of the floodplain forms a less prominent bluff that does not overlook the present course of the Mississippi river. Nevertheless, several cities, including Little Rock and Poplar Bluff (P.B. in Fig. 13.11) are built on the escarpment.

The muddy Mississippi River carries an enormous amount of sediment into the Atlantic Ocean, which has the effect of building the Mississippi Delta outward into the Gulf of Mexico in the shape of a birds-foot. The delta continues to build outward because more sediment accumulates than can be removed by wave action. The location of the delta, however, is not permanent and has shifted locations several times over the past 5000 years. The current birds-foot delta has been building for only 400 to 500 years. Prior to its present location, beginning 900 to 1000 years ago, the delta emptied into what is today Breton Sound. Prior to that, the river emptied into the area of salt domes between the present-day delta and Atchafalaya Bay. One can see in Fig. 13.11 that the delta in the Breton Sound area still retains a bird's foot shape partly protected by a barrier island. The older delta in the salt domes area is more thoroughly rounded and reshaped by wave action. The delta region is also an area of subsidence. Table 13.1 shows that sea level is rising at Grand Isle along the coast south of New Orleans at a rate of 35.8 inches per 100 years.

One area of the Mississippi Embayment where bench-and-slope landscape and inland-facing bedrock cuestas are developed is in the Black Belt region of Alabama and

FIGURE 13.18 Google Earth image looking north-northeast at Bluff Hills, Vicksburg, Mississippi.

Mississippi. This region is shown in its entirety in Fig. 13.19. Individual rock units in the Black Belt exert control on landscape because the layers are relatively thick and because they show strong contrasts in erodability. Here, the inner margin of the Coastal Plain (the Fall Line) swings northward around the southern end of the Piedmont Plateau to form a boundary with the Appalachian and Interior Low Plateaus. The Fall Line, which is distinctive along the Atlantic coast, no longer forms an escarpment that separates crystalline from sedimentary rock. With sedimentary rock on both sides, the boundary between the Coastal Plain and the Appalachian and Interior Low Plateaus becomes indistinct and marked with a line of hills underlain with resistant sandstone referred to as the Fall Line Hills. These hills are shown in Fig. 13.19 in the vicinity of Tuscaloosa, Alabama.

Southwest of the Fall Line Hills a weak limestone known as the Selma Chalk forms the Black Belt Valley near Montgomery, a valley once famous for its large mansions and cotton plantations, and still famous for its rich black soil. Beyond the Black Belt Valley lies the distinctive Pontotoc Ridge—Ripley cuesta followed by a series of less distinct bedrock-controlled valleys and cuestas. The Pontotoc Ridge—Ripley cuesta is an eroded, discontinuous set of hills, elevated 50—200 ft above the Black Belt Valley composed of resistant green, glauconite-bearing sandstone.

Salt domes are present at depth below the continental shelves of New England and the Carolinas. However, they are most abundant in the Gulf region both inland on the Coastal Plain and offshore in the Gulf of Mexico. On land, they are particularly abundant on the Mississippi

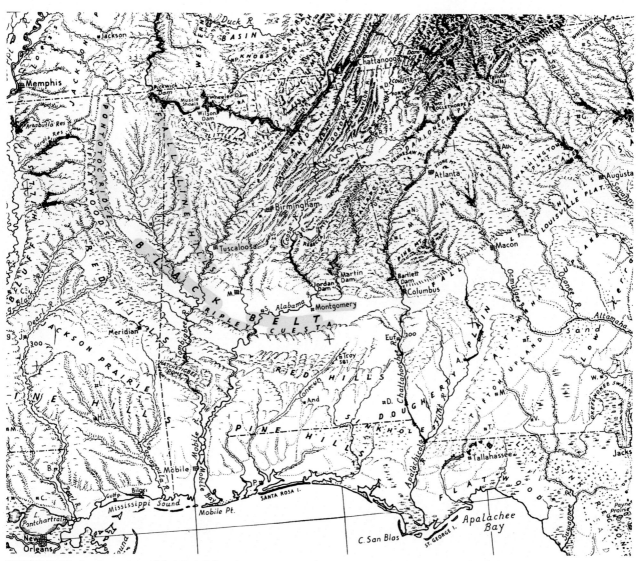

FIGURE 13.19 Landscape map of the Black Belt region.

Delta near New Orleans and Baton Rouge westward to the area surrounding Houston (Fig. 13.11). Thick layers of salt are unstable at depth. Salt is a somewhat unique rock that is low in density, soft, and malleable. It has a tendency to rise buoyantly in the form of a vertical column when under the confining pressure of overlying rocks. As the salt rises, it will fracture, fold, and fault the surrounding and overlying sedimentary layers. The deformation is what creates major oil traps in the Gulf of Mexico. Many of the salt domes along the Gulf coast intersect the surface where they create a circular lake, or a circular upland 1 to 4 miles in diameter and a few feet to 150 ft above surrounding land. Fig. 13.20 shows four of the most visible salt domes on the Louisiana coast. The four domes are located south of New Iberia (N.I. in Fig. 13.11) just north of Atchafalaya Bay.

We now turn our attention to the historical development of the Mississippi Embayment and New Madrid Fault zone. Both have a history that includes an episode of Appalachian—Ouachita Mountain building, and a period of thermal uplift and subsidence associated with a hot spot. The rocks are largely buried beneath unconsolidated sediment; so much of our knowledge is based on seismic and drill hole data.

Faults at depth below the Mississippi Embayment (including the New Madrid fault zone) can be traced to 525 million years ago when a failed rift zone (an aulocogen) known as the Reelfoot Rift produced a large number of extensional normal faults across the region. The faults became inactive and were subsequently covered with sedimentary layers similar to what is depicted in Fig. 4.10. At about 300 million years ago, the area was involved in Appalachian—Ouachita Mountain building events that resulted in reactivation of some of the preexisting faults, the creation of new faults, and the development of a highland. Following uplift, the embayment region eroded and subsided to lowland possibly by 230 million years ago during rifting and the opening of the Atlantic Ocean. The area remained lowland until about 115 to 95 million years ago when it passed over a thermal plume known as the Bermuda hot spot. The addition of heat below the embayment area resulted in thermal isostatic uplift, the

FIGURE 13.20 Google Earth image looking northward at four salt domes along the Louisiana coast south of New Iberia.

reactivation of older preexisting faults, and the creation of mountainous topography of uncertain elevation.

The rocks began to cool by 95 million years ago and erosion coupled with thermal sinking quickly converted the mountains to lowland. An ocean inundated the area by about 80 million years ago and the area has been sinking ever since (Fig. 8.3). Today, the eroded roots of the mountain belt are buried below as much as 8500 feet of younger sedimentary rock, which equates to a slow overall subsidence rate of 0.12 inches per 100 years over an 85-million-year time interval. Constant sinking over a period of tens of millions of years has helped direct rivers into the embayment area, specifically the Mississippi River. During times of high sea level, the Mississippi delta emptied directly into the embayment well north of its present location. The addition of this sediment contributed to the burial of bedrock and development of the floodplain landscape.

Texas

Beyond the swampland of the Mississippi embayment lies another coastal area of nearly continuous barrier islands extending from Galveston to Brownsville, Texas. This area is shown on the landscape map depicted in Fig. 13.12. The barrier islands protect a highly irregular, drowned, coastline consisting of numerous bays similar to, but smaller than, Chesapeake Bay.

Inland, the Texas Coastal Plain, like the Black Belt area, consists of a series tilted layers that become younger toward the coast. Resistant layers produce a series of low-lying, subtle, inland-facing bedrock cuestas that create a mild form of bench-and-slope topography. Several escarpments are highlighted in Fig. 13.12 including the Austin Chalk cuesta, the Bordas—Oakville escarpment, and the White Rock escarpment near Dallas. Also present in this area, although not obvious in Fig. 13.12, are sand terraces that face toward the ocean and which represent old beach deposits. Fig. 13.21 is a cross-section sketch that shows some of the relationships in the San Antonio area.

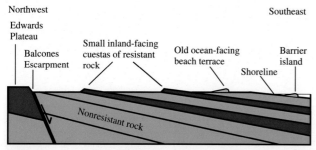

FIGURE 13.21 Cross-section of the Texas coast from San Antonio southeastward showing escarpments and terraces. *Based on Hunt (1974, p. 228).*

The most obvious cuesta is Balcones escarpment between Del Rio and Waco, which separates the Coastal Plain from the Edwards Plateau section of the Great Plains. Unlike many of the bedrock cuestas on the Coastal Plain that face inland away from the ocean, the Balcones escarpment faces the ocean. The escarpment corresponds with a zone of normal faults, active between 25 and 10 million years ago, known as the Balcones fault zone. The Coastal Plain was down-dropped along the fault zone thus creating the escarpment.

In addition to the inactive zone of normal faults that form Balcones escarpment, a zone of active normal faults, downthrown toward the Gulf coast, is present across the Mississippi, Louisiana, and Texas coastline from New Orleans westward to Corpus Christi. The faults are present for up to 100 miles inland on the Coastal Plain and for up to 40 miles offshore in the Gulf of Mexico. Fortunately, the rocks are weak, poorly consolidated, and break easily under even a small amount of stress. There is no chance that faulting will create an earthquake that can do any damage. The primary effect of faulting is subsidence toward the coast.

According to Table 13.1, sea level is rising rapidly along the Texas coastline. Some of this us due to global sea level rise and to faulting, but there is no doubt that ground subsidence due to the withdrawal of groundwater, oil, and gas has also contributed. Tide-gauge measurements in some areas of Texas have recorded relative sea level rises at rates as high as 90 inches per 100 years due primarily to groundwater withdrawal.

Ancient Shorelines of the Coastal Plain

Although sea level has been rising for the past 18,000 years, sea level was higher than present day throughout most of the recent past. The highest sea level stands are recognized at about 125,000 years ago during the most recent interglacial phase, between 5 and 3 million years ago just prior to the beginning of glaciation, approximately 15 million years ago during the mid-Miocene climatic optimum, and 51 million years ago during the Early Eocene climatic optimum (Fig. 8.4).

Sea level high stands on the Coastal Plain have left behind ancient shoreline deposits composed primarily of beach and barrier island sediment. These ancient deposits are separated from younger sediment deposited during lower shoreline elevations by escarpments that represent the edge of the ancient deposit much like the front of the barrier islands we see today. The escarpment is in the form of a rounded, inconspicuous ridge composed of unconsolidated sediment. Such ridges on the Coastal Plain are referred to as scarps. The scarps form a series of stair-stepping escarpments that formed as sea level progressively dropped. Each shoreline scarp faces oceanward.

A schematic example is labeled beach terrace in Fig. 13.21. Shoreline scarps are depositional features that are preserved where sea level was once higher than present-day. The two primary characteristics of the Coastal Plain, an abundance of sand and a gently sloping shelf, are conducive to the preservation of shoreline scarp features.

The inland Coastal Plain, particularly from Virginia to Georgia and along the Texas Gulf coast, steps downward toward the coast across a series of as many as seven scarps that represent shorelines that existed less than 6 million years ago. The physiography of the Coastal Plain changes slightly across each of these scarps. Two scarps are particularly well developed and are located in Figs. 13.9 and 13.10. The Trail Ridge-Orangeburg-Coates-Broad Rock scarp extends discontinuously from Richmond, Virginia, close to the Fall Line, to northern Florida. Relief across the scarp is more than 100 ft in some areas. Sections of this scarp have been recognized as far north as New Jersey and as far south as southern Florida. The scarp represents the frontal edge (the face) of sand ridges and sand dunes that developed along a beach between 6 and 1 million years ago. The scarp may have formed during the short-lived, high sea level stand that occurred just prior to glaciation between 3 and 5 million years ago. In the Carolinas, this scarp forms a terrace that separates the Carolina Sand Hills section of the Coastal Plain from lower, flatter sections closer to the ocean. The transition across the scarp is best seen in the vicinity of Columbia, South Carolina (Fig. 13.9). The Sand Hills themselves are a strip of beach sand dunes that began to form against the Fall Line possibly as early as 35 million years ago.

The Suffolk scarp extends along the eastern side of the Coastal Plain across Virginia and part of North Carolina. This scarp represents the front of an ancient barrier island, much like present-day Daytona Beach (Fig. 13.6). Correlative sediments extend as far south as southern Florida where they have been dated at 110,000 years old. The ancient barrier island probably was in existence during the the high sea level stand from 128,000 to 118,000 years ago. Relief across the scarp is on the order of tens of feet. The Suffolk scarp forms an easily recognized terrace that separates present-day barrier islands and back-barrier tidal flats and marshes to the east, such as the area south of Norfolk shown in Fig. 13.9, including Dismal Swamp and the area surrounding Albemarle and Pamlico Sound, from circa 128,000 to 110,000-year-old tidal flats and marshes to the west that are now elevated 40–50 ft above sea level. Fig. 13.22 is a Google Earth image looking northward at the Suffolk Scarp along the west side of Dismal Swamp. The city of Suffolk, Virginia is built on the scarp.

The various shoreline scarps provide evidence of differential uplift across the Coastal Plain. Differential uplift implies greater uplift (or subsidence) at one location relative to another. The scarps represent ancient shorelines and as such, individual scarp lines of the same age should have formed at the same elevation (at sea level). Regardless of sea level change, if a scarp line presently varies in elevation, then we can assume that there has

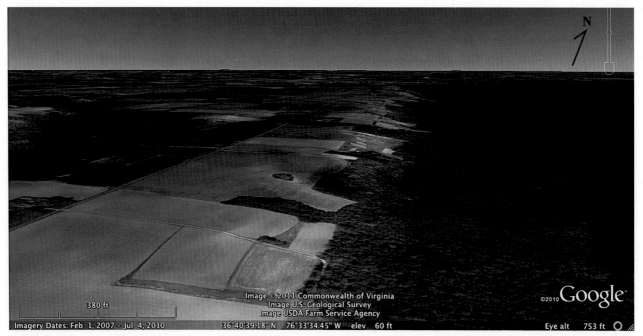

FIGURE 13.22 Google Earth image looking approximately north along the Suffolk scarp. The scarp represents the front of a 110,000 year-old barrier island. The higher elevation to the west is the 110,000-year-old tidal flat. The low-lying, present-day tidal flat to the east is Dismal Swamp.

been vertical displacement (uplift or subsidence) of land along its length. The Trail Ridge-Orangeburg-Coates-Broad Rock scarp varies in elevation by as much as 125 ft. The only explanation is that the land surface must have uplifted or subsided in different areas along the scarp line to produce this relative difference in elevation. The deformation is consistent with broad folding in the vicinity of the Cape Fear arch within the past 6 million years.

Differences in elevation along the Suffolk scarp of 10 to 15 feet suggests that broad deformation of the Coastal Plain has been active in the past 110,000 years. The deformation and uplift is consistent with formation of the scalloped coastline described earlier between Cape Lookout and Cape Romain (Fig. 13.9). A maximum of 15 feet of vertical displacement over the past 110,000 years amounts to an average rate of only 0.16 inches per 100 years. Deformation here is clearly a very slow process. Presumably, it is associated with continued development of the Cape Fear arch.

THE WESTERN MARGIN OF NEARLY FLAT-LYING SEDIMENTARY LAYERS

Areas of the Great Plains, Middle-Southern Rocky Mountains, and Colorado Plateau constitute the western margin of the nearly flat-lying sedimentary layers structural province in the United States (Fig. 11.1). These areas also form the western limit where crystalline shield and interior platform rock successions are widely exposed (Fig. 11.8). The landscape is more complex than other areas of nearly flat-lying sedimentary rock due to the presence of numerous additional structures that include domal mountains, intrusive domes, synclinal down-folds, faults, and volcanic landforms. Fig. 13.23 is an image of the west-central US oriented due north that extends from the Canadian border to the Mexican border. The image is striking because it shows a large, elevated dome structure in central Colorado surrounded on all sides by nearly flat-lying sedimentary rock.

Fig. 13.24 is a landscape map of the region that shows the distribution of structural provinces along with escarpments and lineaments. Nearly flat-lying sedimentary rocks of the Great Plains, Rocky Mountains, and Colorado Plateau (C) are shown in yellow. The area of Wyoming in the Rocky Mountains where nearly flat-lying layers are widespread is known as the Wyoming Basin (W). Note that nearly flat-lying sedimentary layers can be traced through mountain gaps from the Great Plains across the Wyoming Basin (W) to the Colorado Plateau (C).

Other structural provinces in Fig. 13.24 are the crystalline-cored, mid-continent anticlines and domes, normal fault systems, and young volcanic rocks. The

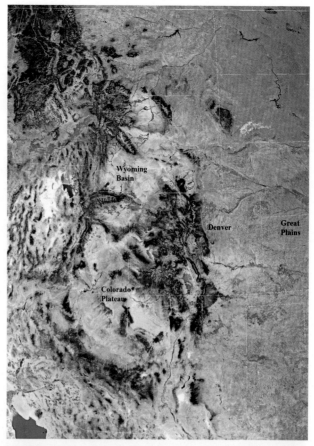

FIGURE 13.23 Google Earth image of the Rocky Mountain region. State boundaries are shown with faint white lines. The Canadian and Mexican borders are shown with faint yellow lines. The figure shows the massive bulge of the Colorado Rockies surrounded by nearly flat-lying sedimentary rocks of the Colorado Plateau and Wyoming Basin.

crystalline dome structure in central Colorado is obvious. Additional crystalline-cored anticlines and domes extend as a discontinuous belt from the Colorado Plateau through the Wyoming Basin to the northwest margin of the Great Plains in Montana. The domes are cored either with ancient crystalline shield rock or with young intrusive rock (c. 20–70 Ma). A few are cored with sedimentary rock. An area of active normal faults in central New Mexico, known as the Rio Grande rift system, separates the Colorado Plateau from the Great Plains. The rift system is outlined with thick lines in Fig. 13.24, which show that it extends northward into the heart of the Colorado Rockies. At least some of the high elevation in the Colorado Rockies can be attributed to normal faulting associated with the Rio Grande Rift system. Young volcanic rocks are widespread, particularly along the margins of the Colorado Plateau and at the western margin of the Great Plains where they are associated with the Jemez Lineament (J). In this chapter, we concentrate our effort on landscape in nearly flat-lying sedimentary rocks.

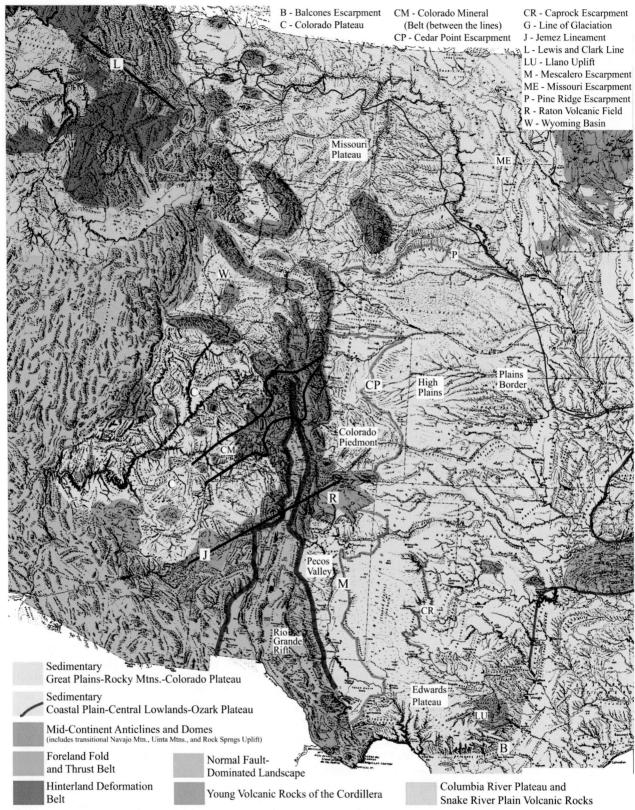

FIGURE 13.24 Regional landscape map showing divisions of the Great Plains, the extent of nearly flat-lying sedimentary layers, the southern limit of continental glaciation (G), and the distribution of other structural provinces and lineaments.

We noted previously that the Western Interior Seaway covered nearly this entire area 80 million years ago (Fig. 8.3). We also know, based on the age of some of the deformed layers, that much of the deformation that initially raised the landscape above sea level took place during Laramide orogeny between 75 and 45 million years ago. This is the orogeny that created the anticlinal structures of the Middle-Southern Rockies and the monoclines of the Colorado Plateau.

The presence of a deformational phase between 75 and 45 million years ago suggests that landscape is no older than 75 to 45 million years. In Chapter 14, we will provide evidence to suggest that landscape across the region is actually much younger. We will suggest that as late as 6 million years ago the entire region was part of an expanded Great Plains. In other words, a drive from Denver to Salt Lake City 6 million years ago would have crossed landscape that more closely resembled the Great Plains than the Rocky Mountains.

Beginning between 10 and 6 million years ago, the entire region was elevated into a broad dome structure centered among the high mountain peaks of Colorado. Evidence discussed in Chapter 14 suggests that uplift at the crest of the dome in central Colorado was on the order of 1640 to 3280 ft (500–1000 m). This was a broad uplift without a lot of visible deformation to the rocks. Flat-lying sedimentary layers were tilted slightly but remained nearly flat-lying. It was this broad domal uplift of previously deformed rock, and its associated erosion, that created the mountains and canyons we see today. This is not to say that mountains and high elevation didn't exist prior to 6 million years ago because there is fossil and oxygen isotope evidence to support such a claim. It simply means that the mountains have been elevated and that overall relief has increased in the past 6 million years. Even if we assume a maximum of 3200 feet of broad domal uplift, that still leaves room in the Colorado Rockies for the existence 11,000-foot mountains prior to 6 million years ago. We next discuss the Great Plains, Wyoming Basin, and Colorado Plateau in sequence.

THE GREAT PLAINS

The Great Plains form a vast area, originally of grassland, that rises gradually toward the Rocky Mountains. The average gradient across a distance of 335 miles from Salina, Kansas at 1240 feet (378 m), to the edge of the High Plains 68 miles east of Denver near Limon, Colorado at 5950 feet (1814 m), is 14 ft/mi (2.67 m/km). Maximum elevation excluding mountains and mesas is nearly 7000 ft between Denver and Colorado Springs. It is the rise in elevation coupled with the near absence of glaciation that separates the Great Plains from the Central Lowlands. The northern part of the Great Plains was glaciated. The line of glaciation (G), shown in Fig. 13.24, follows just south and west of the Missouri River. Fig. 13.24 separates the Great Plains into six subprovinces, two large areas, the High Plains and Missouri Plateau, and five smaller areas, the Colorado Piedmont, Pecos Valley, Edwards Plateau, Plains Border, and Raton (R). The Raton region is volcanic and is discussed in Chapter 17. Several of the subprovinces are separated by escarpments labeled M, CP. and P. Each subprovince is discussed below. The boundary with the Coastal Plain and Central Lowlands is variably an escarpment, a glacial boundary, or is poorly defined. Towns that approximately mark the boundary with the Coastal Plain are Del Rio, San Antonio, Austin, Waco, and Fort Worth, Texas. Towns that approximately mark the boundary with the Central Lowlands are Abilene, Texas; Hutchinson and Salina Kansas; Lincoln and Norfolk Nebraska, Aberdeen, South Dakota; and Jamestown and Minot North Dakota.

The sedimentary rocks that form the surface of the Great Plains are, for the most part, younger than 100 million years. Many of the oldest rock units were deposited at the bottom of the Western Interior Seaway circa 80 million years ago (Fig. 8.3). Younger rock units were derived chiefly from erosion of the Rocky Mountains. These rocks form a wedge across the Great Plains that tapers toward the Central Lowlands. Our discussion will make use of Figs. 13.25–13.28, which are overlapping Raisz landform maps of the entire Great Plains with various features highlighted.

The Missouri Plateau

The Missouri Plateau forms the part of the Great Plains that lies north of the Pine Ridge escarpment (P, in Fig. 13.24) and west of the Missouri escarpment, which forms the eastern slope of the Coteau du Missouri. There is no sharp border with the Wyoming Basin. Most of the nearly flat-lying sedimentary rocks are between 30 and 100 million years old. As seen in Figs. 13.25 and 13.26, the glacial boundary extends across the area, but rather than glacial landscape or one dominated by flat ground, we see a region that is strongly dissected by rivers. Major rivers include the Milk, Musselshell, Missouri, Yellowstone, and Little Missouri. As noted in Chapter 12, the level of dissection is due partly to rivers reoccupying and exhuming older preglacial river valleys.

There are several areas of badlands on the Missouri Plateau including Makoshika State Park near Glendive in eastern Montana, Theodore Roosevelt National Park along the Little Missouri River in western North Dakota, and Toadstool Geologic Park in the Oglala National Grassland near the Pine Ridge escarpment in northwestern Nebraska. The term badlands is used for steep, rough, sparsely vegetated terrain that has been extensively

FIGURE 13.25 Landscape map of the northwest Great Plains. The line labeled G signifies the southern limit of continental glaciation. CP-Cedar Point Escarpment, Pine Ridge Escarpment.

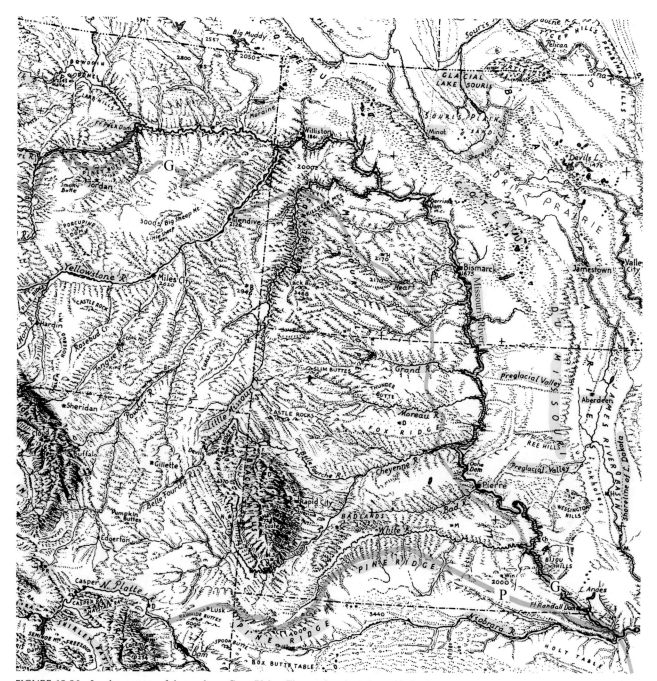

FIGURE 13.26 Landscape map of the northeast Great Plains. The province boundary with the Central Lowlands extends along the eastern base of the Coteau Du Missouri. The line labeled G signifies the southern limit of continental glaciation.

gullied by the carving action of small streams and rivulets. Badlands are found in soft, easily carved rock in arid climates. They are referred to as badlands because cultivation is nearly impossible.

The best-known area is Badlands National Park along the White River a short distance east of the Black Hills in Fig. 13.26. These badlands form within fine-grained, circa 37 to 24-million-year-old volcaniclastic mudstone and claystone (volcanic rocks reworked by streams) of the White River Group. The rock unit not only is famous for hosting badlands topography, but also for its diverse set of mammal fossils that include species similar to present-day rhinoceroses, hippopotamuses, horses, pigs, deer, camels, dogs, and cats. These same rocks form Toadstool Geologic Park in Nebraska.

FIGURE 13.27 Landscape map of the central Great Plains. Thick blue-gray lines are province boundaries. The limit of continental glaciation (labeled G) coincides with the province boundary. The Cedar Point (CP), Mescalero (M) and Pine Ridge (P) escarpments are shown in green.

FIGURE 13.28 Landscape map of the southern Great Plains. Thick blue-gray lines are province boundaries. The Mescalero (M) and Cedar Point (CP) escarpments are shown in green. The Caprock escarpment forms a province boundary with the Central Lowlands. The High Plains extend southward approximately to Odessa and Big Spring.

The High Plains

The High Plains form a young (<17.5 Ma), little dissected, gently east-tilted, flat plateau surface across a large part of the Great Plains from South Dakota to Texas. It is the area that comes closest to being the expansive flat surface that everybody dreams about when they think of the Great Plains. The total extent of the High Plains corresponds closely with the total surface area of an important unconsolidated to semiconsolidated rock unit known as the Ogallala Formation. The Mescalero (M) and Cedar Point (CP) escarpments mark the western boundary of both the High Plains and the Ogallala Formation; the Pine Ridge escarpment (P) defines the northern boundary (Figs. 13.24, 13.27, and 13.28). The eastern limit of the Ogallala Formation separates the Plains Border subprovince from the High Plains. Here, the Ogallala Formation is eroded back (westward) along river valleys to produce a highly irregular High Plains boundary. Farther south, the Caprock escarpment forms a sharp boundary with the Central Lowlands. The southern boundary with the Edwards Plateau subprovince lies between the towns of Odessa and Big Spring, which coincides with the southern termination of the Ogallala Formation. The total present-day extent of the Ogallala Formation is shown in Fig. 13.29 and this can be compared with that of the High Plains in Fig. 13.24.

The Ogallala Formation consists of sand, silt, and gravel transported eastward from the Rocky Mountains via rivers and wind. It is between 17.5 and 5 million years old and was deposited to a maximum thickness of nearly 900 feet. The time of Ogallala deposition was one of stability across the Great Plains. Sediment of the Ogallala Formation filled pre-Ogallala stream valleys and then spread out as an apron across the Great Plains.

Most of Ogallala Formation is unconsolidated except for the top surface layer where sand has been cemented by calcium carbonate into a hard, nearly impermeable rock layer 10 to 30 ft thick. A rock layer of this type is referred to as caliche. It is known as the Caprock layer across most of the region and will be referred to as such in this book. In Kansas, the caliche layer is known as the Algal Limestone. The Caprock layer forms the hard, flat, gently tilted surface that defines the High Plains. This layer protects the underlying unconsolidated sediment from erosion.

Prior to erosion, the Ogallala Formation was more widespread than shown in Fig. 13.29. It extended farther east and probably farther north, but more importantly, it extended farther west, not only to the base of the Rocky Mountains but across the Front Range itself where a few isolated remnant exposures are preserved. The presence of these sediments in the Front Range is evidence that the Great Plains once extended across the mountain landscape. Today, the only location where the Ogallala Formation extends continuously to the foot of the Rocky Mountains is just across the Colorado border near Cheyenne, Wyoming. This part of the High Plains is known as the Cheyenne Table (Fig. 13.27). Here, the High Plains forms a ramp, known as the Gangplank, that connects the Great Plains with the summit area of the Front Range. It is further evidence that the Great Plains once extended across the Middle-Southern Rocky

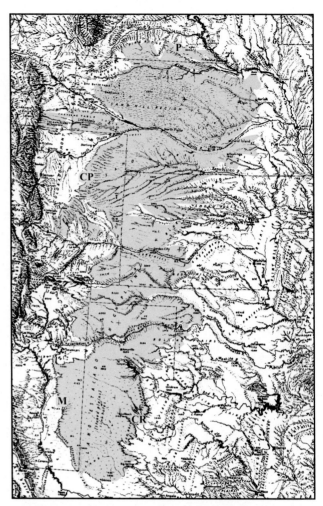

FIGURE 13.29 Regional landscape map showing present-day surface extent of the Ogallala Formation on the Great Plains. Labeled escarpments are CP—Cedar Point, CR—Caprock, M—Mescalero, and P—Pine Ridge.

Mountains prior to broad, domal uplift. The significance of the Gangplank is discussed in Chapter 14.

The High Plains are broadly divided into three areas, the Sand Hill region north of the Platte River in Nebraska, the Central High Plains between the Platte River and Canadian River in Texas, and the Llano Estacado south of the Canadian River (Figs. 13.24, 13.27, and 13.28). These areas are described below.

The Nebraska Sand Hill Region and Ogallala Aquifer

Some people assume that Nebraska is a flat plain across which those of us from the east must travel in order to get to the mountains of the west. However, from a landscape point of view, about 30% of Nebraska is unique. The entire north-central part of the state, as much as 23,600 square miles (61,124 km^2), is covered with the largest sand dune region in the western hemisphere. Some of the dunes are almost 400 feet high and 20 miles long. These dunes are host to one of the largest native grassland regions in North America.

As shown in Fig. 13.26, the Sand Hill region lies east of the Wyoming Rocky Mountains between the Niabrara and Platte Rivers. Dunes develop during periods of drought and are stabilized by vegetation during wet periods. In this area, the protective Ogallala caprock layer has been removed by erosion, thus exposing unconsolidated sand underneath. Wind pushed and shaped the unconsolidated sand into dunes, and at the same time picked up the finer-grained silt and blew it eastward and southward to accumulate on the loess plains of western Nebraska, Kansas, and Iowa. Some of these loess plains are visible in Fig. 13.26 in the vicinity of Hastings. Major episodes of dune formation have occurred in only the past 13,000 years. Dunes have shifted within the past 1000 years. Today, special measures are in place to prevent excessive shifting so that the region does not revert to a Sahara Desert-like landscape. Fig. 13.30 looks east across stabilized dunes near Hyannis, Nebraska.

The Sand Hill region forms an obvious area of sand dunes, but it is not the only dune sand on the Great Plains. Dune sand, generally younger than the Ogallala Formation, covers large areas of eastern Colorado and southwestern Kansas. These sandy areas, including the Sand Hill region, represent the principal recharge areas of the Ogallala aquifer, the largest aquifer by far in the High Plains aquifer system and one of the largest in the world. The Ogallala aquifer forms a giant underground water storage area nearly the size of Lake Huron that stretches the length of the Ogallala Formation from South Dakota to Texas. It is the primary source of water for much of the High Plains region and it supports as much as one-fifth of the wheat, corn, and cattle produced in the United States. Its significance and importance cannot be overstated because, without this source of water to provide irrigation, much of the land on the High Plains would not be suitable for farming.

The top surface of an aquifer is known as the water table. Rainwater and snowmelt enters the Ogallala aquifer principally in dune areas and percolates downward at

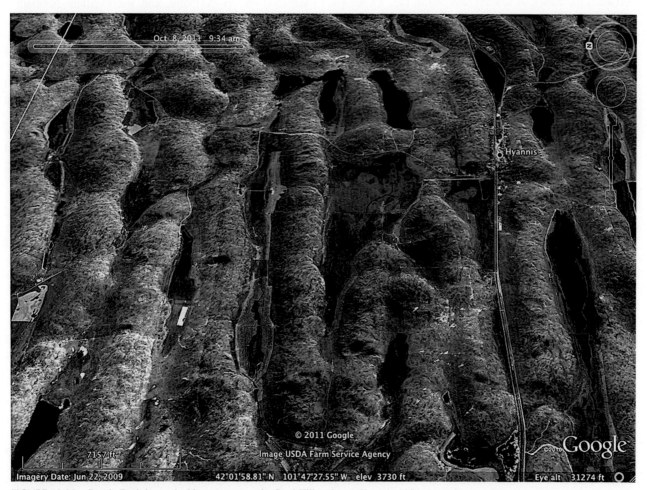

FIGURE 13.30 Google Earth image looking east across the Sand Hill region at Hyannis, Nebraska. Lakes and farm plots are located between the dunes. The white line at upper left is the Grant-Cherry county line.

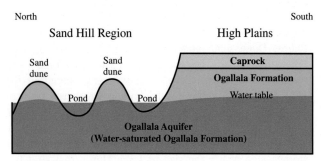

FIGURE 13.31 Cross-section that shows the Ogallala aquifer below the High Plains and the intersection of the aquifer with the ground surface in the Sand Hill region, thus creating ponds.

rates of up to 10 feet/day. The infiltrated water then flows underground generally from west to east at a rate of about 1 foot/day. The thickness of the water-saturated zone varies from more than 1000 feet in Nebraska to less than 100 feet in Texas where recharge is slow. Some interesting features of the Sand Hill region are the hundreds of small ponds and water holes that form in low areas between the dunes. A few are obvious in Fig. 13.30. The ponds exist because the water table (the top of the aquifer) intersects the Earth's surface in low areas between the dunes as shown in Fig. 13.31.

The Central High Plains

The Central High Plains forms the area shown in Fig. 13.27 between the Platte River and Canadian River. Similar with other parts of the High Plains, it forms a flat, gently east-sloping subprovince with several hundred feet of relief primarily along incised river valleys. The Ogallala Formation and younger deposits that include loess and sand dunes underlie the entire region. The western border is sharply defined by what is referred to here as the Cedar Point escarpment (CP), which also forms the eroded edge of the Ogallala Formation (Figs. 13.27 and 13.29). The Arkansas River has its headwaters in the Colorado Rockies. Other rivers that flow eastward across the Central High Plains, including the Republican, Solomon, Smoky Hill, and Cimarron rivers, all have their headwaters at the western edge of the subprovince near the Cedar Point escarpment or in the Raton volcanic field. River dissection, as seen in Figs. 13.27 and 13.29, becomes more intense toward the east such that the Ogallala Formation is eroded from river valleys and the boundary with the Plains Border region is poorly defined.

The Llano Estacado

A glance at Fig. 13.28 revels that there are no rivers south of the Canadian River that cross the southernmost part of the High Plains, the Llano Estacado. The arid climate and lack of major through-going rivers have produced one of the flattest, least dissected treeless areas in the United States. The English translation of Llano Estacado is staked plains. Legend has it that this part of the High Plains is so devoid of landscape markers that early settlers followed stakes set across the region so as not to walk in circles. Fig. 13.32 is a Google Earth image of the Caprock escarpment at the eastern edge of a monotonous expanse of the Llano Estacado. From this image, it is easy to see why this location is known locally as the Break of the Plains. The change in elevation across the escarpment is about 200 feet. At this latitude, the Llano Estacado rises from 2900 feet at the Caprock escarpment to about 4450 feet at the Mescalero escarpment for a gradient of 11 feet/mile (2.1 m/km).

The Colorado Piedmont, Pecos Valley, Plains Border, and Edwards Plateau

The Plains Border, Colorado Piedmont, Pecos Valley, and Edwards Plateau are subregions of the Great Plains that surround the High Plains and the Ogallala Formation. River dissection in the Plains Border subregion east of the Central High Plains has created a series of hills with small east-facing escarpments known in Kansas as the Blue Hills Upland and Smoky Hills (Fig. 13.27). These escarpments step progressively downward to the Central Lowlands exposing a west-dipping sequence of rocks that get progressively older toward the east.

The Colorado Piedmont and Pecos Valley both lie to the west of the High Plains, and both developed as a result of river erosion and removal of the overlying Ogallala Formation. The Colorado Piedmont forms the area in Colorado west of the Cedar Point escarpment (CP) between the Cheyenne Table and the Raton volcanic field (Fig. 13.27). The Arkansas and South Platte Rivers are most responsible for excavating the Colorado Piedmont and creating the Cedar Point escarpment. Elevation drops across the Cedar Point escarpment, but then rises across the Colorado Piedmont to the Rocky Mountains. The piedmont area is more strongly river dissected relative to the High Plains with elevations mostly between 5000 and 6000 feet. The highest elevations are between Denver and Colorado Springs where small mountains top out well above 7000 feet. Denver, the famous mile-high city, is at a somewhat lower official elevation of 5280 feet due to its location on the South Platte River. Rocks exposed in the Colorado Piedmont form a geologic basin (the Denver Basin, Fig. 11.1) filled with sediment derived from erosion of the Rocky Mountains and deposited between about 75 and 40 million years ago.

The Pecos Valley is located in New Mexico and adjacent Texas. It is separated from the Colorado Piedmont by the Raton volcanic field, which reaches elevations

FIGURE 13.32 Google Earth image at Post, Texas, looking west across the Caprock Escarpment at the vast emptiness of the Llano Estacado.

above 9500 feet, and from the Llano Estacado section of the High Plains by the Mescalero escarpment (Fig. 13.28). Elevation drops sharply at the Mescalero escarpment, reaching a low point on the Pecos River and then rising westward. Overall elevation is between 4000 and 6000 feet. We noted earlier that no rivers cross the Llano Estacado. At one time, less than 2.6 million years ago, rivers did flow from the Rocky Mountains across the Llano Estacado, but rapid headward erosion of the Pecos River beheaded those rivers, captured their water, and left the Llano Estacado high and dry. The sequence is depicted in Fig. 13.33. In doing so, the Pecos River removed the Ogallala Formation and cut deeply into highly soluble Permian (299–252 million year old) rock salt and limestone. Dissolution of limestone, not by surface water but by sulfuric acid, has created karst topography that includes the massive Carlsbad Caverns. Sulfuric acid formed when hydrogen-rich water migrated upward from deeply buried oil reservoirs and mixed with groundwater. This unusually strong acid created the Carlsbad Caverns possibly only 4 to 6 million years ago.

In the southernmost part of the Great Plains, the Llano Estacado section of the High Plains merges with the Edwards Plateau in the vicinity of Odessa and Big Spring without an obvious topographic break. The boundary is placed at the southern termination of the Ogallala Formation (Figs. 13.28 and 13.29). Much of the rock on the Edwards Plateau consists of limestone between 145 and 100 million years old. The Edwards Plateau becomes increasingly dissected with entrenched rivers toward the east to the point where erosional mountains are present, such as Riley and Cedar Mountain (see also Fig. 13.5).

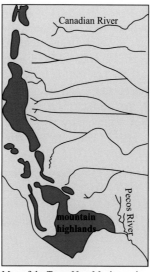

Map of the Texas-New Mexico region prior to about 2 million years ago showing east-flowing rivers crossing the Llano Estacado.

Headward erosion (lengthening) of the Pecos River resulted in capture of east flowing rivers leaving the Llano Estacado dry.

FIGURE 13.33 Headward erosion of the Pecos River, Texas resulting in stream capture. *Based on Thornbury (1965).*

The dissected northeastern part of the plateau is sometimes referred to as the Central Texas uplift.

THE WYOMING BASIN

The Wyoming Basin is truly transitional between the nearly flat-lying sedimentary layers of the Great Plains and the anticline and dome structure of the Middle-

Southern Rocky Mountains. Wyoming is an Indian word that can be translated to several meanings, one of which is large plains. As seen in Fig. 13.24, the Wyoming Basin is an area of wide plains composed of nearly flat-lying layers surrounded on three sides by anticlinal mountains and on the west side by a foreland fold and thrust belt. The nearly flat-lying rocks of the Great Plains extend continuously into and across the Wyoming Basin. In fact, if you were to travel westward through the Wyoming Basin hoping to see the Rocky Mountains, you might be disappointed because it is possible to miss the mountains entirely. The ease with which one could bypass the Rocky Mountains has made the Wyoming Basin a major thoroughfare. The Oregon Trail, the first transcontinental railroad, and Interstate 80, all pass through this region. As shown in Fig. 13.25, the region is not a single basin, but a conglomerate of several basins that include the Green River, Bridger, Washakie, Great Divide, Shirley, Laramie, and Wind River basins, and the outlying Bighorn Basin. Average elevation is between 6000 and 8000 feet, about as high the Colorado Plateau.

The Wyoming Basin, Great Plains, and Colorado Plateau are all underlain with nearly flat-lying sedimentary layers of the interior platform rock succession, but those of the Wyoming Basin are unusual for two reasons. The sedimentary thickness is between 15,000 and 32,000 ft (4573−9756 m), much thicker than is typical of the interior platform rock succession, and the rocks, although nearly flat-lying at the surface, are strongly deformed at depth. The deepest sedimentary rocks were deposited prior to Laramide orogeny. They are relatively thin and typical of the interior platform rock succession except they were folded into synclines and faulted during Laramide orogeny at the same time as development of structures in the surrounding anticlinal mountains. The synclines are complimentary to the anticlines, similar to what is shown in the left-side panel of Fig. 4.2A.

These deformed sedimentary rocks crop out in only a few areas along the margins of surrounding crystalline-cored anticlinal uplifts. For example, beautifully developed folds are exposed along the Bighorn River in the vicinity of Sheep Mountain on the northwest flank of the Bighorn Mountains as seen in Fig. 13.34. Elsewhere, across most of the Wyoming Basin, these deformed rocks are unconformably overlain and hidden beneath weakly deformed and undeformed Late Cretaceous to the Eocene sedimentary layers derived from erosion of surrounding mountains and deposited during the late stages and following Laramide orogeny. These rocks form a continuous surface of nearly flat-lying layers that can be followed continuously from the Great Plains through the Wyoming Basin to the Colorado Plateau.

Sedimentary rocks in some areas of the Wyoming Basin are broadly folded into domes and basins. One of the largest domes is the Rock Springs Uplift centered just east of the town of Rock Springs (Rock Spr in Fig. 13.25). The core of the dome is a topographic basin, known as Baxter Basin, eroded into nonresistant sedimentary rock. The tilted layers surrounding Baxter Basin form bench-and-slope landscape and a series of inward-facing cuestas visible in Fig. 13.25. The Leucite Hills, located just north of the Rock Springs Uplift, are a series of

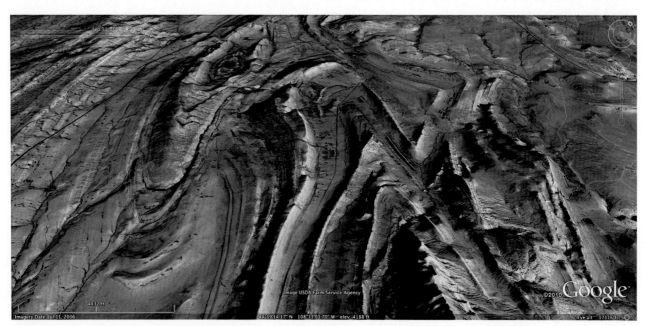

FIGURE 13.34 Google Earth image looking NNW at folds and hogbacks along the northwest margin of the Bighorn Mountains north of Greybull, Wyoming. The northern part of Sheep Mountain anticline is shown at lower right. Note the anticlinal basin at upper left-center.

volcanic flows mostly between 940,000 and 890,000 years old that are now eroded to form isolated mesas and buttes. The rocks contain a variety of relatively unusual minerals including gem minerals.

Most of the basins that surround the Rock Springs Uplift, including the Green River, Bridger, and Washakie, are underlain with nearly flat-lying rocks of Eocene age that cover older, deformed rocks. These are broad structural basins that show considerable topographic relief created by river dissection and the development of outward-facing cuestas. The Washakie Basin, in particular, is nearly surrounded by cuestas up to 600 feet high that include Kinny Rim, Powder Rim, Delaney Rim, and Pine Butte.

East of the Rock Springs Uplift, the Continental Divide passes along the crest of the Sierra Madre and Wind River anticlinal mountains. The Great Divide Basin, host to the Red Desert, lies between these two ranges. It is so named because the Continental Divide bifurcates into a northeastern branch and a southwestern branch, thus enclosing the Great Divide Basin and creating a situation similar with the Great Basin in Nevada where water is trapped with no outlet to the ocean (Fig. 2.2).

Active and dormant sand dunes are present across the region particularly in the Washakie, Great Divide, and Wind River basins. The eastern Wind River Basin is also host to Hell's Half Acre, an area of badlands located just west of Powder River (Fig. 13.25).

UPLIFT OF THE WYOMING BASIN AND NORTHERN GREAT PLAINS

It is well known that the anticlinal mountains of the Middle-Southern Rockies developed initially during Laramide orogeny largely between 75 and 45 million years ago. What is less certain is the timing of uplift of the surrounding Wyoming Basin and adjacent Great Plains. Older rocks in the Wyoming Basin were downfolded into synclines during Laramide orogeny. Yet, today, these basins are at an elevation of around 7000 feet and the adjacent Great Plains is at an elevation between 4000 and 5000 feet. In this section we present evidence for at least two periods of basin uplift, the first between 42 and 37 million years ago at the end of the Laramide orogeny, and the second less than 6 million years ago.

Oxygen isotope data suggests mountains surrounding the Wyoming Basin achieved high elevation primarily between 60 and 50 million years ago (Chapter 14). The Bighorn Mountains and Black Hills, in particular, may have achieved elevation higher than present-day elevation. Conversely, the Wind River Range may have achieved an elevation of only 9200 feet. Surrounding basins, particularly the Wind River and Bighorn Basins, may have been only 1640 feet (500 m) above sea level during this time interval implying substantial basin to mountain peak relief on the order of 11,000 feet across the Bighorn Mountains, and about 7500 feet across the Wind River Range. The studies imply that the Bighorn Mountains and Black Hill have lost elevation (through erosion) since 50 million years ago, while the Wind River Range and basins in the Wyoming Basin and adjacent Great Plains have gained elevation.

A hydrogen isotope study of volcanic glass from the Wyoming Basin and the northern High Plains of western Nebraska found that samples from 36 to 5 million years old all show a gradual increase in hydrogen isotope values from west to east across the region. The recorded increase is similar to the increase seen in modern-day hydrogen isotope values and is interpreted as the drying out of air masses derived from the Gulf of Mexico and Atlantic Ocean as they are lifted over the Middle Rockies. Because a similar gradient is found in all age samples, it was concluded that by 36 million years ago the Wyoming Basin and the adjacent part of the Great Plains had achieved an elevation within 1640 feet (500 m) of present-day elevation and that relief was close to present-day relief. The study correlates uplift with a period of nondeposition across the Wyoming Basin and the western Nebraska Great Plains dated between 42 and 37 million years ago. As with any study of this type, there are uncertainties and assumptions that result in another 1640 feet (500 m) of elevation uncertainty. Thus, by 36 million years ago, relief could have been about the same as today across the Wyoming Basin and the adjacent Great Plains, but elevation was lower by at least 1640 feet and possibly by as much as 3280 feet (1000 m).

Direct evidence for recent uplift and regional doming (within the past 6 million years) includes the eastward tilt of the High Plains across Colorado and Wyoming. By analyzing the present-day tilt of the 17.5 to 5-million-year-old Ogallala Formation, and comparing it with the inferred tilt at the time of deposition, it was found that there has been as much as 2230 feet (680 m) of uplift at the western margin of the Great Plains since the Ogallala Formation was deposited. Approximately 460 feet (140 m) of this uplift has been attributed to erosion-generated isostatic uplift. The remainder (1770 ft, 540 m) must be tectonic uplift. Exactly how long ago tilting began is not well constrained. Tilting could have begun 17.5 million years ago soon after deposition of the Ogallala Formation began, or it could be more recent, perhaps less than 5 million years ago and entirely postdating deposition of the Ogallala Formation.

A study of the deposition history of part of the Ogallala Formation and overlying sediment sheds some light as to when recent tilting of the Great Plains may have occurred. This study focused on sediment deposited from 10 to 2.5 million years ago along the North Platte River from the Wyoming—Nebraska border eastward almost to the confluence with the South Platte River (Fig. 13.27). The data, perhaps not surprisingly, suggest that both tectonics and climate have influenced landscape evolution on the Great Plains.

The study found that sediment from the Ogallala Formation deposited from 10 to 6 million years ago was tilted a minimum of 656 feet (200 m) at the Wyoming—Nebraska border. The Ogallala Formation is primarily a river deposit. When individual stream channels of various ages are compared, they all show approximately the same slope. Such evidence suggests that all of the tilt of the Ogallala Formation occurred following deposition. On the basis of this evidence, tilting, incision, and differential uplift must have occurred after 6 million years ago.

The study also found that there was a period of deposition that occurred from 3.7 to 2.5 million years ago and that these relatively coarse-grained deposits filled and buried previously incised river channels. A history of incision prior to 3.7 million years ago followed by deposition can be correlated with tectonics and climate. Recall from Chapter 8 that the Panama Seaway closed between 6 and 4 million years ago thereby altering ocean currents creating a worldwide cool-down and lower sea level. The lower sea level coupled with tectonic uplift of the Great Plains would likely cause incision and erosion. Recall also from Chapter 8 that the world-wide cool-down was followed by a warm period prior to the beginning of glaciation 2.4 million years ago. On the Great Plains, this warm period was also wet. Such conditions may have caused the switch from incision to deposition and channel filling seen on the Great Plains beginning 3.7 million years ago.

On the basis of the above research, the uplift/deposition/erosion history can be summarized as follows. The Wyoming Basin-northern Great Plains was within 1640 to 3280 feet of present-day elevation by 36 million years ago following Laramide orogeny. Uplift is inferred to have occurred between 42 and 37 million years ago during a period of nondeposition in the Wyoming Basin. Between 36 and 6 million years ago the Wyoming Basin and northern Great Plains underwent periods of deposition that included in the Ogallala Formation. Beginning 6.0 million years ago, deposition was followed by at least 656 feet (200 m) of uplift at the Wyoming—Nebraska border and up to 3280 feet of uplift across the Wyoming Basin and western Great Plains resulting in stream incision and erosion that lasted until 3.7 million years ago when uplift stopped and incised stream channels were filled and buried beneath relatively coarse sediment. Deposition continued until 2.5 million years ago when incision and erosion again became the dominant mechanisms of change.

Sediment deposited between 3.7 and 2.5 million years ago on the northern Great Plains shows no evidence of subsequent tilting or differential uplift. The period of incision beginning 2.5 million years ago, therefore, is likely in response to glaciation and sea level lowering rather than to uplift and tilting. Uplift and tilting since 36 million years ago on the northern Great Plains (and possibly also the Wyoming Basin) is thus bracketed to have occurred between 6 and 3.7 million years ago. This final statement, however, is not entirely consistent with traditional leveling surveys employed over the past 85 years that suggest uplift is active and ongoing. Perhaps uplift recorded in the past 85 years is an isostatic response to stream-channel erosion, or perhaps it is a recent phenomenon that has largely been absent over the greater part of the previous 2.5 million years.

THE COLORADO PLATEAU

Nearly flat-lying sedimentary layers of the interior platform rock succession continue southward from the Wyoming Basin into the Colorado Plateau. Here, in the Four Corners of Colorado, Utah, Arizona, and New Mexico, we enter into one of the most spectacular regions in the world. The Colorado Plateau is home to dozens of national parks, national monuments, state parks, and scenic wonders including the Grand Canyon, Bryce Canyon, Zion, Arches, Canyonlands, and Capital Reef National Parks, the Painted Desert, Monument Valley, and Cedar Breaks. All of these iconic areas are carved into nearly flat-lying sedimentary layers. Most are labeled and highlighted on the landscape map of the region shown in Fig. 13.35. Equally impressive are the exceptionally well-displayed sedimentary structures that include tilted layers, anticlines, and most notably, monoclines. The Colorado Plateau is home to some of the longest and best-developed monoclines in the world. Other features are spectacularly developed as well including crystalline-cored domes, fault and fracture zones, meteor impact structures, and young volcanic landforms.

At an average elevation between 6000 and 7000 feet, the Colorado Plateau is the highest plateau in the country. The general shape is that of a southwesterly tilted topographic bowl with high elevation along its rim and lower elevation near its center. The only through-going river, the Colorado River, enters from the northeast and flows southwestward through the center of the plateau, passing Arches National Park and the Circle Cliffs in Utah, flowing through the Grand Canyon in Arizona, and exiting the plateau at Grand Wash Cliffs near Lake Mead (Fig. 13.35). Tributary rivers include the Green River, which flows southward through Split Mountain and the Book Cliffs to join the Colorado River at the Needles in Canyonlands National Park; the Dirty Devil River, which enters the Colorado River east of the Henry Mountains, the Escalante

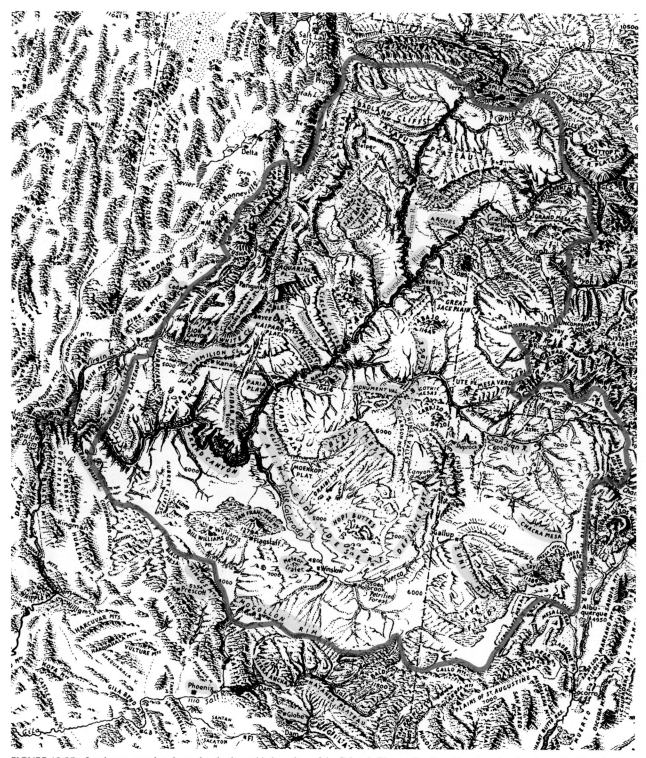

FIGURE 13.35 Landscape map that shows the physiographic boundary of the Colorado Plateau. The Transition Zone lies directly south of the Mogollon Rim.

River, which joins the Colorado between the Circle Cliffs and Straight Cliffs, the San Juan River, which flows westward through northern New Mexico and southern Utah to join the Colorado just north of Navaho Mountain; and the Little Colorado River, which flows northwestward through Arizona to join the Colorado east of Grand Canyon Village. As shown in Fig. 2.3, only two areas of the plateau lie outside the Colorado River system. Rivers in the southeastern corner of the plateau drain to the Rio Grande river system. In the northwestern corner, the Virgin and Sevier Rivers flow in opposite directions off the west flank of the Paunsaugunt Plateau. The Virgin River flows south and west through Zion Canyon to Lake Mead as part of the Colorado River System. The Sevier River flows northward and westward to Sevier Lake as part of the Great Basin river system.

Physiographically, the Colorado Plateau can be divided into a least four areas, the High Plateaus in the northwest, which includes the Paunsaugunt and Aquarius Plateaus; the canyon region along the Colorado and Green Rivers; the Navaho region, which encompasses the entire southeastern part of the plateau including Monument Valley, the Painted Desert, Black Mesa, and Chacra Mesa; and volcanic regions in the south, which include the Hopi Buttes, San Francisco Mountains, and Mt. Taylor (Fig. 13.35). In the following sections, we discuss some of the features that are well displayed in nearly flat-lying sedimentary rocks of the Colorado Plateau.

Incised Meanders

There are two rather amazing features that characterize nearly every tributary of the Colorado River system on the Colorado Plateau. The rivers are deeply incised, and they meander incessantly. Anyone who has visited the Colorado Plateau would agree that it is indeed canyon-country. Rejuvenation, as discussed in Chapter 10, is one process by which rivers can develop these features. Rejuvenation implies that the meandering pattern developed on a flood plain of low relief such as what is seen today along the lower Mississippi River, and that the meander pattern was maintained during deep river incision to form sinuous canyon patterns. River incision can occur due to regional uplift, base level lowering, climate change, or any combination of these. We have already discussed evidence for broad regional uplift within the past 6 million years on the Great Plains and Wyoming Basin. Uplift has also occurred in the past 6 million years on the Colorado Plateau.

Uplift of the Colorado Plateau is intimately tied to the uplift-erosion history of the Middle-Southern Rockies. For this reason, we will hold off on a discussion of uplift history until Chapter 14. In Chapter 14, and in our discussion of the Grand Canyon in Chapter 21, we will discover that the Colorado River did not flow through the Grand Canyon as it does today until at least 6 million years ago. Additionally, we will discover that the Colorado River did not reach the Gulf of California, as it does today, until 5.36 million years ago. Arrival of the Colorado River at the Gulf of California would have lowered base level on the plateau potentially thousands of feet. Coupled with uplift of the plateau, it is likely that the canyon landscape developed largely or entirely within the past 6 million years. Such landscape would include Canyonlands National Park and possibly Arches, Zion, Bryce, and Grand Canyon National Parks. As for the Grand Canyon, in Chapter 21 we will introduce alternative hypotheses in which part or all of the Canyon developed initially as early as 70 million years ago by rivers other than the Colorado River.

Bench-and-Slope Landscape

The Colorado Plateau is host to wide expanses of nearly flat-lying sedimentary layers deformed into both broad, gentle folds and monoclines. Within these folds, there are numerous thick, resistant sandstone and limestone layers that alternate with nonresistant shale-rich layers. The result is well-developed bench-and-slope landscape. The lack of vegetation and absence of thick soil in the dry climate creates landscape in which the edges of benches form angular cliffs of exposed rock. Fig. 13.36 is a schematic cross-section that shows this characteristic feature. Also shown are a mesa and a butte. As a high bench is eroded (stripped) back, sections of the bench become isolated and left behind as erosional leftovers. A mesa is a large erosional leftover. A butte is a small erosional leftover. Both are flat-topped. They represent a remnant of the higher bench surface that once covered the entire area.

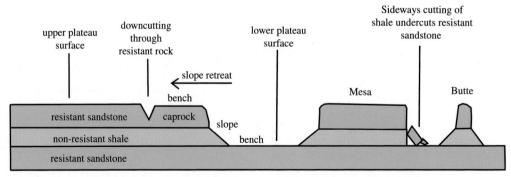

FIGURE 13.36 Schematic cross-section of bench-and-slope topography on the Colorado Plateau showing angular rock walls, a mesa, and a butte.

Two areas where bench-and-slope landscape is well displayed are the Grand Staircase and Monument Valley. The Grand Staircase is a series of five benches that preserve progressively younger rock layers as they step up from the area just north of the Grand Canyon northward to the High Plateaus. The area is part of Grand Staircase-Escalante National Monument. The eroded edge of each bench forms a cliff. Fig. 13.37 looks northwestward from the vicinity of Kanab at all five benches. Three of the most prominent benches, the Vermilion, White, and Pink cliffs, are labeled in Fig. 13.35. Elevation increases from around 4800 ft at the base of the Chocolate Cliffs near Kanab to about 9000 ft above the Pink Cliffs at the southern end of Paunsaugunt Plateau in the High Plateaus. The rock units that form the Grand Staircase range in age from Triassic to Eocene (252—34 million years old). From oldest to youngest, the Triassic Moenkopi Formation and Shinarump Conglomerate form the lowest bench along the Chocolate Cliffs. The Vermilion Cliffs consist of Lower Jurrassic Wingate Sandstone and Kayenta Formation. The Navajo Sandstone, also of Jurassic age, forms the White Cliffs. The Gray Cliffs consist of Upper Cretaceous Straight Cliffs Formation, and the Pink Cliffs are composed dominantly of Eocene Claron Formation. With the exception of the Claron Formation, most of the rock is sandstone with shale, siltstone, and mudstone interlayers. The Claron Formation is mostly limestone. Zion National Park is carved into the White Cliffs. The Pink Cliffs host Bryce Canyon National Park along the east side of Paunsaugunt Plateau. The rock units that form the Grand Staircase also form cliff faces and monoclinal hogback ridges across other parts of the Colorado Plateau.

Fig. 13.38 is a north to south cross-section of the southwestern part of the Colorado Plateau that begins in the Basin and Range north of Cedar City (Fig. 13.35), extends across the Grand Staircase, the Kaibab Plateau, the Grand Canyon, and the San Francisco volcanic field before crossing the Mogollon Rim and ending at Mingus Mountain in the Basin and Range near Prescott, Arizona. Note that the Kaibab Plateau forms a broad anticline known as the Kaibab Uplift (or Kaibab Arch), and that the Grand Canyon cuts across this anticline. Note also

FIGURE 13.37 Google Earth image looking northward at the Grand Staircase and the Paunsaugunt Plateau form near Kanab, Utah. The Chocolate, Vermilion, White, Gray, and Pink Cliffs form the Grand Staircase.

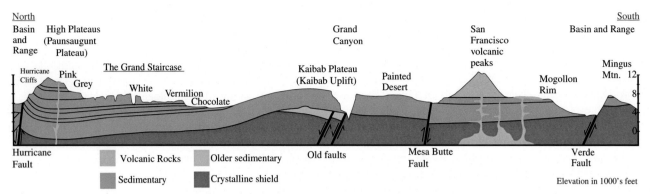

FIGURE 13.38 North to south cross-section from the High Plateaus, across the Grand Staircase, Grand Canyon, and Mogollon Rim to the Basin and Range. *Based on P. Coney and W.J. Breed. Museum of Northern Arizona and the Zion Natural History Association, Zion National Park, Utah, 1975.*

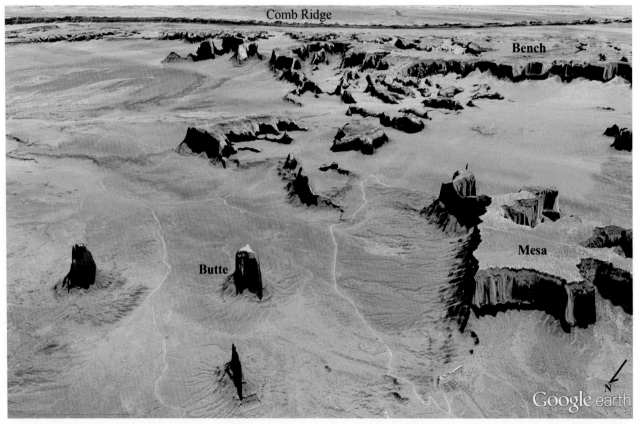

FIGURE 13.39 Google Earth image looking south-southeast at mesas and buttes in Monument Valley, Arizona. Comb Ridge, at top of image, forms a continuous monoclinal hogback ridge for 120 miles. It is part of an eroded monocline that marks the southwest margin of the Monument Upwarp.

that a complimentary syncline underlies the High Plateaus area to the north (left). The Grand Staircase is clearly visible between the Grand Canyon and High Plateaus where progressively younger rock units step up to higher elevation. Given this stair-step arrangement, we can conclude that all of the younger rocks that presently cover the High Plateaus also, at one time, covered the area of the Grand Canyon. These rock units were progressively stripped back to create the Grand Staircase.

Monument Valley is located close to the Four Corners area east of the Grand Staircase. This area, in particular, provides fabulous examples of mesas and buttes. Fig. 13.39 looks south-southeast across Monument Valley in Arizona. Mesas and buttes in the foreground are erosional leftovers of a higher bench that is visible and labeled farther in the distance. Comb Ridge, a monoclinal hogback ridge capped with Navajo sandstone, is visible at the top of the image.

Mogollon Rim

We noted previously that the Colorado Plateau is bowl-shaped with highest elevations along its rim. As seen in the cross-section (Fig. 13.38), the western and southern edges of the plateau form high escarpments that overlook lower topography of the Basin and Range. At Cedar City, the Hurricane Cliffs and Hurricane fault mark the escarpment. Farther south, the Colorado River exits the Grand Canyon by cutting through the Grand Wash Cliffs. Still farther south the Mogollon Rim marks the southern escarpment of the plateau. The prominent Mogollon Rim is obvious in Fig. 13.40, which looks northward across the southwestern margin of the plateau. The Grand Canyon is also visible as are the Grand Wash Cliffs. The Mogollon Rim continues eastward along the southern margin of the plateau where it overlooks a mountainous area of the Basin and Range known as the Transition Zone (Fig. 13.35). This area, which includes Mingus Mountain, Sierra Ancha, and Natanes Plateau, represents an eroded, dissected part of the Colorado Plateau that has been lowered somewhat by normal faults. The Transition Zone, in the area of Prescott and Globe, Arizona, exposes rocks of the ancient North American crystalline shield. Also present are wide areas of basalt mostly less than 10 million years old. Mingus Mountain is capped with young basalt.

Uplifts and Monoclines

Broad folds and monoclines are the primary fold types on the Colorado Plateau. These structures developed between 75 and 45 million years ago at the same time as major crystalline-cored anticlinal uplifts developed in the Middle-Southern Rocky Mountains. Deformation on the Colorado Plateau was less severe than in the Rocky Mountains such that the rocks have remained nearly flat-lying. Many of the largest uplifts on the Colorado Plateau are in the form of wide, elongate domes (known as doubly plunging anticlines) that alternate with wide structural basins. Smaller folds, including anticlines, synclines, and monoclines, are developed within each of the major fold structures, but crystalline rock, in most areas, is not exposed.

In cross-section, the uplifts have the shape of an asymmetric anticline with a gently inclined limb on one side and a monocline on the other. These structures are variably referred to as uplifts, upwarps, or swells. The cross-sectional form is shown in Fig. 13.41. Landscape mimics the structural form in some of the uplifts as shown in Fig. 13.41A. In other uplifts, the center of the dome structure is eroded and breached, creating a central depression with benches and cuestas developed on the gently dipping beds of the uplift, and a hogback along the steep dip of the monocline as shown in Fig. 13.41B. The monoclines

FIGURE 13.40 Google Earth image looking north-northwest at the Mogollon Rim, Arizona, which forms the southwestern rim of the Colorado Plateau. The Transition Zone lies at the southern border of the Colorado Plateau below the Mogollon Rim. The Grand Canyon is visible near the center of the image.

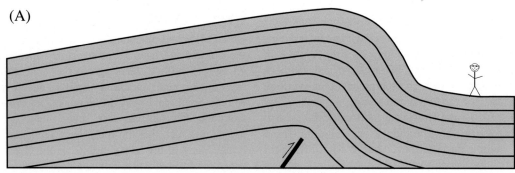

Cross-section of an asymmetric upwarp with a gently inclined limb on left side and a monocline on the right side. Landscape mimics the structural form of the upwarp. A reverse fault typically underlies the monocline, and in some areas pierces the surface.

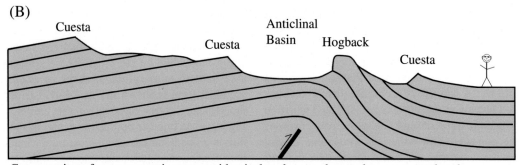

Cross-section of an asymmetric upwarp, identical to the one shown above, except that the center is eroded creating a central depression (anticlinal basin) with benches and cuestas developed in the gently dipping beds and a hogback on the steep dip of the monocline.

FIGURE 13.41 Cross-section sketches of a typical upwarp on the Colorado Plateau, (A) without erosion, and (B) following erosion.

associated with these uplifts are some of the most dramatic structures in the world. Their aggregate length has been estimated at 2500 mi (4023 km). Buried (blind) reverse faults are present at depth below many of the monoclines. Many of these faults originated as normal faults within the North American crystalline shield probably more than 800 million years ago and were later reactivated as reverse faults during Laramide orogeny in a manner similar to that shown Fig. 4.10.

The major uplifts are located and shaded in Fig. 13.42, and the intervening basins are named. The uplifts include the, Kaibab, Echo Cliffs, Circle Cliffs, Miners Mountain, Uncompahgre, Defiance, and Zuni uplifts, the Monument Upwarp, and the San Rafael Swell. Adjacent structural basins include the Uinta, Kaiparowits, San Juan, Piceance, Paradox, and Black Mesa Basins. The White River Uplift is in the Rocky Mountains adjacent to the Colorado Plateau. The Zuni and Uncompahgre Uplifts expose crystalline rock of the North American shield and so are transitional to the crystalline-cored, mid-continent, anticlines and domes structural province. The presence of crystalline rock in these uplifts, as well as at the bottom of the Grand Canyon across the Kaibab Uplift, is because the sedimentary cover rocks, the Interior Platform rock succession, is relatively thin and was removed via eroson.

The Kaibab Uplift (or Kaibab Arch) crosses the deepest and most visited part of the Grand Canyon. The uplift is an asymmetric anticline of the form shown in Fig. 13.41A with a steeply dipping eastern limb marked by the East Kaibab monocline and a gently dipping west limb that, in some areas, includes the smaller West Kaibab monocline near its base. Fig. 13.43 looks northward at the Kaibab uplift from north of the Grand Canyon. In this part of the fold, the resulting landscape mimics the structure of the rock in which the East Kaibab monocline forms a steep monoclinal slope along the eastern flank of the uplift. Landscape across the central crest of the fold is relatively undissected and tilts gently westward with the West Kaibab monocline forming a steep monoclinal slope at the base of the uplift. The uplift is underlain with Permian Kaibab Limestone, the same rock that forms the rim of the Grand Canyon. The Vermilion, White, and Pink Cliffs are visible in the image at upper left. The East Kaibab monocline continues northward for more than 90 miles (150 km) where it is dissected such that landscape resembles the monoclinal hogback ridge

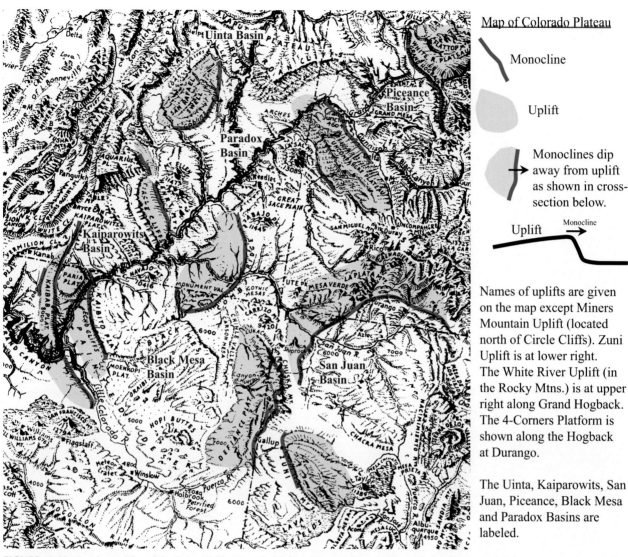

FIGURE 13.42 Landscape Map of the Colorado Plateau and vicinity with monoclines, uplifts, and structural basins labeled. *Based on Kelley (1955) and Davis and Bump (2009).*

shown in Fig. 13.41B. In this area, the monoclinal hogback ridges consist of Cretaceous sandstone that includes the Dakota, Straight Cliffs, and Wahweap Formations. Cottonwood Canyon road follows the hogback ridge for several tens of miles.

Fig. 13.44 looks northwest along the Circle Cliffs Uplift (Ccu) in Utah east of the Paunsaugunt Plateau. The form of the fold is similar to that of the Kaibab Uplift. It is an asymmetric anticline with a gently dipping west limb and a steep eastern limb marked by the Waterpocket Fold (W), an enormous monocline that extends for more than 100 miles (161 km) and which forms part of Capital Reef National Park. Erosion has breached the center of the uplift creating a depression surrounded on three sides by the Circle Cliffs (Cc). Rocks that form the Circle Cliffs include the Triassic-Jurassic Moenkopi Formation, Shinarump Conglomerate, Wingate Sandstone and Kayenta Formation. These are the same rocks that form the Chocolate and Vermilion Cliffs. The bench that forms the Circle Cliffs on the west (left) side of the central depression is dissected with numerous erosional outliers. A section of the Straight Cliffs, not surprisingly composed of the Straight Cliffs Formation, forms a superb bench along the left edge of the image. The Escalante River (Er) in the center of the image, and some of its tributaries, form a meandering pattern that is incised several hundred feet. Fig. 13.45 is a close-up view of the Waterpocket fold that shows the steep dip of the monocline as it passes through the fold. Hogbacks are developed in the Navaho Formation.

The San Rafael Swell, located north of the Circle Cliffs, is another asymmetric, elongate dome with the San

FIGURE 13.43 Google Earth image looking northward at the Kaibab Uplift (Plateau). The location is north of the Grand Canyon looking across the Arizona–Utah border. The East and West Kaibab monoclines are labeled. The Vermilion, White, and Pink Cliffs are visible at upper left.

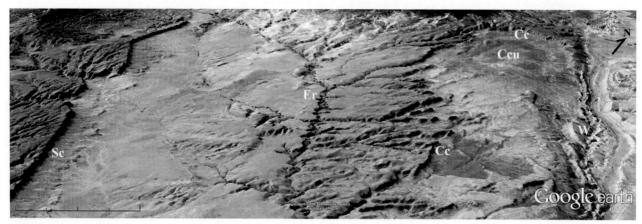

FIGURE 13.44 Google Earth image looking northwest at the Circle Cliffs uplift (Ccu). The steep east-dipping Waterpocket Fold (W) is visible at extreme right. The eroded central part of the uplift (Ccu) and surrounding Circle Cliffs (Cc) are also visible. The Straight Cliffs (Sc) are at left. The Escalante River (Er) shows a deeply incised meandering pattern.

Rafael monocline on its southeast side and gently dipping beds on its northwest side. Fig. 13.46 looks southwestward across the fold structure. Steeply dipping beds of Navaho Formation form a monoclinal hogback ridge along the southeast (left) side. The gently dipping layers on the northwest (right) side form an eroded, dissected bench also partly composed of Navajo Formation. The central part of the fold exposes older rocks disrupted by numerous faults.

One of the largest elongate domes is the Monument Uplift, located in Fig. 13.42 just west of the Four Corners area. The northerly oriented uplift is bordered on its southern and eastern side by Comb Ridge, one of the longest monoclinal hogback ridges on the Colorado Plateau (120 miles). Part of Comb Ridge is visible in Fig. 13.39. The central crest of the Monument Uplift forms a broad, northerly trending, river dissected highland that acts as a drainage divide followed partly by route 261. The southern part of the fold is eroded along its crest to form Monument Valley. The San Juan River, famous for its deeply incised meander patterns known as goosenecks, cuts directly across the central part of the Uplift, and in doing so, cuts through and exposes a beautiful but difficult to access cross-section of the Raplee monocline. This

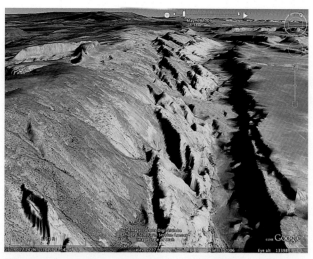

FIGURE 13.45 Google Earth image looking northward along the Waterpocket monocline. Topography partly mimics the structure. The Navajo Formation forms hogbacks. The location is approximately at the W in Fig. 13.44.

Fractures and Impact Features

Strong fracture patterns (joints) within layers of nearly flat-lying sedimentary rock, coupled with the dry climate of the Colorado Plateau, have produced some of the most unusual and picturesque landscape in the world including areas in and around Arches, Zion, and Bryce Canyon National Parks. In this section we discuss these national parks as well as two meteor impact craters, Meteor Crater and Upheaval Dome.

Arches National Park

Arches National Park is located close to the Colorado River near Moab, Utah. Here, we can find the greatest concentration of natural stone arches in the world. There are more than 2000 arches in and around the park, one of which is shown in Fig. 13.48. The Navajo Formation and the overlying Entrada Formation (both sandstone) are the primary rock units in the park. Arches are developed in the Entrada Formation, including one of the longest in the world, Landscape Arch (290 ft).

As shown in Fig. 13.42, the Arches landscape is developed within the Paradox Basin, an area underlain with thick layers of rock salt. Salt layers at depth tend to be unstable because they are less dense than the rock around them, and because they are weak enough to flow when under pressure. The combination of low density and high plasticity causes the salt to rise as a column similar to the igneous intrusion shown in Fig. 4.2A (right side). As the salt rises, the overlying sedimentary rocks are bent, fractured, and faulted into a dome structure. Salt rose very slowly between 300 and 66 million years ago in the area of Arches National Park, and in doing so, bent the overlying Entrada Formation upward to create the Salt Valley anticline. The Entrada Formation, being a brittle rock unit, broke into a series of subparallel, vertical, fractures. Water was then able to percolate downward through the fractures and dissolve some of the salt causing rock layers along the crest of the anticline to slowly collapse along normal faults into a depression to form Salt (anticlinal) Valley.

Fig. 13.49 is a Google Earth image looking northwestward at vertical fractures in the Entrada Formation east of the Salt Valley anticline. The Entrada Formation is dipping gently toward the northeast, away from the Salt Valley anticline. Because of the dry climate and general absence of soil, the best place for vegetation to take root is within the fractures themselves. Weathering and erosion have progressively widened the fractures and, in doing so, have isolated individual vertical walls of rock that are called fins. Several fins can be seen at the top of the image. Arches form where the lower part of a fin is removed by erosion, leaving the upper part intact.

smaller, secondary fold is located on Raplee Ridge along the east side of the Monument Uplift between Comb Ridge and the town of Mexican Hat.

The Echo Cliffs uplift is located just east of the Grand Canyon in close proximity to the Kaibab Uplift. This uplift does not show itself very well because it is breached by the Colorado River. It is significant because it is associated with a long, continuous monoclinal hogback ridge along its northeastern side known as the Echo Cliffs. Fig. 13.47 looks northward along Highway 89 in Arizona at the hogback ridge between The Gap and the Colorado River. Rocks along the ridge dip eastward and are composed of the Jurassic Wingate Sandstone and Kayenta Formation. The Echo Cliffs cross the Colorado River at Lee's Ferry where they form the Vermilion Cliffs as part of the Paria Plateau.

There are two additional monoclinal hogback ridges worth mentioning. The first is the Grand Hogback, northeast of Grand Junction in Colorado. It is a long, continuous monocline that forms the boundary between the White River Uplift (in the Rocky Mountains), and the Piceance Basin on the Colorado Plateau. The second is located in the lower right of Fig. 13.42 near Durango, Colorado. It is known as Basin Mountain or simply as the Hogback. The Zuni and Uncompahgre Uplifts, because they include crystalline rock, are mentioned in Chapter 14.

On the Colorado Plateau, the term, reef, is used for a long, continuous, rocky cliff that acts as a barrier to travel. The classic reef is either a monoclonal hogback ridge or a cliff that defines the edge of a long continuous bench. The Echo Cliffs certainly qualify as do numerous other benches and hogback ridges mentioned above.

FIGURE 13.46 Google Earth image looking southwest at the San Rafael Swell. The well-developed monocline along the southeastern (left) side is partly eroded into hogbacks of Navajo Formation.

FIGURE 13.47 Google Earth image looking northward at a hogback ridge along the Echo Cliffs monocline and Highway 89. The cliffs in the distance are part of the Paria Plateau on the opposite side of the Colorado River.

Sediment and Nearly Flat-Lying Sedimentary Layers Chapter | 13 233

FIGURE 13.48 Photograph of Skyline Arch eroded from a fin, in Arches National Park, Utah.

FIGURE 13.49 Google Earth image looking NNW at a section of Arches National Park, Utah. The Entrada Formation dips gently northeastward (right) away from the crest of the Salt Valley anticline. A strong vertical fracture pattern in the Entrada sandstone is eroded to form fins. Salt Valley (the crest of the anticline) is visible at left.

There are several reasons why arches are so common. The Entrada Formation is not homogeneous sandstone. Instead it is divided into members of different composition. The middle member is resistant sandstone, 200 to 500 feet thick, known as the Slick Rock Sandstone. This member forms most of the fins and the resulting arches. The base of a fin is naturally less resistant to weathering and erosion than the top due to the accumulation of water in shaded areas close to the base. Additionally, the basal member of the Entrada Formation, the Dewey Bridge siltstone/sandstone, is often exposed at the base of a fin and is less resistant than the Slick Rock Member. The combination of excess water and weak rock allows the base of the fin to erode faster than the top. An arch is created and then enlarged as sections of the Slick Rock Sandstone break off. Arches also form higher up within the Slick Rock Sandstone in areas where the rock is strongly fractured or poorly cemented.

In addition to the Salt Valley anticline, there are several other collapsed anticlinal valleys in the Paradox Basin region including the Moab-Spanish Valley where the town of Moab is situated and nearby Castle Valley. The steep western side of Moab Valley is marked by the Moab normal fault. This fault continues northwestward, and is visible in the stream drainage between Highway 191 and the visitor center parking lot at the entrance to Arches. The Moab fault is one of many normal faults in the area.

Zion National Park

Factures have also helped shape Zion National Park. This park is located in the Grand Staircase where the North Fork Virgin River cuts through a bench composed of the Navajo Formation, the 2000+-ft thick sandstone that forms the White Cliffs. The sandstone is fractured in several directions with a dominant joint set oriented just west of north. Fig. 13.50 is a Google Earth image of the canyon looking due north. Notice in this figure that the lower (southern) part of Zion Canyon, near the town of Springdale, forms a valley more than a half mile wide with steep cliffs of Navajo sandstone on both sides. The North Fork Virgin River flows through the valley toward the bottom of the image at a slight angle to the dominant joint set. The valley is wide in this area because the river has managed to cut downward completely through the Navajo sandstone and into relatively weak mudstone and sandstone of the underlying Kayenta Formation. The river is able to meander sideways through the Kayenta Formation, which undercuts the Navajo Sandstone causing collapse of sections of cliffs due to lack of support. Erosion of the Kayenta Formation is aided by groundwater that flows along the contact between the two rock units and seeps out as springs.

Notice in Fig. 13.50 that the North Fork Virgin River and some of its tributaries flow parallel with the fracture set in the upper (northern) part of the canyon. Here, streams have cut downward so swiftly that narrow slot canyons are produced. The canyons are narrow because streams are still in the process of cutting through the Navajo Formation. They have not yet cut into the underlying Kayenta Formation and thus, are trapped by the walls of the Navaho. A famous slot canyon is the Narrows (or Zion Narrows). For 16 miles along the North Fork Virgin River, the canyon is as little as 20 feet wide and bedrock walls on both sides soar vertically upward more than 1000 ft. On a good day one can walk into the Narrows and into one of the most spectacular sights in the world.

Bryce Canyon National Park

Bryce Canyon National Park is a series of amphitheaters (a semicircular area of cliffs) carved by tributaries of the Paria River into the Claron Formation along the east side of the Paunsaugunt Plateau. The Claron Formation forms the Pink Cliffs, the youngest rock unit in the Grand Staircase. It is different than other cliff-forming rock units in the Grand Staircase in that it is predominantly limestone, it is well layered, and it is brightly colored. The lower part is pink, the upper part white, but both parts present patches, streaks, and shades of red, brown, yellow, orange, lavender, and purple. Much of the coloring is due to various amounts of iron oxide and manganese in the rock. Two vertical fracture directions oriented almost at right angles to each other control the landscape along with a nearby fault. Differential erosion along both fracture sets has produced a landscape filled with pinnacles known as hoodoos. As evident in Fig. 13.51, the hoodoos are oddly shaped because some layers of Claron Formation are more resistant to weathering and erosion than other layers.

Meteor Impact Features

Meteor Crater is located in Fig. 13.35 near Flagstaff, Arizona. The crater is young, about 50,000 years old, and still in nearly pristine condition. It is a hole in the ground, 3900 feet (1200 m) in diameter and 570 feet deep (174 m). Rock units are overturned (upside-down) along the rim of the crater, which is elevated 148 feet (45 m) above the surrounding plateau. The center of the crater is filled with rubble left by the impact. The meteorite that created the crater is understood to have been 160 feet (49 m) in diameter and iron-nickel in composition.

Upheaval Dome is an anomaly. This landform is located in Canyonlands National Park between the Green and Colorado Rivers 17 miles north of the confluence of the two rivers. Its location along with a close-up of the

FIGURE 13.50 Google Earth image looking north at Zion Canyon at Springdale, Utah. Note the strong vertical fracture pattern in the White Cliffs of the Navaho Formation. The Narrows are located where the upper part of the North Fork Virgin River flows parallel with the fracture pattern.

impact site and a cross-section are shown in Fig. 13.52. Upheaval Dome is the result of a meteor impact at an uncertain time less than 170 million years ago. The rocks exposed at the surface are not the rocks that were at the surface during the time of impact. The rocks you see today were deeply buried and are now exposed due to erosion. The central part of the impact is a basin about 1 mile in diameter with a small dome at its center.

The pattern results from the presence of fractured, easily eroded rock in the core of an anticlinal structure as depicted in Fig. 10.52C. Not easily seen in Fig. 13.52B is the circular (ring) syncline that completely surrounds the central basin. The two limbs of the syncline form circular cuestas. An inner cuesta (IC) overlooks the central basin. An outer cuesta (OC) looks outward away from the central basin. The inner cuesta can be seen in Fig. 13.52B as a thin sandstone ridge along the lip of the basin. The outer cuesta can also be seen as a sandstone ridge.

SEDIMENTARY-CORED ANTICLINAL AND DOMAL MOUNTAINS

There are several anticlinal mountains in the Colorado Plateau-Rocky Mountain region that are transitional between the nearly flat-lying sedimentary, and the mid-continent anticlines and domes structural provinces. They include the Unita Mountains and Navaho Mountain. The previously discussed Rock Springs Uplift in the Wyoming Basin could be considered a third example.

The Uinta Mountains form a major anticlinal upwarp within the Middle Rocky Mountain physiographic province along the northwestern margin of the Colorado Plateau. In addition to the Ouachita Mountains, it is one of only a few major mountain ranges in the United States that trend east-west. It is the only mountain range in Utah with peaks above 13,000 feet, including Kings Peak, the highest at 13,528 feet (4123 m). There are at least 19 peaks above 13,000 feet.

FIGURE 13.51 Photograph of brightly colored rocks in the Pink Cliffs of Bryce Canyon National Park, Utah.

The Uinta Mountains do not fit well with the nearly flat-lying structural province for two reasons; rocks along the margins of the anticline dip at a relatively high angle, and rocks that form the core of the fold are not part of the interior platform rock succession. The mountains also do not fit well with the mid-continent anticlines and domes structural province because they do not expose crystalline rocks in their core. The rocks that form the core of the fold are circa 975- to 775-million-year-old sedimentary rocks of the Precambrian sedimentary-volcanic rock succession.

The Unita Mountains anticline is large. It runs the length of the mountain (well over 100 miles) and is cut with reverse faults. The primary fold is flat-topped, creating nearly horizontal layers along the summit region (Fig. 18.38). The adjacent Uinta Basin, to the south, forms a large complimentary syncline. The nearly horizontal layers at the summit create a broad mountain landscape heavily scalloped by glaciers into long narrow ridges reminiscent of Glacier National Park except with wide glacial valleys. This rugged landscape transitions into flat-topped mountain peaks surrounded by cirques in the area east of Kings Peak. Rocks are moderately dipping along the flanks of the mountain creating hogbacks and secondary folds including beautifully developed monoclines, anticlines, and synclines at the eastern end of the range. The lower eastern part of the range includes Dinosaur National Monument and several major canyons including Lodore Canyon and Split Mountain Canyon (Fig. 4.18), both of which are breached by the Green River.

The Unita Mountains represent a rare area on the craton and reactivated western craton where interior platform rocks are underlain with Precambrian sedimentary-volcanic rocks rather than directly by rocks of the crystalline shield. Similar with uplifts on the Colorado Plateau, the Unita Mountains were initially uplifted between 65 and 60 million years ago during Laramide orogeny. A second, possibly active episode of uplift may have begun in the past 10 million years.

Navajo Mountain, near the confluence of the San Juan and Colorado Rivers, is a small intrusive dome 10,356 feet high, composed entirely of sedimentary rocks, primarily the Navajo Formation. It is similar to other intrusive domal mountains discussed in Chapter 14 except that erosion has not yet exposed the underlying intrusive rocks.

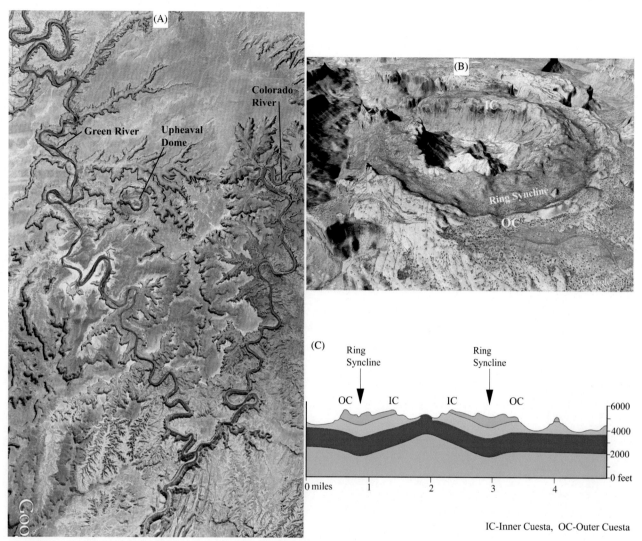

FIGURE 13.52 (A) Google Earth image looking north along the Green and Colorado Rivers at Upheaval Dome. Bowknot Bend (Fig. 10.2B) is seen at upper left. (B) Close-up view of Upheaval Dome looking just west of north. The ring syncline is labeled and completely encircles the impact structure. (C) Cross-section of Upheaval Dome based on Thornbury (1965).

CENTRAL LOWLANDS

The landscape of the Central Lowlands is primarily one of glacial deposition features discussed in Chapter 12. In this section we discuss landscape features that are not glacial in origin. With respect to the geologic time scale (Fig. 1.10) nearly all of the rocks that underlie the Central Lowlands are Cambrian to Pennsylvanian (541–299 Ma) in age. Some of the rocks, therefore, were deposited during Appalachian and Cordilleran orogeny. As shown on the map of structural provinces in Fig. 11.1, these rocks are deformed into broad domes, arches (broad anticlines), and basins. The major structures include the Michigan and Illinois Basins, and the Kankakee, Cincinnati, and Transcontinental Arches, some of which extend into adjacent physiographic provinces. These structures are long-lived. The same rock layers are thicker in basins than they are across arches as shown schematically in Fig. 13.53. This type of relationship suggests that the arches and basins developed at the same time the rock units were being deposited. In other words, these structures developed over the course of hundreds of millions of years during deposition of the interior platform rock succession. They appear to have formed primarily through vertical sagging and arching rather than through horizontal, compressional stress. Slow vertical sagging during deposition would logically result in a thick rock sequence, whereas arching would result in a relatively thin rock sequence or possibly in the development of an

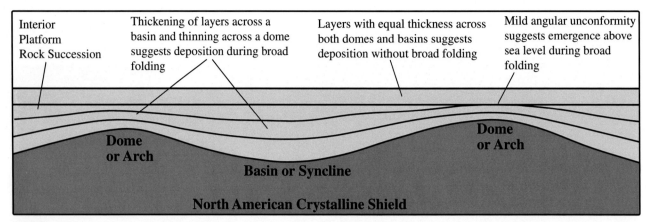

FIGURE 13.53 Deposition and erosion across basins and arches in the Central Lowlands.

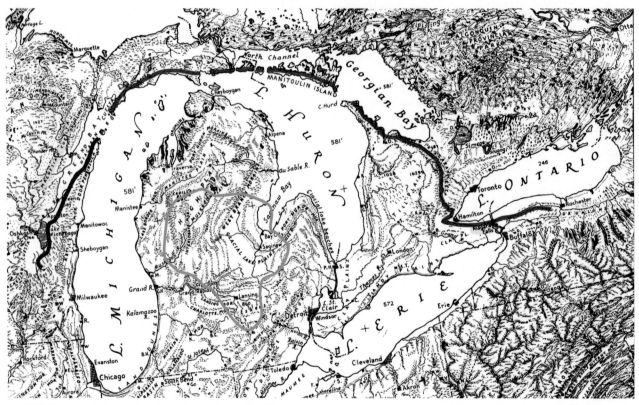

FIGURE 13.54 Landscape map of the Great Lakes region. The orange line encloses the area of youngest rock at the center of the Michigan structural basin (Jurassic and Pennsylvanian age). The blue line locates the Niagara escarpment within Silurian rocks that surround the Michigan Basin. Niagara Falls and the Magnesian cuesta are highlighted in yellow.

unconformity if the arch is elevated above sea level as shown on the right side of Fig. 13.53.

Rocks and structures of the Central Lowlands have little influence on landscape due to burial beneath glacial sediment. A major exception is landscape surrounding the state of Michigan. Rock units associated with the Michigan Basin form a circular bulls-eye pattern. The thick orange line shown in Fig. 13.54 encloses the area of youngest rock at the center of the structural basin (Jurassic and Pennsylvanian age). This line passes through the towns of Jackson, Flint, and Cadillac. The surrounding older rocks form progressively wider circular or semicircular patterns that extend across Lake Michigan and Lake Huron to the contact with rocks of the ancient crystalline shield (the Great Unconformity). The semicircular shape of Lake Michigan and Lake Huron is a direct result of glaciers gouging deep trenches into the circular pattern of soft rocks, particularly of Mississippian, Devonian, and Silurian age. The only obvious bedrock landforms are cuestas of resistant rock layers that face (steep side) outward, away from the center of the Michigan Basin. Three cuestas, the Magnesian, Black River, and Niagara escarpments are developed near Green Bay, Wisconsin. The Magnesian cuesta is developed in Ordovician limestone and is highlighted in yellow in Fig. 13.54. The Black River cuesta, located to the east, is not shown in the figure. The Niagara escarpment (or Niagaran cuesta) is located farther east within Silurian dolomite and is the largest and most continuous of the three escarpments. The trend of the Niagara escarpment follows the curvilinear shape of the Silurian rock unit and is shown with a thick blue line in Fig. 13.54. The escarpment forms a distinctive highland that extends from Door Peninsula in Wisconsin, across the northern shoreline of Lake Michigan to the Manitoulin Islands south of Georgian Bay, and then southeastward to the south side of Lake Ontario where it helps to form Niagara Falls. It then continues eastward to the vicinity of Rochester. It is more than 750 miles long and up to 45 miles wide. It is prominent in several areas including along the east side of Lake Winnebago where it is over 300 feet high and is known as The Ledge.

The Niagara River flows northward to form the US—Canada border from Lake Erie to Lake Ontario, plunging as much as 188 feet at Niagara Falls. There are actually three falls, the large Horseshoe Falls on the Canadian side, and the smaller American and Bridal Veil Falls on the US side. The same resistant Silurian dolomite that forms the Niagara escarpment also forms a bench over which Niagara Falls drops onto weak Ordovician shale. The height of the falls is due in part to the erosional removal of the underlying shale layer at the base of the falls. The falls developed initially at the Niagara escarpment approximately 12,000 years ago following the retreat of glaciers from this area. Erosion of the shale has progressively undercut the overlying Silurian dolomite causing the falls to retreat 7 miles southward (upriver) since its formation, creating Niagara Gorge in its wake. Fig. 13.55 looks southward at the Niagara escarpment, the base of which is shown with a yellow line. Niagara Falls is visible at the top of the image. The upriver retreat of the falls, coupled with development of the Niagara Gorge, is an excellent example of knickpoint retreat as described in Fig. 9.5. The falls continue to retreat but at a much-reduced rate of about 1 foot/year due to diversion and flow control related to hydroelectric power generation.

Comparison of physiographic and glacial maps (Figs. 1.6 and 12.9) shows that the only area not affected by glacial deposition is an arm of the Central Lowlands that lies south of the Kansas and Missouri Rivers. This area,

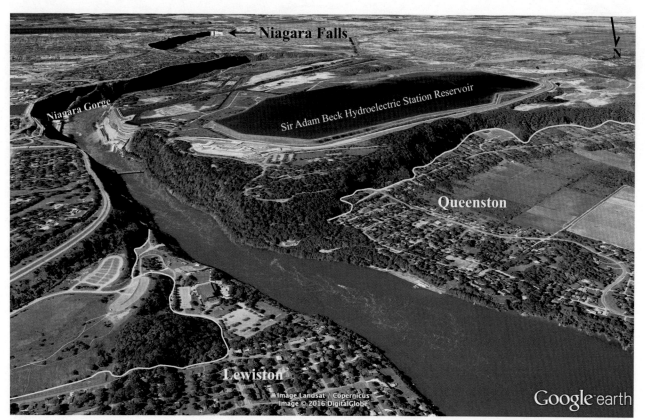

FIGURE 13.55 Google Earth image looking southward at the Niagara escarpment and the Niagara Gorge along the Niagara River. The base of the Niagara escarpment is shown with a yellow line. Niagara Falls is located with an arrow 7 miles up-river at the head of Niagara Gorge. The New York-Canada border follows the river. Lewiston is on the US side.

known as the Osage Plains is shown in Fig. 13.56. The Osage Plains slope eastward from between 2200 and 1200 feet in the west, to about 600 feet in the east. Numerous rivers cross the region from west to east. The regional slope is opposite the regional dip of the rocks, most of which are westward-dipping Pennsylvanian-Permian (323–252 Ma) shale, sandstone, and limestone. In the northern Osage Plains the westward dipping rocks produce

FIGURE 13.56 Landscape map of the Osage Plains section of the Central Lowlands.

a series of small, eroded hills and dissected benches with cuestas that face eastward and step down to older rock toward the east. The most prominent of these are the Flint Hills, Chatauqua Hills, and the Oread, Iola, and Fort Scot Escarpments, all located in southeast Kansas and shown in Fig. 13.56. The Osage Plains south of these escarpments consists of flat and rolling plains partly forested as far west as Oklahoma City and Wichita Falls, and becoming tallgrass prairie farther west. A small surviving remnant of the original prairie can be found at Tallgrass Prairie National Preserve in the Flint Hills of Kansas. Also present are two small mid-continent domes, the Wichita and Arbuckle Mountains, discussed in Chapter 14.

Landscape steps to higher elevation in neighboring provinces on all sides of the Osage Plains except for the southeastern side bordering the Coastal Plain. The border with the Coastal Plain coincides with the city of Fort Worth, Texas where it separates the Western (Upper) Cross Timbers from the Eastern (Lower) Cross Timbers (Fig. 13.56). Both areas are underlain with Cretaceous rocks (145–66 Ma) that unconformably overlie the older Pennsylvanian-Permian rocks. The Western Cross Timbers is the only area on the Osage Plains underlain with Cretaceous rocks. Elevation drops from 653 feet at Fort Worth to 463 feet at Dallas as one crosses onto the Coastal Plain. The Western and Eastern Cross Timbers were both covered with post oak and blackjack oak forest 150 years ago. Today, only a few large, undisturbed tracts remain.

OZARK PLATEAU

The Ozark Plateau is a large domal structure centered in the St. Francois Mountains south of St. Louis where nearly flat-lying sedimentary layers dip gently away from a central crystalline core. In this section we discuss landscape in the sedimentary rocks. The St. Francois crystalline core is discussed in Chapter 14.

The Ozark Plateau is an ancient landscape that has been in existence as a highland for at least the past 265 million years and probably longer. A landscape map of the province is shown in Fig. 13.57. The area lies just south of the southernmost advance of glaciers. The northern boundary of the province along the Missouri and Mississippi Rivers to the vicinity Beckett Mills marks the southern limit of glaciation. The southeastern, southern, and western boundaries are clearly defined against the St. Francois basin, Arkansas River valley, and Cherokee Plains (Osage Plains), respectively.

Rather than uplifted entirely as a large anticlinal bulge, the dome appears to have been uplifted and tilted to the southwest along faults that are exposed along the northeast side of the province and largely hidden along the southeast side. The southwest tilt creates an asymmetric structure, steeper on its east side, which explains why crystalline rocks are exposed on the eastern side of the plateau. Although the overall structure is domal, the sedimentary layers are nearly flat lying. Dip across the dome is typically less than 3° even on the steeper eastern side.

The dome structure is such that younger rocks are exposed with distance from the St. Francois crystalline core. Three age groups of sedimentary rocks form three separate areas of the plateau. Rocks in the Salem Upland (or Salem Plateau) surrounding the St. Francois Mountains are mostly Cambrian and Ordovician (541–444 Ma). Rocks that form the Springfield Plateau are mostly Mississippian (359–323 Ma); and rocks that form the Boston Mountains are mostly Early and Middle Pennsylvanian (323–307 Ma). All of the rocks are part of the Interior Platform with Pennsylvanian rocks of the Boston Mountains derived from erosion of the Appalachian–Ouachita highlands.

The presence of younger rock with distance from the crystalline core creates a series of inward-facing escarpments (facing toward the crystalline core) that separate each of the three areas producing a poorly defined bench-and-slope landscape. There are two major escarpments (cuestas); the Eureka Springs (or Burlington) escarpment separates the Springfield Plateau from the Salem Upland, and the Boston Mountains escarpment separates the Boston Mountains from the Springfield Plateau. Fig. 14.3 is a cross-section that shows both cuestas. Smaller benches are developed within each of the three areas along rock layers of differing resistance. An example is the Crystal escarpment, shown in Fig. 13.57 east of the St. Francois Mountains within the Salem Upland. We will look briefly at each of the three plateau areas and then discuss what is known about the uplift history.

Salem and Springfield Plateaus

The topography of the Salem Upland is one of dissected rolling hills and incised, meandering river valleys up to 400 feet deep. The Springfield Plateau is somewhat flatter but still with incised river valleys up to 300 feet deep. Elevation in both areas is mostly between 1000 and 1500 feet with total relief typically less than 500 feet. The presence of incised, meandering river valleys is reminiscent of rivers on the Colorado Plateau and suggests a phase of recent uplift or base level drop, perhaps associated with glacial advances over the past 2.4 million years. Several of the rivers are dammed along the Missouri–Arkansas border creating large sinuous lakes such as Table Rock Lake, shown in Fig. 13.58 that are ideal for canoeing, fishing, and general recreation.

Limestone and dolostone are the dominant rock types on the Salem and Springfield Plateaus, and given the

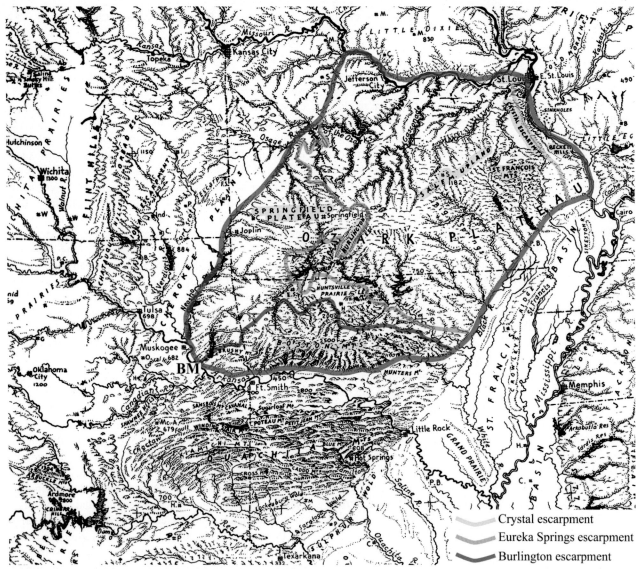

FIGURE 13.57 Landscape map of the Ozark Plateau showing escarpments.

FIGURE 13.58 Google Earth image looking north at Table Rock Lake on the James and White Rivers in southern Missouri just north of the border with Arkansas. Landscape is dissected with incised meanders.

humid climate, karst features are fairly common, including a high concentration of springs. Karst features are, however, less abundant than one might expect. The reason for this is the presence of chert layers and chert nodules within the limestone. Chert is microcrystalline quartz and is immune to dissolution. It weathers out of the limestone and collects in great quantities in soil and along the banks of the many streams in the area. It has been suggested that there is more chert in Missouri than anywhere else in the United States.

The Salem and Springfield Plateaus are separated by the Eureka Springs (or Burlington) escarpment, which is shown with a thick line in Fig. 13.57. A particularly resistant cherty limestone of Mississippian age, known in Arkansas as the Boone Formation, holds up the escarpment and also forms much of the surface area of the Springfield Plateau. Rather than a clear-cut slope, the escarpment forms a sinuous, irregular line of eroded hills along which the cherty limestone has eroded back. Relief across the escarpment is as much as 400 feet near Eureka Springs, Arkansas (E.S., Fig. 13.57) but becomes less, and the cuesta more dissected, toward the north.

Boston Mountains

The Boston Mountains are higher than the Salem and Springfield Plateaus with as much as 1500 feet (457 m) of relief. Topography is that of a dissected plateau with flat-topped summits, steep slopes, and deep, narrow valleys. These are classic erosional mountains. The high point is Buffalo Lookout at 2561 feet (781 m). In contrast to other parts of the Ozark Plateau, the rocks are mostly sandstone and shale. They dip gently southward except along the southern margin of the Boston Mountains where dip steepens as rock strata descends into a syncline that forms the Arkansas River valley and the northern margin of the Ouachita Mountains.

The younger sandstone and shale of the Boston Mountains is separated from the cherty limestone of the Springfield Plateau by the Boston Mountains escarpment, the largest and best defined in the Ozarks with as much as 800 feet of relief. Fig. 13.59 looks west across the highly dissected Springfield Plateau near Mountain View, Arkansas to the topographically higher, flat-topped Boston Mountains. The escarpment is well developed and obvious at this location. Also note the incised channel of the White River at right.

Uplift History

The Ozark dome appears to be a long-lived geologic structure that has been in the process of forming for at least 400 million years and perhaps longer. Similar with some of the basins and arches in the Central Lowlands, the dome structure may have been actively forming during deposition of some of the sedimentary rocks that form the structure. Karst development and the presence of erosion surfaces in some of the older rock units point to periods of emergence circa 345 million years ago. The area emerged permanently above sea level between 307 and 265 million years ago during the final phases of Appalachian–Ouachita orogeny and has been an upland ever since. At that time, the rock layers that presently cover the Boston Mountains and Springfield Plateau probably extended across the entire dome structure. The rock layers have since eroded and retreated to their present

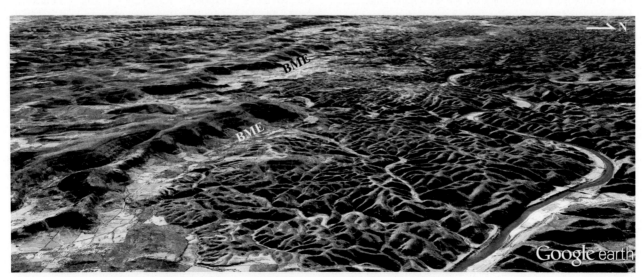

FIGURE 13.59 Google Earth image looking west across the dissected Springfield Plateau near Mountain View, Arkansas at the topographically higher, flat-topped Boston Mountains. BME—Boston Mountains Escarpment. The incised White River is at right.

position creating the Eureka Springs and Boston Mountains escarpments. Overall, the area has slowly grown old or reached quasi-topographic steady-state via climate-driven erosional lowering and deposition punctuated with periods of rejuvenation.

Apatite fission track analysis suggests two significant pulses of exhumation and cooling (rejuvenation) since 265 million years ago. The first occurred between 110 and 85 million years ago associated with the passing of the Bermuda hotspot. The second occurred some time less than 50 million years ago for reasons that are not understood, but could be related to increased internal stresses or thermal isostasy. Given this history, we could expect that changes in climate-driven rates of erosion, changes in sea level, and possible thermal isostatic changes could be responsible for 700 feet or more of elevation change since 265 million years ago. However, with the exception of the passing of the Bermuda hot spot, there is nothing to suggest strong tectonic activity or major reincarnation of the land surface.

Given the potential for rejuvenation, the timing for maximum elevation and relief across the area is not known. The area could have reached maximum elevation 265 million years ago at the end of Ouachita mountain building, or circa 90 million years ago with the passing of the hot spot, or during low sea level stands within the past 2.4 million years. During this entire time, we can suggest that landscape has evolved slowly in a predictable, recognizable way via erosional lowering with periodic rejuvenation. Major changes have been the erosional retreat of the two major escarpments and the unroofing (the exposure) of the St. Francois crystalline core.

THE INTERIOR LOW PLATEAUS

Bench-and-slope topography on the Colorado Plateau can be compared with similar topography developed in a warm, humid climate that has not undergone recent uplift. One such province is the Interior Low Plateaus region of Kentucky, Tennessee, and southern Indiana—Illinois. The Interior Low Plateaus have been tectonically inactive for at least 265 million years implying that landscape has grown old or reached quasi-topographic steady-state over that time span. Climate-driven erosion and deposition with periods of isostatic- and sea level-driven rejuvenation have been the primary agents and mechanisms of change. The climate is warm and humid with a deep soil relative to the Colorado Plateau. The landscape is one of rolling, dissected hills, low-lying escarpments, and incised river valleys controlled largely by the distribution of resistant rock units.

A landscape map that outlines the plateaus region is shown in Fig. 10.60. In sharp contrast to the Colorado Plateau, elevation in the Interior Low Plateaus is mostly between 350 and 1000 feet. It is truly a low plateau. Never the less, the province is sharply defined against the highland region of the Appalachian Plateau to the east, which steps up several hundred feet along the Pottsville escarpment, against weaker rock of the Coastal Plain lowland to the southwest, and against the glacial drift plains of the Central Lowlands northwest of Louisville. One exception is the absence of a distinct physiographic boundary with the Central Lowlands northeast of Louisville. Traditionally, the boundary is placed arbitrarily along the Ohio River. In Fig. 13.60, the entire northern boundary from the Mississippi River eastward is placed at the southern limit of recognizable glacial deposition features. This boundary roughly follows the Ohio River in the area east of Louisville.

A sketch map and two cross-sections that identify major features within the province are presented in Fig. 13.61. The primary landscape-controlling structure is the Cincinnati arch, which is a broad anticline that trends approximately north-south through the eastern part of the province. The location of the arch at the eastern side of the province results in most of the rocks dipping northwestward and becoming younger toward the Illinois Basin, thus forming a stair-step (bench-slope) pattern evident in Fig. 13.61C that consists of four subprovinces separated by three escarpments (cuestas).

Two of the four subprovinces form domal structures along the crest of the Cincinnati Arch, the Nashville Dome and the Lexington (or Jessamine) Dome. The Nashville Dome is eroded to form the Nashville Basin subprovince. It is a classic example of inverted topography as shown in Fig. 13.61C. The central part of the Lexington Dome, by contrast, forms a plain known as the Inner Bluegrass or Lexington Plain subprovince that is more or less continuous with the Central Lowlands across the Ohio River. Both subprovinces expose relatively weak Ordovician (485−444 Ma) shale and limestone, the oldest rocks in the province. Younger, primarily Mississippian limestone, sandstone, and chert (359−323 Ma) forms the Pennyroyal Plateau-Highland Rim subprovince that surrounds both the Nashville Basin and Lexington Plain. Mississippian rocks extend into the Shawnee Hills region, but elsewhere most of the rock that forms the Western Coal Field-Chester Uplands subprovince consists of Pennsylvanian sandstone and shale (359−299 Ma). The rocks are part of the Interior Platform rock succession with part of the Mississippian and much of the Pennsylvanian rocks derived from erosion of the Appalachian−Ouachita Mountains.

A major difference with the Colorado Plateau is the absence of canyons in the Interior Low Plateaus. This is not surprising considering the absence of tectonic uplift. There are, however, areas where rivers are incised several hundred feet, especially in the Western Coal Field-Chester Uplands (Fig. 13.60). This, perhaps, also is not surprising. The incision is at least partly in response to

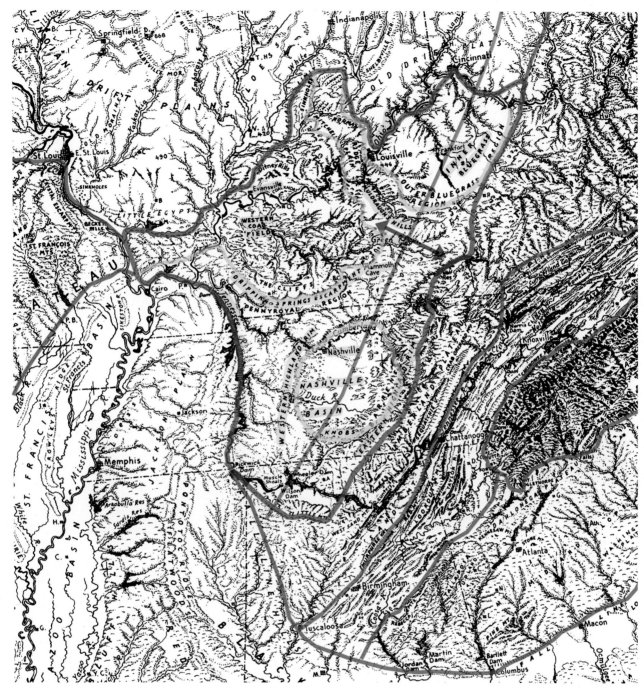

FIGURE 13.60 Landscape map of the Interior Low Plateaus with subregions outlined in green. The crest of the Cincinnati arch is shown with a dark gray line and double arrow. Thick blue-gray lines are province boundaries.

periodic drops in sea level during glacial advances, the most recent of which was the 400-foot drop that occurred between 22,000 and 18,000 years ago. Today, with rising sea level, many of the rivers are undergoing deposition in their channels. Cosmogenic radionuclide dating suggests that the rate of erosion of sandstone-capped ridges in the Mammoth Cave area has apparently been very slow, on the order of 0.0068 in. per 100 years (0.00172 mm/year), suggesting that landscape has approached quasi-topographic steady-state.

A second major difference with the Colorado Plateau is the presence of thick soil with plenty of vegetation that covers most everything. A glance at the cross-sections in Fig. 13.61 shows that, rather than rocky, angular

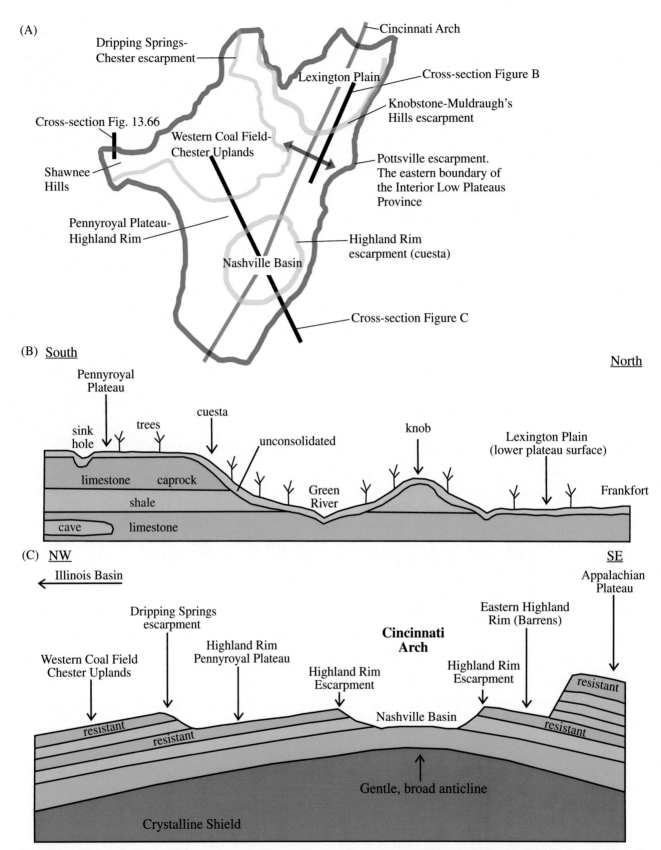

FIGURE 13.61 (A) Sketch map of the Interior Low Plateaus with subregions and structures labeled. (B) North to south cross-section from the Lexington Plain to the Highland Rim. (C) Northwest to southeast cross-section from the Chester Uplands to the Appalachian Plateau.

escarpments as depicted in Fig. 13.36, escarpments (cuestas) on the Interior Low Plateaus are soil-covered with rounded slopes and steep hillsides.

In the following sections, we compare bench-and-slope landscape with landscape on the Colorado Plateau, we introduce the Shawnee Hills as one of the most deformed areas in the central part of the United States, and we discuss Mammoth Cave, the largest cave system in the world.

Bench-and-Slope Landscape

Fig. 13.61C shows bench-and-slope landscape that steps northwestward from the Nashville Basin (or the equivalent Lexington Plain) onto younger rock of the Highland Rim-Pennyroyal Plateau and then onto still younger rock of the Western Coal Field-Chester Uplands. In contrast to the Colorado Plateau where elevation increases across benches of the Grand Staircase, elevation does not change substantially or decreases across benches of the Interior Low Plateaus. Some of the highest elevations in the province are located near Frankfort at the center of the Lexington Dome. The absence of elevation gain is due to the relatively low height of the escarpments that separate the benches, the distance between the benches, and the general northwestward dip of the rock layers, which negates the elevation gain.

The cuestas that separate each bench are well developed in some areas but strongly eroded in others. As shown in Fig. 13.61A, the cuesta that separates the Lexington Plain from the Highland Rim-Pennyroyal Plateau is known as the Knobstone-Muldraugh Hills escarpment. As seen in Fig. 13.62, this particular escarpment is most conspicuous immediately west of Louisville where it forms a step that rises as much as 600 ft above the Lexington Plain. It resembles cuestas that develop on the Grand Staircase except with greater vegetation and human population. Southward, in the vicinity of the Muldraugh Hills, the escarpment becomes so thoroughly dissected and irregular that it becomes difficult to determine where the Lexington Plain ends and the Pennyroyal Plateau begins. Rather than a sharp escarpment, the boundary is marked with scattered isolated hills known as knobs. The knobs are erosional remnants of the Pennyroyal Plateau and, as such, they are the humid equivalent of small mesas and buttes found on the Colorado Plateau. Excellent examples are shown in Fig. 13.63, and these can be compared with the buttes of Fig. 13.39.

As shown in Figs. 13.61A and C, the cuesta that separates the Nashville Basin from the Highland Rim-Pennyroyal plateau is the Highland Rim escarpment. This escarpment is well developed except along its southern margin, which is marked as knobs in Fig. 13.60. The Eastern Highland Rim is known as The Barrens because

FIGURE 13.62 Google Earth image looking southwest across the Knobstone escarpment near New Albany, Indiana. The Ohio River and Louisville, Kentucky are at left.

FIGURE 13.63 Google Earth image looking northward across tree-covered knobs of the eroded Muldraugh Hills escarpment near Lebanon, Kentucky. The Lexington Plain is at top of image.

Mississippian rocks weather to clay-rich, cherty soil that does not support thick vegetation.

The escarpment that separates the Highland Rim-Pennyroyal Plateau from the Western Coal Field-Chester Uplands is known as the Dripping Springs-Chester (or Springville) escarpment (Fig. 13.61C). This escarpment is distinctive in its southern part as seen in Fig. 13.64 but is dissected and less distinctive farther north. Mississippian limestone forms the cap-rock layer across much of the Highland Rim-Pennyroyal Plateau south of the escarpment producing karst features such as sinkholes, some of which are visible in Fig. 13.64. Limestone is present at depth in the Western Coal Fields-Chester Uplands subprovince north of the escarpment below an insoluble caprock layer of Pennsylvanian sandstone. The dissolving limestone in this area is protected from collapse by the overlying insoluble caprock layer, which allows caves, including the Mammoth Cave system, to develop. The Dripping Springs escarpment is so named because water moving through limestone below the caprock layer reaches the surface along the face of the escarpment.

Deformed Rocks of the Shawnee Hills

The Shawnee Hills, in southern Illinois, forms a western outlier of the Western Coal Fields-Chester Uplands region (Fig. 13.60). It is a surprisingly hilly region that rises abruptly from the flat glacial plains of the Central Lowlands. Most of the layers are close to horizontal; however, steeply dipping layers occur along several fault zones that trend through the area. The abundance of faults makes this area one of the most deformed in the central US. The faults are part of the Rough Creek-Pennyrile fault zone, which is an extension of an ancient rift zone known as the Reelfoot Rift. These faults have a history of reactivation as outlined in Fig. 4.10. Similar with the Colorado Plateau, the faults formed initially between 800 and 500 million years ago within the underlying ancient crystalline shield prior to deposition of the overlying sedimentary layers. The faults have periodically reactivated since that time, cutting into and displacing the sedimentary layers. Stress along these buried, reactivated, faults is responsible for earthquakes in the area including the great New Madrid earthquakes of 1811 and 1812.

The Google Earth image and cross-section in Figs. 13.65 and 13.66 show an abrupt boundary between the northeastern part of the Shawnee Hills and the glaciated plains of the Central Lowlands. This boundary corresponds with the inactive Shawneetown fault and with the southern limb of the Eagle Valley syncline. The highlands are due to the presence of a thick, resistant layer of sandstone that forms a cuesta several hundred feet high. The

FIGURE 13.64 Google Earth image looking NNE along the Dripping Springs Escarpment. The many ponds on the Pennyroyal Plateau are sinkholes produced by dissolution of limestone. Caves (including Mammoth Cave) form in the Chester Uplands north of the escarpment because limestone is protected from collapse by an overlying layer of insoluble cap-rock.

cuesta can be followed continuously around to the south side of the syncline where, at one location known as Garden of the Gods, the rocks are eroded into a variety of odd shapes. The Shawneetown fault is present along the north and west sides of the cuesta. This fault has had a complex history of reactivation that includes both south-side up and south-side down displacement, which is not unusual for reactivated faults.

Hicks Dome is also shown in Fig. 13.65. The origin of Hicks Dome is reminiscent of some of the crystalline-cored domes on the Colorado Plateau. It formed as a result of igneous intrusion, which shattered and domed sedimentary rocks above it. In contrast to the crystalline-cored mountains on the Colorado Plateau, Hicks Dome forms a bulls-eye topographic basin composed of resistant and nonresistant sedimentary rock. Notice in the figure that it forms high topography at its center, surrounded by a ring of low topography. A small amount of intrusive rock surrounded by shattered sedimentary rock is present at its center. The intrusive rock is not granitic. It is a rare ultramafic olivine-pyroxine-bearing rock that weathers quickly in the humid climate. These rocks, which are also scattered elsewhere in the Shawnee Hills are thought to be approximately 270 million years old (Permian) and are the only crystalline rocks in the province.

Mammoth Cave

Mammoth Cave is by far the largest cave system in the world that is not underwater. It is located in the Western Coal Field-Chester Uplands region near the boundary with the Pennyroyal Plateau. It is labeled in Fig. 13.60. Fig. 13.67 is a schematic cross section of the area.

Earlier we defined an aquifer as a body of permeable rock or sediment that contains groundwater. The water table is defined as the top surface of an aquifer. Sediment and rock are water saturated below the water table and

FIGURE 13.65 Google Earth image looking east across the Shawnee Hills near Equality, Illinois. A resistant sandstone layer forms a cuesta that surrounds the Eagle Valley syncline on three sides. The Wildcat Hills (w) forms part of the cuesta on the north (left) side of the syncline, Garden of the Gods (G) forms part of the cuesta on the south side. The Shawneetown fault borders the cuesta on the west and north sides. Resistant sandstone also surrounds Hicks Dome to form a domal basin. Lakes periodically occupied nearly all of the flat farmland in the image during glaciation. The Ohio River crosses at the top of image.

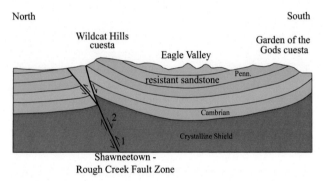

FIGURE 13.66 Cross-section of the Shawneetown-Rough Creek fault zone and Eagle Valley syncline, from Equality, Illinois southward across Garden of the Gods (see Fig. 13.65). Numbers indicate reverse fault (1) motion along the Shawneetown fault followed by normal fault motion (2). The ages of sedimentary rocks are Cambrian through Pennsylvanian (Penn.).

dry above the water table. The dissolution of limestone and the formation of caves typically occur at or just below the water table. The reason for this is that the dissolution of limestone has the effect of neutralizing acidic water. Once neutralized, the water can no longer dissolve limestone. Water within the aquifer has already reacted with limestone and is therefore neutral. The water table is the level at which fresh, acidic rainwater is added to the aquifer, and is therefore the only location where dissolution of limestone can occur. Thus, the elevation of the water table controls the elevation at which active cave formation is occurs.

The water table is not a flat surface. Instead, it mimics the ground surface by rising slightly below hills and dipping slightly below valleys. Water generally flows away from high areas toward low areas such that the water table intersects the surface of major rivers and lakes. For example, in the section on the Nebraska Sand Hills, we described how the water table intersects low areas to form ponds between the dunes (Fig. 13.31). An active water table is what keeps many rivers flowing and many lakes full of water during summer drought.

As suggested in Fig. 13.67, water in the Mammoth Cave system seeps into the ground or descends through vertical cracks (joints and faults) to the water table and then moves horizontally northwestward along nearly horizontal bedding planes eventually to intersect the Green River, less than a half mile from the Mammoth Cave Visitor Center. The elevation of the Green River corresponds roughly with the elevation at the top of the water table, and therefore with the elevation at which cave formation is active. If the Green River cuts downward, then active cave formation will drop to a lower level leaving dry caves at higher elevation. If the Green River rises due to increased water supply or through deposition in its channel, then active cave formation will rise to a higher level and any cave below that level will be flooded.

There are at least 405 miles of interconnected passages in Mammoth Cave, almost double the second largest cave system in the world. The ground surface at Mammoth Cave is at an elevation between 730 and 800 feet. The

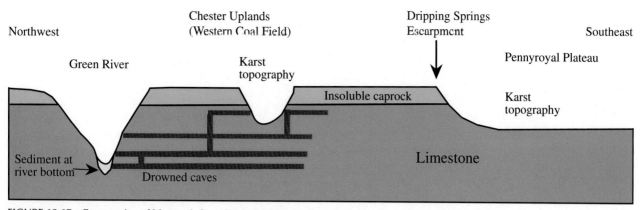

FIGURE 13.67 Cross-section of Mammoth Cave, Kentucky from the Green River southeastward to the Pennyroyal Plateau. *Based on Harris, Tuttle, and Tuttle (1997).*

uppermost cave passages are at an elevation of about 690 feet with passages extending down through at least five levels to the 350-foot elevation and below. The primary passages are above 490 feet. The largest cave is as much as 30 feet wide and almost 100 feet high, but most are much smaller.

It is difficult to date when a cave system first begins to form because it forms very slowly and becomes larger with time. The Mammoth Cave system is still in the process of forming. We have noted that limestone dissolution occurs at or near the water table and that the level of the water table is approximately coincident with the level of the nearby Green River. If we logically assume that a river must incise its channel into bedrock from the top down, we can also assume that cave passages have formed generally from the top down, a conclusion verified in Mammoth Cave by crosscutting relationships at passage intersections. When caves are flooded, they deposit sediment in the cave. The time when this sediment is deposited can be estimated using cosmogenic radionuclide decay ratios. In Chapter 9, we discussed how cosmic rays are used to date how long a surface has been free of water and close enough to the surface to be subject to long-term cosmic ray bombardment. Once sediment is removed from the surface and redeposited in a cave, cosmic ray bombardment ceases to reach the sediment and radioactive isotopes, specifically ^{26}Al and ^{10}Be, that were produced during bombardment, undergo decay at different rates such that the ^{26}Al/^{10}Be ratio decreases with time. The age of sediment can be estimated by measuring the ratio of these two isotopes. The oldest cave sediment, collected between the 557- and the 590-foot elevation (170–180 m), was dated at approximately 3.62 million years ago. This date indicates that caves at this elevation, and all caves at higher elevation, had already formed and had undergone flooding and deposition by this time. Given this date, coupled with additional paleomagnetic, radiometric, and cosmogenic isotope evidence, along with field evidence that correlates cave sediment with river sediment on upland surfaces, we can suggest that Mammoth Cave formation was in progress 5.7 million years ago, and likely began even earlier.

THE APPALACHIAN PLATEAU

The Appalachian Plateau began life as a depositional basin to the rising Appalachian Mountains. Today it is one of the highest provinces in the east with summit elevations above 4000 feet (1220 m) in both its northern and southern parts. The province is large, extending from New York to Alabama. Rivers drain westward to the Mississippi River with the exception of most of New York and parts of Pennsylvania and Alabama, which drain eastward to the Atlantic Ocean (Fig. 2.3).

Fig. 13.68 is a landscape map that outlines most of the Appalachian Plateau from southern New York to northeastern Alabama. The northern part of the province is shown in Fig. 13.8, the southern part in Fig. 13.19. The plateau is wide in its northern part, extending halfway across Ohio, but becomes narrow in its southern part. The regional structure is a broad syncline, the Appalachian Basin, the center of which lies in West Virginia. The central trough of the fold is shown in Fig. 13.68 with a thick line and two inward-facing arrows. There are additional folds and a few faults particularly at the northern and southern ends of the plateau, but they are fewer in number when compared to the Valley and Ridge and for the most part the structure maintains a nearly flat-lying character. Rocks are Permian (299–252 Ma) in the center of the Appalachian Basin, but most of the plateau consists of Pennsylvanian rock (323–299 Ma). Older, primarily Mississippian rock units are present along the peripheries. Devonian rocks (419–359 Ma) underlie much of the plateau in New York and northeastern Pennsylvania. Nearly all rocks are sandstone, shale, conglomerate, and coal derived from erosion of the Appalachian Mountains. The

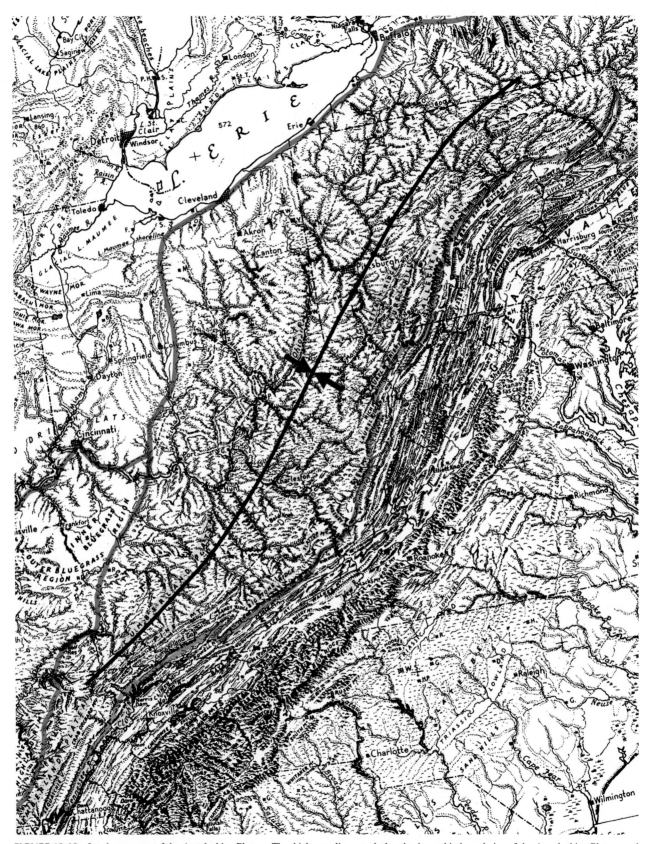

FIGURE 13.68 Landscape map of the Appalachian Plateau. The thick grey lines mark the physiographic boundaries of the Appalachian Plateau and Interior Low Plateaus. The thick black line with arrows marks the trough of the Appalachian Basin.

entire province can be considered part of the interior platform.

The Appalachian Plateau can be divided into two subregions without a clear physiographic distinction. The Alleghany Plateau forms the northern half approximately as far south as the New and Kanawha Rivers (Charleston, West Virginia). The Cumberland Plateau forms the southern half including parts of Virginia, Kentucky, Tennessee, and Alabama. Fig. 13.68 shows a mature degree of river dissection across the entire plateau with a high density of stream valleys between rolling hills and undissected tablelands. The western part of the plateau is more thoroughly worn down, leaving a series of dissected, erosional mountains along the eastern margin. The Catskill Mountains occupy the eastern part of the plateau north of New York City (Fig. 13.8). The Pocono Mountains form a small range in northeastern Pennsylvania (Figs. 13.8 and 13.68). The Alleghany Mountains stretch along the eastern margin of the Alleghany Plateau from Pennsylvania to eastern West Virginia. And the Cumberland Mountains form the eastern edge of the Cumberland Plateau in the Cumberland Gap area where Tennessee, Kentucky, and Virginia meet (Fig. 13.68). The Catskill, Alleghany, and Cumberland Mountains all have peaks above 4000 ft. These are quintessential erosional mountains.

Fig. 13.69 is a shaded relief elevation map that shows the entire Appalachian Plateau and adjacent Appalachian Mountains. The map is rotated 25° counterclockwise from north. The synclinal down-folding of the rocks to form the Appalachian Basin (thus protecting them from erosion relative to surrounding areas) and their strong erosional resistance, helps explain why the province today stands relatively high. All of the erosional mountains are located on the upturned eastern limb of the syncline where they come in contact with tilted, nonresistant rocks of the Valley and Ridge. The contrast in structural and erosional properties between rocks in these two provinces has created the Appalachian Front, the longest and most continuous high escarpment in the United States. The escarpment is located with yellow arrows in Fig. 13.69 and is clearly visible east of the Catskill (c) and Pocono (p) Mountains, near Altoona (A), and north of Chattanooga (C). Local names for the escarpment include the Catskill Front, Allegheny Front, and the Cumberland Escarpment. The more thoroughly eroded western side of the plateau is without an escarpment except for the Tennessee region where Pennsylvanian rocks of the Pottsville escarpment overlook nearly flat-lying rocks of the Interior Low Plateaus. This escarpment is identified in Fig. 13.69 with a blue arrow.

Fig. 13.69 highlights some of the smaller folds on the Appalachian Plateau, which show as ripples in northern Pennsylvania (located with white arrow) and in the area east of Pittsburgh (P) where the folds are identified in Fig. 13.68 as Chestnut Ridge and Laurel Hill. Folds in the southern part of the plateau are tighter and more strongly eroded, creating lineaments such as Sequatchie Valley (s). Fig. 13.69 also shows how the trend of high elevation crosses the Appalachian region. High elevation begins in the North Carolina Blue Ridge (br), crosses into the Valley and Ridge just south of Roanoke (R), and extends across the Appalachian Plateau north of Pittsburgh (P).

The Appalachian Plateau, like the Ozark and Interior Low Plateaus, is an ancient landscape that has not experienced tectonic activity since the final phases of Appalachian mountain building. Deposition during much of the Pennsylvanian and Permian periods was primarily river and lake environments, implying that the area has been emergent (above sea level and undergoing erosion) probably for more than 300 million years. The entire province has grown old in a predictable and recognizable way via slow erosion with periodic rejuvenation due to isostatic and sea level changes.

Allegheny Plateau

The New York area and northeastern Pennsylvania were glaciated as recently as 18,000 years ago. Glaciers carved the eleven Finger Lakes and created the drumlin field south of Rochester (Fig. 13.8). The Finger Lakes, including Lake Seneca and Lake Cayuga, were preglacial river valleys oriented roughly parallel with glacial movement. The parallelism allowed glaciers to carve steep-sided, narrow troughs.

The Catskill Mountains (c, all abbreviations refer to Fig. 13.69) form erosional mountains in eastern New York with two peaks that top 4000 feet and another 33 above 3500 feet. Rocks in the Catskill Mountains are Devonian and are folded into a broad syncline. The eastern upturned lip of the syncline forms the Catskill Front, which is spectacularly displayed in Fig. 13.70. The mountain front forms an escarpment more than 3000 ft above the Hudson Valley, which lies within 200 ft of sea level.

In northeastern Pennsylvania, the Allegheny Plateau loops around the synclinal Wyoming Valley, a northward extension of Valley and Ridge topography centered on the city of Scranton (S). Wyoming Valley is a major producer of anthracite coal, and thus is sometimes referred to as Anthracite Valley. East of Wyoming Valley, at the eastern edge of the Appalachian Plateau, we find the Pocono Mountains (p), which is dissected into low-lying hills and peaks less than 2500 feet in elevation. The Poconos form a southward extension of the Catskill Mountains. They are composed of the same rock but are lower in elevation and separated from the Catskills by the Delaware River lowland. The eastern side of the Poconos forms the continuation of the Catskill Front, which wraps around

A-Altoona, a-Alleghany Mtns., B-Birmingham, br-Blue Ridge, C-Chattanooga, c-Catskill Mtns., Cb-Columbus, Ch-Charleston, Cl-Cleveland, cm-Cumberland Mtns., N-Nashville, n-New/Kanawha River, o-Ohio River, P-Pittsburg, p-Pocono Mtns., R-Roanoke, S-Scranton, s- Sequatchie Valley, v-Valley and Ridge.

FIGURE 13.69 Shaded-relief elevation map of the Appalachian region. Yellow arrows point to the Appalachian Front. White arrow points to folds. Blue arrow points to Pottsville Escarpment. National Atlas of the United States of America, USGS, Nationalmap.gov/small_scale/atlasftp.html.

Appalachian Basin. The hills rise in elevation and show greater river dissection toward the east, reaching their highest elevation close to the Allegheny Front where peaks in the Alleghany Mountains (a) stand above 4000 feet. In Fig. 13.68, the Allegheny Front is identified in Pennsylvania. Southward, the Front corresponds roughly with the eastern side of Laurel Brier, Laurel Ridge, and Yew Mountain, which collectively form part of the Allegheny Mountains. The New River (n) crosses from the Valley and Ridge to the Appalachian Plateau south of Yew Mountain at the same location where high topography also crosses from the Valley and Ridge to the plateau (Fig. 13.69). The Appalachian Front is not distinctive in this area, but regains its distinctive look and relief farther south along Big Stone Ridge (Fig. 13.68). The New River continues across the plateau to Fayetteville, West Virginia, where it forms the New River Gorge, a 2000-foot gorge considered to be the deepest in the eastern US. The smaller Gauley River joins the New River a few miles downstream before reaching Charleston (Ch), and the two joined rivers are renamed the Kanawha River. The Kanawha River, thus, is the downstream continuation of the New River.

The western boundary of the Alleghany Plateau in Ohio is placed roughly at the eastern boundary of the Ohio Shale, a nonresistant Devonian rock unit that extends north to south through the middle of Ohio just east of Columbus (Cb). Here, the boundary with the Central Lowlands is marked by an inconspicuous eastward rise in elevation of no more than about 300 feet where resistant Mississippian rock units replace the Ohio Shale. The rise in elevation is visible in Fig. 13.69 northeast of Columbus. As outlined in Figs. 13.68 and 13.69, the Ohio Shale and correlative shale units continue along the Lake Erie and Lake Ontario shoreline through Cleveland (Cl) into New York State where the boundary with the Central Lowlands is marked by a rise in elevation along a slope with a total elevation gain that can exceed 1000 feet.

Cumberland Plateau

The Cumberland Plateau is a dissected tableland (a high bench) mostly between 1000 and 2000 feet in elevation and almost entirely underlain with a highly resistant caprock layer up to 1500 feet thick composed of Pennsylvanian sandstone, conglomerate, and coal. A shaded relief image of the plateau is shown in Fig. 13.71. The image is rotated 25° clockwise from north and can be compared with Fig. 13.68. A section of the Cumberland Plateau is also shown in Fig. 15.20. The plateau is wide in the north where the New/Kanawha and Ohio Rivers cross a river-dissected landscape. Visible landforms in Fig. 13.71 include Pine Mountain (p) and Cumberland/Stone Mountain (c), both of which show as lineaments created by ridges of dipping rock layers.

Wyoming Valley and continues southward as the Allegheny Front.

The Allegheny Plateau south of Pittsburgh (P), extending to West Virginia, is a river-dissected area of rolling hills underlain with resistant sandstone of Pennsylvanian and Permian age (323–252 Ma). It is the center of the

FIGURE 13.70 Google Earth image looking north at the Catskill Front along Hudson Valley, New York. The Hudson River is at far right. The mountains at distant right are the Taconic, Berkshire, and Green Mountains. The Adirondacks are at distant center-left.

Rocks on Pine Mountain dip southeastward and are underlain by the Pine Mountain thrust, which surfaces along the northwestern base of the mountain. Rocks on Cumberland/Stone Mountain dip in the opposite direction, toward the northwest, such that the Cumberland Mountains (cm), located between the two lineaments, forms a broad syncline of high elevation. The two long ridges are structural outliers of the Valley and Ridge province and are discussed in more detail in Chapter 15. In the early frontier days, these ridges presented a barrier to travel save for a wind gap in Cumberland/Stone Mountain celebrated today as Cumberland Gap National Historic Park, and a water gap in Pine Mountain at Pineville through which the Cumberland River flows. The southeast flank of Cumberland/Stone Mountain forms the Cumberland Escarpment.

The Cumberland Mountains (cm) were originally designated as the area between Pine Mountain and Cumberland/Stone Mountain. It is the highest part of the Cumberland Plateau with Black Mountain rising above 4000 feet, however, areas of high elevation northwest and southwest of Pine Mountain, visible in Fig. 13.71, have also been referred to as part of the Cumberland Mountains. The entire mountain region is thoroughly dissected by rivers. Fig. 13.72 looks south-southwest (SSW) across southeastern Kentucky. The long ridge in the image is Pine Mountain, which separates a dissected part of the Cumberland Plateau on the north (right) side from an equally dissected but topographically higher section of Cumberland Mountains (proper) to the south. The numerous rounded hills and sharp, narrow, deep valleys apparent in this figure form some of the roughest topography in the eastern US. Cumberland/Stone Mountain is barely visible in the upper left of the figure, beyond which lies the Valley and Ridge.

The Cumberland Plateau narrows southward across Tennessee where it becomes a less dissected tableland that stands as much as 1000 feet above the Interior Low Plateaus separated by the Pottsville escarpment (blue arrow, Fig. 13.71), and at least 1000 feet above the Valley and Ridge separated by the Cumberland escarpment (yellow arrow). Fig. 13.73 is a cross-section that extends from the Nashville dome across the northern part of Sequatchie Valley to the Valley and Ridge. Rocks dip gently eastward away from the Nashville dome producing bench-and-slope landscape with two westward-facing cuestas. As you step eastward across each bench, you step upward in elevation onto younger rock units. The Highland Rim cuesta, within the Interior Low Plateaus, separates Ordovician rock (O) in the Nashville dome from resistant Mississippian limestone/chert (M1) on the

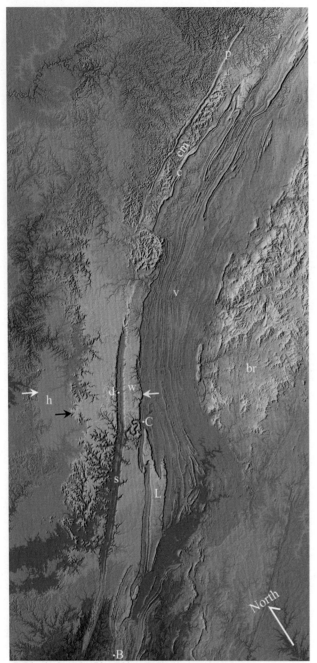

B-Birmingham, br-Blue Ridge, C-Chattanooga,
c-Cumberland/Stone Mtn., cm-Cumberland Mtns., d-Dunlap,
h-Highland Rim, L-Lookout Mtn., p-Pine Mtn., s- Sequatchie Valley,
v-Valley and Ridge, w-Walden Ridge.

FIGURE 13.71 Shaded-relief elevation map (as in Fig. 13.69) of part of the Cumberland Plateau. White arrow points to the Highland Rim cuesta. Blue arrow points to the Pottsville escarpment. Yellow arrow points to Cumberland escarpment.

Eastern Highland Rim. Farther east, the Pottsville Escarpment separates the Eastern Highland Rim from resistant Pennsylvanian rock units (P) that form the Cumberland Plateau. Both benches are visible in Fig. 13.71 (the white and blue arrows) but notice that both rather abruptly lose their identity northward in Kentucky and southward in Alabama.

As shown in Fig. 13.73, Sequatchie Valley is an anticline with thrust faults. In contrast to other areas on the Cumberland Plateau, the resistant Pennsylvanian caprock layer has been eroded from the crest of the anticline exposing weak rock underneath and creating a wide, flat-floored, anticlinal valley with steep, 1000-foot-high escarpment walls that expose the same Ordovician through Pennsylvanian rock sequence present across the Nashville Basin and Highland Rim. The Cumberland Plateau overthrust (1) is an extension of the Sequatchie Valley fault (2) developed along a flat bedding surface (a decollement). In order for fault displacement to occur along a flat layer, the rock must be weak enough to break and slide with minimal resistance. The weak layer in this case is a coal seam. Because it is horizontal and close to the surface, the Cumberland Plateau overthrust is exposed in a few areas where rivers or roads have cut downward through the rock sequence. One such area is along route 8 just north of Dunlap (d, Fig. 13.71). To reach this outcrop from Dunlap, turn north onto Highway 8 at the intersection with Highway 127 and drive toward Nashville. The fault zone is exposed on Highway 8 from 2.3 to 4.8 miles north of the intersection.

The 150 mile long Sequatchie Valley effectively divides the Cumberland Plateau into a northwestern and a southeastern section. The northwestern section becomes dissected toward the south and loses elevation (Fig. 13.71). The southeastern section, known as Walden Ridge (w) in Tennessee, expands in Alabama to become the main part of the Cumberland Plateau. Walden Ridge is deformed into a broad syncline and its eastern upturned lip forms the Cumberland Escarpment (Fig. 13.73).

As we follow Sequatchie Valley northward in Fig. 13.71, we notice it is rather abruptly replaced by mountainous topography known as the Crab Orchard Mountains, whose highest peak stands at 2828 feet, more than 1000 feet above the surrounding plateau. Here, the Pennsylvanian cap-rock layer has not yet been eroded from the top of the anticline thus creating an anticlinal mountain. Similar anticlinal mountains are present in Alabama at the south end of Sequatchie Valley including Wornock Mountain at 1430 feet.

The Tennessee River flows for 75 miles along the western side of the Valley and Ridge to Chattanooga (C) where it cuts the Tennessee River Gorge through Walden Ridge to Sequatchie Valley (Figs. 13.68 and 13.71). South of Chattanooga, Lookout Mountain (L) forms a third physiographic section separated from the rest of the plateau by the Lookout Valley anticline visible in Fig. 13.71 to the west (left) of Lookout Mountain.

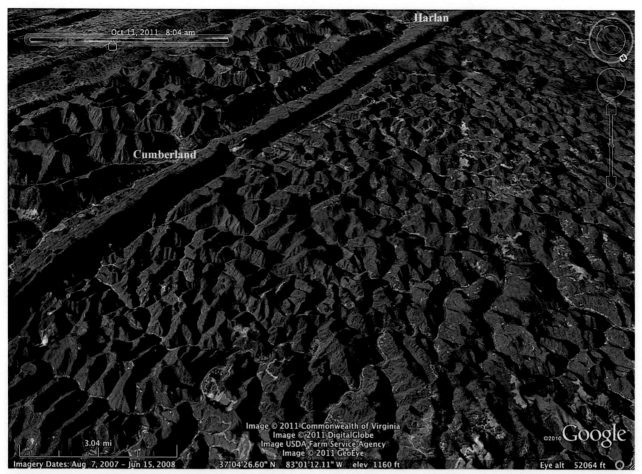

FIGURE 13.72 Google Earth image looking SSW across the dissected Cumberland Plateau in southeastern Kentucky near the border with Virginia and Tennessee. The long ridge between Harlan and Cumberland is Pine Mountain. The Cumberland Plateau is at right, the Cumberland Mountains (proper) are at left, and the Valley and Ridge is at extreme upper left.

Lookout Mountain forms a high tableland underlain with resistant Pennsylvanian rock units similar to Walden Ridge, but is transitional to the Valley and Ridge because the rocks are folded into a series of anticlines and synclines whose axes continue into the Valley and Ridge. This dual relationship suggests that Pennsylvanian rocks, at one time prior to erosion, may have extended across the Valley and Ridge. If this is correct, then erosional stripping of these resistant rocks would have resulted in slow erosional reincarnation of the Valley and Ridge from an Appalachian Plateau-type landscape to the present-day ridge and valley-type landscape. Additional folds are visible along the southeastern side of the Cumberland Plateau as far south as Birmingham (B).

Comparison of the Pottsville and Cumberland Escarpments

The Pottsville and Cumberland escarpments form the western and eastern margins of the Cumberland Plateau across Tennessee. As seen in Fig. 13.71, the Pottsville escarpment (blue arrow) is a dissected, highly irregular boundary marked by a series of knobs (eroded hills) across which elevation rises 800 to 1000 feet. The Cumberland escarpment (yellow arrow), by contrast, forms an abrupt, straight escarpment that stands more than 1000 feet above the Valley and Ridge to form the Appalachian Front, and which is traversed only by the Tennessee River in the vicinity of Chattanooga. The reason for the contrast in physical character between the Pottsville and Cumberland escarpments has to do with geology and location. The Cumberland escarpment developed along a steep, straight geologic boundary that separates weak, tilted, thrust-faulted rock layers of the Valley and Ridge from resistant, nearly flat-lying rock of the Cumberland Plateau. Differential erosion along this geologic boundary has helped create a straight escarpment.

The Pottsville escarpment, by contrast, lies between two areas of nearly flat-lying rocks. Rivers have managed to cut through the resistant sandstone caprock layer at the edge of the Cumberland Plateau and have encountered weak rock underneath. Sideways removal of the weak

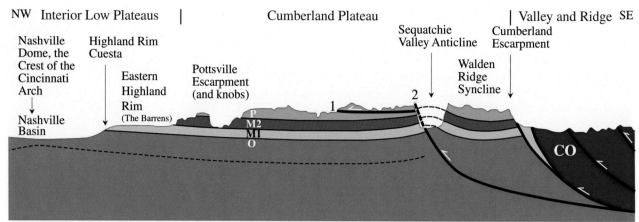

P-Pennsylvanian sandstone/conglomerate/coal (Pottsville, Gizzard, Crab Orchard Mountain, and Crooked Fork).
M2-Mississippian limestone/sandstone/shale (Monteagle, Hartselle, Bangor, and Pennington).
M1-Mississippian chert and cherty limestone (Fort Payne, St. Genevieve, St Louis, and Warsaw).
O-Ordovician limestone and older rocks. **CO**-Cambrian-Ordovician rock in the Valley and Ridge.
1-Cumberland Plateau Overthrust, **2**-Sequatchie Valley Fault.

FIGURE 13.73 Cross-section from the Nashville Basin southeastward to the Valley and Ridge. Names in parentheses below the cross-section are formation names.

rock undercuts and removes the overlying hard sandstone thus creating wide valleys. As valleys merge, they leave behind intervening erosional leftovers in the form of hills (knobs), thus creating the irregular boundary. Contributing to erosion is the large number of rivers that cross the Pottsville escarpment. Additionally, the Pottsville escarpment faces northwest relative to the southeast-facing Cumberland escarpment. It receives less direct rays of the sun and therefore is more likely to retain moisture.

QUESTIONS

1. Beginning with Fig. 13.1A, explain why one area might erode as shown in Fig. 13.1B and another area as shown in Fig. 13.1C. Speculate on how rock type, rock hardness, climate, and surrounding geography might be different in the two areas. Do the same with Fig. 13.1D–F.
2. Explain bench-and-slope landscape by drawing a labeled example in cross-section. What is the meaning of caprock layer?
3. Briefly explain why the dominant structural style in the United States is nearly horizontal sedimentary rocks.
4. Under what circumstances are erosional mountains most likely to form?
5. How can mountains form in flat-lying sedimentary layers?
6. Can hogbacks form within a structure other than a monocline? Explain.
7. Barrier islands are a common feature along the Atlantic coast. What criteria regarding the Atlantic coast are conducive to the construction and maintenance of barrier islands even in the face of rising sea level?
8. Why is sea level rising rapidly at Galveston, the Mississippi Delta, and the Chesapeake Bay?
9. Explain why sea level rise is different at Portland Maine verses Boston Massachusetts.
10. What is the evidence to suggest the existence of the Cape Fear Arch?
11. Why are sedimentary rock layers of the Coastal Plain wedge-shaped?
12. How can sea level fluctuations (also called eustatic rise and fall, or transgression and regression) be recognized from rocks and fossils of the Coastal Plain?
13. How can a geologist in the field discern between a loosely consolidated bedrock cuesta and a beach scarp?
14. Why do the rock units of the Black Belt region exert notable landscape control?
15. How does the Balcones escarpment in Texas differ from other bedrock cuestas such as the Austin Chalk cuesta?
16. What may have caused thermal isostatic uplift and high mountains in the Mississippi Embayment 100 Ma?
17. In spite of active normal faults, why shouldn't the residents of Texas, Louisiana, and Mississippi be worried about a devastating earthquake? What should they be concerned about instead?
18. What is the ground water significance of the Chesapeake Bay Impact Structure?
19. If rivers, at one time, flowed across the Llano Estacado, explain how the Mescalero Escarpment formed?

20. Name and locate several of the basins that form the Wyoming Basin. Which one forms the continental divide?
21. Describe the Ogallala Formation. What is its significance?
22. Why are the High Plains so flat?
23. What is the maximum age for landscape in the Great Plains and Wyoming Basin?
24. How did monoclines of the Colorado Plateau form? Explain with a cross-section.
25. Describe one of the uplifts shown in Fig. 13.42 in as much detail as you can. Use the state geological map (http://mrdata.usgs.gov/geology/state/).
26. Explain the formation of vertical fractures in the Entrada sandstone. Why are they significant to the development of landscape in Arches National Park?
27. Why do arches form? Explain using the following terms: dry climate, vegetation, freeze-thaw, fin, differential weathering.
28. Explain how fractures influence the landscape of Zion and Bryce Canyon National Parks.
29. Referring to Fig. 13.35, how is the northeast border of the Colorado Plateau in Colorado different from the northwest border in Utah. Describe physiographic differences on either side of the border.
30. Describe changes to the cross-section in Fig. 13.36 if the region were to shift to a humid climate.
31. There are several mountains on the Colorado Plateau as indicated in Fig. 13.35 including Mt. Ellen, the La Sal Mountains, and the Navaho Mountains. Speculate on their origin.
32. Define an aulocogen. Name the aulocogen responsible for the New Madrid fault zone.
33. In Figs. 13.25 and 13.26, note the degree of dissection in the Missouri Plateau surrounding the Black Hills. Compare the level of dissection with that of the Llano Estacado (Fig. 13.28), Colorado Plateau (Fig. 13.35), and Appalachian Plateau (Fig. 13.68). Which of these three areas does the Missouri Plateau most resemble? Explain your reasoning. Why is there more dissection west of the Missouri River than to the east in Fig. 13.26?
34. Using Fig. 13.60 as a guide, describe physiographic differences between the Interior Low Plateaus and the Central Lowland along the northwestern border near Evansville and compare these differences with the northeastern border of the Interior Low Plateaus with the Appalachian Plateau.
35. Using Fig. 13.57 as a guide, describe physiographic differences between the Ozark Plateau and Ouachita Mountains.
36. Compare and contrast bench-and-slope landscape of the Interior Low Plateaus with that of the Colorado Plateau.
37. The same sandstone rock unit in Fig. 13.65 underlies the forested areas surrounding the Wildcat Hills (w) and Garden of the Gods (G). Explain why the width of the sandstone at Garden of the Gods (G) is wider.
38. How do erosional mountains differ from plateaus? How do they differ from the "classic" mountain?
39. Explain the stages of river dissection resulting in erosional mountains.
40. Explain why a maturely dissected landscape can be younger in actual years than a less mature landscape.
41. Describe the landscape of the Ozark Plateau.
42. What is chert? How do interbeds of chert in the limestone affect the landscape of the Salem and Springfield Plateaus?
43. What is the minimum age of the Ozark Plateau? How do we know?
44. Under what conditions does unconsolidated sediment control landscape?
45. Name some areas where landscape is controlled largely by unconsolidated sediment.
46. Draw the Adirondack, New England, Piedmont Plateau, Blue Ridge, Valley and Ridge, and Appalachian Plateau physiographic boundaries on Figs. 13.8 and 13.9.
47. Color the entire Mississippi flood plain in Fig. 13.11 yellow. What do Memphis, Vicksburg, Natchez, and Baton Rouge have in common? How is the landscape of the floodplain different from surrounding areas?
48. Why does the Mississippi River in Fig. 13.11 turn sharply at Vicksburg?
49. Note the turn in the Mississippi River south of Natchez in Fig. 13.11. Describe what might happen if the Mississippi were to jump its banks and flow into the Atchafalaya River.
50. Why do the strong rock trends seen in the Ouachita Mountains west of Little Rock in Fig. 13.11 abruptly terminate at the Mississippi floodplain?
51. Draw the boundaries to the Ouachita, Ozark, Interior Low, and Central Lowland physiographic provinces in Fig. 13.11.
52. Why is there sand on the left (west) bank of the Mississippi River in Fig. 13.18?
53. In Fig. 13.12, speculate on the origin of Sandy Hills east of Dallas.
54. Why are there no farms east (right) of the ridge in 13.22?
55. Research Topic: Describe the types of unconsolidated sediments that might be produced in the following environments: river, lake, delta, beach, glacial, shallow marine, tidal flats, swamp, deep marine, volcanic, reef, desert.
56. Explain the origin of Cody Scarp at Apalachee Bay, Florida.

Chapter 14

Crystalline-Cored Mid-Continent Anticlines and Domes

The crystalline-cored, mid-continent anticlines and domes structural province consists of a central anticlinal or domal core of crystalline rock surrounded by sedimentary rock. The typical landscape produced by this combination of rock/sediment type and structural form is a domal mountain. The distribution and character of the structural province is shown in Figs. 11.1 and 11.2. In this chapter, we explore two very different types of crystalline-cored mountains; anticlinal (domal) mountains cored with crystalline shield rock, and domes cored with young granitic rock. In both cases, the surrounding sedimentary rocks are part of the interior platform rock succession (Fig. 11.8). The rocks form part of the craton and reactivated western craton tectonic provinces (Fig. 11.9).

Anticlinal (domal) mountains cored with crystalline shield rock are widespread. This combination of rock type, structure, and landform develops in areas where shield rock and its overlying sedimentary cover are folded and elevated high enough such that sedimentary layers are removed from the crest of the fold and underlying resistant crystalline rocks are exposed to form domal mountains. These structures are transitional to the broad arches and domes previously discussed in Chapter 13. The major distinction is that the folds are tight enough and eroded deep enough to expose shield rock in their core. The folds are large, often forming an entire mountain range. Some of the folds are associated with reverse or thrust faults that accentuate the structure. Major anticlinal and domal structures cored with rocks of the crystalline shield include the Adirondack Mountains of New York, the St. Francois Mountains of Missouri, the Wichita and Arbuckle Mountains of Oklahoma, the Llano Uplift of Texas, the Black Hills of South Dakota, several ranges in the Middle-Southern Rocky Mountains, and the Superior Upland. Crystalline shield rocks in all of these areas possess a complex internal structure of folds, faults, and granitic intrusions that developed, in nearly all cases, more than 1.0 billion years ago during ancient mountain building phases. The anticlinal (domal) structure most responsible for the present-day landscape, however, is far younger than these ancient complex structures.

Intrusive domes are fundamentally different from the previously described anticlinal (domal) mountains. The crystalline rock that forms intrusive domes is granitic in composition and younger than surrounding sedimentary rocks. Rather than folded with sedimentary rocks, granitic magma has forcefully punched its way upward into overlying sedimentary layers such that layers above the intrusion are deformed into a dome structure as shown in Fig. 4.2A (right side). Erosion then removes the sedimentary layers thereby exposing a central core of granitic rock. These structures are relatively small, often forming a single isolated mountain peak or cluster of peaks rather than a large mountain range. They are not necessarily simple structures. They can be complicated with faults, multiple intrusions, volcanic rocks, crystalline shield rocks, or partially covered with young sedimentary layers that were deposited following intrusion. The domes discussed here intrude interior platform sedimentary rocks in the Middle-Southern Rocky Mountains, the eastern Colorado Plateau, and the northwestern Great Plains. All of the intrusions are between 75 and 18 million years old.

We will begin by discussing anticlinal mountains across the United States starting with the Adirondack Mountains. We then characterize intrusive domes and anticlinal mountains in the Great Plains, Middle-Southern Rocky Mountains, and Colorado Plateau. We end the chapter with a discussion of the Superior Upland, which has a domal structural form, but is unusual because it is not mountainous and because crystalline rocks are largely covered with glacial drift.

ADIRONDACK MOUNTAINS

The Adirondack Mountains form a nearly circular dome approximately 125 miles across composed of complexly

FIGURE 14.1 Raisz landform map of the Adirondack Mountains. The blue line separates crystalline shield from surrounding sedimentary rock.

deformed crystalline shield rock 1.3 to 0.95 billion years old. Fig. 14.1 locates the dome in upstate New York west of the Lake Champlain-Hudson River Valley. The entire area is protected as a state park, the largest in the contiguous United States. The blue line in Fig. 14.1 separates the crystalline shield from surrounding sedimentary rock. Note that the surrounding sedimentary rocks form low-lying topography. The western side of the crystalline dome is continuous with the Canadian Shield across a narrow area in the vicinity of Thousand Islands. The eastern side impinges on tilted, thrust faulted sedimentary rocks of the Valley and Ridge. Nearly flat-lying sedimentary layers of the interior platform rock succession dip away from the crystalline dome along the northern and southern flanks of the mountain forming a series of low-lying cuestas. All of the sedimentary rocks are either in fault contact with shield rocks or in depositional contact across the Great Unconformity.

Fig. 14.2 looks north-northeastward at the north-central Adirondacks. The figure reveals a complex structural pattern in the rocks etched by glacial erosion. More than 2000 lakes fill depressions. Many of the curved lakes and valleys follow folds in weak rock layers. Straight valleys, such as Long Lake Valley (L), follow fault or fracture zones. Highlands are composed of relatively resistant rock. The landscape that most resembles the central Adirondacks can be found in the glacially scoured region north of the Great Lakes (Fig. 12.1C). The topography suggests that the entire Adirondack dome underwent erosion beneath a continental glacier. Although glacial erosion features dominate the landscape, glacial depositional features are also present. Most impressive is a system of eskers that trends northeast-southwest from the Saranac Lake area (S) to Stillwater Reservoir (Sr) and Cranberry Lake (C). An esker at Lows Lake-Hitchens Pond is shown in Fig. 12.23.

A close look at Fig. 14.2 shows that the eastern half of the dome maintains a mountainous terrain. There are 2 peaks above 5000 feet and 42 above 4000 feet all located within 20 miles (32 km) of Mt. Marcy (M), the highest at 5344 feet (1629 m). Topographic relief is on the order of 3000 feet. Several of the high peaks, including Mt. Marcy, are composed of a rare intrusive rock known as anorthosite that consists almost entirely of the mineral plagioclase.

The metamorphic and plutonic rocks that form the Adirondack Mountains were involved in the Grenville orogeny approximately one billion years ago. The area was eroded to sea level by 541 million years ago and subsequently covered with sedimentary rock. Evidence for this interpretation is found in the small amount of sedimentary rock that is still preserved in down-dropped fault basins within the dome surrounded by crystalline rock. The faults and associated fracture zones were active about 465 million years ago and are responsible for some of the north to northeast-trending lineaments that were later accentuated by glacial erosion. Long Lake Valley (L) in Fig. 14.2 is an excellent example. Sedimentary rocks continued to be deposited with intermittent periods of uplift and erosion until about 300 million years ago, which is the age of the youngest rock that surrounds the dome. The history of the Adirondacks after 300 million years ago is less well known, but some geologists contend that the area has been above sea level since that time.

Insight to the recent history of the Adirondacks is obtained from apatite fission track and (U-Th)/He dating. The data suggest that present-day surface rocks in the Adirondacks underwent slow cooling from 170 to 130 million years ago followed by a heating event from 130 to 105 million years ago, followed by a more rapid cooling event from 105 million to 95 million years ago. The data suggest that by 130 million years ago the Adirondack landscape was a low-lying flatland or low plateau that had undergone very little uplift or erosion during the 300 to 130 million-year period. Crystalline rocks by 130 million years ago may have already been exposed at the surface or were close to the surface under a thin layer of sedimentary rock.

The heating event from 130 to 105 million years ago correlates with the passage close to the Adirondacks of a thermal plume known as the New England (or Great Meteor) hot spot. This is the same thermal plume, introduced in Chapter 5 that would later produce the New England Seamount chain. Heating associated with the hot spot is understood to be responsible for broad doming and uplift of the Adirondacks, possibly reactivating older

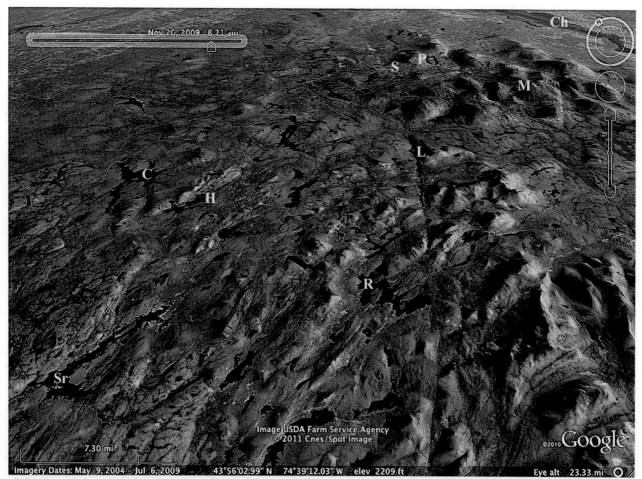

C-Cranberry Lake, Ch-Lake Champlain, H-Hitchins Pond and Lows Lake, L-Long Lake, M-Mt. Marcy, P-Lake Placid, R- Raquette Lake, S-Saranac Lake, Sr-Stillwater Reservoir.

FIGURE 14.2 Google Earth image looking north-northeast across the north-central Adirondack Mountains, New York. Note the change in topography across the Long Lake Lineament (L).

faults. These events, in turn, triggered increased rates of erosion, causing river dissection that continued until about 95 million years ago. The crystalline core likely became widely exposed during this time interval. By 95 million years ago, the hot spot was far enough away for the rocks to cool to a normal geothermal gradient. The mountains likely subsided slightly in response to cooling.

Recent earthquake activity along with geophysical data in the form of gravity and magnetic anomalies, suggest the possibility of recent to active uplift along reactivated fault/fracture zones. Tug Hill Plateau (highlighted in Fig. 14.1) is an elevated area of sedimentary rock surrounded on all sides by what appear to be reactivated faults in the underlying crystalline rock. Discrete changes in elevation across lineaments such the Long Lake Valley lineament (L) also suggest the possibility of recent fault activity.

The study therefore suggests that the Adirondack Mountains, as we know them today, came into being between 130 and 95 million years ago as a result of broad doming and erosion of a plateau surface. The mountains were sculpted, accentuated, and elevated into their present shape by an overall global sea level drop since 95 million years ago and by glaciation during the past 2.4 million years, which covered and scoured the entire mountain dome, and left glacial deposits. Recent and active uplift along discrete reactivated faults could be the result of a combination of compression and glacial isostatic unloading.

ST. FRANCOIS MOUNTAINS

The St. Francois Mountains are located south of St. Louis on the Ozark Plateau where crystalline rock crops out in a faulted, tilted dome surrounded by sedimentary rock of the Salem Upland. The St. Francois Mountains are shown in Fig. 13.55 on a landscape map of the Ozark Plateau and in cross-section in Fig. 14.3. The mountains are not exceptionally high, but they do boast the highest point in Missouri,

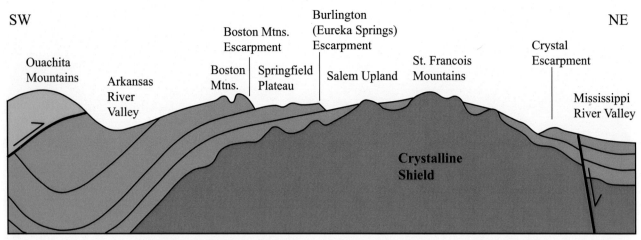

FIGURE 14.3 Schematic cross-section across the Ozark Plateau showing the St. Francois Mountains and the surrounding plateaus. A buttress unconformity surrounds crystalline rock in the St. Francois Mountains.

Tom Sauk Mountain at 1772 feet (540 m). Maximum relief between nearby points is about 700 feet (213 m).

As is typical of many crystalline-cored domes, the crystalline rocks are part of the ancient North American shield. Most are between 1.55 and 1.3 billion years old and underlie Cambrian-Ordovician (541–444 Ma) interior platform sedimentary rocks of the Salem Upland across the Great Unconformity. In addition to granite and a small amount of gabbro, much of what I refer to here as crystalline rock is actually hard, silicic volcanic rock. I have included this rock in the crystalline category because of its age and because it is extensively intruded by granitic rock. One interpretation of these rocks is that, long ago, the area resembled Yellowstone National Park.

An interesting aspect of the St. Francois area is that the erosion surface (the Great Unconformity) between ancient crystalline rock and younger sedimentary rock is not planar. Most erosion surfaces are eroded to a nearly flat surface prior to deposition of overlying rock. What makes the St. Francois unconformity so unusual is that the ancient crystalline landscape was hilly with as much as 2000 feet of relief during deposition of overlying sedimentary layers. The interpretation is that the sedimentary rocks were deposited from a shallow ocean during rising sea level between 541 and 444 million years ago such that ancient crystalline land areas were slowly drowned.

Evidence for this interpretation is the unconformity itself. The type of unconformity present in the St. Francois Mountains and shown in Figs. 14.3 and 14.4 is known as a buttress unconformity because younger sedimentary layers abut and truncate against older crystalline rock. This type of unconformity is explained in Fig. 14.5. The figure suggests that ocean water first covered low areas of crystalline rock while leaving topographically high areas as islands. The ocean would lap up against the islands and deposit a layer of sedimentary rock around their margin perhaps in the form of a beach. As sea level continued to rise, sedimentary layers would progressively cover more and more of the islands. Ultimately, the ocean would drown the islands and a layer of sedimentary rock would cover the entire region.

The situation just described is similar to what is happening today along the coast of Maine where crystalline rock of the Appalachian Mountains trends directly into the ocean. As seen in Fig. 13.13, the Maine coastline is slowly subsiding against rising sea level resulting in an abundance of coastal inlets and islands. As sea level continues to rise, ocean sediment will be deposited at progressively higher levels along the margins of the crystalline islands resulting in a buttress unconformity. Another area where we can find a buttress unconformity is near the bottom of the Grand Canyon. Here, crystalline rock of the inner gorge locally peaks up through the lowest layers of overlying sedimentary rock (Fig. 11.13).

In the case of the St. Francois Mountains, the buttress unconformity has produced some interesting landscape features. River incision (downward erosion) has created patches of crystalline rock surrounded by sedimentary rock. The crystalline rock, because of its hardness, forms highland knobs. The sedimentary rock is weak and forms flat-floored valleys. Note in Fig. 14.3 how crystalline rock sticks up from between layers of sedimentary rock. This type of landscape is well displayed in the vicinity of Ironton, Missouri as shown in Fig. 14.6. Forested highland knobs in this image consist of crystalline rock. Valley farmlands and some of the surrounding forested lowlands consist of nearly flat-lying sedimentary rock.

Rivers, in some instances, cut directly across areas of crystalline rock. Fig. 14.7 shows the East Fork of the Black River (BR) flowing from left to right across the

FIGURE 14.4 Photograph of a buttress unconformity in the St. Francois Mountains, Missouri, where layers of sedimentary rock abut against a knob of older crystalline rock.

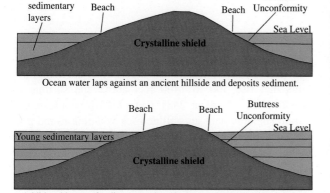

FIGURE 14.5 Origin of a buttress unconformity.

FIGURE 14.6 Google Earth image looking east-southeast at forest-covered, rounded knobs of crystalline rock surrounded by lowland farms and forest-covered areas of sedimentary rock in the St. Francois Mountains near Ironton, Missouri.

center of the image. On the left, the river (BR) flows through a wide, low-lying valley (farmland) underlain with sedimentary rock. In the center of the image, the river enters a gorge of rugged highland topography underlain with crystalline rocks known as the Johnson Shut-ins (JS). On the right side, the river flows back into a wide low-lying valley of sedimentary rock occupied by farmland and the Lower Reservoir (LR).

The interpretation is that the river was in existence and flowing across sedimentary layers prior to exposure of

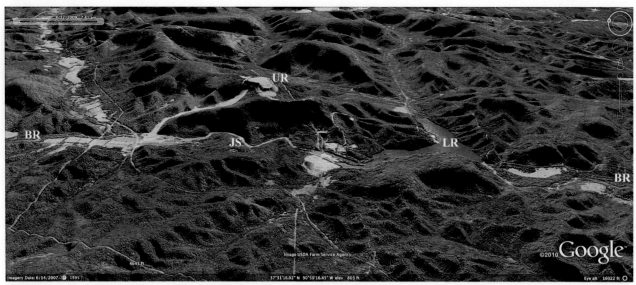

BR-East Fork Black River, JS-Johnson Shut-Ins, LR-Lower Reservoir, UR-Upper Reservoir

FIGURE 14.7 Google Earth image looking east at the East Fork Black River (*BR*) in Missouri. The river flows from left to right (southward) across the photo from sedimentary rock on the left, into crystalline rock at Johnson Shut-ins (*JS*), and back into sedimentary rock on the right occupied by the Lower Reservoir (*LR*). The scar on the hillside above Johnson Shut-ins was created December 14, 2005 when water being pumped into the Upper Reservoir (*UR*) as part of the Taum Sauk Hydroelectric Power Station overflowed and flooded. This image was obtained 1.5 years after the incident.

crystalline rock. Over time, the river cut downward into the slowly emerging crystalline rocks while managing to maintain its original course. Because of the strong contrast in erodability, the river had to cut a gorge through the crystalline rock to maintain a downstream gradient. As a result, the river flows from open valley into a deep gorge and then back into an open valley. River gorges that form in this way are referred to in Missouri as shut-ins. The Johnson Shut-ins are a prime example.

The uplift history of the St. Francois Mountains follows that of the Ozark Plateau discussed in Chapter 13. The area has been a highland since at least 265 million years ago and possibly much longer. As erosion has slowly lowered the landscape, more and more of the buried ancient crystalline hills have been exhumed thus creating the shut-ins. This interpretation would imply that some of the rivers could be as old as the landscape.

WICHITA, ARBUCKLE, AND LLANO STRUCTURAL DOMES

The Arbuckle and Wichita Mountains lie less than 100 miles from each other on the Osage Plains in the Central Lowlands of southwest Oklahoma just west of the Ouachita Mountains and north of Dallas, Texas. The Llano Uplift is located at Llano, Texas, roughly 270 miles south of the Wichita Mountains on the Great Plains northwest of Austin along the dissected eastern margin of the Edwards Plateau. All three areas are located on a Raisz landform map in Fig. 13.12. In this figure, the Llano Uplift corresponds with the Central Mineral Region-Riley Mountain-Cedar Mountain area.

Each of the three areas has a different geologic history but a similar recent landscape history. All three are domal uplifts complicated by faults. They expose crystalline rock in their core and are surrounded by younger sedimentary layers of the interior platform. We will describe all three areas, and then discuss aspects of their landscape history.

Wichita Mountains

The Wichita Mountains form hills and low-lying mountains that rise 500 to 1400 feet above the surrounding plains. Haley Peak, at 2481 feet (756 m), is the highest point. The dominant crystalline rock is a pink to red granite. Also present are a variety of other crystalline rocks including gabbro and anorthosite (as in the Adirondacks), and silicic volcanic rocks. We can consider these rocks to be part of the crystalline shield, but they are unusual in the sense that most are only 530 to 539 million years old, considerably younger than the 1.0- to 3.6- billion year-old crystalline rocks that form the bulk of the North American shield.

The crystalline rocks are surrounded by Cambrian-Early Ordovician sedimentary rocks (541–470 Ma), all of which are deformed along a pair of fracture/fault systems oriented approximately west-northwest and north-south. The dominant west-northwest trending system is responsible for the overall trend of the mountain range. A unique

FIGURE 14.8 Google Earth image looking west at fractured, bare granite exposure in the Wichita Mountains, Oklahoma. The Meers fault is visible as a narrow lineament along the right side of the image. The fault is parallel with the dominant west-northwest fracture orientation of the Wichita Mountains.

aspect to the Wichita Mountains, as shown in Fig. 14.8, is the wide exposure of bare granitic rock. Bedrock is exposed across most of the mountain area except where covered in prairie grasslands of the Wichita Mountains Wildlife Refuge. The west-northwest fracture system is well displayed in this figure.

Somewhat less deformed conglomerate and sandstone of Permian age (299– 252 Ma) surrounds the crystalline rock and the older sedimentary rock. The conglomerate is important because it contains clasts (fragments) derived from erosion of the older sedimentary rocks. The implication is that the older sedimentary rocks were deformed, uplifted into a highland, and eroded just prior to and during deposition of the less deformed Permian conglomerate. It is likely that the fracture/fault systems developed during this deformation.

Another unique aspect to the Wichita Mountains is that it boasts the easternmost active fault to reach the Earth's surface in the United States. Known as the Meers Fault, it forms a 15 mile long lineament oriented parallel with the dominant west-northwest trending fracture system. It is located just north of Lawton, Oklahoma in the northwest part of the range and can be seen along the right side of Fig. 14.8. Trenching along the fault suggests that the last major earthquake occurred between 1100 and 1400 years ago. The Meers fault has likely reactivated an older west-northwest trending fault in the crystalline rock.

Arbuckle Mountains

The Arbuckle Mountains form a low-lying subdued area that perhaps should more appropriately be referred to as an upland or a small dissected plateau. Mountaintops are flat and plateau-like. The highest elevation is 1412 feet (430 m) and there is less than 500 feet of relief between hilltops and the surrounding plains. Most of the mountain area retains its sedimentary cover with crystalline rock exposed primarily in the southeastern part. The primary rock is pink granite between 1.3 and 1.4 billion years old, much older than rocks found in the Wichita Mountains, but about the same age as rock in the St. Francois Mountains. Also, in sharp contrast to both the Wichita and St. Francois granites, the Arbuckle granites are easily eroded relative to the more resistant surrounding limestone. Much of the upland area consists of limestone with karst features that include sinkholes and caves. A small amount of c. 530- to 539-million-year-old granitic rock forms hills in the northwest part of the Arbuckles.

Sedimentary layers in the Arbuckle Mountains range in age from Cambrian to Middle Pennsylvanian (541–307 Ma) and are faulted and deformed into a series of anticlines and synclines. Differential erosion of southward-tilted sedimentary layers along the southern flank of the uplift has created a well-developed sequence of north-facing cuestas. Less deformed rocks, as young as 252 million years old (Upper Permian) are also present, including conglomerate, again suggesting the existence of local highlands during the Permian. An important consideration with respect to landscape history (discussed later) is the presence of nearly flat-lying Cretaceous (113–100 Ma) sandstones in the southern part of the mountains east of Ardmore (Fig. 13.12). These young rocks were deposited directly on top of the ancient granitic rocks.

FIGURE 14.9 Google Earth image looking north across the Llano uplift in Texas. The small hills piercing the surface consist of pink granite. The high peak at center-right is Enchanted Rock.

Llano Uplift

The Llano Uplift corresponds with the Central Mineral Region, Riley Mountain/Cedar Mountain area of the Edwards Plateau (Great Plains) in Fig. 13.12. Although structurally a dome, topographically it is more accurately described as a broad, poorly defined basin with hills and small erosional mountains situated between Balcones Escarpment to the east and undissected Edwards Plateau to the west. The area gains elevation toward the west and northwest from less than 900 feet to about 2000 feet.

The structural dome is cored with rocks of the North American crystalline shield. Originally deposited as sedimentary and volcanic rocks between 1.36 and 1.23 million years ago, the rocks were deformed and metamorphosed to high metamorphic grade primarily between 1.15 and 1.12 billion years ago and intruded between 1.12 and 1.07 billion years ago during Grenville orogeny. Deformed Cambrian to Middle Pennsylvanian sedimentary rock (542–307 Ma) surrounds the crystalline rock, and in a few places it is down faulted into surrounding crystalline rock. Young, nearly flat-lying, Cretaceous (113–100 Ma) sedimentary rocks directly overlie the crystalline rocks on the south side, similar to what is seen in the Arbuckle Mountains. Much of the sedimentary rock is limestone, which stands out as resistant rock in the arid climate. Crystalline rocks are less resistant and tend to occupy small basins between areas of high-standing limestone. Riley Mountain and Cedar Mountain are both limestone mountains located in a fault-bound block surrounded by crystalline rock. Both stand more than 700 feet above the surrounding crystalline lowland. A few small mountains of pink granite stand out as monadnocks (erosional outliers) within the crystalline terrain. The peaks are small with less than 500 feet of relief. The pink granite at Enchanted Rock State Natural Area, shown in Fig. 14.9, is an excellent example.

Landscape Development

The Wichita, Arbuckle, and Llano areas have all undergone similar landscape histories. The presence of deformed rocks older than 307 million years indicates that much of the deformation occurred after 307 million years ago. The presence of circa 299- to 252-million-year-old conglomerate in the Wichita and Arbuckle Mountains, some with clasts of crystalline rock, suggests that both areas, and probably also the Llano area, were deformed, uplifted, and eroded during this timeframe to the point where crystalline rock was exposed at the surface. This phase of uplift and erosion is shown in simplified form (without tight folds or faults) in Fig. 14.10A. Fission track studies suggest that presently exposed crystalline rocks in all three areas were already close to the surface by 252 million years ago.

Following erosion to near flatland, all three areas were covered with Cretaceous rocks associated with the Western Interior Seaway beginning less than 113 million years ago. Deposition continued until at least 75 million years ago. Fission track studies suggest that deposition may have continued in the form of river deposits eroded from the Rocky Mountains until well past 50 million years ago, but these younger rocks have since been eroded from the vicinity of all three uplifts. The presence of Cretaceous rocks directly unconformably above crystalline

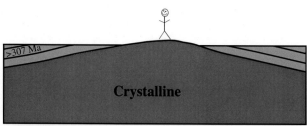

(A) Deformation, uplift, and erosion between 307 and 252 million years ago resulted in exposure of crystalline rock in some areas and deposition of Permian conglomerate in other areas. Deformation in some areas is stronger than what is shown here, and includes faults.

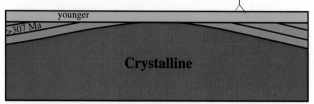

(B) Erosion was followed by deposition of Cretaceous and younger rocks beginning less than 113 million years ago. The rocks were deposited unconformably above the older crystalline and sedimentary rocks.

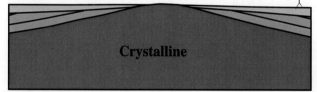

(C) Slight doming and erosion reexposes crystalline rock.

FIGURE 14.10 Sketches that show the sequence of landscape development in the Wichita, Arbuckle, and Llano areas.

rocks as seen in the Arbuckle and Llano areas, and as shown in Fig. 14.10B, indicates that crystalline rocks were exposed at the beginning of Cretaceous deposition less than 113 million years ago. Interestingly, sedimentary layers in the Llano area produced buttress unconformities similar to but far younger than the buttress unconformities seen in the St. Francois Mountains, implying that the Llano area was not eroded to a flatland prior to Cretaceous deposition.

The overall drop in sea level beginning 51 million years ago (Fig. 8.4), coupled with isostatic uplift, resulted in erosion of Cretaceous and younger beds, which re-exposed crystalline rocks in all three areas. Differential erosion since that time has created the present-day landscape shown in Fig. 14.10C. Given the overall history of these areas, it is possible that the landscape that existed 113 million years ago prior to Cretaceous deposition resembled the landscape we see today following erosion of Cretaceous rocks. For example, compare Figs. 14.10A and C.

In the Llano Uplift, it is possible to approximately date the time that crystalline rock was re-exposed at the surface. Clasts of crystalline rock are absent in surrounding sediment dated 2 million years and older, but are present in sediment younger than 2 million years. This suggests that crystalline rock was exposed at the Earth's surface beginning less than 2 million years ago.

WESTERN MARGIN OF CRYSTALLINE-CORED ANTICLINES AND DOMES

Thus far we have surveyed small, isolated, crystalline-cored anticlinal mountains across the interior United States. In the Middle-Southern Rocky Mountains, eastern Colorado Plateau, and northwestern Great Plains region, we have a large area with numerous superb examples of both anticlinal mountains and intrusive domes. This area, shown in Figs. 14.11 and 14.12, marks the western limit of widely exposed crystalline-cored anticlines. In both figures anticlinal mountains cored with ancient shield rock are shown with yellow highlight. Intrusive domes cored with 75- to 18-million-year-old granitic rocks are highlighted in blue. Other areas mentioned in the text are highlighted in green. The physiographic boundary of the Colorado Plateau is shown with a dark blue line. The area enclosed within the thick light red line is the Colorado Mineral Belt, an area of ore deposits that includes silver and gold. The light orange shaded area that includes San Luis Valley is the Rio Grande Rift zone, an area of active normal faults.

For discussion and clarification purposes, we will consider Wyoming and points north to be the Middle Rockies, and Colorado and points south to be the Southern Rockies. All of the yellow highlighted anticlinal mountains shown in Figs. 14.11 and 14.12, with the exception of the Uinta Mountains in northeastern Utah, are cored with crystalline shield rock and flanked with sedimentary rock of the interior platform. The contact between the crystalline core and the overlying sedimentary rocks is the Great Unconformity described in Chapter 11. Shield rocks vary in age from 1.08 to more than 3.0 billion years old. The oldest rocks (>1.8 billion years old) are restricted to the Middle Rockies. Many of the anticlines are accentuated with thrust and reverse faults, and in some areas they are cut by younger normal faults. Some are associated with synclinal valleys, but most are surrounded by nearly flat-lying sedimentary rock. The Uinta Mountains are cored with rocks of the Precambrian sedimentary/volcanic rock succession. This mountain, the highest in Utah with peaks that exceed 13,000 feet, is transitional between the nearly flat-lying structural province and the crystalline-cored province.

FIGURE 14.11 Landscape map of the Middle-Southern Rocky Mountains.

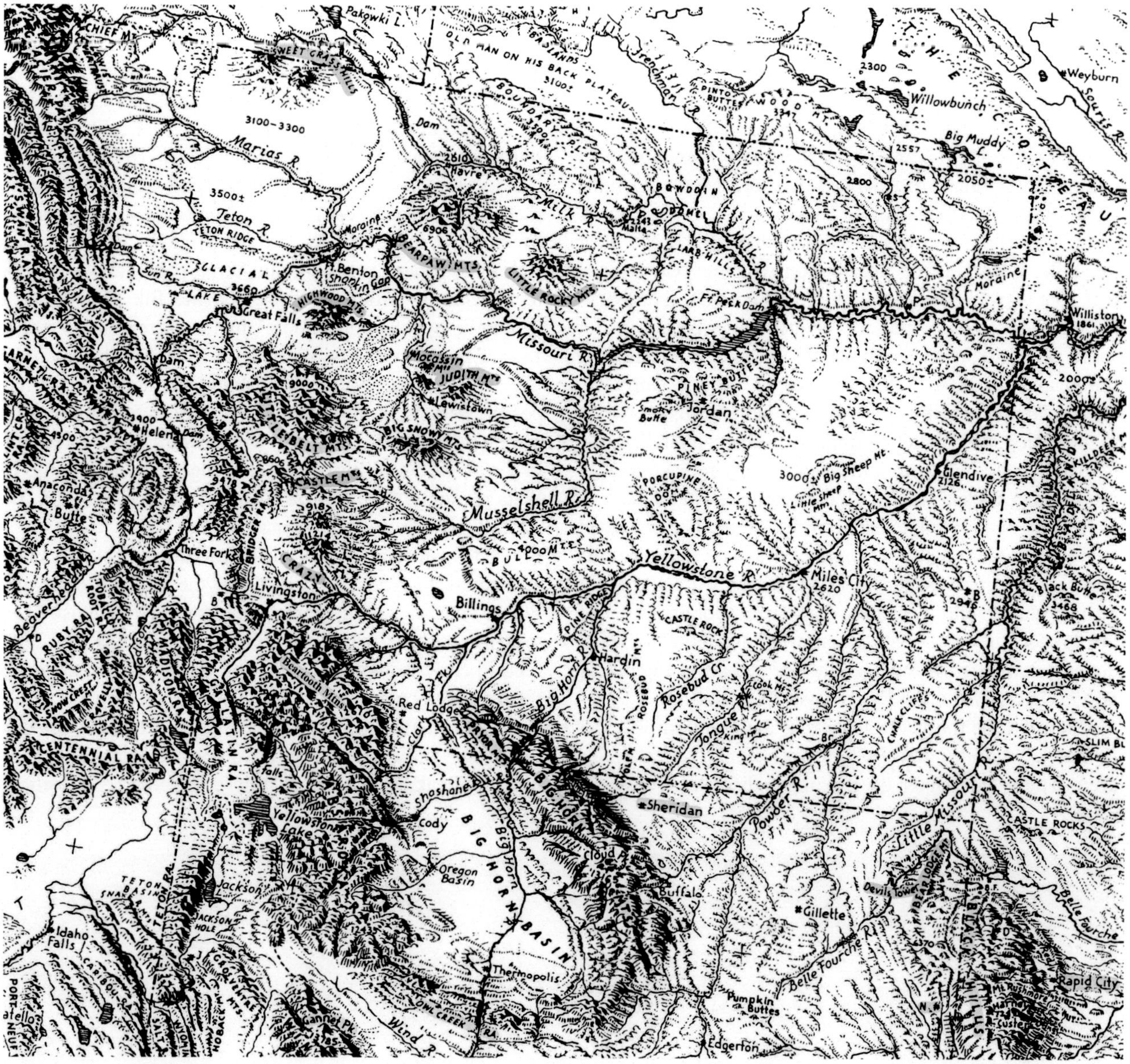

FIGURE 14.12 Landscape map of the northwest Great Plains showing anticlinal mountains (yellow) and intrusive mountains (light blue).

With the exception of Navajo Mountain, all of the blue highlighted mountains shown in Figs. 14.11 and 14.12 are cored with 75- to 18-million-year-old intrusive rocks. Intrusions that do not form distinct domal mountains are not highlighted in the figures. Most of the intrusions are granitic in composition although typically not true granite. Most are fine-grained with scattered large crystals. These intrusions, along with volcanic rocks of similar age, are associated with ore mineralization in the Colorado Mineral Belt. The central core of Navajo Mountain consists of resistant sandstone with unexposed intrusive rock at depth. The Navaho and Uinta Mountains are discussed briefly in Chapter 13.

INTRUSIVE DOMAL MOUNTAINS

The Colorado Mineral Belt is a northeast-trending 300-mile long, 15- to 30-mile wide zone of mining districts that extends from the Four Corners region on the Colorado Plateau across the Southern Rockies to Boulder, Colorado (enclosed within the light red lines in Fig. 14.11). The orientation of the mineral belt is unusual because it cuts obliquely across the topographic grain and seems to have no correlation with regional tectonic or structural elements. Zircon and argon dating indicates that intrusive and volcanic rocks associated with this belt are between 75 and 18 million years old. The main trend and some of the mineralization were created by intrusions that are more than 40 million years old; however, most of the world-class lead, zinc, silver, gold, and molybdenum mining districts are associated with younger intrusions and volcanism concentrated within a bulge in the trend near Salida and Leadville, and in the San Juan Mountains near Ouray. The Climax, Henderson, and Red Mountain mine localities all host intrusive and volcanic rock between 40 and 24 million years old. Notice that the bulge near Leadville occurs at the intersection with the Rio Grande Rift zone. Given this relationship, it is possible that extension associated with the rift opened easy pathways for some of the youngest magma to intrude. Hydrothermal fluids associated with the cooling of shallow intrusive bodies are responsible for emplacement of most of the mineral deposits, which formed by direct crystallization in cracks and voids within the rocks, or by replacement of pre-existing rocks. Much of the mineralization in the volcanic San Juan Mountains occurred along faults associated with caldera collapse, some of it as late as 11 million years ago. Mining began in 1858, reached a peak at the turn of the century, and has continued sporadically into the 21st century. Colorado is renowned for its many abandoned mining camps such as the one shown in Fig. 14.13 nestled in the Mosquito Range east of Leadville.

Many of the intrusions that cross the Southern Rockies in the Colorado Mineral Belt are part of a larger mountain area. These intrusions do not form discrete, individual domes and are not highlighted in Figs. 14.11 and 14.12. Examples of nonhighlighted intrusions include those in the Mount Princeton area in the southern Sawatch Range just north of the San Luis Valley, and numerous intrusions in the mountains east of Breckenridge (B, Fig. 14.11). Intrusive domal mountains in the 40- to 18-million-year-old range in the Southern Rockies and Colorado Plateau include the Henry Mountains, Abajo Peak, and La Sal Mountains in Utah, the San Miguel Mountains in Colorado, where one can find three peaks above 14,000 feet including Mount Wilson, and the Elk Mountains in Colorado, which boast more than a dozen peaks above 11,000 feet cored with young intrusive rock that includes Capitol Peak and Snowmass Mountain (both 14,000-foot peaks), Mount Gunnison, and Whetstone Mountain. Also present are domal mountains with older granitic intrusions in the 50- to 75-million-year-old range that include the La Plata Mountains and Ute Peak in Colorado, and the Carrizo Mountains in Arizona (Fig. 14.11).

The Spanish Peaks form an additional isolated intrusive dome in Colorado located at the edge of the Great Plains near the New Mexico border. West Spanish Peak tops out at 13,626 feet and East Spanish Peak at 12,683 feet. In early frontier days, the two peaks formed an easily visible topographic target for settlers crossing the Great Plains. Intrusive rock forms a large part of East Spanish Peak, but only a small area along the flank of West Spanish Peak. The rock, in this case, consists mostly of

FIGURE 14.13 Photograph of abandoned mining camp in the Mosquito Range near Leadville, Colorado.

FIGURE 14.14 Photograph of a vertical dike with West Spanish Peak in the background.

27 to 21 million-year-old, quartz-poor, granitic intrusions. The peaks are well known for the many dikes that radiate in all directions from the central mountain. A dike is a tabular intrusion (like a tabletop) that is often oriented vertical or near vertical. The largest dikes radiate up to 25 miles from the summit area. Fig. 14.14 is an example of a dike with West Spanish Peak in the background.

As shown in Fig. 14.12, there are numerous small intrusive domal mountains on the Great Plains of Montana. From north to south they include the Sweet Grass Hills, and the Crazy, Bearpaw (or Bears Paw), Little Rocky, Highwood, Moccasin, and Judith Mountains. Elevations are between 6000 and 8000 feet, which is 3000 to 4000 feet above the surrounding Great Plains. The mountains consist of a central area of intrusive rock surrounded with variable amounts of volcanic and sedimentary rock. The intrusive and volcanic rocks are part of the Central Montana alkalic province, which implies they are rich in potassium but poor in quartz and plagioclase relative to true granite. The rocks are 50 to 70 million years old. Intrusive rocks are especially well exposed in the Sweet Grass Hills, Little Rocky, Moccasin and Judith Mountains. Volcanic rocks are abundant in the Bearpaw and Highwood Mountains. The Crazy Mountains consist of a central intrusion surrounded by many dikes that intrude water-laid volcanic material. Intrusive rock is absent in the Big Snowy Mountains but may be present at depth.

A few small intrusions at the north end of the Black Hills have punched into surrounding sedimentary rock to create small dome structures that include Bear Lodge Mountain, Crow Peak, Citadel Rock, and Devils Tower (Fig. 14.11). Bear Lodge Mountain also contains a small amount of ancient (>2500 Ma) crystalline shield rock. These intrusions are mostly between 50 and 62 million years old and are probably related with the larger and more numerous intrusions of similar age that dot the northwestern Great Plains.

Devils Tower, shown in Fig. 14.15, is a vertical rock edifice that rises nearly 1300 feet above the Great Plains. Much of the tower is composed of near vertical five- or six-sided columns of a somewhat rare alkalic igneous rock known as phonolite porphyry, in which white feldspar crystals are visible in a fine-grained groundmass. The tower is interpreted as the eroded remnant of a shallow intrusion around which sedimentary rocks have been eroded. Intrusion at shallow depth allowed the rock to

FIGURE 14.15 Google Earth image looking west in Wyoming at Devils Tower (foreground) and the Missouri Buttes (upper right). The snow-covered Bighorn Mountains are seen in the distance behind Devils Tower.

cool quickly, which created a fine-grained, volcanic-looking groundmass. The columns formed as a result of magma cooling and shrinking in place. Devils Tower is a popular location for rock climbers except in June when it becomes a ceremonial destination for American Indians. The Missouri Buttes form a similar landscape only a few miles to the northwest.

THE SOUTHERN ROCKY MOUNTAINS

The anticlinal structure of rocks in the Middle-Southern Rocky Mountains developed primarily between 75 and 45 million years ago during the Laramide orogeny, the same mountain building phase that formed monoclines and broad uplifts on the Colorado Plateau. In the Middle-Southern Rocky Mountains, much of the deformation occurred between 65 and 50 million years ago. The mountains created at that time, however, are not the same as the mountains we see today. As we shall see, the landscape has evolved considerably since 50 million years ago via erosion, glaciation, normal faulting, and broad domal uplift. The Southern Rockies today form the apex of a broad, regional dome structure that stretches across the Great Plains, the Middle Rocky Mountains, and the Colorado Plateau. With 58 named peaks above 14,000 feet, and hundreds above 13,000 feet, the Southern Rockies are truly the roof of North America. Of particular importance with respect to landscape history is the Rio Grande Rift, which extends along the crest of the dome into the heart of the Southern Rockies as shown in Fig. 14.11.

The crystalline core of each anticline consists of shield rocks, and to a lesser extent the previously described young intrusive rocks. The shield rocks consist of circa 1.8-billion-year-old biotite gneiss, schist, and pegmatitic granite intruded by granitic rocks of three separate ages, 1.7, 1.4, and 1.08 billion years old. The 1.08-billion-year-old granitic rocks form nearly all of Pikes Peak, the southernmost 14,000-foot peak in the Front Range. An example of biotite gneiss and intermingled pegmatitic granite is shown in Fig. 3.3.

In this section, we use Fig. 14.11 to describe the distribution of anticlinal mountains in the Southern Rockies. With the exception of the Mosquito Range south of Leadville, all of the mountains described in this paragraph are anticlinal and cored with crystalline shield rocks. Many of the anticlinal uplifts are associated with reverse and thrust faults, and most were subsequently modified to a certain extent by younger normal faults. Fig. 14.16 is a generalized cross-section across central Colorado that shows two of the largest crystalline-cored anticlinal mountains, the Front Range and Sawatch Range, separated by a complex area of folds and faults that includes South Park and the Mosquito Range. As seen in Fig. 14.11, the crest of the Front Range anticline extends from the North Platte River in Wyoming, to Royal Gorge on the Arkansas River at Canon City, a total distance of 335 miles. The northern part of the Front Range in Wyoming is known as the Laramie Range. The southern part, south of Denver, is known as the Rampart Range. Across the Laramie River in northern Colorado the anticlinal Medicine Bow Mountains extend into Wyoming as a prong emanating from the Front Range. In the south,

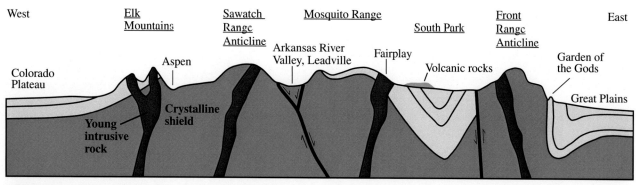

FIGURE 14.16 Cross-section across central Colorado that shows major anticlinal mountains, synclinal valleys, hogbacks at Garden of the Gods, and young intrusive rocks (dark red).

FIGURE 14.17 Photograph looking westward across the South Park synclinal valley from Wilkerson Pass, Colorado.

the anticlinal Wet Mountains are offset westward from the Front Range across the Arkansas River. To the west, South Park, at an elevation between 8500 and 10,000 feet, and North Park at about 8000 feet, both form synclinal valleys (Fig. 14.16). A park in Colorado is a wide, flat-floored, mostly treeless, synclinal valley. South Park as seen from Wilkerson Pass in Fig. 14.17 is particularly wide and flat with volcanic hills in its southern part. Farther west we cross into a line of anticlinal mountains that includes, from north to south, the Sierra Madre, Park, Gore, and Ten Mile/Mosquito Ranges (TM, Fig. 14.11). Still farther west, across the narrow Arkansas River Valley at Leadville, we reach the Sawatch Mountains, a massive anticlinal bulge of crystalline rock only 110 miles long and less than 25 miles wide. The southern terminus of the Sawatch Range is located near Salida where the south-flowing Arkansas River turns eastward. South of Salida, the Sangre de Cristo Mountains extend southward for more than 220 miles well into New Mexico where they become the main frontal range of the Southern Rockies. The Sangre de Cristo Mountains are anticlinal in form, but are especially complex with Laramide-age thrust faults, young intrusive rocks, 29- to 24-million-year-old volcanic rocks, and normal faults. West of the Sangre de Cristo's are a series of valleys that includes the San Luis Valley. These valleys, along with the Arkansas River Valley at Leadville, are part of the Rio Grande Rift system.

The only other major anticlinal mountain region in the Southern Rockies are the Needle Mountains (which

FIGURE 14.18 Google Earth image in Colorado looking northeast at the jagged peaks of the Grenadier Range.

includes the Grenadier Range, Fig. 14.11), located in the Weminuche Wilderness within the primarily volcanic San Juan Mountains of southwest Colorado. The Needle Mountains form a high, remote, rugged crystalline-cored dome well known for its three 14,000-foot peaks, Windom Peak, Sunlight Peak, and Mount Eolus, and its many steep, challenging 13,000-foot ascents, particularly in the Grenadier Range, as shown in Fig. 14.18, where glacial erosion has carved steep spires into quartzite, a highly resistant rock that is rare elsewhere in the Southern Rocky Mountain crystalline shield.

Crystalline-cored anticlinal mountains on the Colorado Plateau include the Uncompahgre Uplift, the Nacimiento (Jemez) Mountains and the Zuni Mountains. The Uncompahgre Uplift is transitional between the broad sedimentary uplifts of the Colorado Plateau and the crystalline-cored uplifts of the Middle-Southern Rockies. Most of the rocks are sedimentary. Shield rocks are exposed in Unaweep Canyon and along the southwest side of the fold beneath a relatively thin cover of Triassic (252−201 Ma) sedimentary rock. The Nacimiento (Jemez) Mountains and the Zuni Mountains form small crystalline-cored uplifts in New Mexico. One could conceivably also consider the Kaibab Plateau as a transitional crystalline-cored anticline due to the presence of crystalline shield rocks at the bottom of the Grand Canyon.

In the following sections, we will take a closer look at the Front Range and Sawatch Range, and then discuss the landscape history of the Southern Rockies. We discuss anticlinal mountains of the Middle Rockies and the northern Great Plains later in the chapter.

The Front Range

In Colorado, the Front Range extends 185 miles from the Arkansas River northward to the Wyoming border. Along this distance, there are six peaks above 14,000 feet mostly in the western part of the range. Farther north, for the next 70 miles in southern Wyoming, the mountain range almost disappears with no peaks reaching 9000 feet. For the final 80 miles, the range remains relatively low with one peak, Laramie Peak, above 10,000 feet. Width varies from less than 20 miles to 45 miles.

Along its eastern flank, eroded, tilted, upturned sedimentary layers form hogbacks that mark the edge of the Great Plains. Known collectively as the Dakota Hogback, different rock units form hogbacks, but the most common are sandstones of the Permian Lyons Formation

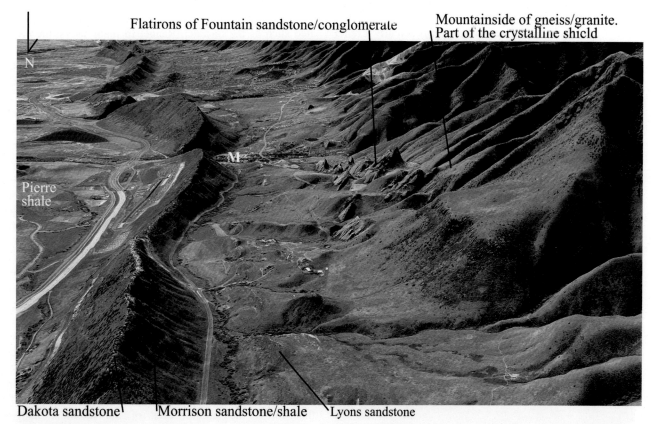

FIGURE 14.19 Google Earth image looking south toward Morrison (M) along the Front Range near Denver, Colorado. The Fountain Formation forms flatirons at Red Rocks Park against a mountainside composed of gneiss of the crystalline shield. The Dakota sandstone caps the hogback at left. The Lyons Formation forms secondary hogbacks.

(c. 275 Ma) and the lower Cretaceous Dakota Formation (c. 145–139 Ma). Fig. 14.19 looks southward at Red Rocks Park along the Front Range near Denver. A large hogback underlain with Dakota sandstone is obvious along the left (east) side of the figure. The lower slope of the hogback consists of Lower Jurassic Morrison Formation where, in 1877, several skeletons of dinosaur bones were unearthed, including the first *Stegosaurus* and *Apatosaurus*, thus making the Morrison Formation, and its namesake, the town of Morrison (M) seen at the top of the image, famous. For this reason, this part of the hogback is known as Dinosaur Ridge. The valley located between Dinosaur Ridge and Red Rocks consists of Permian-Triassic mudstone and sandstone of the Lykins and Lyons Formations. A close look at the lower part of Fig. 14.19 reveals a few low hogbacks of Lyons sandstone. Stratigraphically below the Lyons sandstone the Fountain Formation forms the famous flatirons that rest against 1.8-billion-year-old biotite gneisses of the crystalline shield, which form the mountainside. The contact between these two rocks is the Great Unconformity, celebrated here with a plaque embedded in the rock. A flatiron is similar to a hogback. It is a triangular-shaped resistant bedrock feature resting against a mountainside in which the dip of the flatiron is at a slightly higher angle than the mountain slope. The youngest rocks in the figure form the lowland to the east of Dinosaur Ridge. These rocks consist of Cretaceous Benton and Pierre Shale, which was deposited in the circa 80-million-year-old Cretaceous interior seaway shown in Fig. 8.3A.

The other area along the Front Range where hogbacks are fabulously displayed is Garden of the Gods in Colorado Springs. At this location, the Fountain Formation forms Balanced Rock, Siamese Twins, Steamboat Rock and the Cathedral Spires. Other towers and walls, including the Tower of Babel, Gateway Rocks, and the Kissing Camels, consist of the younger Lyons Formation. A photograph of Garden of the Gods is shown in Fig. 14.20.

When viewing the Front Range in Colorado, one is struck by the unusually flat summit region as seen in Fig. 14.21. As mentioned earlier, there are six peaks above 14,000 feet, but all of them, with the exception of Pikes Peak, are in the western part of the range. Rather than sharp glaciated peaks, the eastern side of the range appears as a river-dissected plateau. One could conceivably walk the entire eastern summit region from southern

Colorado to Wyoming and never rise above 10,000 feet. This wide, relatively flat summit region is actually a relict erosion surface known as the Rocky Mountain erosion surface. Although best observed in the Front Range, it is present in other ranges across the Southern Rocky Mountains. In most areas the erosion surface expresses itself as a tilted, relatively flat, dissected area located either at summit elevations or along mountain flanks. In some areas it is difficult to recognize because it is tilted, occurs at variable elevation, and is partially or wholly destroyed by glacier and water erosion. It appears to be preserved in the Medicine Bow, Sierra Madre, Park, and Gore Ranges among other areas, but is more difficult to recognize in higher ranges that have been heavily glaciated such as the Sawatch Range. In many areas, the erosion surface is recognizable because it is preserved below volcanic and sedimentary rocks deposited directly on the surface. These deposits also help determine the age of the surface as discussed later.

As a result of this erosion surface, the Front Range has a slope-flat-slope profile. Shown in Fig. 14.21, the frontal slope of the Front Range rises abruptly 3000 to 4000 feet above the Colorado Piedmont to an elevation between 8000 and 10,000 feet where, across a wide area, topography levels off to form the Rocky Mountain erosion surface (the flat in the slope-flat-slope profile). The mountain then rises another 4000 to 6000 feet to the summit areas of 14,000-foot peaks such as Pikes Peak, Grays Peak, and Longs Peak.

The presence of the Rocky Mountain erosion surface at high elevation implies a landscape history that is more complex than simple erosional lowering following Laramide mountain building events 75 to 45 million years ago. One clue to understanding the origin and significance of the erosion surface is to follow the surface northward along the trend of the Front Range. If we do this, the first thing we notice is that, without the presence of a high peak such as Pikes Peak, it is the erosion surface itself that forms the summit area of the mountain range (Fig. 14.21). The second thing to realize is that the erosion surface is inclined gently northward such that relief along the frontal slope of the Front Range diminishes from 3000 feet at Denver to zero in the area of Cheyenne,

FIGURE 14.20 Photograph of hogbacks at Garden of the Gods, Colorado Springs, Colorado.

FIGURE 14.21 Google Earth image looking southward at slope-flat-slope topography along the Colorado Front Range south of Denver. The mountains rise abruptly from the Great Plains to a flat on the Rocky Mountain erosion surface, and then rise again to Pikes Peak. The Sangre de Cristo Mountains are seen in the distance at upper right.

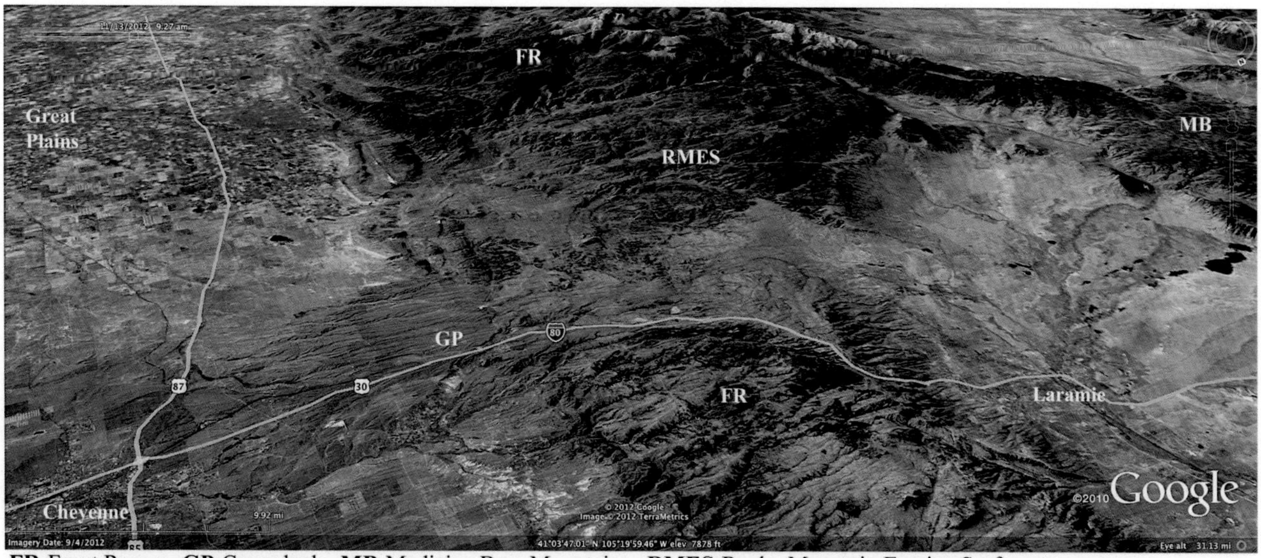

FR-Front Range, **GP**-Gangplank, **MB**-Medicine Bow Mountains, **RMES**-Rocky Mountain Erosion Surface

FIGURE 14.22 Google Earth image looking south-southwest at the Front Range (*FR*) between Cheyenne and Laramie. The Gangplank (*GP*) is an extension of the Great Plains that crosses the summit of the Front Range and continues into the Wyoming Basin. In this image the Rocky Mountain erosion surface (*RMES*) appears to disappear below sediment at the Gangplank.

Wyoming. This relationship is clearly evident in the Google Earth image of the Cheyenne area shown in Fig. 14.22.

Recall from Chapter 13 that the Colorado Piedmont intervenes between the Front Range and the High Plains, and that it represents the location where the High Plains, as represented by the top of the Ogallala Formation, has been eroded (Fig. 14.11). Erosion of the High Plains across the Colorado Piedmont is what creates the steep frontal slope at Denver. Northward, at Cheyenne, the High Plains (capped by the Ogallala Formation, Fig. 13.29) is not eroded, and instead extends all the way to the mountain front at a location known as the Gangplank. Here, the Gangplank, as shown in Fig. 14.22, forms a continuous surface from the Great Plains, across the summit of the Front Range, to the Wyoming Basin. This connection between the Great Plains and Wyoming Basin is a primary piece of evidence to suggest that the Wyoming Basin is a westward extension of the Great Plains. The Gangplank creates easy passage through the Rocky Mountains. It is the route followed by the Oregon Trail, the first transcontinental railroad, and the first coast-to-coast interstate freeway, I-80.

There is evidence that the High Plains land surface was once continuous with the Rocky Mountain erosion surface across the entire Front Range as well as across other ranges in the Southern Rockies. Fig. 14.23 is a profile across the Colorado Piedmont in the vicinity of Colorado Springs. Notice that the surface of the High Plains can be extrapolated to connect with the Rocky Mountain erosion surface. Further evidence is the presence of preserved erosional remnants of Ogallala Formation in the summit region of the Front Range north of Pikes Peak surrounding the town of Divide.

Thus, it appears that at some time following the Laramide orogeny, but prior to the present day, the area we know today as the Southern Rockies was part of the Great Plains. It was a relatively flat region with only a few isolated mountain peaks such as Pikes Peak. The landscape may have resembled what we see today in the Wyoming Basin or along the northwestern fringe of the Great Plains in Montana. At that time it would have been possible to drive across central Colorado and see only a few isolated peaks in the distance. We will discuss landscape evolution of the Southern Rockies later in this chapter.

The southern part of the Front Range is currently experiencing active normal faulting. The Rampart Range fault extends along the front of the range west of Colorado Springs, contributing to the height of the escarpment that separates the Great Plains from the Front Range. The Ute Pass fault follows route 24 northwestward from Colorado Springs to beyond Woodland Park. The down-dropped side of this fault has created the lowland valley at Woodland Park.

Sawatch Mountains

The Sawatch Range is the centerpiece of the Southern Rocky Mountains. Located between Leadville and Aspen it is about 110 miles north to south and less than 25 miles wide. Along its length there are 15 peaks above 14,000

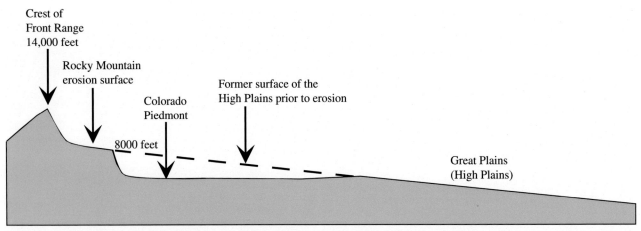

FIGURE 14.23 Profile that shows how the High Plains once connected with the Rocky Mountain erosion surface prior to erosion of the Colorado Piedmont. The mountain has a slope-flat-slope profile. Based on Chronic (1980, p. 20).

Gore Range
Pw-Mount Powell, 13,580'
R-Red Peak, 13,195'

Elk Mountains
C-Castle Peak, 14,272'
F-Fossil Ridge, 12,427'

Ten Mile Range
Q-Quandary Peak, 14,272

Mosquito Range
B-West Buffalo Peak, 13,332'
L-Mount Lincoln, 14,293'
Sh-Mount Sherman, 14,043'

Sawatch Range
E-Mount Elbert, 14,440'
H-Mount Harvard, 14,427'
Hc-Mount of the Holy Cross, 14,011'
P-Mount Princeton, 14,204'
S-Mount Shavano, 14,236'

T-Taylor River
SC-Sangre de Cristo Mountains

\ Active Normal Fault

FIGURE 14.24 Google Earth image of the Sawatch Range and surrounding area oriented due east. The heavy blue lines are active normal faults associated with the Rio Grande Rift system.

feet and more than 300 above 13,000 feet, more than any other range. It includes Mount Elbert, the highest in the Rockies at 14,440 feet. From the Arkansas River Valley looking westward, the range appears massive with wide mountain edifices separated by glacial valleys with moraines that stick out into the Arkansas River Valley where they surround Clear Creek Reservoir, Twin Lakes and Turquoise Lake. The highest peaks are at the eastern margin of the range, creating a steep, straight escarpment down to the Arkansas River. The peaks, although high, are not jagged, and for the most part offer a relatively easy challenge to the summit assuming good summer weather. The western side of the range leads down to Aspen at its northern end and merges with the Elk Mountains across the Taylor River at its southern end.

Fig. 14.24 is a Google Earth image oriented east that looks vertically down on the Sawatch Range and surrounding area. In this figure, the light-colored, dissected areas represent mountain elevations above tree line, at about 11,700 feet in this image. Large areas of the Sawatch Range, the Ten Mile-Mosquito Range, and part of the Gore Range are above tree line. The Continental Divide trends northward along the west side of the Sawatch Range such that all of the 14,000-foot peaks except Mount of the Holy Cross (Hc) are east of the divide (part of the Mississippi River System). The divide turns eastward from the Sawatch Range and separates Quandary Peak (Q) of the Ten Mile Range west of the divide from Mount Lincoln (L) of the Mosquito Range east of the divide, and then extends northeastward to the Front Range before turning westward and northward along the crests of the Park Range and the Sierra Madre.

The structure of the Sawatch Range is anticlinal with a crystalline core that includes not only shield rock but also young intrusive and volcanic rocks. The heavy lines in Fig. 14.24 are active normal faults associated with the Rio Grande Rift zone. Normal faults are present along most of the eastern side of the Sawatch Range, along the east side of the Gore Range at Red Peak (R) and Mount Powell (Pw), on both sides of the Ten Mile Range at Quandary Peak (Q), along the western side of the Mosquito Range at Mount Sherman (Sh), and on the western side and within the southern Mosquito Range at West Buffalo Peak (B). An important aspect of the Sawatch and surrounding ranges is the influence that normal faults have had on the landscape. Geological evidence discussed later suggests that normal faults are responsible for the existence of the upper Arkansas River Valley and most of the Mosquito Range, and are at least partly responsible for the height of peaks, not only in the Sawatch Range, but also in the Mosquito, Ten Mile, Gore, and Sangre de Cristo Ranges. One could argue that normal faults are responsible for more than half of the 14,000-foot peaks in Colorado. The Rio Grande Rift is discussed in Chapter 18, but due to its influence in central Colorado we elaborate on it in the following section.

Rio Grande Rift in Central Colorado

The Arkansas River Valley is interpreted as a normal fault basin, a graben that has been superimposed on what was previously the flank of the Sawatch anticline. Fig. 14.25A shows an ideal crystalline-cored anticlinal mountain where elevation is highest near its center, becomes progressively lower toward its flanks, and reaches sedimentary layers near its base. The Adirondack Mountains are a good

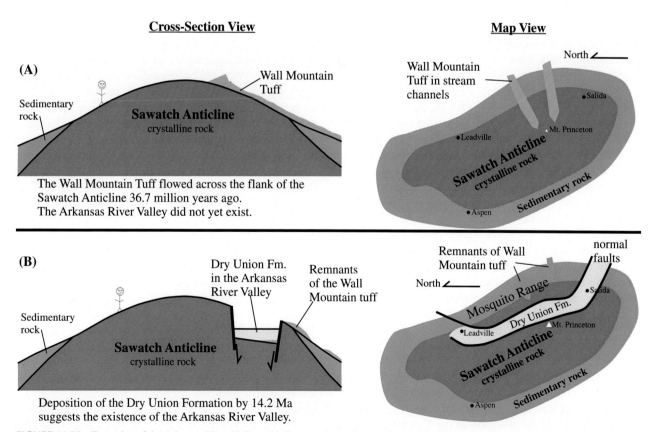

FIGURE 14.25 Formation of the Arkansas River Valley. (A) Cross-section and map prior to valley formation. (B) Cross-section and map following valley formation.

example. There are at least three lines of evidence to suggest that the Arkansas River Valley was carved out of the eastern flank of the Sawatch Range anticline by normal faults as shown in Fig. 14.25B. The first is the presence of active normal faults on both sides of the Arkansas River Valley as shown in Fig. 14.24. The second has to do with the geometry of stream channel deposits that contain fragments of the Wall Mountain Tuff. This volcanic rock, dated at 36.7 million years old, was violently ejected from a volcanic center in the Mount Princeton (P) area of the Sawatch Mountains and carried downstream in river channels. Rather than flowing down the Arkansas River, geologists have traced stream channel deposits eastward directly across the Arkansas River Valley and across the southern part of the Mosquito Range. The logical conclusion is that this part of the Arkansas River Valley, and the southern part of the Mosquito Range, could not have been in existence 36.7 million years ago because, if they were, the stream channels would have been redirected down the axis of the valley. The movement of volcanic fragments across what was then a nonexistent Arkansas River Valley is shown schematically in Fig. 14.25A.

The third line of evidence has to do with the rocks that underlie the Arkansas River Valley. Sedimentary rocks that surround the crystalline core of the Sawatch Range anticline are shown schematically in Fig. 14.25A. If the base of the anticline was located in the Arkansas River Valley, then these same sedimentary rocks should occupy the floor of the valley. Instead, we find the much younger Dry Union Formation, a poorly consolidated rock unit composed of sandstone, mudstone, gravel, and limestone either in fault contact or deposited directly above shield rocks on the floor of the valley as shown in Fig. 14.25B. A logical explanation is that normal faults created the Arkansas River Valley by down-dropping crystalline shield rocks, which were subsequently directly overlain by the Dry Union Formation. Given this explanation, the age of the Dry Union Formation provides an age for when the Arkansas River Valley first came into existence. Interlayered volcanic ash deposits from near Salida indicate that deposition of the Dry Union Formation was occurring between 14.2 and 8.5 million years ago. On this basis, we can conclude that normal faulting associated with the down-dropping of the of the Arkansas River Valley was in progress by 14.2 million years ago as depicted in Fig. 14.25B. These normal faults not only dropped the Arkansas River Valley, they also likely resulted in uplift of the bordering Sawatch and Mosquito Ranges.

Now, let us turn our attention to the effect that normal faults have had on the elevation of the Ten Mile-northern Mosquito, Gore, and Park Ranges. As seen in Fig. 14.11, the Park Range lies north of the Gore Range and north of the area shown in Fig. 14.24. On the basis of their outcrop pattern (crystalline shield rock in their core surrounded by Paleozoic and younger rock of the interior platform) all four ranges appear to be crystalline-cored anticlinal mountains. What is noticeable is that range elevation and relief decrease northward as active normal faults die out. Fig. 14.24 shows that active normal faults are prominent on both sides of the Ten Mile Range and on the east side of the Gore Range. The Ten Mile and Gore Ranges are correspondingly high and rugged. There is one peak above 14,000 feet in the Ten Mile Range, and several in the adjacent northern Mosquito Range. The Gore Range boasts peaks above 13,000 feet along with rough, glacial-carved terrain. Normal faults die out north of the Gore Range such that they do not affect the Park Range to any great degree. The Park Range is correspondingly less rugged with wide summit areas and only a few peaks above 12,000 feet. Here, the landscape more closely approximates that of a typical anticlinal mountain without the complicating effect of normal faulting.

Field data from the southern Gore Range suggests that normal faulting was active as early as 27 million years ago. (U-Th)/He data suggest a period of rapid exhumation between 20 and 13 million years ago, presumably brought on by accelerated normal fault displacement. A second period of exhumation began about 10 million years ago, presumably associated with active normal faults. Because the normal faults are both recent and active, we can presume that high topography, steep relief, and exposure of some of the Precambrian crystalline core in the southern Gore Range is a direct result of normal faulting. By extension, we could argue that high topography and rugged relief in the Ten Mile-northern Mosquito Range is also a direct result of active normal faulting, and that the absence of normal faults in the Park Range may explain the rather subdued topography. As for the Sawatch Range, normal faults along its eastern flank have presumably uplifted the eastern side of the mountain range creating, or increasing, the number of 14,000-foot peaks along its eastern side, and creating the steep escarpment leading down to the Arkansas River Valley.

Normal faults associated with the Rio Grande Rift system have shaped landscape in other parts of the Southern Rockies. The San Luis Valley, like the Arkansas River Valley, is a down-dropped basin separated from the Sangre de Cristo Range by an active normal fault that extends along the western base of the range for more than 160 miles. Similar with the Arkansas River Valley, geophysical evidence suggests that part of the San Luis Valley was a highland section of the Sangre de Cristo Mountains that was down-dropped due to normal faults. We might also assume that the height of some of the lofty Sangre de Cristo peaks is a manifestation of recent normal faulting. There are 10 peaks above 14,000 feet in the

Sangre de Cristo's. Apatite fission track dates and cross-cutting relationships suggest 2 periods of normal fault activity. The first began about 19 million years ago and may have lasted until 12 million years ago. The second began about one million years ago, is active, and is largely responsible for the steep slopes and great height along the west side of the Sangre de Cristo Range.

In conclusion, we can state that the central Colorado Rockies consist of two anticlinal culminations, the Front Range and the Sawatch Range, separated by a complex synclinal/anticlinal region where normal faults have helped shape and, in some cases, completely reincarnated the landscape. The development of the upper Arkansas River Valley during the past 14.2 million years shows how quickly normal faults can completely reincarnate landscape. Normal faults carved the Arkansas River Valley and the southern Mosquito Range out of the Sawatch Range anticline, and they appear to be partly responsible for the lofty heights of the present-day Gore, Ten Mile, Mosquito, Sawatch, and Sangre de Cristo Ranges. We now turn our attention to a discussion of the origin and significance of the Rocky Mountain erosion surface.

Landscape History of the Southern Rocky Mountains and Colorado Plateau

Given the existence of the Rocky Mountain erosion surface and its connection with the Great Plains, how do we interpret the landscape history of the Southern Rocky Mountains? Recall from Chapter 8 that, based on the distribution of marine sedimentary rocks, this entire region was near or below sea level 80 million years ago (Fig. 8.3). Thus, the landscape we see today must be younger than 80 million years. Based on the age of deformed rocks verses undeformed rocks, and on the age of sediment eroded from mountains and deposited into adjacent basins, we can say with strong certainty that the rocks were deformed and uplifted into a mountain range between 75 and 45 million years ago. We have indicated previously that deformation and uplift associated with this mountain building event is known as the Laramide orogeny. All geologists generally agree with this part of history. However, the elevation and relief of the mountain ranges that existed 45 million years ago and the evolution of landscape since 45 million years ago are not universally agreed upon. In this section we discuss one possible interpretation and some of the evidence used to support it.

Clues as to how landscape developed during and since the Laramide orogeny can be found in the erosional and depositional history of the region. Fig. 14.26 provides a simplified summary interpretation of mountain evolution. The Laramide orogeny created most of the anticlines,

(A) Mountain building (Laramide orogeny) between 75 and 45 million years ago creates the anticlinal structure and high mountains in the Southern Rockies.

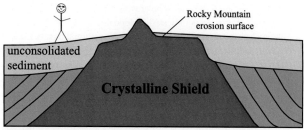

(B) The mountains are lowered, beveled, and partly buried in their own debris with only a few peaks, such as Pikes Peak, protruding above the erosion surface. The beveled shoulder of the mountain is the Rocky Mountain erosion surface, which began to form as early as 50 million years ago and was incised between 42 and 37 million years ago.

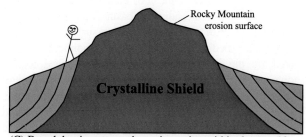

(C) Broad doming across the entire region within the past 10 million years resulted in erosion of unconsolidated sediment from around the buried, beveled mountain. The mountain and the Rocky Mountain erosion surface are exhumed and elevated, thus creating the present day landscape.

FIGURE 14.26 Series of cross-section sketches that show the evolution of the Southern Rockies since 75 million years ago.

synclines, monoclines, reverse, and thrust faults we see today across the Middle-Southern Rockies and across the Colorado Plateau. In a sense, this mountain building event provided the canvas for subsequent landscape development. There is no doubt that the Laramide orogeny created mountains in the Southern Rockies such as illustrated in Fig. 14.26A. Recall from Fig. 8.4 that the time period of Laramide orogeny, from 75 to 45 million years ago, was warm, potentially creating optimal conditions for rapid erosion. Much of the sediment eroded from interior mountains was not carried great distances, but instead was deposited in adjacent basins, or spread-out like an apron on the Great Plains. As mountains were lowered and adjacent basins filled with sediment, the mountain front would retreat, leaving behind a flattened, beveled, low-lying, slightly

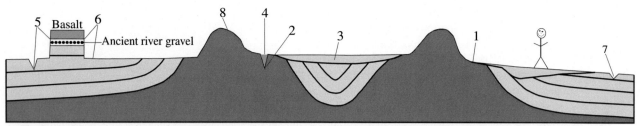

1 - The Rocky Mountain erosion surface can be projected to the top of the Denver Basin where sediment is dated at 50 million years old.
2 - Deep river incision into the Rocky Mountain erosion surface between 42 and 37 million years ago signifies a period of strong erosion and the generation of relief. The erosion surface was incised to depths that exceed 1970 feet (600 m).
3 - A period of erosion is indicated in depositional basins by the absence of rock in the age range from 40 to 37 million years.
4 - Deep river incision (2) and erosion of depositional basins (3) ended by 37 million years ago when some of the paleovalleys began to fill with sediment of the White River Group (37-24 Ma) and the Wall Mountain (volcanic) tuff (36.7 Ma).
5 - The difference in elevation between ancient river gravel below the basalt and the present-day elevation of a nearby modern river channel provides an estimate of the depth of river incision since the basalt was deposited.
6 - The difference in elevation between ancient river gravel below the basalt and the present-day elevation of the surrounding ground surface provides an estimate of erosional stripping since the basalt was deposited.
7 - The difference in elevation between the top of the Ogallala Group and the present-day elevation of a nearby modern river channel provides an estimate of the depth of river incision since the Ogallala Group was deposited (since about 5 or 6 million years ago).
8 - Apatite thermochronology provides an estimate of the thickness of overburden that has been removed.

FIGURE 14.27 Summary cross-section that explains criteria used to interpret landscape evolution in the Southern Rocky Mountains. Numbers are keyed to the text. Based on Rosenberg, et al. (2014).

tilted bedrock surface at the foot of the mountain roughly at the elevation of the adjacent basin and partly covered with a thin layer of river gravel. Such a surface in geology is known as a pediment. In our case, the pediment is represented by the Rocky Mountain erosion surface and is shown in Fig. 14.26B as the beveled shoulder of a mountain thinly buried in debris. It would have been at this time, beginning as early as 50 million years ago, when it would have been possible to travel across central Colorado and see only a few mountain peaks. The Colorado Rocky Mountains may have remained buried in their own debris and essentially part of the Great Plains until as late as 6 million years ago when broad uplift and erosion exhumed the buried, beveled mountains thereby creating the present-day landscape as shown in Fig. 14.26C.

Given the summary explanation above, let us now provide a more detailed explanation. Fig. 14.27 is a cross-section that attempts to explain the type of information gathered and how it is used to interpret landscape evolution. Numbers given in the following discussion are keyed to this figure. The figure is not to scale, nor it is an actual cross-section of any specific location in the Rocky Mountains.

The Rocky Mountain erosion surface had already begun to form across the Front Range by 50 million years ago because the erosion surface can be projected to the top of the Denver Basin where sediment is dated at 50 million years (#1). The erosion surface continued to develop across the Rocky Mountains and by 42 million years ago the mountains were largely buried and part of an expanded Great Plains. Fig. 14.28 shows the location of large lakes and depositional basins that were in existence circa 50 million years ago superimposed on present-day landscape. Note the presence of north-flowing rivers (opposite their present-day flow direction). The modern Grand Canyon and possibly the Colorado River had not yet formed.

Late in the development of the erosion surface, after about 42 million years ago, there is evidence for widespread river incision. Keep in mind that during its formation, the Rocky Mountain erosion surface would remain flat (with little relief) and partially covered in river gravel. Deep river incision across the erosion surface signifies a period of strong erosion, the development of paleovalleys, and the generation of relief. In northern Colorado-southern Wyoming, the erosion surface was incised to depths that exceed 1970 feet (600 m, #2). At the same time, many of the depositional basins shown in Fig. 14.27 underwent erosion. A period of erosion is indicated in these basins by the absence of rock in the age range from 40 to 37 million years (#3). Deep incision and erosion of the basins ended by 37 million years ago when erosion switched to deposition as indicated by the deposition of rocks of the White River Group, dated 37 to 24 million years old, and the Wall Mountain (volcanic) Tuff, dated 36.7 million years old (#4).

Some significant events occurred in the southern Rockies from 37 to 10 million years ago, but this was not a time of deep river incision or widespread erosion. The Wall Mountain Tuff signaled the beginning of a massive volcanic period emanating from the San Juan Mountains that reached a climax between 29 and 26 million years

FIGURE 14.28 Map that shows the location of large lakes and depositional basins (shaded areas), and river drainage directions (thick blue lines with arrows) circa 50 million years ago superimposed on a present-day landscape map. The Gore Range is highlighted in yellow. Based on Cather et al. (2012).

ago. Elevation of the San Juan's may have exceeded 16,400 feet (5000 m) at that time and ash from volcanic explosions likely covered all of Colorado. The Chuska erg, a large Sahara-like desert sand deposit, covered nearly all of the Colorado Plateau from 26 to at least 16 million years ago. Also by this time, normal faults were active in the Rio Grande Rift system. Between 12 and 8 million years ago, basalt flowed across large areas of the Colorado Plateau and the Southern Rockies. But irrespective of these changes, by 10 million years ago the Rockies north of the San Juan Mountains may still have resembled Figs. 14.26B and 14.28 with large basins, low-relief, low-lying mostly buried mountains, regions covered in volcanic rock, possibly north-flowing rivers, and only local exposure of the Rocky Mountain erosion surface. The Colorado River was flowing across Colorado by this time but was not flowing through the Grand Canyon area or to an ocean. Instead, it flowed into one or more depositional basins west of the Rockies.

The past 10 million years, and particularly the past 6 million years, have produced profound landscape changes across the region including the erosion of rock layers, the generation of modern-day elevation and relief, normal faulting, incision of canyons including Canyonlands and the Grand Canyon, basalt lava flows, and the development of the modern Colorado River system. By using a combination of stratigraphic correlation, apatite fission track and (U-Th)/He dating techniques, it is possible to determine how much rock thickness has been removed (stripped) from the land surface and the depth to which rivers have incised into the landscape since 10 million years ago. Erosional bedrock stripping involves removal of rock layers such as in the evolution of bench-and-slope landscape (Fig. 13.2). We are perhaps lucky that there was an outpouring of basalt between 8 and 12 million years ago because these rocks are easily dated and they can be used to determine the amount of river incision and erosional stripping that has occurred since they were deposited.

Basalt will flow into and fill river channels, which buries and preserves the underlying river gravel and prevents the river from reoccupying the channel. The river is forced to establish itself along the margin of the basalt flow. Subsequent river erosion will cut into and remove the surrounding sedimentary layers and leave the basalt flow as a highland mesa. The difference in elevation between ancient river gravel buried beneath the basalt mesa and a nearby modern river channel, or between the ancient river gravel and the surrounding ground surface, provides respectively, an estimate of the depth of river incision (#5) and the depth of erosional bedrock stripping (#6) since the basalt was deposited. Grand Mesa, near Grand Junction Colorado, is an extreme example of this type of topographic inversion. Grand Mesa is capped with circa 10-million-year-old basalt, yet stands as much as 5400 feet above the nearby Gunnison River. The relationships indicate that more than 5000 feet of sedimentary rock has been removed from around Grand Mesa via river downcutting and layer stripping since the basalt was deposited. Similar information can be obtained on the Great Plains based on the amount of river incision that has occurred following deposition of the Ogallala Formation 5 or 6 million years ago (#7). In the absence of such features, apatite fission track and (U-Th)/He theromochronology can be used to provide information regarding the depth from which rock has been exhumed to the surface over the past 10 million years. This procedure indirectly determines the amount of overburden that has been removed (#8).

Analyses of this type suggest that over the past 10 million years the Colorado River has incised 3600 to 4900 feet (1100−1500 m) across the central Colorado Rockies whereas rivers in southern Wyoming have incised less than half as much (#5). The analyses suggest further that 3280 feet (1000 m) of overburden has been removed by erosional stripping across the highest peaks in the southern Rockies in the past 10 million years, and over the same time period there was as much as 6560 feet (2000 m) of river incision in the Canyonlands region of the Colorado Plateau. Adding these layer thicknesses to the existing land surface will result in elevations that are substantially higher than present-day elevations. However, we must also add isostatic subsidence due to the weight of the additional overburden. When these calculations are done, we can suggest that elevation across the Colorado Rockies and Colorado Plateau 10 million years ago was on the order of 1500 to 3500 feet less than present-day elevation. Although there is a great deal of uncertainty attached to this estimate, it does imply that reasonably high elevation was maintained during and since the Laramide orogeny in the Southern Rockies and Colorado Plateau even if relief was not maintained. It also implies that between 1500 and 3500 feet of uplift has occurred across the region in the past 10 million years. This uplift, coupled with glaciation, is responsible for exhumation of the present-day Southern Rockies, the incision of deep canyons, the generation of modern-day relief, and the establishment of numerous (if not all) 14,000-foot peaks. Cosmogenic ^{10}Be measurements suggest that highlands underlain with crystalline rock are eroding at much slower rates than basins underlain with sedimentary rock. Such measurements imply that basins are getting deeper relative to mountain heights. In other words, mountain exhumation and the generation of relief across the Southern Rockies continues to the present day.

Cause of Accelerated Erosion in the Southern Rocky Mountains and Colorado Plateau

Although the details and timing of erosion are debatable, we can conclude that the Southern Rocky Mountain landscape developed as a result of at least three separate erosional periods. The first occurred during Laramide orogeny from 75 to 45 million years ago including initial development of the Rocky Mountain erosion surface. The second was between 42 and 37 million years ago when the Rocky Mountain erosion surface and geologic basins underwent incision and erosion. The third occurred during the past 10 million years, and particularly in the past 6 million years when buried anticlinal uplifts and the Rocky Mountain erosion surface were exhumed, and canyons were formed across the Colorado Plateau. All three accelerated incision periods are the result of some combination of uplift, climate change, and base level drop.

The Laramide uplift was clearly tectonic in nature and likely resulted in high overall elevation on the order of 10,000 feet or so, although rapid erosion at the same time might have prevented high relief. The cause of the 42- to 37-million-year-ago erosion event is less certain, but data from the Middle Rockies (discussed later) suggests that erosion was associated with uplift. This uplift, however, apparently still left the Southern Rockies and the Colorado Plateau 1500 to 3500 feet (450–1100 m) below present-day elevation. Given the available data, we can suggest that this elevation deficiency was made up within the past 10 (or possibly 6) million years by broad uplift of a giant dome structure centered in the Colorado Rockies that encompasses the Colorado Plateau, the western Great Plains, and the Middle Rockies. Broad doming occurred apparently without strong deformation in the sense that nearly flat-lying sedimentary rocks have remained nearly flat-lying. Broad uplift is likely ongoing, and is ultimately responsible for the erosional exhumation of the Southern Rockies as described in Fig. 14.26.

Broad tectonic uplift within the past 10 million years coupled with base level changes brought on by changes to river patterns, is the likely cause of canyon cutting on the Colorado Plateau. The drainage pattern on the plateau prior to 10 million years ago may have resembled that shown in Fig. 14.28 with large lakes and rivers that were not integrated (connected). It is only in the past 8 million years that the Green River has flowed into the Colorado River, and only in the past 6 million years that the Colorado River has flowed through the Grand Canyon and reached the Pacific Ocean. As high elevation rivers become connected with low elevation rivers, base level would fall, thereby favoring incision and canyon cutting.

Indirect evidence for broad uplift in the past 10 million years includes the presence of high elevation across the entire region (Fig. 1.7). The Colorado Plateau, western Great Plains, and Wyoming Basin together form the highest nonmountainous region in the country. The Colorado Rockies are arguably the highest area in North America. Although other parts of North America have high peaks, including Denali (Mt. McKinley), the highest at 20,310 feet, there are only 40 peaks that rise above 14,000 feet in Alaska, Canada, and Mexico combined. Colorado has 58 peaks above 14,000 feet, all within a relatively small area. Additional evidence includes the many hot springs in the Colorado Rockies (as many as 84). The existence of so many hot springs indicates that rocks below the mountain topography are hot and, therefore, buoyant enough to isostatically rise as a broad dome. Such a possibility is supported by measurements of crustal thickness, which varies from 23 to 35 miles (38–56 km). This thickness, alone, is not enough to support the height of the mountains. The height must be partly supported by thermal isostasy. Additional evidence is provided by geophysical studies, which indicate that the mantle below the crust is relatively warm and buoyant. Some of the near-surface heat may have been introduced 40 to 18 million years ago during intrusion associated with the Colorado Mineral belt, or more recently with development of the Rio Grande rift system. Finally, the high rates of river incision and the high apatite fission track and (U-Th)/He derived exhumation rates across the Southern Rockies and Colorado Plateau are consistent with accelerated tectonic and isostatic uplift over the past 6 to 10 million years.

First There Is a Mountain

Donovan P. Leitch wrote the lyrics "first there is a mountain, then there is no mountain, then there is" (from There is a Mountain, 1967). Although not meant to fit the history of the Southern Rocky Mountains, it does seem appropriate. Given the scenario outlined in Fig. 14.26, we can suggest at least two iterations of mountains since 80 million years ago. The first were the Laramide Mountains when the Front Range and Sawatch anticlinal mountains (among others) were produced. These mountains could have initially been high and rugged ($\pm$ 10,000 feet), but given that sea level was high (Fig. 8.3) and that the Rocky Mountain erosion surface had already begun to form by 50 million years ago, it is possible they were rather quickly eroded to low-relief although not low elevation. These are not the mountains we see today. There are two major differences between these mountains and the present-day mountains. The first is the presence of the Rocky Mountain erosion surface, which did not exist at high elevation in the Laramide incarnation of the mountain range. The second is the ubiquitous glacial sculpturing that began 2.4 million years ago. Many of us are quite happy with both of these creations. The existence of a relatively flat erosion surface at high elevation allows

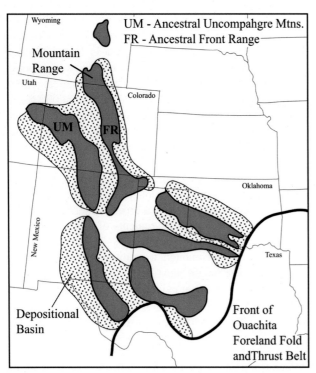

FIGURE 14.29 Extent of the Ancestral Rocky Mountains and associated depositional basins.

Colorado to be one of the premier locations in the world for mountain biking. It is one of the few places where one can ride for great distance at high elevation without too much worry of overly steep slopes and a quick trip over your handlebars. Glaciers, although certainly not unique to the Rockies, carved many of the spires and cliffs that make all high mountains so picturesque.

Amazingly, the Laramide orogeny was not the first event to create the Rocky Mountains. An earlier iteration known as the Ancestral Rocky Mountains was in existence roughly 300 million years ago. As seen in Fig. 14.29, the Ancestral Rockies stretched from Wyoming to Oklahoma and Texas. The Arbuckle and Wichita Mountains, and the Llano Uplift, are remnants of the Ancestral Rockies. Similar to the Laramide event, deformation included anticlines, synclines, high-angle reverse faults, exposure of ancient shield rock, and the development of deep depositional basins. Differential uplift (uplift plus subsidence) was as much as 20,000 feet, generating mountains that may have been on the order of 10,000 feet.

Evidence for the existence of the Ancestral Rockies is based on the presence of eroded residue preserved as sedimentary rock in nearby depositional basins. If you travel through Colorado, you will likely notice that many areas consist of red sandstone, shale and conglomerate. Two such reddish formations are the Fountain and Lyons Formations. These are the same formations that form the flatirons at Red Rocks Park and some of the hogbacks at Garden of the Gods. In fact, if you chance to visit the Great Unconformity at Red Rocks, you will likely notice that the Fountain Formation contains clasts (pebbles and cobbles) that look identical to the underlying crystalline shield gneisses. These clasts were derived directly from erosion of the shield rocks. The age of the Fountain Formation (between 323 and 299 Ma), therefore, indicates when shield rock was exposed at the surface and undergoing erosion. The Fountain Formation and correlative rocks of Pennsylvanian-Permian age (323–252 Ma) are the erosional remnants of the Ancestral Rockies. We know there were mountains at this time based on the sheer volume of red rocks. The rocks were deposited in basins along the margins and eventually above the eroded stub of the mountains. Some of these basins were later elevated into mountains during Laramide orogeny. Rocks surrounding Vail, for example, consist in large part of Pennsylvanian-Permian red sandstone and shale. Similar rocks form the summit areas of numerous 14,000-foot peaks that include Maroon Peak, North Maroon Peak, Pyramid Peak, and Castle Peak in the Elk Mountains, Mount Bross in the Mosquito Range, and Crestone Peak, Humboldt Peak, and Kit Carson Mountain in the Sangre de Cristo Mountains. One could suggest that it is these rocks from which Colorado (Spanish for "colored red") got its name. Early Spanish explorers named the Rio Colorado ("river colored red") for the red silt carried by the Colorado River. They also named the Sangre de Cristo ("Blood of Christ") Mountains possibly for the red hues visible at sunset.

The location of the now vanished Ancestral Rockies can be determined based on the age of sedimentary rocks that depositionally overlie crystalline shield rocks. In order for shield rocks to be exposed at the surface during the Ancestral Rockies uplift, the overlying sedimentary rocks must first be removed. At the time of the Ancestral Rockies, only the oldest rocks of the interior platform were in existence (the Cambrian–Mississippian rocks between 541 and 323 Ma). We can gauge where ancestral mountains once existed by looking for areas where the Cambrian–Mississippian rock succession is missing and where Pennsylvanian through Cretaceous rocks are in direct contact above the crystalline rocks. We have seen at Red Rocks Park that the Pennsylvanian-Permian Fountain Formation is in direct contact with shield rocks of the Front Range (Fig. 14.19). In this case, the Cambrian–Mississippian rocks were removed via erosion and shield rocks were exposed at the surface prior to deposition of the Fountain Formation. We can infer from this relationship that an ancestral mountain was in existence at this location and that the mountain was eroded and eventually covered with rocks of the Fountain Formation. A second ancestral mountain range was at the location of the present-day Uncompahgre Plateau where

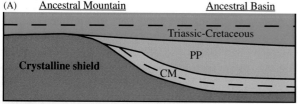

Relationships on the left side of the diagram indicate that crystalline shield rocks and the Cambrian-Mississippian rocks (CM) were uplifted and eroded to create a crystalline-cored anticlinal mountain (Ancestral Rockies). The Pennsylvanian-Permian (PP) rocks represent detritus eroded from the mountain and deposited in a basin on the right side of the diagram. As the mountain was eroded to lowland, late PP deposition transgressed across the eroded mountain flank and was deposited unconformably above tilted CM rocks and above shield rocks. The mountain remained a highland until Triassic rocks were deposited across the entire area.

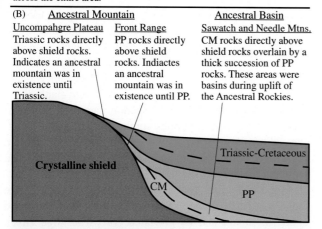

FIGURE 14.30 (A) Cross-section showing stratigraphic relationships across the Southern Rocky Mountains following erosion of the Ancestral Rockies but prior to Laramide orogeny. (B) Cross-section showing present-day (post-Laramide) stratigraphic relationships surrounding the Uncompahgre Plateau, Front Range, Sawatch and Needle Mountains.

Triassic rocks (252–201 Ma) directly overlie shield rocks. The absence of even Pennsylvanian-Permian rocks on the Uncompahgre Plateau suggests that this area remained a highland until the Triassic.

Conversely, we can gauge where depositional basins were located by looking for areas where the Cambrian–Mississippian rock succession is preserved and the overlying Pennsylvanian-Permian rocks are especially thick. When we look at the rocks surrounding the crystalline cores of the Sawatch and Needle Mountains, we find that both preserve the Cambrian–Mississippian rock sequence as well a thick accumulation of Pennsylvanian-Permian red rocks. Both areas must have been depositional basins during the rise of the Ancestral Rockies. The depositional relationships that led to the discovery of the Ancestral Rockies and associated depositional basins are illustrated in Fig. 14.30.

Using the above relationships it is possible to map the location of the Ancestral Rockies and the adjacent depositional basins as shown in Fig. 14.29. The two mountain ranges in Colorado are known as the Ancestral Front Range and the Ancestral Uncompahgre Mountains due to their coincidence with the present-day Front Range and Uncompahgre Plateau. The deep trough between these ancient mountains collected almost 13,000 feet (4000 m) of sediment, which is now exposed surrounding the Sawatch Range and Needle Mountains. The Ouachita Mountains were also undergoing uplift at this time, and their proximity to the Ancestral Rockies suggests the two mountain building episodes are related.

If we go farther back in time, there is evidence for uplift of an ancient Front Range long before the Ancestral Rockies. The age and distribution of faults within the crystalline shield suggests that an ancient Front Range may have been in existence sometime between 1.4 and 1.1 billion years ago. Faults originally associated with this most ancient of mountain building events were likely reactivated during Ancestral Rocky and Laramide mountain building events, and possibly again in the form of the Rampart Range and Ute Pass normal faults. Thus, we can reasonably say that there have been at least four separate iterations of the Southern Rocky Mountains.

ANTICLINAL MOUNTAINS OF THE MIDDLE ROCKIES

Turning now to the Middle (Wyoming) Rocky Mountains and adjacent Great Plains, the major crystalline-cored anticlinal mountains are the Wind River Range, Bighorn Mountains, Beartooth Mountains, and Black Hills. There are, in addition, several smaller anticlinal mountains that include the Laramie Range, Sierra Madre, Medicine Bow, Ferris, and Shirley Mountains, and the Casper, Owl Creek, Granite, and Rattlesnake Ranges (Fig. 14.11). All of these mountains host crystalline shield rocks that are more than 2.5 billion years old, much older than the oldest rocks in the Southern Rockies. Young shield rocks (<1.8 Ga), similar to those in the Southern Rockies, are present the Sierra Madre, Medicine Bow, southern Laramie Mountains, and in the Black Hills, but elsewhere these rocks are absent. Young intrusive rocks (<75 Ma) are rare in these ranges except in the Rattlesnake Range and the northern Black Hills. The Little Belt Mountains contain both circa 2.5-billion-year-old crystalline rock and young intrusive rock in their core, but shield rocks otherwise are absent in the isolated mountains of the northwest Great Plains (Fig. 14.12).

Although the structure is similar across the Middle and Southern Rockies, the landscape of the Middle Rockies is distinctly different from that to the south. In Wyoming, there is more basin than mountain. If we look for a solution as to why numerous basins occupy the

Middle Rocky Mountains, and why the Southern Rockies are mountainous, we can suggest that broad domal uplift over the past 10 million years was a factor. The Southern Rockies are at the apex of this broad uplift such that most of the sediment that had buried the mountains has since been removed. The Middle Rockies have not been uplifted to the same extent and, therefore, are still largely buried in their own debris. Other factors may have played a role as well. Perhaps the Middle Rockies were more deeply buried than the Colorado Rockies. Or, perhaps the Wyoming Basin is an area where there are fewer anticlinal upwarps to be uncovered.

The intervening synclinal basins are important because, as seen in the interpretation of the Ancestral Rockies, they provide a plethora of information that can be used to interpret uplift of the surrounding mountains. The age and composition of sediment deposited during Laramide orogeny, for example, can indicate how long ago mountains were uplifted and when ancient crystalline rock was exposed at the surface. Certain basin deposits, such as volcanic debris, carbonate cement, and freshwater shells, provide hydrogen and oxygen isotope information that can constrain paleoelevation and paleorelief.

Erosion surfaces are present in some of the major ranges, but it is uncertain if these are the same age as the Rocky Mountain erosion surface. All of the ranges were deformed and uplifted during Laramide orogeny, but individual landscape histories are not fully documented or universally accepted. In the following sections, we briefly survey the Wind River, Bighorn, and Beartooth Mountains, and the Black Hills, and discuss at least one interpretation of landscape development. The structure of each mountain region is that of an elongate dome (a doubly plunging anticline) with their long axes oriented approximately northwest-southeast. Thrust faults accentuate the anticlinal structure in each of the major ranges except for the Black Hills.

Wind River Range

The Wind River Range is a crystalline-cored anticlinal mountain approximately 125 miles long and 40 miles wide located in the northern part of the Wyoming Basin (Fig. 14.11). It forms an outlying mountain range on the Yellowstone Plateau. Crystalline rocks are primarily 2.6 to 3.1 billion years old and are part of the crystalline shield. Most of the rock is granite and gneiss, but gold- and iron-bearing metasedimentary rocks are present surrounding South Pass. The mountain range is one of the highest, most glaciated, and most remote in the contiguous United States. There are 32 peaks above 13,000 feet and many more above 12,000 feet. The highest peaks are in the northern part of the range including Gannett Peak, the highest in Wyoming at 13,810 feet. Nearly the entire range was glaciated. Glaciation was so strong that a small icecap developed in the north-central part of the range. Glacial lobes moved down valley to create the wide Green River Valley in the northern part of the range as well as moraine-damned lakes such as New Fork, Willow, Freemont, Boulder, and Bull Lakes along mountain flanks. The moraine that surrounds Willow Lake is shown in Fig. 12.6. Today, there are 25 named glaciers and a spectacular array of glacial features that include the Cirque of Towers, a popular climbing area in the southern part of the range. In most areas there are no roads within 15 miles of the glacial-carved summit region, which also forms the Continental Divide. The central northeastern flank of the range forms the Wind River Indian Reservation, home to the Eastern Shoshone and the Northern Arapaho tribes.

Uplift of the Wind River Range and initial development of the anticlinal structure was enhanced with displacement on major thrust faults, particularly the Wind River thrust. This fault is mostly buried beneath younger sediment. It is located along the southwestern flank of the mountain based on seismic reflection data. The fault displaced shield rocks 13 miles southwestward above younger sedimentary rocks. It was active prior to 49 million years ago based on the age of the oldest undeformed sediment that covers the fault.

Fig. 14.31 looks southeastward along the crest of the range. One can immediately notice the anticlinal shape of the mountain and its glaciated core. The large Green River glacial valley (Gr) is visible at the northern (front) of the range. Although the summit region is rugged and remote, there is a well-developed erosion surface at the 9000 to 13,000 foot elevation on both flanks of the mountain that allows for relatively easy foot travel. The erosion surface is labeled e and pe, and is present on several visible mountain peaks that are also labeled in Fig. 14.31. On the northeast (left) flank, the erosion surface is well displayed on Horse Ridge (H) and Goat Flat (G). On the southwest side, the erosion surface is fairly continuous along the entire flank of the mountain. It slopes downward from well over 11,000 feet in the north (front) of the image where it is marked by several flat-topped summit peaks that include Squaretop (S) and Big Sheep Mountain (B), to less than 10,000 feet in the south (upper center-right) where it is continuous with a pediment surface (labeled pe) beveled into the side of the mountain, which slopes downward to the level of the adjacent Green River Basin. Fig. 14.32 is a photograph looking southward from Green Lakes Valley that shows the erosion surface at the summit of Squaretop Mountain. There are several additional flat-topped mountains farther south in the summit area that also appear to preserve part of the erosion surface including Jackson Peak (13,517 feet), Downs Mountain (13,349 feet), and Desolation Peak (13,155 feet). None are readily visible in Fig. 14.31, however the

e-Erosion surface. pe-Pediment erosion surface on the southwest flank. Additional areas that show the erosion surface are B-Big Sheep Mountain, G-Goat Flat, H-Horse Ridge, L-Lost Eagle Peak, O-Osborn Mountain, S-Squaretop Mountain, Sh-Shale Mountain, T-Three Waters Mountain. Also Gr-Green River Lakes. Arrows point to the southern edge of the pediment surface. A northeast-southwest cross-section line (Figure 14.33) is shown.

FIGURE 14.31 Google Earth image looking southeast along the crest of the Wind River Range showing the erosion surface.

existence of these peaks suggests that the erosion surface was continuous across the summit area prior to glacial erosion. The erosion surface does appear to project across the northern nose of the range as marked by Osborn (O), Shale (Sh), and Three Waters Mountains (T).

If you trace the erosion surface at Horse Ridge (H) or Goat Flat (G) back to the base of the Wind River Range (to the left side of Fig. 14.31), you will notice that the erosion surface traces directly below hogbacks and cuestas of Cambrian-Ordovician (541–444 Ma) sedimentary rock. The contact between these sedimentary rocks and the underlying shield rocks is the Great Unconformity. The implication is that the erosion surface formed initially along the Great Unconformity.

Fig. 14.33 is a cross-section across the southern part of the range that shows the inferred relationships with glacial erosion removed. The cross-section is oriented as if you are looking southeastward, the same look direction as Fig. 14.31, so that northeast is on the left. In Fig. 14.33, the summit area along the northeastern (left) side of the cross-section forms a relatively smooth erosion surface that corresponds roughly with the Great Unconformity.

The pediment surface on the southwest side appears to be the one location along the flanks of the mountain where the erosion surface has developed well below the Great Unconformity. In the southernmost part of the range, shown with a short blue arrow in Fig. 14.31, the pediment surface is buried beneath Upper Eocene–Lower Oligocene (c. 37–28 Ma) rocks of the Green River Basin. This relationship indicates that the erosion surface formed prior to deposition of these rocks. The Upper Eocene–Lower Oligocene rocks are eroded from the top of the erosion surface farther north in the area of the long blue arrow creating a small escarpment that is visible both in the image and in cross-section. Still farther north, the pediment surface (pe) appears to be continuous with, and part of the erosion surface (e) preserved on Squaretop (S) and Big Sheep Mountain (B). Given these relationships, we can presume that the erosion surface (e), and the pediment surface (pe), had formed across the entire mountain range prior to 37 million years ago.

Apatite fission-track ages from crystalline rock in the core of the mountain suggest a period of rapid cooling between 62 and 57 million years ago during Laramide orogeny associated with activity on the Wind River thrust

FIGURE 14.32 Photograph looking southward over the Green River Lakes, Wyoming. Peaks from left to right are Lost Eagle Peak, White Rock, Squaretop Mountain, and unnamed Peak 11,545. The flat surfaces along mountaintops appear to trace out a broad anticline with White Rock tilted east (left) and Squaretop Mountain, and Peak 11,545 tilted west (right).

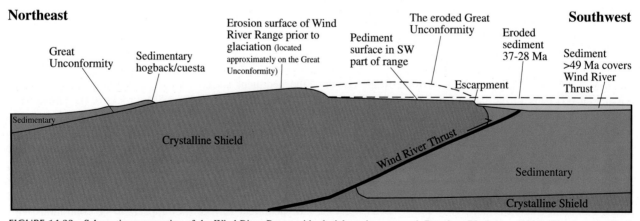

FIGURE 14.33 Schematic cross-section of the Wind River Range with glacial erosion removed. Based on Blackstone (1993). The cross-section is viewed toward the southeast to be consistent with Figure 14.31 (northeast is on the left side).

and with initial uplift-erosion of the anticlinal mountain. Clasts of crystalline rock in the surrounding sedimentary rocks indicate that the ancient crystalline core was exposed at the surface by 51 million years ago. Oxygen isotope data suggest that the surrounding Green River and Wind River Basins were within about 1640 feet (500 m) of sea level around 51 million years ago, and that the mountain range was at an elevation of about 9200 feet (2800 m). The data therefore imply that overall elevation of both the basin and the range following Laramide orogeny was 3000 to 4000 feet lower than present day, but that the difference in elevation (the relief) between the basin and the range was similar to present day. The erosion surface (e) began to form along the folded Great Unconformity beginning 51 million years ago when ancient crystalline shield rock was first exposed at the surface. The idea is that the soft sedimentary rocks were progressively stripped from the hard surface of the shield in manner similar to that shown in Figs. 10.3D and 10.3E.

The entire region appears to have undergone the same uplift-erosion stage experienced in the Southern Rocky Mountains between 42 and 37 million years ago resulting in 1500 to 2000 feet (457–610 m) of broad uplift (without strong deformation) of both the mountain and adjacent basins. Given that the time period from 51 to 37 million years ago was relatively warm, we can assume that there were no glaciers. Thus, by 37 million years ago, most of the range may have assumed the shape of a broad, weakly dissected, low-relief dome whose summit area corresponded roughly with the surface of the Great Unconformity as shown along the northeast (left) side of Fig. 14.33. The one exception would be the southwestern flank where erosion beveled into the side of the mountain to create a pediment surface at the level of the surrounding Green River Basin. The pediment surface was then buried in sediment associated with the Green River Basin from 37 to 28 million years ago.

There must have been another 1500 to 2000 feet of broad uplift of both the mountain and surrounding basins within the past 10 million years in order for landscape to reach its present-day elevation. This final phase of uplift was accompanied by normal faulting but was otherwise without strong deformation. The 37- to 28-million-year-old Green River Basin sedimentary rocks that had covered and buried the pediment surface were stripped off the top of the pediment surface in most areas creating an escarpment similar to, but smaller than the Colorado Front Range escarpment at Denver. The escarpment is visible at the long blue arrow in Fig. 14.31 and is labeled in the cross-section, Fig. 14.33. The southern edge of the pediment surface, located with a short blue arrow in Fig. 14.31, remains buried below these sedimentary rocks such that the ground surface merges with the Green River Basin, a situation similar in form to the Gangplank. Given that the pediment surface tilts southward, and that it appears to be continuous with the high elevation erosion surface in the northern part of the range, we can suggest that differential uplift in the past 10 million years was greater in the north, perhaps due to its proximity to the Yellowstone Plateau. The overall broad, low-relief dome shape of the mountain depicted in Fig. 14.33, with remnants of the Great Unconformity forming the erosion surface at high elevation, could have persisted until glacial erosion within the past 3 million years gouged and shaped the mountain summit region into the spectacular u-shaped valleys, cirques, mountain cliffs, and spires we see today.

Thus, the overall timing and development of the Wind River Range mirrors that of the Colorado Front Range.

An interesting aspect to the history of the Wind River Range is that zircon fission track ages are on the order of 600 million years old or more. Recall from Chapter 9 that zircon tracks are erased above 310°C and retained below about 210°C. The presence of such old zircon track ages indicates that the presently exposed crystalline rocks have not been buried more than 5 to 7 miles (7–11 km) since at least 600 million years ago. The area, over that time span prior to Laramide orogeny, must have been a lowland region of slow erosion that periodically sank below a shallow sea.

Beartooth Mountains

The Beartooth Mountains form a rugged, heavily glaciated region at the northern edge of the Yellowstone Plateau (Fig. 14.12). The highest 55 peaks in Montana are all in the Beartooth Mountains including 26 peaks above 12,000 feet and, of course Granite Peak, the highest at 12,804 feet. The Beartooth Mountains are rectangular in shape, extending 75 miles northwest/southeast from mountain front to mountain front and 45 miles across from the Yellowstone River east of Gardiner to the mountain front near Nye. Seen in its entirety in Fig. 14.34 it forms a high, glacier-carved anticlinal plateau bound along steep mountain flanks that form 4000-plus foot escarpments on three sides overlooking the wide Yellowstone River Valley (Yr) in the northwest at Pray, the Great Plains to the northeast at Nye, and the Clarks Fork Yellowstone River (CFYr) in the Bighorn Basin to the southeast. To the southwest, the Beartooths merge with the Absaroka Mountains. Practically the entire range forms the Absaroka-Beartooth Wilderness, which borders Yellowstone National Park.

Most of the range, including its entire northeastern side, consists of ancient crystalline shield rock greater than 2.5 billion years old. Most of the rock is granitic and gneissic, but also present along the mountain front west of Nye is the Stillwater mafic-ultramafic layered intrusive complex, a 2.7-billion-year-old complex with deposits of chromium, gold, silver, nickel and copper, and which is actively mined for its rich deposits of platinum group metals.

The northeastern side of the range rises abruptly along a bold 4000- to 5000-foot escarpment that is deeply incised with long, u-shaped glacial valleys easily visible in Fig. 14.34. This side of the range is truly plateau-like with flat mountain summits and small plateau-like tracts between steep-walled, glacial-carved valleys. The valleys cross the mountain summit region, which appears to be an elevated, dissected erosion surface much like that of the Wind River Range. Mt. Hague (H), Silver Run Peak (Sr), the namesake Beartooth Mountain (B), and Chalice Peak (C) are examples of flat-topped mountains in the summit

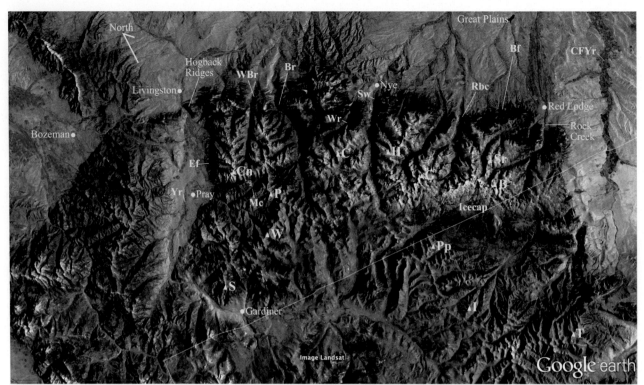

B-Beartooth Mtn., Bf-Beartooth thrust, Br-Boulder River, C-Chalice Peak, CFYr-Clarks Fork Yellowstone River, Cn-Mount Cowen, Ef-Emigrant normal fault, G-Granite Peak, H-Mount Hague, I-Indian Peak, Mc-Mill Creek, P-The Pyramid, Pp-Pilot Peak, Rbc-Rosebud Creek, S-Sheep Mtn., Sr-Silver Run Peak, Sw-Stillwater River, T-Trout Peak, W-Mount Wallace, WBr-West Boulder River, Wr-West Fork Stillwater River, Yr-Yellowstone River.

FIGURE 14.34 Google Earth image looking east of north at the Beartooth Mountains. The thin white line is the Montana—Wyoming border.

area. Many other peaks, including Granite Peak (G) and Mount Cowen (Cn), are horns with glacial-carved cirques on all sides. The remoteness and steep approaches make for challenging mountaineering. The area southwest of Beartooth Mountain (B) was an icecap during glaciation that scraped clean and flattened the landscape creating continuous rock exposure and many trout-filled lakes, some of which developed along glacial-gouged linear fractures and faults within the underlying shield rock.

The crystalline core was uplifted and folded into an asymmetric anticline along the Beartooth thrust, which extends along the northeast and southwest escarpments west of Red Lodge. A major difference with the Wind River Range is that the Beartooth thrust displaced the crystalline core toward the east or northeast, not toward the southwest. Sedimentary rocks along and below the thrust are turned up and tilted steeply eastward to form a series of hogbacks and ridges along the margin of the range. These ridges can be seen crossing Yellowstone River Valley in Fig. 14.34. The Beartooths also form an escarpment overlooking Yellowstone Valley, here marked by the Emigrant normal fault, which has been active within the past 15,000 years. The gouging of weak rock along an inactive normal fault is responsible for the cross-valleys marked by the West Fork Stillwater River (Wr) and Mill Creek (Mc). Most of the rock west of and including The Pyramid (P) and Pilot Peak (Pp) consists of circa 49-million-year-old Absaroka volcanic rocks. Also present are areas of young intrusive rocks and a few areas of sedimentary rock. Shield rock is present in this area, most notably at Sheep Mountain (S).

On the basis of apatite (U-Th)/He and fission track data, the landscape history is similar with that of the Wind River Range. Northeastward thrusting, anticlinal folding, and uplift took place during Laramide orogeny between 61 and 52 million years ago. A plateau-like, low-relief domal mountain shape formed and was maintained over the next several million years subject to river incision as sedimentary layers were progressively stripped from the hard crystalline core

Just as in the Wind River Range, the erosion surface likely initiated along the Great Unconformity that separates the shield rocks from overlying platform sedimentary rocks. A second period of uplift, without significant deformation, began between 15 and 5 million years ago and remains active. Uplift is associated with normal faults and possibly with thermal isostasy associated with the Yellowstone hot spot. This uplift likely created or enhanced the escarpments that partially surround the mountain range, especially along the active Emigrant normal fault. We can also assume that the mountain interior maintained its domal, low-relief, plateau-like topography until glaciers literally ripped into it, thus creating the spectacular scenery we see today.

Bighorn Mountains

The Bighorn Mountains form a massive range isolated from the Yellowstone Plateau and the Rocky Mountains by the Bighorn Basin. It is 150 miles long measured northwest to southeast from the Bighorn River to the Badwater River, but mountainous terrain extends northward into the Pryor Mountains of Montana, and curves westward to become the Owl Creek Range in Wyoming (Figs. 14.11 and 14.12). Much of the range consists of unglaciated, low-relief, rounded peaks below 9000 feet, but the central 45 miles or so rises above 10,000 feet and shows clear signs of glaciation with jagged, vertical cliffs. There are two peaks above 13,000 feet and an additional seven above 12,000 feet. The highest is Cloud Peak at 13,167 feet.

The Bighorn Mountains possess an elongate domal structure with tilted sedimentary rocks along its flanks and ancient crystalline shield rock in its central glaciated core. Shield rocks are of two dominant types. Circa 2.85-billion-year-old granitic rocks occupy the north-central summit area, and circa 2.95-billion-year-old gneisses occupy the south-central summit. Shield rocks are surrounded on all sides by Cambrian-Ordovician (541–444 Ma) and younger dolomite, limestone, sandstone, and shale of the interior platform rock succession. Unlike the Wind River and Beartooth Mountains, sedimentary rocks occupy unglaciated areas across the northern and southern summit areas of the mountain. The crystalline-sedimentary contact forms the Great Unconformity.

Topography of the range mimics the anticlinal structure. The mountain has the overall geometry of an elongated dome, highest at its center and gradually losing both fold amplitude and elevation toward the north and south. With few exceptions, all peaks above 10,000 feet are in the central core of crystalline rock. A buried thrust fault along the eastern side of the range with about 6 miles (10 km) of eastward displacement accentuates the anticlinal structure creating an asymmetry with steeper beds along the east side. The tilted Cambrian-Ordovician rocks and some of the younger rocks are resistant enough to form hogbacks along the eastern flank, and cuestas along the western flank, both of which are visible in Fig. 14.12. A second concealed thrust fault present in the northwest part of the range east of Cody displaces rocks westward, but this fault dies out southward into folds. Both faults are located several miles from the range front and both are buried beneath young sediment of the Great Plains. The fault east of Cody forms the approximate southwest limit of strongly folded rock in the eastern Bighorn Basin, an example of which is shown in Fig. 13.34. A monocline situated along the range front in the area east-northeast of Greybull creates local areas of steep dip and an escarpment up to 5000 feet high along the mountainside.

Easterly and northerly striking reverse and normal faults cross the mountain and in some areas create lineaments, especially in rocks of the crystalline shield. A strong north-northeast-trending lineament crosses the central part of the range defined by North Paint Rock Creek, Lake Elsa, Kearny Creek, and Kearny Lake Reservoir. This lineament follows an ancient fault in crystalline rock that partially separates granitic rock in the north from older gneisses in the south. The fault is not active, but it may have been active during Laramide orogeny because it forms a topographic boundary between peaks below 11,500 feet to the north and peaks above 12,000 feet to the south. The change in elevation is clearly visible in Fig. 14.35, which looks eastward up North Paint Rock Creek.

Based on (U-Th)/He apatite ages from the crystalline core, and on the composition of rock fragments eroded from the Bighorn Mountains and deposited in adjacent basins, deformation and mountain uplift associated with the Laramide orogeny began 65 million years ago and was active 56 million years ago when crystalline rock was exposed and eroded from the core of the mountain. Large variation in oxygen isotopes in the adjacent Great Plains have been interpreted to suggest that by 56 million years ago the mountains could have been above 14,000 feet. Thus, in the case of the Bighorn Mountains, it appears that most of the uplift is associated with Laramide orogeny. However, it remains highly probable given the amount of regional doming in the Colorado Rockies and surrounding

FIGURE 14.35 Google Earth image looking northeast at the Bighorn Mountains along North Paint Rock Creek. The creek follows an inactive fault zone that separates peaks above 12,000 feet on the right (south) including Cloud Peak (center-right) and peaks below 11,500 feet on the left (north).

FIGURE 14.36 Google Earth image looking northeastward along the crest of the Bighorn Mountains east of Sheridan. There is a marked change in landscape between areas underlain with granitic rocks of the crystalline shield (right) and areas underlain with sedimentary rocks of the interior platform (left). The contact (the Great Unconformity) is shown with a thick white line. The thin white line is route 14. The image is approximately 28 miles right to left.

areas over the past 10 million years that the Bighorn Mountains and surrounding basins have also experienced broad uplift (without deformation) on the order of 1000 feet or more in the past 10 million years.

The presence of unglaciated sedimentary rock across the summit area of the mountain provides a glimpse at what the Wind River and Beartooth Mountains may have looked like and how they may have evolved prior to and following the erosional stripping of sedimentary rocks from the crest of the anticlinal mountain and prior to glaciation. Fig. 14.36 looks northeastward across the crest of the range west of Sheridan. The heavy white line marks the boundary (the Great Unconformity) between shield rocks to the south (right) and sedimentary rocks to the north. The shield rocks display smooth, domal-shaped topography without a great deal of relief. The sedimentary rocks, by contrast, are cut by streams into a series of ridges that show greater local relief. Thus, prior to glaciation, one would have expected a transition from rough to smooth domal topography in the Wind River and Beartooth Mountains as overlying sedimentary rocks were stripped from across the mountain and crystalline rocks were progressively exposed.

The Black Hills

Fig. 14.11 locates the Black Hills on the Missouri Plateau of the Great Plains just north of the Pine Ridge

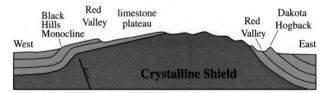

FIGURE 14.37 Cross-section of the Black Hills dome. Note the limestone plateau, Red Valley, and Dakota Hogback.

escarpment between the White and Little Missouri Rivers. The Black Hills form an elliptical dome, approximately 125 miles northwest to southeast and 65 miles wide. The dome is lower in elevation than the Middle Rocky Mountains and unglaciated with peaks mostly between 5000 and 6500 feet. The highest is Black Elk Peak (formerly Harney Peak), which, at 7233 feet, rises some 4000 feet above the surrounding Great Plains.

The cross-section in Fig. 14.37 revels an asymmetric structure in which the east side of the dome dips more steeply than the west. The asymmetry causes rocks of the crystalline shield to be widely exposed on the eastern side. The dome shape itself results in concentric rings of progressively younger sedimentary layers of the interior platform rock succession to be exposed in the form of cuestas, hogbacks, and valleys that surround the crystalline core. The youngest rocks merge with nearly flat-lying layers of the Great Plains. Fig. 14.38 is a Google Earth

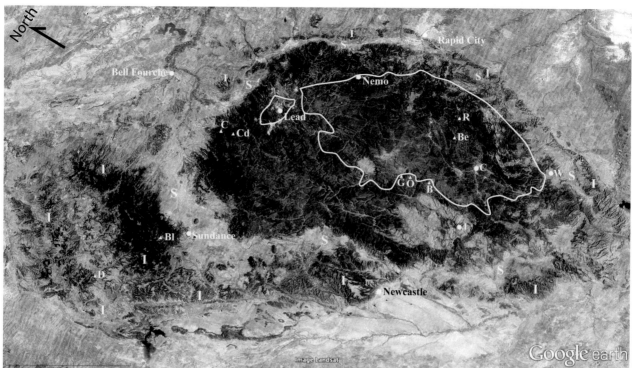

I-Areas of Inyan Kara Group hogbacks, cuestas, and hills	**Peaks**			**Places**
S-Areas of Spearfish Formation and Red Valley	B-Bear Mountain	C-Crow Peak	G-Green Mountain	C-Custer
White lines enclose areas of crystalline shield	Be-Black Elk Peak	Cd-Citadel Peak	O-Odakota Mtn.	J-Jewell Cave
	Bl-Bear Lodge Mtn.	D-Devils Tower	R-Mt. Rushmore	W-Wind Cave

FIGURE 14.38 Google Earth image looking northeast at the Black Hills. Black Elk Peak (*Be*) is formerly known as Harney Peak.

image oriented northeastward that shows some of the geology and localities in the Black Hills. The first thing to notice is that sedimentary layers, mostly limestone, form a gently tilted, stream-dissected plateau along the entire west side of the mountain. Fig. 14.37 shows that the plateau rises to the crest of the dome. Odakota Mountain (O, 7210 feet), Bear Mountain (B, 7166 feet), and Green Mountain (G, 7164 feet), the second, third, and fourth highest peaks in the Black Hills, all form cuestas on limestone that look eastward from the top of the dome across rocks of the crystalline shield. Although not well expressed in the landscape, the westward, the edge of the Black Hills is marked by the Black Hills monocline. As on the Colorado Plateau, this monocline likely formed during Laramide orogeny along a pre-existing fracture or fault in underlying crystalline rock, similar to what is shown in Fig. 4.10. The monocline is shown in Fig. 14.37.

Sedimentary layers on the east side of the dome are variably folded, but overall the rocks dip more steeply than on the west side. Of special interest is the lower Cretaceous Inyan Kara Group, which contains a resistant sandstone-conglomerate layer that forms variably a hogback, cuesta, or hills around the entire Black Hills. It is located in Fig. 14.38 with the letter I. A hogback composed of Inyan Kara sandstone cuts directly through Rapid City where it is known as the Dakota Hogback or Hogback Ridge. Perhaps, not surprisingly, the Inyan Kara Group is the same age as the Dakota Group, the rock that caps many hogback ridges along the east side of the Colorado Front Range. The Morrison Formation underlies both rock units. A second rock unit, known as the Spearfish Formation is located in Fig. 14.38 with the letter S. This older (Permian-Triassic) rock unit consists of red shale and siltstone that erodes to form Red Valley (also known as The Racetrack), a continuous valley of red soil that encircles the dome between the mountain and the Dakota hogback/cuesta.

The sedimentary layers are in contact with rocks of the crystalline shield across the Great Unconformity. The oldest rocks are schist and gneissic granite between 2.5 and 2.7 billion years old, best seen along the east side of Bear Mountain directly below sedimentary rocks, or near the town of Nemo. The vast majority of shield rock, however, consists of 2.2-to 1.85-billion-year-old schist and quartzite intruded by 1.7-billion-year-old Harney Peak granite. Mount Rushmore National Memorial was carved into Harney Peak granite with older schist present at the base. At the town of Lead, crystalline rocks are host to the famous Homestake gold mine, which was in continuous operation for more than 125 years before closing operations in 2002. The mine remains open for tours. Also present in the northern part of the dome, within both sedimentary and shield rocks, are circa 60- to 50-million-year-old intrusive domes discussed earlier in the chapter. Some of these are located in Fig. 14.38.

The Black Hills followed a Laramide uplift history that began between 66 and 64 million years ago based on the age of material eroded from the Black Hills and deposited in the adjacent Great Plains to the west. The timing of events is not well constrained beyond initial uplift, but there is disputed evidence that crystalline rocks were exposed between 58 and 54 million years ago. In any case, it appears that the Black Hills has remained a highland region overlooking the Great Plains since initial uplift. There is little evidence of an erosion surface similar to what is found in the Colorado Front Range. Rock units younger than 40 million years old are tilted, indicating periods of uplift following the Laramide orogeny. Considering the history of the Great Plains, we can suggest that some of that uplift has occurred within the past 10 million years. Leveling surveys suggest uplift is active today.

An interesting feature in the Black Hills is the presence of Jewel Cave National Monument and Wind Cave National Park. Both are found in the Mississippian (359–323 Ma) Madison Limestone, and both are labeled in Fig. 14.38. Excluding a mostly drowned cave system in Mexico, Jewel Cave at 179 miles, and Wind Cave at 143 miles of passageways, are the second and fifth longest cave systems in the world. Jewel Cave was named for the many calcite crystals that line some of the walls. Wind Cave is famous for the wind that blows into and out of it at various times of the day dependent on whether air pressure in the cave is lower or higher than air pressure outside the cave. Wind Cave may be one of the oldest cave systems in the world, having begun to form as early as 320 million years ago. It also may have the most compact concentration of passages in the world, as the entire cave network fits under just one square mile of land.

WATER GAPS IN THE ROCKY MOUNTAINS

A water gap forms where a river flows directly through a mountain or ridge rather than around. The flow of Rapid Creek through the Dakota hogback at Rapid City is an example. There are several water gaps in the Rocky Mountains, particularly in the Middle Rockies. Major water gaps are present where the Bighorn River cuts across the Sheep Mountain anticline north of Greybull, Wyoming, and again where it cuts through the north end of the Bighorn Mountains south of Hardin, Montana (Fig. 14.12). Additional water gaps include Wind River Canyon at the east end of the Owl Creek Range, Devil's Gate along the Sweetwater River at the east end of the Granite Range southwest of Casper, Lodore Canyon and Split Mountain Canyon at the east end of the Uinta Mountains, and several in the Front Range including Royal Gorge on the Arkansas River near Canon City (Fig. 14.11). Water gaps are also present on the Colorado Plateau. For example, the San Juan River cuts directly across the Raplee Anticline in the Monument Uplift region, and the Colorado River cuts across the Kaibab Uplift to form part of the Grand Canyon.

The existence of a water gap is perhaps a bit surprising because rivers normally go-with-the-flow so to speak. In other words a river will seek the easiest route to the sea and, in this case, that should have been around the mountain. The explanation for some of the rivers in the Rocky Mountain region is that the rivers are younger than the Laramide Mountains, but older than the present-day landscape, and that they were superposed onto the present-day landscape. Recall that many of the anticlinal Laramide mountains were buried beneath younger nearly flat-lying sedimentary rocks prior to being exhumed within the past 10 to 6 million years. A superposed river is one that establishes a channel in overlying low-relief, nearly flat-lying rock layers, and then maintains its channel as it cuts downward into the slowly exhuming crystalline-cored anticlinal mountain. The process is shown schematically in Fig. 14.39. Note that an unconformity exists between the nearly flat-lying rocks and the older deformed rocks.

There are other mechanisms that can produce water gaps and at least one other mechanism may be responsible for some of the water gaps in the Rocky Mountains. I could discuss other mechanisms at this time but would rather present it as a question at the end of the chapter for you to research. We will revisit this problem when we discuss the Appalachian fold and thrust belt (Chapter 15).

SUPERIOR UPLAND

The Superior Upland is one of two areas where rocks of the Canadian Shield extend directly and continuously into the United States. The other location is the Adirondack Mountains. Fig. 14.40 is a Raisz landform map that

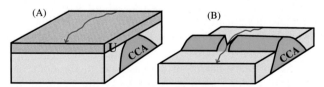

FIGURE 14.39 Diagram that shows a water gap formed by river superposition. (A) The river establishes itself on low-relief, nearly flat-lying rock layers above an unconformity (*U*). (B) The river maintains its channel as it cuts downward through a crystalline-cored anticline (*CCA*) to form a water gap.

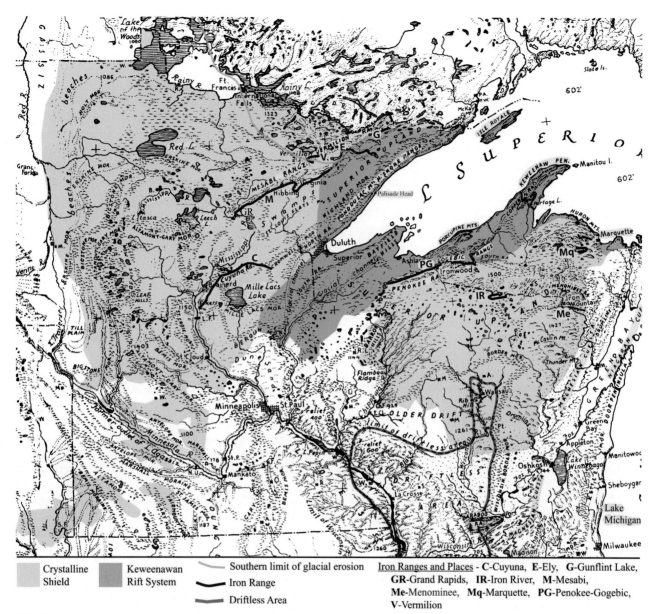

FIGURE 14.40 Raisz landform map of the Superior Upland region.

divides the Superior Upland into three areas, a western area in Minnesota, a central area surrounding Lake Superior, and an eastern area in Wisconsin and Michigan. The western and eastern areas, shown in green, represent ancient, eroded mountain belts where rocks of the North American crystalline shield are exposed. Most of the rock is schist, gneiss, and granite, but an unusual amount of sedimentary rock that includes iron formations is also present. All of the rock is between 3.6 and 1.6 billion years old. Rocks in the central area, shown in orange, are part of the Keweenawan (or Mid-Continent) Rift system. These rocks include sandstone, basalt, and gabbro, and represent a 1.1-billion-year-old failed attempt by the North American shield to rift (separate) into two pieces. The rocks are part of the Precambrian sedimentary/volcanic rock succession.

The western (Minnesota) crystalline shield is bordered on its west and south sides by younger rock of the Central Lowlands, on its east side by younger rock of the Keweenawan Rift system, and on its north side by the vast Canadian Shield. We can infer from this pattern that the Minnesota crystalline shield is exposed across a broad anticlinal arch with younger rock on all but the north flank. The arch continues southwestward as the Transcontinental arch but the level of erosion is not deep enough to expose crystalline rock. The trend of the

Transcontinental arch is shown in Fig. 11.1. A close look at Figs. 14.40 and 12.18 reveals that the eastern (Wisconsin–Michigan) part of the crystalline shield is detached from the Canadian Shield. It forms a broad dome surrounded on its northern side by younger rock of the Keweenawan Rift system, and on its west, south, and eastern sides by younger sedimentary rock of the Central Lowlands. The intervening Keweenawan Rift system forms a broad syncline whose axis trends northeast-southwest through the center of Lake Superior. Thus, from west to east, the overall structure of the Superior Upland consists of an broad anticline, a syncline, and a dome. The entire surface extent of both the Canadian Shield and Keweenawan Rift system is shown in Fig. 12.18.

Although the overall structure of the Superior Upland is domal, the mode of uplift-subsidence is different from the Middle-Southern Rockies in that uplift appears to have occurred slowly and periodically over the course of hundreds of millions of years. The Keweenawan trough had already formed by 1.06 billion years ago, but we can be sure that additional broad folding has occurred because sedimentary rocks younger than 541 million years old are deformed across structures such as the Transcontinental arch. Based on Fig. 9.2, we can also be sure that the region is currently rising due to isostatic rebound following the retreat and melting of glaciers.

Although the bedrock of the Superior Upland is distinctive, the primary landscape features in the province are glacial. The glacial landscape is, in fact, so similar to the Central Lowlands that it is not possible to draw a physiographic boundary between the two provinces based on glacial features. A boundary line could conceivably be drawn between areas dominated by glacial erosion and areas dominated by glacial deposition. However, this line separates only the northern fringe of Minnesota, and therefore is not particularly useful as a physiographic boundary. It is the rocks that produce the primary landscape differences. In areas where rocks are reasonably well exposed, such as in Wisconsin, Michigan, and eastern Minnesota, deformed rocks of the Superior Upland produce three landforms that are not prominent in the Central Lowlands. Within the shield there are long narrow ridges composed of resistant iron formation rocks, and hills underlain with quartzite. Within the Keweenawan Rift system there are well-formed cuestas composed of tilted volcanic rock, sandstone, and conglomerate, especially surrounding Lake Superior. These features are absent in western Minnesota where the thickness of glacial sediment (>100 feet) is enough to eliminate any landscape influence that bedrock may have. Rather than a true physiographic boundary, we will locate the western boundary with the Central Lowlands based on the type of bedrock present at depth below the glacial deposits.

Knowledge of the composition and structure of buried bedrock in Minnesota is based on gravity and magnetic studies and on drilling spearheaded by the Minnesota Geological Survey. These studies indicate that ancient, crystalline rocks of the Canadian Shield extend below the cover of glacial sediment as far west as the Red River along the Minnesota–North Dakota boundary as shown in Fig. 14.40. Also shown in the figure are a few isolated areas in southern Minnesota and South Dakota where ancient crystalline rocks crop out at the surface. Altitudes in the Superior Upland are mostly between 1000 and 2000 feet with topographic relief rarely more than 600 feet. The elevation of Lake Superior is 602 feet. Glacial features were discussed in Chapter 12. In this section we discuss bedrock geology and bedrock landscape features.

Geologic Overview

Shield rocks of the Superior Upland are of interest because they record as many as three orogenic episodes, are host to wide areas of iron formation rocks, and include some of the oldest crystalline, sedimentary, and volcanic rocks in the country. In the following sections, we will look briefly at the distribution of these rocks and their landscape characteristics. A general geologic map of the area surrounding Lake Superior is given in Fig. 14.41 in order to facilitate discussion. Shield rocks are subdivided into an older Superior Province, a younger Penokean Province, and the still younger Barron Quartzite. The Superior Province is further subdivided into the Minnesota River Valley Subprovince, and a subprovince composed of granitoid gneiss, granite, greenstone, and metasedimentary rocks (shown as a single rock unit). The Penokean Province is divided into three rock units, a volcanic arc terrane, a fold and thrust belt, and the Animikie Basin. Iron formations occur primarily in the Penokean Province and are shown with a thick blue line. There are two ancient suture zones. The Great Lakes Tectonic Zone separates the two rock units in the Superior Province, and the Malmo Discontinuity-Niagara Fault separates the volcanic arc terrane from the fold and thrust belt within the Penokean Province. The Keweenawan Rift system is of interest because it represents the rifting of the interior crystalline shield at a time when the Grenville orogeny was occurring along the eastern seaboard. The rift system is divided into three rock groups in Fig. 14.41, the volcanic series, the Duluth Intrusive Complex, and the sedimentary series. Uncolored areas are either part of Canada or are covered with nearly flat-lying interior platform sedimentary rocks of the Central Lowlands. Not shown are a few areas of Cretaceous rock that cover ancient shield rocks primarily southeast of Hibbing and near St. Cloud. The following sections briefly describe the geology.

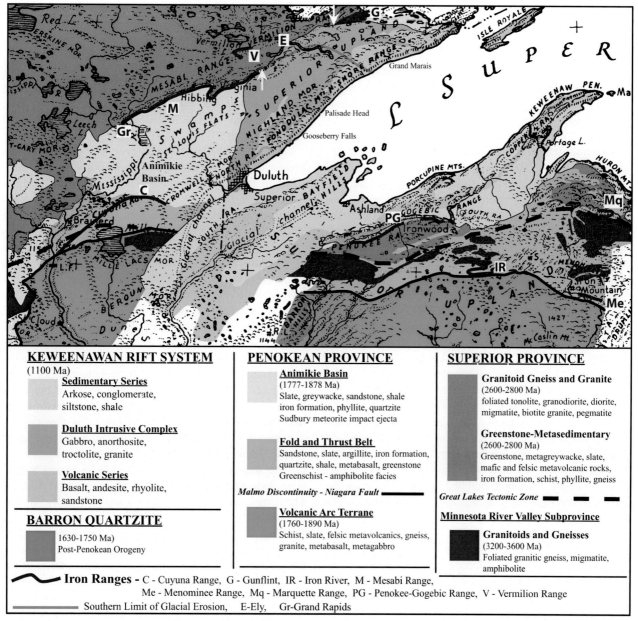

FIGURE 14.41 General geology of the area surrounding Lake Superior superimposed on a Raisz landform map. The two white arrows show where the Duluth Gabbro separates the Mesabi Iron Range (*M*) from the Gunflint Range (*G*). Uncolored areas in the United States at the south and east edges of the map are rocks of the interior platform rock succession.

Superior Province

Most of the truly crystalline rocks that form the Superior Upland are 2.6 billion years old and older, and are part of the Superior Province. The oldest rocks form the Minnesota River Valley subprovince, shown in red in Fig. 14.41. Included are granitic gneisses between 3.6 and 3.2 billion years old. These rocks represent some of the oldest examples of continental crust in the world. They form the very core of the North American crystalline shield. An example is the famous (in geological circles) Morton Gneiss, which is dated at 3.524 billion years old. A good place to see the Morton Gneiss and associated rocks is within an outlier of the Superior Upland province along the Minnesota River between Montevideo and Franklin (Fig. 14.40). Specific locations are on the Hambdecka Trail at the namesake town of Morton where it was quarried and used as a building stone, and at Redwood Falls. Although not well documented, the rocks appear to be part of a protracted and sporadic period of intrusion and metamorphism that began more than 3.5 billion years ago and culminated between 3.14 and

3.05 billion years ago, possibly with the formation of the first continental supercontinent (a small, early version of Pangea).

Also present and more widespread in the Superior Province are rocks between 2.8 and 2.6 billion years old. In this age group, we find two distinct rock types, greenstone-metasedimentary rocks, and granitoid gneiss and granite rocks (shown as a single unit in Fig. 14.41). Greenstone is basalt that has been metamorphosed at low to moderate temperature. The metasedimentary rocks are primarily schist and schistose meta-sandstone. The greenstone-metasedimentary rocks are interpreted as remnants of ancient ocean basins, island volcanic arc terranes, and possibly accretionary prisms. They are well exposed in the zone of glacial erosion along route 169 between Lake Vermilion and Ely (E) where most are metamorphosed at high temperature. Included in the sequence is the Soudan Iron Formation in the Vermilion Range (V), the only iron producing formation of this age in United States. The Soudan mine closed in 1962 and now hosts Soudan Underground State Park. Located on the south shore of Lake Vermilion near the town of Soudan, park visitors can tour the underground mine as well as an active underground physics laboratory.

The granitoid gneiss and granite rocks intrude the greenstone-metasedimentary belt. These rocks likely represent the batholithic roots of an island volcanic arc complex perhaps similar in origin to the Sierra Nevada and Idaho Batholiths. The best exposures are in the Gunflint Lake area (G) east of Lake Vermilion. These rocks underwent a major mountain building event known as the Kenoran or Algoman orogeny that culminated between 2.6 and 2.5 billion years ago when this terrane collided with the Minnesota River Valley subprovince along the Great Lakes tectonic zone (shown as a thick dashed line in Fig. 14.41). There is evidence that this collision also produced a supercontinent.

Penokean Province

Rocks of the Penokean Province are between 2.2 and 1.76 billion years old and are associated with the circa 1.88- to 1.83-billion-year-old Penokean orogeny, which is named after the Penokee Range near Ironwood, Michigan. The Penokean orogeny was one of several orogenic events that occurred worldwide about 1.8 billion years ago. The end result was yet another pre-Pangean supercontinent. The rocks are part of the crystalline shield, but are uncharacteristic in the sense that, in addition to granite, schist and gneiss, there are wide areas of unmetamorphosed and weakly metamorphosed rocks. Collectively, the rocks record a complete orogenic belt created from the collision of a volcanic arc terrane (shown in dark green in Fig. 14.41) with a continental rock terrane (shown as the fold and thrust belt) along the Malmo Discontinuity-Niagara fault (shown as a thick black line). The volcanic arc terrane consists of schist, slate, gneiss, granite, and metavolcanic rocks. Rocks north of the fault, within the fold and thrust belt and the Animikie Basin, consist of slate, quartzite, sandstone, shale, basaltic, and silicic metavolcanic rocks, and iron formations. These rocks become less deformed and less metamorphosed toward the north.

The fold and thrust belt is best exposed in northern Wisconsin and Michigan, but it does not control landscape to the same extent as glacial deposition features. Nevertheless, this belt includes the foreland-hinterland transition. If you travel north from the Menominee Range (Me) across the Marquette Range (Mq) to Lake Superior (eastern edge of Fig. 14.41), you would notice that rocks of the fold and thrust belt are strongly deformed and metamorphosed near the Menominee Range, but become less deformed and less metamorphosed to the north. The primary structures across this area are thrust faults that have pushed the rocks northward. These rocks are also exposed in the Penokee and Gogebic Ranges surrounding Ironwood. Notice that there are inliers and domes of older rocks in the fold and thrust belt, including rocks of the Minnesota River Valley subprovince south of the Great Lakes Tectonic zone. These older rocks were incorporated into the younger fold and thrust belt. The Great Lakes Tectonic zone was locally reactivated and then buried during Penokean orogeny.

The Animikie Basin is a foreland basin in which rocks were deposited both during and following thrust faulting. These rocks are weakly deformed and essentially unmetamorphosed, especially in the Mesabi Iron Range (M) near Hibbing. Deformed, weakly metamorphosed sandstone and shale are spectacularly exposed along with rocks of the Keweenawan rift system on the St Louis River at Jay Cooke State Park, located in the southern part of the basin 16 miles southwest of Duluth. Animikie Group rocks are also present in the Gunflint lake region near the Canada—Minnesota border north of Lake Superior (northern edge of Fig. 14.41). Here, the rocks contain breccia derived from the Sudbury meteorite impact event (discussed later).

Iron Formations

Iron formations, because of their resistance to erosion, form narrow upland ridges with less than 600 feet of relief. An example is the Marquette Iron Range (Mq) south of Marquette, Michigan as seen in Fig. 14.42. These ridges form one of the primary bedrock landscape features of the crystalline shield. The iron ranges are identified with a dark blue line and are labeled in Figs. 14.40 and 14.41. They include the Cuyuna (C), Mesabi (M),

FIGURE 14.42 Google Earth image looking east at the Marquette Iron Range near Ishpeming, Upper Peninsula, Michigan. An active iron mine is visible at upper right.

Vermilion (V), and Gunflint (G) Ranges in Minnesota, and the Penokee-Gogebic (PG), Menominee (Me), and Marquette (Mq) Ranges in Wisconsin and Michigan. These ranges constitute the largest iron deposits in the United States and our only domestic source of iron, although only the Mesabi and Marquette Ranges host active mines. With the exception of the Soudan iron formation (S) in the Vermilion Range near Ely (E), all of the iron ranges are in either the Penokean fold and thrust belt or the Animikie Basin.

Iron formations are typically banded with dark, inch-thick layers rich in iron oxide (hematite) that alternate with lighter chert-rich layers. The iron formations in the Mesabi (M) and Gunflint (G) regions are deformed but virtually unmetamorphosed. Those in other ranges are deformed and weakly metamorphosed. All but one of the major iron formations are between 1.88 and 1.85 billion years old based on the dating of volcanic ash layers and detrital zircon. As noted earlier, the Soudan iron formation is on the order of 2.7 to 2.6 billion years old. The Mesabi Range (M) is home to some of the largest open-pit iron mines in the world. Several of the mines are visible as grey patches in Fig. 12.16. There is also a triple divide located in the middle of an iron mine in the Mesabi Range. Surrounding this triple divide, the Red River flows north along the Minnesota–North Dakota border to the Hudson Bay, the Mississippi River flows from its birthplace near Lake Itasca and Leech Lake to the Gulf of Mexico, and the Great Lakes drain eastward to the Gulf of St. Lawrence (Fig. 2.2).

Sudbury Meteorite Impact Event

Sedimentary rocks deposited 1.85 billion years ago contain ejecta from the 1.85-billion-year-old Sudbury meteorite impact event. This particular event was no small potatoes! The meteorite was likely more than 5 miles in diameter at impact, forming a crater as much as 81 miles wide. It is the third largest and fourth oldest known impact site in the world, on par with the slightly larger Chicxulub impact crater in the Yucatan Peninsula that extinguished the dinosaurs 66 million years ago and created profound devastation across the globe. The Sudbury impact crater (SI) is located in Fig. 12.18.

The most obvious evidence of the Sudbury event can be found in the form of breccia located directly above iron formations across Michigan and Minnesota. Breccia is a rock composed of randomly oriented blocks of (in this case) iron formation and chert. Individual blocks vary in size from 10 feet across to less than an inch. Apparently these blocks were ripped from the underlying iron formation by landslides and tsunamis associated with the event. The breccia is best seen along the Gunflint Trail (Route 12) west of Gunflint Lake nearly 400 miles from the impact site. In Minnesota, the breccia is generally less than 25 feet thick, but in Michigan, closer to the impact site, it is as much as 100 feet thick. A thin layer of accretionary lapilli can be found at the top of the breccia. Accretionary lapilli are small airborne pellets that form by accretion of fine ash around water droplets or larger solid particles. The Sudbury event is important because it essentially ended iron formation deposition across the Superior region.

Barron and Baraboo Quartzite

Areas of highly resistant red quartzite, along with iron formation ranges, form some of the most noticeable landscape within the crystalline shield. Red quartzite is abundant in two areas of Wisconsin, the Barron Hills south of Duluth as shown in Fig. 14.41, and the Baraboo Range north of Madison as highlighted in Fig. 14.40. The rocks are between 1.75 and 1.63 billion years old based on the age of the youngest underlying rock and the age of metamorphism. The quartzite at Barron Hills sits between older crystalline shield rocks to the east, and young, nearly flat-lying interior platform sedimentary rocks of the Central Lowlands to the west. Although the rock itself is not well exposed, the resistant quartzite produces a hilly, forest-covered landscape that rises 400 to 600 feet above the surrounding area. The Baraboo Range forms an outlier of crystalline rock surrounded on all sides by sedimentary interior platform rock of the Central Lowlands. The quartzite is folded into a syncline such that it forms an oval ring of highlands approximately 30 miles long, 10 miles wide, and 400 to 600 feet above surrounding areas. Fig. 14.43 looks westward across the southern part of the syncline in the Baraboo Range at Devil's Lake (D).

FIGURE 14.43 Google Earth image looking westward at the South Baraboo Range and Devil's Lake (D).

Keweenawan Rift System

The Keweenawan (or Mid-Continent) Rift system does not fit very well with any of the structural provinces. It contains sedimentary, volcanic, and crystalline rocks so it cannot be defined based on rock type. It was shaped initially by normal faults so it could logically fit in Chapter 18, but normal faults do not control the landscape. The landscape instead, is a reflection of the most recent structure, a synclinal trough between a crystalline-cored anticline on one side, and a dome on the other. It is included in this chapter because of its close association with these structures.

The rift system is exposed in a down-folded synclinal trough, oriented northeast-southwest through the center of Lake Superior. The rocks represent a failed attempt by the North American continent to separate (rift) into two pieces. Fig. 14.41 shows three major rock units, the Volcanic Series, the Duluth Intrusive Complex, and the Sedimentary Series. The Volcanic Series is the oldest and most voluminous with a huge thickness on the order of 12.5 miles (20 km) that was deposited during normal faulting. Most of the rock consists of basalt but sandstone, conglomerate, and silicic volcanic rocks such as rhyolite are also present. The volcanism produced some of the purist copper deposits in the world. The Duluth Intrusive Complex intrudes the Volcanic Series. It forms much of the rock on the northwest side of Lake Superior as well as small intrusions elsewhere. In addition to gabbro (composed of pyroxene and plagioclase), the Duluth Complex includes anorthosite (composed of plagioclase), troctolite (composed of olivine and plagioclase), and a small amount of granite. Notice in Fig. 14.41 that the Duluth Intrusive Complex separates iron formation rocks in the Mesabi Range (M) from those at Gunflint Lake (G) across a distance of 50 miles (indicated by the two white arrows). At the time of intrusion, approximately 1.1 billion years ago, the two iron formations were deeply buried and were continuous with each other. Intrusion of the Duluth gabbro engulfed and destroyed the part of the iron formation that connected the two. Truncation of the iron mines is also visible in Fig. 12.16. The Sedimentary Series consists of red sandstone (arkose), conglomerate, siltstone, and shale, and is more than 3.7 miles thick (6 km). Most of the rocks were deposited unconformably above the other two Keweenawan rock units, and faults are uncommon. Such a relationship implies that most of the Sedimentary Series was deposited in a slowly subsiding depositional basin after volcanism, intrusion, and normal faulting had ceased.

Radiometric dating indicates that the attempted rift event was short-lived. Rifting began 1.109 billion years ago with normal faulting and deposition of the Volcanic Series. Rifting reached a climax between 1.099 and 1.095 billion years ago with intrusion of the Duluth Complex, and had largely ended by 1.086 billion years ago without going to completion. Following the failed rift event, the basin continued to subside and receive deposits of the Sedimentary Series until at least 1.060 billion years ago when subsidence was interrupted with periods of mild compressional folding and reverse faulting, possibly associated with the Grenville orogeny. Subsidence created the Lake Superior syncline, but the syncline continued to develop post-541 million years ago with concomitant development of the Transcontinental arch, Michigan Basin, and other broad structures that deform sedimentary interior platform rocks of the Central Lowlands.

The syncline is asymmetric. It dips greater than 35 degrees on the southeast side and about 15 degrees on the northwest side. Fig. 14.44 is a simplified cross-section across the Lake Superior syncline with normal and reverse faults not shown. Erosion of the synclinal structure has produced a series of cuestas on both sides of Lake Superior. Along the northwestern shoreline, northwest-facing cuestas composed of rhyolite and basalt of the Volcanic Series, and of Duluth Gabbro, form Isle Royale National Park and the Fond du Lac and North Shore Ranges (also known as the Sawtooth Mountains), which includes the highest point in Minnesota (Eagle Mountain, 2301 feet). Thick flows of rhyolite form Palisade Head (Fig. 14.41), which rises more than 300 feet above the lake. Southeast-facing cuestas along the southeastern shoreline consist of the same rock with the addition of conglomerate and hard sandstone of the Sedimentary Series. The Porcupine Mountains are one of the more prominent ranges along the southeastern shoreline. As seen looking eastward in Fig. 14.45, the range is only 12 miles long and 6 miles wide. It is distinguished by the presence of a prominent southeast-facing cuesta composed of basalt and interlayered conglomerate. Lake of the Clouds (LC) sits at the base of the cuesta. The

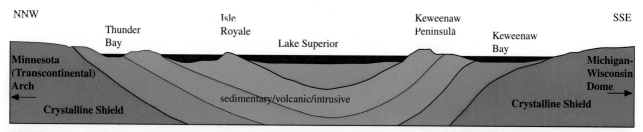

FIGURE 14.44 Cross-section sketch of the Lake Superior syncline with cuestas on both sides of Lake Superior. The syncline trends through the center of Lake Superior between the Transcontinental arch to the west and the Michigan–Wisconsin dome to the east.

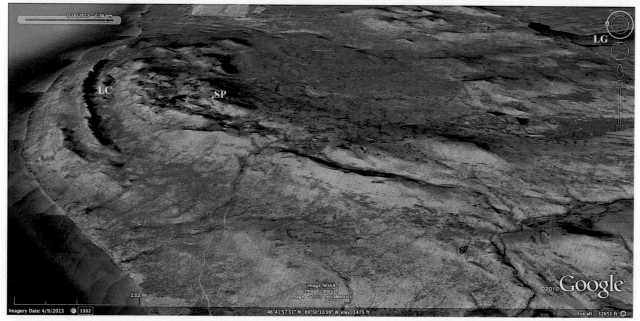

FIGURE 14.45 Google Earth image looking east across the Porcupine Mountains. Highland areas are underlain with resistant rhyolite, conglomerate, basalt, and sandstone. *LC*-Lake of the Clouds; *LG*-Lake Gogebic; *SP*-Summit Peak.

rugged highland areas seen in the image are underlain with resistant silicic volcanic rock (rhyolite), conglomerate, basalt, and sandstone. Summit Peak (SP, 1960 feet) is more than 1000 feet above Lake Superior. Lowland areas are underlain with nonresistant sandstone and shale. Well-developed cuestas are also present on Michigan's Keweenaw Peninsula as shown in Fig. 14.46. Additional exposures of Keweenawan Rift system rocks can be found at Goosebury Falls State Park along the Lake Superior shoreline northeast of Duluth, and at Interstate State Park on the St. Croix River south of Duluth. At Gooseberry Falls, the edges of individual lava flows have eroded to create waterfalls.

The entire exposed extent of the Keweenawan Rift system is shown in Fig. 12.18. There are, in addition, two arms of the rift system that extend at depth below the interior platform rock sequence, neither of which is exposed. One arm extends southwestward from Lake Superior across the southeast corner of Nebraska into Kansas and Oklahoma. A second, less-developed arm extends across central Michigan and possibly southward through western Ohio to northern Alabama. Extension associated with the Keweenawan Rift system may be responsible for intrusion the circa 1.08-billion-year-old Pikes Peak batholith in the Front Range of the Colorado Rocky Mountains.

The origin of the rift system is similar in some respects to that of the Triassic Rift Basins in the Appalachian Mountains described in Chapter 18, in which extension and normal faulting opened narrow basins that were subsequently filled with volcanic, intrusive, and sedimentary rock. In addition to age, significant differences between the Keweenawan rift basin and the Triassic basins are the sheer volume of volcanic rocks, the widespread presence of intrusive rocks, and the absence of border faults in the Keweenawan system. Contacts between Keweenawan rift rocks and those of the surrounding crystalline shield are depositional (unconformities rather than

FIGURE 14.46 Google Earth image looking eastward at southeast-facing cuestas on the Keweenaw Peninsula.

faults) because passive subsidence and deposition of the Sedimentary Series continued long after volcanism and normal faulting had ceased. The great volume of volcanic rocks in the Keweenawan system suggests the rift may have been associated with a short-lived hot spot mantle plume.

QUESTIONS

1. If you look closely at the Long Lake lineament (L in Fig. 14.2) you will notice that topography is significantly higher to the immediate east of the lineament than to the west. Suggest several hypotheses that might account for this landscape pattern.
2. Use tracing paper to trace the crystalline rock and the sedimentary layers in Fig. 14.4. Label the buttress unconformity.
3. What is the nature of the treeless area at the top of the mountain labeled UR across the river from Johnson Shut-ins in Fig. 14.7? Why does the treeless area extend down the mountain to the river? Research this by flying to Johnson Shut-ins State Park, MO in Google Earth.
4. Use a yellow pencil to color all lowland areas (farmland and lakes) in Fig. 14.7. Assuming these areas are underlain with sedimentary rock, and the highlands with crystalline rock, are there any locations, other than Johnson Shut-ins, where a river or stream appears to flow from sedimentary rock into crystalline rock and back into sedimentary rock? Do this with additional images downloaded from Google Earth.
5. Speculate as to why the Meers fault in Fig. 14.8 is parallel with the dominant fracture orientation in the Wichita Mountains.
6. Note in Fig. 14.9 that the pink granite is visible in a series of hills that form a circular (oval) shape. Note also that the ground is darker to the right of the granite than to the left. Explain the outcrop pattern and the change in ground shade.
7. Describe the landscape history of the Wichita Mountains.
8. Describe significant differences between the two types of crystalline-cored anticlinal (domal) mountains (shield and intrusive) as described in this chapter. With respect to Fig. 14.22, describe the sequence of events that created the Rocky Mountain erosion surface and the Gangplank.
9. In Google Earth, fly to Sundance, Wyoming. Locate Green Mountain about 2.5 miles to the east and describe the landform. Where are you located in the regional sense? Speculate as to the origin of Green Mountain.
10. What are the rock successions that make up most of the Middle and Southern Rocky Mountains?
11. How is the interior platform different in the Middle-Southern Rockies relative to other areas?
12. Provide a hypothetical example to show how sediment in a depositional basin can be used to determine 1)

how long ago an adjacent mountain first comes into existence, and 2) when crystalline rock is first exposed. Provide simple sketches with your explanation and suggest how the sediment could be dated.
13. Fig. 14.26 shows the evolution of the Southern Rockies since 75 million years ago. Describe some additional events that occurred between 50 and 10 million years ago.
14. Describe evidence to suggest that the present-day Colorado Front Range was, at one time, mostly buried beneath young, unconsolidated sediment?
15. In what way is the crystalline core of the Middle Rocky Mountains different from the Southern Rocky Mountains?
16. Make a list of anticlinal mountains in the Middle-Southern Rockies cored with shield rock. Which of these have been modified by normal faults?
17. Summarize some of the evidence that suggests the Southern Rockies have been elevated within the past 10 million years.
18. In Google Earth, fly to Pinedale, Wyoming. Locate the erosion surface along the western flank of the Wind River Range. How far does the erosion surface extend along the western flank of the range? What is the relationship between the erosion surface and the adjacent basin? Is there evidence in the northern part of the range to suggest that the erosion surface crosses to the eastern side of the range?
19. In one paragraph, summarize the landscape evolution of the Wind River Range. What happened between 62 and 57 million years ago? What is the inferred elevation of the mountain and adjacent basins 51 million years ago? When did the Wind River thrust become inactive and how do we know? During what time period did the erosion surface form and how do we know? On what surface did the erosion surface form? What happened between 42 and 37 million years ago and between 37 and 28 million years ago? What processes within the past 10 million years contributed to the present-day landscape?
20. What is a superposed stream? Could the East Fork Black River (BR, in Fig. 14.7) be considered a superposed stream? Why or why not? Fly to Greybull, Wyoming and then fly about 9 miles north along the Bighorn River to the Sheep Mountain anticline. Compare this area with the East Fork Black River area. How are they similar, how are they different?
21. A water gap can form if a river first establishes its course within overlying unconsolidated material and then maintains its course as it cuts downward into a pre-existing, but buried, mountain or ridge of resistant rock. Can you think of other possibilities that could produce a water gap? As a hint, consider the flow of rivers prior to Laramide uplift.
22. Research wind gap. What is a wind gap? Where do they occur? How do they form?
23. Given an average river incision rate of 0.5 inches per 100 years, what is the total incision in feet and in meters over a 10 million-year period?
24. Research the history of one of the crystalline-cored intrusive domes shown in Fig. 14.11. Do the same for one of the domes shown in Fig. 14.12.
25. Within the Middle Rocky Mountains, describe one method used to determine how long ago crystalline shield rock first became exposed at the surface.
26. Locate all unconformities in Fig. 14.30 and indicate the age of rocks on both sides. When did each unconformity form? When did mountain uplift occur? Explain your answers.
27. In Colorado, if we find Tertiary (Paleogene) rocks deposited directly above shield rocks, explain how or why this relationship can or cannot be used to gauge the location of the Ancestral Rocky Mountains.
28. What are the Missouri Buttes? Where are they located? How did they form?
29. In the Black Hills, is the Spearfish Formation older or younger than the Inyan Kara Group? Explain your answer.
30. On what basis is the Superior Upland separated from the Central Lowlands? Why is it difficult to place a western boundary between these two physiographic provinces?
31. Trace the cuestas in Fig. 14.46. How many are present? What is their origin?
32. Note the orientation of the Mesabi and Vermilion Ranges in Fig. 14.40. Speculate as to why these ranges survived flattening during glaciation.
33. Research the Keweenawan (Mid-Continent) rift system. What is its extent? How was it traced beneath the surface? How did it form and how long ago?
34. Research the Morton Gneiss, the origin of iron formations, or the Sudbury meteor impact site.
35. How many potential pre-Pangean supercontinents are recorded in the rocks of the Superior Uplands, and when did they exist?
36. Describe the history of the Penokean orogeny.
37. What special name is given to the contact that separates crystalline shield rocks of the Superior Upland from rocks of the Central Lowlands.

Chapter 15

Foreland Fold and Thrust Belts

From the perspective of the geologist, the most common and obvious structure within a compressional mountain system is the thrust fault. Thrust faults are present in sedimentary, crystalline, and volcanic rocks along the entire length of a mountain system and across its entire width in both the foreland and hinterland. Thrust faults permeate the Appalachian-Ouachita and Cordilleran Mountain systems. This chapter describes landscape and geology that developed in sedimentary rocks at the front of these orogenic belts where dominantly miogeoclinal and foreland basin rocks have been thrust toward the continental interior onto the interior platform (Fig. 11.2). Chapter 16, describes structures and landscape within hinterland belts at the rear of each mountain system where native North American and accreted terrane crystalline rocks are involved. Together, these two chapters form the Appalachian-Ouachita and Cordilleran Mountain systems group of structural provinces (Table 11.1).

The Valley and Ridge is the most conspicuous foreland fold and thrust belt in the United States. The southern (Tennessee) Valley and Ridge, in particular, shows the characteristic landscape, which is one of narrow, linear mountain ridges and valleys. In addition to the Valley and Ridge, this combination of rock type, rock structure, and resulting landscape is found in the Ouachita Mountains, the Marathon Basin region of West Texas, the Northern Rocky Mountains of Montana and Idaho, and in an area known as the Idaho-Wyoming thrust belt in the Middle Rocky Mountains. These areas are located in Figs. 11.1 and 11.3. As noted in Chapter 5, the Valley and Ridge, Ouachita, and Marathon fold and thrust belts formed at about the same time and were once continuous as part of the greater Appalachian-Ouachita thrust system. Similarly, the fold and thrust belts in the Northern and Middle Rocky Mountains formed at about the same time and were once continuous as part of the Cordilleran thrust system. Following a discussion on the structural form of fold and thrust belts, we will look first at the Cordilleran fold and thrust belt, followed by discussion of the Appalachian-Ouachita thrust belt.

STRUCTURAL FORM OF FORELAND THRUST FAULTS

The formation and resulting geometry of a simple foreland thrust fault is illustrated in Fig. 15.1. The fault is shown with a thick line. Rocks above the thrust fault are referred to collectively as the upper plate. Rocks below the thrust fault are referred to as the lower plate. As illustrated in Fig. 15.1, foreland thrust faults typically have a flat-ramp-flat geometry. The lower and upper flats are parallel with rock layers. The ramp cuts across rock layers at an angle between 20 and 45 degrees. Rock layers on the upper plate that begin movement along the lower flat are bent parallel with the ramp and then bent back to horizontal as they encounter the upper flat. Rock layers on the upper plate that begin movement in the ramp area become permanently bent once they encounter the upper flat. Rock layers on the upper plate that began movement along the upper flat remain horizontal throughout thrust faulting. Rock layers on the lower plate ideally remain undeformed and horizontal throughout thrusting as shown in the figure, but this is not always the case.

Flats typically form in weak or weakened rock layers such as shale, in which sliding is relatively easy. Each flat can be referred to as a décollement, which means to unglue, or to detach from. Ramps cut across strong rock layers that resist sliding. The thrust fault must break initially along the ramp for sliding to occur. Ramping and subsequent displacement have the effect of stacking older sedimentary layers above younger ones. The stacking, in turn, results in burial and overall shortening of the land surface. It is quite possible for foreland thrust faults to shorten an area originally 400 miles wide to one that is less than 200 miles wide.

Faulting and stacking begins at depth, and with increasing displacement, the fault grows and climbs by stair stepping (ramp-flat-ramp) up the layers until it reaches the surface, usually along a ramp. The stair-step geometry is illustrated in Fig. 15.2. It is not unusual for rock to be transported 20 miles or more along a single thrust fault that extends along strike for more than 100

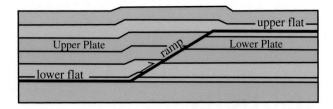
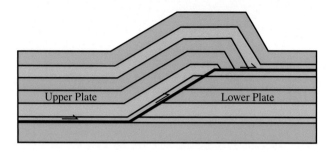
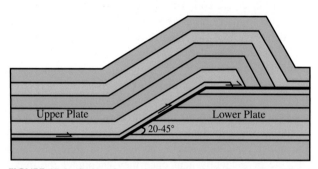

FIGURE 15.1 Series of cross-sections that show the process of thrust faulting and development of an anticline above a ramp.

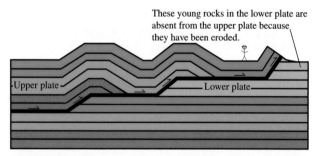

FIGURE 15.2 Schematic cross-section showing how a thrust fault stair-steps to the surface.

miles. Major thrust faults can extend the length of a mountain belt and can carry rock hundreds of miles. Note in Fig. 15.2 that stair stepping produces an anticline above ramps and a syncline along flats between ramps. Ramps can be continuous over great distances, thus producing the characteristic landscape of long linear (anticlinal) mountain ridges, and intervening narrow synclinal valleys. Additionally, the thrust fault itself is a zone of weakened, broken rock that will often erode to form a linear valley wherever it intersects the surface.

COMPARISON WITH THE CRYSTALLINE-CORED ANTICLINAL STRUCTURE

The foreland fold and thrust belt and the crystalline-cored mid-continent anticlines and domes both develop via compressional stress. Their structural development and resulting landscape are different for at least two reasons. The first is the thickness of sedimentary rocks. The second is the involvement of the crystalline shield. The crystalline-cored, anticlinal structure is prominent in the interior of the continent where a relatively thin sequence of interior platform rock overlies the crystalline shield. Shield rocks close to the surface become involved in the deformation and are subsequently exhumed to form the core of a mountain.

The fold and thrust belt, by contrast, forms in the miogeocline closer to the continental margin where the thickness of sedimentary rocks can exceed 30,000 feet and where planar layering provides ready surfaces along which a thrust fault can develop. Rocks of the ancient crystalline shield are too deeply buried to become involved in the deformation. The result is that the mountains consist of sedimentary rather than crystalline rock.

CORDILLERAN FOLD AND THRUST BELT

There are significant differences between the Cordilleran and the Appalachian-Ouachita fold and thrust belts. Primary among these are: (1) the age of the Cordilleran belt, which is considerably younger than the Appalachian-Ouachita belt, and (2) the extent to which each area has been modified by younger tectonic processes. Tectonic modification is extensive in the Cordilleran belt but nearly absent in the exposed part of the Appalachian-Ouachita belt.

Foreland thrust faulting in the western part of the Cordillera had begun by about 173 million years ago (Jurassic) following tectonic accretion in the hinterland and initial building of the tectonic wedge. Thrust faults migrated eastward over the next 100-plus million years, culminating between 115 and 52 million years ago with the Sevier orogeny. In its heyday, 55 million years ago, the linear ridge and valley landscape created by the thrust belt stretched across Montana, Idaho, Utah, Nevada, and California. The line shown in Fig. 15.3 represents the frontal (eastern-most) location of the thrust system in the United States. The eastern front crosses the Snake River Plain and then continues southward along the western edge of the Colorado Plateau into the Basin and Range. The linear ridge and valley landscape associated with the thrust belt likely stretched westward from this line for 100 miles or more. Total shortening across central Nevada due to thrusting is as much as 150 miles (240 km).

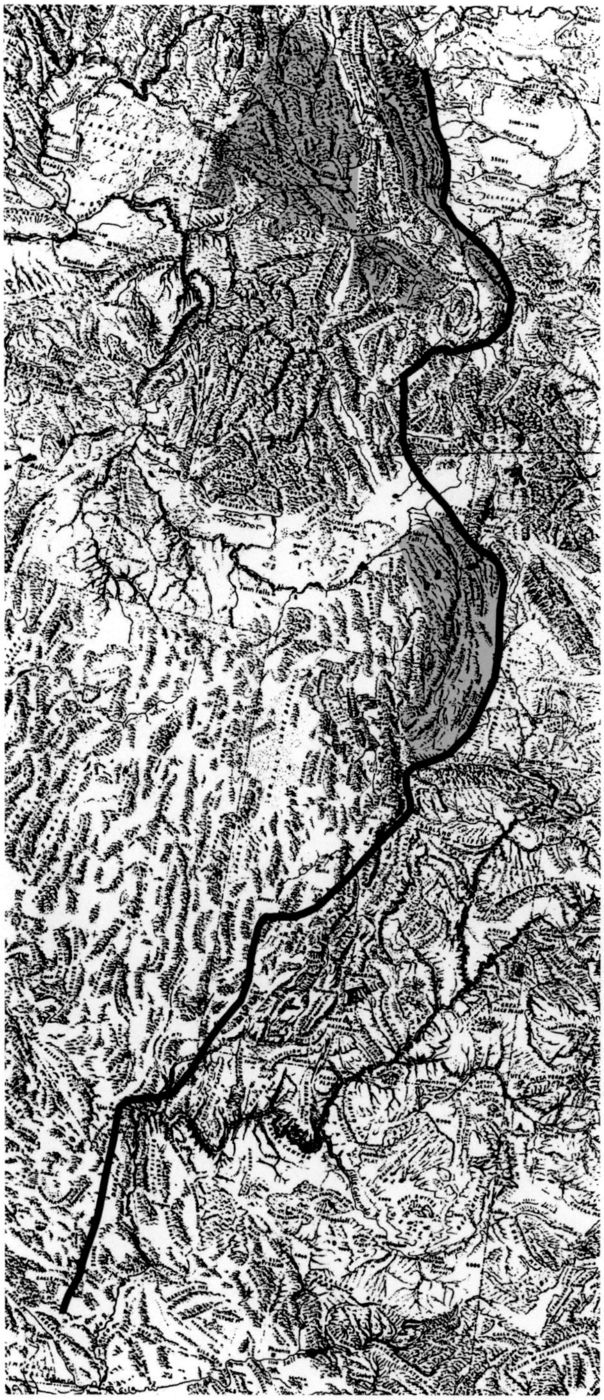

FIGURE 15.3 Landscape map that shows the frontal (eastern-most) trace of the Cordilleran foreland fold and thrust belt (heavy dark line). Thrust faults are present in rocks along this line and to the west, however much of the landscape has undergone tectonic reincarnation. Areas where fold and thrust landscape is preserved are shaded areas.

The demise of the thrust belt landscape began when younger tectonic processes, most notably normal faults and volcanism associated with the Basin and Range and Snake River Plain, created an entirely new landscape. Tectonic reincarnation began soon after the fold and thrust belt was established, but it wasn't until about 17 million years ago when the pace of volcanism and normal faulting accelerated, and wholesale destruction of the older landscape began. Volcanism and normal faulting remain active across large parts of the Cordillera. The older fold and thrust landscape is preserved in only a few areas discussed later. In other areas, we can locate and map thrust faults in the rock layers, but these structures no longer control the landscape.

Partly as a result of tectonic reincarnation, the Northern Rocky Mountains have some of the most varied landscape in the country. As seen in Fig. 15.4, the area is host to no fewer than four structural provinces; the foreland fold and thrust belt discussed here, the hinterland deformation belt discussed in Chapter 16, young volcanic rocks discussed in Chapter 17, and normal fault landscapes discussed in Chapter 18. Each structural province results in different and distinct landscape. According to Fig. 15.4, the fold and thrust linear ridge and valley landscape survived reincarnation in two areas of the Northern Rockies and one area of the Middle Rockies. But even these areas have undergone significant tectonic modification primarily along normal faults that have infiltrated and influenced the landscape. The two areas in the Northern Rockies will be referred to as the eastern and western segments, respectively. The fold and thrust landscape in the Middle Rocky Mountains occurs along the mutual borders of Utah, Idaho, and Wyoming where it is known as the Idaho-Wyoming fold and thrust belt.

Fig. 15.5 is a Google Earth image of Idaho and adjacent Montana that identifies mountain ranges. The area between the two yellow lines is the area where active normal faults are numerous and where tectonic reincarnation has largely overprinted the fold and thrust landscape. Active normal faults outside the yellow lines are shown with a thick blue line such as those in the Idaho-Wyoming thrust belt (Wy). A lineament, not obvious in Fig. 15.4, but visible in Fig. 15.5, is located between the two white arrows. It is identified by a change in the orientation of mountain ridges such as the Garnet Range (Gt) and the Reservation Divide Mountains (Rd), and by its parallelism with river valleys. The lineament corresponds with a zone of concentrated right-lateral strike-slip and normal faults known as the Lewis and Clark Line (labeled LC in Fig. 15.5, and L in Fig. 15.4). This fault zone is believed to have originated during the Precambrian (pre-541 Ma) within the underlying North American crystalline shield, and has periodically reactivated. An active normal fault is present within the lineament north of Missoula (M). Interstate 90 follows this lineament through Missoula to Spokane, Washington (S). A secondary lineament extends along the Clark Fork River north of the Coeur d'Alene Mountains (Cd) to Lake Pend Oreille (p).

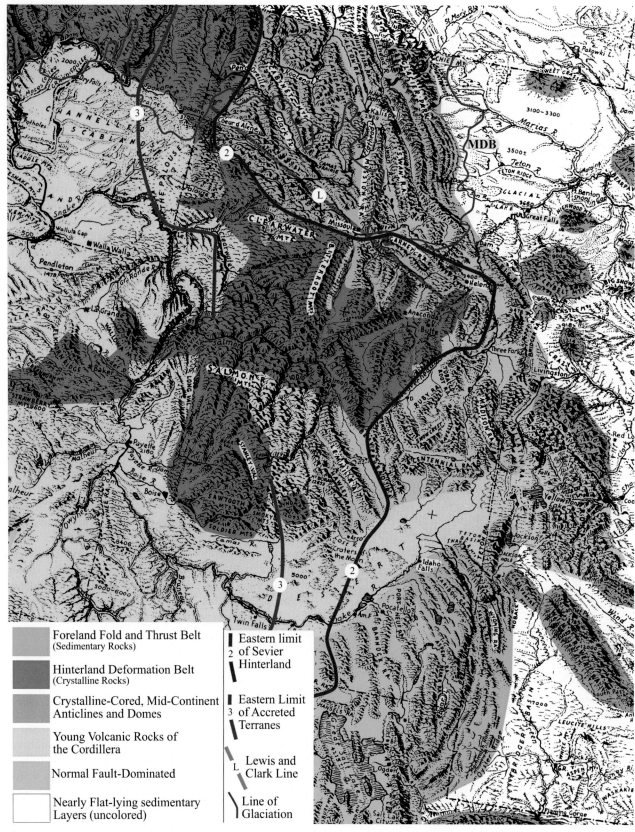

FIGURE 15.4 Landscape map of the Northern Rocky Mountains that shows the distribution of structural provinces.

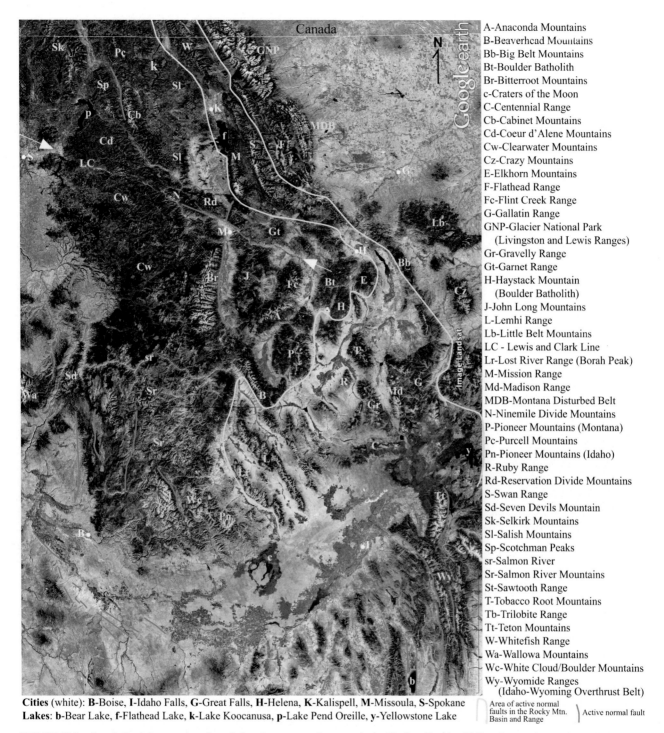

FIGURE 15.5 Google Earth image oriented north that shows mountain ranges in the Northern Rockies. Yellow lines enclose the area of active normal faults known as the Rocky Mountain Basin and Range discussed in Chapter 18. Active normal faults outside the yellow lines are shown with a thick blue line. The Lewis and Clark line (LC) is located between the two white arrows.

Glaciation affected both the Northern and Middle Rocky Mountains. The Cordilleran Ice sheet covered the entire eastern fold and thrust segment of the Northern Rockies, and part of the western segment as far south as Lake Pend Oreille (p) and Kalispell (K). The thin blue line in Fig. 15.4 marks the approximate southern limit of the Cordilleran Ice sheet. Alpine glaciation has continued to affect high elevation areas.

Northern Rocky Mountains

The landscape expression of the Northern Rocky Mountain fold and thrust belt is located primarily north of the Lewis and Clark Line (L, Fig. 15.4) where it is separated by a zone of active normal faults into a western and eastern segment. South of the Lewis and Clark Line, the fold and thrust belt is disrupted by crystalline, primarily granitic rock, metamorphism, and by active and inactive normal faults to the point where the linear mountain trend has largely been obliterated. The foreland-hinterland boundary (line 2 in Figs. 15.4 and 11.9) is placed at the first occurrence of crystalline rock such that sedimentary rocks, and part of the fold and thrust belt, extend south of the boundary. Many of the normal faults in this area have north-northwest trends similar in orientation with the trend of thrust faults. This similarity preserves some of the original linear landscape even though down-dropped normal-fault basins typically are wider than fold and thrust valleys. In some instances, normal faults reactivate older thrust faults. Because of their strong influence on landscape, the area of active normal faults in the Northern Rockies shown in Figs. 15.4 and 15.5, is referred to as the Rocky Mountain Basin and Range.

Eastern Segment of the Northern Rocky Mountains

The eastern segment forms the Rocky Mountain Front with the Great Plains. Two subareas are shown in Figs. 15.4 and 15.5, Glacier National Park (GNP), located just south of the Canadian border, and the Montana Disturbed Belt (MDB, also known as the Sawtooth Range) located farther south, which includes the Flathead Range (F). There are five peaks above 9000 feet in the Montana Disturbed Belt, which forms part of the Great Bear and Bob Marshall Wilderness areas. In Glacier National Park, there are six peaks above 10,000 feet and at least 55 additional peaks above 9000 feet. The continental divide extends through both areas, reaching a triple divide at Triple Divide Peak in Glacier National Park where water can begin its journey to the Atlantic, Pacific, or Arctic Oceans. The area is known for its incredibly beautiful glacier-carved sedimentary peaks, its many hiking trails, and its remote backcountry, but because the rock is sedimentary and easily broken, it does not offer ideal rock climbing opportunities. Thrusting along the frontal zone is Late Cretaceous to Eocene in age (75–52 Ma) based on the age of deformed rock units, cross-cutting relationships, radiometric dating, and foreland basin deposits.

Thrust ramps in the Montana Disturbed Belt are numerous and closely spaced resulting in a sequence of stacked thrust sheets and some of the best developed ridge and valley landscape in the Cordillera. A Google Earth image and schematic cross-section of the structure and resulting landscape are shown in Fig. 15.6. Erosion has preferentially attacked and partly removed elevated rocks in ramp areas along the crest of anticlines producing an asymmetric shape to the mountain with steep slopes on the eastern side and relatively gentle dip-slopes on the western side. A dip slope is where the dip of the rock layer is parallel with the inclination of the slope. Thrust faults occupy narrow intervening valleys. One possible explanation for the many closely spaced thrust faults is the thin layering of the Paleozoic-Mesozoic miogeoclinal sedimentary rocks that are involved in the deformation.

Thrust faults similar to those in the Montana Disturbed Belt, but less numerous and with less total displacement, extend northward into the foothills region of the Rocky Mountains along the western edge of the Great Plains east of Glacier National Park, but these faults have little in the way of topographic expression. West of the foothills region, the Rocky Mountains rise abruptly at Glacier National Park, but linear ridge and valley landscape is not well developed due to a combination of rock type and rock structure. Rather than a series of thin, stacked thrust sheets as seen in the Montana Disturbed Belt, landscape in Glacier National Park is shaped by a single large, synclinally folded thrust sheet that underlies the entire region. The thrust sheet was carried northeastward on the Lewis thrust, one of the largest in the Cordillera. The Lewis thrust extends northward from the US–Canada border for 140 miles (225 km) and southward into the United States for another 141 miles (227 km) through the Flathead Range (F, Fig. 15.5) west of the Disturbed Belt. Rocks have been pushed northeastward along this thrust for more than 62 miles (100 km). A schematic cross-section and Google Earth image showing the folded thrust sheet is presented in Fig. 15.7. Note the correspondence of high elevation with structurally high areas along the eastern and western flanks of the syncline. The Lewis Range occupies the eastern flank and the Livingston Range occupies the western flank north of Lake McDonald. These mountains are separated by a low area at Flattop Mountain (f), visible in Fig. 15.7A, which forms the trough of the fold. The Lewis thrust crops out along the eastern and southern margins of Glacier National Park at the foot of the Lewis Range. In Fig. 15.7B, a segment of the thrust sheet can be seen on Chief Mountain where it occurs as an erosional remnant detached from the main part of the thrust sheet. Detached erosional remnants of thrust sheets, such as Chief Mountain, are known as klippen (plural) or as a klippe (singular).

One of the most significant aspects of the Lewis thrust is that it carries sedimentary layers that are significantly older than sedimentary rock in the Montana Disturbed Belt. Most of the rocks are between 1470 and 1370 million years old and are known as the Belt Supergroup

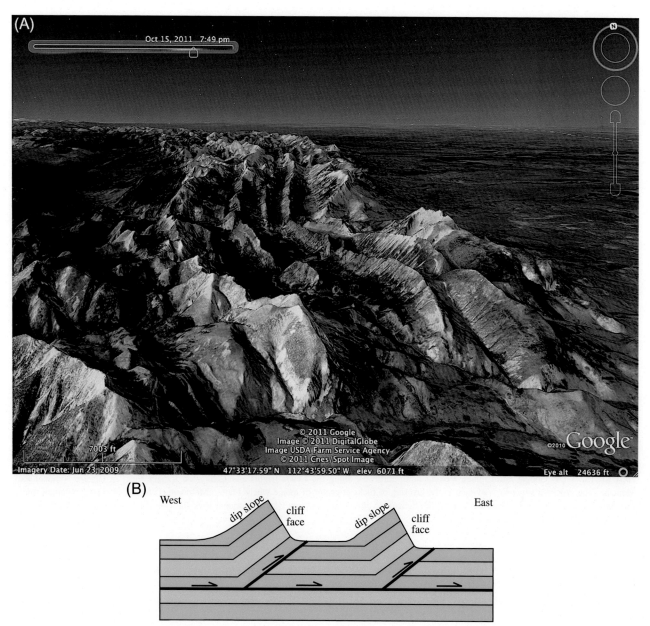

FIGURE 15.6 (A) Google Earth image looking north along the Montana Disturbed belt, Montana. (B) Schematic cross-section that shows the general structure of the Montana Disturbed belt.

(Purcell Supergroup in Canada). The rocks are part of the Precambrian sedimentary/volcanic rock succession and are as old or older than some of the rocks that form the North American crystalline shield. Much of the rock unit consists of argillite, siltstone, quartzite, quartz sandstone, limestone, and dolostone metamorphosed at low temperature. Argillite is a hard, indurated mudstone. The Belt Supergroup at Glacier National Park is about 21,000 feet thick, but the rocks are more than double that thickness farther west. The Lewis thrust transports these rocks over rocks that are, in some areas, less than 100 million years old.

A major rock unit within the Belt Supergroup, and present throughout the central part of the Glacier National Park as well as in the Flathead Range, is the Helena Formation (previously known as the Siyeh Limestone). This rock unit is significant for several reasons. It is as much as 3380 feet thick and composed primarily of thick-beds of limestone and dolostone. As such, it is a fairly resistant cliff-former. It forms steep summit cliffs on Mount Stimson and Mount Jackson, and steep cliffs just below the summits of Mount Siyeh, Mount Cleveland, Kintla Peak, and Mount Merritt; all six of the 10,000-foot peaks in the park. It is present along most of the drive on

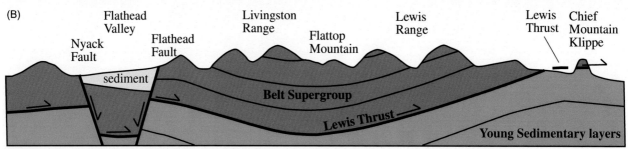

FIGURE 15.7 (A) Google Earth image looking north-northwest along the Lewis (right) and Livingston Ranges (left) in Glacier National Park, Montana. The synclinal structure is evident in the layering of rocks at center of image where Flattop Mountain (f) forms a lowland synclinal trough. The line in the distance is the Canadian border. (B) Schematic cross-section across Glacier National Park showing the Lewis thrust and the synclinal structure. Based on Kiver and Harris (1999, p. 553).

Going-to-the-Sun Highway from Granite Park to Logan Pass and beyond. It forms most of the meadow at Hanging Gardens below Reynolds Mountain, and forms steep cliffs below the summit of Going-to-the-Sun Mountain. The rock itself is interesting because the upper part contains stromatolitic limestone, a strongly laminated rock formed by the growth of blue-green algae and preserved as some of the oldest fossils anywhere. The middle and lower parts of the formation contain a texture known as molar-tooth structure, which consists of closely spaced, vertical and inclined calcite veins that are often crumpled, squashed, and broken. Additionally, when viewing cliff faces of Helena Formation, you might see a dark band sandwiched between two thin white bands. The dark band is a slightly metamorphosed layer of intrusive rock (a sill) composed primarily of dark granitic rock known as diorite. The two white bands result from reaction of the limestone with the hot magma. The only other nonsedimentary rock in the park is the Purcell Lava, which can be seen at Granite Park in rocks that overlie the Helena Formation (there is no granite at Granite Park).

Rocks composed of thin-bedded argillite and siltstone are considerably less resistant to erosion than the Helena Formation. Reynolds Mountain (9125 feet), shown in Fig. 15.8, literally looks like it is crumbling to the ground. This mountain is composed of argillite-rich and siltstone-rich formations that overlie the Helena Formation. These same weak rocks cap some of the high peaks including Mount Cleveland, Kintla Peak, and Mount Merritt. It is amazing that these weak rocks can form any mountain at all. These peaks likely would not be so high without the presence of underlying Helena Formation to form a strong, resistant base.

Areas where thrust faults are closely spaced such that they overlap like roof tiles are known as imbricate zones. They are common features in fold and thrust belts. The Montana Disturbed Belt, for example, forms an imbricate zone on a large scale as depicted schematically in

FIGURE 15.8 Photograph of nonresistant rock of the Belt Supergroup that forms Reynolds Mountain (9125 feet) in Glacier National Park, Montana. These rocks overlie the Helena Formation.

Fig. 15.6B. Smaller versions of these zones are present along and below the Lewis thrust especially in the southern and eastern part of the park. Imbricate thrust faults form rings around Eaglehead Mountain, the Cloudcroft Peaks, and Battlement Mountain. They are also present below the summits of Mad Wolf, Rising Wolf, White Calf, Curly Bear, and Bearhead Mountain. Two imbricate thrusts form the Chief Mountain klippe.

Western Segment of the Northern Rocky Mountains

The western segment of the fold and thrust belt extends west of Flathead Lake across northern Montana and Idaho to the foreland-hinterland boundary (line 2 in Figs. 15.4 and 11.9) located in Idaho near Lake Pend Oreille and Coeur d'Alene. Most of the mountains in the thrust belt are in the 6000 to 8000 foot range with rounded, forest-covered peaks. The only range with peaks above 8000 feet are the Cabinet Mountains (Cb, Fig. 15.5), but the high elevation is likely due to the presence of an active normal fault along its western base. The higher ranges show evidence of alpine glaciation, but these features are strongly developed only in the Cabinet Mountains (Cb), the nearby Scotchman Peaks (Sp), and the US Purcell Mountains (Pc, Fig. 15.5).

The western segment, as a whole, is deformed into a broad, composite, anticline known as the Purcell anticlinorium (an anticlinorium is a large composite anticline with many sub-anticlines and synclines). The region does not display a particularly strong linear landscape for several reasons. First, the area is underlain with thick-bedded rocks of the Belt Supergroup. Consequently, the thrust faults also carry thick sequences of rock. Second, the thrust faults are widely spaced relative to those in the Montana Disturbed Belt with major thrust faults present only in the Salish Mountains (Sl, Fig. 15.5), along the west side of the Whitefish Range (W), and in the vicinity of the Cabinet Mountains (Cb). Third, the fold and thrust landscape is disrupted by younger, mostly inactive normal faults, some of which are parallel with and reactivate thrust faults, whereas others cut at an angle across thrust faults. The age of thrusting in this area is not well constrained, but likely began in the Jurassic as early as 173 million years ago. Overall, this area is subdued in its mountain linearity relative to the better-developed eastern segment of the fold and thrust belt.

Lake Koocanusa (k, Fig. 15.5) is a long, narrow lake that extends for 93 miles along the Kootenai River beginning south of Mt. Fisher in Canada and continuing into Montana. It is a man-made lake, created in 1974 with the building of Libby Dam. It is mentioned here because of its unique name. It was named Lake Koocanusa after KOOtenai, CANada, and USA.

The Rocky Mountain Trench

Fig. 15.9 is a Google Earth image oriented eastward that shows both the Canadian and the US Northern Rocky Mountains. Flathead Lake and Lake Pend Oreille are clearly visible on the US side. Also visible is a remarkably long, straight valley known as the Rocky Mountain Trench that cuts the Canadian Rockies lengthwise nearly in half. In this usage, the word trench refers to a long, continuous valley. It is not a subduction zone. A second long valley, the Purcell trench, extends northward from Coeur d'Alene, Idaho, into Canada, where it merges with the Rocky Mountain trench. Both valleys are underlain with a combination of thrust faults and younger normal faults, and both have been widened and deepened by glaciers. The Cordilleran fold and thrust belt in Canada to the east of the Rocky Mountain Trench is not strongly modified with normal faults and thus shows excellent linear ridge and valley landscape, particularly in the vicinity of Banff and Jasper. The Rocky Mountain Trench continues into the United States within the zone of active normal faults through Flathead Lake to Missoula and perhaps southward along the east side of the Bitterroot Mountains where another active normal fault is present as seen in Figs. 15.4, and 15.5, however, the trench, in this area, is not particularly well developed.

The Idaho-Wyoming Fold and Thrust Belt

The Idaho-Wyoming fold and thrust belt is shown in Fig. 15.4 south of Idaho Falls and is labeled Wyomide Ranges (Wy) in Fig. 15.5, which shows the area as far south as Bear Lake (b). It is a small surviving island of linear ridge and valley landscape that projects south from the Yellowstone Plateau. This area is similar to the Montana Disturbed Belt in that there are at least eight

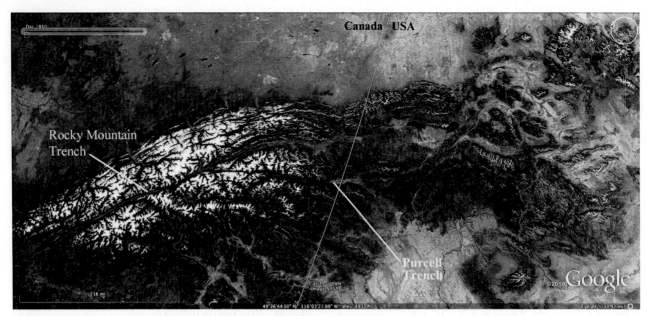

FIGURE 15.9 Google Earth image looking eastward at the Rocky Mountain and Purcell trenches in Canada.

closely spaced thrust faults that have carried thick piles of miogeoclinal sedimentary rock eastward. The rocks are well layered and mostly Devonian to Cretaceous in age (419–66 Ma). The thrust belt is surrounded by a variety of structural provinces that include normal fault block mountains of the Teton Range, nearly flat-lying sedimentary layers of the Wyoming Basin, anticlinal mountains that include the Wind River and Uinta Mountains, and areas of volcanic rock.

As seen in Fig. 15.10 looking north, this part of the thrust belt shows excellent linear ridge and valley landscape with typical asymmetric ridges in which the dip slope is on the west side and the steep slope is on the east. Crosscutting relationships indicate that thrusting and initial landscape formation occurred between 90 and 55 million years ago. These mountains boast three peaks above 11,000 feet, Wyoming Peak (11,383 feet), Mount Coffin (11,247 feet) and Triple Peak (11,132 feet), all in the Wyoming Range (W), and at least 18 additional peaks above 10,000 feet including no fewer than 10 in the Salt River Range (S), which lies west of Grays River Valley (G).

The fold and thrust belt is not without tectonic modification. Basaltic volcanic rocks are present in the area between Idaho Falls and Bear Lake, and active normal faults are present across the region as illustrated in Fig. 15.5. The normal faults are oriented northward, roughly parallel with older thrust faults, and thus, in some cases, they enhance the linear ridge and valley landscape. Normal faults are present on both sides of Bear Lake (b), implying a graben structure. A normal fault active in the past 15,000 years separates the Wyoming Range (W, Fig. 15.10) from Grays River Valley (G). It is likely that displacement on this fault is responsible for elevating the western margin of the range above 11,000 feet and for tilting rocks gently eastward creating wide, flat, tilted summit areas such as those on Mount Coffin. The normal fault may also be responsible for the down-dropping and eastward tilting of Grays River Valley (G) such that the Grays River flows along the eastern edge of the valley creating a steep 3500-foot escarpment along the base of the Wyoming Range that leads up to the 11,000-foot peaks.

OVERVIEW: APPALACHIAN-OUACHITA FOLD AND THRUST BELT

The Appalachian fold and thrust belt corresponds roughly with the physiographic Valley and Ridge province. It extends the length of the US Appalachians from Lake Champlain, Vermont, to Tuscaloosa, Alabama. Similar rocks, structure, and landscape are present in the Ouachita Mountains west of the Mississippi River, and in the Marathon region of southwest Texas. These areas collectively form the Appalachian-Ouachita foreland fold and thrust belt. In contrast to the Cordilleran fold and thrust belt, this is an ancient belt, having formed prior to 265 million years ago. Geologic, geophysical, and drilling data indicate that the three now separate regions were continuous 265 million years ago when mountain building events ended. Intervening areas on the Coastal Plain, and the entire Marathon region, were eroded and covered with younger rock of the Atlantic miogeocline primarily during high sea level stands over the past 145 million years. The Marathon region has since been uncovered. Most of the

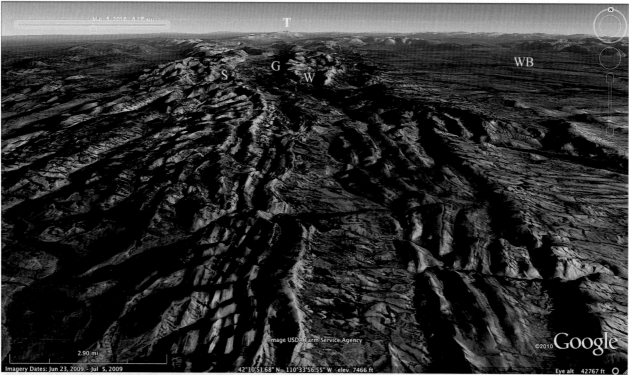

G-Greys River Valley, S-Salt River Range, T-Teton Mountains, W-Wyoming Range, WB-Wyoming Basin

FIGURE 15.10 Google Earth image looking north across the Idaho-Wyoming fold and thrust belt from southwestern Wyoming. An active normal fault forms a steep 3500-foot escarpment between Grays River Valley (G) and the Wyoming Range (W).

Valley and Ridge and Ouachitas, by contrast, have existed as a highland since their initial formation.

Partial burial beneath rocks of the Coastal Plain has been the only major tectonic event to affect the Appalachian-Ouachita foreland fold and thrust belt since 265 million years ago. There have been changes in elevation and relief due to sea level and isostatic variation in the exposed thrust belt, but no major tectonic modification. The structure we see today is identical with the structure that existed some 265 million years ago. The primary change to the landscape has been erosional lowering and sculpting due to the presence of tilted interlayered hard and soft rock. Today, the region forms an erosion-controlled landscape of long, linear resistant sandstone ridges and nonresistant limestone/shale valleys. Elevation rarely exceeds 4000 feet. Here we see a clear distinction between an ancient Appalachian-Ouachita fold and thrust system that has grown old over time via erosion, and a relatively young Cordilleran fold and thrust system whose landscape, in many areas, is no longer recognizable due to recent tectonic reincarnation.

Rocks in the Appalachian-Ouachita foreland fold and thrust belt are primarily miogeoclinal, but foreland basin rocks derived from ancient, now vanished hinterland mountains, are also present. All of the rocks are sedimentary, from 541 to 265 million years old, and much thicker than interior platform rocks of the same age on the craton to the west.

VALLEY AND RIDGE FOLD AND THRUST BELT

The Appalachian fold and thrust belt closely mimics the physiographic Valley and Ridge as seen in Fig. 11.3. The structure and landscape of the fold and thrust belt, however, varies considerably from south to north. The southern part from, Alabama to Roanoke, Virginia, is a classic fold and thrust belt in which rocks have been transported west-northwestward more than 200 miles along 10 or more major thrust faults that are long, continuous, and closely spaced with typical flat-ramp-flat geometry. However, as we travel northward across the central fold and thrust belt from Roanoke to Pennsylvania, we begin to depart from the classic fold and thrust architecture. Long continuous thrust faults give way to folds. Ridges and valleys are maintained, but they form a zigzag pattern that follows fold patterns rather than thrust faults. Mississippian and Pennsylvanian rocks are involved in all of the deformation in both the southern and central regions, which implies that thrust faulting and folding occurred during the last of the great Appalachian orogenic events listed in Chapter 5, the Alleghany orogeny from 335 to 265 million years ago (also known as the Alleghanian orogeny).

Changes in both the structure and landscape become drastic as we travel across the northern fold and thrust belt along the New England-New York border. Here, the fold and thrust belt is narrow, occupying only the Lake Champlain-Hudson River valley, and it is older, having formed in the Ordovician during the first of the great Appalachian orogenic events, the Taconic orogeny from 480 to 446 million years ago. This part of the thrust belt does not show classic linear ridge and valley landscape. In the following discussion, we first introduce the Great Valley, and then look briefly at the northern and central sections of the fold and thrust belt. We end with a detailed look at the classic fold and thrust belt of Tennessee.

The Great Valley

The Great Valley (or Great Appalachian Valley) is the Appalachian counterpart to the Rocky Mountain Trench. It is a long, continuous valley that extends along the eastern margin of the Valley and Ridge from Vermont to Alabama. It sits directly adjacent to the Blue Ridge and to the Berkshire and Green Mountains of western Massachusetts and Vermont.

It is not a single river valley, but rather, a composite of several valleys. It begins in western Vermont in the Lake Champlain valley where it encompasses the entire Valley and Ridge. It extends southward along the Hudson River Valley, but prior to reaching New York City, it turns southwestward through eastern Pennsylvania and northern Virginia to Roanoke where it forms the Shenandoah River Valley. South of Roanoke it is interrupted with a series of short, discontinuous ridges, but is still distinctive at the western foot of the Blue Ridge. From Tennessee southward, it occupies essentially the entire Valley and Ridge but is interrupted with low-lying linear ridges. Fig. 15.11 is a Google Earth image looking northeast along the Valley and Ridge from Georgia to Pennsylvania. The Great Valley is obvious along the east side of the Valley and Ridge against the Blue Ridge. Also visible are the thrust-faulted linear ridges of southern Virginia (SVVR) and Tennessee (TVR), the transitional, folded and faulted ridges of northern Virginia (NVVR), and the folded, zigzag ridges of Pennsylvania. The Valley and Ridge and Great Valley are shown in their entirety on Raisz landform maps, Figs. 13.8, 13.9, and 13.60.

The Great Valley is underlain with thrust faults along most of its length. The presence of a valley is due to

C-Chattanooga
D-Dalton
G-Great Valley
K-Knoxville
P-Pine Mountain
R-Roanoke,
S-Sequatchie Valley,
NVVR-Northern Virginia Valley & Ridge
SVVR-Southern Virginia Valley & Ridge
TVR-Tennessee Valley & Ridge

FIGURE 15.11 Google Earth image oriented northeastward along the Appalachian Valley and Ridge and Great Valley.

weak rock, particularly limestone, which is easily dissolved in the humid climate. The dissolution of limestone has produced karst topography, particularly north of Roanoke where one can find numerous cave systems including Crystal Cave, Indian Echo Cave, and the Baker, Shenandoah, Massanutten, and Dixie Caverns. In most areas of karst topography, such as Florida and the Interior Low Plateaus, limestone layers are approximately flat-lying. Karst topography in the Valley and Ridge is unusual because it is developed in dipping layers.

Northern Appalachian Fold and Thrust Belt

The northern Appalachian fold and thrust belt does not conform to typical ridge and valley landscape. It is older than other parts of the thrust belt and there are few ridges. Nonresistant rock, eroded to form the Great Valley, occupies most of the region. Fig. 15.12 is a Google Earth image looking just south of east that shows the landscape north of Saratoga Springs, New York (S). The thrust belt occupies the Champlain Valley between the Catskill and Adirondack Mountains on the west, and the Green Mountains and Green Mountain foothills (GMf) on the east. A mountain range, the Taconic Mountains (outlined with a yellow line), sits in the middle of the Great Valley separated from the Green Mountains by the Valley of Vermont at Rutland (R) and Manchester (M). Lake Champlain (LC) is only 98 feet above sea level, and most of the surrounding valley is less than 500 feet in elevation. Equinox Mountain (E, 3850 feet) is the highest point in the Taconic Mountains.

A major thrust fault, the Champlain thrust, extends along the east side of Lake Champlain (LC) and continues southward to the area east of Lake George (LG) where major displacement switches to other thrust faults. The Champlain thrust transported resistant Cambrian (541–485 Ma) dolomite and quartzite as much as 60 miles westward above weak Ordovician (485–444 Ma) shale, from which glaciers gouged Lake Champlain. The resistant Cambrian rocks dip gently eastward to form a series of cuestas that rise up to 300 feet above the lake. The arrow in Fig. 15.12 points to the location of one such cuesta.

From St. Albans City (A) southward, the miogeoclinal rocks that form the thrust belt are folded into a large composite syncline known as the Middlebury synclinorium. There is little expression of the synclin in the landscape

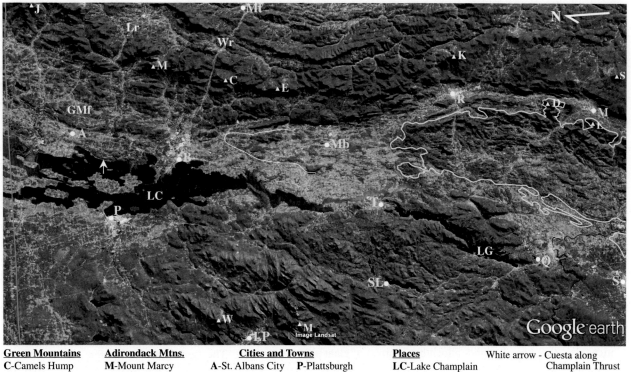

Green Mountains	Adirondack Mtns.	Cities and Towns		Places	
C-Camels Hump	M-Mount Marcy	A-St. Albans City	P-Plattsburgh	LC-Lake Champlain	White arrow - Cuesta along Champlain Thrust
E-Mt. Ellen	W-Whiteface Mtn.	B-Burlington	Q-Queensbury	LG-Lake George	
J-Jay Peak		LP-Lake Placid	R-Rutland	GMf -Green Mtn foothills	White line - Outlines resistant hills of Quartzite
K-Killington Peak	Taconic Mountains	M-Manchester	S-Saratoga Springs		
M-Mount Mansfield	D-Dorset Peak	Mb-Middlebury	SL-Schroon Lake	Lr-Lamoille River	Yellow line - Encloses Taconic Allochthons
S-Stratton Mountain	E-Equinox Mtn.	Mt-Montpelier	T-Ticonderoga	Wr-Winooski River	

FIGURE 15.12 Google Earth image looking eastward across the Champlain Valley, New York–Vermont. The thin line at left is the Canadian border.

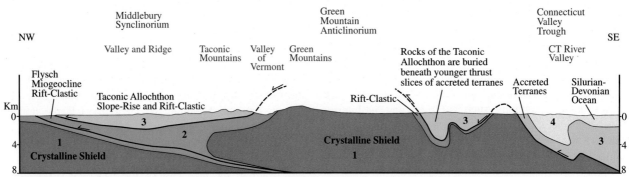

1-Crystalline shield. 2-Miogeocline, underlying rift-clastic rocks, and foredeep basin rocks (flysch) derived from erosion of the now vanished Taconic Mtns. 3-Slope-rise and rift-clastic rocks are exposed in the Taconic Allochthon on the northeast side. Accreted terranes are exposed on the southeast side. The Taconic Allochthon was thrust westward from the eastern side of the Green Mountain anticlinorium along thrust faults that are now buried beneath accreted terranes. 4-Silurian-Devonian rocks that unconformably cover older Taconic-age rocks.

FIGURE 15.13 Simplified cross-section across southern Vermont from the Valley and Ridge to the Connecticut River. The cross-section is 64 miles wide and based on cross-section I-I' of the Geologic Map of Vermont (2011).

except surrounding Middlebury (Mb) where highly resistant quartzite forms a series of hills that trace out the shape of the fold (shown with a white line in Fig. 15.12).

South of Middlebury, in the middle of the Great Valley, the Taconic Mountains form a group of giant thrust slices known as the Taconic Allochthon that sit like an island in the trough of the Middlebury synclinorium directly on top of the folded, faulted miogeoclinal rock assemblage. The extent of the allochthon is enclosed within the yellow line, which also outlines the Taconic Mountains. A simplified cross-section shown in Fig. 15.13 shows the relationship between the allochthon (#3, left side of figure) and surrounding rocks. The allochthon is not part of the miogeocline or part of the fold and thrust belt. It is part of the hinterland that was transported westward during Taconic orogeny along multiple thrust faults a distance of more than 70 miles and placed above the deformed miogeocline and associated foredeep basin rocks (#2).

The rocks that form the allochthon are a combination of slope-rise rocks that are the same age as rocks of the miogeocline (Cambrian-Ordovician) and rift-clastic rocks that are part of the underlying Precambrian sedimentary/volcanic rock succession. Some of the rocks were metamorphosed prior to and during transport, and these have resisted erosion to form the present-day Taconic Mountains. Notice in Fig. 15.13 that the Valley of Vermont forms a lowland of non-resistant miogeoclinal rock that separates the Taconic and Green Mountains. Notice also that the Taconic allochthon forms a giant klippe, detached from its thrust belt, which lies east of the Green Mountains. The thrust belt is not well exposed east of the Green Mountains. It has, instead, been overrun and buried beneath younger thrust faults associated with the emplacement of accreted terranes (#3, right side of Fig. 15.13). The Taconic Allochthon extends as far south as Poughkeepsie, New York, but similar thrust-faulted rocks extend discontinuously southward through Manhattan Island to eastern Pennsylvania.

Geology of the Southern Hudson River Valley

The Great Valley occupies most of the fold and thrust belt in the southern Hudson River Valley. Fig. 15.14 is a Google Earth image of southeastern New York looking northeastward from the vicinity of Port Jarvis (PJ) toward Poughkeepsie (Pk). The area extends from the Appalachian Plateau (Catskill Mountains) across the Valley and Ridge (Great Valley) to the Reading Prong section of the New England Highlands (RP). Fig. 15.15 is a west to east cross-section across the area with an exaggerated vertical scale to show relative rock hardness. From a landscape perspective, the area is controlled by rock hardness. From a geological perspective, it provides a tale of ancient, now vanished mountains. We will discuss geological history with reference to Figs. 15.14 and 15.15. Numbers refer to rock units listed in Fig. 15.15.

Monticello (Mc) is located in the southern (lower) part of the Catskill Mountains where surrounding elevation is between 1,200 and 1,600 feet. The landscape surrounding Monticello (Mc) can be described as a weakly dissected plateau relative to the topographically higher and more strongly dissected northern Catskill Mountains seen in the distance. The landscape is held up by resistant, nearly flat-lying sandstone and conglomerate known as Catskill molasse (6). The term molasse refers to nonmarine conglomerate, sandstone, and shale produced from the erosion of mountains and deposited in rivers and lakes. The Catskill molasse (6) is Middle and Upper Devonian (393–359 Ma), and represents the eroded residue of the now vanished Acadian Mountains (mountains that formed during Acadian orogeny). The rocks decrease in thickness

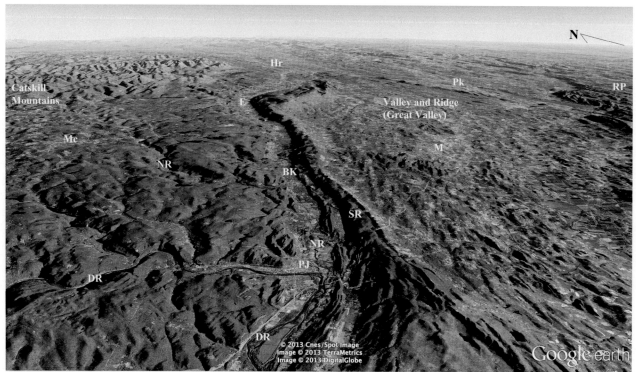

BK-Basher Kill, DR-Delawere River, E-Ellensville, Hr-Hudson River, M-Middletown, Mc-Monticello, NR-Neversink River, Pk-Poughkeepsie, PJ-Port Jarvis, RP-Reading Prong, SR-Shawangunk Ridge

FIGURE 15.14 Google Earth image looking northeast across the Appalachian Plateau and Great Valley in southeastern New York.

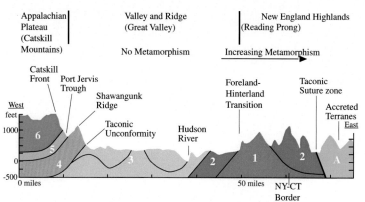

FIGURE 15.15 Cross-section across the area shown in Fig. 15.14.

and grain size toward the west, implying that the mountains were located to the east. It can be seen in cross-section that the rocks begin to tilt westward at the eastern edge of the Catskill Mountains to form the Catskill Front. Some of the west-dipping rock layers are visible at the bottom center of Fig. 15.14 near Port Jervis (PJ). The west dip allows us to see progressively older rock as we travel eastward across the area.

The narrow valley at Basher Kill (BK) is known as the Port Jervis Trough. It forms lowland, 300 to 600 feet in elevation that marks the beginning of the Valley and Ridge. A single river did not carve this valley. Roundout Creek flows north from Ellenville (E) to the Hudson River (HR). Basher Kill (BK) flows south to the Neversink River (NR) and the Delaware River (DR). Although Roundout Creek and Basher Kill flow in opposite directions, they are connected across the drainage divide to form the Hudson and Delaware Canal. Built between 1825 and 1829, the canal was used to transport anthracite coal from Pennsylvania to towns along the Hudson River including New York City. The trough itself exists due to the presence of thin layers of nonresistant, westward-tilted limestone and shale known as the Taconic Overlap sequence (5). These rocks were laid down largely during the Early Devonian between 419 and 393 million years ago in a miogeocline along a passive continental margin similar to the east coast of today. They represent a quiet time following Taconic orogeny but prior to Acadian orogeny.

The Catskill Front is partly hidden behind the narrow but high Shawangunk Ridge (SR), the westernmost ridge in the Valley and Ridge. The ridge consists of resistant, westward-tilted sandstone, and conglomerate known as Taconic molasse (4). The ridge is more than 2200 feet high at its northern end where visitors can find world-class rock climbing. These rocks form the older Taconic equivalent of Catskill molasse (6). They were deposited during the Upper Silurian between 433 and 419 million years ago and represent the eroded residue of the now vanished Taconic Mountains (mountains that formed during Taconic orogeny).

So here, in the space of a few miles, we have traversed across the erosional debris of two ancient, now vanished mountain ranges (4 and 6), separated by the remains of an ancient shallow sea (5) that covered the eroded residue of one mountain range (4) and was itself covered by residue from erosion of the younger mountain range (6). Ironically, it is the eroded remnants of these mountains that now form the highest elevations across the region. Elevation in the Great Valley east of Shawangunk Ridge is less than 1000 feet and drops to sea level at the Hudson River, which forms a major estuary to the Atlantic Ocean.

Rocks in the Great Valley consist mostly of deformed Middle and Upper Ordovician (470–444 Ma) Austin Glen shale, which is one of several rock units that form Taconic flysch (3). The term flysch refers to deep marine fine-grained sandstone, mudstone, volcanic, and sedimentary rocks produced from the erosion of a volcanic arc and surrounding highlands, and deposited either in a subduction trench or in the oceanic foreland during the early stages of mountain building. The Taconic flysch (3) dates the time the Taconic Mountains began to rise. Farther eastward in the cross-section, below Taconic flysch (3), we find limestone, dolostone, and sandstone that form the original Cambrian-Early Ordovician miogeocline (2, c. 570–470 Ma) that existed prior to Taconic orogeny. These same rocks extend northward to form the miogeocline in the Champlain Valley.

If we continue our journey eastward, we enter the Reading Prong where the level of deformation in miogeoclinal rock (2) increases, metamorphism begins, and underlying crystalline shield rocks (1) are exposed. Included within the shield rocks are granitic gneisses and schist. We have now crossed into the hinterland where farther east even the miogeoclinal rocks are strongly metamorphosed to marble and schist. If we travel east of Danbury, Connecticut, we would cross the Taconic suture zone and enter an area composed of schistose rock and gneiss that represents the Shelbourne Falls island volcanic arc (unit A), one of several volcanic accreted terranes that collided with North America during Taconic orogeny.

The story that can be garnered from these rocks is outlined in Fig. 15.16. The upper panel shows the situation during Middle and Late Ordovician deposition of Taconic flysch (3). By this time, the Shelbourne Falls island volcanic arc (A) had already collided, thereby raising the Taconic Mountains. Shield rocks (1) and the eastern part of the miogeoline (2) were both involved in hinterland deformation, metamorphism, and the growing of the tectonic wedge. The area of Fig. 15.14 is located at the

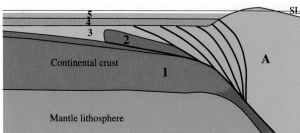

FIGURE 15.16 Schematic representation of the geologic history of southeastern New York keyed to Figs. 15.14 and 15.15. (A) Landscape during deposition of Middle and Upper Ordovician Taconic flysch (unit 3). (B) Landscape during late Silurian-Early Devonian deposition of the Taconic overlap miogeocline (unit 5). The rise of distant Acadian Mountains in the Middle and Late Devonian will eventually shed Catskill molasse (unit 6) across the miogeocline (unit 5) creating a landscape of rivers and lakes. SL-Sea level.

western side of Fig. 15.16. At this time, it was a foreland basin receiving sediment (3) from erosion of the mountains. The start of the Taconic orogeny at any one location can be determined by dating Taconic flysch (3) deposited directly above rocks of the miogeocline (2).

The lower panel in Fig. 15.16 shows the situation during Late Silurian-Early Devonian deposition of the Taconic overlap sequence (5). Notice in Fig. 15.15 that there is a time gap (the Taconic unconformity) between deposition of Taconic flysch (3) and Taconic molasse (4). Taconic flysch (3) shows stronger deformation below the unconformity than Taconic molasse (4) above the unconformity. These relationships imply that Taconic flysch (3) became incorporated into the growing, advancing tectonic wedge during mountain building, and underwent deformation, uplift, and erosion during Taconic orogeny. Taconic molasse (4) represents deposition across a landscape of rivers and lakes during erosion of distant Taconic Mountains after deformation and mountain building had ended. The Taconic unconformity between these two rock units thus dates the end of the Taconic orogeny. The miogeocline (5) represents a quiet time prior to Acadian orogeny. The miogeocline (5) would eventually be covered with Catskill molasse (unit 6) derived from erosion of distant Acadian Mountains, but this final depositional stage is not shown in Fig. 15.16. The size and thickness of Catskill molasse (6) suggests that the now vanished Acadian Mountains were larger than the previously vanished Taconic Mountains.

Central Appalachian Fold and Thrust Belt

The fold and thrust belt widens considerably in the vicinity of Allentown, Pennsylvania where we begin to see something that resembles ridge and valley landscape. Major thrust faults are largely absent. Instead, folds dominate the structure. It is possible and even probable that thrust faults exist at depth, but they are rare at the surface. Perhaps this is our good fortune because it is the manner in which the folds are eroded that makes the Pennsylvania Valley and Ridge rather geologically unique. Erosion of alternating resistant and nonresistant rocks has created a truly amazing zigzag pattern of bedrock ridges and valleys, as is clearly evident in Fig. 15.17 where the Alleghany Front, Pocono Mountains, Blue Ridge, and the Reading Prong each form a boundary with the Valley and Ridge. The ridges and valleys are controlled by rock hardness resulting in numerous examples of anticlinal valleys and synclinal mountains, such as the example shown in Fig. 15.18. Note in Fig. 15.17 that the Susquehanna River (Sr) forms numerous water gaps as it cuts directly through ridges, most of which are between 1800 and 2500 feet in elevation. Water gaps were introduced in Chapter 14 as a location where a river forms a narrow gorge by cutting directly through a ridge or mountain. A water gap is present (barely visible) on Fishing Creek in Fig. 15.18 between Mill Hall and Cedar Springs. We will discuss the origin of Appalachian water gaps later in this chapter.

A-Allentown, **H**-Harrisburg, **R**-Reading, **S**-Scranton, **SC**-State College, **Sr**-Susquehanna River, **W**-Williamsport

FIGURE 15.17 Google Earth image looking northward across the Pennsylvania Valley and Ridge. The Hamburg allochthon is enclosed within the yellow line.

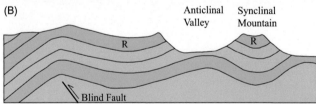

FIGURE 15.18 (A) Google Earth image looking westward along Nittany Valley, Pennsylvania. Rock layers dip away from the center of the valley creating an anticlinal valley. Fishing Creek forms a water gap between Mill Hall and Cedar Springs. (B) Schematic cross-section showing anticlinal valleys and synclinal mountains. R signifies resistant rock.

A significant difference between the Pennsylvania Valley and Ridge and areas farther north is the age of rocks and the age of deformation. Mississippian and Pennsylvanian rocks (359–299 Ma) are folded in the central Appalachians. This folding must have occurred after the rocks were deposited, which implies deformation during Alleghany orogeny, much younger than what is found in the north. Taconic-age deformation is present, but is restricted to the Great Valley along the southeast margin of the Pennsylvania fold belt where, in Fig. 15.17, we find the Hamburg Allochthon (enclosed within the yellow line). These thrust slices are Taconic in age with an origin that mirrors the Taconic Allochthon. The major difference is that the Hamburg Allochthon consists dominantly of sedimentary rock and therefore does not form resistant mountains.

The Valley and Ridge begins to transition into a classic fold and thrust belt south of Pennsylvania where thrust faults become more numerous and where deformation is entirely associated with Allegany orogeny. In this region, Taconic-age deformation is found only in the hinterland. We will next take a detailed look at the Tennessee Valley and Ridge, one of the best developed and best-studied foreland fold and thrust regions in the world.

Southern Appalachian (Tennessee) Fold and Thrust Belt

The southern Appalachian fold and thrust belt extends across the Valley and Ridge to the base of the Blue Ridge Mountains. The miogeoclinal rocks of the fold and thrust belt record near continuous Cambrian through Pennsylvanian deposition, which implies they were not deformed during Taconic and Acadian orogeny. The Cambrian-Ordovician rocks represent the pre-Taconic miogeocline as well as foreland basin rocks derived from erosion of the distant, ancient Taconic Mountains. The Silurian, Devonian, and Mississippian rocks record stable marine deposition distant from, and apparently not influenced by, the rise of the ancient Acadian Mountains. Pennsylvanian conglomerate, sandstone, and coal, deposited in rivers and lakes, represents the erosional debris of a third mountain range, the Alleghany Mountains. Following deposition of these rocks, folding and thrust faulting associated with the Alleghany orogeny migrated across the Tennessee Valley and Ridge creating the present-day structural form. A study of fault gauge indicates that thrusting was active by 280 million years ago.

Thrust faults dip east, therefore most of the sedimentary layers also dip east, producing long asymmetric ridges with steep slopes on their west side and gentle dip-slopes on their east side. The structure and resulting landscape is illustrated in the Google Earth image and schematic cross-section shown in Fig. 15.19. The landscape mirrors that of the Montana Disturbed Belt with major differences being elevation, relief, and age of deformation. After 55 million years of erosional modification, the Montana Disturbed Belt still maintains landscape that mimics its underlying structure in which topographic high areas correlate with ramp areas (Fig. 15.6B). Rocks in the southern Valley and Ridge, by contrast, have undergone hundreds of millions of years of erosion, resulting in topographic high areas that do not necessarily correlate with ramp areas (Fig. 15.19B). High areas instead correlate with hard sandstone, and intervening valleys with weak, thinly layered sandstone, limestone, and shale. The tilted, alternating layers create the strong linear topography seen in Fig. 15.19A. Anticlinal valleys and synclinal mountains are common. Presumably, the southern Valley and Ridge was once similar in grandeur to the Montana Disturbed Belt, but erosion has since reduced the mountain heights.

Although characterized by numerous linear ridges, the Tennessee Valley and Ridge forms a southwestward-sloping valley traversed by the Tennessee River and its tributaries. We can consider the entire region to be part of the Great Valley. Fig. 15.20 is a Google Earth image oriented north across the region with superposed geological features. With few exceptions, maximum elevation of ridge crests decreases southwestward from 2500 feet northeast of Kingsport (Kp), to less than 1200 feet southwest of Knoxville (K). Relief between ridge and valley is between 200 and 800 feet throughout the area with the exception of a few of the most prominent ridges. Clinch Mountain (clm) is consistently more than 1000 feet above surrounding valleys. Faults on the Cumberland Plateau and Valley and Ridge are shown with thin white lines and are numbered in yellow. Symbols are listed in Table 15.1. There are 10 or more major thrust faults across the Tennessee Valley and Ridge at any one location. Thrust faults, on average, are spaced 3 to 5 miles apart. Minimum displacement across all faults is close to 200 miles, implying that the area currently occupied by the Valley and Ridge was once 200 miles wider than it is today.

The Great Smoky (26) and Holston Mountain (11) thrust faults both bring older (>541 Ma), more resistant sedimentary and crystalline rocks to the surface. These faults, therefore, mark the geologic boundary that separates the foreland fold and thrust belt from the hinterland deformation belt (Figs. 11.6 and 11.9). This geologic boundary corresponds closely, but not exactly, with the physiographic boundary between the Valley and Ridge and Blue Ridge.

The foreland fold and thrust belt is widest in the area surrounding Thorn Hill (T, just above the center of Fig. 15.20). From Pine Mountain pm to the base of the Blue Ridge, the thrust belt is 66 miles (106 km) wide. Pine Mountain is physiographically part of the Cumberland Plateau, but geologically part of the foreland fold and thrust belt. It also has historical significance. The Pine Mountain thrust (1) was the focus of a classic study in the 1930s in which the flat-ramp-flat thrust geometry was well documented. If we include Pine Mountain, we can divide the Tennessee thrust belt into three parts. A western area that extends from the Pine Mountain thrust (1) to the Wallen Valley thrust (2). A central area of closely spaced thrust faults that lies between the Wallen Valley thrust (2) and the Rocky Valley/Carter Valley thrust (7). And an eastern area that extends from the Rocky Valley/Carter Valley thrust (7) to the Holston Mountain-Great Smoky thrust (11, 26). The approximate width of each area is 20 miles (32 km), 15 miles (24 km), and 31 miles (50 km), respectively.

Our discussion will concentrate on the western and central part of the thrust belt. We will use two figures in order to better understand thrust geometry. Fig. 15.21 is a Google Earth image that looks east across the western and central sections of the thrust belt. Thrust faults are shown with thin white lines and are numbered. The eastern part of this figure is shown in more detail in Fig. 15.19A. Fig. 15.22 is a cross-section that extends from the Pine Mountain thrust (1) to the Saltville thrust (6). The cross-section is a generalized composite of two

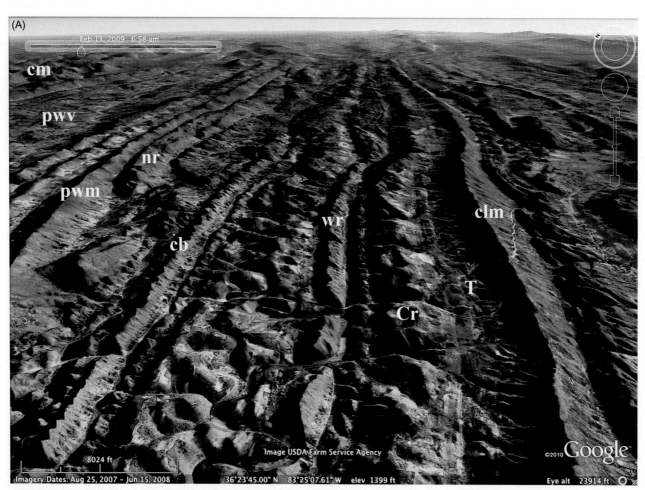

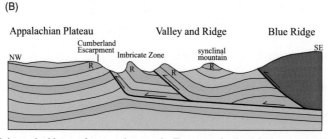

cm-Cumberland/Stone Mountain, clm-Clinch Mountain, cb-Comby Ridge, Cr-Copper Ridge, nr-Newman Ridge, pwm-Powell Mountain, pwv-Powell River Valley, wr-War Ridge. T-Thorn Hill

FIGURE 15.19 (A) Google Earth image looking northeastward across the Tennessee foreland fold and thrust belt (Valley and Ridge) from just south of the Virginia border. The Cumberland front is seen at upper left corner. (B) Schematic cross-section of an erosion-controlled fold and thrust belt showing an imbricate zone and a synclinal mountain. R signifies resistant rock.

cross-sections that are located in Figs. 15.20 and 15.21 with two black lines.

The first thing you might notice in cross-section is that thrust faults are stacked one above the other like overlapping roof tiles. This pattern is an example of an imbricate zone and is typical of fold and thrust belts worldwide. The thrust faults bring to the surface primarily a Cambrian (541–485 Ma) and Ordovician (485–444 Ma) miogeoclinal rock sequence up to 14,000 feet thick that repeats itself along each thrust. These rock units thicken toward the east and are much thicker than interior platform rocks of the same age in the continental interior. The Pine Mountain thrust involves rock as young as Pennsylvanian (323–299 Ma). Such involvement indicates that thrust faulting was active during Alleghany orogeny following deposition of these rock units.

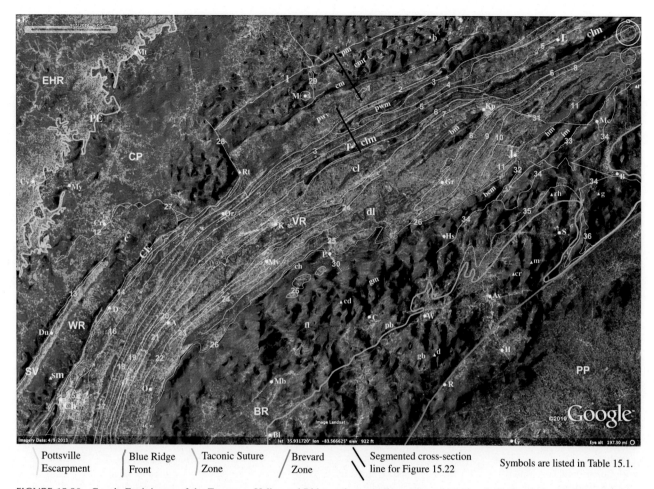

FIGURE 15.20 Google Earth image of the Tennessee Valley and Ridge and surrounding region oriented north. Faults on the Cumberland Plateau, Valley and Ridge, and the Grandfather Mountain area of the Blue Ridge are shown with thin white lines and are numbered in yellow. Faults in other parts of the Blue Ridge and Piedmont Plateau are not shown. All symbols are keyed to Table 15.1.

Geologic mapping and drilling indicate that the Cambrian Rome formation is the oldest rock unit cut by any of the thrust faults in this part of the Valley and Ridge. In the cross-section, the Hunter Valley (3), Clinchport (4), Copper Creek (5), and Saltville (6) thrusts all carry Rome Formation to the surface. On the basis of these observations, we can conclude that the thrust faults root (or merge) into a single floor thrust near the base of the Rome Formation. This floor thrust is shown as the basal décollement in Fig. 15.22. Rocks below the floor thrust are largely undeformed across the length of the cross-section. Fig. 15.22 can be compared with the idealized fold and thrust belt shown in Fig. 11.6.

One oddity within the thrust belt is the circular basin that surrounds the town of Middlesboro (M, Fig. 15.21). This area was shown to be a meteorite impact feature as far back as the 1960s based on the presence of circular normal faults surrounding the basin, intensely deformed and folded rocks, and shattered and shocked quartz grains.

In the next two sections, we will look in more detail at the thrust belt along cross-section line Fig. 15.22.

The Western Thrust Belt

The western thrust belt is not typical of the Valley and Ridge. More than half of it lies within the physiographic Cumberland Plateau. Rather than forming a thrust imbricate zone of linear ridges, the western thrust belt, as seen in Figs. 15.21 and 15.22, is characterized by the Middlesboro syncline and Powell Valley anticline. Pine Mountain (pm) and Cumberland/Stone Mountain (cm) are the only prominent linear ridges. Both extend for more than 100 miles and reach an elevation above 3000 feet. Both mountains are strike ridges as is typical of nearly all ridges (and valleys) in the Valley and Ridge. A strike ridge is one in which the trend of the ridge-top follows the strike of the underlying resistant rock unit. Both mountains are asymmetric. Rocks on Pine Mountain (pm)

TABLE 15.1 Explanation to Fig. 15.20

Mountain Areas and Lakes	Towns	Thrust Faults
b-Black Mountain	A-Athens	1-Pine Mountain
bm-Bays Mountain	Av-Asheville	2-Wallen Valley
bcm-Buffalo and Cherokee Mtns	B-Boone	3-Hunter Valley
c-Crab Orchard Mountains	Bl-Blairsville	4-Clinchport
cd-Clingmans Dome	C-Cherokee	5-Copper Creek
ch-Chilhowee Mountains	Ch-Chattanooga	6-Saltville
cl-Cherokee Lake	Cr-Crossville	7-Rocky Valley-Carter Valley
cr-Craggy Dome	Cv-Cookeville	8-Pulaski
clm-Clinch Mountain	D-Dayton	9-Dunham Ridge
cm-Cumberland/Stone Mountain	Du-Dunlap	10-Spurgeon
cmt-Cumberland Mountains	E-Edmonton	11-Holston Mountain
d-Devil's Courthouse	Gr-Greeneville	12-Cumberland Plateau Overthrust
dl-Douglas Lake	G-Greenville	13-Sequatchie Valley Thrust
fl-Fontana Lake	H-Hendersonville	14-Rockwood and Chattanooga
g-Grandfather Mountain	HS-Hot Springs	15-Missionary Ridge
gb-Great Balsam Mountains	J-Johnson City	16-Kingston
gm-Great Smoky Mountains	K-Knoxville	17-White Oak Mountain
hm-Holston Mountain	Kp-Kingsport	18-Copper Creek
im-Iron Mountain	L-Lebanon	19-Beaver Valley
m-Mount Mitchell (Black Mtns)	M-Middlesboro	20-Saltville
pb-Plott Balsam Mountains	Mb-Marble	21-Knoxville
pm-Pine Mountain	Mc-Mountain City	22-Chestuee
pwm-Powell Mountain	Mc-Montecello	23-Dumplin Valley
pwv-Powell Valley	Mv-Maryville	24-Wildwood
r-Mount Rogers	My-Monterey	25-Guess Creek
rh-Roan High Knob (Roan Mtns)	O-Ocoee	26-Great Smoky
sm-Signal Mountain	Or-Oak Ridge	
	P-Pigeon Forge	**Transverse Faults**
Physiographic Areas	R-Rosman	27-Emory River
BR-Blue Ridge	Rt-Rocky Top	28-Jacksboro
CE-Cumberland Escarpment	S-Spruce Pine	29-Rocky Face
CP-Cumberland Plateau	T-Thorn Hill	30-Pigeon Forge
EHR-Eastern Highland Rim	W-Waynesville	31-Cross Mountain
PE-Pottsville Escarpment		
PP-Piedmont Plateau	**Grandfather Mtn. Area Faults**	
SV-Sequatchie Valley	32-Buffalo Mountain	
VR-Valley and Ridge	33-Iron Mountain	
WR-Walden Ridge	34-Unaka Mtn-Stone Mtn-Linville Falls	
	35-Fries	
	36-Tablerock	

dip toward the southeast creating a steep northwest-facing slope and a gentle southeast-facing dip slope. These rocks are well exposed along highway 23 at Pound Gap, 65 miles northeast of Pineville (Pv). Cumberland/Stone Mountain (cm) is asymmetric in the opposite direction. The steep southeastern face of Cumberland/Stone Mountain forms the Cumberland Front, the physiographic boundary of the Appalachian Plateau. Together, the dip of rocks in the two mountain ridges defines the Middlesboro Syncline.

Only two thrust faults reach the surface, the Pine Mountain thrust (1), and the Wallen Valley thrust (2). The Pine Mountain thrust is unusual in that it ramps upward from the Rome Formation to the Chattanooga Shale (part of the thin Silurian-Devonian layer in Fig. 15.22) where it forms a thrust flat below the Middlesboro syncline. Rocks above the thrust flat are horizontal and are the same resistant Pennsylvanian rocks found elsewhere on the Cumberland Plateau. These resistant rocks at the center of the Middlesboro syncline form the Cumberland Mountains (cmt, Fig. 15.21) and some of the highest elevations on the plateau including Black Mountain at 4145 feet (b, Fig. 15.20). A third fault, the Bales Thrust, ramps upward from the Rome Formation to the Chattanooga Shale, and in doing so, folds the Pine Mountain thrust above it to create both the

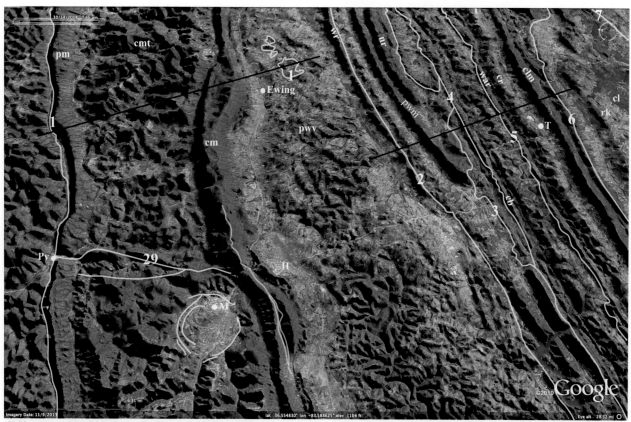

H-Harrogate, M-Middlesboro, Pv-Pineville, T-Thorn Hill.
cb-Comby Ridge, cl-Cherokee Lake, clm-Clinch Mountain, cm-Cumberland/Stone Mountain, cmt-Cumberland Mountains, cr-Copper Ridge, nr-Newman Ridge, pm-Pine Mountain, pwm-Powell Mountain, pwv-Powell Valley, rk-Richland Knobs, war-War Ridge, wr-Wallen Ridge.
1-Pine Mountain Thrust, 2-Wallen Valley Thrust, 3-Hunter Valley Thrust, 4-Clinchport Thrust, 5-Copper Creek Thrust, 6-Saltville Thrust, 7-Rocky Valley/Carter Valley Thrust, 29-Rocky Face Fault.

FIGURE 15.21 Google Earth image of the Thorn Hill area looking east. White lines are faults. The two black lines show the location of cross-section Fig. 15.22.

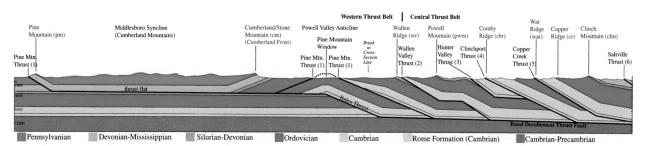

FIGURE 15.22 Composite cross-section of the Thorn Hill area. Location is shown in Figs. 15.20 and 15.21 with two thick black lines. Based on Harris and Mixon (1970), Mixon and Harris (1971), and Mitra (1988).

Powell Valley anticline and Middlesboro syncline. The Bales Thrust does not reach the surface.

Resistant Pennsylvanian rocks must have been displaced to high elevation in the Powell Valley anticline above the Bales Thrust, but these rocks have since been removed by erosion thereby exposing weak Cambrian-Ordovician rocks underneath. The weak rocks, in turn, were eroded to create an anticlinal valley. It is possible that prior to erosional removal of higher parts of the thrust belt, some of the thrust faults in the central thrust belt may also have carried Pennsylvanian rocks. If these rocks were nearly flat-lying like those in the Middlesboro syncline, then landscape across much of the Valley and Ridge would have more closely resembled that of the

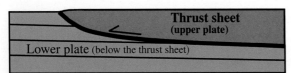

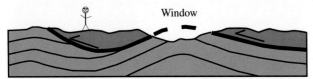
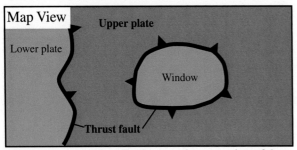

FIGURE 15.23 Cross-section sketches and a map that shows the formation of a window through a thrust sheet.

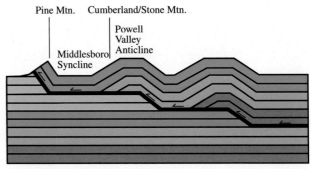

FIGURE 15.24 Schematic cross-section showing how the Middlesboro syncline and Powell Valley anticline are associated with underlying thrust ramps. This figure is inverted from Fig. 15.2.

Cumberland Plateau. The present-day Valley and Ridge landscape would have emerged only after slow erosional removal of these rocks.

Notice in Fig. 15.22 that the Pine Mountain thrust breaches the surface along the crest of the Powell Valley anticline creating the Pine Mountain Window. Fig. 15.23 explains how a window might form. Windows are common in the southern Appalachians. They provide constraints on the geometry of thrust faults. In this case, they indicate that the Pine Mountain thrust is folded as part of the Powell Valley anticline. Maximum displacement on the Pine Mountain thrust has been estimated to be 13.25 miles (21.3 km). This is less than some faults in the Valley and Ridge, but more than other faults on the Cumberland Plateau.

The overall geometry of the western thrust belt resembles that shown in Fig. 15.24. Notice that an anticline forms above a thrust ramp and a syncline above a thrust flat between two ramps. The primary difference between this schematic cross-section and the one shown in Fig. 15.22 is that erosion in Fig. 15.22 has reduced the anticline to form Powell Valley and left the Middlesboro syncline to form the Cumberland Mountains. The geometry of the folded Pine Mountain thrust can be compared to the geometry of the much younger Lewis thrust in Glacier National Park as shown in Fig. 15.7B.

The Central Thrust Belt

The heart of the Appalachian foreland fold and thrust belt, and the location where ridge and valley landscape is best developed, is in the central imbricated part from the Wallen Valley thrust (2) to the Saltville Thrust (6). Each thrust fault repeats the Cambrian-Ordovician rock sequence, capped in a few areas with Silurian, Devonian, and Mississippian rocks. This area shows a close correlation of rock hardness with landscape. Highland ridges are invariably composed of sandstone and, in a few places, hard dolostone. Valleys are underlain with shale, limestone, dolostone, and fine-grained sandstone. As shown in Fig. 15.22, most of the rock units dip toward the southeast creating asymmetric strike ridges. Whereas the western (Pine Mountain) thrust belt could be likened to Glacier National Park, the central thrust belt is more akin to the Montana Disturbed Belt (Fig. 15.6B).

The most prominent ridges within the central imbricate thrust zone are those capped with hard Silurian (444–419 Ma) Clinch and Rockwood sandstone. Comparison of Figs. 15.21 and 15.22 show that these rocks cap Clinch Mountain (clm), Powell Mountain (pwm), and Wallen Ridge (wr). The shape of Powell Mountain (Pwm) follows the bending of rock layers across a fold prior to being cut off by the Hunter Valley Thrust (3). The fold is visible in Fig. 15.22. The prominent Newman Ridge (nr) forms a similar fold, in this case capped with younger Mississippian sandstone.

The Rome Formation consists primarily of shale and sandstone. Some of these sandstone units form low-lying resistant ridges such as Comby Ridge (Cb) and War

Ridge (war). Above the Rome Formation, there are only a few ridge formers within the thick Cambrian and Ordovician rock sequence, and the ridges that do form tend to be low and rounded. One example is Copper Ridge (cr), held up by the Copper Ridge-Chepultepec dolomite. Richland Knobs (rk), located near Cherokee Lake (cl) in Fig. 15.21 consists of folded Copper Ridge Dolomite. Highway 32 (the Dixie Highway) crosses the central thrust belt from Middlesboro (M) to Cherokee Lake (cl). In doing so, it crosses the Wallen Valley (2), Hunter Valley (3), Clinchport (4), Copper Creek (5), and Saltville (6) thrust faults. This road is visible as a faint white line through the center of Fig. 15.21. There are several locations along this road where one can view rocks and possibly thrust faults.

If we were to continue our journey eastward to the Douglas Lake-Bays Mountains area (dl and bm, Fig. 15.20), we would find a lowland region of folded rocks generally without thrust faults similar to what we see in the Great Valley of southeastern New York (Fig. 15.14). This area consists of Middle Ordovician Sever shale and Bays sandstone/conglomerate (470–458 Ma), a sequence derived from erosion of the ancient, now vanished Taconic Mountains. The rocks are similar to the Austin Glen/Shawangunk sequence seen in New York. (#3 and 4, Fig. 15.15). Thus, although the Taconic orogeny did not affect the southern Appalachian foreland, the presence of Sevier and Bays rocks in the foreland indicate the existence of distant Taconic Mountains.

Fault Zones on the Cumberland Plateau

Nearly flat-lying sedimentary rocks on the Cumberland Plateau are broken in places by faults that represent the northwest limit of deformation associated with the Tennessee Valley and Ridge fold and thrust belt. Two fault types are present. The first are thrust faults oriented parallel with the strike of nearby rock layers. Examples include the Sequatchie Valley and Pine Mountain thrusts (13 and 1 in Fig. 15.20). The second are transverse strike-slip faults oriented nearly perpendicular to thrust faults. Examples include the Emory River (27), Jacksboro (28), and Rocky Face (29) faults.

These two fault types were active at the same time between 300 and 265 million years ago. In some cases, the strike-slip faults acted as tear faults that allowed one side of a thrust fault to move farther than the other side. For example, synchronous fault activity along the Jacksboro strike-slip fault (28) and the Pine Mountain thrust fault (1) has allowed the Pine Mountain thrust to push farther to the northwest than the Rockwood and Chattanooga faults (14).

Distribution of Appalachian Foreland Deformation

The Tennessee fold and thrust belt was built during Alleghany orogeny, which climaxed with the collision of Africa and Eurasia with North America to form Pangea and the present-day Appalachian Mountains. It is therefore surprising that the thrust belt does not extend northward into Pennsylvania where we have to stretch the definition of a foreland fold and thrust belt just to include it. Perhaps more surprising is that the effects of Alleghany orogeny disappear entirely in the northern Appalachian fold and thrust belt. Thrust faults break and transport rocks many miles. Folds, on the other hand, squeeze rocks, which usually implies far less total displacement. On this basis, we can suggest that Africa–North America collision was head-on and most intense in the southern Appalachians, but was just a glancing blow in the foreland farther north.

OUACHITA FOLD AND THRUST BELT

The Ouachita fold and thrust belt is located in Arkansas and Oklahoma and is surrounded on all sides by nearly flat-lying sedimentary rocks. The Coastal Plain lies to the south and east, the Central Lowlands to the west, and the Ozark Plateau to the north. A Raisz landform map of the region is shown in Fig. 15.25.

The Ouachita landscape has similarities to both the Tennessee fold and thrust belt and the Pennsylvania fold belt. There are major thrust faults that produce long linear ridges, and there are folds, some of which produce a zig-zag pattern while others show dome and basin pattern.

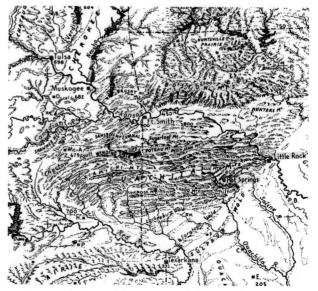

FIGURE 15.25 Raisz landform map of the Ouachita Mountains.

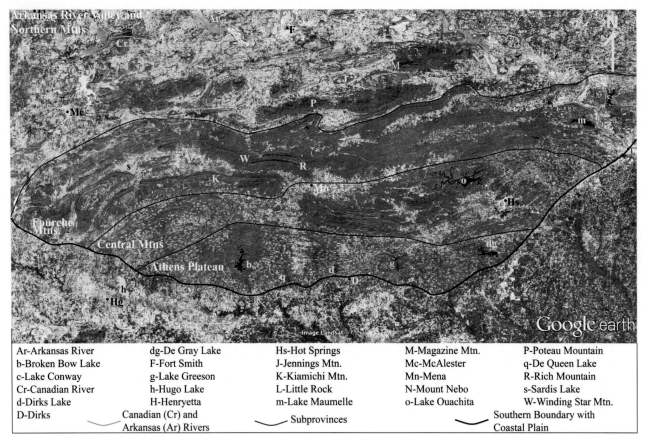

FIGURE 15.26 Google Earth image of the Ouachita Mountains oriented north. The Choctaw and Ross Creek thrust faults follow the subprovince boundary that separates the Arkansas River Valley–Northern Mountains from the Fourche Mountains.

Fig. 15.26 is a Google Earth image of the Ouachita Mountains oriented north. Dark blue lines subdivide the Ouachita region into four areas. From north to south they are the Arkansas River Valley–Northern Mountains, the Fourche Mountains, the Central Mountains, and the Athens Plateau. The light blue lines trace the Canadian River (CR) and the Arkansas River (AR). The black line is the approximate boundary with younger, nearly flat-lying rocks of the Coastal Plain. Following an introduction to the rocks, we will look briefly at each of the four sub-regions.

Almost all of the rocks are sandstone, shale, chert (microcrystalline quartz), and a light-colored, often translucent and gritty form of chert known as novaculite. In sharp contrast to the Valley and Ridge, limestone is largely absent. Rocks in the Arkansas River Valley–Northern Mountains are Early to Middle Pennsylvanian (323 to 307 Ma). Southward, the rocks range in age from Ordovician to Middle Pennsylvanian (485–307 Ma). Similar with the Valley and Ridge, the rocks underwent deformation and mountain building between 335 and 265 million years ago, and have undergone erosion ever since, resulting in landscape controlled by rock hardness. Chert and novaculite are particularly resistant to erosion and are strong ridge formers. The two most widespread formations are the Bigfork Chert (circa 465 million years old) and the Arkansas Novaculite (circa 359 million years old).

A major feature of the Ouachita Mountains is the alternation of resistant layers of sandstone, chert, and novaculite with less resistant layers of sandstone and shale. The alternating layers have created a landscape of closely spaced ridges and valleys that include anticlinal valleys and synclinal mountains. A second major feature is the abundance of beautifully developed quartz crystals. The crystals formed within quartz veins primarily from deposition out of hot water during the final stages of mountain building between 285 and 245 million years ago. The belt of quartz veins extends across the center of the mountain range from eastern-most Oklahoma to Little Rock, Arkansas. Some of the finest crystals are found in the Crystal Mountains south of Mount Ida, Arkansas.

Arkansas River Valley–Northern Mountains

The Arkansas and Canadian River Valleys lie north of the fold and thrust belt proper where elevation is

FIGURE 15.27 Google Earth image looking west at a syncline centered on flat-topped Jennings Mountain (*J*), and an anticline (elongate dome) centered near Washburn, Arkansas (*W*). Fort Smith (*FS*) is at upper right.

between 300 and 700 feet. The southern part of the valley is marked by a series of isolated mountains sometimes referred to collectively as the Northern Mountains. These mountains can be considered outliers of the Ouachita Mountains. Between 315 and 307 million years ago, during active thrust faulting and mountain building in the south, this area acted as a basin into which sediment from erosion of the uplifting thrust belt was deposited. The rocks were broadly folded and locally faulted soon after deposition and have undergone erosion ever since. The result is an erosion-controlled landscape of curved ridges, anticlinal valleys, and flat-topped, synclinal mountains.

Fig. 15.27 is a Google Earth image that looks west at folded layers south of Fort Smith (FS). Jennings Mountain (J) forms a synclinal mountain around which broadly folded rock layers form a series of benches and outward-facing cuestas that step to higher elevation toward the 750-foot flat-topped center of the mountain. There are numerous flat-topped, synclinal mountains in this area in addition to Jennings Mountain, including Carter, Spring, and Huckleberry Mountains, Mount Nebo (N, Fig. 15.26), the Poteau Mountains (P) where one can find Oklahoma High Top at 2377 feet, and Magazine Mountain (M), which includes Signal Hill, the highest point in Arkansas at 2753 feet. All of these synclinal mountains are capped with nearly flat-lying, resistant, 315 to 307 million year old sandstone. Fig. 15.28 looks northwest at flat-topped Mount Nebo. Many of these flat-topped mountain summits can reasonably be regarded as mesas, tablelands, or small plateaus. The fold on the right side of Fig. 15.27 near Washburn (W) forms a faulted anticlinal valley.

FIGURE 15.28 Google Earth image looking northwest at flat-topped Jones Mountain (*j*), Mount Nebo (*n*), and Spring Mountain (*s*), sitting in the trough of a syncline, Arkansas. Huckleberry and Carter Mountain are seen in the distance.

The Fourche Mountains

South of the Northern Mountains, the Ouachita Mountains fold and thrust belt proper begins with a belt of linear, thrust-faulted ridges that are sometimes referred to collectively as the Fourche Mountains. These mountains mark the northern limit of strong deformation associated with the Ouachita fold and thrust belt. The boundary with the Northern Mountains is placed at the northernmost occurrence of major thrust faults, the Choctaw fault in the west, and the Ross Creek fault in the east. The landscape, as evident in Fig. 15.26, consists of a series of long, closely spaced ridges and valleys with short oval-shaped ridges of folded rock that form dome and basin structures. Southward-dipping hard sandstone forms most of the ridge crests creating steep slopes on the north side and more moderately inclined dip-slopes on the south side similar to the Tennessee Valley and Ridge shown in Fig. 15.19A. An area of folded, zigzag ridges directly east of Sardis Lake (s, Fig. 15.26) consists of Bigfork Chert and Arkansas Novaculite. The large fold on the west side of Sardis Lake is a structural and topographic basin composed of sandstone, shale, minor chert, and limestone surrounded by resistant ridges of sandstone that form highlands.

The Ouachita Mountains, and the Fourche Mountains in particular, are one of the few east-west oriented mountain ranges in the United States, and because of this, there are visible contrasts in vegetation. South-facing slopes are

FIGURE 15.29 Google Earth image of the area surrounding Lake Ouachita (*o*) oriented north. The image shows the outcrop pattern of the Arkansas Novaculite. The Zigzag Mountains lie east of Hot Springs (*HS*). The Crystal Mountains (*C*) lie south of Mount Ida (*MI*).

warm, sunny, and dry. They boast large areas of pine forest. North-facing slopes are wet and cool with plenty of hardwood forest. Major ridges include Winding Star Mountain in Oklahoma (W, 2451 feet), and Rich Mountain in Arkansas (R, 2681 feet). These ridges mark an approximate drainage divide between the Arkansas River to the north and the Red River to the south. As shown in Fig. 15.26, the Arkansas River cuts through the eastern end of these ridges in the vicinity of Little Rock.

The Central Mountains

The primary landscape feature of the Central Mountains is an amazing pattern of fold ridges similar to the Pennsylvania Valley and Ridge but more tightly spaced. Also present are some of the oldest rocks in the Ouachitas, some dating back to the Ordovician (485–444 Ma).

The Central Mountains are referred to as the Novaculite Uplift due to the abundance of folded, resistant layers of novaculite. The Bigfork Chert and the Arkansas Novaculite are particularly abundant in the area of zigzag ridges surrounding Lake Ouachita (o, Fig. 15.26). The outcrop pattern of the Arkansas Novaculite is shown in Fig. 15.29. The zigzag ridge crests located east of Hot Springs (HS) are known quite appropriately as the Zigzag Mountains. It is in these mountains where one can find Hot Springs National Park. Novaculite ridges in the lower left of the image include the Caddo, Missouri, and Cossatot Mountains. Several of the ridge crests top 2000 feet in elevation. The Crystal Mountains (C), home to several quartz mines, is also located in the figure.

Athens Plateau

South of the Central Mountains, we drop onto the Athens Plateau at an elevation of less than 1000 feet. The landscape here is relatively flat, but the rocks are just as folded and faulted as those in the north. Resistant layers create low-lying, tightly spaced ridges across the region. Several rivers flow southward directly across the low-lying ridges creating rapids, waterfalls, and excellent whitewater rafting. Some of the rivers have been dammed, creating lakes such as Broken Bow (b), De Queen (q), Dirks (d), Greeson (g), and De Gray (dg) Lakes (Fig. 15.26). The damming of the rivers has allowed lake water to expand into low areas thus accentuating the low-lying ridge and valley landscape. De Queen Lake, shown in Fig. 10.4, is an example. Average relief between ridge and valley in this figure is less than 200 feet.

The low-lying topography and cross-cutting rivers suggest that the Athens Plateau was at one time covered with nearly flat-lying sedimentary layers of the Coastal Plain. Rocks on the Coastal Plain in the immediate vicinity are between 126 and 100 million years old (Cretaceous), suggesting that the Athens Plateau was worn down and covered with rocks of the Coastal Plain some time after 126 million years ago. The area was re-exposed less than 100 million years ago implying that the landscape is much younger than topographically higher areas in the Ouachita Mountains to the north. Over time, we would expect erosional lowering to increase relief by preferentially deepening lowland valleys in nonresistant rock and leaving resistant rock to form highland ridges.

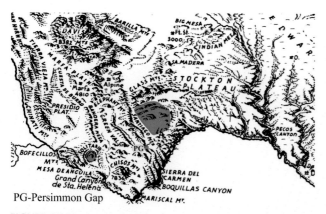

FIGURE 15.30 Landscape map of Marathon Basin region. Areas of fold and thrust landscape within the Marathon Basin, Solitario, and at Persimmon Gap are highlighted.

MARATHON BASIN FOLD AND THRUST BELT

Rocks of the Ouachita fold and thrust belt are present in several areas of southwest Texas along the transitional boundary between the Great Plains and Basin and Range. The three largest areas are the Marathon Basin, the Solitario, and Persimmon Gap. All three are highlighted on a landscape map shown in Fig. 15.30. In this section, we concentrate on the Marathon Basin where fold and thrust landscape is well developed. The Solitario is volcanic in origin and is discussed in Chapter 17. Persimmon Gap is more strongly influenced by normal faults and is discussed in Chapter 18.

The Marathon Basin is located in a pivotal area both from a geological and a landscape point of view. It is part of the Appalachian-Ouachita mountain belt, yet it is located at the edge of the Cordilleran belt. It is a structural dome and a physiographic basin and, therefore, similar to the Llano uplift. But, unlike the Llano Uplift, the central part of the basin contains no crystalline rock. The basin instead exposes thrust-faulted and folded sedimentary layers that create a linear landscape similar to what is found in the Ouachita Mountains. The rocks are of the same type, age, and structure as in the Ouachitas including the presence of novaculite and chert, which form many of the ridges in the western part of the basin. Drilling beneath the Coastal Plain confirms the connection with the Ouachitas. Clearly, the Appalachian Mountain system at one time extended continuously westward across the Gulf region where it must have interacted with rocks that later became part of the Cordilleran Mountain system. The influence of the Cordillera is evident. The northwestern margin of the Marathon basin is flanked with 35 million year old volcanic rocks of the Trans-Pecos volcanic field. The southwestern margin is flanked with normal faults associated with the Basin and Range.

Fig. 15.31A is a Google Earth image that looks southeastward across the eastern part of the Marathon Basin. The town of Marathon (M) is located to the north, just below the lower right edge of the image. One can clearly see the upturned edges of folded layers in the bottom and middle right of the image. Look closely at the folded layers at right center and how they disappear beneath nearly flat-lying layers of the Great Plains (GP). Note also how the landscape changes its morphology as a result of the change in rock structure. The nearly flat-lying layers at this location are approximately the same age as rocks that partly cover the Athens Plateau. Fig. 15.31B is a close-up view where one of the tilted layers of older rock is covered with younger nearly flat-lying layers. Can you find this location in Fig. 15.31A? The contact is an unconformity. The rock layers below the unconformity were thrust-faulted, folded, uplifted, and eroded prior to deposition of the younger nearly flat-lying layers.

The young nearly flat-lying sedimentary rocks at one time covered the entire Marathon region. The reason for exposure of the fold and thrust belt today is because the nearly flat-lying layers have been folded into a broad dome and eroded, thereby re-exposing the fold and thrust belt underneath. It is a situation that mirrors the history of the Llano Uplift shown in Fig. 14.10 although there are some important differences. Crystalline rocks are not exposed in the Marathon Basin, and although both areas were deformed between 335 and 252 million years ago, deformation in the Llano area was relatively mild. Rocks in the Llano Uplift are not part of the Appalachian-Ouachita foreland fold and thrust belt. The Marathon region, on the other hand, is definitely part of the thrust belt.

The unconformity displayed in Fig. 15.31B shows the same relationship that we would see at the eastern and western margins of the Ouachita fold and thrust belt as well as at the southwestern margin of the Valley and Ridge in Alabama. In all three areas, older faulted and folded sedimentary layers disappear beneath younger, nearly flat-lying, sedimentary layers. Unfortunately, the unconformable contact is not so well exposed in these areas relative to the Marathon Basin.

WATER GAPS IN THE VALLEY AND RIDGE AND OUACHITA MOUNTAINS

There are numerous examples in the Valley and Ridge and Ouachita Mountains where rivers cut directly through ridges to form water gaps. In the Valley and Ridge, we have seen water gaps on the Susquehanna River and Fishing Creek (Figs. 15.17 and 15.18A). They are also present along several rivers in Tennessee. Examples in the

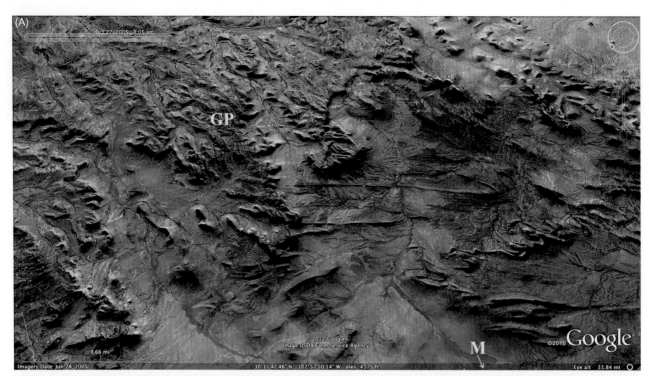

FIGURE 15.31 (A) Google Earth image looking southeastward across the east side of the Marathon Basin. Folded resistant ridges form the Marathon fold and thrust belt at lower right. Younger, nearly flat-lying sedimentary rocks cover the Great Plains (*GP*) at upper left. Note the change in landscape where the fold ridges disappear below the nearly flat-lying rocks. The town of Marathon (*M*) lies just below the image. (B) Google Earth image looking northeastward at the angular unconformity visible in A.

Ouachita Mountains include the Arkansas River near Little Rock, and the many rivers that traverse the Athens Plateau. Perhaps the most famous is the Delaware Water Gap on the Delaware River north of Allentown, Pennsylvania.

There are several methods by which a water gap can form. In Chapter 14 we described superposed rivers that were let down across an unconformity onto a pre-existing structural form (Fig. 14.39). This same process was active on the Athens Plateau. It has also been suggested that superposed rivers created some of the water gaps in the Valley and Ridge, but this is difficult to prove because there are no flat-lying rock layers in close proximity to water gaps. A related possibility is that some of the rivers established themselves, not in flat-lying rock, but on a flat surface underlain with deformed shale or deep soil. An example of a nearly flat surface underlain with deformed shale is the Great Valley in southeastern New York shown in Fig. 15.14. A river could establish itself in any direction within the deformed shale, and then maintain its course as underlying resistant ridges are encountered.

Many rivers, especially those in the southern Valley and Ridge, form straight paths in valleys for some distance, and then abruptly cross a ridge at nearly a right angle. One might expect meanders to cross ridges and valleys in a more haphazard manner if they were simply let down onto deformed rock. One possible solution is to suggest that ridges oblique to the river course were too strong for the river to cross, thereby deflecting the river back into the valley until it found a ridge or a location where it could cross, preferably one oriented at 90 degrees to the river course so that the full power of the river is utilized in cutting the water gap. One could modify this proposal by suggesting that rivers exploited areas of fractured, weak rock along resistant ridges. The idea is that two tributary rivers flowing in opposite directions will channel themselves into areas on a ridge that are strongly fractured and, therefore, weaker than other points on the ridge. Eventually, as the weak part of the ridge wears down, water from a nearby major river is pirated and diverted through the gap. The sequence is depicted schematically in Fig. 15.32A. The presence of fractured rock within many water gaps, including the Delaware Water Gap, lends credence to this possibility. Notice also that this sequence produces a perpendicular path through the ridge.

A completely different possibility is that the rivers are antecedent. An antecedent river first establishes itself in flat-lying rock, and then maintains its channel during active deformation as the flat-lying rock is deformed into a fold or fault. This particular proposal seems unlikely for the Valley and Ridge because the fold and thrust belt is hundreds of millions of years old. It would require that rivers not change their course during that entire time.

Also present, particularly in the Valley and Ridge, are wind gaps. A wind gap is a low-lying mountain pass through which a river may have at one time passed, but is no longer present. An excellent example is Cumberland Gap, which crosses Cumberland/Stone Mountain on the Appalachian Plateau at the edge of the Valley and Ridge in Tennessee (located between Middlesboro, M, and Harrogate, H, in Fig. 15.21). Cumberland Gap forms a

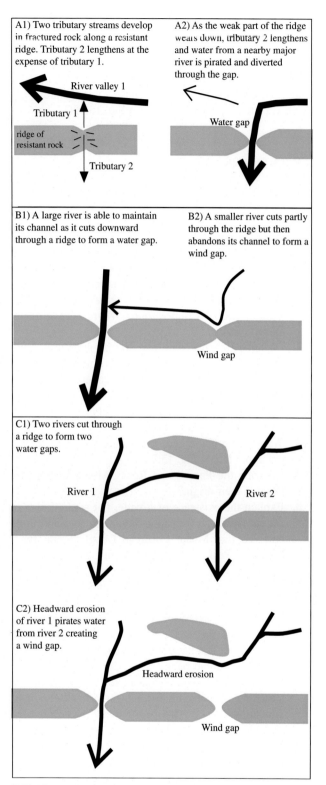

FIGURE 15.32 Schematic maps showing (A) formation of a water gap by headward erosion of a river through a fractured, weakened, section of a ridge; (B) formation of a wind gap where a smaller river is defeated in its attempt to maintain its channel through an emerging ridge; and (C) formation of a wind gap by headward erosion and river piracy.

famous gateway to the west first used by migrating animals and later by migrating settlers. A wind gap can form if a river cuts partly across a ridge prior to abandoning its channel as depicted in Fig. 15.32B. This process has likely occurred in the Valley and Ridge, including Cumberland Gap, but one must be careful because a wind gap can also form via stream piracy as shown in Fig. 15.32C.

QUESTIONS

1. What type of rock, rock structure, and landscape is typical of a foreland fold and thrust belt?
2. Calculate the amount of shortening that has occurred in the bottom panel of Fig. 15.1. Measure the amount of separation (in inches or cm) between two formerly adjacent layers and use the equation $(L'-L)/L$ where L' is the present-day length of the cross-section, and L is the original length prior to thrust faulting.
3. What is the Belt Supergroup? Where does it occur? What is its age?
4. Describe the rock type and special structures found in the Helena Formation. Why is this formation an important contributor to the scenery of Glacier National Park.
5. Describe the rock types you might see at Granite Park in Glacier National Park.
6. What is the name and origin of the major valley in the upper left of Fig. 15.7A?
7. Describe the Purcell and Rocky Mountain trenches. What is their origin?
8. Explain the existence of a wide, flat summit area on Mount Coffin in the Idaho-Wyoming fold and thrust belt. To find Mount Coffin in Google Earth, fly to Wyoming Peak, WY. Mount Coffin is the next peak to the north. Use a cross-section sketch to show Mount Coffin and the Grays River Valley immediately following thrust faulting, and another sketch to show the summit area and valley following normal faulting. Based on your two sketches, estimate the amount of vertical displacement that has occurred along the normal fault?
9. Describe tectonic reincarnation along the Cordilleran fold and thrust south of Salt Lake City. What are the structures responsible for reincarnation? Are these structures active? What can you say regarding when tectonic reincarnation began?
10. What surface geologic evidence could you look for in order to determine if the Valley and Ridge, Ouachita, and Marathon regions were once a continuous fold and thrust belt? As helpful hint, let us assume you know the rock types and geological history of all three areas.
11. Describe the Great Valley? What is its origin?
12. Some areas of the Appalachian-Ouachita fold and thrust belt have remained emergent since their initial formation. Other areas, such as the Athens Plateau and Marathon Basin, were covered with sediment from a shallow ocean at some time between 135 and 80 million years ago. Explain how the landscape in these two areas differs from areas that have remained emergent. Specifically, why are emergent areas in the Appalachian-Ouachita fold and thrust belt at higher elevation with greater relief than areas on the Athens Plateau and Marathon Basin?
13. With reference to Fig. 15.15, answer the following. What does the age of the miogeoclinal rock (2) tell us regarding when the Taconic orogeny occurred? Describe the sequence of events that created the unconformity between Taconic Flysch (3) and Taconic molasse (4). Describe the geologic setting during deposition of Catskill molasse (6); where were the Acadian Mountains? Was Catskill molasse deformed during Acadian orogeny?
14. Would you consider the Green Mountain anticlinorium to be a tectonic window? Explain.
15. What evidence is there that the southeastern side of Fig. 15.13 was deformed during Acadian orogeny? Was it also affected by Taconic orogeny? How do you know?
16. If the Llano Uplift was not involved in foreland fold and thrust faulting, and the Marathon region was involved, locate these areas on a copy of Fig. 1.3 and draw the inferred front (the northern and westernmost extent) of the foreland fold and thrust belt from the Valley and Ridge to the Ouachitas and then to the Marathon region.
17. Using a copy of Fig. 15.31A, trace the fold structure in the fold and thrust landscape, and trace the boundary with nearly flat-lying sedimentary rocks.
18. What is novaculite and where does it occur?
19. Why are north-facing slopes in the Ouachitas wet and cool relative to south-facing slopes?
20. Describe the landscape history of the Marathon Basin.
21. Fly to the following areas, describe the landscape, and determine if the structure is an anticline or syncline. Scranton, PA, Oriole, PA, Shamokin, PA, Strasburg, VA, Powells Crossroads, TN, East Signal Peak, OK. How do you distinguish an anticline from a syncline based on landscape alone? What human activity is occurring at Shamokin? What is the name of the river that flows past Strasburg? Suggest hypotheses to explain way the river at Strasburg meanders along the side of the mountain? Name the physiographic province at Powells Crossroads and at East Signal Peak.

Chapter 16

Hinterland Deformation Belts

We noted in Chapter 11 that the term hinterland refers to the central core of a collisional mountain belt where crystalline rocks dominate, but where volcanic and sedimentary rocks are also present. The rocks are crystalline due to intrusion and the metamorphism of sedimentary and volcanic rocks during the mountain building process. The dominance of resistant crystalline rocks creates a somewhat random, unpredictable landscape of rugged highlands and rolling hills. In this chapter, we look at rocks and landscape within hinterland regions of the Appalachian and Cordilleran Mountain belts where crystalline rocks are widespread, and where the structural form that developed during mountain building is well preserved. We will also look briefly at the precursor to the Appalachian orogeny, the Grenville orogeny, which took place in the eastern and southeastern United States between 1.3 and 1.0 billion years ago.

Deformation, volcanism, igneous intrusion, and metamorphism during Appalachian orogeny took place between 480 and 265 million years ago resulting in a large, nearly continuous area of crystalline rock that includes the Blue Ridge, Piedmont Plateau, and New England Highlands. Since that time, the only major tectonic event has been the opening of the Atlantic Ocean, which resulted in subsidence and burial of part of the mountain belt beneath sedimentary and volcanic rocks of the Coastal Plain and Triassic Lowlands. Rocks across most of the Blue Ridge, Piedmont Plateau, and New England Highlands were unaffected by this tectonic modification and still retain the hinterland structural form that developed more than 265 million years ago. These areas form the Appalachian hinterland landscape shown in Fig. 11.1. Although the structural form is ancient, the present-day landscape associated with these rocks is far younger, and shaped by erosion, isostasy, and sea level change. We will discuss the physiographic character of the hinterland Appalachians, the role of isostasy on recent landscape history, and the origin of the Fall Line and Blue Ridge escarpment.

Deformation, volcanism, igneous intrusion, and metamorphism in the Cordillera began less than 400 million years ago and ended 45 million years ago except along the Pacific coast where orogeny remains active. In contrast to the Appalachian hinterland, crystalline rocks are not widely exposed, and large areas of the Cordilleran hinterland have undergone tectonic reincarnation via younger normal faulting and volcanism. We will restrict our discussion to the northern Cordilleran hinterland where crystalline rocks are abundant, and where the structural form that developed prior to 45 million years ago is still reasonably well preserved. These areas include the North Cascade Mountains, the Northern Rocky Mountains including the Idaho Batholith, and part of the Columbia Plateau as shown in Fig. 11.1.

ROCKS WITHIN HINTERLAND DEFORMATION BELTS

Crystalline rocks that form the hinterland deformation belt structural province originate both as accreted terranes, and as native North American rocks. A major characteristic of these rocks is that they were deformed, metamorphosed, and intruded during Appalachian and Cordilleran orogeny. As such, intrusive granitic rocks also play a prominent role. Native North American rocks include the crystalline shield, and deformed metamorphic equivalents of the Precambrian sedimentary/volcanic rock succession, slope/rise rocks, and in a few areas, rocks of the miogeocline. These rocks form the narrow Appalachian native North American belt shown in Fig. 11.9, which encompasses the western Blue Ridge and the Green Mountains of Vermont. Native North American rocks in the Northern Rocky Mountain hinterland of Idaho and Montana are primarily metamorphic equivalents of the Precambrian sedimentary/volcanic rock succession.

Accreted terrane rocks are represented by metamorphosed and intruded island volcanic arcs, ocean basins, and microcontinents. In the Appalachians, these areas include part of the Blue Ridge, and most of the Piedmont Plateau and New England Highlands. In the Cordillera, they include the northern Rocky Mountains and North Cascades in Washington, and the Wallowa and Aldrich-Strawberry Mountains on the Columbia Plateau in Oregon.

Granitic intrusions are widespread. The largest intrusion, the Idaho Batholith, intrudes across the boundary that separates accreted terranes from native North American rocks (Line 3, Fig. 11.8). Common rocks throughout the hinterland are slate, schist, marble, gneiss, greenstone, and amphibolite.

Hinterland deformation is more complex than what is seen in foreland fold and thrust belts. Thrust faults are pervasive and are deep enough to carry crystalline rocks to the surface. Strike-slip faults are common, and in some areas normal faults are present. In many areas there are multiple generations of folds, faults, metamorphism, and intrusion. Layering and differential erosion of layers takes a less prominent role in these dominantly crystalline rocks, and therefore, they create rugged highland regions with somewhat unpredictable landscape characteristics such as the North Cascades, or low-lying rolling hills such as the Piedmont Plateau. We will begin with a discussion of the Appalachian highlands.

APPALACHIAN MOUNTAINS

Formation of the US Appalachian Mountains culminated between 335 and 265 million years ago during Alleghany (or Alleghanian) orogeny with collision of North America, Africa, Europe, and South America to form the supercontinent Pangea. During final collision, the Appalachians may have reached elevations of 20,000 feet or more. Since that time, rifting and subsidence has buried a large part of the Appalachian hinterland beneath younger rock of the Coastal Plain and Triassic rift basins (Fig. 5.2). The remaining exposed hinterland regions have undergone only erosion-deposition without substantial tectonic modification. The Piedmont Plateau is now reduced to lowland such that only the Blue Ridge and New England Highlands retain their mountainous character and even these are low, rounded mountains without the jagged peaks that characterize many of the younger mountain ranges of the Cordillera. The Blue Ridge Mountains, Piedmont Plateau, and New England Highlands together form the largest continuous area of crystalline rock in the country. The rocks are dominantly metamorphic, including an abundance of metamorphosed granitic rock (granitic gneiss), but sedimentary rocks and low-temperature metamorphic rocks are common particularly in the western part of the Blue Ridge, western Vermont, northern Maine, and a few other parts of New England and the Piedmont. The structural forms are complex with an assortment of thrust faults, strike-slip faults, and folds that are evident both at the scale of an outcrop and at the scale of the entire mountain range. Resistant rock units follow the strike of major faults and together they create a regional northeastern physiographic grain (trend) that is obvious on any map of the mountain belt.

A poorly understood aspect of the Appalachians is that it is still a mountainous region some 265 million years after mountain building events ended. The presence of mountainous topography after all this time suggests that overall erosional lowering has been interrupted with punctuated thermal isostatic changes, sea level changes, and climate change, all three of which can cause elevation (and relief) gain or loss. Such changes imply that the Appalachians were likely lower at times in the past. One possible time period was between 110 and 34 million years ago when sea level was higher than it is today (Fig. 8.3). We will have more to say about the ups and downs of the mountain range later in this chapter. For now we can conclude that the erosional landscape we see today is controlled by a combination of rock hardness, structure, and thermal isostatic changes punctuated with sea level and climate changes.

Physiographic Overview of the Blue Ridge

The Blue Ridge is the quintessential mountain range in the eastern United States. From its southernmost peak, Mount Oglethorpe (3288 feet) in northern Georgia, to one of its northernmost peaks, Long Mountain (1583 feet) near Harrisburg, Pennsylvania, the Blue Ridge stands out against the lower, flatter landscape of the Great Valley and Piedmont Plateau. The Blue Ridge is nearly 600 miles long with an overall shape of a teardrop. It is more than 70 miles wide in the south but tapers to less than 5 miles wide in the north. The entire physiographic province and most of the Piedmont Plateau are outlined in a Raisz landform map shown in Fig. 16.1, and in a shaded relief elevation map shown in Fig. 13.69.

The physiographic character of the Blue Ridge varies considerably from north to south. In different areas it is a narrow ridge, a rugged mountain landscape, or a plateau. Physical changes are most pronounced north and south of Roanoke, Virginia. North of Roanoke, the Blue Ridge is a series of ridges less than 12 miles across. South of Roanoke, the Blue Ridge is a plateau, never less than 12 miles across, with mountainous regions particularly along its western and eastern sides.

At the city of Roanoke, the Blue Ridge almost disappears into the Roanoke Basin, which occupies a position where one would expect to see the crest of the mountain range. We will take a brief look at the geology of the Blue Ridge and then describe its physiographic character beginning at Roanoke.

Geologic Overview of the Blue Ridge

Although traditionally considered part of the hinterland, the rocks that form the Blue Ridge are actually transitional between the foreland fold and thrust belt and the

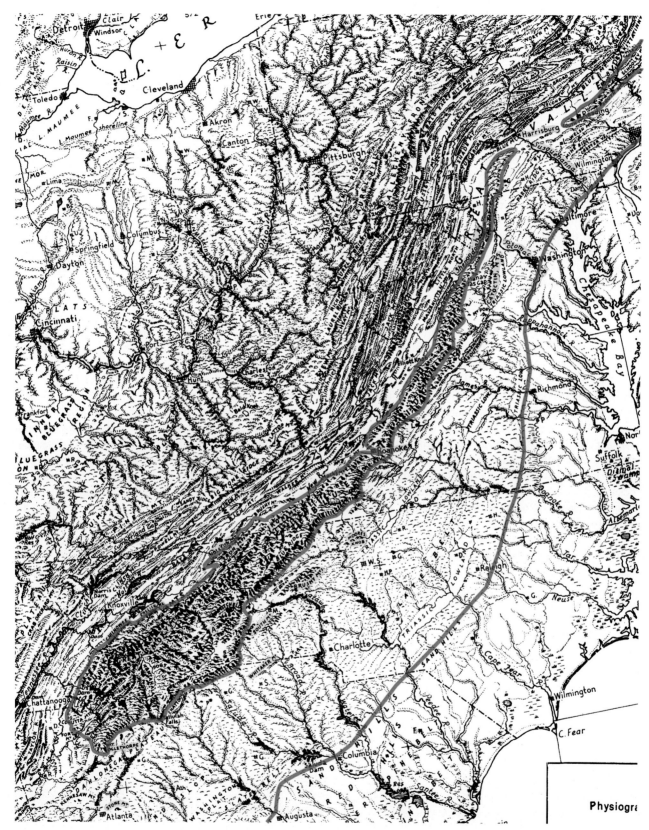

FIGURE 16.1 Landscape map of the Blue Ridge. Thick lines trace the Fall Line, the Blue Ridge, and the Reading Prong.

hinterland deformation belt. Foreland rocks, as we have seen, are sedimentary and remain unmetamorphosed during the mountain building process. Hinterland rocks are sedimentary and volcanic in origin, but they undergo metamorphism and intrusion during the mountain building process, creating crystalline rocks. Although rocks in the Blue Ridge were metamorphosed during Appalachian orogeny, the intensity of metamorphism is so low in some areas, especially along the western margin of the range, that the rocks still retain much of their original sedimentary and volcanic character. It is only south of Roanoke in the eastern part of the Blue Ridge where the intensity of Appalachian-age metamorphism is high. The overall structure of the Blue Ridge mirrors that of the Valley and Ridge with thrust faults in the south, including the Great Smoky thrust, which forms the approximate boundary with the Valley and Ridge, and folds in the north, including a large composite anticline known as the Blue Ridge anticlinorium. The Great Smoky thrust forms part of line 2 in Fig. 11.8 and Fig. 11.9.

Sedimentary and weakly metamorphosed rocks in the western part of the Blue Ridge were brought to the surface at the same time and along the same set of thrust faults and folds that developed in the Valley and Ridge. The difference is that thrust faults in the Blue Ridge have penetrated to greater depth and are more deeply eroded than those in the Valley and Ridge, resulting in older rocks reaching the surface. One of the deepest rocks to reach the surface in the Valley and Ridge is the Rome Formation of Cambrian age (541 to 485 million years old). Nearly all rocks in the Blue Ridge, by contrast, are older than the Rome Formation. Rocks of the crystalline shield, the Precambrian sedimentary/volcanic rock succession, and the lowest part of the miogeocline are represented. As a group, these rocks are more resistant than the younger miogeoclinal rocks found in the Valley and Ridge, and thus they form highlands. Part of the physiographic distinction with the Valley and Ridge, therefore, is simply the resistant nature of rocks that form the Blue Ridge.

Fig. 16.2 is an along-strike cross-sectional representation from Georgia to Pennsylvania that provides the names of rock units below the Rome Formation. These are the rocks, along with accreted terranes, that form nearly all of the Blue Ridge. Not all of these rocks are strongly metamorphosed. For the purpose of discussion, we will equate the term low-grade with low-temperature or weak metamorphism, and the term high-grade with high-temperature or strong metamorphism. For most of its length, it is the youngest rocks below the Rome Formation, the Shady Dolomite and especially the Chilhowee Group, that form the westernmost part of the Blue Ridge. Both rock units are Cambrian in age, and together they form the base of the miogeocline directly below the Rome Formation. The grade of metamorphism is low enough in both rocks such that they retain their sedimentary character. The Shady Dolomite is not particularly resistant or widespread. The Chilhowee Group, on the other hand, consists of sandstone and quartzite that is both highly resistant and widespread. As seen in Figs. 16.1 and 13.69, the Blue Ridge rises abruptly from the Great Valley along ridge crests that extend the length of the boundary from Georgia to Pennsylvania. Most of these ridge crests are underlain with resistant Chilhowee Group rocks.

Rocks below the Chilhowee Group and above the crystalline shield are part of the Precambrian sedimentary/volcanic rock succession. North of Roanoke, the Chilhowee Group is underlain with circa 570- to 555 -million-year-old weakly metamorphosed basalt (greenstone) of the Catoctin Formation, which in turn overlies granitic gneisses of the crystalline shield. These three rocks, the Chilhowee, Catoctin, and crystalline shield, form nearly all of the Blue Ridge north of Roanoke. Only the shield rocks are strongly metamorphosed, but it is important to understand that metamorphism of these rocks occurred during Grenville orogeny, one billion years ago, and not during the more recent Appalachian orogeny. These shield rocks, therefore, are not hinterland in the traditional sense of undergoing metamorphism during Appalachian orogeny. Never the less, they give the Blue Ridge a crystalline flavor, and therefore a hinterland character.

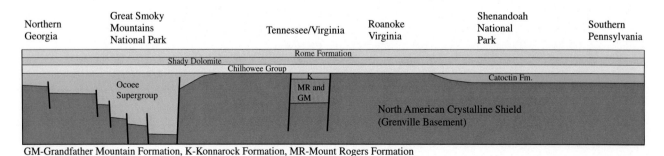

FIGURE 16.2 Schematic along-strike cross-sectional representation that shows relative thickness of rock units below the Rome Formation. Rocks below the Rome Formation are found only in the Blue Ridge. Based on Hatcher et al., 2007.

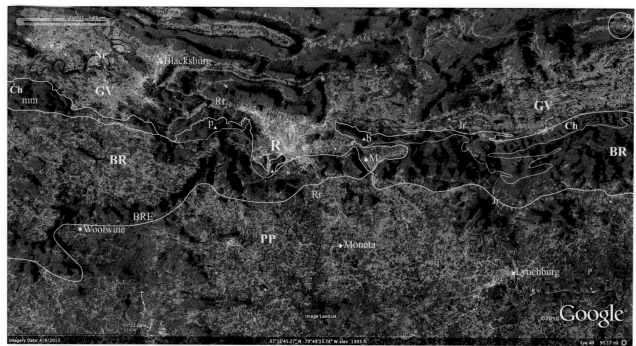

b-Blue Knob, **BR**-Blue Ridge, **BRE**-Blue Ridge Escarpment, **GV**-Great Valley, **Jr**-James River, **M**-Montvale, **mm**-Macks Mountain, **Nr**-New River, **p**-Poor Mountain, **PP**-Piedmont Plateau, **r**-Roanoke Mountain, **R**-Roanoke (and Roanoke Basin), **Rr**-Roanoke River

FIGURE 16.3 Google Earth image looking northwestward at the Roanoke, Virginia area. The Blue Ridge is enclosed between the two white lines. Note that the Blue Ridge is nearly absent near Montvale (*M*). The area between the white line on the western side (at top of image) and the yellow line consists of Chilhowee sandstone/quartzite. Crystalline shield and metamorphic accreted terrane rocks form the Blue Ridge between the yellow line and the white line on the eastern side (lower-center of image).

Chilhowee Group rocks extend south of Roanoke to form ridges along nearly the entire boundary with the Valley and Ridge. In some areas, these rocks sit directly above rocks of the crystalline shield creating a strong contrast in metamorphic intensity across the Great Unconformity. In other areas, a thick sequence of rift-related Precambrian sedimentary-volcanic rocks intervenes. Such rocks include metasandstone-conglomerate, and metavolcanic rocks of the Mount Rogers and Grandfather Mountain Formations near the Tennessee-Virginia border, and metasandstone-conglomerate of the Ocoee Supergroup, which forms nearly all of Great Smoky Mountains National Park (Fig. 16.2). In these rocks, the intensity of Appalachian-age metamorphism increases from low-grade at the western edge of the Blue Ridge, to high-grade in the interior Blue Ridge. A gradual increase in metamorphic intensity is well displayed in Great Smoky Mountains National Park where rocks of the Ocoee Supergroup are low-grade on the western side of the mountain, and high-grade on the eastern side.

The Blue Ridge at Roanoke

Fig. 16.3 is a Google Earth image looking just west of north that shows landscape surrounding Roanoke (R). The Blue Ridge physiographic province lies between the two white lines. Notice that the Blue Ridge changes from a narrow ridge northeast of Roanoke (right side) to a wide plateau in the southwest. Chilhowee Group sandstone-quartzite forms all of the rock between the northern (top) white line and the yellow line. These areas form highlands that mark the boundary with the Valley and Ridge. Areas within the Blue Ridge south of the yellow line consist of crystalline shield, Precambrian sedimentary/volcanic, and accreted terrane rock successions. Lowland areas in the Valley and Ridge consist of nonresistant Cambrian-Ordovician sandstone/shale/limestone that includes the Rome Formation. Ridges consist of younger, more resistant, dominantly Devonian sandstone. The Roanoke Basin (R) and a smaller basin centered at Montvale (M) both exist because they are underlain with infolded, weak sandstone/shale of the Rome Formation. The Blue Ridge all but disappears around the Roanoke and Montvale Basins because of the presence of these rocks.

As seen in Figs. 1.7 and 13.69, Roanoke is the location where the locus of high elevation in the eastern US begins to shift westward from the Blue Ridge, across the Valley and Ridge, to the Appalachian Plateau of Pennsylvania and New York. This shift coincides with a similar westward shift in the Mississippi River drainage

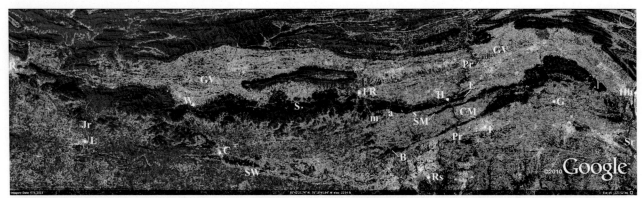

a-Ashby Gap, **B**-Bull Run Mountain, **C**-Charlottesville, **CM**-Catoctin Mountain, **E**-Elk Ridge, **F**-Frederick, **FR**-Front Royal, **G**-Gettysburg, **GV**-Great Valley, **H**-Harper's Ferry, **Hb**-Harrisburg, **Jr**-James River, l-Long Mtn., **L**-Lynchburg, **m**-Manassas Gap, **Pr**-Potomac River, **R**-Roanoke, **Rs**-Reston, **S**-Stony Man, s-Snickers Gap, **SM**-South Mountain, **Sr**-Susquehanna River, **SW**-Southwest Mountain, **W**-Waynesboro

FIGURE 16.4   Google Earth image looking northwestward at the northern Blue Ridge.

divide as shown in Fig. 2.3. The Roanoke River (Rr, Fig. 16.3) traverses Roanoke Basin on its way to the Atlantic Ocean becoming the southernmost river to flow completely across the Blue Ridge.

The Blue Ridge North of Roanoke

Fig. 16.4 shows the extension of the Blue Ridge northward to Pennsylvania. In this figure, one can easily follow a continuous series of small ridges from Roanoke (R, left edge) to Harpers Ferry (H) in which the narrow Blue Ridge stands out in forest-covered hills verses the partly deforested farmland of the Great Valley and Piedmont Plateau. In this part of the Blue Ridge only a few peaks rise above 4000 feet, the northernmost being Stony Man (S, 4011 feet) in Shenandoah National Park. North of Stony Man (S), the Blue Ridge loses elevation such that only a few peaks top out above 2000 feet north of Front Royal (FR). Chilhowee sandstone/quartzite and Catoctin metavolcanic rocks form most of the resistant ridge crests save for a few areas of resistant shield rock.

The westward shift of the Mississippi River drainage divide at Roanoke implies that rivers must cross the Blue Ridge on their way to the Atlantic Ocean. In addition to the Roanoke River, the James River (Jr), which flows through Lynchburg (L), and the Potomac River (Pr) at Harpers Ferry (H), both form water gaps through the Blue Ridge. The Shenandoah River follows the western edge of the Blue Ridge from Front Royal (FR) to Harpers Ferry where it joins the Potomac River. Wind gaps such as Manassas Gap (m), Ashby Gap (a), and Snickers Gap (s) formed via stream piracy by the Shenandoah River as depicted in Fig. 16.5. These gaps are labeled but barely visible in Fig. 16.4.

In the vicinity of Harpers Ferry (H), the Potomac River forms prominent water gaps through three successive ridges labeled Elk Ridge (E), South Mountain (SM), and Catoctin Mountain (CM) in Fig. 16.4. Elk Ridge is more properly known as Blue Ridge, but the local name is used here to avoid confusion. Some workers show all three ridges as part of the Blue Ridge. Others show only Elk Ridge and South Mountain as part of the Blue Ridge, and this is how it is shown in Fig. 16.1.

The geology at Harpers Ferry is shown in Fig. 16.6. It is clear from this figure that all three ridges exist due to the presence of resistant Chilhowee sandstone/quartzite and Catoctin metavolcanic rocks (unit 2). The three ridges collectively form part of the Blue Ridge anticlinorium, a giant anticline whose crest is traced with a dashed yellow line. Elk Ridge (E) and South Mountain (SM) form the west limb, separated by a thrust fault. Catoctin Mountain (CM) forms part of the east limb. Granite and granitic gneiss of the crystalline shield (unit 1) is present in the center of the anticlinorium and west of the thrust fault. Surprisingly, the shield rocks form nonresistant lowlands. Note the presence of Cambrian Frederick Limestone (unit 3) on the Piedmont Plateau east of Catoctin Mountain (CM). These rocks are part of the unmetamorphosed miogeocline, a rock sequence more characteristic of the Valley and Ridge. Rocks of the Triassic Rift valleys (unit 4) are also present and are discussed in Chapter 18.

South Mountain (SM) and Catoctin Mountain (CM) merge northward into a single, wide ridge that forms the northern terminus of the Blue Ridge just south of Harrisburg, Pennsylvania (Hb, Fig. 16.4). In this area, most of the Blue Ridge consists of Chilhowee Group rocks, but Catoctin metavolcanic rocks and crystalline shield rocks are also present. Catoctin metavolcanic rocks help define the Blue Ridge anticlinorium. They are present on the west limb as far south as Waynesboro (W, Fig. 16.4), and on the east limb of the anticlinorium as far south as Lynchburg (L). Bull Run Mountain (B), the Southwest Mountains (SW), and Stony Man (S) all consist of Catoctin Formation. Both limbs of the anticlinorium

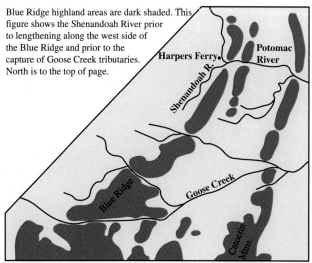

Blue Ridge highland areas are dark shaded. This figure shows the Shenandoah River prior to lengthening along the west side of the Blue Ridge and prior to the capture of Goose Creek tributaries. North is to the top of page.

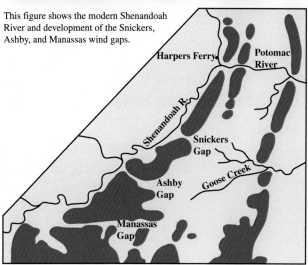

This figure shows the modern Shenandoah River and development of the Snickers, Ashby, and Manassas wind gaps.

FIGURE 16.5 Sketch map that shows development of wind gaps via progressive piracy by the Shenandoah River. *Modified from Thornbury (1965).*

are clearly visible in Fig. 16.4 due to the presence of resistant Catoctin and Chilhowee Group rocks. As shown in Fig. 16.2, the Catoctin Formation is restricted to the northern Blue Ridge.

The Blue Ridge South of Roanoke

If we return to Fig. 16.3, we immediately see that the Blue Ridge physiographic province becomes wider southwest of Roanoke. Here, the Blue Ridge extends from Macks Mountain (mm) on its western side, to the Blue Ridge escarpment (BRE) on its eastern side. Fig. 16.7 shows the entire Blue Ridge south of Roanoke. For about 70 miles, from Roanoke (R) to Galax (G), the Blue Ridge forms a broad, relatively flat, dissected plateau where deep forest is replaced with open farmland. The plateau is mostly between 2300 and 2800 feet in elevation with no peaks above 4000 feet. With the exception of Willis Ridge (wr) in the middle of the plateau, most of the 3000-plus foot peaks are located along the western border, including Poplar Camp Mountain (p) and Macks Mountain (mm). The plateau continues southwestward for another 50 miles to Boone (B), rising to between 2800 and 3200 feet and becoming more thoroughly dissected. Mountains are numerous along the margins, especially on the western side where one can find Mt. Rogers (r, 5729 feet).

Southwest of Boone, the central plateau rises to form the Roan Mountains (rm), the northernmost mountains with 6000-foot peaks. Of the more than 60 peaks above 6000 feet in the Blue Ridge, the Carolina Mountain Club recognizes 40 peaks within six mountain ranges with enough prominence (relief of 200 feet or more to a saddle between one peak and another) and enough distance (at least 0.75 miles between peaks) to be considered a separate high point. Three of the peaks are in the Roan Mountains.

South of the Roan Mountains the central plateau reappears in the form of the Asheville Basin (ab) with arms that extend from Asheville (Av) to Hendersonville (H) and Canton (Ct). Here the plateau is between 2000 and 2200 feet in elevation and is surrounded by mountains. Northeast of Asheville, one peak in the Great Craggy Mountains (Craggy Dome) and 10 peaks in the nearby Black Mountains rise above 6000 feet, including Mt. Mitchell (m), the highest in the Appalachians at 6684 feet.

Southwest of Asheville, the central plateau area is replaced by the highest and most rugged region in the Blue Ridge. Three adjacent mountain ranges have peaks that rise above 6000 feet. The Great Smoky Mountains boast 12 peaks including Clingmans Dome (c). The Plott Balsams boast four peaks, including Waterrock Knob (w), and the Great Balsam Mountains boast 10 peaks, including Richland Balsam (r). Here, it is possible to walk for at least 35 miles across all three mountains and never descend below 3000 feet. The mountain landscape is one of long, continuous ridges with high, rounded, often-treeless knobs known as balds.

The plateau area reappears southwest of the Great Smoky Mountains surrounded by mountains that form the southern terminus of the Blue Ridge. There are several peaks above 4000 feet in this area, which is better known for its large damned lakes including Chatuge (cl), Hiwassee (h), Nottely (n), and Blue Ridge lakes (b). Here, the plateau is mostly at an elevation between 1800 and 2100 feet. The southern termination of the Blue Ridge is in the vicinity of Mount Oglethorpe (o), which marked the end of the Appalachian Trail from 1937, when the trail was first completed, to 1958 when the trail was moved to Springer Mountain (s).

The western border with the Valley and Ridge is sharp and sinuous. In many areas the border is marked with

CM-Cotactin Mountain, E-Elk Ridge, SM-South Mountain. 1-Granite and Granite Gneiss (North American crystalline shield), 2-Catoctin Fm. and Chilhowee Group (includes older rock near Frederick), 3-Frederick Limestone, 4-Rocks of the Triassic rift valley. Thrust Fault (teeth on upper plate). Normal Fault (ticks on down side).

FIGURE 16.6 Google Earth image looking northward at the geology in the vicinity of Harpers Ferry. The crest of the Blue Ridge Anticlinorium is traced with a dashed yellow line.

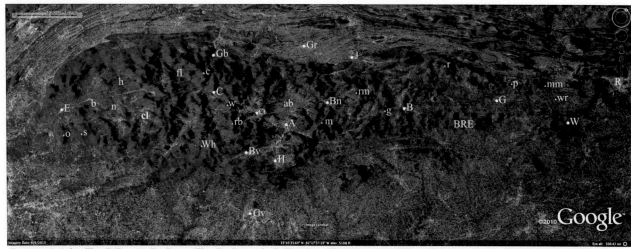

Towns - A-Asheville, B-Boone, Bn-Burnsville, Bv-Brevard, C-Cherokee, Ct-Canton, E-Ellijay, Fl-Fontana Lake, G-Galax, Gb-Gatlinburg, Gr-Greeneville, Gv-Greenville, H-Hendersonville, J-Johnson City, R-Roanoke, W-Woolwine. **Physical Features -** ab-Asheville Basin, b-Blue Ridge Lake, c-Clingmans Dome (Great Smoky Mtns.), cl-Chatuge Lake, g-Grandfather Mountain, h-Hiwassee Lake, m-Mt. Mitchell (Black Mtns.), mm-Macks Mountain, n-Nottely Lake, o-Mt. Oglethorpe, p-Poplar Camp Mtn., r-Mt. Rogers, rb-Richland Balsam (Great Balsam Mtns.), rm-Roan Mountain, s-Springer Mtn., w-Waterrock Knob (Plot Balsam Mtns.), Wh-Whiteside Mountain, wr-Willis Ridge

FIGURE 16.7 Google Earth image looking north-northwest that shows the Blue Ridge south of Roanoke (R).

major thrust faults, specifically the Great Smoky and Holston Mountain faults (26 and 11 in Fig. 15.20), which bring resistant Chilhowee Group sandstone, quartzite, and conglomerate to the surface. Numerous mountain ranges along the border are composed principally of Chilhowee Group rocks including the Chilhowee, Holston, English, Buffalo-Cherokee, Rich, Iron, Stone, and Unaka Mountains. Several of these are highlighted in Figs. 16.1 and 15.20. Similar to the Roanoke Basin, the sinuous nature of the border is due to folded inliers of soft Valley and Ridge rock that form small basins between the hard sandstone ridges. The valleys near Johnson City (J, Fig. 16.7) are an example.

The Blue Ridge escarpment (BRE) marks the boundary that separates the southern Blue Ridge from rolling hills and flatland of the Piedmont Plateau (PP). Known also as the Blue Ridge Front, this escarpment forms a continuous ridge overlooking the Piedmont Plateau with more than 2000 feet of relief. The escarpment is physiographically distinct and sharply defined southwest of Roanoke as seen in Fig. 16.3 near Woolwine. It can be followed southward at least to Hendersonville (H) where it becomes dissected and less distinct. It is not well developed southwest of Hendersonville. What is unusual is that rock units in the Blue Ridge near Woolwine can be followed northeastward across the escarpment into the Piedmont Plateau without a break, implying that the escarpment is not controlled by rock type or by a fault. We discuss the origin of the Blue Ridge escarpment later in this chapter.

The geology of the southern Blue Ridge is more complex than in the north. Appalachian-age metamorphism is weak along the western side of the Blue Ridge, but intense in the east where we find native North American rocks and accreted terranes. We will look first at the level of erosion in the Appalachians, and then look in more detail at specific areas that include the Great Smoky Mountains and the Balsam Mountains.

Level of Exhumation Across the Great Smoky Mountains

The Appalachian Mountains were, at one time some 265 million years ago, a mountain range as magnificent as the Himalayas. We do not know if the Appalachians ever achieved the height of the Himalayas, but without a doubt they were much higher than they are today. Most of their topographic lowering has been through the process of erosion.

In a world without isostatic compensation, we could erode the Himalayan peaks from 26,000 feet in elevation down to 6000 feet by removing 20,000 feet of rock (overburden) from the highest part of the mountain range. But,

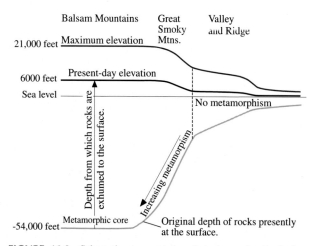

FIGURE 16.8 Schematic representation that shows the depth from which rocks were exhumed to the surface and the corresponding metamorphism assuming a total of 75,000 feet of erosion.

if you think about isostasy, you would know that for every 5 feet of elevation removed from a mountain, the mountain will isostatically uplift approximately 4 feet. Given our overly simple Himalayan example, we would actually have to remove 100,000 feet of rock in order to lower the mountain from 26,000 feet to 6000 feet. The Appalachians may not have been this extreme, but it is easy to imagine that anywhere from 50,000 to 80,000 feet of rock (overburden) has been removed from the core of the mountain range.

Fig. 16.8 is a schematic representation that shows the cumulative depth from which rocks presently at the surface have been exhumed. The top blue line represents the maximum elevation of the mountain, taken here as 21,000 feet. The middle black line represents the maximum present-day elevation, taken as 6000 feet. The lower orange line is the maximum depth from which rocks presently at the surface were exhumed. This depth is shown as 54,000 feet below sea level (75,000 feet total depth). Also shown is the increase from low-grade to high-grade metamorphism from the Valley and Ridge, across the Great Smoky Mountains, to the Balsam Mountains. The metamorphic core represents the part of the mountain belt where the most deeply buried rocks have been exhumed to the surface.

Even though the relationships in Fig. 16.8 are highly simplified, they do reflect what is seen when traveling across the Great Smoky Mountains. Rocks in the Valley and Ridge at the margin of the Great Smoky Mountains are unmetamorphosed. As one progresses across the mountain, one encounters progressively more strongly metamorphosed rock until high-temperature conditions are reached on the eastern flank where we enter the metamorphic core. According to this simplified example, the high-grade rocks have undergone metamorphism at a depth of 75,000 feet below the mountain. However, if you think about it further, 75,000 feet (22.86 km) is not all that deep considering that the continental crust averages from 18 to 30 miles thick (30–48 km), and becomes thicker during mountain building. The combination of isostatic uplift and erosion removes the excess crustal thickness brought on by mountain building, and in doing so, exhumes what were once deeply buried rocks.

The Appalachians have undergone a long and complex history that includes multiple mountain building events. Such a history implies that the location of highest peaks, highest rates of erosion, highest rates of uplift/subsidence, metamorphism, and intrusion have all migrated with time and with each orogenic event. The result is that there is no unique metamorphic core. Instead, there is what can be described as the largest area of intensely metamorphosed rock in the United States; one that extends rather continuously along nearly the entire length of the mountain belt from the southern Blue Ridge and Piedmont Plateau northward to the New England Highlands. We can think of this area as a composite metamorphic core.

The Great Smoky Mountains

The Great Smoky Mountains are located on the western side of the Blue Ridge along the Tennessee-North Carolina border. They extend southwest-northeast for about 55 miles, and if we include the Western Foothills region, they are more than 25 miles wide. Most of the area forms Great Smoky Mountains National Park. Fig. 16.9 looks west-northwest along the southern part of the range. Immediately obvious is a strong lineament, indicated with black arrows, that extends from the bottom-center of the image, through Gatlinburg (G), at the top of the image roughly parallel with the mountain range. The lineament marks the main trace of the Gatlinburg fault. It separates low ridges and valleys of the Western Foothills region from the high ridge that forms the Great Smoky Mountains proper. Rocks along the fault are fractured and broken, which implies the fault is post-metamorphic. The crushed rock produces the topographic lineament, which expresses itself in the form of stream valleys. Little River Road and route 73 follow the

B-Bryson City, C-Cherokee, Cd-Clingmans Dome, c-Cades Cove, cm-Chilhowee Mtn., F-Fontana Dam,
F-Fontana Lake, G-Gatlinburg, Gb-Gregory Bold, L-Lake Santeetlah, LT-Little Tennessee River, m-Miller Cove,
P-Pigeon Forge, S-Silers Bald, T-Townsand (Tuckaleechee Cove), Tm-Thunderhead Mtn., Wd-Walland, w-Wear Cove

FIGURE 16.9 Google Earth image looking west-northwest along the southern part of the Great Smoky Mountains, Tennessee. The Great Smoky thrust is offset at two locations along transverse faults, one of which is at Pigeon Forge. The Gatlinburg fault is marked by a lineament located with arrows. The Gatlinburg fault displaces the older Greenbrier thrust, which roughly parallels the Gatlinburg fault. Place name symbols are shown with different colors to maximize contrast.

lineament for some distance southwest and northeast of Gatlinburg respectively. The fact that the lineament is nearly straight implies that the fault is steeply dipping (Fig. 4.11). Fault zone fabrics suggest it is a reverse fault with a component of right-lateral strike-slip displacement. Its age is unknown. It could potentially be younger than 265 million years.

The high ridge of the Great Smoky Mountains extends from Gregory Bold (Gb) to Clingmans Dome (Cd) in Fig. 16.9 where it forms the Tennessee-North Carolina border. It is slightly asymmetrical, steeper on its northwest (left) side, with several satellite ridges that extend down the more gentle southeastern side. The steep northwest flank of the range is visible between Gatlinburg (G) and Clingmans Dome. Fig. 16.10, looks southeastward from Pigeon Forge (PF) across Gatlinburg (G) to Mount Le Conte, the only recognized 6000-foot peak entirely in the state of Tennessee. There is 5304 feet of relief across 6 miles between Gatlinburg (1289 feet) and Mount Le Conte (6593 feet).

The Great Smoky thrust fault, shown in Fig. 16.9 with a white line, extends along the base of Chilhowee Mountain (cm). It is one of the largest thrust faults in the Appalachians with an estimated displacement of more than 200 miles. The fault separates weak limestone and shale in the Valley and Ridge from hard sandstone of the Chilhowee Group, which forms Chilhowee Mountain. Notice that the Great Smoky thrust is displaced near Pigeon Forge (P) by a transverse strike-slip fault that cuts off Chilhowee Mountain (shown with a blue line). The mountain abruptly ends because the hard rocks that form the mountain are displaced. Rocks are fractured, broken, and eroded along the transverse fault to form a transverse valley that connects Pigeon Forge with Gatlinburg.

Stratigraphically below the Chilhowee Group, nearly all of the rocks across the entire Great Smoky Mountains (including the Western Foothills) are part of the Ocoee Supergroup, a giant rock unit more than 50,000 feet thick, composed of metasandstone, metaconglomerate, graphitic slate, phyllite, and schist. The rocks were deposited in a large normal-fault basin shortly after 640 million years ago during rifting and initial opening of the Iapetus Ocean (the precursor to the Atlantic Ocean) following Grenville orogeny. The rocks are part of the Precambrian sedimentary/volcanic rock succession. Fig. 16.2 shows that the Ocoee Supergroup is mostly restricted to the Great Smoky Mountains. The rocks thin southwestward in Georgia and abruptly terminate northeastward in northern Tennessee.

Initial deformation and metamorphism in the Great Smoky Mountains, as well as in other areas of the Blue Ridge and across most of the Piedmont Plateau, occurred during Taconic orogeny beginning some 480 million years ago. Rocks of the Blue Ridge and Piedmont Plateau subsequently underwent additional faulting, folding, metamorphism, and intrusion during later orogeny resulting in crustal thickening at the rear of the tectonic wedge. The Great Smoky thrust initiated during Alleghany orogeny between 305 and 270 million years ago when the thickened wedge began to push into the continental interior. The thrust fault was active after rocks in the Blue Ridge

FIGURE 16.10 Google Earth image looking southeastward at the northwestern front of the Great Smoky Mountains in Tennessee. Pigeon Forge (*PF*), Gatlinburg (*G*) and several 6000-foot high peaks are named. Gatlinburg is at an elevation of 1289 feet.

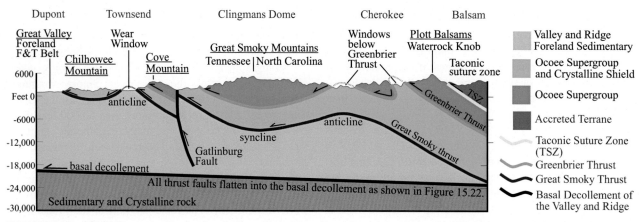

FIGURE 16.11 Schematic cross-section showing the Great Smoky Mountains sitting in a broad syncline. Also shown are widows below eroded parts of the Great Smoky and Greenbrier thrust faults.

and Piedmont Plateau had already been deformed and metamorphosed, but prior to and during foreland thrust faulting in the Valley and Ridge. The 200 miles of displacement on the Great Smoky thrust, combined with about 200 miles of displacement on faults in the Valley and Ridge, suggest that rocks in the Blue Ridge and Piedmont Plateau were carried a distance of more than 400 miles during Alleghany orogeny.

The Great Smoky thrust became inactive and was broadly folded toward the end of the Alleghany orogeny. The broad folds trend roughly parallel with the length of the mountain range. Fig. 16.11 is a schematic cross-section across the range. The Great Smoky Mountains sit in the trough of a broad syncline above the folded Great Smoky thrust. Anticlines are present on both sides of the syncline whose crests pass close to the towns of Townsend (T) and Cherokee (C). The Great Smoky Mountains thus form a large synclinal mountain. Erosion of the broad anticline in the vicinity of Townsend has created windows similar to what we saw along the Pine Mountain thrust where rocks underneath the thrust sheet are exposed (Figs. 15.22 and 15.23). Rocks within these windows consist of nonresistant Cambrian-Ordovician limestone correlative with limestone in the miogeoclinal fold and thrust belt of the Valley and Ridge. These rocks erode to form flat-floored basins known as coves. The three largest coves: Wear (w), Tuckaleechee (T), and Cades Cove (c), each form windows below the Great Smoky thrust. The Wear Cove window is shown in cross-section in Fig. 16.11. A fourth cove, Miller Cove, is present at Walland (Wd) in Fig. 16.9. This cove is not a tectonic window. It is a location where the Great Smoky thrust has carried relatively weak rocks of the Shady Dolomite and Rome Formation to the surface, thus creating a valley. All four coves result from erosion of weak rock.

The Great Smoky thrust is the westernmost fault to bring crystalline rock to the surface, and as such it forms the geologic boundary between the foreland fold and thrust belt and the hinterland deformation belt. The transition from sedimentary rock of the Valley and Ridge to crystalline rock in the Blue Ridge, as previously stated, is subtle and transitional. A drive across the mountain range on highway 441 is a drive from virtually unmetamorphosed rock near Gatlinburg (G), to intensely metamorphosed rock near Cherokee (C). It is a wonderful opportunity to see how the intensity of metamorphism gradually increases as one enters the core of the mountain belt.

Another thrust fault, the Greenbrier thrust straddles the boundary between the Western Foothills and the Great Smoky Mountains extending semi-parallel with the Gatlinburg fault. It is shown with a yellow line in Fig. 16.9 and is offset by the Gatlinburg fault. Ocoee Supergroup rocks are present on both sides of the fault. The Greenbrier thrust shows nothing in the way of fractured and broken rock. It has no obvious topographic expression and there are few fault zone fabrics of any kind. The primary evidence for its existence is a change in rock type across the contact (even though both sides are part of the Ocoee Supergroup). The absence of fault zone fabrics indicates that displacement took place prior to metamorphism. The fault is understood to have initiated during Taconic orogeny between 480 and 446 million years ago.

Nearly all of the rocks across the mountain range are part of the Ocoee Supergroup. However, by the time we reach valleys occupied by the towns of Cherokee (C, Fig. 16.9) and Bryson City (B), we cross, of all things, the Greenbrier thrust. Similar with coves on the northwest side of the mountain that form windows surrounded by the Great Smoky thrust, these valleys also form windows surrounded by the Greenbrier thrust. The windows are exposed along the eroded crests of anticlines. Two windows are shown in cross-section, Fig. 16.11. Rocks exposed in these windows are not those of the Ocoee Supergroup. They are granitic gneisses of the crystalline shield. These rocks were metamorphosed initially during Grenville orogeny 1.3 to 1.0 billion years ago, and subsequently re-deformed and re-metamorphosed during Taconic and younger Appalachian events. One might expect that these thoroughly crystalline rocks would form resistant highlands. But, similar with crystalline rock near Harpers Ferry (Fig. 16.6), they instead form lowland valleys. The surrounding highlands are composed of massive schist of the Ocoee Supergroup, which acts as the more resistant rock. The Blue Ridge Parkway, a 469-mile long scenic tour that connects Great Smoky Mountains National Park with Shenandoah National Park, begins (ends) at Cherokee and takes us across the Balsam Mountains to Asheville and beyond.

The Balsam Mountains

The Plott Balsams and the Great Balsam Mountains, although not as well known as the Great Smoky Mountains, are every bit as high with a total of 14 recognized peaks above 6000 feet. In this region, all of the rocks are gneiss and schist and are strongly deformed and metamorphosed. We are in the metamorphic core of the Appalachians. Fig. 16.12 looks north-northeast across the Plot Balsams (w), the Great Balsam Mountains (rb), and the Asheville Basin (A). The Plott Balsams form a narrow mountain range no more than 20 miles long and 7 miles wide composed of a single northeast-oriented ridge. The highest peak is Waterrock Knob (w) at 6292 feet. Rocks along the crest of the Plott Balsams consist of strongly metamorphosed schist of the Ocoee Supergroup and gneiss of the crystalline shield. The Taconic suture zone, here represented by the Hayesville thrust, is located on the southeastern flank of the Plott Balsams. Recall from Chapter 11 that the Taconic suture zone is a major tectonic boundary that separates North American rock from accreted terranes along the length of the Appalachians. It is here where we pass into the accreted terranes rock succession. Northward, the suture zone winds its way through the Asheville Basin as shown in Fig. 15.20. Accreted terranes in the Plott Balsams east of the suture zone were emplaced during Taconic orogeny and were variably redeformed and remetamorphosed during later orogeny.

The Great Balsam Mountains are part of a massive mountainous and in some areas plateau-like area about 38 miles northeast-southwest and 24 miles wide. The highest peak is Richland Balsam (rb) at 6410 feet. The rocks are primarily gneiss and schist associated with several accreted terranes. Of note are areas of Devonian (c. 380 Ma) metamorphosed granitic intrusive rocks such as those located at Looking Glass Rock (lg) and nearby Cedar Rock Mountain and John Rock. Looking Glass Rock (lg),

A-Asheville, **Bn**-Burnsville (Laurel Creek lineament), **Bv**-Brevard (Brevard fault zone), **C**-Cherokee, **cm**-Mt. Cammerer (Great Smoky Mtns.), **Ct**-Canton (Swannanoa lineament), **Gr**-Greeneville, **H**-Hendersonville, **Hs**-Hot Springs, **J**-Johnson City, **lg**-Looking Glass Rock (Great Balsam Mtns.), **m**-Mt. Mitchell (Black Mtns.), **O**-Old Fort, **rb**-Richland Balsam (Great Balsam Mtns.), **S**-Sylva, **w**-Waterrock Knob (Plott Balsam Mtns.)

FIGURE 16.12 Google Earth image looking north-northeastward across the Plot Balsams, the Great Balsam Mountains, and the Asheville Basin. Arrows point to lineaments.

a well-known destination for rock climbers, is a 3976-foot pinnacle with a sheer rock wall that rises 600 feet from its base. It was so named for the way sunlight reflects off the rock face.

Located near the towns of Cashiers and Highlands about 21 miles south-southwest of Richland Balsam, there are a series of knobs with well exposed cliff faces within high-grade rocks of the Tugaloo accreted terrane, the same accreted terrane that forms the city of Asheville and Mt. Mitchell. Most spectacular among these is Whiteside Mountain, a 4880-foot ridge with up to 1000 feet of exposed rock along a series of south-facing cliffs that extend for more than 1.5 miles. These are probably the highest sheer cliffs east of the Mississippi River. Nearby knobs with high cliff faces include Blackrock Mountain, Rocky Mountain, Laurel Knob, and Bald Rock Mountain. Most of the knobs consist of granite and foliated granite gneiss whose smooth surfaces and few cracks create difficult, challenging rock climbs. Vertical grooves on some of the cliff faces, such as at Laurel Knob, were made by running water. Schist, gneiss, and amphibolite of the Ashe Metamorphic Suite forms the upper cliffs of Whiteside Mountain offering somewhat easier climbing conditions. These rocks are intruded by the granitic rock that forms the base of the mountain.

Asheville Basin

The Asheville Basin is a wide, relatively flat area mostly between 2000 and 2250 feet in elevation centered on the town of Asheville (A, Fig. 16.12). It is surrounded by mountains and incised by the French Broad River, which flows through Asheville. Fig. 16.13 is a profile across the basin that begins in the Valley and Ridge near Greeneville, Tennessee (Gr), passes just south of Asheville (A) and just north of Hendersonville (H), and ends on the Piedmont Plateau a short distance north of Greenville, South Carolina (Gv). These towns are located in Fig. 16.7 and all but Greenville (Gv) are located in Fig. 16.12. The profile line parallels the French Broad River, crossing it seven times.

In this profile we see some characteristic features of the southern Blue Ridge. Mountain areas are present primarily along the northwest and southeast margins, both of which form escarpments that look out over the Valley and Ridge and Piedmont Plateau, respectively. The Asheville Basin is incised 100 to 300 feet by the French Broad River, with incision increasing downstream to more than 1000 feet south of the profile line where the river crosses the mountainous northwest margin of the Blue Ridge. The deep incision suggests that the river has recently cut downward into the Blue Ridge. On the southeast side, the North Saluda Reservoir follows a steep fault zone of crushed rock that produces a deep linear valley.

A close look at Fig. 16.12 shows three well-developed lineaments. All three are indicated with white arrows. The best known of these is the Brevard fault zone, located on the right side of Fig. 16.12, which forms a continuous lineation more than 375 miles long from Alabama through Atlanta to Brevard (Bv) and continuing through Old Fort (O), North Carolina. The lineament forms the southeast flank of the Great Balsam Mountains. The origin and history of this lineament is not well known. An abundance of shear fabrics and broken rock clearly show that it is a fault zone. It is not a suture zone because similar rock of the Tugaloo accreted terrane is present on both sides. It may have initially been a thrust fault active during Taconic orogeny. Although not active today, there is clear evidence of multiple phases of reactivation. It was active as a strike-slip fault during the Late Devonian-Early Mississippian Neoacadian orogeny (385–345 Ma) and during the Pennsylvanian-Early Permian Alleghany orogeny (335–265 Ma). Following Alleghany orogeny, it may have been reactivated as a high-angle fault with vertical motion.

The Swannanoa lineament can be traced more-or-less continuously from Fontana Lake, through Cherokee (C),

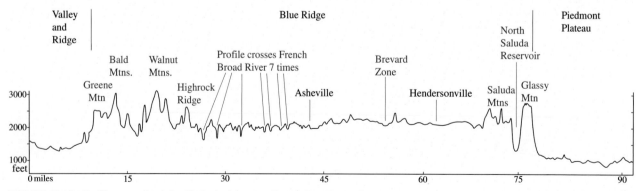

FIGURE 16.13 Profile across the Asheville basin from near Greeneville, TN (Valley and Ridge), passing just south of Asheville and just north of Hendersonville, to near Greenville, SC (Piedmont Plateau). The profile shows mountains at the margins of the Blue Ridge. The incised French Broad River crosses the profile seven times.

through Canton (Ct) and Asheville (A), to Old Fort (O) where it intersects the Brevard fault zone. It roughly borders the south side of the Asheville Basin. The Laurel Creek lineament can be traced for 45 miles through the town of Burnsville (Bn), roughly bordering the north side of the Asheville Basin. The origin of these lineaments is not entirely clear. Both cut across the regional trend of rock units and structures, which implies they developed following Appalachian orogeny. Fieldwork indicates they are associated with zones of intense fracture and with small-scale normal fault displacement. Several earthquakes have occurred in the vicinity of these lineaments including a strong earthquake in 1916 centered near Asheville (A) and one in 2005 centered near Hot Springs (Hs). The concentration of earthquakes and the presence of small-scale normal faults suggest they are active fault structures. One interpretation is that the lineaments result from fracturing and normal faulting associated with active isostatic uplift and broad doming of the southern Blue Ridge.

The Grandfather Mountain Area

Grandfather Mountain (g in Figs. 15.20 and 16.7) is located along a dissected part of the Blue Ridge escarpment between Asheville and Boone in an area where we find all of the 6000-foot peaks that lie outside the Great Smoky-Balsam Mountain region. Fig. 16.14 is a closer view of the area looking northeast. The mountain forms a small, steep ridge, 8 miles southwest-northeast and less than 5 miles wide. The high point, Calloway Peak (g, 5946 feet, top right-center of figure), forms the northeastern end of an impressive array of rocky knobs, pinnacles, and cliffs that rise 4000 feet in 4 miles on its southeastern flank, and 2000 feet in 1.5 miles on its northwestern flank. Calloway Peak also forms the Eastern Continental divide. Water draining the northern slopes is part of the Mississippi River system; water draining the southern slopes is part of the Atlantic Seaboard river system (Fig. 2.3).

The Linville River has its headwaters on the southern slopes of Grandfather Mountain and is one of several rivers that drop steeply down the Blue Ridge escarpment. At Linville Falls overlook (Lf), the river begins to incise into the escarpment to create a rocky gorge more than 1500 feet deep before entering Lake James and the Catawba River at the edge of the Piedmont Plateau. It is this gorge that separates Wisemans view scenic overlook (w) from Hawksbill Mountain (h). The gorge is likely the result of headward erosion and stream piracy as the river cut westward through the escarpment into the Blue Ridge upland.

The Grandfather Mountain area is structurally complex and geologically unique. It includes the largest tectonic window in the Appalachians, the Grandfather Mountain window. It is host to the oldest known rocks in the Appalachians, and to what were probably the most deeply buried rocks. One of the best exposures of a ductile fault zone can be found at Linville Falls (Lf). There is a second tectonic window, the Mountain City window, and at least one duplex structure (explained below). Similar to windows elsewhere in the Appalachians, the Grandfather Mountain and Mountain City windows are exposed across large, eroded anticlinal flexures. The white lines in Fig. 16.14 are faults that divide the area into 9 tectonic units. The smaller figure is a reduced tectonic map of the same area that shows the 9 fault-bound rock units. The faults are also shown in Fig. 15.20. Each rock unit consists of several formations and all contain additional faults. With the exception of unit 9, all of the tectonic units consist of native North American rock successions. Unit 9 is an accreted terrane. Below is a short introduction to the nine rock units followed by a summary of the geologic structure.

Rock unit 1 consists of Rome Formation and younger generally nonresistant sandstone, limestone, and shale typical of the miogeocline and the Valley and Ridge. Rock units 2, 3, and 4 consist mostly of resistant Chilhowee Group sandstone-conglomerate. These rocks form highlands that define the northwest margin of the Blue Ridge including Holston (hm), Buffalo-Cherokee (bcm), Rich (rm), Iron (im), Stone (sm) and Unaka Mountains (um). The Holston Mountain thrust (11) forms the geologic boundary between the Valley and Ridge and Blue Ridge, and is the northern continuation of the Great Smoky thrust. Nonresistant rocks typical of the miogeocline and the Valley and Ridge extend across the thrust southeast of Johnson City (J) where they form all of the lowland valleys in units 2 and 4.

Rock unit 5 consists of circa 1.1 billion-year old Grenville crystalline shield gneisses along with the 747- to 735-million-year-old Crossnore Plutonic suite, which consists of granite and the Bakersville Gabbro. The Crossnore and Bakersville rocks are considered to be an intrusive part of the Precambrian sedimentary/volcanic rock succession. In addition to rocks of the crystalline shield and the Crossnore Plutonic suite, unit 6 contains a rift-clastic sandstone-conglomerate known as the Grandfather Mountain Formation, a rock that is approximately the same age or slightly older than the Crossnore plutonic rocks. It is these rocks, along with Bakersville Gabbro, that form the many cliffs along the summit ridge of Grandfather Mountain. Rock unit 7 consists of our old friends, the Chilhowee Group and Shady dolomite. Chilhowee Group rocks form cliffs at Wisemans view scenic overlook (w) above the Linville River.

Rock units 1 through 7 escaped metamorphism until Alleghany orogeny when all but unit 1 was metamorphosed primarily at low grade. Unit 8 is the westernmost

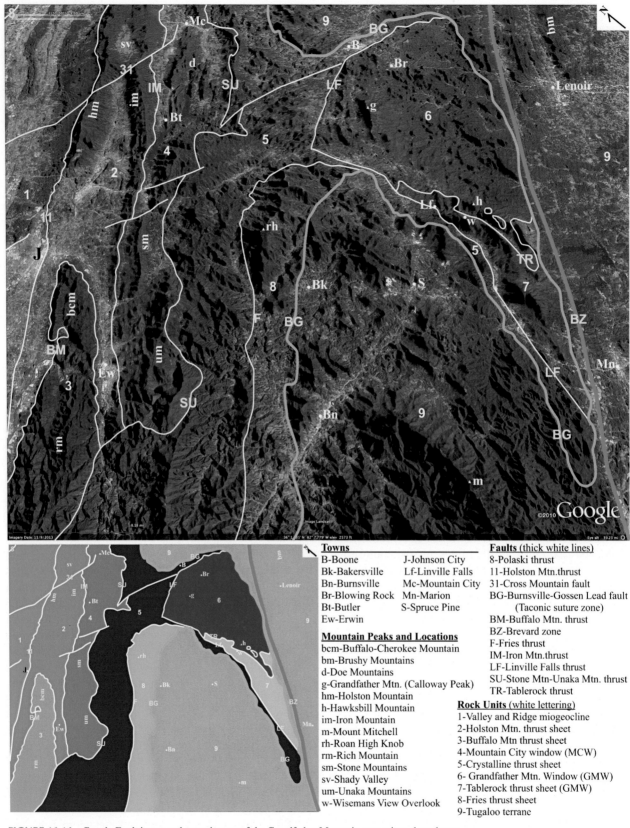

FIGURE 16.14 Google Earth image and tectonic map of the Grandfather Mountain area oriented northeast.

rock unit with strong Appalachian-age metamorphism. Radiometric dating suggests these rocks were metamorphosed during Taconic orogeny and again at low grade during Alleghany orogeny. Unit 8 consists of Bakersville metagabbro and a variety of crystalline shield gneisses and granitic rocks collectively referred to as the Pumpkin Patch metamorphic suite. These rocks form 6000-foot peaks at Roan Mountain (rh). Included within unit 8 is what could be the oldest rock formation in the Appalachians, the circa 1.8 billion-year-old Carvers Gap pyroxene granulite gneiss. This ancient rock unit can be viewed on Roan Mountain along highway 143 approximately 0.75 miles from Carvers Gap on the Tennessee side.

The final rock formation, unit 9, is an accreted terrane composed of strongly metamorphosed schist, gneiss, amphibolite, and ultramafic rocks known as the Tugaloo terrane. These rocks form 6000-foot peaks in the Black Mountains (Mt. Mitchell, m) and the Great Craggy Mountains. The deformation history of this rock unit mirrors that of unit 8 with strong Taconic metamorphism followed by weak Alleghany metamorphism. One significant aspect to unit 9 is the presence of a metamorphic rock known as eclogite. These rocks likely represent the most deeply buried rocks preserved in the Appalachians. They formed within the Taconic subduction zone at a depth of more than 35 miles and at temperatures as high as 790°C. You can visit these rocks at the intersection of Tater Hill Road and Loafers Joy Road in Silverstone, North Carolina, or in Bakersville about 5 or 6 tenths of a mile north on Redwood Road (Rt. 1217) from its intersection with North Mitchell Avenue (Rt. 1211).

Fig. 16.15 is a simplified northwest to southeast cross-section of the Grandfather Mountain area that shows the stacking sequence of the nine rock units. There are several things to notice. The Holston Mountain (11) and Iron Mountain (IM) thrust faults both underlie unit 2 implying they are the same thrust fault. Similarly, the Stone Mountain-Unaka Mountain thrust (SU) and the Linville Falls thrust (LF) both underlie unit 5 implying they are the same thrust fault. This thrust cuts across the Holston Mountain-Iron Mountain (11, IM) and the Buffalo Mountain (BM) thrust faults implying that it is younger. Rock units 6 and 7 together form the Grandfather Mountain window (GMW). Unit 7 (the Tablerock thrust sheet) is a small fault sliver in the Grandfather Mountain window wedged between units 5 and 6. Unit 7 pinches out toward the northwest. Unit 4 forms the Mountain City window (MCW) where numerous internal thrust faults are shown in cross-section. In Chapter 15 we referred to a series of closely spaced thrust faults as an imbricate zone. A duplex is where the imbricate zone is sandwiched between a lower (floor) thrust and an upper (roof) thrust. A buried duplex is shown on the southeastern side of unit 4. This duplex is emergent across an anticlinal flexure where the roof thrust (the Iron Mountain thrust, IM, in this case) has been eroded. The emergent duplex can be seen in the Doe Mountains (d) of unit 4 located between Mountain City (Mc) and Butler (Bt, Fig. 16.14) where a series of ridges are capped with resistant Chilhowee sandstone, and intervening valleys are underlain with Shady dolomite, and to a lesser extent, Rome shale and fine-grained sandstone. As you cross each ridge on highway 167 or on any of the side roads, you would find that the rock units are repeated numerous times across more than a dozen thrust faults.

The well-exposed Linville Falls thrust (LF) can be viewed by walking 0.5 miles from the Linville Falls parking lot (Lf) to the overlook above the main falls. Rocks along the trail consist of strongly sheared and deformed granitic rock of the Crossnore plutonic suite (unit 5). By the time you reach the overlook you will have crossed the fault and are standing on deformed Chilhowee quartzite (unit 7). To view the thrust fault, cross the stonewall at the overlook and walk upstream about 300 feet to an overhanging rock outcrop on the left-hand side of the smaller upper falls. The

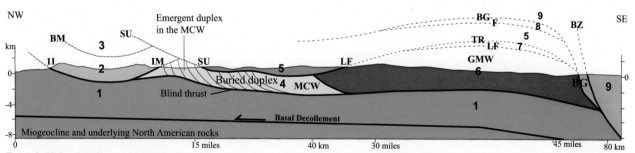

Tectonic Units: **1**-Valley and Ridge (miogeocline), **2**-Holston Mtn. thrust sheet, **3**-Buffalo Mtn thrust sheet, **4**-Mountain City window (MCW) **5**-Crystalline thrust sheet, **6**- Grandfather Mtn. Window (GMW), **7**-Tablerock thrust sheet (GMW), **8**-Fries thrust sheet, **9**-Tugaloo terrane.
Thrust Faults: **11**-Holston Mtn. thrust, **BG**-Burnsville-Gossen Lead fault (Taconic Suture zone), **BM**-Buffalo Mtn. thrust, **BZ**-Brevard Zone, **F**-Fries thrust, **IM**-Iron Mtn. thrust, **LF**-Linville Falls thrust, **SU**-Stone Mtn.-Unaka Mtn. thrust, **TR**-Tablerock thrust.

FIGURE 16.15 Cross-section oriented southwest-northeast (left to right) across Grandfather Mountain (g) in Fig. 16.14. Based on Bryant (1970) and Hatcher et al. (2006).

Linville Falls thrust is exposed near the base of the outcrop immediately below the upper falls. Here, the thrust separates sheared Crossnore granitic rock (unit 5) from underlying, highly deformed Chilhowee quartzite (unit 7). This is a ductile fault that formed under high temperature conditions such that the rocks were metamorphosed at the same time they underwent faulting. An 18-inch thick layered, lineated, fine-grained rock known as mylonite separates the two coarser-grained rocks. Mylonitization (shearing resulting in a finer-grained rock) is not confined to the contact but instead extends into the Crossnore rocks for more than one mile.

The Taconic suture zone is represented by the Burnsville-Gossen Lead fault (BG). This major tectonic suture zone wraps around the window on three sides along the flanks of the anticlinal flexure creating the shape of a foot. The fault was reactivated during post-Taconic orogenic events and therefore no longer represents the original suture zone boundary.

Recall from our earlier discussion that units 8 and 9 reached high metamorphic grade during Taconic orogeny. The fact that they directly overlie low-grade rock units indicates that units 8 and 9 reached high-grade metamorphism prior to being thrust onto the other rock units. These rocks must have been more deeply buried, and located far to the east, during Taconic orogeny relative to the other rock units. All of the rock units were much later pushed one above the other and metamorphosed at low grade during Alleghany orogeny to create the present-day structural set-up. Other than scattered circa 380 million-year-old intrusions, the effect of the intervening Neoacadian orogeny is not well documented in this area.

Piedmont Plateau

The Piedmont Plateau, perhaps because it is closer to the ocean, has been nearly completely eroded to a low-lying relatively flat plateau of broad, rolling hills. Elevation in its eastern part is mostly less than 500 feet with little relief (<50 feet) beyond the immediate confines of river channels. Higher elevation is present on the western side especially where isolated mountains of particularly resistant rock protrude above the surroundings. These mountains include Stone Mountain in Georgia, Kings Mountain, South Mountain, Moores Knob, and Brushy Mountain in North Carolina, Whitaker Mountain in South Carolina, and Big Cobbler in Virginia. A few of these are highlighted in Fig. 16.1. Each of these mountains can be referred to as a monadnock, which is defined as a mountain or rocky mass that has resisted erosion and stands isolated, surrounded by flatland. A Google Earth image of two monadnocks near Kings Mountain, North Carolina is shown in Fig. 16.16. These mountains are erosional remnants of a slowly retreating Blue Ridge plateau.

Most of the crystalline rock is schist and gneiss that represent deformed and metamorphosed accreted terranes. Structural patterns in the rock are highly complex, however, with the exception of stream channels, road cuts,

FIGURE 16.16 Google Earth image looking northwestward at two monadnocks on the Piedmont Plateau near Kings Mountain, North Carolina; King's Pinnacle (1625 feet) at left, and Crowder's Mountain (1706 feet) at right. Both stand 700 to 900 feet above the plateau. The Blue Ridge escarpment can be seen in the distance at the skyline.

and a few summit areas, bedrock is poorly exposed due to millions of years of weathering in a dominantly humid climate that has locally produced soil horizons up to 300 feet thick. The result is that there is little in the way of bedrock control on landscape. As is visible in Fig. 16.1, there are narrow Triassic lowland valleys that represent normal-fault rift basins of sedimentary and volcanic rock that formed during the breakup of Pangea. These basins are also present in New England and are described in Chapter 18.

The Blue Ridge Escarpment

The Blue Ridge escarpment is a major east-facing scarp that separates the Blue Ridge Mountains from low-lying hills of the Piedmont Plateau. It extends for more than 300 miles from northern Georgia to southern Maryland and averages between 1000 and 2000 feet in relief. It is best developed and clearly visible south of Roanoke both on landscape maps (Fig. 16.1) and in the Google Earth image shown in Fig. 16.17. Along its southern part it separates rivers that drain into the Gulf of Mexico from rivers that drain eastward directly into the Atlantic Ocean.

The origin of the Blue Ridge escarpment is a bit of a mystery. It is not the result of differential erosion because similar crystalline rock units are present on both sides. The Taconic suture zone crosses the escarpment and in some areas the escarpment is developed in relatively non-resistant rock. The escarpment does not appear to be the result of recent faulting as there is no evidence for a fault near its base and no evidence that nearby mapped faults have been active in the past 50 million years.

Apatite (U-Th)/He and apatite fission track dating techniques have been applied to rocks on both sides of the escarpment in the North Carolina-Virginia area in order to look for trends in exhumation rates that may shed some light on the origin of the escarpment. Given an estimated geothermal gradient of 25°C per kilometer of depth, the apatite fission track clock starts at about 3.2 km (1.95 miles) depth and the apatite (U-Th)/He clock starts between 1.6 and 2.0 km (1.0–1.2 miles) depth. These dating techniques, therefore, indicate how long a rock has been within 1 or 2 miles of the surface.

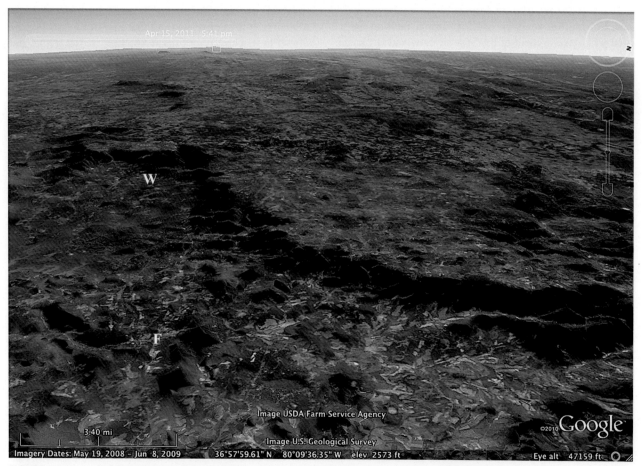

FIGURE 16.17 Google Earth image looking west along the Blue Ridge escarpment from just south of Roanoke, Virginia. F-Ferrum, W-Woolwine. The view crosses into North Carolina southwest of Woolwine.

Calculated apatite fission track dates range from 152 to 111 million years. Apatite helium dates range from 204 to 68 million years. These rather old dates suggest that the rate of exhumation in this part of the Appalachians has been slow for at least the past 150 million years. For example, one mile of exhumation in 100 million years is equivalent to about 0.06 inches per 100 years. However, we have already indicated that such dates from the Appalachians are problematic, and given the fact that some of the apatite helium ages are older than fission track ages from the same area, any conclusion regarding absolute numbers and rates must be viewed with caution.

Data from this study, however, do show some significant trends that can be used to infer a possible origin for the escarpment. Both the fission track and helium ages are older in the Blue Ridge than in the Piedmont. Apatite fission track ages range from 152 to 129 million years ago (Ma) in the Blue Ridge, compared with 127 to 111 Ma in the Piedmont. Apatite helium ages show a stronger separation with ages between 204 and 122 Ma in the Blue Ridge compared with 106 to 68 Ma in the Piedmont. On the basis of these trends, it has been suggested that the escarpment formed originally between 237 and 174 million years ago during normal faulting associated with development of the Triassic Lowland rift basins. The normal faults created an escarpment much like the escarpment that forms along normal faults in the Basin and Range. This escarpment originally was far to the east of where it is today and, over time, has retreated to its present position via erosion. Ages are younger in the Piedmont because, as the escarpment erodes and retreats, rocks directly below the escarpment are brought closer to the surface faster than rocks that remain below the Blue Ridge. The relationships are shown in Fig. 16.18.

Thus, one explanation for the escarpment is that it formed initially on the up-thrown block of a normal fault that, over time, has retreated westward such that the width of the Piedmont has grown at the expense of the Blue Ridge. What is unusual is that the escarpment has persisted as a topographic feature even though it has undergone erosional retreat for upwards of 200 million years. One suggestion for the persistence of the escarpment is that it is due to microclimate. The idea is that moisture becomes trapped on the Piedmont Plateau below the escarpment. This entrapment results in concentrated precipitation along the escarpment face, which in turn focuses preferential erosion on the escarpment face causing retreat without destruction of the escarpment itself. The escarpment, over time, will continue to retreat westward at the expense of the Blue Ridge.

The Fall Line

The Fall Line is a 100- to 300-foot drop in elevation where crystalline rock of the Piedmont Plateau and New England Highlands meets sedimentary rock of the Coastal Plain. The Fall line extends from Cape Cod to the southern end of the Piedmont Plateau near Montgomery, Alabama. It is marked with a thick line in Fig. 16.1 and is shown in its entirety in Figs. 13.8, 13.9, 13.10, 13.11, and 13.19. It is the easternmost of three major escarpments in the eastern US, the others being the Appalachian Front and the Blue Ridge escarpment. Most of the rivers in the eastern US flow eastward across the Fall Line to the Atlantic Ocean. The abrupt change in rock type across the Fall Line creates a knickpoint that disrupts the rivers ability to maintain a smooth gradient to the sea. Rivers are forced to cut downward into crystalline rock in order to meet the lower topography of the Coastal Plain. In

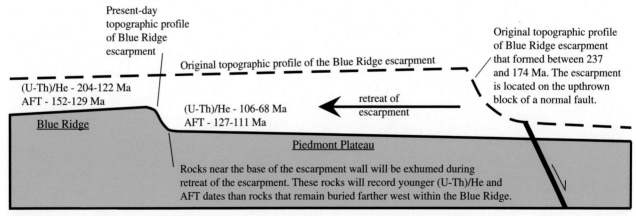

FIGURE 16.18 Topographic profile across the Blue Ridge escarpment in the North Carolina-Virginia area with apatite fission track (AFT) and (U-Th)/He ages. As the escarpment erodes and retreats, rocks directly below the escarpment in the Piedmont are brought closer to the surface faster than rocks that remain below the Blue Ridge. The result is relatively younger ages on the Piedmont. Based on Spotila et al. (2004).

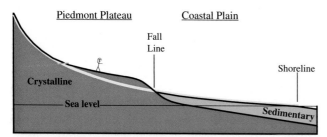

FIGURE 16.19 Cross-section sketch across the Fall Line. A stream seeks a smooth gradient to the sea along the yellow line. Hard crystalline rock of the Piedmont Plateau resists erosion and sits higher than weak sedimentary rock of the Coastal Plain. The change in rock type disrupts the rivers ability to maintain a smooth gradient. The river is forced to cut downward (toward the yellow line) in order to meet the lower topography of the Coastal Plain. In doing so, the river forms gorges, rapids, and waterfalls at the Fall Line.

doing so, rivers form gorges, rapids and waterfalls. Once a river reaches the lower gradient of the Coastal Plain, it slows down and deposits material creating a wide, navigable waterway that extends all the way to the Atlantic Ocean. The relationships are depicted in Fig. 16.19. The Fall Line is most conspicuous between Washington D.C. and Philadelphia, where the Delaware Bay and an arm of the Chesapeake Bay extend inland all the way to the Fall Line. A mini-version of a fall line forms shut-ins in the St. Francois Mountains as discussed in Chapter 14.

In the early days of colonization, before there were widespread railroads and automobiles, most freight was moved by ship. The ships could navigate up-river as far as the Fall Line but no further because they could not navigate through rapids. The rapids and waterfalls also provided water-based power for growing industries. The result was the development of a string of major cities along the Fall Line that includes Philadelphia, Baltimore, Washington DC, Richmond, Raleigh, and Augusta (Fig. 16.1). A few of the rivers, including the Roanoke, Neuse, Cape Fear, and Peedee, have cut downward into crystalline rock at the Fall Line producing rapids for up to 20 miles into what would normally be regarded as part of the Coastal Plain. In a sense, this stripping of sedimentary layers is expanding the Piedmont eastward at the expense of the Coastal Plain.

New England Highlands

The New England Highlands form the northern glaciated extension of the Blue Ridge and Piedmont Plateau. Landscape in northern New England (Vermont, New Hampshire, and Maine) consists of linear ranges and isolated highlands, 3000 to 6000 feet in elevation, with lots of rock exposure, a feature indicative of glacial erosion. Southern New England is one of rolling highlands, 1000 to 3000 feet in elevation, with low relief, less rock

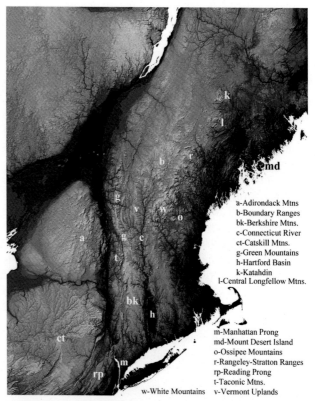

FIGURE 16.20 Elevation map of New England. National Atlas of the United States of America, USGS, Nationalmap.gov/small_scale/atlas ftp.html. Blue is low elevation, brown is high elevation.

exposure, and many glacial depositional features. A Raisz landform map of the area is shown in Fig. 13.8.

Landscape will be discussed with reference to three major highland trends, the Green Mountain, the Boundary Ranges, and the White Mountain trends. Each trend includes several disconnected or semi-connected highland areas that together create a linear arraignment of highlands. Highland areas are labeled in the shaded relief elevation map shown in Fig. 16.20. The Green Mountain trend extends for 450 miles from western Vermont to eastern Pennsylvania. It includes the Green (g), Berkshire (bk), Taconic (t), Manhattan Prong (m), and Reading Prong (rp) highlands. The Boundary Ranges trend extends along the Maine-Canada and New Hampshire-Canada border as the Boundary Ranges (b) and then southward to include the Vermont Uplands (v), a distance of 200 miles. The White Mountain trend includes the White Mountains (w), which form a dome 50 miles in diameter in north-central New Hampshire with two linear highland extensions. The first is a 55-mile extension northward to intersect the Boundary Ranges at the New Hampshire-Canada border. The second is a 165-mile extension to the northeast to include the Rangeley-Stratton (r), Central Longfellow (l), and Katahdin (k) highlands. The White

Mountain trend is easily the highest and most rugged. There are 67 peaks above 4000 feet in New England and 62 are in the White Mountain trend. The remaining 5 reside in the Green Mountain trend.

Three lowland trends are prominently displayed in Fig. 16.20. The first is the Great Valley (Valley and Ridge) located west of the Green Mountains (g) and discussed in Chapter 15. The second is a narrow corridor along the Connecticut River (c) that widens in central Connecticut to become a Triassic Lowland rift basin known as the Hartford Basin (h), discussed in Chapter 18. The third is along a wide swath of coastline from Manhattan (m) to eastern Massachusetts extending northward to inland Maine where it encroaches on the White (w), Rangeley-Stratton (r), central Longfellow (l), and Katahdin (k) highlands.

Deformation and metamorphism across the region is intense and primarily the product of Taconic (480–446 Ma) and Acadian (421–387 Ma) orogeny with contributions that correlate with Salinic (446–423 Ma) and Neoacadian (385–345 Ma) orogeny. Strong Alleghany (335–265 Ma) orogenic effects are nearly absent in New England, restricted to a small area in eastern Connecticut, Rhode Island, and eastern Massachusetts. Most of the rock is crystalline including high-grade schist and gneiss, and there are several generations of granitic intrusions. The Taconic suture zone extends along the east side of the Green Mountain trend from Manhattan to western Vermont implying that all rock successions in the Green Mountain trend are North American, and that all rocks east of the suture zone, including those in the White Mountain and Boundary Ranges trends, are either accreted terranes, or are composed of rocks that were deposited onto or intruded into accreted terranes. The approximate trace of the suture zone is shown with a black line in Fig. 16.21, which looks northward over the Reading (rp) and Manhattan (M) Prongs near the southern end of the Green Mountain trend. Note the subtle change in topography across the suture zone near White Plains (WP) due to the presence of less resistant schist within accreted terranes to the east. A large section of Ordovician oceanic lithosphere (serpentinite), probably related to the Taconic suture zone, covers central Staten Island (located off the image south of Manhattan).

On the basis of radiometric dating we can say that the New England metamorphic core has stepped eastward during successive orogeny. North American rocks in the

Place Abbreviations: *AP*-Appalachian Plateau, *Bk*-Berkshire Mtns., *Ct*-Catskill Mtns., *e*-East River, *GV* - Great Valley, *Hr*-Hudson River, *M*-Manhattan Prong, *p*-Palisades diabase sill, *rp*-Reading Prong, *SR*-Shawangunk Ridge, *TR*-Triassic Rift Basin (Newark Basin), *w*-Watchung basalt flow. **Cities: D**-Danbury, **Mc**-Monticello, **Ms**-Morristown, **Pk**-Poughkeepsie, **Pj**-Port Jervis, **S**-Sparta, **Sp**-Stony Point, **Wp**-White Plains, **Y**-Yonkers. ⎯⎯⎯ Approximate trace of the Taconic Suture zone.

FIGURE 16.21 Google Earth image looking just east of north across the Manhattan and Reading Prongs, New Jersey, New York, Connecticut. The black line shows the trend of the Taconic suture zone, which continues northward along the east side of the Berkshire and Green Mountains.

Green Mountain trend of western New England were deformed and metamorphosed during Taconic orogeny with the grade of metamorphism increasing from low-grade in the western part of the Green Mountain trend, to high-grade along the eastern part near the suture zone. Rocks east of the suture zone were strongly deformed, metamorphosed, and intruded during Acadian orogeny, and in some areas during Taconic, Salinic, and Neoacadian orogeny, creating a complex geologic history. Acadian orogeny is primarily responsible for a prominent area of high-grade gneiss, schist, granitic gneiss, and granite that extends across eastern Vermont, New Hampshire, southern Maine, and parts of eastern Connecticut and Massachusetts. This high-grade metamorphic belt includes the Vermont Upland (v, Fig. 16.20), the White Mountains (w), part of the Rangeley-Stratton Ranges (r), and areas south of the White Mountains. Acadian orogeny is also responsible for low-grade metamorphism across central and northern Maine from Bangor northward, eastward, and westward to include the Boundary Ranges (b), the Central Longfellow (l) and Katahdin (k) areas, as well as areas north of Katahdin and the northern Maine coastline including Mount Desert Island (md, Acadia National Park). Included within the low-grade terrain are intrusive rocks present primarily from the Rangeley-Stratton Ranges (r) to Katahdin (k), and southeastward from Katahdin to the northern Maine coast to include Mount Desert Island (md).

There are several exceptions to the dominance of crystalline rock across New England. The first is the area of low-grade metamorphism across central and northern Maine. The grade of metamorphism is so low, particularly north of Katahdin (k), that many of the rocks retain their sedimentary character. The second area is the previously mentioned Hartford Basin (h), a Triassic rift basin composed of red sandstone, shale, and basalt. A third area surrounding and east of Providence, Rhode Island consists of sedimentary rock deposited during the Middle and Late Pennsylvanian (315–299 Ma), long after Acadian orogeny. We next discuss each of the three highland trends.

The Green Mountain Highland Trend

With respect to all of the highland areas in New England, those in the Green Mountain trend are the most continuous. The Green Mountains (g, Fig. 16.20) extend the entire length of western Vermont along the eastern flank of the Great Valley where the Taconic Mountains (t) form an additional westward extension. The Green Mountains continue southward into Massachusetts where they are known as the Berkshire Mountains (bk). Farther south, in Connecticut and Eastern New York, the Berkshire Mountains merge with the Taconic Mountains. In Connecticut, the highland trend diverges in two directions. The Manhattan Prong (m) extends southward, ending at Manhattan Island. The Reading Prong (r) extends southwestward through Reading, Pennsylvania almost to the Susquehanna River. The two prongs are visible as highland areas in Fig. 16.21.

Elevation in the Green Mountain trend decreases southward. The Green Mountains boast five peaks above 4000 feet and many above 3000 feet. All five are located between the two g symbols in Fig. 16.20. Mount Mansfield (4393 feet) is the highest and farthest north. Killington Peak (4235 feet) is farthest south. A few peaks, such as Mount Mansfield and Camels Hump (4088 feet), are above tree line providing panoramic views from summit areas. Although peak elevations are not particularly high, the terrain is steep with long slopes that offer some of the best skiing in the east. There are at least 10 peaks above 3000 feet in the Taconic Mountains and three in the northern part of the Berkshires (bk) in Massachusetts. Farther south there are only three peaks above 2000 feet in Connecticut and none above 1500 feet on the Reading Prong.

The Green Mountain trend, taken in its entirety, forms the topographic continuation of the Blue Ridge Mountains. The physical connection with the Blue Ridge is seen in Fig. 15.17 where the northernmost mountains of the Blue Ridge and southernmost mountains of the Reading Prong are separated only by the Susquehanna River. Both the Blue Ridge and the Green Mountain trends contain water gaps. The Delaware, Ramapo, and Hudson Rivers all cut water gaps through the Reading Prong (Fig. 13.8). In Vermont, the Winooski and Lamoille Rivers traverse through the Green Mountains (Fig. 15.12). Several of these water gaps are visible in Fig. 16.20.

The Green Mountain trend is not only the topographic continuation of the Blue Ridge, it is also the geologic continuation; the rocks, structure, and metamorphism are similar. Most of the Blue Ridge consists of North American rock successions, specifically the crystalline shield, the Precambrian sedimentary/volcanic, and the miogeocline. These same rock successions are found in the mountains that form the Green Mountain trend. The rocks in both mountain trends show a general increase in metamorphism from low-grade in the west, to high-grade in the east, and both areas were deformed and metamorphosed during Taconic orogeny. Similar with the northern Blue Ridge, the most prominent structure from a landscape perspective is a large composite, post-Taconic anticline known as the Green Mountain anticlinorium whose age of formation is not well constrained. The Green Mountain anticlinorium is the along-strike equivalent of the Blue Ridge anticlinorium discussed previously in this chapter. North American shield rocks, primarily gneiss and granitic gneiss, are exposed in the core of the anticlinorium from the Reading Prong northward through the Berkshire Mountains to the central Green Mountains of

Vermont. In contrast with the northern Blue Ridge where shield rocks form lowland areas in the core of the Blue Ridge anticlinorium, similar rocks in the core of the Green Mountain anticlinorium form highlands. Killington Peak is underlain with shield rocks. The other four peaks above 4000 feet (Mount Abraham, Ellen, Mansfield, and Camels Hump) all lie north of any exposure of shield rock and instead are underlain with low- and high-grade schist of the Precambrian sedimentary/volcanic rock succession. The Green Mountain anticlinorium shows stronger metamorphism relative to the Blue Ridge anticlinorium, particularly in its eastern part.

The Valley of Vermont separates the Green Mountains from the Taconic Mountains. An oblique view of the Taconic Mountains is shown in Fig. 16.22 looking north. The highest peak is Equinox Mountain (e, 3840 feet), which overlooks the town of Manchester (M). Although not high by western standards, Equinox Mountain rises more than 3000 feet from the Valley of Vermont (vv) in less than 2.5 miles creating a bold backdrop to the town of Manchester. Dorset Mountain (d, 3760 feet), the second-highest peak in the Taconic Range, is home to the largest underground marble quarry in the world. It has been producing high quality white marble for our national cemeteries and national buildings since 1902. Although separated from the Green Mountains by the Valley of Vermont, the Taconic Mountains merge with the Berkshire Mountains (or Hills) across narrow valleys in Massachusetts and Connecticut.

As noted in Chapter 15, the Taconic Mountains are unusual because they sit in the trough of the Middlebury synclinorium in the middle of the Great Valley. The rocks are part of the Taconic allochthon. They are composed of thrust slices of sedimentary and metamorphic slope-rise and rift-clastic rock derived from the eastern margin of North America and transported westward more than 70 miles to the Great Valley. The rocks form highlands because they are more resistant than the miogeoclinal rocks that are present in both the Great Valley and in the Valley of Vermont.

The Boundary Ranges Highland Trend

Fig. 16.23 is a Google Earth image of northern New England oriented north in which the shading accentuates the highland trends. The Green Mountain trend (g) through Vermont is clearly visible. Also visible are the Vermont Uplands (v), which continue along the New Hampshire-Canada and Maine-Canada border as the Boundary Ranges (b) to form the Boundary Ranges highland trend. This highland trend boasts numerous peaks above 3000 feet, but none above 4000 feet. High points include East Mountain (3439 feet) in the Vermont Uplands, Stub Hill (3627 feet) in the Boundary Ranges of northernmost New Hampshire, and Snow Mountain (3960 feet) in Maine. Most of the rock consists of Silurian and Devonian granitic gneiss and schist metamorphosed during Acadian orogeny. With one possible exception, the

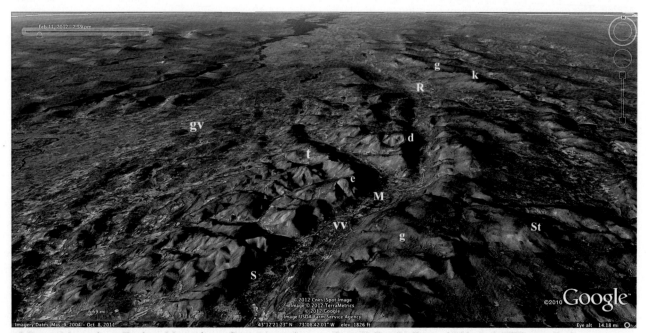

d-Dorset Mountain, **e**-Equinox Mountain, **g**-Green Mountains, **gv**-Great Valley (Valley and Ridge), **k**-Killington Peak **M**-Manchester, **R**-Rutland, **S**-Shaftsbury, **St**-Stratton Ski Area, **t**-Taconic Mountains, **vv**-Valley of Vermont

FIGURE 16.22 Google Earth image looking north at the Taconic Mountains (t), western Vermont. The Valley of Vermont (vv) is visible from Rutland (R) to Shaftsbury (S).

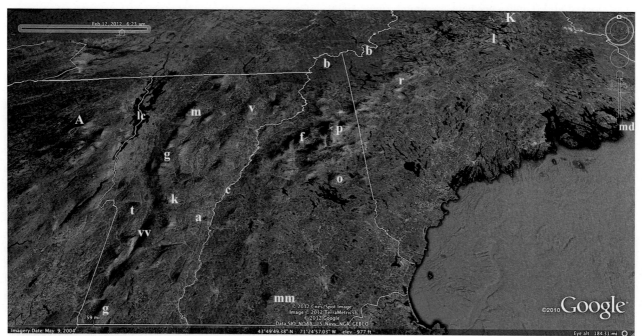

A-Adirondack Mountains, a-Mount Ascutney, b-Boundary Ranges, c-Connecticut River, g-Green Mountains, K-Katahdin, k-Killington Peak, f-Franconia Range (White Mtns.), l-Central Longfellow Mountains, lc-Lake Champlain, m-Mt. Mansfield, md-Mount Desert Island, mm-Mt. Monadnock, o-Ossipee Mountains, p-Presidential Range (White Mtns.), r-Rangeley-Stratton Ranges, t-Taconic Mountains, vv-Valley of Vermont, v-Vermont Upland.

FIGURE 16.23 Google Earth image looking north across northern New England. The Taconic suture zone is located along the east side of the Green Mountains (g).

rocks are part of the accreted terrane succession. Snow Mountain and several nearby 3000-foot peaks in the Boundary Ranges, including Kibby Mountain, are underlain with Precambrian gneiss known as the Chain Lakes Massif. The origin of this gneiss is uncertain. High-grade metamorphism is Grenville in age, which suggests the rocks may be an isolated part of the North American crystalline shield.

The White Mountain Highland Trend

Rising from the hills of central New Hampshire are the White Mountains where one can find the highest and most massive peaks in New England. As seen in Fig. 16.23, the White Mountains (p and f) form a circular domal welt of 4000- and 5000-foot peaks, with a northward linear extension and a more obvious northeastward extension. The highest elevations are in the northern and northeastern part of the dome within the Presidential (p) and Franconia (f) Ranges. Five 5000-foot peaks are in the Presidential Range (p) including Mount Washington (6288 feet), the only peak above 6000 feet. Two additional 5000-foot peaks are in the Franconia Range (f), leaving only one (Mount Katahdin) not in the White Mountains. Numerous 4000-foot peaks form the remainder of the dome area.

All five 5000-foot peaks in the Presidential Range are clustered along a continuous ridge top within 4.5 miles of Mount Washington creating the largest unbroken summit area above tree line in the eastern US. In this area, tree line is at 4800 feet. The peaks have rounded summit areas connected with ridges that were carved and narrowed during a period of alpine glaciation following the retreat of continental glaciers. Perhaps the most prominent alpine glacial feature is the Great Gulf, a deep amphitheater-shaped cirque on the north side of Mount Washington with walls that plunge more than 1250 feet to Spaulding Lake over a horizontal distance of 2000 feet. Tuckerman Ravine, a glacial cirque on the southeast flank of Mount Washington, is a popular challenge for spring skiers who ascend the steep wall on foot. The absence of trees on all of the 5000-foot peaks allows visitors phenomenal 360-degree views of surrounding landscape, but at the same time, creates potentially fierce weather conditions. The Mount Washington observatory, at the mountain summit, has been recording weather conditions since 1932. Below zero temperatures and wind speeds in excess of 100 mph are common. Maximum daily temperature rarely climbs above 70°F even in summer, and average wind speed on any given day is between 24 and 46 mph. On April 12, 1934, wind at the observatory was clocked at 231 miles per hour, becoming (until 1996) the highest surface wind speed ever recorded. The mountain summit is accessible by foot, road, or cog railroad.

Highlands extend northward and northeastward from the White Mountains to form two separate highland trends. The northward extension north of the Presidential Range (p, Fig. 16.23) in New Hampshire includes two 4000-foot peaks (Mount Waumbek and Mount Cabot) and several 3000-foot peaks, all considered to be part of the White Mountains. The northward extension ends just south of the Canadian border in New Hampshire where 3000-foot peaks of the White Mountains merge with 2000- and 3000-foot peaks of the Boundary Ranges (b). The northeastward extension is more prominent and includes two additional ranges with 4000-foot peaks, the Rangeley-Stratton Ranges (r, and sometimes referred to as the High Peaks and Bigelow Ranges), where there are 10 peaks above 4000 feet, and the Katahdin area (k) where there are three peaks above 4000 feet including Mount Katahdin (Baxter Peak), the only 5000-foot peak in New England not in the White Mountains. There is one additional 4000-foot peak in Maine, Old Speck Mountain, located in the northeastern highland trend just over the border with New Hampshire in an area that is considered to be part of the White Mountains.

The Rangeley-Stratton region (r) has the highest concentration of 4000-foot peaks anywhere in New England outside of the White Mountains. The highest peak is also a ski area, Sugarloaf Mountain (4240 feet). Several of the high peaks top out above tree line, again providing panoramic views. Still farther northeast, along the same line, we reach another area of high ground known as the Central Longfellow Mountains (l). An interesting aspect to the naming of mountains is that the state legislature named the entire collection of mountains in Maine the Longfellow Mountains, after the poet and educator who was born in Portland. The Central Longfellow Mountains (l) are a small specific group of peaks located 30 miles southwest of Katahdin with White Cap Mountain, at 3654 feet, the highest.

The northeastward extension of the White Mountain highland trend ends in the Katahdin area (k). At 5268 feet, the summit of this highland region, known as Baxter Peak or Mount Katahdin, marks the end of the Appalachian Trail. The Katahdin highland is amazingly isolated, being surrounded by lowland areas and giant lakes for miles in all directions. Fig. 16.24A looks northward across the region. The Katahdin highland forms a ring of peaks that surrounds the high summit area on the east, north and west sides, and which includes several 3000-foot peaks and one 4000-foot peak, North Brother (4151 feet). The highland area ends northward in the vicinity of The Traveler (t), another 3000-foot peak. Relief between the Katahdin summit (Baxter Peak) and some of the surrounding lakes exceeds 4000 feet. Relief between the summit and the West Branch Penobscot River (pr) is 4668 feet over a horizontal distance of 5.25 miles. The entire area, and much of the surrounding area, is part of Baxter State Park, an independently funded park that is not part of the state park system. The park is purposely kept in its forever wild state as much as possible. Visitation is limited. Those wishing to visit during peak summer season should call for reservations many months in advance.

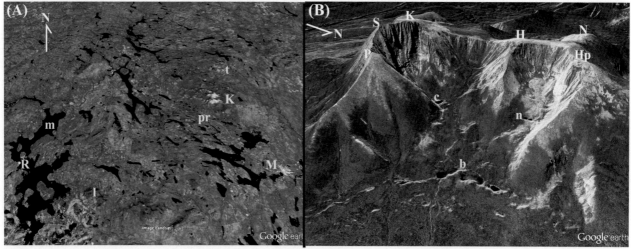

b-Basin Ponds, **c**-Chimney Pond, **H**-Hamlin Peak, **Hp**-Howe Peaks, **l**-Central Longfellow Mtns., **K**-Katahdin (Baxter Peak), **M**-Millinocket, **m**-Moosehead Lake, **N**-NW Plateau, **n**-North Basin, **P**-Pamola Peak, **pr**-west West Branch Penobscot River, **R**-Rockwood, **S**-South Peak, **t**-The Traveler

FIGURE 16.24 (A) Google Earth image looking north across Maine at Mount Katahdin and surrounding lakes. (B) Google Earth image of the Katahdin summit area looking west-southwest. The Knife Edge lies between Pamola Peak (P) and Mount Katahdin (K).

Fig. 16.24B looks southwestward at the summit area revealing glacially carved cirques along the steep eastern face. Of the several summit peaks, only Hamlin Peak (H) qualifies as a separate 4000-foot peak. The glacier-carved chasm located between Mount Katahdin (K) and Hamlin Peak (H), and occupied by Chimney Pond (c), is known as the Great Basin. Glacial cirques have helped create the famous Knife Edge, a mile-long trek on a narrow, potentially dangerous rocky ridge between Katahdin (K) and Pamola Peak (P, 4919 feet). Some consider the Knife Edge to be the most spectacular maintained hiking trail in the country. The west face (not seen in this image) offers a less-steep ascent. The rocky terrain and steep ascent from any direction combine to create a challenging climb to the summit, but a must do for those wishing to complete the Appalachian Trail. The summit area itself is rounded with gentle slopes and a wide area above the 3800-foot tree line.

The origin of the White Mountain trend appears to largely be due to rock resistance and glaciers, which gouged surrounding lowland areas and accentuated relief. The dome shape of the White Mountains is attributed to intrusion and differential erosion of granitic rocks of various ages that surround Mount Washington. These rocks disrupt layering and are resistant enough to form knobs. Undeformed Jurassic (201–145 Ma) granite forms the summit areas of the two 5000-foot peaks in the Franconia Range (Mount Lafayette and Mount Lincoln), as well as the bulk of the central dome area including Mount Carrigain (4680 feet). These rocks are post-orogenic and possibly associated with the breakup of Pangea. They form part of the White Mountain Batholith. A variety of additional deformed and undeformed Ordovician to Pennsylvanian granitic intrusions (454 to 320 million years old) forms much of the western and southern parts of the dome with Devonian intrusions most abundant. Bucking this trend are the 5000-plus-foot peaks in the Presidential Range (Washington, Jefferson, Adams, Monroe, and Madison). These summit areas consist of Silurian-Early Devonian metasedimentary and metavolcanic schist (444–393 Ma). The rocks represent accreted terranes and the erosional debris of accreted terranes metamorphosed during Acadian orogeny.

The origin of the two highland extensions of White Mountains is at least partly associated with the presence of resistant rock, but some areas of weak rock also forms highlands. Both highland trends host intrusive rocks of various ages. Several 4000-foot peaks in the Rangeley-Stratton region such as Saddleback Mountain and Mount Redington, are underlain with hard Devonian granitic rocks. Sugarloaf Mountain and Crocker Mountain are underlain with Devonian gabbro. Other peaks, such as Spaulding Mountain and both peaks on Mount Bigelow are composed of less resistant schist and low-grade metamorphic rocks. White Cap Mountain, in the Central Longfellow Mountains, is underlain with rather nonresistant mudstone metamorphosed to low grade. Perhaps some of these areas were mantled with hard rock that has since eroded.

The origin of the Katahdin highlands is problematic when one considers the huge expanse of lowland surrounding the mountain massif. The area was obviously a highland prior to glaciation. The entire Katahdin massif shown in Fig. 16.24B, is underlain with resistant Devonian granite (c. 400 Ma). But this same granite also forms lowland areas to the immediate southwest of Katahdin. The Katahdin highland appears to have been a monadnock prior the glaciation as surrounding areas were eroded to lowland. Relief was subsequently accentuated as glaciers gouged lowland areas surrounding the mountain. The Traveler (t, Fig. 16.24A) forms another monadnock highland area underlain with resistant Devonian rhyolite volcanic rock. Surrounding these areas, the wide expanse of wilderness in northern Maine is one of giant lakes underlain with weak dominantly low-grade and sedimentary rocks with few highland areas above 1500 feet.

Isolated Mountain Peaks

Outside of the previously discussed highland areas, New England is an uneven hilly landscape transected by the Connecticut River with as few as 5 peaks above 3000 feet, four in New Hampshire, and one in Vermont. Southward in Massachusetts, only one peak exceeds 2000 feet, Mount Wachusett. Here, we discuss two of the isolated 3000-foot peaks and one area of 2000-foot peaks. All three appear as highlands due to the presence of resistant rock. Mount Ascutney (a, Fig. 16.23, 3130 feet), located in southeastern Vermont near the Connecticut River, forms a distinctive round mountain shape underlain with Permian-Triassic granitic rock. Mount Monadnock (mm, 3149 feet) in southern New Hampshire, is a narrow, isolated, bald mountain peak underlain with accreted terrane metamorphic rocks. This mountain represents the type location for a monadnock. The Ossipee Mountains (o), located just south of the White Mountains and just north of Lake Winnipesaukee, forms a nearly circular range 10 miles in diameter visible in Fig. 16.20 with Mount Shaw the highest peak at 2990 feet. The Ossipee Mountains are underlain with some of the youngest intrusive and volcanic rocks in the state, including rhyolite, basalt, and granite. The rocks represent the inner guts of a volcano that formed in the Cretaceous between 130 and 100 million years ago when the Great Meteor (New England) Hot Spot passed below New Hampshire. The volcano collapsed into a caldera during eruption and was subsequently eroded leaving behind a ring dike (a circular intrusion) and other features that outline the remnants of a volcano.

Acadia National Park

As noted in Chapter 12, Maine is a glaciated, drowned coastline with many deep inlets and islands. It is also host to Acadia National Park, the only national park in the northeastern US and the oldest national park east of the Mississippi River. Most of Acadia is located on Mount Desert Island (md, Fig. 16.23), a glacially streamlined island that includes the highest point on the Atlantic seaboard, Cadillac Mountain (1528 feet). The area is geologically interesting because it records aspects of three microcontinent collisions, Ganderia, Avalon, and Meguma. The term microcontinent implies an accreted terrane with a continental, rather than oceanic origin. Fig. 16.25 is a tectonic map of Acadia and the surrounding region that we can use to outline the sequence of events.

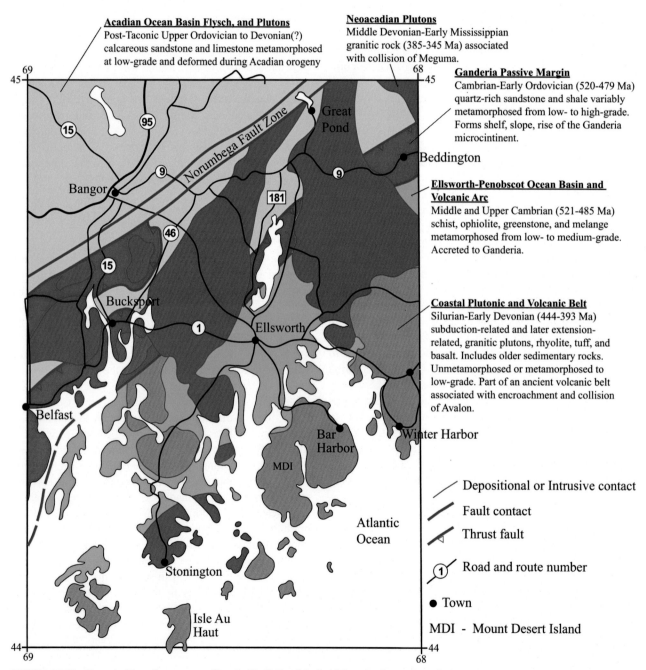

FIGURE 16.25 Tectonic Map of area surrounding Acadia National Park, Maine. Acadia consists of Mount Desert Island, the Schoodic Peninsula south of Winter Harbor, Isle Au Haut, and a few additional small islands.

The Ganderia microcontinent was accreted to the edge of North America during Taconic and Salinic orogeny in the Ordovician-Silurian between about 460 and 423 million years ago. This microcontinent, and associated rock, forms all of coastal Maine. Two rock units that were once part of Ganderia are shown in Fig. 16.25. The Ganderia passive margin represents the outer shelf, slope, and rise of the micocontinent. The rocks are variably metamorphosed sandstone, shale, and schist, and are exposed surrounding the towns of Beddington and Bucksport, and along the coast near Belfast. The second rock unit is represented by the Ellsworth-Penobscot ocean basin and volcanic arc. These rocks are oceanic in origin and were accreted to Ganderia while Ganderia was drifting across the ocean prior to collision with North America. The rocks were already part of Ganderia when the microcontinent collided. They include an assortment of schist, greenstone, and sliced sections of oceanic crust characteristic of a subduction-suture zone. These rocks crop out surrounding the town of Ellsworth and on Mount Desert Island at the bridge that connects the island with the mainland.

The Acadian Ocean Basin Flysch and Plutons rock unit consists of calcareous sandstone, limestone, and a variety of other rocks deformed and metamorphosed primarily to low-grade during Acadian orogeny. The rocks were deposited in an ocean setting following collision of Ganderia, but prior to and during collision of Avalon. These rocks are exposed in the area surrounding Bangor.

Avalon was accreted to the edge of North America during Acadian orogeny between 421 and 385 million years ago. The microcontinent itself lies offshore and is not exposed in this area, but it is exposed in the area surrounding Boston. The rock unit associated with Avalon is the Coastal Plutonic and Volcanic belt. This rock unit formed a volcanic arc on the continental margin of North America during Silurian-Early Devonian (444−393 Ma) prior to collision with Avalon. The volcanic arc developed above a subduction zone that dipped west below North America and which was associated with the encroachment of Avalon, which was approaching North America from the east. The Coastal Plutonic and Volcanic belt remained active in the late Silurian-Devonian as an extensional volcanic belt during and following accretion of Avalon with North America. This period of volcanism developed into explosive calderas not unlike those presently at Yellowstone National Park. Upper Silurian caldera-forming plutonic and volcanic rocks form nearly all of Acadia National Park. The rocks are primarily granite, but rhyolite, tuff, basalt, and gabbro are also present, all of which are unmetamorphosed or metamorphosed to low grade.

Meguma was accreted to the edge of North America in the late Devonian-early Mississippian (385−345 Ma) during Neoacadian orogeny. This microcontinent is not exposed in the United States, but is well exposed in Nova Scotia, Canada. The collision created the Neoacadian plutons, a group of granitic intrusive rocks, which are exposed between Beddington and Ellsworth, between Bucksport and Ellsworth, and along the coast surrounding Stonington.

Also shown in the figure is one of the youngest faults in Maine. The Norumbega Fault was likely active as a right-lateral strike-slip fault in mid-Devonian (c. 388 Ma) and was reactivated intermittently during the late Paleozoic and Mesozoic. Fission-track dates on the west side of the fault zone are between 113 and 89 million years old. Those east of the fault zone are between 159 and 140 million years old. This age discontinuity has been interpreted to suggest final reactivation with vertical, east-side-down displacement during the Late Cretaceous, at or soon after 89 million years ago.

Erosional History of the Appalachian Mountains

The erosional history of the Appalachian Mountains during the 265 million years following mountain building is not entirely clear. One way to gain a better understanding of this history is to look at the age and thickness of sediment that has been eroded from the mountain. The idea is to correlate periods of high sediment accumulation with periods of regional uplift. One has to be careful here because changes in climate or sea level could also cause changes in sediment accumulation rates. In the case of the Appalachians, we have a good enough understanding of climate and sea level history to take these changes into consideration. When we do this, we find that isostatic uplift and subsidence are major influences on landscape history. Isostatic adjustment results from many factors, one of which is the temperature of rock below the mountain. An increase in temperature, due perhaps to upwelling of hot material from deep within the Earth, or from a granitic intrusion, will lower the density of rock and result in isostatic uplift in the form of a broad, anticlinal bulge. We have already discussed numerous examples of this phenomenon such as the Adirondacks, the Mississippi Embayment, and the Colorado Plateau/Middle-Southern Rocky Mountains. The location of warm rock beneath the mountain can migrate over time, which implies that the crest of an anticlinal bulge, and therefore, the location of high topography in the mountain belt, can also migrate.

Large basins off the eastern US coastline, such as Baltimore Canyon and the Carolina Trough, trap sediment eroded from the Appalachian Mountains. Both troughs extend parallel to the coastline along the shelf-slope break. The Baltimore Canyon extends from New Jersey southward to Maryland. The Carolina trough parallels the North and South Carolina coastlines. Fossils are used to date the sediment, which is as much as 180 million years old. These troughs, particularly the Baltimore Canyon Trough, records erosion primarily from New England and the central Appalachians. Sediment thicknesses suggest that more than 4.3 miles (7 km) of crust has been removed from the central and northern Appalachian Mountains in the past 180 million years (equivalent to 0.15 inches/100 years). The data indicate high rates of sediment accumulation at about 155, 130, 85, and 15 million years ago. The data, however, must be interpreted with the understanding that exhumation did not occur uniformly along the length of the Appalachians, but rather was more likely concentrated in one area or another at different times. One must also keep in mind that sediment reaching these oceanic troughs could be derived from anywhere within the Atlantic Seaboard river system, which today, includes the Piedmont Plateau, most of the Coastal Plain, and the central and northern parts of the Appalachian Plateau, Valley and Ridge, and Blue Ridge. Most of the southern Blue Ridge, including all 40 of the 6000-foot peaks, is not presently part of the Atlantic Seaboard river system. These areas drain westward to the Mississippi River as shown in Fig. 2.3. Obviously, the size and locations of river systems have varied over the past 180 million years, so it is unclear how much, if any of the erosion of the southern Blue Ridge is recorded by sediment deposited off the Atlantic coastline.

Another clue to landscape history are the sedimentary rocks on the Coastal Plain that overlie the eroded stump of the Appalachians. The age and composition of these rocks indicates how long ago this part of the Appalachian Mountain chain had eroded and subsided to sea level. The oldest exposed oceanic rocks suggest that part of the Appalachian chain had already eroded and sunk below sea level by 130 million years ago. Sedimentary rocks on the Coastal Plain generally become younger toward the ocean suggesting an overall retreat of the sea (a relative lowering of sea level) since 85 Ma.

Given the above information, we can suggest a scenario for the erosional history of the Appalachian landscape. The Appalachians likely reached their highest elevation during the culmination of mountain building some 265 million years ago. Following mountain building, part of the mountain belt may have already eroded to lowland by 237 million years ago prior to the rifting event that produced the Triassic rift basins and the Atlantic Ocean. Actual separation of North America from Africa and Europe occurred between 237 and 174 million years ago in a part of the mountain belt that is currently below the Coastal Plain (east of the present day Appalachians). The rifting event likely produced elevation gain in the exposed part of the Appalachians through thermal isostatic uplift although elevations were nowhere near the height attained during the earlier continental collision. The first pulse of sedimentation at 155 Ma may be related to the late stages of this elevation gain. Additional uplift in the New England Appalachians may have been associated with intrusion of the White Mountains batholith.

It is typical for a region to undergo thermal cooling and slow subsidence following a rifting event. This phenomenon, coupled with erosion, would have again worn the Appalachians to lowland perhaps by 140 million years ago. The pulse of sedimentation at 130 million years ago was relatively small and thus, probably does not signal a significant gain in elevation. A high sea level stand began about 130 million years ago and lasted until 34 Ma (Fig. 8.3). It was during this time interval that elevations were likely lower than today and possibly the lowest in the history of the Appalachians. Much of the Coastal Plain and perhaps part of the Piedmont Plateau was below sea level. Because of high sea level, the 85 million year spike in erosion/sedimentation rates probably correlates with thermal isostatic uplift but not with significant elevation gain. Any uplift generated would instead have been neutralized by increased rates of erosion and sedimentation. The Appalachian landscape at this time likely was low-lying with isolated highlands restricted to the vicinity of the Blue Ridge and New England Highlands.

Elevations likely remained subdued and below present-day elevation for the next 60 to 65 million years as sea level slowly retreated thereby exposing the Coastal Plain and marginally increasing overall elevation. The final relatively strong pulse of sedimentation began about 15 million years ago and has continued to the present-day. This pulse is correlated with regional bulging of the Appalachians, possibly due again to thermal isostatic changes brought on by the upward flow of warm rock from the mantle. Circumstantial evidence for thermal bulging is seen in Fig. 13.67 where areas of high elevation trend obliquely across the structural grain of the Appalachians from the Blue Ridge to the Appalachian Plateau. This somewhat odd trend could correspond with the axis of a thermal bulge.

Thus, present-day elevation and relief in the Appalachian Mountains is interpreted as the result of uplift that is only 15 million years old. If this is the case, then the Appalachian landscape we see today, although shaped by 265 million years of erosion, may actually be relatively young. In a sense, we can say that the Appalachians are an old mountain range that has found the fountain of youth and has reverted back to a younger stage. It has undergone rejuvenation. The Roanoke, James, and Potomac Rivers, among others, form water gaps through the northern Blue Ridge, and additional water gaps are found in the New England Appalachians. It is possible that these rivers were already in existence long before 15 million years ago and were able to maintain their channel by down-cutting through buried structures during regional uplift. The more recent glaciation would have, at times, lowered sea level substantially enough to cause river incision and, therefore, increase overall relief thus contributing to the present-day landscape.

This analysis does not directly apply to the southern Appalachian Mountains because these areas drain to the Gulf of Mexico. There have been several studies on the rate of erosion (exhumation) primarily in the southern Appalachians using apatite and zircon fission track data, cosmogenic ^{10}Be data, and apatite and zircon (U-Th)/He data. The data have not been entirely robust. Often there is considerable spread in ages obtained from nearby samples resulting in uncertainty as to how to interpret the data. However, some of this uncertainty can be mitigated using complex modeling techniques.

These modeling techniques suggest slow steady erosion along mountaintops over the past 200 million years of 0.079 in/100 years. Calculated erosion rates appear to have been more variable in valleys where, from 180 to 120 million years ago, rates may have averaged 0.138 in/100 years. Greater erosion in valleys relative to mountaintops suggests that streams were incising their channels thereby generating relief. The data suggest that as much as 3000 feet of relief could have been generated from 180 to 120 million years ago, a time period that includes both the 155 and 130 million-year intervals of rapid offshore sediment deposition. The data seem to support the idea that the Appalachian Mountains have undergone periodic gains in elevation and especially relief since its heyday 265 million years ago.

Given the data that we have to work with, it is fair to say that the Appalachians Mountains today are higher than they were during certain periods of the past, and perhaps higher today (or with greater relief) than they have been for most of the past 100 million years. One interesting aspect of these analyses is that it is likely the southern Blue Ridge and perhaps other parts of the Appalachians have remained above sea level since mountain building events began more than 450 million years ago. Thus, in spite of its mountainous character, the Blue Ridge could form some of the oldest continually exposed landscape in the United States.

Relict Erosion Surfaces in the Southern Blue Ridge

An analysis in support of recent uplift for the southern Appalachians has been conducted in the Blue Ridge Mountains west of Highlands, North Carolina. The analysis involves knickpoint migration on the Cullasaja River. Recall from Chapter 9 that a knickpoint is any location along a stream channel where the gradient is steep

FIGURE 16.26 Google Earth image looking southwestward at relict, low relief landscape (lr), several knickpoints (k), and high relief (hr) below knickpoints. Location is near Highlands, NC (35.04753 N, 83.27156 W). Background topography has been washed from the image.

relative to the gradient above and below (Fig. 9.5). Once a knickpoint is established, it will slowly migrate upstream. In this analysis, geologists identified numerous knickpoints, some of which form waterfalls. They showed that the topographically highest set of knickpoints separate a low-relief relict landscape above, from a more rugged, actively incising landscape below. Both landscapes and three of the many knickpoints described in the area are shown in Fig. 16.26. Notice that landscape above the knickpoints is relatively flat and populated with fields and houses. Terrain below the knickpoints is steep and entirely wooded. Field observation and LiDAR (light detection and radar) elevation data confirm that local relief, hillslope steepness, stream channel steepness, and frequency of landslides are all significantly greater in the rugged area below the knickpoints, and that mean soil thickness is less.

The magnitude of river incision below a knickpoint can be estimated by reconstructing the original low-relief stream profile and comparing it to the present-day incised profile. The results for the Highlands area are shown in Fig. 16.27. Based on this reconstruction, the present-day topographic relief between the drainage divide (the high point) at the top of the river profile and the mouth (the low point) of the Cullasaja River where it empties into the Little Tennessee River near the town of Franklin, is 790 m (2591 feet). Given that the estimated topographic relief along the reconstructed relict profile prior to incision is 300 ± 25 m (984 ± 82 feet), the analysis implies that incision and knickpoint migration has added roughly 490 m (1607 feet) of relief to the landscape.

Based on characteristic rates of incision, it is estimated that knickpoint migration began sometime between 17.6 and 4.6 million years ago. The low end of this estimated time range predates continental glaciation, therefore sea level changes due to glaciation were eliminated as a

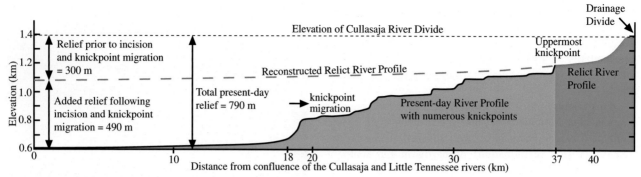

FIGURE 16.27 Longitudial stream profile of the Cullasaja River showing the reconstructed relict profile prior to uplift and erosion, and the present-day profile with numerous knickpoints. *Redrawn from Gallen and others (2013).*

possible cause. A study of the geology indicates no correlation between knickpoint location and bedrock contacts; therefore bedrock control was also eliminated as a possible cause. Climate change was eliminated based on the small size of the Cullasaja River basin. In such a small area, any change in climate likely would affect the entire basin. The conclusion was that incision and knickpoint migration were most likely initiated due to a change in base level in the form of surface uplift. There is no evidence of active faulting; therefore, uplift was more likely regional in nature, affecting the entire southern Appalachians or perhaps the entire Appalachians. Although the estimated timing for knickpoint migration is not well constrained, it is within the time frame of the sediment accumulation pulse that began 15 million years ago, implying a possible correlation.

THE NORTHERN ROCKY MOUNTAINS AND NORTH CASCADES

Hinterland regions form at the rear of foreland fold and thrust belts where thrust faults are able to cut deep enough into the crust to carry crystalline rock to the surface. It is obvious from Fig. 11.1 that the Appalachian hinterland deformation belt extends the length of the mountain system along the eastern margin of the Valley and Ridge fold and thrust belt. The situation is different in the Cordillera because crystalline rock is not as widespread or as continuous as it is in the Appalachians. One possible reason for this is that Cordilleran mountain building events are relatively young and rocks have not yet eroded deep enough to expose crystalline rock. Given the level of erosion, it is likely that crystalline rock has never been exposed as a continuous belt along the length of the Cordillera in the same manner it is today in the Appalachians. Another important factor has been the nearly complete disruption, burial, and tectonic reincarnation of landscape that has resulted from younger volcanism, normal faulting, and deposition. The part of the US Cordilleran hinterland least affected by these later tectonic events is the far northern region of Idaho and Washington. Here, the transition from foreland fold and thrust belt to hinterland deformation belt remains reasonably well intact.

Areas of the Pacific Northwest where hinterland crystalline rock and structures are preserved is shown on a landscape map in Fig. 16.28. One similarity with the Appalachians is that many of the metamorphic rocks have undergone multiple fold and fault generations. One difference is that there is significantly more granitic rock in the Cordilleran hinterland. Granitic rocks, shown in red in Fig. 16.28, form the Idaho and Boulder Batholiths and the Challis intrusive rocks. These rocks appear to have been associated with ancient subduction zones. The lack of layering in these rocks produces a monotonous, random landscape that is shaped in some areas by normal faults. The figure also shows active volcanoes and areas of volcanic and sedimentary rock within the northern Washington hinterland. Volcanic and sedimentary rocks in other areas are not shown. The short, thick, blue lines across the entire figure represent active faults. Most are normal faults, but thrust faults are present in the Seattle area and in the vicinity of the Saddle Mountains in central Washington. The eastern limit of accreted terranes is shown with a thick magenta line. This line separates native North American rock successions from accreted terranes. The contact is buried beneath young basalt across the Columbia Plateau and intruded by granitic rocks of the Idaho Batholith. We will begin our survey in southern Idaho and work our way northward.

Southern Idaho

The Idaho Batholith forms one of the largest areas of massive granitic rock in the United States covering most of the red-shaded area in Fig. 16.28. The rocks vary widely in age from 180 to 45 million years old although most are between 100 and 65 million years old. The

FIGURE 16.28 Landscape map that shows areas of crystalline rock in the Pacific Northwest.

FIGURE 16.29 Google Earth image looking west at a plateau-like surface within the Salmon Mountains that is deeply incised by the Salmon River.

largest continuous part of the Idaho Batholith is its southern part, known as the Atlanta lobe, covering most of the Sawtooth Mountains northward across the Salmon River to include the Salmon River Mountains. Also known as the River of No Return, the Salmon River flows for 130 miles through an area of few roads between the town of Salmon and the Seven Devils Mountains. As seen in Fig. 16.29, the Salmon River Mountains surrounding the river is one of wide, weakly dissected, plateau-like surfaces with incised river valleys. Elevation in the highlands varies from 6000 to more than 8000 feet with relief to the Salmon River that exceeds 5000 feet in some areas such as south of Elk City. The topography is similar to, but less evolved than the erosional mountains of the Appalachian Plateau. One interpretation is that the area was a low-lying, nearly flat, erosion surface that formed as long ago as 50 million years. The present day landscape developed less than 5 million years ago when the area was broadly uplifted into a plateau and incised. The landscape is an example of erosional mountains developed on a plateau of massive crystalline rock.

Part of the Atlanta lobe is shaped by active normal faults associated with the Rocky Mountain Basin and Range. Three active faults are shown within the lobe in Fig.16.28. All three produce lineaments. The western one passes along the west side of Lake Cascade. The central one passes through Deadwood Lake. The most obvious and most active of the three is the easternmost, which separates the northeastern flank of the Sawtooth Mountains from Stanley Valley.

The Sawtooth Mountains are distinctly different from the Salmon River Mountains both in their geology and landscape. These mountains easily form the most rugged area in Idaho. They are higher in elevation and strongly glaciated (eroded) with sharp serrated ridges and peaks from which the mountains receive their name. There are 33 peaks above 10,000 feet and more than 100 sub-peaks and towers above 9000 feet, many with remote, rugged approaches and very difficult ascents. The strong rock and steep cliffs provide a superb playground for mountaineering and backcountry rock climbing. Fig. 16.30 shows the view from the summit of Decker Peak (10,709 feet). The Sawtooth Mountains are highest in the northeast along the steep, precipitous eastern slope that leads down across the active normal fault to Stanley Valley. There is no doubt that displacement along this fault is at least partly responsible for the high elevation. Relief from the Salmon River on the Stanley Valley floor, to the top of the highest peak, Thompson Peak (10,756 feet), is 4400 feet across about 5 miles. This main ridge overlooking Stanley Valley forms the Sawtooth Range, but the mountain system stretches westward into rugged, remote areas of the Sawtooth Wilderness. The rocks are different from those in the Salmon River Mountains. The mountain front that overlooks Stanley Valley consists mostly of young (52 to 39 million-year-old) granitic rock associated with the Challis volcanic field of approximately the same age (Chapter 17). The Challis volcanic rocks occupy the uncolored area in Fig. 16.28 located due east of Stanley Valley. Challis granitic rocks form Mount Cramer, Decker Peak, and Williams Peak. Older granitic rocks that are part of the Idaho Batholith and which are similar to those in the Salmon River Mountains are present at the summits of Horstmann and Barron Peaks as well as across wilderness areas to the west. Thompson Peak and Mount Regan sit in a small area of high-grade Precambrian rocks

FIGURE 16.30 Photograph that shows the Idaho Sawtooth Range from the summit of Decker Peak.

that were metamorphosed or re-metamorphosed during Cordilleran orogeny.

Central Idaho, Montana, and Oregon

A belt of metamorphic rocks separates the Atlanta lobe from the Bitterroot section of the Idaho Batholith. The metamorphic rocks form a complex pattern that likely includes crystalline shield, the Belt Supergroup, and younger rocks metamorphosed during Cordilleran orogeny. These rocks form the northern Beaverhead Mountains east of Salmon where there are numerous 10,000-foot peaks, the Ruby Range farther east, and the Tobacco Root Range where again there are 10,000-foot peaks. The latter two ranges are bound partly along active normal faults. The most prominent mountain area is the central Bitterroot Range located north of the Beaverhead Mountains and south of Missoula, composed mostly of Idaho Batholith granitic rocks (Fig. 16.28). The crest of this range and that of the northern Beaverhead Mountains forms the sinuous border between Idaho and Montana. A normal fault is present along the eastern flank of the Bitterroot Range and is probably responsible for the wide Bitterroot River valley located between the central Bitterroot and Sapphire Ranges.

The fault, although considered active, has shown no activity in at least 750,000 years. Consequently, the Bitterroot Range boasts only one peak, Trapper Peak, above 10,000 feet. Although not as high as other ranges in Montana, there are those who argue that the central Bitterroot Range is the most rugged and scenic, referring to it as the Montana Alps. The ruggedness is likely the result of the resistant, massive nature of the underlying granitic rock. As seen in Fig. 16.31, many of the valleys on the eastern flank of the range are remarkably straight. The reason for this is unclear, but it may have resulted from fractures in the granite that were later excavated by glaciers. This area is also unique in that it is part of the Selway-Bitterroot Wilderness, which lies just north of the Frank Church-River of No Return Wilderness. Together, these two areas cover an area larger than the state of Connecticut. The Clearwater Mountains west of the Bitterroot Range are less rugged with peaks between 6000 and 9000 feet. This area, composed mostly of Idaho Batholith and some metamorphic rock, appears to be a northward extension of the dissected plateau surface found in the Salmon River Mountains except that the level of dissection is greater and there are isolated areas of glacier-carved peaks.

FIGURE 16.31 Google Earth image looking east across the Bitterroot Range at straight glacial valleys. The Idaho-Montana border snakes along the summit area. An active normal fault separates the range from Bitterroot River Valley at Darby. Trapper Peak is out of view south (right) of Darby.

Granitic rocks farther east in the Anaconda-Butte-Helena area of Montana are part of an 81- to 70- million-year-old volcanic-intrusive association known as the Boulder Batholith and the Elkhorn Mountains volcanic suite. The presence of granitic and volcanic rocks of approximately the same age, both here and in the Challis belt, suggest a level of erosion great enough to expose underlying intrusive rock, but not enough to completely remove volcanic rock.

Stretching westward into the belt of accreted terranes in western Idaho and central Oregon are several mountainous, glaciated areas that contain crystalline rock, including the Seven Devils, Wallowa, Elkhorn, and Aldrich-Strawberry Mountains (Fig. 16.28). These mountains exist as islands surrounded by Columbia River Flood Basalt. Peaks in the Strawberry, Elkhorn, Wallowa, and Seven Devils Mountains rise above 9000 feet. The Wallowa Mountains boast at least 31 peaks above 9000 feet and are known as the Oregon Alps. Included are circa 160 to 90 million-year-old granitic rocks in the Wallowa Mountains and Elkhorn Ridge, as well as variably metamorphosed sedimentary and volcanic rocks, some only weakly metamorphosed. The uplift history of this area is poorly constrained. Faulted, folded, and tilted basalt flows along the margins of some of the mountains suggest that uplift has occurred within the past 12 to 10 million years. Faults active during the past 1.6 million years are especially common surrounding the higher Wallowa Mountains and Elkhorn Ridge as well as the Blue Mountains. Most are normal faults, but others show strike-slip displacement. These faults imply active uplift.

Northern Washington

In contrast to other parts of the US Cordillera, a belt of mountains extends uninterrupted from the foreland region at Glacier National Park across the hinterland region to the Puget Sound of northern Washington. Fig. 16.32 is a Google Earth image across northern Washington that shows five primary mountain areas. The North Cascade Mountains occupy the western third of the area west of Winthrop (W). North of Winthrop, the North Cascades are more-or-less continuous eastward with the Okanogan Range across the Chewuch River (ch). Farther east, the Okanogan Highlands are located between the Okanogan (o) and Sanpoil Rivers (s), the Kettle River Range is located between the Sanpoil and Columbia Rivers (cr),

B-Bellingham, E-Everett, S-Seattle, Sp-Spokane, T-Tacoma, W-Winthrop, b-Mt. Baker, c-Chewuch River, lc-Lake Chelan, cr-Columbia River, g-Glacier Peak, k-Kootenay River, o-Okanogan River, r-Ross Lake, s-Sanpoli River, st-Mt. Stuart

FIGURE 16.32 Google Earth image of northern Washington and Idaho oriented north. The state boundary is shown with a thin white line. The Canadian border is shown with a thin yellow line.

and the greater Selkirk Mountains between the Columbia River and the north-flowing Kootenay River (k). As seen in Figs. 16.28 and 16.32, there are strong topographic differences east and west of the Okanogan River (o). The area east of the Okanogan River is one of broad, rounded, relatively low-lying mountains. There is one peak (Mount Bonaparte) above 7000 feet in the US Okanogan Highlands, five peaks above 7000 feet in the Kettle River Range, and many peaks above 7000 feet but none above 8000 feet in the US Selkirk Mountains. The area west of the Okanogan River (o) is topographically higher and more rugged with greater relief. There are at least ten peaks above 9000 feet in the North Cascade Mountains and several peaks above 8000 feet in the Okanogan Range. We will look first at the North Cascade-Okanogan Mountains, and then discuss the Okanogan Highlands, Kettle River Range, and Selkirk Range. We end this section with a short discussion on the San Juan Islands.

North Cascade Mountains and Okanogan Range

The North Cascade Mountains stand out relative to the other nearby ranges with respect to both elevation and relief. It is a vertical landscape that extends northward into Canada from Snoqualmie Pass on interstate 90, and is arguably the most rugged region in the contiguous United States. Most impressive are the number of glaciated cliffs and bare rock faces above tree line that create 5000—7000 feet of steep relief between mountain top and valley bottom. Few roads traverse this great land, which is a Mecca for alpine mountaineers. Fig. 16.33 is a photograph from above Washington Pass that captures some of the rugged beauty.

The area consists of accreted terranes amalgamated primarily between 160 and 65 million years ago and intruded with granitic rock. Metamorphic and intrusive rocks dominate, but sedimentary rocks are present across a large area known as the Methow Basin. Strong deformation and metamorphism peaked in the Cretaceous 90 million years ago and the area likely has been a highland since at least that time. Following metamorphism, between 55 and 50 million years ago and possibly earlier, the area underwent strike-slip and normal faulting along major high-angle faults that include the Straight Creek and Ross Lake faults. Right lateral strike-slip displacement on the Straight Creek fault is estimated to be more than 100 miles based on offset marker beds. Granitic rocks vary in age from more than 90 million years old to about 2.5 million years old. Most of the intrusives are subduction-related. Older granitic rocks are deformed and metamorphosed. Younger intrusions are weakly deformed and unmetamorphosed. These intrusions invade and obliterate sections of older fault zones including parts of the Straight Creek and Ross Lake faults. There are two active volcanoes, Mount Baker (10,786 feet) and Glacier Peak

FIGURE 16.33 Photograph of the North Cascades looking west from above Washington Pass and the North Cascades Highway.

(10,525 feet), both of which are part of the Cascadia volcanic arc system (Chapter 19). Excluding these, there are at least 10 peaks above 9000 feet including Mount Stuart (9420 feet), Bonanza Peak (9516 feet), Mount Shuksan (9135 feet), Mount Fernow (9254 feet), and Goode Mountain (9204 feet), most of which require technical skill to climb.

Fig. 16.34 shows the geology on a landscape map oriented north. The heavy lines are faults, some of which are labeled. Thin yellow lines are the only through-going roads. The northern road is route 20, the North Cascades Highway, which passes through North Cascades National Park. Fig. 16.35 is a Google Earth image of the same area looking southward from the Canadian border to the latitude of Glacier Peak. Faults (heavy lines) in this figure are dashed where obliterated by intrusions. Major faults in both figures serve to separate the North Cascades into domains. West of the Straight Creek fault (SF), thrust slices composed of sedimentary rock and moderately metamorphosed sedimentary/volcanic rock occupy the Northwest Cascades thrust system. A subpart of this domain, south of the Devils Mountain fault (DF), consists of a subduction mélange zone (an ancient accretionary prism) and granitic intrusions. The northern part hosts the active Mount Baker volcano.

East of the Straight Creek fault (SF), the Cascade Crystalline Core forms the heart of the North Cascades. The rocks are dominantly gneiss and granite, and they are uniformly hard enough to produce a rather random landscape of jagged mountain peaks (Fig. 3.9). Here we find most of the 9000-foot peaks. The hard crystalline rocks give way to dominantly sedimentary rocks in the Methow Basin east of the Ross Lake fault (RF). The Methow Basin is mountainous. It is (was) a depositional basin in the geological sense. The area is slightly lower in elevation and relief than the crystalline core, but is still incredibly rugged due in part to the additional presence of metamorphic, intrusive, and volcanic rocks. High peaks include Jack Mountain (9070 feet), Robinson Mountain (8731 feet) and Mount Lago (8750 feet), each of which is partly composed of rocks other than sedimentary. Relief is extreme at Jack Mountain where the west flank forms a wall of rock that rises some 7464 feet from the Ross Lake shoreline to the summit area across a distance of only three miles. Sedimentary rocks of the Methow Basin extend eastward to the Pasayten fault (PF). Farther east

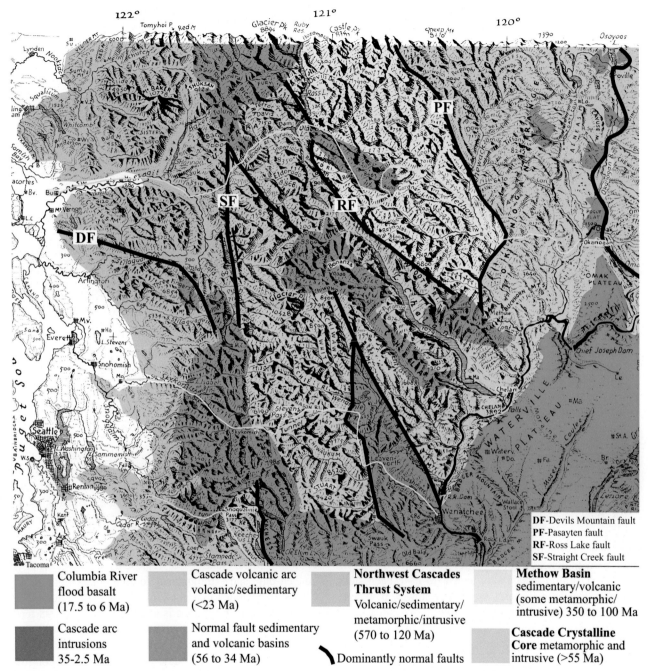

FIGURE 16.34 Detailed landscape map of the North Cascades with geology. The thin yellow lines are the only three roads that traverse the region. From north to south, they are Route 20 (North Cascades Highway), Route 2 over Stevens Pass, and Interstate 90 through Snoqualmie Pass.

we find granitic rocks that are more than 35 million years old covering most of the Okanogan Range.

The reason for the present-day North Cascade landscape is likely due to a combination of recent and rapid uplift, and climate. There are several lines of evidence to suggest recent uplift. The presence of batholithic rocks as young as 2.5 million years old is an indication of recent uplift, as these rocks must have recently been exhumed to the surface. Another line of evidence is the absence of young volcanic rock material beyond the general confines of Mt. Baker and Glacier Peak, the two active volcanoes. The absence of volcanic rock implies relatively rapid erosion and, therefore, the existence of both elevation and relief. The absence of Columbia River basalt along the crest of the North Cascades constitutes yet another line of evidence. Large volumes of Columbia River basalt were

Okanogan Range   Methow Basin   Cascade   Northwest Cascades
                                   Crystalline Core   thrust system

b-Bonanza Peak, **C**-Concrete, **c**-Chewuch River, **E**-Everett, **f**-Mount Fernow, **g**-Goode Mountain, **gp**-Glacier Peak, **j**-Jack Mountain, **lc**-Lake Chelan, **mb**-Mount Baker, **l**-Mount Lego, **r**-Robinson Mountains, **rl**-Ross Lake, **S**-Stehekin, **s**-Mount Shuksan, **W**-Winthrop
DF-Devils Mountain fault, **PF**-Pasayten fault, **RF**-Ross Lake fault, **SF**-Straight Creek fault

FIGURE 16.35 Google Earth image of the northern part of the North Cascade Mountains looking south from the Canadian border. Heavy lines are faults (dashed where obliterated by intrusions).

erupted between 16.8 and 15.6 million years ago. The absence of these rocks along the crest of the North Cascades suggests that topography was high enough to block or deflect lava as it flowed westward.

A final line of evidence are apatite (U-Th)/He and fission-track ages that suggest rapid exhumation rates of 2 to 4 inches per 100 years between 12 and 4 million years ago. Exhumation rates prior to this time, perhaps as far back as 55 million years ago, appear to have been slow, on the order of 0.4 inches/100 years or slower. If we correlate rapid exhumation with rapid uplift we can suggest that the region was at lower elevation with less relief prior to 12 million years ago when it underwent its most recent uplift phase.

The climate in the North Cascades can be summarized as wet and cold. The west side receives 80 to as much as 250 inches of precipitation per year, much of it in the form of snow. In 1980, there were more than 700 glaciers in the North Cascades, more than anywhere in the lower 48 states. These glaciers have excavated valleys and carved around mountain peaks contributing immensely to the scenic beauty and relief of the range. Although uplift may have slowed over the past 4 million years, one possible reason for steep relief is the combination of high elevation, high precipitation, ongoing uplift, widespread alpine glaciation, strong rock, and periodic sea level (base level) drop during continental glacial advances. Rather than producing typical U-shaped glacial valleys, this combination has produced deep glacial V-shaped troughs lined with strong rock walls that put an exclamation point on some of the most rugged and vertical landscape in the United States.

A unique feature of the North Cascades is the presence of long, linear lakes that represent old river valleys gouged-out and straightened during glaciation. Lake Chelan is 55 miles long, one mile wide, and almost 1500 feet deep. The lake surface is at an elevation of 1098 feet, which implies that the bottom of the lake is 400 feet below sea level. This was a natural lake, damned by glacial moraine deposits. Today, a man-made dam controls lake level. The town of Stehekin sits at the northern end of this lake and can be reached by foot, small plane, or boat, but not by road. Ross Lake is a dammed reservoir along route 20 near the border with Canada. It is 23 miles long, 1.5 miles wide, and 540 feet deep. The northern end of the lake is in Canada. Both lakes are visible in Figs. 16.32, 16.34 and 16.35.

The Okanogan Range is a north-south-oriented highland located between the Chewuch and Okanogan Rivers. Most of the rock is granitic, with areas of schist and gneiss. The area is lower in elevation than the North Cascades. High peaks include Windy Peak (8338 feet) and Tiffany Mountain (8250 feet). Lower elevations along mountain flanks are forested, especially along the eastern flank. Mountaintops are free of forest, but without significant bare rock exposure. The northwest-facing slopes on the highest peaks are scalloped and very steep as a result of glacial gouging, but there are few if any active glaciers. The difference in precipitation between this area and the North Cascades is stark. The Okanogan Range is located in the rain shadow of the Cascade Mountains, and receives only 11 to 35 inches of precipitation per year, primarily in the winter. The overall absence of glaciers, the lower elevation, and the considerably drier climate has resulted in terrain significantly less rugged than the North Cascades. Volcanic rocks in the age range of 30 to 50 million years occupy a few of the 6000-foot peaks, including Chickadee Ridge. The preservation of these surface rocks at high elevation suggests that the Okanogan Range has not undergone the degree of deep erosion that is evident in the North Cascades. Apatite (U-Th)/He cooling ages suggest a slow exhumation rate of less than 0.4 inches/100 years throughout most of the Eocene (56–33.9 Ma). Much of the present-day relief has been attributed to glaciation and subsequent stream erosion without significant uplift.

Okanogan Highlands, Kettle River Range, and Selkirk Range

The Okanogan Highlands, Kettle River Range, and Selkirk Range form a series of north-south oriented mountains that are lower in elevation and far less rugged than the North Cascades. Overall landscape is subdued with long slopes partly covered in forest and few rocky cliffs. The climate is dry. Only the Selkirk Mountains and the northern part of the Kettle River Range show obvious signs of past alpine glaciation. Granitic rocks, some correlative with the Idaho Batholith, are widespread across all three mountain ranges as are areas of Cambrian-Ordovician gneiss and schist. Also present, principally along the Sanpoli River (s), are areas of circa 56- to 39-million-year-old volcanic rocks. As shown in Fig. 16.28, the Okanogan Highlands, the Kettle River Range, and the western Selkirk Mountains lie in the area of accreted terranes. The central and eastern Selkirk Mountains host North American rock successions where one can find Belt Supergroup rocks metamorphosed during Cordilleran orogeny in addition to intrusive rock.

On the basis of their lower elevation and relief, we can suggest that none of three mountain regions have undergone the same degree or recency of uplift as the North Cascades, at least over the past 12 million years. The structural form in each range is that of a north-south-oriented elongate, crystalline-cored dome surrounded by inactive normal faults. It is a special type of

FIGURE 16.36 Schematic cross-section of a dome structure developed via normal faulting in a metamorphic core complex.

structural form known as a metamorphic core complex. There are three core complexes, one in each mountain range, the Okanogan, Kettle River, and Priest River (Selkirk) core complexes. A simplified sketch of a metamorphic core complex is shown in Fig. 16.36. Unlike the active normal faults that border some of the mountains in the Northern Rockies to the east, these normal faults were active long ago, primarily between about 56 and 39 million years ago during volcanism along the Sanpoli River. Because normal faulting and doming represent the final tectonic deformation, it is these structures that exert landscape control. The origin and discussion of a metamorphic core complex is best left for the chapter on normal fault systems (Chapter 18), but in general, as normal faults stretch and remove near surface rock layers, deeply buried rocks undergo various degrees of metamorphism and intrusion and rise isostatically to form an elongate central dome. The mountain shape mimics the shape of the central dome.

Different styles of deformation and climate between the North Cascades and the metamorphic core complexes to the east have resulted in different topography. The North Cascades experienced differential uplift along high-angle faults followed by broad uplift within the past 12 million years coupled with extensive glaciation in a wet climate to produce a spectacular high elevation, high relief landscape. Exposure of crystalline rock occurred principally through uplift and erosion. It is an example largely of erosional exhumation as discussed in Chapter 9.

Crystalline rock in the Okanogan Highlands, Kettle River Range, and Selkirk Mountains underwent normal faulting and broad isostatic uplift between 56 and 39 million years ago to form elongate dome structures where the mountain shape mimics the structural form. Since that time, the mountains have undergone erosion in a relatively dry climate without extensive glaciation and without significant uplift. Initial exposure of the central crystalline dome, in this case, is largely the result of tectonic exhumation along normal faults rather than through erosional exhumation.

Fig. 16.37 is a schematic sketch that illustrates the contrasting geology. The suggestion here is that the presence of vertical faults in the North Cascades may have allowed crustal blocks to move up or down independent of each other potentially creating isolated areas of high peaks. An intact dome structure, on the other hand, would distribute broad uplift across the entire area, thus creating low-lying, subdued mountains such as what is seen east of the Okanogan River.

San Juan Islands

The San Juan Islands are located in the Puget Sound just off the southeastern coastline of Vancouver Island less than 70 miles north of Seattle. There are at least 172 islands in the group many of which are very small. Fig. 16.38 shows the main area of large islands labeled with white lettering. Only San Juan (s), Orcas (o), Lopez (l), and Shaw Islands (w) are accessible by the Washington State Ferry out of Anacortes (A). The islands are surprisingly sunny for most of the year with only

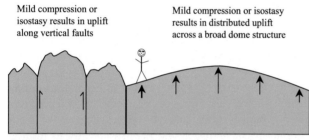

FIGURE 16.37 Cross-section sketch that shows how the style of uplift might vary dependent on the presence or absence of preexisting vertical zones of weakness.

about half the total precipitation that falls on Seattle. Landscape is one of flat, fertile farmland composed of glacial drift with forested hilly areas. Elevation is mostly less than 500 feet with a few peaks above 1000 feet and only one peak above 2000 feet, Mount Constitution (C, 2411 feet), on Orcas Island. Some of the best rock exposure is along the coastline, but additional rock exposure is found inland.

The rocks consist of oceanic accreted terranes thrust onto the edge of the North American continent primarily between 95 and 85 million years ago as part of the mid-Cretaceous San Juan Islands—northwest Cascades thrust system. Major thrust faults are shown in Fig. 16.38 with yellow lines. Rock units are labeled with black lettering. The rock units form an imbricate thrust stack that appears to have been thrust from southeast to northwest such that rock units in the northwest are overlain along thrust faults by rock units toward the southeast. The Fidalgo ophiolite (*F*) forms the structural top of the thrust stack. These rocks form most of Lopez (l), Blakely (b), Decatur (d), Cypress (c), Guemes (g), and Fidalgo (f) Islands. The Fidalgo ophiolite (*F*) represents a section of oceanic lithosphere composed of serpentinite, peridotite, gabbro, diorite, and sedimentary breccia. The thrust stack that underlies the Fidalgo ophiolite is best observed on Orcas Island (o) where several rock units are exposed. From south to north (i.e., from the top downward) the thrust stack on Orcas Island consists of Constitution siltstone-sandstone ocean trench and ocean floor deposits (*C*), the

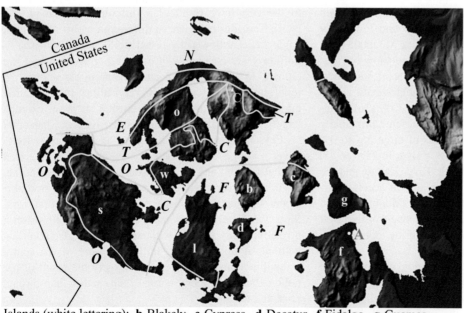

FIGURE 16.38 Shaded relief map of the San Juan Islands oriented north. Yellow lines represent thrust faults.

Islands (white lettering): **b**-Blakely, **c**-Cypress, **d**-Decatur, **f**-Fidalgo, **g**-Guemes
 l-Lopez, **o**-Orcas, **s**-San Juan, **w**-Shaw. **A**-Anacortes, **C**-Mount Constitution
Rock Uints (black italics): *C*-Constitution Fm., *E*-East Sound Gp., *F*-Fidalgo ophiolite
 N-Nanaimo Group, *O*-Orcas Chert, *T*-Turtleback Igneous Complex.

Orcas Chert and pillow basalt (O), the Turtleback Igneous Complex (T), and the East Sound Group island volcanic arc sequence (E, andesite-dacite pyroclastic rocks, with limestone and shale). Mount Constitution (C), the highest point in the San Juan Islands, is part of the Constitution Formation. Nanaimo Group rocks (N), at the north end of the island, are basin rocks composed of marine and nonmarine sandstone, conglomerate, and shale deposited during and following thrusting. San Juan Island (s) consists of Constitution Formation (C) inland and along the eastern coastline, and Orcas Chert (O) along the western and northern coastlines. The rocks have been weakly metamorphosed under conditions of high pressure.

THE GRENVILLE FRONT

The Grenville orogeny was a bit like the Appalachian/Ouachita orogeny in that it affected both the east coast of North America and the Gulf coast. Both events began with continent-arc collision and ended with continent-continent collision and the formation of a supercontinent. The Appalachian/Ouachita orogeny had begun by 480 million years ago, and culminated about 265 million years ago with the formation of the supercontinent Pangea. The Grenville orogeny began 1.3 billion years ago and culminated about 1.0 billion years ago with formation of the supercontinent Rodinia or proto-Pangea. High-grade metamorphic shield rocks associated with the Grenville orogeny are found in the Adirondack Mountains, the Llano Uplift, and in the Appalachian Mountains along the Green Mountain-Blue Ridge Mountain trend. One additional area where Grenville rocks are exposed is in the west Texas area of the Carrizo, Baylor, and Sierra Diablo Mountains near Van Horn, and in the Franklin Mountains near El Paso. The Van Horn area is especially interesting because it straddles the Grenville Front. The tectonic front of a compressional orogenic belt is defined as the location within the continental interior where deformation associated with orogeny becomes minor or insignificant. The Appalachian-Ouachita and Grenville fronts are shown in Fig. 16.39 along with the location of the Van Horn area (V) and Franklin Mountains (F). The Appalachian-Ouachita Front corresponds approximately with the boundary between the Valley and Ridge-Ouachita Mountains (the foreland fold and thrust belt) and the Appalachian Plateau-Ozark Plateau (the foreland basin). The Appalachian-Ouachita Front must be inferred across the Mississippi Embayment based on drill core and geophysical evidence. West of the Ouachita Mountains, the front must again be inferred across southern Texas to the northern margin of the Marathon Basin (M) and the area north of the Solitario (S).

The Grenville Front was defined in Canada where it marks the western limit of metamorphism and deformation associated with the Grenville orogeny. Unlike the Appalachian Front where foreland sedimentary rocks are exposed, the Grenville Front in Canada is so deeply eroded that all foreland sedimentary rocks have been removed. Only crystalline rocks are exposed. Rocks east of the front record the metamorphic effects of Grenville orogeny. Crystalline rocks west of the front record primarily older orogenic events in which the effects of Grenville orogeny are absent or weak.

Recall from Chapter 15 that Appalachian rocks are exposed in the Marathon Basin a short distance south of Van Horn, making west Texas the only location where aspects of Grenville, Appalachian, and Cordilleran orogeny are in close proximity.

In the United States, the Grenville Front is buried beneath sedimentary rocks of the Interior Platform rock succession, but can be located based on drill hole, magnetic, gravity, and chemical data. As shown in Fig. 16.39, the Grenville Front is traced southward from the Great Lakes area to the Mississippi Embayment. The Front cannot be traced through the deep sedimentary trough of the Mississippi Embayment where it is inferred to take a sharp turn, or be offset along a strike-slip fault, in order to connect with the Llano Front, which is the western continuation of the Grenville Front. A strike-slip fault is shown in Fig. 16.39.

The Van Horn area is exceptional in that it actually straddles the Llano (Grenville) Front. It is the only location where foreland sedimentary rocks related to the Grenville orogeny are exposed at the Earth's surface. Here, it is possible to walk from the deformed crystalline core of the Grenville orogenic belt, across a narrow fold and thrust belt, to the undeformed sedimentary foreland. The traverse is similar in style to walking from the Blue Ridge-Piedmont crystalline core of the Appalachian Mountains across the Valley and Ridge fold and thrust belt to the undeformed foreland on the Appalachian Plateau. The difference is that the Grenville landscape has largely been replaced by Basin and Range topography. From the Van Horn area, the Grenville Front extends westward to the Rio Grande but is not exposed. Both the Grenville and Appalachian-Ouachita Fronts become impossible to follow across the border into Mexico due to Cordilleran deformation that includes normal faults. Notice in Fig. 16.39 that the Adirondack Mountains, the Appalachian belt, and the Llano Uplift are all located within the Grenville metamorphic core well to the east of the Grenville Front.

The Franklin Mountains form a north-south range located north of the Grenville Front. The rocks of the

384 PART | II Structural Provinces

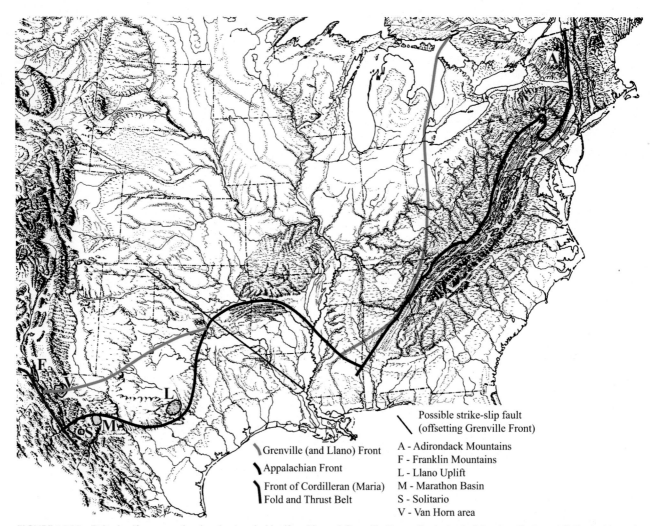

FIGURE 16.39 Raisz landform map showing the Appalachian/Ouachita and Grenville Fronts. Rocks in the Marathon Basin and Solitario (shown in yellow) are associated with Appalachian/Ouachita orogeny. Rocks in the Franklin Mountains, Llano Uplift, and Van Horn area (also shown in yellow) are associated with the Grenville orogeny.

Franklin Mountains are unaffected by compressional Grenville-age deformation-metamorphism, however, they are intruded by the Red Bluff Granite dated at 1.12 billion years old. The Red Bluff Granite forms a large area along the northeastern side of the mountain where it intrudes a Precambrian sedimentary-volcanic rock succession composed of volcanic breccia, mudstone, quartzite, rhyolite, and limestone. Limestone close to the granite is metamorphosed to marble as a result of heat and hot fluids that emanated directly from the cooling magma. This is a form of contact metamorphism common anywhere granitic magma intrudes cold reactive rocks at shallow depth. The entire rock sequence is exposed along route 375, which crosses the mountain. This is the final location where Grenville-age rocks are found. Older crystalline shield rocks in New Mexico and in other parts of the Cordillera are entirely unaffected by Grenville orogeny. Let us now turn our attention to the Van Horn area.

Van Horn Area

Fig. 16.40 is a Google Earth image that shows the geology of the Van Horn area looking north-northwest. The rocks crop out on the up-thrown side of the East Baylor Mountains-Carrizo Mountain fault (EBC), a large normal fault active in the past 750,000 years. Visible in the distance are the Franklin Mountains, the Guadalupe Mountains, the Guadalupe Dune Field, and the White Sands Dune Field.

Rock units in Fig. 16.40 are given numbers with 1 the oldest and 5 the youngest. Precambrian rocks involved in the Grenville orogeny are numbered 1, 2, and 3. Units 4 and 5 are younger, unconformably overlying interior platform rocks that bury and hide Grenville-age rocks. There are two major boundary zones: the Streeruwitz thrust (ST) and the Llano Front (LF). The Streeruwitz thrust is a post-metamorphic fault that acts as the foreland-hinterland boundary perhaps

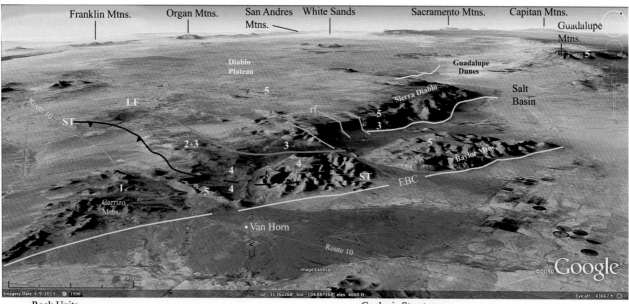

FIGURE 16.40 Google Earth image looking northwest at the Van Horn area.

similar in style to the Great Smoky thrust in the southern Appalachians. It separates metamorphic rocks of the Carrizo Mountain group (unit 1) from a fold and thrust belt composed of low-grade and unmetamorphosed rocks of the Allamoore, Tumbledown, and Hazel Formations (units 2 and 3). The Llano Front (LF) marks the location where the fold and thrust belt dies out into weakly deformed and undeformed rocks of the Hazel Formation (unit 3). The rocks are completely undeformed in the Sierra Diablo only a few miles north of the Llano Front.

The Streeruwitz thrust is poorly exposed, but scattered outcrop can be seen on both sides of the fault along side roads that lead north from route 10. Light gray areas in Fig. 16.40 are mines and quarries. The thrust is shown as a dashed line at its eastern end because the fault trace and all associated rocks (units 1, 2, and 3) disappear beneath younger rock of units 4 and 5.

The Carrizo Mountain Group (unit 1) consists of sandy schist (metasandstone), amphibolite, metavolcanic rocks, quartzite, marble, phyllite, and minor intrusions. The rocks are metamorphosed to high-grade within the Carrizo Mountains, but metamorphism decreases to low grade toward the Streeruwitz thrust (ST), a situation similar to the Great Smoky Mountains and the Great Smoky thrust. The high-grade rocks represent part of the Grenville crystalline core. Radiometric dating and field work suggests the rocks were deposited between 1.39 and 1.327 billion years ago in a shallow water continental rift setting, and deformed and metamorphosed soon thereafter, possibly prior to 1.325 billion years ago. Such a scenario implies very early Grenville-age metamorphism. The rocks are exposed along route 10 and along side roads that lead south. If you take the time to look at these rocks, you may find areas where rocks are intensely sheared, strongly foliated, and lineated. These rocks are mylonites. They represent shear zones (ductile fault zones) that probably developed during the late stages of metamorphism. Following the strongest phases of metamorphism, the entire Carizzo Mountain Group was thrust northwestward along the Streeruwitz thrust onto low-grade and unmetamorphosed rocks of the Allamoore, Tumbledown, and Hazel Formations (units 2 and 3).

The Grenville fold and thrust belt in Fig. 16.40 is located between the Streeruwitz thrust (ST) and the Llano Front (LF). Although the Hazel Formation (unit 3) is present in the fold and thrust belt, most of the rock in the belt consists of Allamoore and Tumbledown Formations (unit 2). The Allamoore Formation consists of cherty limestone, tuff, basalt, and phyllite. Some of the volcanic rocks are dated at 1.256 to 1.253 billion years old. The cherty limestone contains mound-like fossil layers known as stromatolites. These structures exist today in warm, shallow marine environments such as Shark Bay,

Australia. Given this similarity, we can suggest that the Allamoore was deposited in a warm, shallow marine environment apparently in close proximity to volcanic sources. Limestone in the Franklin Mountains also contains stromatolites and is interpreted as correlative with the Allamoore.

The Tumbledown Formation consists of sandstone, basalt, and tuff dated at 1.243 billion years old. These rocks are extraordinary in that you can find megablocks (large clasts) of carbonate rock imbedded within layers of volcanic rock. Some of the megablocks are the size of a house and their composition indicates they were derived from the Allamoore Formation. They are interpreted as submarine slumps or gravity slides known as olistostromes. The Tumbledown Formation is interpreted to have been deposited below sea level during a time when strike-slip and normal faults were active. The megablocks are assumed to have slid off the up thrown side of submarine fault blocks.

Three northwest-trending inactive faults are shown in orange and labeled rf in Fig. 16.40. The faults cut the southern Sierra Diablo ridge. These faults are representative of several northwest-trending high-angle faults in the area. One suggestion is that these are ancient faults active during deposition of the Tumbledown Formation. The faults also must have been active following deposition of the Tumbledown Formation because they cut upward through younger Permian rocks (unit 5). Unlike the currently active range-front normal faults (shown in yellow), these faults show no evidence of recent activity.

The Hazel Formation (unit 3) consists of sandstone, siltstone, and conglomerate that sit with angular unconformity above the Tumbledown Formation. The unconformity is marked with a thick conglomerate at the base of the Hazel Formation containing boulders derived directly from the Allamoore and Tumbledown formations. Although deformed and overthrust within the fold and thrust belt, the Hazel Formation shows no Grenville-age deformation in the Sierra Diablo north of the Llano front. The Hazel Formation is interpreted as a syndeformational rock unit that was deposited during and following development of the fold and thrust belt similar to foreland basin rocks found in the Appalachian foreland (Chapter 15).

Dating of granite and rhyolite boulders within the conglamerate suggests that the Hazel Formation is younger than 1.123 billion years. This age, coupled with the dating of low-grade mylonitic rock in the fold and thrust belt at 1.035 billion years, suggests that the Hazel Formation was deposited between 1.123 and 1.03 billion years ago when the Streeruwitz thrust and the fold and thrust belt were both active. The Hazel Formation may have continued to be deposited following thrust faulting. The rock unit is well exposed at the base of Sierra Diablo where it is directly unconformably overlain with Permian cliff-forming limestone (5). Permian limestone (unit 5) is the most widespread of the rock units (4 and 5) that cover and hide Grenville-age rock units. It forms the cliffs of Sierra Diablo, the Diablo Plateau, the Guadalupe Mountains, and part of the Baylor Mountains. Unconsolidated alluvium (basin-fill stream sediment) covers most of the valley region surrounding Van Horn.

In summary, Grenville-age metamorphism affected the Carrizo Mountain Group (unit 1) circa 1.325 billion years ago. Following metamorphism, the Allamoore and Tumbledown formations (unit 2) were deposited in a foreland basin setting affected by strike-slip and normal faults circa 1.256 to 1.243 billion years ago. The rocks were deformed between 1.123 and 1.03 billion years ago during development of the Streeruwitz thrust and the fold and thrust belt. The Hazel Formation was deposited during and following thrust faulting.

QUESTIONS

1. Name and briefly describe the different types of crystalline rocks and their location in the United States.
2. In Google Earth, fly to Woolwine, Virginia. What is the relief across the Blue Ridge escarpment at this location? How far south can you follow the escarpment? What is the relief across the escarpment at Lowgap, NC? How does the escarpment change between Lowgap and Millers Creek, NC? How does the escarpment change between Millers Creek and Dahlonega, GA? What happens to the Blue Ridge Mountains and the escarpment between Woolwine and Syria, Virginia?
3. What is the relief between the Blue Ridge and the Piedmont at the northern end of the Blue Ridge near Dillsburg, Pennsylvania?
4. What is the relief between Mt. Katahdin and Upper Togue Pond directly south of the mountain?
5. What are some of the differences, in terms of rock type and relief, between the Fall Line and the Blue Ridge escarpment?
6. Explain differences in the rocks and topography between the Green Mountains of Vermont and the White Mountains of New Hampshire.
7. What is the origin of the many islands along the coast of Maine?
8. Referring to Fig. 16.28, what rock type or feature lies directly south of the Northern Rocky Mountain-Cascade crystalline belt in Washington?
9. What is the origin of Lake Chelan and Ross Lake?
10. What is the Fall Line? Where is it located? What is its significance with respect to American history?

11. What is a monadnock? Where is the Mount Monadnock namesake? Could Katahdin be considered a monadnock, why or why not?
12. Speculate on the origin of the line of lakes in Fig. 16.23 that stretches south and east of the White Mountains across the New Hampshire-Maine border.
13. How is the border between the Blue Ridge and Valley and Ridge different from the border between the Blue Ridge and Piedmont Plateau?
14. If we assume zero uplift at the confluence of the Cullasaja and Little Tennessee Rivers, and we assume that uplift began 15 million years ago, and that added relief (1607 feet or 490 m) is approximately equivalent to total uplift, what is the rate of uplift over the 15 million year period in mm per year and inches per 100 years? What is the rate of incision over the 15 million year period in inches per 100 years? Given the same assumptions, what is the rate of knickpoint migration between the 18 and 37 km markers in mm per year and in inches per 100 years?
15. If you consider that there are a large number of tributary valleys that feed the Cullasaja River on its way to the Little Tennessee River, why might you expect the rate of erosion and knickpoint migration to progressively decrease with distance upstream?
16. Speculate on the origin of water gaps in the Green Mountains
17. Why does the Blue Ridge almost disappear at Roanoke?
18. In the Appalachians, how does one distinguish between gneiss that belongs to the North American crystalline shield and gneiss found in accreted terranes?
19. What is the evidence for recent uplift in the North Cascades?
20. What is the Methow Basin? Describe its topography.
21. What type of structure forms the landscape of the Kettle River Range?
22. List differences between mid-continent, crystalline-cored domes discussed in Chapter 14, Crystalline-Cored Mid-Continent Anticlines and Domes and domes in the Okanogan Highlands, Kettle River Range, and Selkirk Mountains. Include differences in the types and ages of rocks involved, the structural form, and mode of uplift.
23. What is the Grenville Front? What does it look like in the Van Horn, TX area and in Canada?

Chapter 17

Young Volcanic Rocks of the Cordillera

Volcanic rocks have unique attributes as well as attributes of both sedimentary and crystalline rock. A unique attribute is that they form a distinctive set of landforms that includes major volcanoes such as those in the Cascade Mountains, lava plateaus and plains such as on the Columbia Plateau and Snake River Plain, and smaller volcanic features that include cones, lava fields, domes, calderas, and buttes. Another unique attribute is that they can potentially reincarnate landscape in one person's lifetime. Volcanic rocks, particularly basaltic lava flows, are similar to sedimentary rocks in that they are layered and originally nearly flat-lying. However, they do not produce well-developed bench and slope landscape because each layer is of similar composition, and because the layers are less continuous, irregular in outline, and less planar in comparison with sedimentary rocks. Volcanic rocks are similar to crystalline rocks in that they are easily dated using radiometric methods.

In this chapter, we are concerned with volcanic rocks in the Cordillera that are less than 70 million years old. The general distribution of these rocks is shown in Fig. 17.1. The chapter begins with an introduction to the rock types and volcanic landforms found in the Cordillera. It is followed with a discussion of two of the largest volcanic provinces, the Columbia Plateau and Snake River Plain. We end the chapter with short descriptions of volcanic fields across the United States. The discussion is separated into rocks that are 70 to 20 million years old, and those that are less than 20 million years. Excluded are young volcanic rocks of the Central-Southern Cascade Mountains, which are more properly considered with respect to the larger Cascadia volcanic arc system (Chapter 19). Widespread volcanism discussed in this chapter, and normal faulting discussed in Chapter 18, constitute the two primary forms of post-orogenic tectonic reincarnation in the Cordillera (Table 11.1). These two forms of landscape reincarnation are contemporaneous and often associated with each other.

MAGMA TYPES AND COMMON VOLCANIC LANDFORMS

Previously we distinguished between volcanic rocks of basaltic and silicic composition. In this section, we distinguish between basalt, andesite, rhyolite, and tuff, and also introduce volcanic landform types characteristic of the Cordillera.

Basaltic magma is a dark, heavy liquid derived from partial melting in the Earth's mantle. It tends to be nonexplosive and flows easily. The Hawaiian basalt flows are an example. An area of many flows and volcanic landforms constitutes a volcanic field. Some of the examples described in this chapter are the San Francisco, Raton-Clayton, and Mount Taylor fields. Common landforms within a basaltic volcanic field are shown in Fig. 17.2. They include lava flows, cinder cones, maars, fissures, and basalt-capped mesas. Not shown are larger landforms that include shield volcanoes and basalt (or lava) plateaus.

A cinder cone (also known as a scoria cone) is a small, well-shaped volcano with steep sides that consists almost entirely of small, dark, vesicular, loose fragments of basalt known as scoria, the basaltic equivalent of pumice. Typically, a cinder cone is a few thousand feet wide and less than 1000 feet high with a central crater that is large relative to its overall size. As lava pools in the large central crater, gas bubbles rise and pop in a manner vaguely similar to boiling honey. Lava is shot or sprayed into the air, landing around the central vent to create the volcanic shape, and cooling to form scoria. Many cinder cones erupt only once. Some are associated with a lava flow. Cinder cones litter basaltic lava fields often numbering in the hundreds, and are often referred to as buttes.

Maars are also common in basaltic lava fields. A maar is a hole in the ground that forms as a result of a water-charged explosion. As hot magma moves upward, it encounters groundwater, which flashes to steam creating an explosion that blows a hole in the ground. Cinder cones and maars are often aligned along narrow cracks in the ground known as fissures. Large volumes of magma can erupt along fissures that can be miles to tens of miles long.

FIGURE 17.1 Landscape map showing volcanic areas. Some areas include intermingled sedimentary, intrusive, and metamorphic rocks.

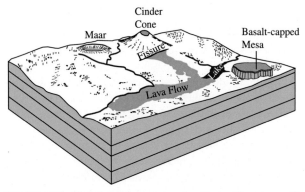

FIGURE 17.2 Sketch showing common basaltic landform types. The lava flow fills the valley blocking and deflecting rivers. Based on Hamblin and Christensen (2001).

Cinder cones, maars, and fissures within a volcanic field are often concentrated in a relatively small area referred to as the volcanic center. The abundance of warm rock causes the volcanic center to isostatically uplift into a broad dome creating what is known as a shield volcano. Lava will flow freely from the volcanic center into surrounding valleys at the base of the dome. As younger flows cover older flows, the volcanic dome builds into a broad, gentle-sloped, shield-shaped edifice that is much wider than it is tall. Areas near the central vent where lava has extruded may sag due to loss of support thereby creating small depressions. As more lava extrudes, several depressions may coalesce, perhaps along a fissure or along several intersecting fissures, to form a large semi-circular depression known as a caldera. In this case, the caldera is not explosive. It is a semi-circular sag (a depression) in the ground where magma has extruded to the surface. Thus, a shield volcano can be described as a very large, very broad, nonexplosive, shield-shaped volcano composed principally of basalt, with fissures, cinder cones, maars, many overlapping lava flows, and possibly a central caldera. It forms as a result of large, centralized extrusions of basaltic magma. Shield volcanoes can be tens of miles wide. Kilauea and Mauna Loa in Hawaii are shield volcanoes. The above sea level part of Mauna Loa is between 60 and 75 miles wide and 13,679 feet tall with a summit caldera 9.3 miles across.

Large volumes of basalt can erupt continuously for years, or discontinuously over millions of years, to create a series of flows tens of miles to hundreds of miles long piled one above another. Eruptions can be on such a massive scale and of such long duration that they can completely bury and reincarnate pre-existing landscape to form a basalt (or lava) plateau. A succession of lava flows of this type can be thousands of feet thick and is known as flood basalt. The basalt that covers most of the Columbia Plateau is an example.

A single lava flow will fill a river valley, which either blocks the river to form a lake, or displaces the river to the margin of the lava flow (Fig. 17.2). Waterfalls form where the river flows over the steep front face of a lava pile. Young flows, generally less than a few thousand years old, will appear as dark objects in satellite images due to the absence of vegetation and soil. As basalt fills lowland areas, it covers young sedimentary rocks. Basalt is relatively more resistant; therefore, subsequent erosion and topographic lowering are concentrated in the area of soft sedimentary rocks that surround the basalt flow. Over time, erosion will lower the surrounding sedimentary landscape, but leave the basalt flow to form a flat-topped highland mesa with basalt serving as the resistant caprock layer, thus inverting topography. Undercutting of weak sedimentary rock at the margins of the basalt mesa results in numerous landslides. An example is Grand Mesa, Colorado.

The term silicic has been used in this book to refer to three types of volcanic rocks that often occur together, andesite, rhyolite, and tuff. Rhyolite is a relatively low density, light-colored liquid derived primarily from partial melting of continental crust. It is the volcanic equivalent of granite. It tends to be gas-rich and does not flow very easily resulting in massive, short-lived, explosive eruptions. Andesite is intermediate between basalt and rhyolite in terms of composition, color, density, and flow characteristics. It is less fluid and more explosive than basalt, but generally more fluid and less explosive than rhyolite. It can be derived through differentiation of basaltic magma. Differentiation is any process by which magma changes composition. One such process is the partial crystallization, below the Earth's surface, of mantle-derived basaltic magma. In this case, silica-poor minerals crystallize first, leaving behind a more silica-rich liquid that erupts later to form andesite. Andesite is significant because it is found primarily in volcanoes that are situated above subduction zones. Keep in mind that a subduction zone has continuously existed off the west coast of the United States for more than 300 million years. Thus, the age and distribution of andesite across the Cordillera will, to a first approximation, tell us when and where subduction was occurring. Tuff is a name used for magma that was explosively ejected into the air, deposited as volcanic ash, and solidified into rock. If ash is hot when it settles out of the air, it can be compressed and fused to form welded tuff.

Landforms typical of silicic volcanic rocks include short, thick lava flows, lava domes, ash layers, ignimbrite and volcanic breccia layers, calderas, and composite (strato) volcanoes. When silicic magma reaches the surface, rather than flow freely away from a vent, it can pile up to form a mound. Once formed, a mound can freeze in place to form a lava dome. Alternatively, the lava dome can grow higher, tip over, and create a short, thick lava flow. A third possibility is for the lava dome to tip over and blow up to create a pyroclastic flow. A pyroclastic flow is not lava. It is a high velocity, high temperature, gas-rich volcanic cloud of suspended rock and magma particles that hugs the ground following a violent eruption. You do not want to be in the way of one of these because it will kill you in short order. The resulting rock is composed of a mixture of pumice, welded tuff, and many rock fragments, and is referred to as an ash-flow tuff or ignimbrite. A pyroclastic flow can enter a river channel and flow for many miles down-river creating a chaotic-looking deposit of mixed volcanic material and sediment referred to as volcanic breccia. A breccia is any rock in which angular blocks of any size are encased in a finer-grained matrix. In this case, most of the blocks are volcanic, so we can refer to it as a volcanic breccia. Similar to a basaltic lava flow, volcanic breccia can choke and displace a river.

Silicic calderas are large volcanic landforms, on the order of 15 to 50 miles across. There are three active calderas in the United States, Yellowstone, Valles, and Long Valley. All three are discussed later in this chapter. Silicic calderas are explosive and therefore different from the caldera at the top of a shield volcano. Typical caldera formation is as follows. As silicic magma within a magma chamber rises toward the surface, it domes and stretches overlying rock layers to create a series of radial and circular (ring) fractures. These fractures act as conduits for magma and gaseous clouds to escape, potentially marking the beginning of a massive, very dangerous eruption that could result in the ejection of huge amounts of volcanic material. Once the magma chamber is depleted, the overlying land area will sink into the void along the ring fractures, which now act as normal faults, creating a semi-circular depression. Additional pyroclastic eruptions and possibly a lake may fill or partially fill the central depression. A resurgent caldera is one in which all or part of the caldera subsequently rises into a dome possibly due to the arrival at depth of another injection of magma. The result is a central roundish dome area surrounded by a circular valley marked by ring faults that outline the margins of the caldera. Dozens of calderas were active in the Cordillera from 38 to 27 million years ago during a particularly violent volcanic period referred to as the ignimbrite flare-up discussed later in the chapter.

A composite volcano (also known as a stratovolcano or stratocone) is tall, often thousands of feet high, with steep sides built by a combination of explosive pyroclastic flows and silicic lava flows. It is the classic volcano that everybody instantly recognizes, and the one most typically associated with a subduction zone. A central vent, known as a crater, may be present at the top of the volcano. Silicic lava emanating from the central vent will flow a short distance down the side of the volcano only to freeze before reaching the valley. Successive flows build up one above the other, thus building the volcano to great height. The lava is of variable composition. Andesite is common, but rhyolite and even basalt may be present. Explosive eruptions create pyroclastic flows that gain speed as they avalanche down the side of the volcano often entering a river to create a volcanic breccia and mudflow. An explosive eruption can blow the top off a volcano thereby lowering its summit elevation as it did with Mount St. Helens in 1980. Alternatively, the entire volcano can collapse into the void left after the underlying magma chamber is depleted, thus creating a depression, a caldera, where the volcano once stood. Crater Lake is an example. In terms of mass, a straovolcano is large, but much smaller than a shield volcano. Most of the Cascade volcanoes are composite volcanoes. Mount Rainier, the largest Cascade volcano, is 15 miles wide and 14,417 feet tall with a summit crater about 0.25 miles wide.

In addition to the above volcanic landforms, two shallow intrusive features often create landforms in volcanic fields. They are diatremes and volcanic necks. A diatreme is a near vertical cone-shaped, brecciated pipe of magma that originates from deep below the Earth's surface. Brecciation occurs either through forceful internal gas-charged intrusion, or by steam-charged explosion, such as within the feeder pipe below a maar. A volcanic neck is a landform that forms from differential erosion of a cone-shaped intrusive feeder pipe. It can be thought of as the eroded throat or the plumbing system of a volcano. Rocks and magma in a volcanic neck often become brecciated as a result of a water-charged maar explosion, in which case, they could be called a diatreme.

COLUMBIA PLATEAU

The Columbia Plateau is located between the Cascade Mountains to the west and the Northern Rocky Mountains to the north and east. It is a vast area of dominantly basaltic rock more-or-less continuous with basaltic rock of the Snake River Plain. In the Google Earth image shown in Fig. 17.3, the Columbia Plateau is divided into the

B-Bend, **BM**-Blue Mtns., **Bn**-Burns, **BR**-Basin and Range, **c**-Craters of the Moon, **CB**-Columbia Basin, **CM**-Cascade Mtns., **cr**-Columbia River, **h**-Hells Canyon, **HP**-High Lava Plains, **IB**-Idaho Batholith, **OU**-Owyhee Upland, **p**-Picture Gorge, **S**-Selkirk Mtns., **SR**-Snake River Plain, **sr**-Snake River, **st**-Steens Mtns., **Tf**-Twin Falls, **W**-Willamette Valley, **Y**-Yellowstone Caldera, **y**-Yakama Fold Belt

FIGURE 17.3 Google Earth image of the Columbia Plateau and Snake River Plain oriented north. State boundaries and the international boundary with Canada are barely visible. Snow-capped volcanic peaks are visible in the Cascade Mountains.

Columbia Basin (CB), Blue Mountains (BM), and High Lava Plains (HP). These areas are discussed following an introduction to the primary rock unit that forms the Columbia Plateau, the Columbia River flood basalt.

Columbia River Flood Basalt

The Columbia Plateau is covered almost entirely with nearly flat-lying lava flows of the Columbia River flood basalt. The term flood basalt refers to repeated nonexplosive outpourings of lava massive enough to completely bury (flood) pre-existing landscape. These rocks, together with volcanic rocks of the Snake River Plain and Cascade volcanic arc system, cover more than half of Washington and Oregon and about one third of Idaho. The outpouring of basalt was great enough to bury and reincarnate what was once a mountainous terrain similar to the surrounding Rocky Mountains.

A landform map of the Columbia Plateau is shown in Fig. 17.4. The dark shade shows the approximate area covered by Columbia River flood basalt regardless of physiographic province. The lighter shade shows volcanic rocks only on the Columbia Plateau and Snake River Plain that are about the same age or younger than the flood basalt. Additional volcanic rocks associated with the Cascade Mountains and Basin and Range are not

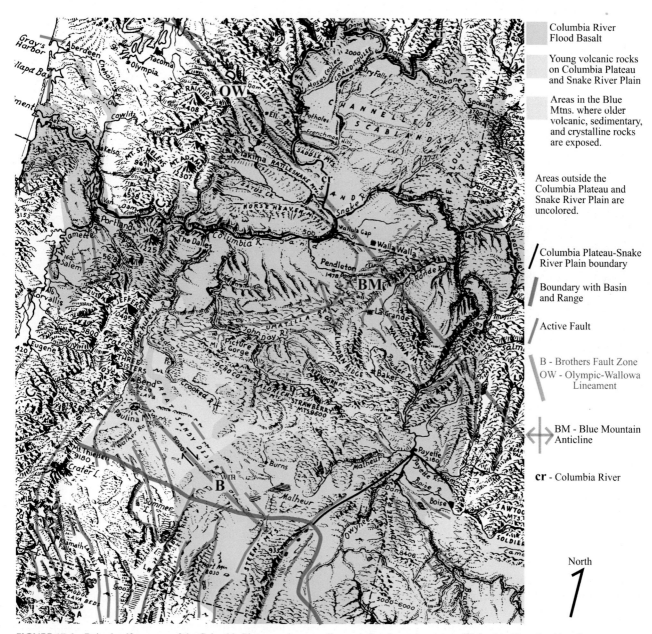

FIGURE 17.4 Raisz landform map of the Columbia Plateau and surrounding area that shows the extent of Columbia River flood basalt.

shown. Light blue shaded areas in the Blue Mountains of central Oregon are underlain with older volcanic, sedimentary, and crystalline rocks. Faults that have been active in the past 1.6 million years are shown with thick, light-blue lines. The physiographic boundaries of the Basin and Range and the Snake River Plain are also shown. Flood basalt extends beyond the confines of the Columbia Plateau, most notably into the Steens Mountain area of the Basin and Range, the Owyhee Upland of the Snake River Plain southwest of Boise, Idaho, in the Willamette Valley near Salem, Oregon, and along the Columbia River to the Pacific Ocean.

The Columbia River flood basalt has been extensively studied and radiometrically dated. Eruptions of lava began as early as 17.5 million years ago and ended 6 million years ago. Today, the volume of basalt is large enough to cover an area equivalent to 275 by 275 miles to a thickness of 3800 feet, an area slightly larger than the entire state of Washington. Maximum thickness is 16,000 feet along the Oregon-Washington border west of Walla Walla, Washington. The weight of the basalt has caused the Walla Walla area to subside 4000 to 10,000 feet since 17 million years ago. The subsidence, coupled with the addition of the basalt itself, has completely reincarnated this part of the Rocky Mountains producing a relatively flat plateau surface 500 to 2000 feet above sea level from what was once a mountainous terrain.

More than 98% of the basalt was extruded between 16.6 and 14.5 million years ago creating more than 300 individual lava flows, some several hundred feet thick. If we assume that a basalt thickness of 3800 feet was achieved in 2.1 million years, then the average rate of accumulation across an area slightly larger than the state of Washington is equivalent to 2.17 inches per 100 years.

The source area for most of the flood basalt is exposed as a series of dikes located southeast of Walla Walla in the area east and south of the Wallowa Mountains (Fig. 17.4). Lava from this source area buried and overfilled mountainous topography in the Columbia Basin, lapped up against the Cascade and Rocky Mountains, and covered most or all of the Blue Mountains. Some of the flows are among the longest on Earth, flowing more than 300 miles through the Columbia River channel all the way to the Pacific Ocean. In doing so, the lava repeatedly choked, damned, and displaced the Columbia River northward. Today, the remnant lava flows have created more than 70 waterfalls in the Columbia River Gorge between The Dalles and Portland, Oregon, including Multnomah Falls, which at 620 feet is one of the tallest in the United States. A second vent located near Picture Gorge in the Blue Mountains is responsible for lava that covers the Ochoco and Aldrich Mountains as well as part of the High Lava Plains near the Crooked River (Fig. 17.4).

Lava from a third smaller source area near the northern part of Steens Mountain created the large area of Columbia River Basalt surrounding Steens Mountain in southeast Oregon. Here, the basalt overlies and is partly intermingled with andesite/rhyolite and ignimbrite associated with the Cascade volcanic arc. This area is shown as part of the Basin and Range in Fig. 17.4 because normal faults, rather than the volcanism, exert primary control on the landscape. Steens Mountain is the largest normal fault-block mountain in the northern Basin and Range, reaching an elevation of 9738 feet.

Columbia Basin

The Columbia Basin (CB, Fig. 17.3) is centered in eastern Washington State. It forms the lowest part of the Columbia Plateau with elevation that varies from less than 500 feet to about 2500 feet. It is a structural basin nearly completely covered with basalt and down-warped due to the weight of the basalt. The basalt layers are nearly flat lying except in the Yakama fold belt (discussed later). The relatively young age of the basalt, its resistance to erosion, and the arid (rain shadow) climate have allowed the area to remain relatively undissected except in the vicinity of major rivers such as the Columbia and Snake Rivers, which have locally cut canyons into the volcanic landscape that are 1000–2000 feet deep. The most unusual and spectacular feature in the Columbia Basin are the Channeled Scablands discussed next.

Channeled Scablands

The Channeled Scablands constitute a large area of bare basalt in the Columbia Basin eroded by massive floodwaters that poured through the region primarily between 18,200 and 14,700 years ago when ice dams holding back a large glacial lake known as Lake Missoula suddenly failed. The scablands (c) are visible as scars in a Google Earth image of the Columbia Basin shown in Fig. 17.5. Lake Missoula formed in valleys surrounding Missoula, Montana. Evidence for the lake is seen along mountainsides surrounding Missoula where wave-cut shorelines are well preserved. Fig. 17.6 is a landscape map that shows Lake Missoula and some of the downstream floodwater lakes. Floodwaters would cross the scablands and enter and overflow the Columbia River channel creating a series of downstream lakes that include Lake Allison, which flooded the Willamette Valley to a depth of more than 400 feet.

Lake Missoula formed when the Cordilleran Ice Sheet advanced far enough southward to dam the Clark Fork River, a tributary of the Columbia River. The lake grew to the size of Lake Erie and Lake Ontario combined, 250 miles long and 1000–2000 feet deep. Apparently there

A-Mt. Adams, b-Banks Lake, ch-Lake Chelan, cr-Columbia River, c-Channeled Scablands, D-The Dalles, d-Dry Falls, f-Franklin D. Roosevelt Lake, g-Grand Coulee, G-Glacier Peak, K-Kennewick, L-Lewiston, m-Moses Coulee, R-Mt. Rainier, S-Spokane, sr-Snake River, W-Walla Walla, Y-Yakama, y-Yakama Fold Belt

FIGURE 17.5 Google Earth image looking north across the Columbia Basin. Thin white lines are state boundaries.

were many renditions of the lake. As the lake grew, the ice dam would periodically rupture, releasing a wall of water estimated at its maximum to be several hundred feet high, that poured through the area at 30 to 50 miles per hour in a sudden torrent removing vegetation and soil, and creating the scabland landscape. Maximum discharge is estimated to have been on the order of 10 to 30 million m^3/s. To put that number in perspective, average discharge of the Columbia River today is 7504 m^3/s. There may have been as many as 100 floods over the 3500 years that Lake Missoula was in existence.

The number and timing of floods is based on lake deposits (varves), the dating of interlayered volcanic ash deposits, and radiocarbon dating of wood fragments found in flood deposits. The data indicate that floods were separated by periods that lasted from 60 years to less than 10 years. The size of the floods decreased over time, but the frequency increased. There is also evidence for earlier periods of flooding. Deeply buried flood deposits dated from 200,000 to 400,000 years old, and one that is more than 780,000 years old, suggest that floods were associated with earlier glacial advances.

As floodwater moved across the Scablands, it would coalesce into river channels, eroding them into much larger channels. These now mostly dry flood channels are known as coulees. It is the interlocking network of coulees, scoured bare rock surfaces, and gravel bars that form the scablands (c) visible in Fig. 17.5. Two of the largest coulees are Grand Coulee (g)—50 miles long, 1000 feet deep, and between 0.5 and 4 miles wide—and the smaller Moses Coulee (m). A Google Earth image of Moses Coulee, shown in Fig. 17.7, depicts a deep, wide canyon floor with steep canyon walls composed of slightly tilted layers of basalt. Grand Coulee was cut initially by the Columbia River at a time when the Cordilleran Ice Sheet temporally displaced the river from its present channel (Fig. 17.6). Subsequent floodwaters gouged and enlarged the channel. Note in Fig. 17.5 that the northern part of Grand Coulee (g) is occupied by Banks Lake (b) as part of the Columbia Basin Project and the building of Grand Coulee Dam. The dam itself is on the Columbia River. Franklin D. Roosevelt Lake (f) is located behind the dam.

In addition to coulees, the scablands contain dry waterfalls, plunge pools, and giant streambed ripples up to 50 feet high. Many of the lakes presently on the floor of the coulees formed initially as plunge pools below waterfalls that no longer exist. A most impressive ancient waterfall is Dry Falls (d) in Grand Coulee. Flood waters plucked large

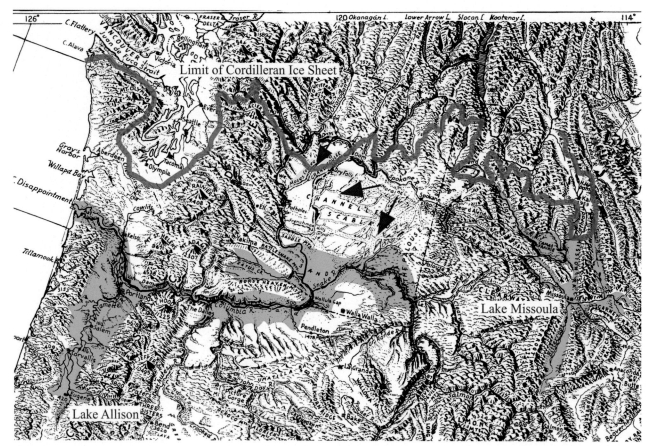

FIGURE 17.6 Landscape map that shows Glacial Lake Missoula and associated floodwater lakes. Arrows point in the direction of floodwater flow. The thick line is the southern limit of the Cordilleran Ice sheet.

FIGURE 17.7 Google Earth image looking south-southwestward at Moses Coulee and surrounding Columbia River basalt near Palisades, Washington.

blocks of basalt to create one of the largest falls known to have existed. During a flood, water would have dropped 400 feet across a shear cliff nearly 3.5 miles across.

Yakama Fold Belt

The Yakama fold belt is a series of actively forming anticlinal ridges and synclinal valleys that deform basalt layers in the Columbia Basin near Yakima, Washington. The fold belt is located in Figs. 17.3 and 17.5 with the letter y. Several of the anticlinal ridges are highlighted in Fig. 17.4, including Frenchman Hills, Saddle Mountains, and Horse Heaven Mountains.

Notice in Fig. 17.4 that the Columbia River (cr) steps eastward away from the edge of the flood basalt at several locations as it crosses the fold belt. It is possible that the growing anticlinal ridges were able to divert the river. As shown in Fig. 17.8, the river cuts a water gap through the Saddle Mountains, but appears to deflect around Umtanum Ridge. Such differences may in part reflect different rates and timing of anticlinal growth. Thinning of basalt layers above the anticlines suggests the folds were active about 16 to 15 million years ago at the time basalt flowed across the area.

The anticlinal mountains rise to as much as 2000 feet above synclinal valleys and appear to have undergone little erosion in the dry climate. One of the smallest ridges, Frenchman Hills, rises just 500 feet above its surroundings and has a fault along its northern side that has been active in the past 1.6 million years. The north side of Saddle Mountain is also faulted with activity within the past 130,000 years. Additional hidden faults are present at depth below some of the other anticlinal ridges. The absence of erosion and the presence of active faults suggest that the folds are active and that the ridges are in the process of forming. The rate of uplift along the anticlines has been estimated to be between 0.12 and 0.24 inches per 100 years (0.03–0.06 mm/year).

The water gap at Saddle Mountain is different from the ones we discussed earlier in the Appalachians, Ouachitas, and Middle-Southern Rocky Mountains. In the case of Saddle Mountain, the river was already in existence (antecedent) when the anticline began to grow. In the previous examples, the rivers were let down (superposed) across an unconformity onto an already existing anticlinal structure.

Blue Mountains

Intervening between the Columbia Basin and the High Lava Plains is a set of closely spaced mountains, plateaus, and intervening valleys known collectively as the Blue Mountains. The mountains are located in Fig. 17.3 (BM). Individual mountains are labeled on the landscape map shown in Fig. 17.4, and on a Google Earth image oriented north in Fig. 17.9. As seen in Fig. 17.4, the region is partly covered with Columbia River flood basalt (dark green shade), but there is significant exposure of older

FIGURE 17.8 Google Earth image looking northwest at a series of active anticlinal ridges in the Yakima fold belt near Yakima, Washington.

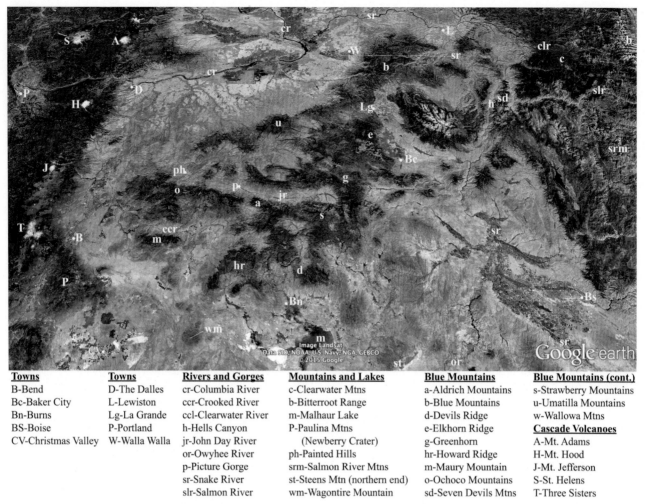

Towns	Towns	Rivers and Gorges	Mountains and Lakes	Blue Mountains	Blue Mountains (cont.)
B-Bend	D-The Dalles	cr-Columbia River	c-Clearwater Mtns	a-Aldrich Mountains	s-Strawberry Mountains
Bc-Baker City	L-Lewiston	ccr-Crooked River	b-Bitterroot Range	b-Blue Mountains	u-Umatilla Mountains
Bn-Burns	Lg-La Grande	ccl-Clearwater River	m-Malhaur Lake	d-Devils Ridge	w-Wallowa Mtns
BS-Boise	P-Portland	h-Hells Canyon	P-Paulina Mtns	e-Elkhorn Ridge	**Cascade Volcanoes**
CV-Christmas Valley	W-Walla Walla	jr-John Day River	(Newberry Crater)	g-Greenhorn	A-Mt. Adams
		or-Owyhee River	ph-Painted Hills	hr-Howard Ridge	H-Mt. Hood
		p-Picture Gorge	srm-Salmon River Mtns	m-Maury Mountain	J-Mt. Jefferson
		sr-Snake River	st-Steens Mtn (northern end)	o-Ochoco Mountains	S-St. Helens
		slr-Salmon River	wm-Wagontire Mountain	sd-Seven Devils Mtns	T-Three Sisters

FIGURE 17.9 Google Earth image, oriented north showing the Blue Mountains and part of the High Lava Plains, Oregon. State boundaries are shown with a faint white line.

rock units (light blue shading) as well as younger mostly volcanic rock (light green shade).

The area, on average, lies between 2000 and 5000 feet in elevation. Peaks are in the 6000- to 10,000-foot range except in the Umatilla Range where they are less than 6000 feet. The highest is Sacajawea Peak (9843 feet) in the Wallowa Mountains. Relief between peak and valley varies from about 3000 feet to as much as 6500 feet in the Wallowa Mountains. Several of the mountains appear to have been level plateau surfaces subsequently uplifted into an anticlinal bulge. Folded basalt layers indicate that at least some of the uplift has occurred within the past 15 million years.

The low-lying Umatilla Range (u, Fig. 17.9) and the Blue Mountains proper (b) form a broad basalt plateau surface that has been warped upward along the Blue Mountain anticline, a structure that follows the crest of these mountains and is possibly actively growing (Fig. 17.4). Erosion along the crest of the fold has exposed older underlying volcanic and sedimentary rock.

Other ranges, including the Ochoco (o), Aldrich (a) and Strawberry Mountains (s) also appear to be uplifted, dissected plateau surfaces. They are asymmetric mountains with a gentle southern slope and steep, dissected northern slope that faces the John Day River valley (jr). The apparent concentration of erosion on the north slope of mountains could be due to their proximity to the John Day River, or it could be climate-related (wetter and less sun on the north slope). Flood basalt sits at 6926 feet on Lookout Mountain in the Ochoco Range (o) suggesting that flood basalt, at one time, covered most of the range, and that uplift and erosion occurred following basalt deposition.

North of the Ochoco (o) and Aldrich Mountains (a), in Fig. 17.9, the John Day River valley (jr) hosts a succession of volcanic ash beds (tuff), thick soil horizons, and lake and stream sediment that was deposited mostly between 36 and 18 million years ago, prior to extrusion of flood basalt. The beds are unique in that they contain a great variety of plant and mammal fossils including dogs, rodents, and horses. Three separate sites near the towns of Dayville, Mitchell, and Fossil constitute John Day Fossil Beds National Monument. Included are Picture Gorge (p), where one can find pictographs (ancient paintings) on rock walls, and the Painted Hills (ph), where the rocks present a fabulous display of color.

Elkhorn Ridge (e) is a warped, anticlinal plateau surface that gains elevation from 3000 feet in the north near La Grande (Lg), to 9000 feet in the south near Baker City (Bc). Flood basalt surrounds Elkhorn Ridge and is present along its anticlinal crest to an elevation of 6500 feet, implying that at least this much of the ridge was once covered with basalt, and also suggesting that the anticline has formed within the past 15 million years (the approximate age of the basalt). Flood basalt is absent in the high, glaciated peaks of the southern Elkhorn Ridge where older granitic, volcanic, and sedimentary rocks are exposed. It is possible, therefore, that this part of ridge was a highland prior to 15 million years ago and that it was never covered with flood basalt.

The Wallowa Mountains (w) form the highest mountain range with at least 31 peaks above 9000 feet. The mountain was extensively glaciated, creating numerous sharp ridges and peaks that expose older granitic and sedimentary rock. Perhaps surprisingly, there is evidence to suggest that the mountain was mostly or entirely covered with flood basalt by about 15 million years ago. Erosional remnants of flood basalt can be found along several high ridge crests, and on mountain peaks including Aneroid Mountain (9707 feet). The presence of basalt at such high elevation suggests that the mountain was at low elevation 15 million years ago, and that it has seen significant uplift in the past 15 million years. Fig. 17.4 shows that it is surrounded by active normal faults, implying that uplift is active and ongoing.

Hells Canyon (h), on the Snake River, separates the Wallowa Mountains (w) from the Seven Devils Mountains (sd). The Seven Devils is a high mountain range with six peaks above 9000 feet including He Devil (9405 feet) and his bride She Devil (9404 feet). The mountain consists mostly of deformed and metamorphosed volcanic and sedimentary accreted terrane rock with granitic intrusions. Here too, there are erosional remnants of basalt along glaciated ridge crests as high as 8000 feet. We can suggest, therefore, that similar to the Wallowa Mountains, most or all of the Seven Devils Mountains were buried beneath basalt prior to uplift beginning less than 15 million years ago.

Hells Canyon is deeper than the Grand Canyon. Total elevation change is on the order of 8000 feet over a horizontal distance of 5.5 miles from He Devil to the Snake River. Fig. 17.10 is an image of Hells Canyon looking

FIGURE 17.10 Google Earth image looking southward up the Snake River at Hell's Canyon on the Idaho-Oregon border. The plateau area at right (west) consists of layered Columbia River basalt. The dissected high peaks along the upper left skyline are part of the Seven Devils Mountains volcanic accreted terrane.

southward that shows significant landscape differences between a low-relief terrain underlain with tilted Columbia River basalt on the west (right), and the steep, rugged, glaciated, crystalline terrain of the Seven Devils Mountains to the east. Notice that the tilted Columbia basalt layers would project over the Seven Devils Mountains in support of the suggestion that the mountains were once completely covered with basalt. It appears that the Snake River was able to keep pace with recent and active uplift of the Wallowa-Seven Devils area thus creating one of the deepest canyons in the country. Geological mapping suggests that the Snake River eroded its channel partly along a series of normal faults. The US Geological Survey does not show these faults as active.

In summary, landscape evidence suggests that the Blue Mountain region was a relatively flat area with no more than 1500–2000 feet of relief prior to complete or nearly complete burial beneath flood basalt by about 15 million years ago. The basalt layers created an extensive plateau across the area that was subsequently warped into anticlinal folds, locally cut by normal faults, and glaciated. The overall increase in elevation from west to east, coupled with a greater number of active faults toward the Wallowa and Seven Devils Mountains, suggest that uplift was most intense in the east. Geological evidence suggests deformation and mountain uplift began as late 5 or 6 million years ago and is ongoing, particularly in the vicinity of the Wallowa Mountains. The uplift history of the Blue Mountains can be contrasted with that of the Columbia Basin, which, with the exception of the Yakama Fold Belt, has remained largely undeformed or has subsided over the past 15 million years. Fig. 9.2 implies minor uplift in both areas over the past 100 years.

Olympic-Wallowa Lineament

The Olympic-Wallowa lineament, labeled OW in Fig. 17.4, cuts obliquely across the Columbia Basin, Yakama fold belt, and the Blue Mountains, continuing northwestward to the north side of the Olympic Mountains and southeastward into the Idaho Batholith. Recall from Chapter 4 that a lineament is any visible scar or alignment of features that can be followed across the Earth's surface. The Olympic-Wallowa lineament, although not obvious in Fig. 17.4, aligns the margins of mountains such as the Olympic and Wallowa Mountains with several other less obvious features that include the confluence of the Snake and Columbia Rivers. Erwin Raisz, the man who drew the landscape maps used in this book, was the first to identify the Olympic-Wallowa lineament.

The origin of the lineament is not clear. It is parallel with the more prominent Brothers Fault Zone (B, Fig. 17.4) described later, and with the Lewis and Clark line described in Chapter 15. All three lineaments are associated with zones of active normal and right-lateral strike-slip faults. They may have a similar origin in which all three are reactivated, ancient, buried fault zones.

High Lava Plains

As seen in Fig. 17.4, the Columbia Plateau region south of the Blue Mountains between Bend and Burns is occupied by the Great Sandy Desert, a largely semi-arid, flat region of sagebrush, farm and grazing fields 4000 to 5000 feet above sea level and underlain with nearly flat-lying basaltic lava flows, ash/pumice deposits, and cinder cones mostly less than 10 million years old. The sand by which the desert gets its name is largely disintegrated pumice. The drainage divide that separates the Columbia River watershed from the Great Basin watershed loops across the Great Sandy Desert such that the town of Burns and Malheur Lake are part of the Great Basin, and the Crooked, Owyhee, and Malheur Rivers are part of the Columbia River system.

The primary structure on the High Lava Plains is the second of two parallel lineaments, the Brothers fault zone, labeled B in Fig. 17.4. This lineament is marked with a well-developed set of active normal and strike-slip faults, the largest of which are shown in Fig. 17.4. Also present along the lineament are more than 100 explosive, silicic volcanic centers. The silicic rocks become progressively younger from the southeast, where they are approximately 10.5 million years old, to the northwest where they are less than 2000 years old. They are young enough that they have not been strongly modified by erosion in the dry climate. The volcanic centers form small mountains and hills (buttes) that rise above the Great Sandy Desert. Several are drawn but not labeled in Fig. 17.4 including Wagontire Mountain (shown as wm in Figs. 17.4 and 17.9).

The youngest volcanic feature in the Brothers fault zone is the Newberry volcanic field located at Newberry National Volcanic Monument in the Paulina Mountains south of Bend (Fig. 17.4, and P in Fig. 17.9). The Newberry volcanic field is somewhat enigmatic in that it is transitional between explosive, silicic composite volcanoes, and nonexplosive, basaltic shield volcanoes. It can be considered a shield volcano with basaltic flows that extend northward for more than 45 miles and southward for at least 20 miles. The city of Bend is built on lava from this volcano. It can also be considered a composite volcano with an explosive past. It is located near the margins of both the Brothers fault zone and the Cascade volcanic arc. The oldest volcanic rocks are about 400,000 years old. The youngest are only 1300 years old. There have been well over 20 eruptions in the past 12,000 years. Hot springs and a sharp increase in temperature with

b-Big Obsidian Flow, BM-Blue Mountains, c-China Hat, cp-Central Pumice Cone, e-East Butte, el-East Lake, GSD-Great Sandy Desert, l-Inter Lake Flow, L-Little Crater, LM-Lava Mountain (East Lava Field), p-Paulina Peak, pl-Paulina Lake, pm-Pine Mountain,

FIGURE 17.11 Google Earth image looking east at the Newberry Caldera and Great Sandy Desert, Oregon.

depth suggest it is an active volcano with magma at depth.

Fig. 17.11 looks east at the Newberry volcanic field and the Great Sandy Desert. This image shows that the central part of the volcanic field is a caldera 5 miles across partly occupied by two lakes. The caldera formed 75,000 years ago during the last of three major explosive events. Evidence for this eruption is exposed as tuff in the walls of Paulina Creek Falls located just west of Paulina Lake (pl). As seen in Fig. 17.11, cliffs mark the edge of the caldera on all but the west side. Circular (ring) normal faults are present at the base of the cliffs along which the central caldera has collapsed. Paulina Peak (p) forms the highest point on the caldera rim at 7989 feet, about 1600 feet above the lakes.

The most recent wave of activity in the Newberry volcanic field began 7000 years ago when lava spewed from a series of fissures and cinder cones within and north of the caldera. Silicic volcanism within the caldera created the Central Pumice cone (cp, Fig. 17.11), Little Crater (L), and two pumice/obsidian lava flows, the Inter Lake flow (I) and the Big Obsidian flow (b). This volcanism separated what was once a single lake into two lakes. The pumice/obsidian lava flows are rhyolitic in composition and therefore resistant to flow. As lava flowed slowly downhill from the caldera wall, it developed into lobes with beautiful ripple-like ramp structures. The flows consist roughly of 90% pumice and 10% obsidian. The Big Obsidian flow (b) on the south side of Paulina Lake is the largest of the pumice/obsidian flows and the youngest. It erupted 1300 years ago covering 1.7 square miles. Also visible in Fig. 17.11 is the flat terrain of the Great Sandy Desert (GSD), the Blue Mountains (BM), the circa 13,000-year-old Lava Mountain basalt flow (Lm), and three silicic volcanic centers, Pine Mountain (pm), China Hat (c), and East Butte (e).

Several basaltic lava flows less than 7000 years old are present northwest of the Newberry Caldera. At Lava Cast Forest, just 5 miles north of the caldera, basaltic lava flows engulfed trees to create tree molds. Farther northwest, on the outskirts of Bend, Lava Butte, shown in Fig. 17.12 looking due west, is a perfect location to show landform features associated with a basaltic lava flow. Lava Butte, at lower left, is a beautifully shaped cinder cone with a large lava flow emanating from fissures at its base. Notice at the top of the image that the lava flow blocked and redirected the Deschutes River. Can you locate where the river was initially blocked? Over time, the Deschutes River will strip away surrounding rock layers leaving the flow largely intact to create a basalt-capped mesa.

There are many additional, young, well-preserved volcanic landforms that surround the Newberry volcanic field. Hole-In-The-Ground, located 22 miles to the south, is a superbly developed maar where groundwater came in contact with rising magma to cause an explosion. There are three young lava flows within 30 miles east of Hole-In-The-Ground, one of which, Lava Mountain (LM), is visible in Fig. 17.11. There are spectacular cross-sectional views of a variety of volcanic rocks at Cove Palisades, north of Bend, where the Crooked, Metolius, and Deschutes Rivers meet. Also near Bend, there are views

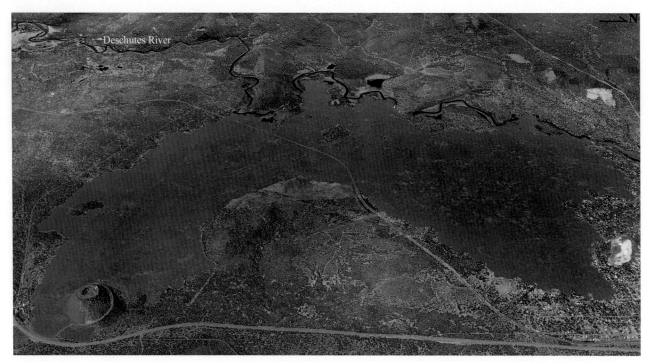

FIGURE 17.12   Google Earth image looking west at Lava Butte (cinder cone) and the redirected Deschutes River in Oregon.

of volcanic tuff at nearby Smith Rock State Park, where you can also do some fabulous rock climbing. The tuff is part of the John Day Formation and was derived from collapse of a large caldera approximately 30 million years ago located southeast of Smith Rock. If this were not enough volcanic excitement, there are, of course, Crater Lake, the Three Sisters, and other Cascade volcanoes nearby.

The landscape south of the Great Sandy Desert, including Steens Mountain, Summer Lake, and Klamath Falls, although largely underlain with young volcanic rock, is strongly influenced by normal faults, and therefore shown in Fig. 17.4 part of the Basin and Range. Clearly, the boundary is transitional and somewhat arbitrary. Many of the faults in this part of the Basin and Range are active and trend directly into the active Brothers fault zone. Given this relationship, we could alternatively extend the Basin and Range northward to the vicinity of Bend to include the Brothers Fault zone.

SNAKE RIVER PLAIN

The Snake River Plain forms the southeastern extension of the Columbia Plateau without a distinct boundary. A landscape map of the region is shown in Fig. 17.13. Included within the Snake River Plain physiographic province is the geologically linked Yellowstone Volcanic field, which forms a broad plateau at the eastern end of the province, and the Owyhee Upland, which occupies the area surrounding the Owyhee and Bruneau Rivers, including the Owyhee Mountains. The most conspicuous feature is the Snake River, which flows south off the Yellowstone Plateau to the east side of the Teton Mountains, and then westward and northwestward through the northern Wyoming Range before turning southwestward toward Idaho Falls where it continues along the length of the Snake River Plain. River elevation varies from 4700 feet at Idaho Falls to about 2200 feet west of Boise.

The Snake River Plain proper is a broad, flat, downwarp surrounded by highlands of the Basin and Range to the south, the Idaho batholith and Northern Rockies to the north, the Owyhee Upland and High Lava Plains to the west, and the Yellowstone Caldera to the east. It is composed entirely of nearly flat-lying basaltic lava flows, small volcanic cones, explosive rhyolite calderas, and river and lake sediment. The western Snake River Plain, west of Twin Falls, is a deep graben filled with volcanic and sedimentary rock that developed along normal faults active between 11 and 9 million years ago. The eastern Snake River Plain is a downwarp due to isostatic sinking under the weight of the volcanic and sediment load. Basin and Range normal faults extend into the area, especially in the west, but most are covered with younger lava flows.

Basaltic and silicic volcanism began about 16.5 million years ago in the Owyhee Upland at about the same time as massive outpourings of flood basalt in the Columbia Basin and Steens Mountain areas. Volcanism has migrated eastward with time and remains active in areas such as the

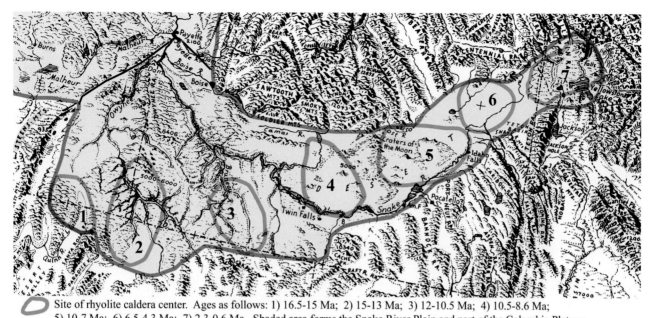

Site of rhyolite caldera center. Ages as follows: 1) 16.5-15 Ma; 2) 15-13 Ma; 3) 12-10.5 Ma; 4) 10.5-8.6 Ma; 5) 10-7 Ma; 6) 6.5-4.3 Ma; 7) 2.3-0.6 Ma. Shaded area forms the Snake River Plain and part of the Columbia Plateau.

FIGURE 17.13 Landscape map with the Snake River Plain and part of the Columbia Plateau shaded. The thin black line shows the boundary with the Columbia Plateau. Caldera locations based on Smith and Siegel (2000).

Yellowstone Caldera and Craters of the Moon. Surface rocks across the Snake River Plain are quite young. Basalt less than 2.6 million years old covers part of the western Snake River Plain along with river and lake sediments associated with floods and the remnants of a large glacial lake known as Lake Idaho. Older basaltic and rhyolitic rocks are present in the Owyhee Upland and at the margins of the plain. The eastern Snake River Plain is covered almost entirely with basalt flows that are less than 2.6 million years old. The average thickness of volcanic rock is not well known but is probably on the order of 5000 feet. Maximum thickness could be more than 20,000 feet in some areas.

The most recent basalt flows originated from fissures and from central vent areas composed of cinder cones and craters aligned along fissures. The lava fields appear to have displaced the Snake River to the southern edge of the Snake River Plain. There are as many as eight basaltic lava fields in the eastern Snake River Plain where basalt is so young and fresh (mostly less than 10,000 years old) that it stands out as black or gray blotches due to lack of vegetation and soil cover. The largest and most obvious is Craters of the Moon National Monument and Preserve, labeled c in Fig. 17.14. Two others are visible to the east and one (the Wapi field) to the south. At Craters of the Moon, one can find at least 60 lava flows that erupted from a single source area known as the Great Rift. There were eight eruptive episodes between 15,000 and 2000 years ago, each separated by 2000 years of quiet. The last eruption occurred about 2000 years ago, so we should expect another eruption soon. The 60 or so lava flows at Craters of the Moon are only the youngest flows to have erupted. Lighter colored areas within and surrounding these young volcanic rocks represent older volcanic rocks that erupted 30,000–500,000 years ago. These older flows do not stand out in satellite images because they are covered with vegetation and soil.

The Great Rift is an alignment of cinder cones and fissures 50 miles long and 1 to 9 miles wide. It not only is responsible for the young flows at Craters of the Moon, it is also responsible for two additional young volcanic fields, the Kings Bowl and Wapi. The Wapi field is visible in Fig. 17.14 directly south of the Craters of the Moon. The Kings Bowl field lies near the northern margin of the Wapi field, but is too small to be easily visible in the image. All three last erupted about 2000 years ago. Fissure eruptions are particularly interesting because the Columbia River flood basalt erupted along similar, albeit much larger fissures. Fig. 17.15 is an image looking northwestward at a fissure along the Great Rift that created lava flows in the Kings Bowl. The northern edge of the Wapi field (w) and part of the Craters of the Moon (c) are also visible. Kings Bowl (k) is a 260-foot long, 120-foot wide, and 100-foot deep explosion crater (maar) along the fissure that was created when hot magma came in contact with ground water.

Volcanism on the Snake River Plain is bimodal (2 distinct compositions) consisting both of basaltic lava flows

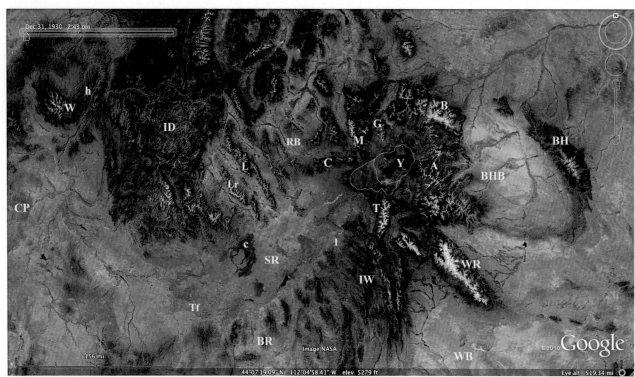

A-Absaroka Mountains, B-Beartooth Mountains, BH-Bighorn Mountains, BHB-Bighorn Basin, BR-Basin and Range, c-Craters of the Moon, C-Centennial Range, CP-Columbia Plateau (High Lava Plains), G-Gallatin Range, h-Hells Canyon, Seven Devils Mtn. I-Idaho Falls, ID-Idaho Batholith, IW-Idaho-Wyoming Fold and Thrust belt, L-Lemhi Range, Lr-Lost River Range, M-Madison Range, RB-Rocky Mtn. Basin and Range, SR-Snake River Plain, T-Teton Mountains, TF-Twin Falls, W-Wallowa Mtns., WB-Wyoming Basin, WR-Wind River Range, Y-Yellowstone.

FIGURE 17.14 Google Earth image looking northward across the Snake River Plain. The inner white circle is the Yellowstone caldera, active from 640,000 years ago to present. The combined inner and outer circles form the composite Yellowstone caldera, active from 2.053 million years ago to present.

FIGURE 17.15 Google Earth image looking northwestward at a large fissure along the Great Rift responsible for lava flows at Kings Bowl volcanic field. Kings Bowl (k) is a maar. The Wapi (w) and Craters of the Moon (c) volcanic fields are visible.

and explosive rhyolite calderas. Fig. 17.13 shows the location and age of seven rhyolite caldera centers beginning in the Owyhee Upland and continuing the length of the Snake River Plain to the active Yellowstone Caldera.

The calderas are partly covered with younger basalt layers, especially in the eastern Snake River Plain, but are identified in drill holes at the margins of the plain. Note in Fig. 17.13 that the calderas become younger toward the east. A similar progression in age is seen in the Hawaiian Island chain. In both cases, the position of active volcanism corresponds with movement of a tectonic plate over a stationary or nearly stationary hot spot (Fig. 5.14). In the case of Hawaii, the moving tectonic plate is the Pacific (oceanic) plate. In the case of the Snake River Plain, the moving tectonic plate is the North American (continental) plate. Those who live in Hawaii know that volcanic eruptions associated with the Hawaiian hot spot are not especially explosive. Basaltic magma generated in the mantle can easily penetrate the thin oceanic crust and reach the surface virtually unchanged. Thus the Hawaiian Islands are composed mostly of nonexplosive basaltic lava flows.

Basaltic magma generated in the mantle below the continental crust of the Snake River Plain has a much harder time reaching the surface due in part to the thickness of the crust and to buoyancy contrasts between the magma and the low-density continental crust. Under these conditions, some of the basalt reaches the surface to create the thick piles of lava previously described, and some of it ponds and cools at depth near the base of the crust. The cooling basaltic magma releases heat that can melt continental crust, thus creating rhyolite magma that rises to the surface to form explosive calderas in combination with outpourings of basaltic magma. Since the formation of the first rhyolite caldera in the Owyhee Upland, active volcanism has migrated eastward at a rate of 8.2 feet per 100 years to its present position below Yellowstone National Park. Each caldera in the chain essentially represents an ancient Yellowstone National Park. Given enough time, the Yellowstone hot spot, and active volcanism, will migrate across southeastern Montana into North Dakota, and eventually into Canada.

The eastern Snake River Plain contains what is probably the largest aquifer west of the continental divide. It is the sole source of drinking water for most of the population across the region. Water for the aquifer originates from precipitation and snowmelt in the mountains north and east of the Snake River Plain. Once in the aquifer, groundwater moves from northeast to southwest through rubble zones at contacts between buried basalt flows. There are two large natural discharge areas (springs); the first is near American Falls Reservoir, the second larger one is at Thousand Springs. Collectively, these springs greatly increase the volume of water in the Snake River.

Owyhee Upland

The area located between the High Lava Plains and the Snake River Plain, including the upper reaches of the Owyhee and Bruneau Rivers in Oregon, Idaho and Nevada, does not fit perfectly within either the Columbia Plateau or Snake River Plain physiographic provinces. It is a region without well-defined boundaries that is transitional between the two physiographic provinces. We will refer to this area as the Owyhee Upland and show it in Fig. 1.6 as part of the Snake River Plain. The area is shown in its entirety in Fig. 17.13.

Structurally, the Owyhee Upland is best described as a domal uplift. Elevation increases from less than 3000 feet near the confluence of the Snake and Owyhee Rivers to more than 5000 feet in the upper reaches of the Owyhee and Bruneau Rivers. Both rivers cut incised channels several hundred to one thousand feet deep through the terrain. The Owyhee River in particular cuts a gorge through the Owyhee Range, which reaches a maximum elevation in its eastern part of 8407 feet at Cinnabar Mountain (Hayden Peak).

Volcanic rocks across the upland are transitional between those of the Columbia Plateau and the Snake River Plain. Included are rocks correlative with Columbia River flood basalt, silicic volcanic centers, and basaltic rock less than 10 million years old. Basalt flows at Jordan Craters volcanic field, located west of the Owyhee Range, are less than 3200 years old. Erosion in the Owyhee Mountains has progressed enough to expose underlying granitic and metamorphic rock in the vicinity of Cinnabar Mountain.

Notice in Fig. 17.13 that the earliest of seven hot spot caldera centers is situated on or near the Owyhee Upland. Active hot spot calderas, such as the Yellowstone Caldera, form topographically high plateau areas due to thermal isostatic uplift associated with warm rock and molten magma at depth. These areas typically sink once they pass the hot spot and the rocks begin to cool. The calderas in the Owyhee Uplift are unusual because rather than sinking, they have apparently undergone uplift. We can presume that uplift is younger than the 12 to 10.5-million-year age of caldera number 3. Given the relationships, we can presume further that uplift is recent (or active) and most likely tectonic in nature, similar with uplift in the Blue Mountains and Yakama fold belt.

Yellowstone Plateau Volcanic Field

The Yellowstone Plateau is located directly above the Yellowstone hot spot. It hosts the largest active volcanic system in North America and the world's largest hydrothermal system. There are at least 180 geysers, mud pools, and fumaroles associated with the system including the famous Old Faithful geyser. Yellowstone Plateau is a nearly circular welt of high topography that rises 3000 feet above the Snake River Plain. Cooling magma less than 7 miles below the surface provides thermal isostatic buoyancy that elevates the region to between 7000 and 8000 feet above sea level and also provides heat and energy for the famous geysers and hot springs. The center of the plateau is an active caldera partly occupied by Yellowstone Lake, one of the highest lakes in the country at 7732 feet. Magma movement at depth causes the land surface in some areas to act as if it were alive. The ground bulges as magma moves underneath, and subsides when magma moves away. It is almost as if the Earth is breathing.

As seen in Fig. 17.14, the area is part of what is known as the Yellowstone crescent of high terrain, a locus of high elevation and high seismic activity that includes normal fault mountains such as the Teton (T), Madison (M), Gallatin (G), Centennial (C), Lemhi (L), and Lost River (Lr) Ranges, anticlinal mountains such as the Wind River Range (WR) and Beartooth Mountains (B), foreland fold and thrust belt landscape such as the Idaho-Wyoming overthrust belt (IW), nearly flat-lying sedimentary rocks of the Wyoming (WB) and Big Horn Basins (BHB), and older volcanic rocks of the Absaroka Range (A). All of the ranges have peaks above 10,000 feet and most have peaks above 11,000 feet.

The Yellowstone plateau has been the site of three massive explosive eruptions 2,053, 1,292, and 0.640 million years ago, creating the Huckleberry Ridge, Mesa Falls, and Lava Creek tuffs respectively, and a huge composite caldera within and west of national park boundaries as shown in Fig. 17.14. The first and third of these explosive eruptions are two of the largest known in Earth history, depositing ash as far east as Missouri. The third and last of the major eruptions formed the modern day Yellowstone Caldera. Rhyolite and basalt flows preceded and followed each of the massive eruptions.

Fig. 17.16 shows some of the relationships along the northeastern edge of the caldera looking northwest. The most recent volcanic activity occurred between 165,000 and 70,000 years ago when more than 20 rhyolite lava domes and flows of the Central Plateau rhyolite member (r) filled and overflowed the Yellowstone caldera floor, covering most of the area between Yellowstone Lake and the Idaho-Wyoming border. The flows damned streams, creating lakes that include Yellowstone Lake. The steep edge of the Central Plateau rhyolite member (r) is visible near the center of the image. The hills to the north consist of Lava Creek tuff (L), deposited during the last of the great eruptions 640,000 years ago, along with older rocks of the Absaroka volcanic series. Most of the brightly colored stone in the Grand Canyon of the Yellowstone (GC) consists of interlayered rhyolite-tuff and lava deposited approximately 481,000 years ago as part of the Upper Basin Member flows and tuff (u), and subsequently altered by the ubiquitous hot waters of the region.

ORIGIN OF VOLCANISM ON THE COLUMBIA PLATEAU AND HIGH LAVA PLAINS

There is support for the idea that the Yellowstone hot spot is responsible for the massive outpouring of Columbia River flood basalt and possibly also with volcanism on the High Lava Plains even though the volcanism in these areas is not in line with the track of the hot spot. One idea is that the initial outpouring of magma was deflected northward and westward from the Snake River tract to the Columbia Plateau when it first neared the surface 16.5 million years ago due to differences in the structure and composition of the deep crust beneath the two areas. As shown in Fig. 11.5, deep crust below the Snake River Plain forms part of the North American crystalline shield, which is typically strong, rigid, and thick. Deep crust below the Columbia Plateau, on the other hand, consists of accreted terranes that were added to the edge of the North American crystalline shield. These rocks form

g-Grebe Lake, **GC**-Grand Canyon of the Yellowstone, **L**-Lava Creek Tuff & Absaroka volcanic rock (640,000 yrs. old)
r-Central Plateau Member Rhyolite (165,000-70,000 yrs. old), **u**-Upper Basin Member flows and tuff (481,000 yrs. old)

FIGURE 17.16 Google Earth image looking northwest at the northeastern edge of the Yellowstone caldera, Wyoming.

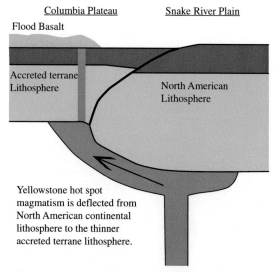

FIGURE 17.17 Cross-section sketch that shows deflection of rising magma to the edge of the North American continental lithosphere below the Columbia Plateau. Based on Winter (2010).

crust that is thinner, weaker, younger, and probably with through-going fractures and faults that allow easy access for magma to reach the surface. Thus, rather than moving vertically through continental crust, the magma was deflected to the edge of the continental crust into accreted terranes where it reached the surface to form the Columbia River flood basalts. The situation is depicted in Fig. 17.17. Evidence in support of this interpretation is the presence a major dike swarm (the primary source area for the flood basalt) along the edge of the continental shield southeast of Walla Walla, Washington.

A Yellowstone hot spot origin for volcanism on the High Lava Plains is more difficult to accept because even though silicic volcanism originates in about the same area (the Owyhee Plateau) as Snake River Plain volcanism, the age trend is for volcanism to become younger toward the west in a direction opposite that of the hot spot trend. One argument is that the High Lava Plains does indeed result from the Yellowstone hot spot. The idea is that hot material flowed westward in what can be called a counter-flow channel, caused by the sinking of the Juan de Fuca plate along the Cascadia subduction zone.

CORDILLERAN VOLCANIC AREAS 70 TO 20 MILLION YEARS OLD

In this section, and in the section that follows, we discuss some of the more significant volcanic fields in the Cordillera. Each is located and named in Fig. 17.18. Rocks that are 70 to 20 million years old are shown in yellow; those less than 20 million years are shown in blue. Many of the large volcanic fields between 70 and 20 million years old are mountainous regions composed of andesite, andesite breccia, rhyolite, ash-flow tuff, and volcaniclastic sedimentary rocks (derived from the erosion of volcanic rocks). Associated landforms are stratovolcanoes, calderas, and thick piles of ash-flow tuff. Collectively, these areas are referred to as the ignimbrite flare-up. They are the product of the sinking of the Farallon tectonic plate. Recall from Chapter 5 that the Farallon plate subducted beneath western North America at such a shallow angle that it underplated the Cordilleran crust all the way to Wyoming and Colorado as shown in Fig. 5.13. Slow sinking of the slab created conditions condusive to the melting of crustal rock and the creation of silicic volcanism.

Also discussed in this section are basalt fields with an origin not associated with subduction, and rhyolite fields in California that are probably associated with subduction. Each of the volcanic areas discussed here have been modified and reincarnated to various degrees by younger normal faults, and in one case, strike-slip faults.

Northern Great Plains

In Chapter 14 we noted the existence of domal mountains along the northwest fringe of the Great Plains in Montana with a central core of intrusive rock between 70 and 50 million years old. Volcanic rocks of approximately the same age occur in association with several of these crystalline domes, including the Bears Paw, Highwood, Castle, and Crazy Mountains (Figs. 14.12 and 17.18). The volcanic-intrusive association implies that erosion was sufficient in some areas to expose underlying granitic rock, but not so much that all of the volcanic rocks were removed. On the Great Plains, the volcanic rocks consist of potassium-rich, quartz-poor basalt, and pyroclastic breccia (ignimbrite). These rocks, along with associated intrusive rocks, form part of an igneous belt known as the Central Montana alkalic province.

In addition to volcanic rocks in the northwestern Great Plains, Fig. 17.18 shows volcanic rocks to the west within the Northern Rocky Mountains. These rocks are variably between 50 and 81 million years old and are of several possible origins. Some are associated with the Central Montana alkalic province. Others may be part of the Absaroka volcanic field discussed later. Volcanic rocks surrounding Butte, Montana are between 81 and 70 million years old and are part of the Elkhorn Mountains volcanic suite and associated Boulder Batholith granitic rocks. The Elkhorn Mountains volcanic rocks are andesitic to basaltic in composition and associated with an ancient subduction zone.

North and South Table Mountain

Fig. 17.19 looks south-southeastward at two mesas on the Great Plains that overlook the town of Golden at the foot of the Colorado Rocky Mountains. Known as North

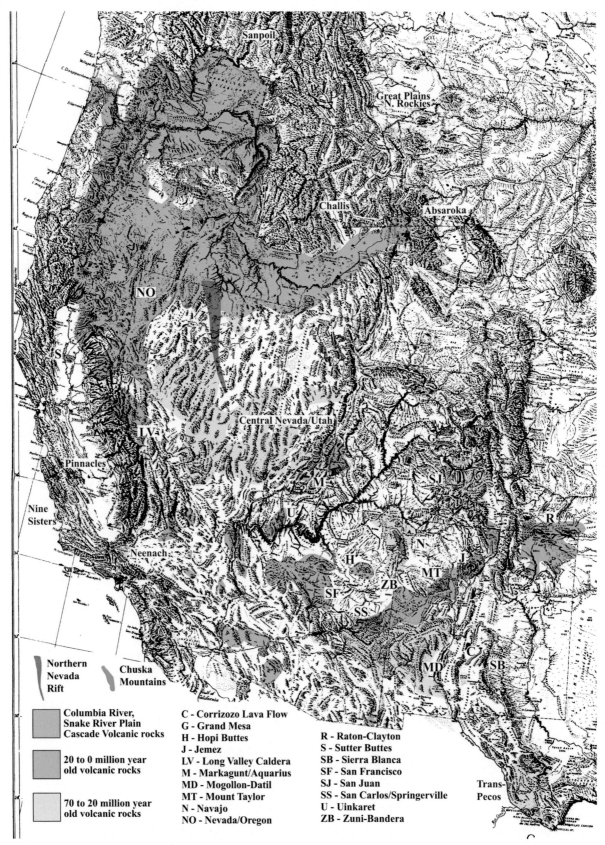

FIGURE 17.18 Landscape map that shows volcanic areas in the Cordillera.

G-Golden, n-North Table Mtn., r-Ralston Pluton, s-South Table Mtn.

FIGURE 17.19 Google Earth image looking south-southeastward at North and South Table Mountain and the Ralston pluton near Golden, Colorado. Also visible are hogbacks capped with Dakota Group rocks.

Table Mountain and South Table Mountain, the two mesas are capped with potassium-rich basalt known as shoshonite. The source region for the basalt is the Ralston pluton located 3.5 miles to the north near Upper Long Lake, and visible in Fig. 17.19. The magma followed a fracture zone to the surface provided by the Golden Fault. At least four flows emanated from the source region between 65 and 63 million years ago, and flowed into what was then a lowland valley. All four flows reached the area of North Table Mountain, but only the last two flowed as far as South Table Mountain. The flows were originally continuous from the Ralston pluton to South Table Mountain, but erosion has since removed sedimentary and volcanic rocks from around the mesas and inverted topography by leaving the two separate areas (North and South Table Mountain) as remnant highland mesas. The youngest lava flow forms a resistant caprock layer on both mesas.

Ignimbrite Flare-Up

The ignimbrite flare-up refers to a time between 56 and 18 million years ago that is perhaps unmatched in US geologic history in terms of the sheer number and magnitude of violent volcanic eruptions. This eruptive period was especially intense in the mid-Tertiary (mid- to late-Paleogene) between 38 and 27 million years ago with more than a dozen eruptions on a scale comparable to those at Yellowstone. Volcanism is associated with the sinking of the Farallon slab and is similar in origin to subduction zone volcanism. The Farallon slab underwent metamorphism as it sank into the mantle, releasing fluids that caused melting and the generation of basaltic magma in the upper mantle above the slab. The basaltic magma rose to the base of the continental crust where it cooled, releasing heat that melted some of the overlying crustal rock, thus creating silicic magma that rose to the surface. The oldest fields are those in the northwest including the Sanpoil, Challis, and Absaroka. Volcanism during the mid-Tertiary height of the flare-up was widespread and includes the central Nevada-Utah fields, the Mogollon-Datil field (MD) of southwestern New Mexico-southeastern Arizona, the San Juan Mountains (SJ) of southwestern Colorado, and the Trans-Pecos region of west Texas. Volcanism in all of these areas began primarily with andesite lava flows and volcanic breccia emanating from composite volcanoes, but later evolved into more than 200 separate rhyolite caldera source areas, each capable of multiple violent eruptions, and each dominated with deposits of ash-flow tuff. With so many explosive calderas, it is likely that by 27 million years ago, large areas of the west, including most of Nevada, Colorado, Arizona, and New Mexico, were buried in volcanic ash tens of feet to hundreds of feet thick. Subsequent uplift, erosion, and normal faulting has converted some of these volcanic piles to mountainous areas, and in doing so, has uncovered isolated areas of associated intrusive rocks.

Sanpoil and Challis Volcanic Fields

A belt of volcanic rocks known as the Sanpoil and Challis magmatic arc complexes occurs discontinuously from northeastern Washington to southern Idaho (Fig. 17.18). These rocks erupted from 56 to 39 million years ago and are dominantly andesite and rhyolite-tuff with small areas of granitic intrusions. Most of the volcanism occurred between 54 to 44 million years ago. The Sanpoil volcanic field, in Washington, occurs in normal fault basins along the Sanpoil River on the west side of the Kettle River Range surrounded by granitic rock. In southern Idaho, the main belt of Challis volcanic rocks forms lowland valleys and mountainous areas north of Craters of the Moon near Borah and Hyndman Peaks (Lost River Range and Pioneer Mountains). These rocks extend into the Salmon River Mountains west of Salmon where they are disrupted along normal faults and are interspersed with sedimentary, granitic, and metamorphic rocks. The level of subsequent deformation and erosion is such that these rocks do not form particularly distinctive volcanic landscape.

Absaroka Volcanic Field

The Absaroka volcanic field occupies the mountainous eastern and northern parts of Yellowstone National Park extending from the eastern side of Yellowstone Lake eastward and southward across the entire Absaroka Range (A) shown in Fig. 17.14, and northward across the Gallatin Range (G). The entire mountainous area is an eroded remnant of silicic, explosive volcanism that may have begun as early as 55 million years ago, culminated with massive eruptions between 49.5 and 47.5 million years ago, and was extinct by 43 million years ago. These volcanic rocks are far older than the hot spot-generated volcanic rocks that form Yellowstone Caldera. The two disparate volcanic fields are intermingled along the eastern shores of Yellowstone Lake.

The Absaroka Mountains are one of the largest, highest, most rugged, and most remote highland regions in the United States. There are 47 peaks above 12,000 feet including Francs Peak at 13,158 feet. The area is similar in some respects to the Wind River Range in the sense that it appears to be a high plateau elevated into a dome, and strongly glaciated and eroded into a mountainous terrain. The rocks and landscape are very different from the Wind River Range. Being volcanic and sedimentary rather than part of the ancient crystalline shield, the landscape is one of wide glacial valleys with long slopes that lead to narrow ridges and sharp mountaintops. Cliff faces of exposed rock, and the many bedrock lakes that characterize the Wind River Range, are largely absent. Relatively flat, dissected plateau surfaces are best preserved in the southwestern part of the range between about 10,500 and 11,000 feet in the area directly east of the northern Teton Mountains. Only a few flat summit areas exist at the 12,000-foot level that may represent a remnant of the elevated plateau surface. One example is Spar Mountain in the southern part of the range.

During its heyday, the Absaroka field probably resembled a concentrated version of the Cascade Mountains with as many as 12 composite (strato) volcanoes including Mt. Washburn (10,248 feet) within Yellowstone National Park and Sunlight Peak (11,927 feet) east of the Park, and possibly several calderas. The area was one of explosive eruptions, pyroclastic flows, and lahars (mudflows). Volcanic rocks, and sedimentary rocks eroded from the volcanic rocks, piled to a maximum thickness of 5000 feet and a volume of 7000 mi^3 (29,177 km^3). Included are andesite, ash-flow tuff, volcanic breccia, and various sedimentary rocks including conglomerate. The abundance of breccia and conglomerate attests to the explosive nature of the volcanic field. Additionally, there are small isolated areas of intrusive granitic rock, some of which represent the eroded guts of ancient volcanic centers.

Following extinction, the volcanic field underwent a long period of erosion likely creating a wide, relatively flat, river-dissected plateau surface with isolated highlands. The present-day landscape evolved probably within only the past 3 million years due to thermal isostatic uplift on the Yellowstone hot spot and strong glacial erosion. Given the above scenario, one could argue that the landscape is one of erosional mountains.

The volcanic rocks encase some spectacular fossils including an area within Yellowstone National Park

FIGURE 17.20 Photograph of a 12-foot-high tree now fossilized to silica (chert) on Specimen Ridge, Yellowstone National Park, Wyoming.

known as Specimen Ridge, located on the north side of Amethyst Mountain. Here, volcanic ash and mudflows have buried and fossilized as many as 27 separate forests. Fig. 17.20 is a photograph of a petrified tree 12 feet tall that is now chert due to the influx of silica-rich fluids.

The Gallatin Range (G, Fig. 17.14) extends northward from Yellowstone National Park to the vicinity of Bozeman, Montana. The mountain is part of the Rocky Mountain Basin and Range with active normal faults on its southern side. The southern (Yellowstone) part of the range is highest, most rugged and most strongly glaciated, probably due to a combination of active normal faults, its proximity to the Yellowstone Caldera, and the presence of resistant crystalline rocks in the form of Absaroka intrusive rocks and ancient, crystalline shield gneisses. Also present in the southern part are sedimentary rocks and both the Absaroka and Yellowstone volcanic rocks. The central and northern part of the range consists dominantly of Absaroka volcanic rocks that include the Gallatin Petrified Forest, similar in origin to Specimen Ridge. The landscape appears to be that of a dissected plateau similar to the Absaroka Range, but partly uplifted along normal faults. There are at least 16 peaks above 10,000 feet scattered along the length of the range, but mostly in the high southern part underlain with crystalline rocks. Rounded and flat summit areas are separated by steep glacier-carved cirque faces, narrow ridges, and wide glacial valleys.

San Juan Mountains and Surrounding Areas

The San Juan Mountains, in southwestern Colorado, occupy a mountainous area of silicic volcanic rocks similar to the Absaroka field except with higher mountain peaks and less overall glacial erosion. Smaller volcanic fields in Colorado, active at the same time as the San Juan region, and located in Fig. 17.18, include the Elk, West Elk, Rabbit Ears, Never Summer, and Sawatch Mountains, and the Thirtynine Mile volcanic area in South Park. Volcanism began primarily with andesitic eruptions prior to 32 million years ago, and continued with explosive caldera eruptions of ash-flow tuff until 23 million years ago. Erosion has left the San Juan Mountains, the West Elk Mountains, and the Thirtynine Mile volcanic areas as the largest remnants. The latter two areas and the outer margins of the San Juan Mountains consist mostly of older andesite lava flows and volcanic breccia. The central San Juan Mountains are dominated by somewhat younger ash-flow tuff. Following explosive volcanic activity, the San Juan Mountains underwent uplift and erosion with considerable uplift within the past 6 million years as part of the Middle-Southern Rocky Mountains-Colorado Plateau uplift phase discussed in Chapter 14. This uplift likely is primarily responsible for elevating 14 peaks above 14,000 feet. Glacial erosion over the past 3 million years has put the final touches on the present-day landscape.

Volcanic rocks in the San Juan Mountains are younger than those of the Absaroka Range, and in many areas, are not as deeply eroded. Volcanic rocks can still be found at the summits of several 14,000-foot peaks including Mount Wilson, Uncompahgre, Wetterhorn, San Luis, Sunshine, and Redcloud peaks, as well as Grizzly Peak (13,995 feet) in the Sawatch Range. The level of erosion in several other areas of Colorado is such that volcanic rocks have already been removed, and underlying associated intrusions are now exposed. The summits of Wilson, Gladstone, and San Miguel peaks in the western San Juan Mountains, Capital Peak and Snowmass Mountain in the northern Elk Mountains, and Mount Princeton, Mount Antero, and Tabegauche Peak in the southern Sawatch Range (all 14,000-foot peaks) each consists of intrusive rocks of approximately the same age as the volcanic rocks.

Fig. 17.21 is a Google Earth image looking north across the western San Juan Mountains where no fewer than 18 calderas are recognized. One of the largest is the La Garita caldera surrounding the town of Creede (C). The La Garita is the source for the 27.6-million-year-old Fish Canyon tuff, one of the largest ash-flow sheets in the world. The amount of material that was ejected from this

C-Creede, c-Chief Mountain, e-Mount Eolus, I-Ironton, L-Lake City, O-Ouray, R-Ridgeway, r-Red Mountain, rc-Redcloud Peak, S-Silverton, s-Sunlight Peak, sf-Mount Sneffels, sl-Slumgullion Slide, sm-San Miguel Peak, sn-Sunshine Peak, ss-Snowshoe Mountain, T-Telluride, t-Trinity Peaks, w-Windom Peak

FIGURE 17.21 Google Earth image looking north across the western San Juan Mountains, Colorado. The white line shows the boundary with crystalline shield rock of the Needle Mountains.

eruption alone created enough ash to bury the entire state of Colorado to a depth of 61 feet (5000 km^3, 1200 mi^3). The Fish Canyon tuff is widespread, particularly northeast and east of Creede. One well exposed area is tucked away in a remote part of the San Juan Mountains east of Creede. The Wheeler Geologic area is a spectacularly unique location of odd-shaped pinnacles, domes, and arches made of Fish Canyon tuff and carved by water. The area, however, is a challenge to reach. One would have to walk more than 7 miles along a footpath, or drive more than 14 miles along a nearly impassable 4-wheel drive road just to get a well-deserved glimpse of the area.

The shape of the La Garita caldera is not well preserved primarily because there are at least six younger, smaller calderas nested within the larger caldera. Collectively, these calderas contributed to a truly explosive period of volcanism in the San Juan's between 28.8 and 26.9 million years ago when at least 12 calderas were active. The Creede Caldera is one of the best preserved. The remnants of this caldera form a circular resurgent dome clearly visible in Fig. 17.21 as Snowshoe Mountain (ss). Nearly the entire dome consists of Snowshoe Mountain Tuff, which likely erupted from the Creede caldera circa 26.9 million years ago. The surrounding circular valley represents the collapsed margin of the caldera now buried in stream deposits of the Rio Grande and its tributaries.

The La Garita area is one of two locations that host large clusters of calderas. The second is in the western San Juan Mountains between Lake City (L) and Silverton (S). Much of the area consists of three calderas, the Uncompahgre, San Juan, and Silverton, which were active from about 28.4 to 27.6 million years ago. The Silverton is the youngest. It forms a circular mountain pattern visible in Fig. 17.21 between Silverton (S), Ironton (I), and Eureka (E, a mining ghost town). The circular mountain pattern results from erosion along circular ring faults that form the surrounding valleys. Route 550 follows the ring complex between Silverton (S) and Ironton (I). Most of the rock that forms the central mountain area within the caldera consists of andesitic flows associated with older calderas.

The youngest caldera in the San Juan Mountains is the 23-million-year-old Lake City caldera, which forms the circular mountain region south of Lake City. This area, which includes two 14,000-foot peaks, Sunshine (sn) and Redcloud (rc), consists dominantly of the Sunshine Peak Tuff. Red Mountain (r), also within the caldera, consists of shallow intrusive rock that at one time formed the neck of an ancient, eroded volcano.

Landslides are common in volcanic areas due to steep terrain and relatively weak, altered rock. A landslide is defined as movement of rock, rock debris, or earth down a slope under the force of gravity. Movement can be rapid such that an entire hillside fails in a matter of seconds, or slow in which the hillside creeps downslope at rates that vary from inches to tens of feet per year. A particularly well-known landslide is the Slumgullion Slide, part of which is active today. The slide is located along the eastern rim of the Lake City Caldera just east of the town of Lake City. It is labeled sl in Fig. 17.21 and shown in more detail in Fig. 17.22. Downslope movement blocked the Lake Fork

FIGURE 17.22 Google Earth image looking northward at the Slumgullion Slide. The active slide is lightly forested. The inactive slide includes areas marked o and o'. Route 149 is highlighted.

of the Gunnison River and created Lake San Cristobal (LSC). Elevation change from the top of the landslide on Mesa Seco to the bottom at Lake San Cristobal is 3200 feet. The total length is 4.23 miles (6.8 km).

Fig. 17.22 shows the steep, 800-foot headwall scarp (HW) from which the slide originated. The north (left) base of the scarp is unstable and is actively sliding as indicated by the absence of vegetation relative to the right side where dark vegetation indicates an older slide area (shown with an o in Fig. 17.22). Below the headwall scarp, the actively moving part of the landslide is visible based on its lighter vegetation and its thickness at the toe (where the active slide ends). The active slide is 2.4 miles long. If you visit the active landslide, in addition to bare ground, you can see tilted trees, trees with curved trunks, and trees that have been split in half due to differential movement across a fault. Multiple surveys over the past 50 years, coupled with photogrammetric and GPS studies, indicate downslope movement at rates that vary from 3 to 23 feet per year (0.1−0.75 inches/day). Radiocarbon dating indicates that the active slide began about 300 years ago. From a structural point of view, the headwall part of the active slide shows normal faults, the central part shows strike-slip faults in which individual sections move downslope as rigid blocks, and the toe is characterized by thrust faults. The slide itself consists of a colorful assortment of yellow, red, brown, and purple fine-grained sand, silt, and clay derived from hydrothermally altered volcanic rocks, primarily ash-flow tuff. The landslide initiated due to weak, fractured rock material, the steep slope, and periodic high groundwater levels.

A section of inactive slide material, marked in dark vegetation with an o' in Fig. 17.22, is located part way down the active slide on the southwest (right) side. Radiocarbon dating of this material records the earliest movement on the Slumgullion slide, between 1300 and 1000 years ago. This episode blocked Slumgullion Creek (SC) but was not large enough to block the Lake Fork of the Gunnison River.

On the basis of radiocarbon dating, the slide that blocked the Lake Fork of the Gunnison River, and damned Lake San Cristobal (LSC), occurred between 900 and 800 years ago. The geometry of this slide is clearly visible in Fig. 17.22 below the active slide. It appears that landslide material butted against the side of Red Mountain and then spread out in both an upstream and downstream direction. The upstream arm forms part of the shoreline of Lake San Cristobal. The toe of the slide in the downstream direction is indicated in Fig. 17.22. This part of the landslide is now stable and is crossed by route 149. The Lake Fork River has breached the landslide and created an 80-foot waterfall, Argenta Falls, across a resistant layer of bedrock.

Ground failure and the movement of Earth material downslope tend to create irregular, bumpy, hummocky, and bulging ground below a steep headwall scarp. This type of topography is present along the hillside above and east (to the right) of Lake City, suggesting the presence of numerous older landslides as indicated in Fig. 17.22.

Let us now return to our discussion of calderas. With the amount of material ejected during each volcanic eruption, one would expect to find evidence of choked streams and floods. One of the oldest explosive caldera deposits, the 36.7-million-year-old Wall Mountain Tuff, is associated with massive flooding. We have already mentioned this tuff in Chapter 14 during our discussion of the Arkansas River Valley. The source caldera of the Wall Mountain Tuff has been eroded, but based on its map pattern, it probably originated in the vicinity of what is now intrusive rock of the Mount Princeton (P) area in the southern Sawatch Range. Evidence of flooding is seen in the form of the Castle Rock Conglomerate, which overlies the Wall Mountain Tuff at Castlewood Canyon State Park on the Great Plains near Castle Rock. Here, large angular (flood) boulders of Wall Mountain Tuff are imbedded in the conglomerate along the canyon walls. The Wall Mountain Tuff is also well exposed at Florissant Fossil Beds National Monument in Florissant. This is the site of an ancient lake that formed 34.1 million years ago following a massive volcano-induced mudflow that dammed a river channel. The lake that formed behind the dam was at least one mile wide and 12 miles long. Periodic volcanic explosions would send enough ash through the air and into the lake to kill hundreds of living things. More than 1700 species of plants, insects, and animals were fossilized in exquisite detail at the bottom of the lake as a result of these explosions.

Central Nevada-Utah Volcanic Fields

Calderas in the Central Nevada-Utah volcanic fields are difficult to recognize due to reincarnation via younger Basin and Range normal faulting. The caldera shape has, for the most part, been altered or destroyed. The calderas, instead, are recognized by virtue of their rock deposits of andesite, andesite breccia, and tuff. Mountains in central Nevada with significant exposure of silicic rock include the Desatoya, Shoshone, Toiyabe, Toquima, Kawich, and Reveille Ranges, and Stonewall Mountain (Figs. 17.18 and 18.7). The belt extends eastward to the Tushar Mountains of Utah with calderas from 36 to 18 million years old. The volcanic eruptions associated with this belt were no less impressive than those in the San Juan Mountains. In central Nevada alone, there are 11 exposed calderas that created as many as 8 super eruptions on the order of the Yellowstone eruptions, some of which rivaled the La Garita caldera eruption. The total volume of ash is on the order of 25,000 km^3 (6000 mi^3) enough to bury the entire state of Colorado in more than 300 feet of ash.

Mogollon-Datil Volcanic Field

Calderas of the Mogollon-Datil field are not easily recognizable due to subsequent erosion, normal faulting, and burial beneath younger volcanic rock. Today, this area forms mountainous topography along with variably dissected plateaus and basins in the Basin and Range. The Mogollon-Datil field extends discontinuously from west of Socorro, New Mexico to east of Globe, Arizona and southward to the Organ Mountains north of El Paso, Texas (Figs. 17.18 and 18.9). The field was active primarily between 40 and 24 million years ago with no fewer than 10 calderas creating more than 10,000 km^3 (2400 mi^3) of rock. Volcanism was similar to the San Juan area, beginning with andesitic volcanoes and ending with calderas and massive explosions of tuff. Areas with significant exposure of volcanic rock include the Gila, Mogollon, Tularosa, Gallo, Datil, and San Mateo Mountains, the Sierra Mimbres, and the Nantanes Plateau. These mountains surround the Plains of San Augustin, a wide, flat basin that contained a large lake during the glaciation. Remnant caldera concentrations are present southwest of Socorro in the Chupadera, Magdalena, and San Mateo Mountains, and in the Mogollon Mountains east of Glenwood, New Mexico. An extension of this volcanic field occurs in the Peloncillo and Chiricahua Mountains along the Arizona-New Mexico border southward to the Mexican border. The northern part merges with younger, dominantly basaltic rock of the San Carlos-Springerville and Zuni-Bandera fields discussed later in this chapter. The volcanic rocks are offset along normal faults, few of which appear to be active.

Trans-Pecos Volcanic Field

The Trans-Pecos area is traditionally considered to be the part of Texas that lies west of the Pecos River. Much of the region is part of the Chihuahuan Desert, which lies mostly in Mexico and is the largest desert in North America, larger than the state of California. Volcanic rocks are abundant in the southern part of the Trans-Pecos as highlighted in Fig. 17.18. The volcanic area includes Big Bend National Park, Big Bend Ranch State Park, and the Davis Mountains. Although a large majority of the volcanic rocks are part of the mid-Tertiary ignimbrite flare-up, some of the younger basaltic volcanism is probably associated with the Rio Grande rift. The Rio Grande Rift is an area of active normal faults, connected with, but separate from the Basin and Range. It is discussed in Chapter 18. There are 12 recognized calderas in the Trans-Pecos field, all between 38 and 27 million years old with total ash volume on the order of 3000 km^3 (720 mi^3). Basaltic volcanism is as young as 18 million years old. The volcanic landscape is well preserved in spite of normal faulting and erosion, which has exposed underlying intrusive rock.

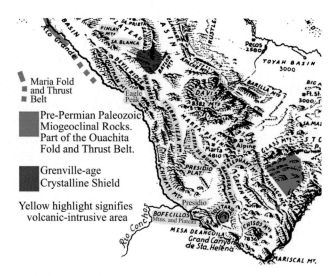

FIGURE 17.23 Raisz landform map of the Big Bend area, Texas. Volcanic areas are highlighted in yellow.

Fig. 17.23 is a Raisz landscape map of the Trans-Pecos with volcanic-intrusive areas highlighted in yellow. Areas with calderas include the Quitman, Davis, Chinati, Bofecillos, and Chisos Mountains, Eagle Peak, and the Solitario. Additional volcanic-intrusive areas not shown in Fig. 17.23 include Black Mesa, the Bofecillos Plateau, the Cienega Mountains, Crossen Mesa, McKinney Hills, Mine Point Mesa, Rosillos Mountains, Santiago Peak, YE Mesa, and the Christmas Mountains. The volcanic-intrusive rocks form mountains and hills due to their resistance to erosion and their young age.

The Chisos Mountains form the primary mountain area in Big Bend National Park. Fig. 17.24 is a Google Earth image of the Chisos Mountains looking east. Here, we find a rugged terrain with several peaks that circle a central lowland known as The Basin (b). A secondary circular mountain area surrounds Pine Canyon (P). Peaks exceed 7000 feet in both areas, the highest being Emory Peak (e) at 7827 feet. The rocks are a mixture of sedimentary, volcanic, and intrusive. The oldest sedimentary rock is the Chisos Formation, consisting of conglomerate, sandstone, and siltstone with a few interlayers of rhyolite, tuff, and basalt. The Chisos Formation is between 47 and 33 million years ago based on radiometric ages of interlayered volcanic rocks. These rocks are relatively nonresistant and are found primarily in lowland areas that surround mountain peaks. They form the substratum to younger volcanic rocks. The presence of these rocks indicates that the area now occupied by the Chisos Mountains was as lowland 33 million years ago.

One of the largest landforms shown in Fig. 17.24 is the Pine Canyon caldera, which forms the circular mountain region surrounding Pine Canyon (P) with peaks above 7000 feet. The Pine Canyon caldera underwent a short-

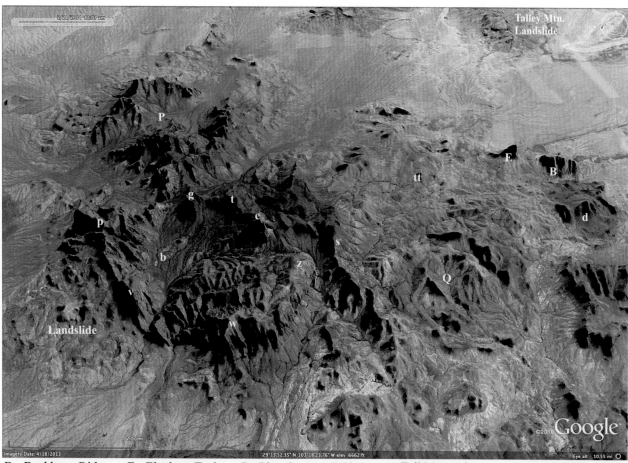

B - Backbone Ridge	E - Elephant Tusk	P - Pine Canyon Caldera	t - Toll Mountain	v - Vernon Bailey Peak
b - The Basin	e - Emory Peak	p - Pulliam Peak	tt - Tortuga Mountain	w - Ward Mountain
d - Dominguez Mtn.	g - Casa Grande	Q - Sierra Quemada	s - South Rim	z - Peak 7147

FIGURE 17.24 Google Earth image of the Chisos Mountains, Texas, looking east.

lived explosive period about 32 million years ago when several eruptions produced ash-flow tuff, volcanic breccia, and rhyolite that today forms the South Rim Formation. The Pine Canyon caldera consists almost entirely of South Rim Formation as does South Rim (s), Toll Mountain (t), Casa Grande (g), and Emory Peak (e). Mountain peaks north and west of The Basin (b), including Pulliam Peak (p), Vernon Bailey Peak (v), Ward Mountain (w), and Peak 7147 (z) consist of granitic rock that intruded prior to and subsequent to Pine Canyon caldera eruption and collapse. The granitic rocks are fine-grained with imbedded large crystals such that they resemble volcanic rocks. Their fine-grained nature suggests they intruded within a mile or two of the surface and cooled rather quickly. They are exposed due to erosion. Typical of volcanic landscapes, landslides are common. As shown in Fig. 17.24, a large landslide occupies the north (left) slope of Pulliam (p) and Vernon Bailey Peaks (v). Landslide material also occupies part of The Basin (b) on the north slopes of Emory Peak (e) and Toll Mountain (t).

Sierra Quemada (Q) forms a large circular area with peaks above 5000 feet occupied by circa 31- to 28-million-year-old intrusive rocks of variable composition that includes fine-grained granite and gabbro surrounded by rocks of the Chisos Formation. Sierra Quemada has been interpreted as a caldera, but other workers have questioned this interpretation because evidence for a large eruption is inconclusive. These workers refer to the landform as an intrusive structural dome. A volcanic breccia is present near the center of the dome and on its western rim where one can find giant inclusions, more than 300 feet wide, of Cretaceous and Paleozoic sedimentary rocks.

A small circular structure, Dominguez Mountain (d), is visible along the right side of Fig. 17.24. This structure is interpreted as the remnants of an extinct, eroded volcano. The top of the mountain consists of basaltic and andesitic lava flows surrounded at lower elevation by fine-grained granitic and gabbroic intrusive rock. The most interesting aspect is along the southwest flank where one can find hundreds of dikes of highly variable composition including dikes that pass upward into lava flows. The age of volcanic rocks on Dominguez Mountain is not well constrained but is probably between 31 and 28 million years. Nearby Elephant Tusk (E), Backbone Ridge (B), and Tortuga Mountain (tt) are cored with intrusive rocks.

The Bofecillos Mountains and adjacent plateau form a large circa 32- to 18-million-year-old volcanic field along the Rio Grande within Big Bend Ranch State Park. The Bofecillos Plateau is composed of basaltic lava flows that emanated from fissures and filled low areas thus creating a flat landscape between 4200 and 4500 feet, now incised by streams. The volcanism is some of the youngest in Texas and is likely associated with an early period of normal faulting along the Rio Grande Rift. The adjacent Bofecillos Mountains form a rugged terrain with jagged peaks and deep canyons composed of rhyolite lava, ash-flow tuff, and intrusive rocks. There are at least eight extinct volcanoes now visible as circular domes or calderas across the landscape, some with intrusive rock. Rhyolite lava, ash-flow tuff, and intrusive rock are resistant enough to form ridges and high areas. Oso Mountain, the highest point in the state park at 5135 feet, consists mostly of intrusive rock. It lies at the rim of a circular basin known as Bofecillos Vent, a probable caldera.

The Solitario is a beautifully developed circular landform located in Big Bend Ranch State Park. It is not a caldera or a structural dome. It is both! Along its margins it traces out a nearly perfect circular dome 7.5 miles in diameter defined by Cretaceous rocks that dip radially away from the center in all directions. The dome shape is obvious in an overhead view oriented north as shown in Fig. 17.25. Road access into the dome is from the north side only, the only side where streams flow into the dome. Within the dome, streams flow off a central high area and exit the structure through three major canyons, Righthand Shutup (Rhs), Lefthand Shutup (Lhs), and Lower Shutup (Ls), none of which are drivable.

The Solitario, like most volcanic areas in Big Bend, had a short-lived intrusive/volcanic eruptive history from 36.0 to 35.0 million years ago. The dome structure was created 36 million years ago when a large intrusion pushed overlying Cretaceous rocks into the shape of a dome. A major volcanic eruption occurred 35.4 million years ago that culminated with collapse of the dome to form a caldera. Today, the caldera is not active and the intrusive body remains buried.

Because of the lack of vegetative cover, major rock units in the Solitario can be distinguished in Fig. 17.25 based on their appearance in reflected sunlight. The rock units are numbered for easy reference. The medium-dark horseshoe-shaped area near the center of the dome labeled unit 1 is underlain with Paleozoic miogeoclinal sedimentary rocks that form part of the Ouachita fold and thrust belt. These are the same rocks that crop out in the Marathon Basin discussed in Chapter 15. Included within the sequence are resistant beds of Caballos Novaculite and Maravillas Chert, both of which form many of the ridges visible in Fig. 17.25. Sandstone, limestone, and shale are also present.

Unit 2 forms a slightly lighter-shaded area that surrounds unit 1. This rock unit consists of alternating layers of the lower Cretaceous conglomerate, sandstone, shale, and limestone between 126 and 100 million years old. The rocks dip radially away from the central part of the dome in all directions, thus defining the dome structure. A ring dike composed of fine-grained granitic rock nearly encircles the dome within the lower part of the Cretaceous sequence. The contact between Paleozoic rocks (unit 1) and Cretaceous rocks (unit 2) is an angular unconformity, similar to the unconformity shown in Fig. 15.31B, marked by the Shutup Conglomerate at the base of unit 2. The angular nature of the unconformity is visible both in the eastern part of Fig. 17.25 between the unit 2 and unit 1 designations, and in the western part of the dome immediately below the unit 1 designation.

Unit 3 is the teardrop-shaped highland area distinguished by its light shading. Although not obvious in the figure, unit 3 is a fault block completely surrounded by a steeply dipping circular ring fault that formed when the rock unit dropped as a coherent block into the caldera following eruption. We will refer to unit 3 as the collapse block. It is composed of several different rock units. Most of the block consists of the same Cretaceous formations that form unit 2. The Caballos Novaculite and Maravillas Chert (unit 1) form the north end of the block. Intrusive rocks and lava flows form some of the darker areas visible within the collapse block. The most interesting aspect

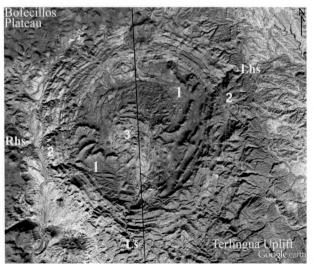

Ls-Lower Shutup, Lhs-Lefthand Shutup, Rhs-Righthand Shutup. 1-Paleozoic rocks, 2-Lower Cretaceous Rocks, 3-Collapse Block.

FIGURE 17.25 Google Earth image of the Solitario, Texas, oriented north.

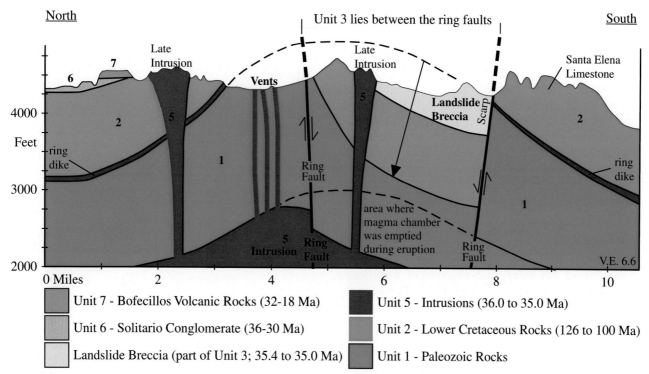

FIGURE 17.26 Schematic north to south cross-section of the Solitario along the thin black line shown in Fig. 17.25. Based on Henry et al., 1997.

of the collapse block is that the entire southern half consists of landslide breccia that we will describe shortly.

Fig. 17.26 is a north to south cross-section along the thin black line shown in Fig. 17.25 that shows the landscape and general rock structure across the Solitario. The cross-section can be used to illustrate an interpretation on the origin of the Solitario. The sequence of events consists of magma intrusion pushing upward to form a dome structure at about 36 million years ago. Volcanism 35.4 million years ago began with eruption of ash-flow tuff along vents within Paleozoic rocks (unit 1) surrounding the caldera. These vents are shown schematically in cross-section as three thick lines. The eruption partially emptied the magma chamber below the caldera resulting in caldera collapse. In Fig. 17.26, the area below the lower dashed line represents the part of the magma chamber that was emptied during eruption. Following eruption, unit 3 (the collapse black) dropped as a single coherent unit into the space previously occupied by the magma chamber. The block was hinged on the north side such that the south side fell a much greater distance, thus tilting the block southward and exposing Paleozoic rocks at its north end. A high, steep fault scarp was created at the southern end of the block that is labeled scarp in cross-section. The upper dashed line shows where the contact between the Cretaceous and Paleozoic rocks was located prior to collapse. The arrow connects the pre-eruption Cretaceous-Paleozoic rock contact with the inferred present-day location of the contact following collapse. The arrow shows how far down the collapse block (unit 3) dropped.

The cross-section suggests that the block dropped well over 1000 feet. Most of this displacement is likely to have occurred during a single catastrophic faulting event immediately following eruption, leaving behind a steep, highly unstable fault scarp along the south wall (labeled scarp). Further sinking of the caldera and the unstable nature of the fault scarp wall created numerous landslides that, today, are preserved in the form of the landslide breccia. The breccia is coarse-grained and rather chaotic (nonlayered) in its southern part where one can find giant blocks of units 1 and 2 more than 600 feet in diameter. Also present in the landslide breccia are fluvial (stream) deposits, lava flows, and ash-flow tuff that formed following the main eruption and collapse. As shown in cross-section, the breccia covers Cretaceous rocks across the entire southern half of the caldera. The entire cycle of formation and collapse took less than 1 million years. From this example alone we see how quickly volcanism can reincarnate an area.

Located in the Alpine-Fort Davis area, the Davis Mountains field is the largest volcanic-intrusive part of the Trans-Pecos. With two peaks above 8000 feet and many above 7000 feet, it forms the second-highest mountain range in Texas (second to the Guadalupe Mountains, which extend into New Mexico). The mountains capture up to 18 inches of precipitation, which is twice the precipitation of surrounding lowlands, creating a separate desert ecosystem that includes emory oak, live oak, piñon pine, and alligator juniper. Volcanism occurred between 38 and 32 million years ago in the form of volcanoes, calderas, and fissure vents. Calderas are located at Paradise Mountain, El Muerto Peak, and at Buckhorn Mountain, and one volcano is known at Paisano Peak. Most of the rock consists of rhyolite lava flows and tuff with little of the andesite found in other areas related to the mid-Tertiary ignimbrite flare-up. Intrusive rocks are present in several areas including the summit of Baldy Peak (Mt. Livermore), the highest peak at 8381 feet.

Sierra Blanca

Sierra Blanca (SB, Fig. 17.18) is an old eroded volcano located about 128 miles south-southeast of Albuquerque, New Mexico near White Sands National Monument. The summit is part of the Mescalero Apache Indian Reservation and requires a permit to access. The volcano was active between 38 and 26 million years ago at about the same time as activity in the nearby Mogollon-Datil field (MD), and so is included in the ignimbrite flare-up. The level of erosion is such that the mountain summit is an intrusive dome surrounded by volcanic rocks. The rocks are mostly andesitic in composition, alkali-rich and quartz-poor. Basalt, rhyolite, and volcanic breccia are also present. Most of the intrusions were emplaced near the end of volcanism, but a few occurred prior to volcanism. A series of dikes crosses Highway 380 along the northeast side of the mountain near Capitan. Volcanic rocks are restricted to the Sierra Blanca area, but intrusive rocks (intrusive domes) are more widespread especially to the north where they form most of the peaks in the Jacarilla, Gallinas, and Capitan Mountains. Sierra Blanca is located at the eastern margin of the Rio Grande rift system. The great height of the mountain (11,977 feet) is due to uplift along normal faults on the west flank that have been active in the past 15,000 years.

Navaho Volcanic Field and Shiprock

The Navajo volcanic field (N, Fig. 17.18) consists of more than 80 relatively small intrusions and volcanic features scattered widely across the nearly flat-lying sedimentary layers of the Four Corners area of the Colorado Plateau. K-Ar ages indicate that volcanism occurred during the latter part of the ignimbrite flare-up between 28 and 19 million years ago. Different parts of the field show different levels of erosion. Most of the landforms are shallow intrusive dikes and volcanic necks, but less eroded lava flows, and at least one maar crater (the Narbona Pass volcano), are preserved in the Chuska Mountains. The volcanic necks can be described as diatremes because

FIGURE 17.27 Google Earth image looking eastward at Shiprock volcanic neck and radiating dikes on the Colorado Plateau, New Mexico, near the Four Corners area.

they were brecciated during water-charged explosions that created the maars at the surface.

The most recognizable landform is Shiprock, located north of the Chuska Mountains and shown in Fig. 17.27. Shiprock is a brecciated volcanic neck (a diatreme) that rises some 1700 feet from the desert floor. A series of six vertical dikes radiate like giant curved walls from the central neck, some of which are visible in Fig. 17.27. The presently exposed Shiprock volcanic neck and associated dikes likely developed some 2500 to 3300 feet below the surface as part of a maar, and are exposed due to erosional stripping of overlying and surrounding volcanic and sedimentary rocks. There are numerous additional diatremes in the Navaho field including Mules Ear, Mitten Rock, Agathla Peak, and Green Knobs, whose surface expression may also have been maars prior to erosion.

The rocks that form the Navajo field, including Shiprock are unusual. The two principal rocks are minette and an altered ultramafic microbreccia. Minette is a quartz-absent, dark-colored rock that consists of potassium and magnesium-rich minerals. In the Navaho field, the rocks consist of mica, pyroxene, alkali feldspar, apatite, and in some areas, olivine. The microbreccia is greenish with olivine and other minerals in a fine-grained matrix. The mineralogy of these rocks suggests they originated from deep below the Colorado Plateau and likely reached the surface along fractures in the underlying North American crystalline shield. They are unrelated to subduction.

Also present in this area are remnants of a giant sand dune field known as the Chuska erg, in existence between about 34 and 27 million years ago. The largest remnant is in the Chuska Mountains within the Navaho volcanic field shown in Fig. 17.18, where deposits reach a thickness of 1750 feet (533 m). During its peak, some 30 million years ago, the Chuska erg may have stretched westward along the Utah-Arizona border to the Colorado River, southward to the Mogollon-Datil volcanic field, and northeastward to the San Juan volcanic field. With the exception of the Chuska Mountains, much of the evidence for the Chuska erg has since been eroded.

Pinnacles, Neenach, and Nine Sisters

A few small volcanic fields are present across the southern California strike-slip structural province (Fig. 17.18). Pinnacles National Park in the Gabilan Mountains in the Coast Ranges east of Salinas Valley represents part of a circa 23-million-year-old rhyolitic volcanic field. The area is famous for its massive monoliths, spires, and sheer-walled cliffs exploited by rock climbers. This small volcanic area is located west of the San Andreas Fault. The other half of the volcanic field is located near Neenach, which is east of the San Andreas fault in the Mojave Desert 192 miles to the south. The volcanic field was split in half by displacement on the San Andreas Fault.

The Nine Sisters (or the Morros) form a linear array of extinct volcanic peaks in the Coast Ranges that extends for 16 miles from the Pacific Ocean through San Luis Obispo. The volcanoes were active between 28 and 20 million years ago, but have since been eroded to form volcanic necks composed of fine-grained intrusive rocks of granitic composition. Five of the largest volcanic necks are labeled in a Google Earth image shown in Fig. 17.28. The surrounding valleys, hills, and ranges shown in the figure consists dominantly of older (Mesozoic) subduction-related melange and ultramafic rocks.

The Nine Sisters, the Pinnacles, and the Neenach field, could each be associated with subduction of the Juan de Fuca plate beneath the North American plate at a time when the Juan de Fuca plate extended farther south than today. The linear orientation of the Nine Sisters suggests that magma may have found an easy path to the surface along a fracture or fault zone.

CORDILLERAN VOLCANIC AREAS YOUNGER THAN 20 MILLION YEARS

Volcanic rocks younger than 20 million years old include the previously discussed Columbia Plateau and Snake River Plain, as well as rocks associated with the Cascadia volcanic arc system, to be discussed in Chapter 19. Additional volcanic rocks of this age are abundant around the margins of the Colorado Plateau, and in the northern California-Nevada-Oregon area within the Basin and Range and Sierra Nevada. Their distribution is shown in blue in Fig. 17.18. The larger more predominant fields are low plateaus and mesas composed of basaltic lava flows, cinder cones, fissures, and maars. But, situated within these fields are high stratovolcanoes and large calderas of silicic rock, principally andesite and rhyolite. These fields are largely unmodified by normal faults.

Uinkaret and Markagunt Volcanic Fields

Volcanic rocks on the Colorado Plateau in the Grand Canyon–High Plateaus region include the Uinkaret field (U) near the rim of the western Grand Canyon in Arizona, and the Markagunt field (M) in the High

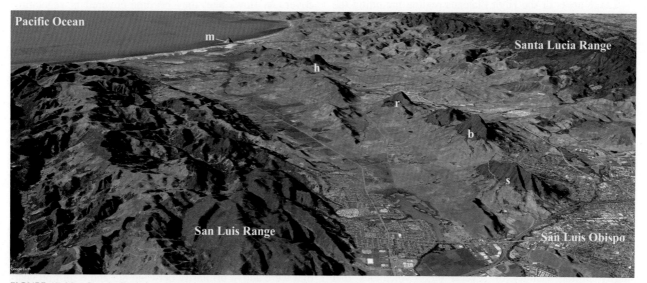

FIGURE 17.28 Google Earth image looking north-northwest near San Luis Obispo, California. The Nine Sisters form a linear array of volcanic necks that extends for 16 miles through the center of the image from the Pacific coast through San Luis Obispo. The five labeled volcanic necks are Morro Rock (m, 578 feet), Hollister Peak (h, 1409 feet), Cerro Romualdo (r, 1307 feet), Bishop Peak (b, 1559 feet), and Cerro San Luis Obispo (s, 1292 feet), Morro Rock, on the Pacific coast, is home to the peregrine falcon. Its shape is the result of quarrying, which continued until 1969.

420 PART | II Structural Provinces

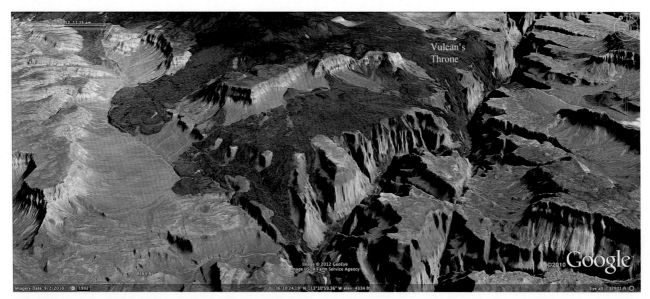

FIGURE 17.29 Google Earth image looking northward at basalt flows of the Uinkaret volcanic field in the western Grand Canyon. A few of the young basalt flows are seen spilling over the rim near Vulcan's Throne. Arrows point at the approximate trace of the active Hurricane normal fault. The active Toroweep normal fault lies just east of Vulcan's Throne.

Plateaus region of Utah. The Uinkaret field contains at least 213 volcanic cones and associated basaltic lava flows. The oldest are between 3.7 and 2.6 million years old and are located in the Mount Trumbull Wilderness. The youngest is only 1000 years old. An amazing site is in the Grand Canyon south of Mount Trumbull where basalt flows less than 723,000 years old are seen frozen in time tumbling down the canyon walls of the Colorado River. A Google Earth image of this site is shown in Fig. 17.29. Vulcan's Throne, a 700-foot tall cinder cone, sits on the rim of the Canyon.

Large areas of the High Plateaus region, including the Markagunt and Aquarius Plateaus, are covered with young undissected basaltic lava flows that form a flat, sloping cap-rock surface on part of both plateaus. There are at least 50 cinder cones and basaltic lava flows, mostly from 5.3 million years old to less than 10,000 years old. These rocks partly cover older volcanic rocks associated with the central Nevada-Utah ignimbrite fields.

San Francisco Volcanic Field

The San Francisco volcanic field (SF) is located at the southern margin of the Colorado Plateau near Flagstaff, Arizona. It is a large field with more than 600 volcanoes, cinder cones, and lava flows, all less than 6 million years old. Cinder cones and lava flows are mostly basaltic in composition. Silicic rock (andesite-rhyolite) is present primarily in the vicinity of several large volcanoes that include San Francisco Mountain (Humphreys Peak, 12,633 feet), Kendrick Peak (10,422 feet), Bill Williams Mountain (9256 feet), and Sitgreaves Mountain (9389 feet). Three of these volcanoes are visible in Fig. 17.30, which looks northwestward across the volcanic field. San Francisco Mountain was built between 1 million and 93,000 years ago. Its high point is Humphreys Peak, the highest point in Arizona. The youngest volcano in Arizona, Sunset Crater (S), is also visible in the image. This small cinder cone last erupted in 1085 A.D. The dark spot at upper right in the image is SP Crater and lava flow, the age of which is not well constrained between 71,000 and 4000 years old. Young lava flows often stand out as black blotches due to lack of vegetation and soil cover. As flows age, they become covered in vegetation, and consequently reflect lighter shades. One can clearly see different shades in the foreground of Fig. 17.30, nearly all of which is covered in basalt. Given this generalization, the deep black color of the SP flow suggests it is on the young side of its age estimate. The San Francisco field is thought to have developed above an ancient fault or fracture zone along which magma was able to reach the surface. Although no eruptions have occurred for almost 1000 years, it is likely that the field is still active.

Hopi Buttes Volcanic Field

About 80 miles east of San Francisco Mountain, the remnants of another volcanic field, the Hopi Buttes (H), forms a circular area of maars, diatremes, mesas, and buttes on the Colorado Plateau. There are about 300 volcanic centers that were active primarily between 8.5 and 6 million years ago with the most recent activity about 4.5 million years ago. Most of the rock is basalt. The eastern part of the field shows the least erosion, preserving surface landforms such

FIGURE 17.30 Google Earth image looking northwestward across the San Francisco volcanic field. S-Sunset Crater, SP-SP Crater.

as maar craters and basaltic lava flows. Erosion of sedimentary layers surrounding the more resistant lava flows has created lava-capped mesas that stand 300 to 500 feet above the surrounding plateau. The western part of the volcanic field is more deeply eroded. It preserves intrusive diatremes (volcanic necks) and more thoroughly dissected mesas. Thus, the level of erosion has created a cross-sectional view of volcanic features. By walking westward, it is possible to walk from examples of surface volcanoes to examples that show the interior of a volcano.

Grand Mesa

Grand Mesa is located at the northeast edge of the Colorado Plateau east of Grand Junction, Colorado. It is reportedly the largest mesa (flat-topped mountain) in the world. The mesa is Y-shaped with the base oriented east. The northwest top of the Y is shown in Fig. 3.10. The mesa is surrounded on three sides by major river valleys. Plateau Valley lies to the north, the Gunnison River to the south, and the confluence of the Gunnison and Colorado Rivers to the west. The basalt that forms the resistant caprock layer of the mesa emanated from two source areas in the east, and flowed westward into lowland valleys. The basalt is as thick as 620 feet in the east and as thin as 50 feet in the west. $^{40}Ar/^{39}Ar$ dates on several different flows range from 9.45 to 10.99 million years. Basalt-capped plateaus of similar age are present in areas north of Grand Mesa including Battlement Mesa and Flat Top Mountain as shown in Fig. 17.18. The Colorado River lies north of Battlement Mesa.

Grand Mesa is perhaps the most extreme example of topographic inversion in the United States. The surface is a flat tableland tilted westward and capped with basalt. It stands as high as 11,224 feet in the east and as low as 9630 feet in the west. There is 5400 feet of relief between the western end of Grand Mesa and the Gunnison River 11 miles to the southwest. Given that basalt initially flowed into lowlands, such relief indicates that more than 5000 feet of rock has been removed from around Grand Mesa via river downcutting and layer stripping since the basalt was deposited roughly 10 million years ago.

Grand Mesa is almost entirely surrounded by landslide material. Most impressive are giant slump blocks called Toreva blocks that have slid off the cliff face and rotated backward along a very weak claystone rock layer as depicted in Fig. 17.31A. Each slumping Toreva block creates an escarpment (a fault scarp) up to 500 feet high. Water is trapped at the base of the escarpment creating numerous lakes. Fig. 17.31B looks southwest at the Mesa Lakes region near the town of Mesa. The Grand Mesa tableland extends from lower left to top right in the image. There are several slump blocks below the tableland, each forming a ridge more than a mile long. Freeze-thaw activity, erosion, and steady downslope creep of material has destroyed some of the older slide blocks to create an irregular landslide surface part of which is seen in the upper right of the figure.

(A) 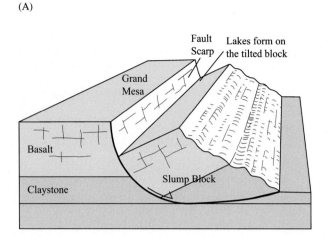 (B)

FIGURE 17.31 (A) Sketch that shows formation of Toreva slump blocks on Grand Mesa, Colorado. Redrawn from Figure 22 of Yeend (1969). (B) Google Earth image looking southwest along the northern edge of Grand Mesa at three rotated slump blocks in the foreground, and several slumps and slides in the background. Intrusive domes of the La Sal Mountains are seen in the distance.

Slumping and subsequent erosion are so pervasive that the entire east side of Grand Mesa has been reduced to landslide debris. Here, the only remnant of the intact basalt tableland is a narrow, knife-edged ridge known as Crag Crest. A wide landslide area of hills and interspersed lakes is present on both sides of the ridge. The nearby Battlement Mesa has also been nearly completely reduced to landslide debris, leaving only a small area of basalt caprock to form the mesa. It is unclear how long ago slumping began, but it has been active throughout the recent continental glaciation, and is active today.

Jemez Lineament

The Jemez lineament is a chain of volcanic centers that extend from southern Arizona east of Phoenix, across northwestern New Mexico to the Colorado border, and then eastward to the Oklahoma border (Fig. 17.18). Beginning in Arizona, the lineament includes the San Carlos-Springerville (SS), Mount Taylor (MT), Zuni-Bandera (ZB), Jemez (J), and Raton-Clayton (R) volcanic fields. The lineament cuts obliquely across several structural provinces. It begins in the Basin and Range, extends along the southeast margin of the Colorado Plateau, crosses the Rio Grande Rift and Southern Rocky Mountains, and ends on the Great Plains. Most of the lineament is shown in Fig. 18.9.

The lineament boasts numerous cinder cones, lava flows, and a few high, extinct volcanoes. Volcanism began sometime prior to 13 million years ago and remains active. Much of the volcanism has occurred in the past 4 million years. Basalt is the dominant rock, but explosive silicic volcanism is an important feature. Because the lineament parallels the track of the Yellowstone hot spot (Fig. 17.18), some have suggested a hot spot origin. However, unlike the Snake River Plain, a characteristic age progression is absent. Seismic evidence has revealed discontinuities in the lower crust that suggest instead that magma reached the surface along an ancient, circa 1.65- to 1.68-billion-year-old fault zone. We next discuss the volcanic fields along the lineament from southwest to northeast.

Volcanic Fields in the Southern Part of the Jemez Lineament

The San Carlos, Springerville, Zuni-Bandera, and Mount Taylor volcanic fields form the southern part of the Jemez lineament roughly between San Carlos, Arizona, and Albuquerque, New Mexico. These fields overlap, in part, with the older Mogollon-Datil field (Fig. 17.18). The southernmost (San Carlos) field contains basaltic lava flows that erupted between 7 million and 500,000 years ago. The most prominent feature is Peridot Mesa, located between the towns of Peridot and San Carlos. The basalt that forms the caprock layer on the mesa is unusual because it contains nodules (inclusions) of a mantle-derived rock rich in olivine. Apparently, the basalt magma carried the olivine nodules to the surface in a gas-charged diatreme that originated deep in the mantle. The olivine in these nodules is special because it is a beautiful gem-quality, yellowish-green variety known as peridot, from which the mesa gets its name. Peridot Mesa is not accessible to the public without permission.

The Springerville field, located in Arizona near the New Mexico border, includes as many as 400 cinder cones and associated lava flows mostly between 2.1 million and 300,000 years old. Nearly all of the small hills between the towns of Greer and Show Low are cinder cones. Most of the rock is basaltic in composition. The

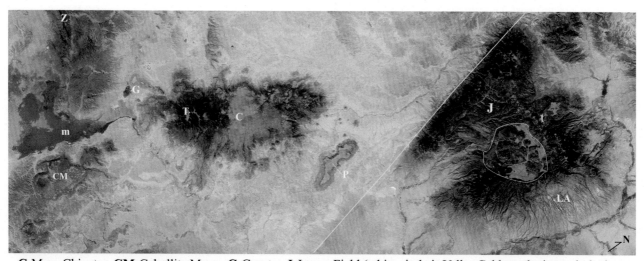

C-Mesa Chivato, **CM**-Cebollita Mesa, **G**-Grants, **J**-Jemez Field (white circle is Valles Caldera, the inner dashed circle is the resurgent dome), **LA**-Los Alamos, **m**-McCarthys Flow, **P**-Mesa Prieta, **T**-Mount Taylor, **t**-Toledo Caldera, **Z**-Zuni Mountains. Active normal faults of the Grande Rift zone lie to the right of the yellow line.

FIGURE 17.32 Google Earth image looking northwest at McCarthys Flow (m) on the left side, the Mount Taylor field (T), and the Jemez field (J) at right. Young basalt flows, such as McCartys Flow and the small flow at Grants (G), show black reflectance. Older flows form basalt-capped mesas such as Mesa Prieta (P), Cebollita Mesa (CM), and Mesa Chivato (C). The yellow line marks the western limit of active normal faults associated with the Rio Grande Rift zone. The Jemez field hosts the Valles Caldera (white circle) within the Rio Grande rift.

most prominent feature is the 11,408-foot volcano, Mount Baldy, where rhyolite and andesite compositions dominate. This extinct volcano straddles the area between the Jemez lineament and the Mogollon-Datil field. It was active between 12 and 8 million years ago, and therefore, it is older than the Springerville field and younger than the Mogollon-Datil field.

The Zuni-Bandera field surrounds the southeastern and eastern side of the Zuni Mountains in New Mexico. This area is labeled lava fields in Fig. 17.18. Cinder cones are present, but most of the field consists of basaltic flows that erupted in the past few million years. The youngest flow, known as McCartys Flow, is only 3000 years old. It flowed northward along the east side of the Zuni Mountains for about 35 miles, ending near the town of McCartys. There are several additional flows in the area, all less than 20,000 years old. Fig. 17.32 looks northwest at McCarthys Flow (m) on the left side, the Mount Taylor field (T) at center, and the Jemez field (J) at right. The city of Grants (G) is located adjacent to another young lava flow.

The Mount Taylor volcanic field is dominated by Mount Taylor (T), a 11,306-foot extinct volcano composed of several volcanic centers, each of which extruded lava and ash of different composition. The volcanic field was active between 3.9 and 1.25 million years ago. The Mount Taylor volcano was built upward by numerous lava domes and flows of primarily andesite, but also dacite and rhyolite. Many of the flows froze before leaving the side of the mountain. Toward the end of its active history, basaltic lava was extruded from vents distributed around the base of the Mt. Taylor cone. Erosion of surrounding sedimentary rock has transformed these flows into basalt-capped mesas elevated 1000 to 2000 feet above surrounding land surfaces. The largest is Mesa Chivato (C), visible in Fig. 17.32 along the northeast (right) side of Mt. Taylor where one can find numerous cinder cones, such as Cerro Chivato, and maar craters that form small lakes such as Laguna Reyes. The mesas are partly surrounded by landslide deposits. Remnant basalt/andesite flows also form Cebolitta Mesa (CM) and Mesa Prieta (P).

Jemez Volcanic Field and Valles Caldera

There is one area in particular along the Jemez lineament where explosive rhyolite volcanism has occurred at a scale on par with those of the Yellowstone eruptions. The area is known as the Jemez volcanic field (J in Figs. 17.18 and 17.32). It lies just west of Santa Fe at the western edge of the Rio Grande Rift, and is approximately 50 miles across. Active normal faults associated with the rift zone are concentrated in the area to the east (right) of the yellow line in Fig. 17.32, which itself marks an active normal fault.

Much of the Jemez field consists of a single large volcano with a central crater 7 to 12 miles across known as the Valles Caldera (or Sierra de los Valles). An older, smaller, and partly destroyed crater, known as the Toledo Caldera, is best preserved at the northeast margin of Valles Caldera. The Jemez field has shown nearly continuous volcanism for the past 13 million years with more than half of the volcanic rocks accumulating between 10 and 7 million years ago. Fig. 17.32 shows the Jemez field with the Valles Caldera outlined with a white line, and the Toledo Caldera labeled (t). An inner resurgent dome is outlined with a dashed white line. The Valles Caldera formed via two explosive eruptions approximately 1.45 and 1.12 million years ago. Both explosions produced thick deposits of pumice and ash known today as the Bandelier Tuff. The second explosion was the largest and is responsible for producing the final shape of the caldera as well as the greatest thickness of Bandelier Tuff. Excellent exposures of Bandelier Tuff can be seen at nearby Bandelier National Monument about 6 miles east of Los Alamos (LA). In addition to tuff, Bandelier Monument preserves evidence of human presence that dates back as far as 11,000 years including the ancient Pueblo People from 1150 to 1600. Ceremonial structures and cliff dwellings are preserved in the soft tuff layers.

Originally, it was thought that the smaller Toledo Caldera formed during the 1.45-million-year explosion. However, based on the distribution, thickness, and flow direction of associated volcanic deposits, more recent analysis suggests instead that volcanic debris of this age was ejected from the Valles Caldera. It is now understood that the Toledo Caldera formed a short time prior to 1.45 million years ago.

The resurgent dome, outlined with a dashed white line in Fig. 17.32, developed within 100,000 years following caldera collapse. The round, dome-shaped protrusions within the caldera surrounding the resurgent dome are lava domes of obsidian and rhyolite that pierced the surface and then froze beginning some time after the 1.12-million-year explosion, but prior to about 530,000 years ago. Formation of the lava domes was followed by almost 500,000 years of dormancy that ended between 57,000 and 40,000 years ago with the creation of additional ash deposits and rhyolite lava domes at the southern (lower left) edge of caldera, including the Battleship Rock ignimbrite and the Banco Bonito obsidian dome.

Although there has been no volcanic activity since at least 40,000 years ago, seismic imaging suggests the presence of a magma body within four miles of the surface implying that the region remains active. A drilling project to look into the feasibility of geothermal energy reached temperatures of 340°C at a depth of 2 miles (3.2 km). Although these are very warm temperatures, it was determined that the reservoir was too small to support a commercial project. The Valles Caldera became a government-owned national preserve in 2000.

Raton-Clayton Volcanic Field

The Raton-Clayton volcanic field occupies a large area composed of basalt-capped mesas and cinder cones along the Colorado-New Mexico border, extending from the western edge of the Great Plains eastward to just over the Oklahoma border at Black Mesa, the highest point in Oklahoma. The entire volcanic area can be referred to as the Raton Mesas or the Mesa de Maya. It is the easternmost recently active volcanic field in the United States, and is shown in its entirety in Figs. 17.18 and 13.27. Volcanism was active from about 9 million years ago to about 45,000 years ago. Fig. 17.33 is a Google Earth image oriented north that extends from Culebra Peak (C, 14,053 feet) in the Sangre de Cristo Mountains at the western (left) edge of the figure, across the Park Plateau (P), to the western half of the Raton-Clayton volcanic field.

The mesas are highest and oldest in the northwest part of the volcanic field near Trinidad, Colorado (T) where they reach an elevation of 9500 feet, 3500 feet above the surrounding Great Plains. These mesas are visible in Fig. 17.33. The mesas are tilted eastward such that they are only a few hundred feet above the Great Plains in the eastern part of the volcanic field. Decreasing elevation and relief toward the east suggests uplift of the western Great Plains and eastward tilting within the past 9 million years. Uplift is also suggested by the presence of the highly dissected Park Plateau (P). This area is underlain with sandstone and siltstone derived from erosion of the Sangre de Cristo Mountains. Strong dissection evident across the plateau suggests that streams are downcutting and eroding in response to uplift.

Although basalt is by far the dominant rock, several isolated peaks visible in Fig. 17.33 that include Laughlin Peak (L) and Green Mountain (G) consist of dacite (a composition intermediate between andesite and rhyolite). These volcanoes erupted between 9 and 5 million years ago. Also present is a relatively large volcano, Sierra Grande (S), composed of 4-million-year-old andesite. The most prominent and accessible basaltic cinder cone is Capulin (Cp) in New Mexico, which is designated a national monument. Capulin last erupted about 55,000 years ago. It is one of about 120 cinder cones in the field. These rocks, like other rocks of the Jemez lineament, probably reached the surface along an ancient fracture or fault zone. None are directly related to subduction. Also visible at the northern end of Park Plateau are the Spanish Peaks (Sp).

Carrizozo Lava Flow

Located in the Rio Grande rift zone just north of White Sands National Monument in New Mexico, the Carrizozo lava flow (C, Fig. 17.18) is one of the longest and best preserved basaltic flows of the past 10,000 years. The flow shows up brilliantly in Google Earth images as a long black mark in the New Mexico desert. Except for its very northern end where a short second flow is present, it is a single lava flow that, over the course of about 30

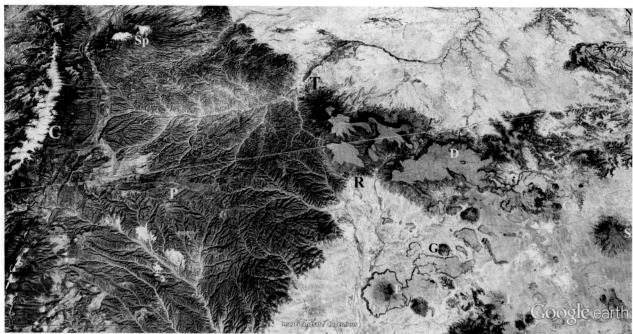

C-Culebra Peak, **Cp**-Capulin Mountain, **D**-Dale Mountain, **E**-Eagle Tail Mountain, **G**-Green Mountain, **L**-Laughlin Peak, **P**-Park Plateau, **S**-Sierra Grande, **Sp**-Spanish Peaks. Towns: **R**-Raton, **T**-Trinidad.

FIGURE 17.33 Google Earth image looking northward across the Sangre de Cristo Mountains (C), the Park Plateau (P), and the western half of the Raton-Clayton volcanic field. The Spanish Peaks (*Sp*) are visible.

years, slowly and continuously emanated from Little Black Peak, a small cinder cone at its northern end. The flow is between 5200 and 4800 years old, and is more than 45 miles (73 km) long. Route 380 crosses the flow providing magnificent views and access.

Northern Nevada Rift Zone

The Northern Nevada rift zone is a conspicuous area of rhyolite and basalt located in the northern part of the Basin and Range. It is shown in Fig. 17.18 as a narrow zone that extends for about 300 miles from the Nevada-Oregon border southward through central Nevada. The rift zone is defined by an alignment of volcanic flows, dikes, normal faults, and gold-silver deposits between 18.6 and 13.6 million years old that cuts across the grain of present-day ranges without obvious topographic expression. Most of the rocks are between 16.5 and 15 million years old. The rift is further identified by a strong magnetic signature that indicates the presence of a deep, linear, normal fault (graben-like) structure filled with silicic and basaltic volcanic rocks. This structure is not visible at the surface.

The rift is interpreted as a product of local stresses imposed during the passing of the Yellowstone hot spot. The hot spot created a bulge that caused the area to stretch and crack, creating a linear fracture (a trough) that filled with magma, some of which reached the surface. The volcanism was short-lived and had largely ended by 15 million years ago, presumably when this part of the North American plate had migrated sufficiently far from the hot spot. The fact that the rift zone cuts obliquely across present-day mountain ranges indicates that it developed prior to the initiation of normal faults responsible for the present-day landscape.

The Northwest Basin and Range and Northern Sierra Nevada

West of the Northern Nevada rift, we can find a great variety of volcanic rocks spread out over a large area of the Basin and Range and northern Sierra Nevada. This region, labeled NO in Fig. 17.18, includes basalt, rhyolite, andesite, and tuff of all ages from 20 million years to the present, as well as some volcanic rocks that are older than 20 million years. Much of the basalt is associated with the Columbia River flood basalt. Silicic rocks and basalt too young to be part of the flood basalt are more likely related to subduction associated with the Cascadia volcanic arc system. Overall, the landscape is more strongly influenced by Basin and Range normal faulting than by the volcanic rocks themselves.

Within the Sierra Nevada, the Lovejoy Basalt consists of a series of flows that erupted from a vent at Thompson Peak in the Diamond Mountains west of Honey Lake, and flowed westward into the Sacramento Valley near Lake Oreville. This is the largest single volcanic unit in California. It is 15.4 million years old and is understood to be associated with the Columbia River flood basalt and Yellowstone hot spot. There are numerous additional locations in the Sierra Nevada where andesite and rhyolite form small volcanic domes, ignimbrites, and volcanic breccia. These rocks have filled ancient streambeds thus displacing stream water to the side of the channel as shown in Fig. 17.2. The rocks are between 14 and 6 million years old and are associated with the Cascadia volcanic arc system when the Juan de Fuca plate extended farther south than its present position. Volcanism ended when the San Andreas Fault grew longer and replaced the subduction zone with a strike-slip system.

Long Valley Caldera and the Inyo-Mono Craters

There are many volcanic fields interspersed with sedimentary and crystalline rock along the boundary zone between the Sierra Nevada and Basin and Range. This area, labeled LV in Fig. 17.18, includes the circa 80,000 year-old basalt flows that form 5- and 6-sided polygonal columns at Devil's Postpile west of Mammoth Lakes, as well as the massive, silicic lava dome (volcano) that is Mammoth Mountain, cinder cones and lava flows near Big Pine that include Red Mountain and Crater Mountain, the giant Long Valley Caldera, and the Inyo-Mono Craters. Most are visible in Fig. 17.34, which looks southward along western flank of the Sierra Nevada. Volcanism in this area began about 3.8 million years ago with basalt and andesite eruptions that, over time, became more silicic. Glass Mountain (G) is composed of rhyolite lava domes, lava flows, and pyroclastic flows 2.1 to 0.8 million years old.

Mammoth Mountain, near the rim of the Long Valley caldera, was built between 230,000 and 57,000 years ago by successive rhyolitic and dacitic lava domes and flows that chemically appear unrelated with the Long Valley caldera. The youngest volcanic activity near Mammoth Mountain took place between 8900 and 8500 years ago with the formation of Red Cones, two basaltic cinder cones located within the Sierra Nevada 6.5 miles southwest of Mammoth Lakes. Water-charged (phreatic) explosions took place along the north side of the mountain only 700 years ago.

The Long Valley caldera (LV) is visible in Fig. 17.34 as a 9- by 19-mile elliptical depression located between Mammoth Mountain (M) and Glass Mountain (G). The

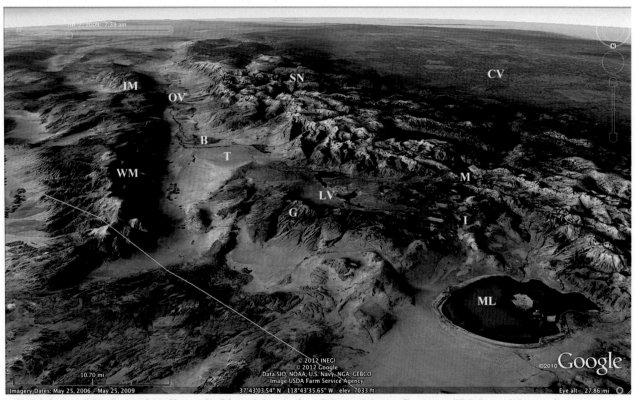

B-Bishop, **CV**-Central Valley, **G**-Glass Mountain, **I**-Inyo-Mono Craters, **IM**-Inyo Mountains, **LV**-Long Valley Caldera, **M**-Mammoth Mountain, **ML**-Mono Lake, **OV**-Owens Valley, **SN**-Sierra Nevada, **T**-Volcanic Tableland, **WM**-White Mountains.

FIGURE 17.34 Google Earth image looking south along the east side of the Sierra Nevada at the Long Valley Caldera-Mono Lake region. The white line is the California-Nevada border.

caldera is one of three major calderas in the United States. It formed via violent explosion 758,900 years ago. The eruption is understood to have lasted at least 90 hours. It created a huge ash fall that swept as far east as Kansas. Pyroclastic flows and volcanic mudflows filled and then swept over the caldera wall covering an area of more than 850 square miles (2200 km^2) to a depth of as much as 656 feet (200 m). These rocks, known today as the Bishop tuff, form the Volcanic Tableland (T) to the southeast, and the area between the caldera and Mono Lake (ML) to the north. Within 100,000 years following eruption and caldera collapse, rising magma resulted in uplift of the caldera floor in the form of a resurgent dome. Obsidian and rhyolite lava flows associated with resurgence now cover the caldera floor along Highway 395 south of Crestview. Additional rhyolite, and in some cases basaltic lava flows and cinder cones, were extruded onto the caldera floor roughly 500,000, 300,000, and 100,000 years ago.

The Long Valley caldera remains active and dangerous. In the United States, it is second only to the Yellowstone caldera in terms of its capacity to create a large volcanic eruption. The caldera floor has risen more than 31 inches since 1978 suggesting that magma as little as 1.8 miles below the surface is on the move. Additional evidence of magma movement includes a swarm of earthquakes beneath Mammoth Mountain in 1989 that lasted more than 6 months, and the discovery of trees that were suffocated by exceptionally high concentrations of CO_2 emitted into the soil by cooling underground magma. The 1989 seismic event is now thought to be associated with the emplacement of dikes at depth beneath Mammoth Mountain. The US Geological Survey, other government agencies, and universities continue to monitor this potentially explosive situation.

The most recent volcanic activity in the region has been along the Inyo-Mono Craters volcanic chain, which extends northward for 30 miles beginning just west of the Long Valley caldera and ending at Mono Lake. The volcanic chain is visible as small hills labeled IM in Fig. 17.34. It consists of small explosion craters (maars), silicic lava domes, and obsidian flows. The southern part of the chain forms the Inyo Craters, the northern part forms the Mono Craters. Volcanic activity along the chain began as early as 50,000 years ago, but most of the activity and all of the surface features have formed in the past 6000 years. Eruptions between 700 and 500 years ago created several features located just north of Mammoth Mountain that include the Inyo Crater Lakes, two explosion craters atop Deer Mountain, and several rhyolite/obsidian domes including South and North Deadman Domes, Glass Creek Dome, and Obsidian Dome. Several of these landforms are shown in Fig. 17.35.

Spectacular tufa towers dot the shoreline at Mono Lake. Tufa towers are tall columns of calcium carbonate (limestone) that form below lake level through chemical reaction of spring water with saline lake water. The towers are exposed because the lake is much smaller, and lake level much lower, than in the past. Lava and craters on the north and east shores of the island erupted as recently at 170 years ago. Other features are also quite young. Black Point, now exposed on the northwest shore of Mono Lake, is a basaltic cone that formed below lake level 13,300 years ago. Of several islands on Mono Lake, Paoha Island is a domed mass of lake-bottom sediment uplifted between 230 and 100 years ago by an intrusion at depth. Lava on Negit Island erupted between 2000 and 200 years ago.

Sutter Buttes

We end our discussion of young volcanic rocks with what is probably the most famous volcanic landmark in the Central Valley of California, the Sutter (Marysville) Buttes, located in the vicinity of Yuba City (S in Fig. 17.18). These volcanic rocks form a series of small andesite and rhyolite lava domes. The volcanoes were active between 2.5 and 1.5 million years ago as part of the Cascade subduction system when the Juan de Fuca plate extended farther south than its present position. The buttes formed via subduction just like other volcanoes in the Cascade chain. However, since that time, the southern boundary of the subduction zone has migrated northward causing volcanism to cease in this area. Today the buttes stand as an isolated relict landscape that can be seen for miles rising from the flat valley floor.

FIGURE 17.35 Google Earth image looking north-northwest at (from left) the Inyo Crater Lakes, Deer Mountain, and South Deadman rhyolite/obsidian dome. The Inyo Crater Lakes and South Deadman Dome formed between 700 and 500 years ago. Deer Mountain consists of older volcanic material.

QUESTIONS

1. The Appendix contains an uncolored version of Fig. 17.1. Color this map and describe the distribution of volcanic rocks across the United States.
2. Why are there almost no volcanic rocks in the Interior Plains and Plateaus?
3. What are the characteristics of the following volcanic types: basaltic lava, andesite lava, rhyolite lava, ash-flow tuff? What is ignimbrite?
4. Describe and comment on the origin of the following volcanic landforms, lava dome, caldera, resurgent caldera, basalt-capped mesa, fissure, maar, cinder cone, composite volcano, shield volcano,
5. Using Fig. 17.13 and Google Earth, calculate and compare the velocity and direction of North American plate movement based on the Yellowstone hot spot track between caldera centers numbers 1 and 3, and between numbers 3 and 7. Use the center of each caldera area and average the age ranges of each caldera.
6. Speculate as to why the Columbia River flows around the margin of the Columbia Plateau rather than through the center?
7. Based on Fig. 17.5 alone, how might one infer the flow direction of floodwaters across the Channeled Scablands?
8. In Google Earth, fly to Dry Falls, Washington. Describe what you see. What are the dimensions of the Falls and its height as measured in Google Earth?
9. Trace all of the anticlinal ridges visible in Fig. 17.8. Given that these are actively forming folds what can you infer about the present-day direction of maximum stress in this part of the Earth's crust? What type of faults might you expect to be associated with these folds?
10. In Fig. 17.8, what is the relationship between the Columbia River and Saddle Mountain.
11. Suggest possible hypotheses to explain why the Columbia River cuts through Saddle Mountain but is deflected eastward around ridges north of Rattlesnake Mountain as seen in Figs. 17.4 and 17.8.
12. In Google Earth, fly to Pinnacles National Park, CA and then to Neenach, CA. Use the Google Earth ruler to measure the distance between the two. Given that volcanic rocks in both areas are 23 million years old and that the two areas were once adjacent to each other, calculate the rate of displacement in feet per 100 years, miles per million years, kilometers per million years, and mm/year along the San Andreas fault.
13. Trace the Hurricane fault across Fig. 17.29. How is the fault expressed in the landscape?
14. What is odd about the age progression of volcanic rocks in the Brothers Fault zone relative with the Yellowstone hot spot track? Can you explain the age progression?
15. In Google Earth fly to Jordan Craters, Oregon. What are you looking at? Describe this feature as completely as possible including its size. Speculate on direction of flow, source area, and origin of Cow Lakes, Batch Lake, and the lighter and darker areas near Batch Lake.
16. What are some differences between the two age groups of volcanic rocks discussed in this chapter 70 to 20 Ma and less than 20 Ma? Name the common rock types and landscape features in each group.
17. What is different (structurally) about the Central Nevada volcanic field relative to other fields?
18. What evidence could you look for to determine the age of the Northern Nevada Rift and whether it pre-dates or post-dates Basin and Range normal faulting.
19. What is the age (or ages) of volcanic rocks at Yellowstone National Park? Provide specific examples.
20. In Google Earth, fly to Creede, Colorado. Zoom out and notice the circular valley surrounding Creede. What is the origin of this valley? Do the same by flying to Silverton, Colorado.
21. Describe the Slumgullion Slide.
22. Describe the San Francisco Volcanic field.
23. Research Specimen Ridge in the Absaroka Volcanic Field. What is special about this area?
24. Why hasn't the geothermal potential of the Valles Caldera been utilized?
25. Research any of the following and write a few paragraphs; Mammoth Mountain, Long Valley Caldera, Devil's Postpile, Inyo-Mono Craters, Mogollon-Datil volcanic field, San Juan volcanic field.
26. Name the three volcanoes in Fig. 17.7.
27. What is the origin of Mesa Chivato (C, Fig. 17.32).
28. Explain the difference between an antecedent and superposed river with respect to a water gap.

Chapter 18

Normal Fault Systems

Active and recently active normal faults form perhaps the most conspicuous and widespread landscape in the Cordillera. Not only are normal faults ubiquitous across the Basin and Range physiographic province, they have infiltrated every state in the Cordillera. Normal faults cut the Grand Canyon region of the Colorado Plateau, they cut into the heart of the Colorado Rockies, and they cut lava flows and volcanic features on the Snake River Plain, Columbia Plateau, Cascade Range, and Basin and Range. Normal faults are partly or wholly responsible for some of the highest and most spectacular mountains in the country including the Sawatch Range in Colorado, the Sierra Nevada of California, the Wasatch Range in Utah, and the Teton Range in Wyoming.

Normal faults create a distinctive landscape that manifests itself in the form of small, tilted mountain blocks usually arranged in parallel orientation that alternate with wide, flat-floored, often tilted valleys. The physiographic province that most exemplifies normal fault landscape is the Basin and Range. In this chapter, we discuss the Basin and Range physiographic province and some aspects of its infiltration into surrounding provinces. We will discuss two active arms of Basin and Range topography: the Rio Grande Rift in New Mexico that extends northward into the Southern Rocky Mountains, and the Rocky Mountain Basin and Range, which extends northward through the Northern Rocky Mountains. We will also look in detail at two active normal fault mountains, the Teton Range and the Wasatch Range. There are no active normal faults east of the Great Plains, however, we will discuss an area of ancient normal faults in the Appalachian Mountains known as the Triassic Lowlands (or Triassic Rift Basins) where topography is primarily erosion-controlled rather than structure-controlled.

STRUCTURAL CHARACTER AND TERMINOLOGY OF NORMAL FAULTS

Normal faulting involves the downward dropping of one rock relative to another under the force of gravity. The weight of the rock is the driving force behind normal faulting. In order for normal faults to occur, there must be some form of extension of the earths crust to provide space for the down-dropped block, or there must be removal of underlying rock to allow overlying rock to collapse into a hole such as a caldera. With respect to landscape development, extension of the earths crust via tensional stress is the more important mechanism.

A common method by which tensional stress is generated in rock is illustrated in Fig. 18.1A. In this figure, opposing horizontal tectonic forces pull the rock apart. Tectonic forces of this type, brought on by plate movement, are common in the Earth's crust and are perhaps the only forces capable of generating long-lived tensional stress in rock. Anticlinal folding by vertical tectonic forces or by horizontal compressional tectonic stress, as shown in Fig. 18.1B, is also capable of producing tensional stress in rock, but only in the crest of the fold where the total amount of extension is small. Under these conditions, fractures (joint sets) typically develop rather than normal faults, and consequently normal fault landscapes of this type are uncommon. In this section, we discuss some of the more characteristic features of normal faults including horst and graben structure, tilted fault blocks, half-grabens, detachments, rollover anticlines, rotations, and flexural rebound.

Horst and Graben Structure

Rocks, as we all know, are rigid and brittle near the surface. If tectonic forces act to pull them apart, then they will break into blocks separated by steeply dipping faults as shown in Fig. 18.2. Blocks that have the shape of a downward-tapered wedge will slide downward along fault planes under the force of gravity. The adjacent block will remain elevated thus producing the most distinctive set of landforms associated with normal faults, a basin and a range. In geology, we refer to the down dropped block as a graben and the up-thrown block as a horst. In this terminology, the basin is the landform and the graben is the structural form irrespective of whether or not the graben creates a topographic basin.

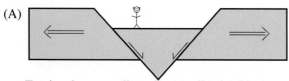

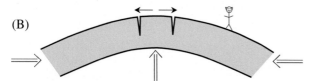

FIGURE 18.1 Mechanisms by which extension occurs.

Returning to Fig. 18.2, did you notice that extension and normal faulting results in an increase in surface area? Compare the width of the cross-sections in Figs. 18.2A and 18.2D. This is an important consequence of normal faulting. A second important consequence is that the process of normal faulting results in tectonic exhumation. Notice that rock 1 is deeply buried in Fig. 18.2B, but has reached the surface in Fig. 18.2D without any erosion. We will also discover that normal faults are not always steeply dipping and that isostasy and associated volcanism play an important role in landscape development.

Tilted Fault Blocks, Half-Grabens, and Flexural Rebound

Fig. 18.3 is a schematic representation of normal faulting across the southern Basin and Range without significant erosion. Ranges are bordered on one or both sides by normal faults that dip steeply near the surface. If two adjacent faults dip in opposite directions then horst and graben structure is formed, potentially creating symmetric fault-block mountains that alternate with flat-floored basins. The blocks may tilt if fault displacement is greater on one side of the mountain than the other. If surface faults dip in the same direction, half-grabens are formed, producing tilted, asymmetric mountains and basins as shown in Fig. 18.3. In half-graben geometry, or when there is a fault only on one side, the rising mountain tilts as a rigid block as shown in Fig. 18.4 such that elevation rises steeply to its highest point above the fault zone and then descends along a gentle slope to a hinge zone where displacement goes to zero. Sediment thickness in adjacent valleys tends to be wedge-shaped, thick in the area close to the fault where the basin is deepest and progressively thinner with distance from the fault.

(A) Land surface prior to normal faulting.

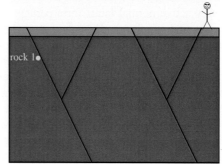

(B) Fractures prior to normal faulting. Rock 1 is deeply buried.

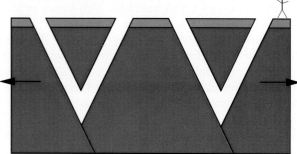

(C) Extension creates space and additional land surface.

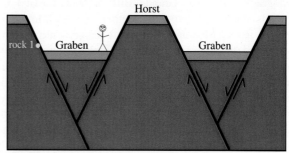

(D) Blocks drop down under force of gravity to fill the void created by extension. Horst and graben structure and range and basin topography is created. Rock 1 reaches the surface without erosion.

FIGURE 18.2 Schematic set of cross-sections that shows the mechanics of near-surface normal faulting.

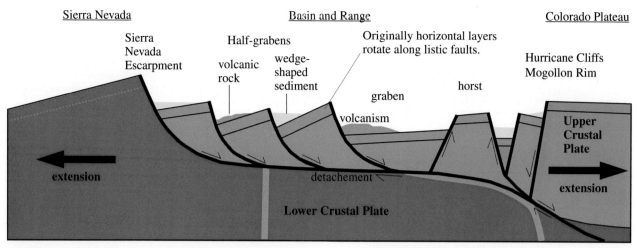

FIGURE 18.3 Schematic cross-section of Basin and Range landscape.

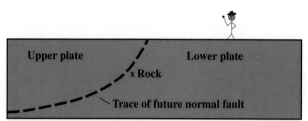

Fault-block prior to faulting. Rock x in the lower plate will undergo tectonic exhumation.

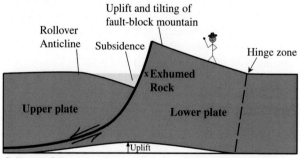

Collapse of the upper plate produces a rollover anticline and a deep asymmetric basin. Flexural and isostatic rebound of the lower plate results in uplift of a tilted, internally rigid fault-block mountain. Uplift is greatest along the normal fault and goes to zero at the hinge zone. The hinge zone intersects the Earth's surface as a line (a lineament). Rock x is exhumed to the surface. The crust rises isostatically due to tectonic thining. Both the mountain and the basin are uplifted.

FIGURE 18.4 Flexural rebound and isostatic unloading along a listric normal fault.

Although it is obvious that extension allows the basin to drop down under the force of gravity, perhaps less obvious is the simultaneous uplift of the mountain block. Uplift occurs due to the flexural rebound and isostatic unloading that occurs when part of the crust breaks allowing one side to act independent of the other. The process is depicted in Fig. 18.4 and is similar to what was discussed regarding the North Cascade Mountains in Fig. 16.37.

The end result of normal faulting is the formation of a fault-block mountain. A fault-block mountain is an isolated, internally intact, rigid mass of Earth bordered on one or both sides by a normal fault. Ideally, there is no associated internal deformation within the mountain in the form of folds or faults. Fault-block mountains tend to be relatively small (50–75 miles long, 10–25 miles wide). When present in sequence, they are oriented parallel with each other separated by somewhat wider flat-floored, tilted valleys that are usually in horst-graben or half-graben structure.

Detachments

The large-scale process of extension is more complex than formation of simple horst and graben or half-graben geometry. Although steep at the surface, many normal faults curve and flatten at depth into a single, horizontal fault known as a detachment into which all overlying faults merge. A horizontal detachment surface is shown in Fig. 18.3. A curved fault that is steep near the surface and horizontal at depth is known as a listric fault. Given the presence of a horizontal detachment fault, the crustal plate above the listric fault is pulled horizontally in one direction while the crustal plate below the fault is stationary or pulled horizontally in the opposite direction. The result is crustal delamination (the splitting of layers) in which the lower crust is pulled out from below the upper crust. The upper crustal plate stretches and breaks, producing horsts, grabens, and half-grabens as shown in Fig. 18.3. The

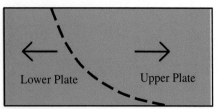
Before extension. The future normal fault is dashed

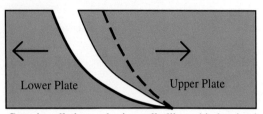
Crust is pulled apart horizontally like a ship leaving harbor. A gap is created between the upper plate and lower plate. A future secondary fault is dashed.

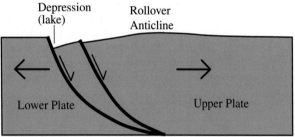
Upper plate collapses into the gap to form a rollover anticline and a depression against the lower plate.

FIGURE 18.5 Development of a rollover anticline.

lower crustal plate below the fault may stretch a bit but remains largely intact. The removal of weight from above the lower plate may cause the lower plate to isostatically rise into a dome structure as explained later.

Fault-Block Rotation and Rollover Anticlines

Horizontal crustal delamination coupled with listric geometry creates two additional characteristics of normal faults. The first involves the fault blocks themselves, which tend to rotate (tilt backward) as they slide down the fault surface. For example, the originally horizontal sedimentary layers in the half-grabens shown in Fig. 18.3 have rotated toward the vertical. They will continue to rotate as they slide down the listric fault surface.

The second characteristic is visible in Fig. 18.4. It involves development of a rollover anticline on the upper plate of a listric normal fault. The sequence is shown in Fig. 18.5. As rocks on the upper plate are pulled horizontally away from a listric fault, they will fold, fault, and collapse into the void that was created between the fault surface and retreating upper plate. This action produces a bulge in the landscape (a rollover anticline) located a short distance from the fault. It also produces a low area at the base of the fault where the rocks have collapsed. It is not uncommon for lakes to form above the collapsed depression adjacent to the mountain front.

THE BASIN AND RANGE

One of the most striking aspects of the physiography of North America is the wide desert area of the Basin and Range province centered in the state of Nevada with its alternating mountain ranges and flat-floored valleys. Recent earthquakes and volcanism indicate that the Basin and Range is an actively forming landscape. It is a structure-dominated landscape that has been under tensional stress for at least the past 35 million years. Most of the present-day landscape, however, has developed only during the past 17 million years, and in some areas during the past 10 million years. Sedimentary, volcanic, and crystalline rocks are present, but these have secondary landscape influence to the much stronger influence of normal faults. Normal faults overprint a long history compressional tectonics that includes landscape previously associated with the Cordilleran foreland fold and thrust belt (Chapter 15). Fig. 2.3 shows that most of the Basin and Range lies within the Great Basin river system. The northwestern part lies within the Columbia River system, and the southern part lies within the Colorado and Rio Grande River systems. The location of the Basin and Range within the Cascade-Klamath-Sierra Nevada rain shadow creates a dry desert and steppe climate in which rocks and recent faults are well exposed (Figs. 6.1 and 6.5).

Physiographic Limit

The signature basin-and-range landscape is clearly evident on the shaded relief map of the Cordillera shown in Fig. 18.6 where one can count more than 150 fault-block mountains across 8 states. It is the largest physiographic province in the Cordillera; too large to show on a single landscape map, and instead, most of the area is shown in a series of four overlapping maps. Fig. 18.7 shows the northern and central regions across the state of Nevada. Fig. 18.8 shows the southern continuation across California and Arizona. Fig. 18.9 shows the Rio Grande rift in New Mexico and Colorado; and Fig. 18.10 shows the continuation of the Rio Grande Rift and Basin and Range across West Texas. The physiographic Basin and Range is shaded in each figure. A number of additional features are shown on these maps, all of which are discussed below. The Basin and Range is shown in its entirety in Fig. 1.6.

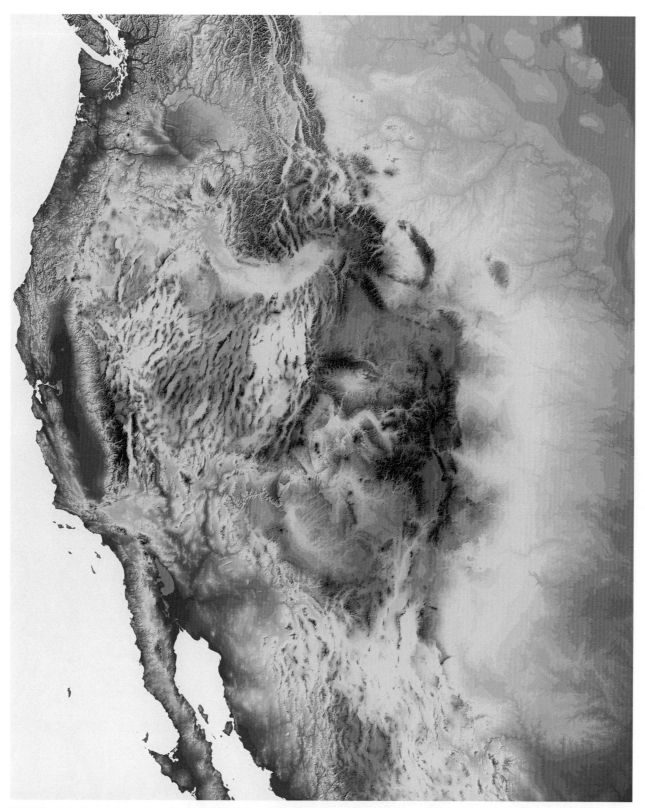

FIGURE 18.6 Shaded-relief elevation map of the Cordillera. Obtained from the National Atlas of the United States of America, USGS, Nationalmap.gov/small_scale/atlasftp.html. Low elevation shown in blue, high elevation shown in brown.

434 PART | II Structural Provinces

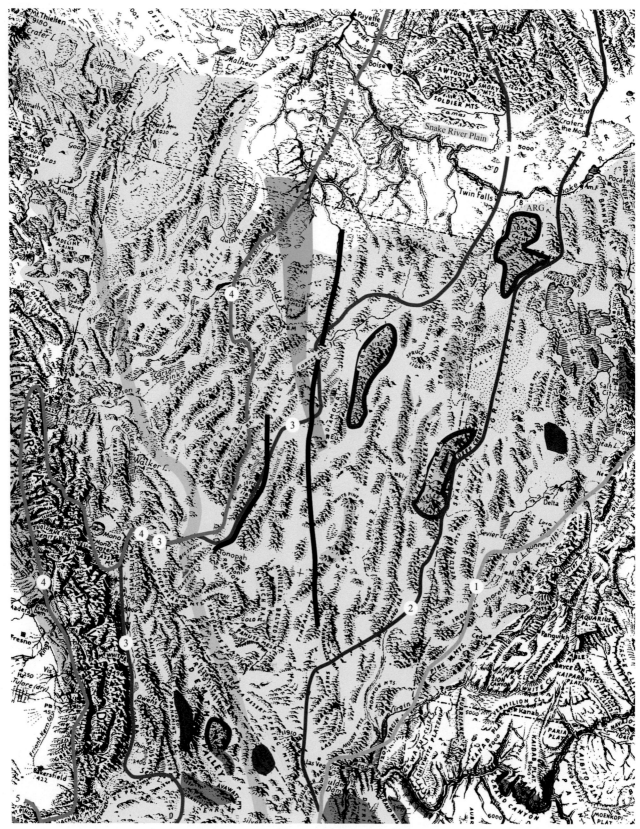

FIGURE 18.7 Landscape map of the Nevada-Utah Basin and Range. See Fig. 18.8 for explanation.

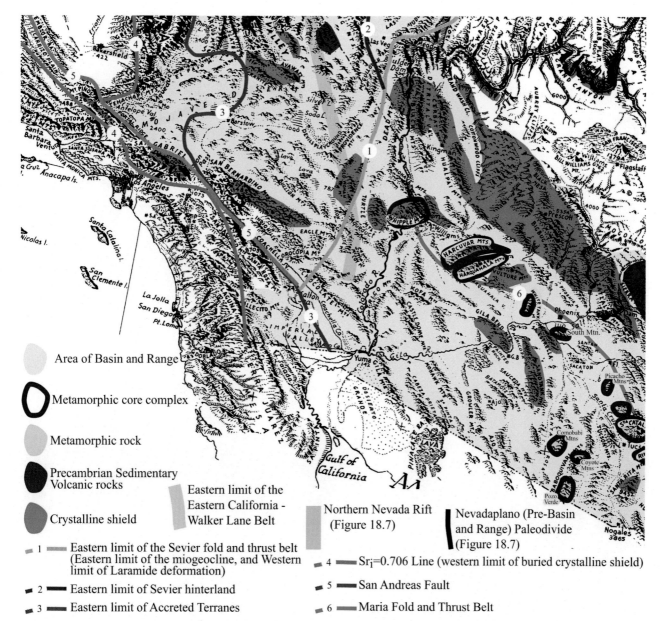

FIGURE 18.8 Landscape map of the southern California–Arizona Basin and Range. The primary source for Figs. 18.7, 18.8, 18.9, and 18.10 is the Geologic Map of North America. Lines 1 and 2 from DeCelles (2004) and DeCelles et al. (2006). Line 3 based on Crafford (2008). The $Sr_i = 0.706$ line is from Armstrong et al. (1987) and Kistler (1990). Metamorphic areas are from Miller and Gans (1989) and Ernst (1990). Metamorphic core complexes are from Armstrong (1982) and Coney (1980). Some metamorphic areas within the Precambrian sedimentary/volcanic succession in Idaho are not shown.

The physiographic boundary of the Basin and Range is drawn where Basin and Range topography is no longer visible across the landscape. The boundary is abrupt against the Sierra Nevada in the west and against the Colorado Plateau in the east, but gradational in the north against volcanic rocks of the Cascade Mountains, Columbia Plateau, and Snake River Plain. The southern border extends into Mexico. As implied in Fig. 18.3, the boundary with the Sierra Nevada marks the location where the upper crustal plate (the Basin and Range) has detached from the lower plate (the Sierra Nevada). Flexural and isostatic uplift of the Sierra Nevada, coupled with faulting and collapse of rocks in the Basin and Range, has produced a massive fault-block mountain, the Sierra Nevada, with a gentle west-tilted slope on its west side that leads down to the Central Valley of California, and a steep escarpment on its east side that leads down to the active Southern Sierra Nevada Frontal fault. Here we find some of the greatest relief anywhere in the United States. The eastern front of the southern Sierra Nevada, capped

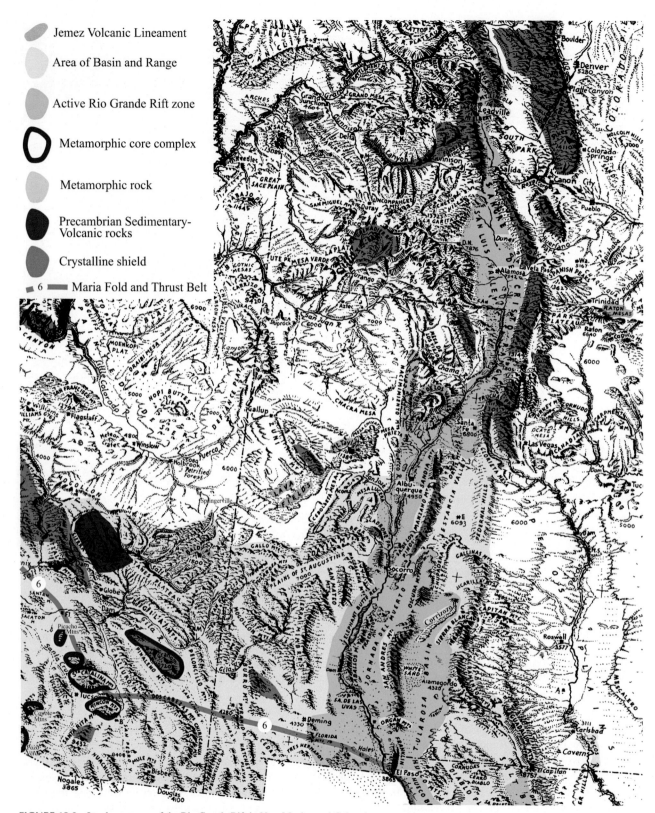

FIGURE 18.9 Landscape map of the Rio Grande Rift in New Mexico and Colorado.

with Mt. Whitney and several other 14,000-foot peaks, towers nearly 11,000 feet above Owens Valley as is evident in the Google Earth image shown in Fig. 18.11. Along the eastern boundary, the Colorado Plateau has managed to remain largely intact in spite of its position on the upper plate of the detachment fault (Fig. 18.3). It too forms an escarpment with Basin and Range topography, marked by the Mogollon Rim, Grand Wash Cliffs, and Hurricane Cliffs as seen in Figs. 13.38 and 13.40. Directly north of the Colorado Plateau, the Wasatch Mountains are a large normal fault block mountain that traditionally forms the eastern boundary of the Basin and Range. It is shown as such in Fig. 18.7. The northeastern corner of the Basin and Range is shown in Fig. 18.34.

Fig. 11.3 shows that most of the Basin and Range physiographic province lies within the normal fault-dominated structural province discussed in this chapter. One major exception is western Nevada and adjacent California, which lies in the California strike-slip structural province. This province includes Death Valley, the White Mountains, Pyramid Lake, Lake Tahoe, and the Sierra Nevada. It is discussed in Chapter 20. The boundary that separates the two structural provinces is both structural and physiographic. Fig. 18.6 shows very clearly the break in trend of mountain ranges that define the boundary. Mountain ranges in central Nevada show a strong north-northeasterly trend. Those in California are oriented north-northwest. The structural and physiographic boundary is highlighted in Figs. 18.7 and 18.8 with a thick tan line. The area to the west of this line is known as the Eastern California-Walker Lane belt. Similar to the Brothers Fault zone, it is an area where strike-slip faults, in addition to normal faults, become the dominant landscape-forming influence.

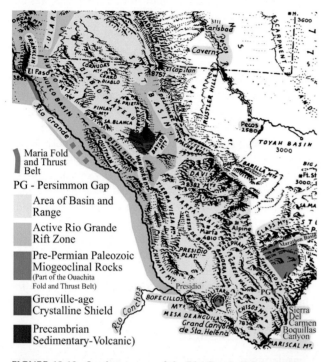

FIGURE 18.10 Landscape map of the Rio Grande Rift in the Trans-Pecos region of West Texas.

FIGURE 18.11 Google Earth image looking southward at Owens Valley. The Sierra Nevada escarpment is at right. The Inyo Mountains are at left. The 1872 Owens Valley earthquake rupture (OVF) forms a linear scar that extends south of the town of Independence (IN).

Expansion into Surrounding Areas

Normal faults are strong landscape-formers, therefore, expansion of these faults into surrounding physiographic provinces has had a strong effect on landscape. The US Geological Survey in conjunction with state geological surveys, has surveyed all known active faults in the country and placed them on a website with a Google Earth link. Fault displacement tends to be episodic, therefore; the definition of active is relative. The US Geological Survey provides five maps each based on the age of last known movement. Two of the maps are shown in Fig. 18.12. The maps show all types of active faults in addition to normal faults. However, the majority are normal faults. Major exceptions include dominantly strike-slip faults along the California coast where reverse and thrust faults are also present, especially near Los Angeles; and dominantly thrust faults off the Oregon-Washington coast associated with the Cascadia subduction zone. The website (given in the figure caption) provides detailed information on each fault.

Fig. 18.12A shows fault activity over the past 1.6 million years. The first thing to notice is that normal faulting is active across nearly all of the Basin and Range with the exception of a few small pockets. The largest area with only a few active faults extends south of the Colorado Plateau from southwestern New Mexico, across Arizona, to the Eastern California-Walker Lane Belt in the eastern Mojave Desert. Two additional fairly large pockets are present. One is located in southern Oregon just across the California border. The other is in the Great Salt Lake Desert of northwestern Utah. Small pockets are present in the western corner of the Mojave Desert in California, and in southwestern Nevada.

Active normal faults extend beyond the Basin and Range in several directions. Normal faults are beginning to chop away and lower the western margin of the Colorado Plateau, particularly in the vicinity of the Grand Canyon. One such example is the Hurricane fault, shown in Fig. 17.29. Active normal and strike-slip faults in the Eastern California-Walker Lane Belt merge southward in southern California with the San Andreas Fault zone. In the northwestern Basin and Range, active normal faults extend to the vicinity of Mt. Jefferson and Mt. Hood in the Oregon Cascade Mountains, they cross the northern Sierra Nevada, and they extend across the western Snake River Plain to the Blue Mountains region along the Idaho-Oregon border. A mixture of normal, strike-slip, and reverse faults are present along the Columbia River to the Willamette Valley near the cities of Salem and Portland, and in the Puget Sound area near Seattle.

Two additional well-developed arms of active normal faults are visible in Fig. 18.12A, both of which are discussed later in this chapter. The first is the Rio Grande rift, shown in Figs. 18.9 and 18.10. It is a fairly compact, isolated active fault zone that extends from the Texas-New Mexico Basin and Range northward along the east side of the Colorado Plateau into the heart of the Colorado Rocky Mountains. The second is the Rocky Mountain Basin and Range, which extends northward from Utah across western Wyoming, Idaho and western Montana just west of the eastern front of the Cordilleran fold and thrust belt. Some of the normal faults in this area have reactivated older thrust faults.

Fig. 18.12B shows fault activity over the past 15,000 years. Particularly active areas include the eastern, northern, and western Basin and Range, the Rio Grande rift, the Rocky Mountain Basin and Range surrounding Yellowstone National Park and the eastern Snake River Plain, the southeastern Cascade Mountains, the San Andreas system near San Francisco and Los Angeles, the offshore Cascadia subduction zone, and faults near Portland and Seattle. Most of these faults are capable of generating destructive earthquakes. Two individual faults shown in Fig. 18.12B are worth mentioning; the Meers strike-slip fault in Oklahoma, and the Cheraw normal fault on the High Plains of southeastern Colorado. These two faults are the eastern-most active faults to reach the surface in the United States.

Landscape Characteristics

Considering that most of the Basin and Range forms the Great Basin, that it includes Death Valley, and that the eastern and western margins are marked with high mountains and escarpments, one may not immediately realize that nearly the entire province is elevated well above sea level. As seen in Fig. 18.6, a region of high elevation forms a broad north to south arch across central-eastern Nevada flanked on both sides by lower elevation.

Within the area of high elevation there are several ranges with peaks above 11,000 feet including the Toiyabe, Toquima, Ruby, Schell Creek, and Spring Mountains (Fig. 18.7). The Deep Creek Range has two peaks above 12,000 feet, and the Snake Range has four, including Wheeler Peak (13,068 feet). Intervening basins are mostly between 5000 and 7000 feet except near the Spring Mountains where they drop below 4000 feet. Only a few peaks rise above 10,000 feet across the lower elevation west and east flanks of the Basin and Range in western Nevada and western Utah where valleys are between 4000 and 5000 feet. These areas, during the glaciation, were occupied by Lake Lahontan and Lake Bonneville as shown in Fig. 12.27.

Elevation rises again within the Walker Lane belt where White Mountain Peak, in the White Mountains of eastern California, reaches 14,246 feet, and seven

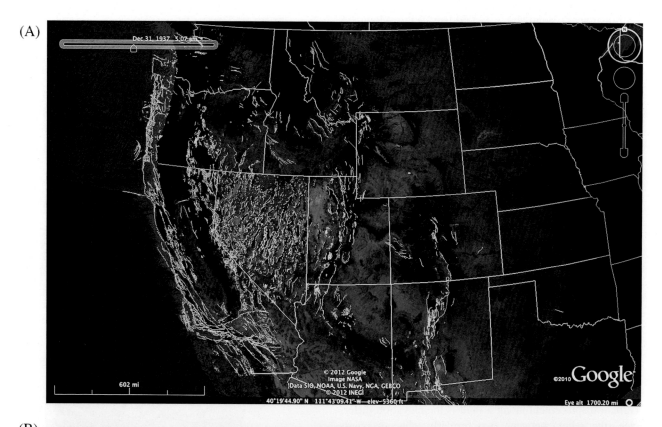
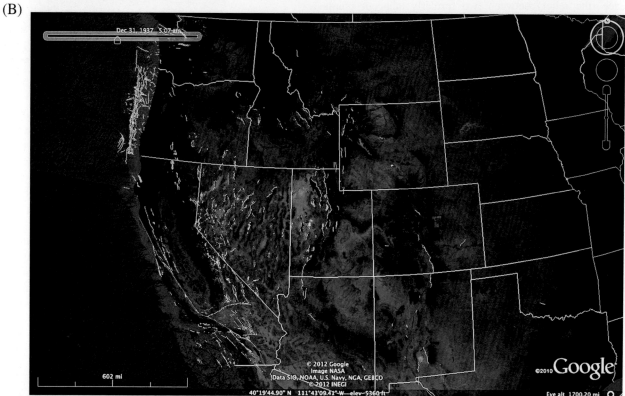

FIGURE 18.12 (A) Google Earth image showing faults active in the past 1.6 million years. (B) Google Earth image showing faults active in the past 15,000 years. Obtained from the US Geological Survey Quaternary fault and fold database for the United States, https://earthquake.usgs.gov/hazards/qfaults/.

additional peaks stand above 13,000 feet. White Mountain Peak is the only 14,000-footer not located in the Sierra Nevada, Cascade Range, or Colorado Rocky Mountains. Several additional ranges in eastern California boast peaks above 11,000 feet including the Panamint, Inyo, and Wassuk Ranges. Elevation in intervening valleys varies considerably from more than 6000 feet to below sea level at Death Valley.

The lowest overall elevations are in the southern Basin and Range from the Mojave Desert eastward to New Mexico (Fig. 18.8). With few exceptions, most peaks are below 8000 feet, and many are in the 4000 to 6000 foot range. In Fig. 18.6, some of the ranges look as though they have been buried in sand. Intervening basins vary in elevation from 4000 feet to -234 feet at the Salton Sea. With the exception of the Mojave Desert, this area is without active faults (Fig. 18.12A). In a later section, we will link lower peak elevation and relief to relative inactivity along bounding normal faults. Elevation rises again in the vicinity of the active Rio Grande rift, particularly toward Colorado where several peaks rise above 12,000 feet and basins are above 6000 feet (Fig. 18.9).

Vertical Displacement

Vertical displacement along a fault is the elevation difference (the relief) between two formerly adjacent points that are now offset by the fault. It includes both upward and downward motion. Total vertical displacement along some of the faults in the Basin and Range is between 26,000 and 33,000 feet. Averaged over 35 million years, this implies between 0.89 and 1.13 inches of vertical movement every 100 years. However, none of the faults have been active over the entire 35 million year period. Faults that have been active for 17 million years or less have vertical displacement rates on the order of 2 to 4 inches per 100 years.

The reason we do not have 26,000 to 33,000 feet of relief between mountain and valley is because valleys are constantly filled with sediment derived from erosion of uplifted mountains. Some valleys contain more than 20,000 feet of sediment. Subsiding valleys that do not accumulate sediment fast enough will sink below sea level. An example is Death Valley, which at 282 feet below sea level is the lowest point in North America. Death Valley was occupied by Lake Manly about 15,000 years ago at the end of the glaciation. Today the valley is bone dry with summer temperatures well above 100°F (38°C). Consequently there is little erosion of surrounding mountain blocks except during thunderstorms.

Horizontal Extension

In Fig. 18.2, we pointed out that extension and normal faulting result in an increase in surface area. The amount

FIGURE 18.13 Landscape map showing the inferred location of the West Coast prior to Basin and Range extension. Based on Moores and Twiss (1995, p. 91).

of increased surface area in the Basin and Range is not trivial. Total east-west extension in the state of Nevada is on the order of 155 miles. If all of this extension occurred in the past 35 million years, it would amount to an average of 28 inches of increased land width every 100 years. This, again, is an underestimate because large parts of Nevada were not active until sometime after 17 million years ago. Actual measured rates of extension vary from 19.7 inches to 6.6 feet per 100 years. Fig. 18.13 shows the presumed location of the Pacific coastline prior to Basin and Range extension. Most of the extension has occurred across the state of Nevada, which has essentially doubled in width.

Crustal Thinning and Volcanism

Extension and delamination in the Basin and Range have thinned the crust to between 17 and 24 miles thick in most areas. This is thinner than the average crustal thickness of 25 to 32 miles. The thinnest crust is located in northern Nevada and adjacent Utah, and in southern

Arizona and parts of adjacent California and New Mexico. The thickest crust is 28 miles near the Belted Range in south-central Nevada. As a result of thin crust, the warm mantle asthenosphere has risen close to the surface, which, in turn, has resulted in a geothermal gradient (the increase in temperature with depth in the Earth) that is approximately three times normal. This makes the Basin and Range a prime location for development of geothermal energy. The warm interior allows the entire region to rise isostatically thus creating basin floors between 3000 and 7000 feet in most areas.

Associated with a warm interior is lots of volcanism. Volcanic rocks less than 45 million years old are ubiquitous across the Basin and Range where they are intermingled with generally older sedimentary rocks (Fig. 17.1). As discussed in Chapter 17, there are large areas of explosive silicic volcanic rocks between 43 and 18 million years old in Nevada, Utah, Arizona, and New Mexico, as well as younger, dominantly basaltic volcanism scattered across the region.

Metamorphic Core Complexes

Although the Basin and Range consists dominantly of sedimentary and volcanic rock, crystalline rocks, exposed through a combination of tectonic exhumation and uplift/erosion, are also common. Figs. 18.7, 18.8, and 18.9 show the general distribution of crystalline shield rock and rocks metamorphosed during Cordilleran orogeny. Not shown are plutonic rocks that intruded during Cordilleran orogeny and which are abundant along the Nevada-California border, in the Mojave Desert, and across Arizona. Nearly all nonvolcanic mountain ranges in the Basin and Range across southern California and Arizona expose crystalline rock of one type or another.

Perhaps the most significant exposures are the metamorphic core complexes. When seen from a distance, a metamorphic core complex is a domal mountain with a central crystalline core surrounded by younger sedimentary and volcanic rock. These structures appear similar to the crystalline-cored, mid-continent anticlines and domes discussed in Chapter 14, but there are significant differences. The mid-continent anticlinal structures developed primarily through compressional uplift and erosional exhumation of cold crystalline rock. A fault may occur along the margin of the structure but there is little, if any, internal deformation and no metamorphism or intrusion. The crystalline rock is part of the North American shield, and the sedimentary rock is typically that of the interior platform. The contact that separates these rocks is the Great Unconformity.

A metamorphic core complex, on the other hand, is the product of extreme lithospheric extension, isostatic adjustment, and tectonic exhumation. The amount of extension is up to four times the original width. In contrast to a mid-continent anticlinal structure, a core complex is associated with internal deformation, metamorphism, and intrusion in the underlying crystalline rock, and with listric normal faulting in the overlying sedimentary/volcanic rocks. The contact that separates crystalline rock from overlying sedimentary rock is a low-angle normal fault, a detachment.

The process of core complex development is illustrated and explained in Fig. 18.14. Lower plate crystalline rocks consist variably of ancient crystalline shield rocks, rocks metamorphosed during the Sevier thrust event or during the extension process, and young intrusive granitic rock. Upper plate rocks are dominantly sedimentary and volcanic although ancient shield rocks may also be present. The detachment fault that separates the upper and lower plates forms initially at depth at elevated temperature creating a mylonite, a fine-grained, strongly foliated and lineated fault rock that forms in a ductile fault zone. Lower plate rocks close to the detachment fault may undergo metamorphism, which creates a gently dipping planar fabric (known as a foliation) that is often mylonitic. Sedimentary layers on the upper plate are stretched, thinned, and rotated along listric normal faults during extension but remain largely unmetamorphosed. As sedimentary layers on the upper plate are thinned, the crust becomes thinner and warmer. The crust weighs less, and therefore rises isostatically to the point where the mylonite zone and lower plate crystalline rocks reach the surface in the form of a dome. As the dome continues to uplift, it will form some of the highest topography in the immediate area. Uplift is aided by the buoyant rise of magma, which may intrude the lower plate. As the mylonite zone approaches the surface it cools and is overprinted with brittle fault zone fabrics. Note that exposure of lower plate crystalline rock is primarily the result of tectonic processes (normal faulting) rather than erosional processes. Exposure of once deeply buried rocks within a core complex is a classic form of tectonic exhumation.

There are more than 25 individual metamorphic core complexes stretching from Canada to Mexico. Most are in the United States. Four are shown in Fig. 18.7, the Albion-Raft River-Grouse Creek, Ruby, Snake Range-Deep Creek Range, and the Death Valley complexes. The greatest density is in the southern Basin and Range of Arizona where Fig. 18.8 locates and names more than ten. There are at least seven additional complexes within the US north of the Snake River Plain. The Clearwater, Bitterroot, Anaconda, and Pioneer core complexes are present in the Rocky Mountain Basin and Range. The Okanogan, Kettle, and Priest River (Selkirk) core complexes are located farther north in Washington and Idaho and were briefly introduced in Chapter 16. We will discuss the timing of core complex formation in the next section.

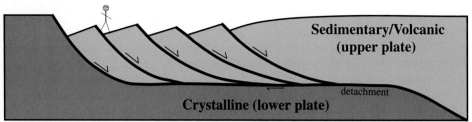

(A) The crust delaminates as it spreads apart creating high-angle, brittle normal faults at the surface in upper plate sedimentary/volcanic rocks. The high-angle faults bend at depth and merge into a single horizontal detachment fault that separates the upper and lower plates. Lower plate rocks consist of crystalline shield and young granitic rock. These rocks undergo metamorphism and possibly intrusion as the crust thins. The detachment fault at depth deforms in a ductile manner at elevated temperature creating a fine-grained, lineated fault rock known as mylonite.

(B) Fault blocks on the upper plate rotate and spread apart with additional extension. Tectonic thinning of the upper plate coupled with heat in the lower plate causes the lower plate to rise isostatically into a dome structure.

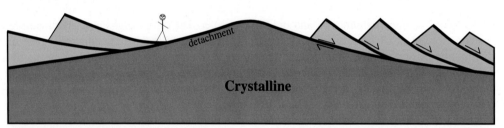

(C) Rocks are stripped off the top of the dome exposing the detachment fault and lower plate crystalline rocks at the surface. As it rises toward the surface, the detachment fault evolves from a ductile structure to one characterized by brittle deformation.

FIGURE 18.14   Schematic cross-section that shows the origin of a metamorphic core complex with no erosion. Based on Lister and Davis (1989) and Van der Pluijm and Marshak (2004, pp. 392–393).

Timing of Normal Faulting

The timing of faulting virtually anywhere can best be determined by comparing the depositional or intrusive age of rocks deformed by a particular fault with those that are undeformed. It should be obvious to you by now that a particular normal fault must have been active after deposition or intrusion of deformed rocks, but before deposition or intrusion of undeformed rocks. A second method is to determine the depositional age of rocks derived from erosion of fault-generated mountains. The age of these rocks (or sediment) indicates the time when the mountain was in existence. Some examples are given in Figure 20.25. A third method is to dig a trench across the fault trace so that cross-cutting relationships within unconsolidated sediment are clearly visible. This method allows geologists to date very recent fault activity from several hundred to several million years old. Faulting can also be dated indirectly by determining periods of rapid exhumation using $^{40}Ar/^{39}Ar$, U-Th/He, and fission track methods as outlined in Chapter 9. It is also possible to date metamorphic and mylonitic rocks that formed during development of a metamorphic core complex.

When field analysis is performed across the Cordillera, it becomes apparent that normal faulting has been a dynamic process that has migrated, expanded, and reactivated over time. There are areas that were active 10, 20, even 50 million years ago, but are no longer active. Conversely, there are areas that were active 30 to 40 million years ago that became inactive and then active again. There is, however, a body of field evidence to suggest that normal faulting and development of present-day Basin and Range landscape began, in earnest, about 17 million years ago and has continued to the present-day. In northern Nevada, there is evidence for two distinct periods of normal faulting, an early period between 17 and 10 million years ago, and a later period beginning less than 10 million years ago. The orientation of these two episodes is similar, but basins in some areas that formed during the early phase of normal faulting were uplifted into a mountain range during the second, younger phase. The older phase shows the greatest displacement, but the younger phase, because it remains active, has a strong influence on present-day landscape.

All of the metamorphic core complexes south of the Snake River Plain were active between 32 and 13 million years ago implying that initiation of core complex formation pre-dates the development of widespread Basin and Range topography. Core complexes are numerous in Arizona, but sparse in Nevada. Given their distribution, one could suggest that core complex formation had a strong effect on landscape in Arizona prior to 17 million years ago, but less of an effect in Nevada. Exactly what the landscape looked like in central Nevada prior to 17 million years ago is discussed in the section on the Nevadaplano. Metamorphic core complexes in the Rocky Mountains north of the Snake River Plain were active between 56 and 39 million years ago.

Normal Fault Activity Verses Erosion

Let us now turn our attention to landscape evolution and how the look of a normal fault mountain block changes depending on the recency of fault activity. First, we must realize that Basin and Range topography is evolving in a dominantly arid climate where there are few perennial rivers. Erosion occurs, not so much by wind, but during infrequent thunderstorms, and the progress is slow. Fault scarps that formed 50 to 100 years ago remain well exposed today. The Pleasant Valley earthquake scarp along the west side of the Tobin Range (shown as Mt. Tobin in Fig. 18.7) is one of several in the Basin and Range that are visible in Google Earth imagery. As seen in Fig. 18.15, the scarp is visible discontinuously along the base of the mountain. The scarp formed during a series of three large earthquakes within seven hours on October 2, 1915. In a humid climate, this fault scarp would be eroded and hidden in thick vegetation.

Much of the Basin and Range lies within the Great Basin such that any precipitation that does fall will have no outlet to the sea (Fig. 2.2). Thunderstorm spawned streams will wash loose sediment through mountain canyons as far as the valley floor at the base of the mountain where it will accumulate. Excess water collects in a low part of the basin where it evaporates forming a saltpan or dry lakebed known as a playa. The net effect of erosion is the transfer of sediment from the mountain to the adjacent basin. Thus, the mountain is eroded and simultaneously buried in its

FIGURE 18.15 Google Earth image looking southeast at a well-preserved fault scarp along the southern Tobin Range. The fault scarp formed during a series of three earthquakes on October 2, 1915.

own debris. The process is similar to what has been proposed for the crystalline-cored domes of the Middle-Southern Rocky Mountains (Fig. 14.26). This self-burial is an important characteristic of the Basin and Range.

The extent to which a mountain is buried in its own debris is dependent on the rate of mountain uplift relative to the rate of erosion and burial. The rate of uplift, in turn, depends on the level of activity along the range-front fault. The term range-front fault refers to a fault at the base of a mountain located roughly at the boundary between the mountain-front and adjacent valley. It is the fault that separates the basin from the range. The active fault shown in Fig. 18.15 is an example. Recall from Fig. 4.11 that a steeply dipping plane (a normal fault in this instance) will produce a relatively straight line (a lineament) across the Earth's surface. Also keep in mind that, because of the dry climate, the rate of uplift across an active fault will likely outpace the rate of erosion. Thus, if we have constant displacement along a steeply dipping, active, range-front normal fault, the net effect of that displacement will be to maintain a relatively straight and abrupt transition from valley floor to mountain front. This type of transition is seen in the southern Tobin Range (Fig. 18.15) and in the southern Sierra Nevada (Fig. 18.11). A mountain block with an active range-front fault on both sides is shown in cross-section and map view in Fig 18.16A. Fault scarps on both sides of the mountain are exposed because the average rate of fault displacement is greater than the rate of erosion. The high dip angle of the fault produces a sharp, straight mountain-front.

A sharp, straight mountain-front is not maintained once the fault becomes inactive. Instead, the mountain-front is eroded back while, at the same time, the valley is filled with sediment. The mountain and the range-front fault are progressively buried in their own erosional debris creating an increasingly sinuous (curvy) mountain-front as depicted in Fig. 18.16B. As erosion lowers the mountain, and the mountain-front retreats, a gently tilted erosion surface known as a pediment surface is cut into the mountainside at about the same elevation as the valley floor. Canyon streams sweep sediment out onto the pediment surface in the form of an alluvial fan. The term alluvium (or alluvial) refers to any sediment deposited by a stream. The term fan refers to the shape of the deposit, which develops as stream water spreads out into the basin.

As the mountain continues to be eroded and buried, alluvial fans grow larger and coalesce to form a continuous alluvial apron on the pediment surface known as a bajada. Burial is complete when a bajada on one side of a mountain begins to merge with a bajada on the other side. At this point, only isolated peaks protrude above the alluvium as shown in Fig. 18.16C. Notice in Fig. 18.16C that sediment in cross-section forms a steers-head geometry—thick on the down-thrown (valley) side of the fault and thin on

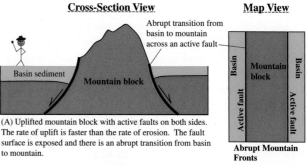
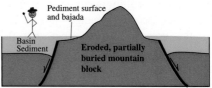
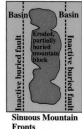

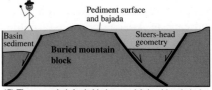
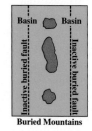

FIGURE 18.16 Schematic cross-sections (left side) and maps (right side) that show the progressive erosion and burial of mountain blocks in the Basin and Range. (A) Active faulting. (B and C) Inactive faulting and progressive erosion-deposition.

the upthrown (mountain) side where the pediment surface is developed. Notice also that the mountain shape still exists in cross-section although it is buried beneath sediment. Finally, notice that although the structure is different, the erosional process and the resulting slope-flat-slope mountain profile is identical to what happened in the Southern Rocky Mountains 50 million years ago except that mountains in the Basin and Range have not been exhumed.

Different mountains in the Basin and Range are at different stages of erosion and self-burial depending on the recentness and magnitude of activity along range-front faults relative to the rate of erosion and burial. By comparing active mountain fronts with inactive ones at various stages of burial, we can see, at the scale of an individual mountain, how a normal fault landscape slowly grows old.

Straight mountain fronts are fairly easy to find in the seismically active northern Basin and Range. The Cortez Range, located east of the Tobin Range in Fig. 18.7, and shown in Fig. 18.17, exemplifies a tilted fault-block mountain with an active range-front fault on its western side. In this figure we can compare the sharp, active, range-front boundary on the west side of the mountain with the sinuous, eroded eastern side where no active fault exists. Note also that the highest peaks are along the extreme western side of the range directly above the active fault. One can use the Cortez Range as a template to interpret other tilted mountain ranges.

Examples of eroded and mostly buried range-fronts can be found in the southern Basin and Range region where active normal faults are absent. Fig. 18.18 is a Google Earth image of Sierra Estrella located just south of Phoenix, Arizona (Fig. 18.8). The mountain has a sinuous, hembayed range-front on both sides suggesting a period of erosion without fault activity. An alluvial fan is located with the letter a. Alluvial fans on both sides of Sierra Estrella coalesce to form a bajada, which on the north side is truncated by the Gila River.

If fault activity remains dormant then we would expect this mountain to be progressively lowered and buried in its own debris. This final stage of mountain evolution is depicted in Fig. 18.19, which shows the mostly buried Rosamond Hills in the Mojave Desert northeast of Rosamond, California.

FIGURE 18.17 Google Earth image looking northeastward at the Cortez Mountains, north-central Nevada. The mountain block is tilted eastward (to the right) along an active normal, range-front fault on its western side. The active fault maintains a sharp boundary with the valley to the west. Compare the mountain shape in this image with the sketch shown in Fig. 18.4. The white area in the valley is a dry lakebed (playa). The range-front fault was active 2500 to 3000 years ago.

FIGURE 18.18 Google Earth image looking northeast at sinuous mountain fronts along Sierra Estrella near Phoenix, Arizona. Letter a denotes alluvial fan.

FIGURE 18.19   Google Earth image looking west at buried mountains (the Rosamond Hills) in the western Mojave Desert northeast of Rosamond, California. There are no known active faults in this area. A large playa is at left. Part of Edwards Air Force base is at lower right corner.

Although it is possible to predict the recency of active faulting along range-fronts, it is important to realize that the actual time required for a complete cycle will vary depending on the rate of fault activity verses the rate of climate-driven weathering and erosion. It is possible for one landform to be numerically younger than another but still be farther along in the erosion cycle. Additionally, the cycle can be interrupted at any time by renewed fault activity.

The Nevadaplano

If Basin and Range landscape developed sometime after 17 million years ago, what did the landscape look like 20 or 40 million years ago? In Chapter 15 we noted that the Cordilleran fold and thrust belt once extended the length of the US Cordillera from Montana to Arizona. We also noted that the Idaho-Wyoming thrust belt south of the Teton Mountains stands as an island of relict landscape surrounded by areas of younger volcanism, deformation, and deposition. Thus, we can suggest that prior to reincarnation into Basin and Range landscape, the eastern part of the Basin and Range was characterized by landscape that resembled the Idaho-Wyoming thrust belt. It was a landscape of long, narrow mountain ridges separated by long narrow valleys.

Evidence for this earlier, now vanished, landscape is seen in individual fault blocks (horsts) that contain thrust faulted and folded sedimentary layers typical of a fold and thrust belt. A schematic cross-section that shows the internal structure of a fault block is presented in Fig. 18.20. The landscape associated with these structures has since been completely reincarnated by normal faults. We might also presume that 20 million years ago, just prior to extensive Basin and Range extension, much of the region was blanketed in a layer of volcanic ash expelled from the many calderas that were in existence across Nevada and surrounding states as part of the ignimbrite flare-up discussed in Chapter 17.

If we conclude that the eastern Basin and Range was a landscape of linear mountains and valleys created by foreland fold and thrust faulting, what did the landscape look like in the western Basin and Range prior to 17 million years ago? Here, the calderas and their ignimbrite deposits show some critical relationships. The most significant relationship is the fact that pyroclastic material ejected from calderas in central Nevada can be followed along ancient river valleys all the way to the western slope of the Sierra Nevada. These are not present-day river valleys, but rather they are deposits of volcanic breccia left by ancient, now vanished rivers. The presence of these river deposits on the western slope of the Sierra Nevada has at least three implications. The first is that the great eastern escarpment of the Sierra Nevada, the one shown in Fig. 18.11, could not have been in existence at the time of the ignimbrite flare-up (from 36 to 18 million years ago) because it would have blocked rivers from reaching the western slope. Secondly, if rivers were flowing from central Nevada directly to the Central Valley of California, then central Nevada must have been topographically higher than the Sierra Nevada. Third, the presence of west-flowing river sediment confirms that present-day Basin and Range topography could not have been in existence because some of the many fault-block mountains would have blocked or redirected rivers.

Certain clay minerals and certain types of silicate clasts found within stream channel volcanic breccia are capable of incorporating water in their chemical structure. This water can be extracted and used for oxygen and hydrogen isotope analysis, which in turn gives information on the relative change in elevation across an area. When these studies are combined with field and paleobotanical studies, they indicate that both the Sierra Nevada and the western Basin and Range were at high elevation in what was probably a warm, dry, rain shadow climate. The evidence suggests that the entire region was part of a high elevation low relief plateau that was in existence from well before 40 million years ago to at least 17 million years ago. Average elevation of the plateau is not known with certainty and may have varied somewhat over time. Steep isotopic gradients are recorded in sediment between 27 and 23 million years old suggesting that elevation may have reached 11,000 to 13,500 feet in central Nevada at that time. Geologists refer to this now vanished plateau as the Nevadaplano or altiplano.

Thus, on the basis of field and isotopic data, we can conclude that the landscape that existed prior to 17 million years ago consisted of high mountain ridges in the eastern Basin and Range (the Sevier thrust belt), and a high plateau in the western Basin and Range (the Nevadaplano) that sloped gently toward an ocean located in the present-day Central Valley of California. Fig. 18.21 is a cross-section that depicts landscape circa 25 million years ago. Although some extension would have already occurred in the form of core complex formation, the state of Nevada, 17 million years ago, would still have been about half as wide as it is today. Also, keep in mind that the plateau would have been blanketed with volcanic deposits. Note the location of the Nevadaplano paleo-drainage divide in the figure. Rivers would have flowed westward from this divide to the paleo-Pacific Ocean. As shown with thick black lines in Fig. 18.7, the paleodivide is inferred to have extended from the vicinity of the Tuscarora Mountains southward to an uncertain location between the Monitor and Pahranagat Ranges. The demise of the Nevadaplano coincides with development of the modern-day Basin and Range beginning 16 to 17 million years ago.

Cause of Basin and Range Extension

As mentioned in Chapter 5, the cause of Basin and Range extension has been attributed to a combination of lithospheric heating associated with the ignimbrite flare-up and Yellowstone hot spot, coupled with slight divergence of the Pacific plate relative to the North American plate allowing the underplated Farallon plate to act as a mechanical couple that has dragged and stretched the weakened North American plate westward (Fig. 5.13).

Basin and Range Geology

The Basin and Range contains all rock successions except the Atlantic miogeocline, a feature evident in Fig. 18.22. The great variety and age of rocks provide a wealth of information on the history of the Cordillera that dates back several hundred million years. This part of Cordilleran history is recorded only in the rocks. It is no longer present in the landscape. We touched on some of

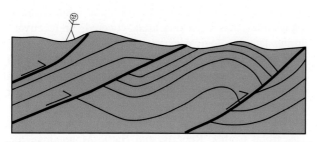

FIGURE 18.20 Schematic cross-section that shows the internal fold and thrust belt structure in normal fault block mountains of the Basin and Range.

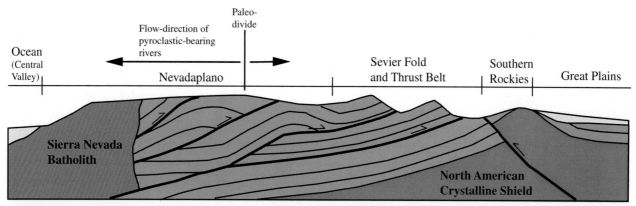

FIGURE 18.21 Cross-section that shows how the Cordilleran landscape may have looked 25 million years ago. A high elevation, low relief plateau, the Nevadaplano, occupies the western Basin and Range. Valley and ridge-type mountains occupy the Sevier fold and thrust belt in the eastern Basin and Range. The Nevadaplano and fold and thrust belt are covered with volcanic ash.

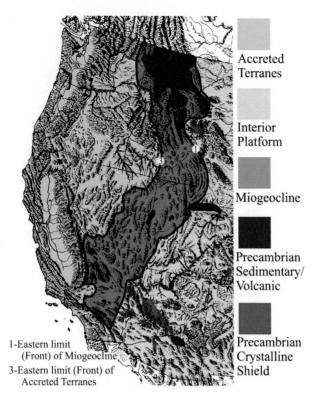

FIGURE 18.22 Landscape map that shows the present-day distribution of rock successions in the Cordillera. This is the same map shown in Fig. 11.8 except without the batholiths.

this history with our discussion of the Nevadaplano. Here, we go back deeper in time.

In this section, we first explain the significance of the six geological boundary lines shown and labeled in Figs. 18.7, 18.8, and 18.9, and then discuss geology associated with each of these lines. Some of the lines were introduced in Chapter 11. Line 1 marks the front of the Sevier fold and thrust belt. It represents the series of thrust faults that separate miogeoclinal rocks thrust westward during Sevier orogeny, from interior platform and crystalline shield rocks deformed into anticlines, domes, and monoclines during Laramide orogeny. Line 1, therefore, not only marks the front of the Cordilleran foreland fold and thrust belt, it also marks the eastern limit of the Cordilleran miogeocline, and the western limit of strong Laramide cratonic deformation. The line, as shown in Figs. 18.7 and 18.8, extends along the eastern edge of the Basin and Range in Utah, and then southward through the southern Basin and Range such that all landscape associated with the thrust belt has been reincarnated. This same line 1 is shown in Figs. 11.8, 11.9, 15.3, and 18.22.

Line 2 forms the foreland-hinterland transition. It separates sedimentary rocks of the Sevier foreland fold and thrust belt from crystalline rocks of the native North American hinterland belt. As such, it forms the front of Native North American Belt. It is the same line 2 as shown in Fig. 11.9. In contrast to the Appalachians, where a continuous belt of North American crystalline rock intervenes between the miogeocline and the accreted terrane belt, the Cordilleran hinterland is marked with an abundance of sedimentary rock that extends across both the hinterland and the accreted terrane belt. In the absence of a well-defined belt of crystalline rock, we define line 2 based on the first appearance of rocks metamorphosed or intruded during Cordilleran orogeny. The largest exposures of metamorphic rocks are shown in light blue in Figs. 18.7 and 18.8. Areas of intrusive rock are not shown. These crystalline rocks were not brought to the surface solely via thrust faulting and erosion as in the Appalachians. Most crop out discontinuously as isolated anticlinal culminations, granitic intrusions, or normal fault exhumed metamorphic core complexes, and all are surrounded by sedimentary rock. In Figs. 18.7 and 18.8, line 2 merges with line 1 south of the Grand Canyon due to the proximity of Cordilleran intrusive rock at the front of the Sevier fold and thrust belt. A similar relationship occurs in southeastern Montana as depicted in Fig. 11.9.

Line 3 separates North American rock successions from accreted terranes. As such, it forms the front (the eastern limit) of accreted terranes in the Cordillera. The same line 3 is shown in Figs. 11.8, 11.9, and 18.22. On the ground, this line represents a collage of different geologic contacts in which the original suture zone, in many areas, has been obscured, covered, or reactivated during later geological events. However, across most of Nevada, this line represents a single thrust fault known as the Roberts Mountains thrust. This fault, like other thrust faults in the Basin and Range, has been chopped and partially obscured by subsequent normal faulting. It is presently exposed in several mountain blocks that include the Tuscarora, Cortez, Roberts, and Battle Mountains. Rock composition and fossil similarities indicate that rocks in the thrust sheet were originally deposited along the North American continental slope/rise, partly on oceanic crust and partly on thinned, extended continental crust. A few rocks of possible ocean lithosphere origin (serpentinite) are present at the base of the thrust sheet, which is the basis for designating the Roberts Mountains thrust as line 3, the easternmost occurrence of accreted terranes.

Line 4 forms the strontium-initial 0.706 line, which marks both the westernmost extent of buried North American rock successions below the accreted terrane belt, and the approximate buried western limit of the miogeocline. The origin of this line and how it was determined is discussed in the following section. Line 5 is the San Andreas Fault. Line 6 is the Maria fold and thrust belt discussed below.

If we were to travel across central Nevada, we would find geological evidence for at least three orogenic cycles now hidden within the normal fault block-mountains of the Basin and Range. Fig. 18.23 is a conceptual cross-section that defines orogenic cycles based on the presence of unconformities labeled 1, 2, and 3. From oldest to youngest the orogenic cycles are the Antler, Sonoma, and Sevier. Geological principles require that thrusting is younger than any rock within or below the thrust sheet, about the same age as sediment shed from rising mountains, and older than unconformably overlying rock sequences.

Fig. 18.23 shows that the Antler orogeny is the oldest of the three events. The Antler orogeny is associated with activity on the Roberts Mountain thrust, which carries the Roberts Mountains allochthon in its upper plate. Recall from Chapter 15 that an allochthon is a large slab of rock that has moved a great distance. The Roberts Mountain allochthon consists of Late Devonian and older continental margin and oceanic rock carried some 45 miles eastward on the Roberts Mountains thrust and placed above Cordilleran miogeoclinal rocks of the same age. Rocks that form part of the allochthon are present in the previously cited mountains directly above the Roberts Mountains thrust. Notice that the Antler foredeep clastic wedge covers the Roberts Mountains thrust to create unconformity 1. This unconformity marks the end of the Antler orogeny. The Antler foredeep clastic wedge represents the eroded remains of the Antler Mountains deposited during and following thrusting. Overlying the Antler foredeep clastic wedge, the Antler overlap sequence represents a return to miogeoclinal deposition on a passive continental shelf following complete erosion of the Antler Mountains. The above relationships suggest that the Antler orogeny occurred during Late Devonian–Early Mississippian approximately 360 to 347 million years ago.

The cause of the Antler orogeny is uncertain. A common interpretation is that a volcanic arc collided with North America and bulldozed the Roberts Mountains allochthon onto the North American miogeocline. An alternative or contributing possibility is that orogeny in the Appalachian Mountains caused the Cordilleran miogeocline to slide below the allochthon, with or without a push by a volcanic arc in the west.

The Sonoma orogeny has a history similar to that of the Antler orogeny. In this case, the Golconda allochthon was transported 35 to 40 miles along the Golconda thrust partly over the Roberts Mountains allochthon and partly above the Antler Overlap sequence. The Golconda allochthon consists of Late Devonian-Permian (c. 360 −252 million years old) chert, sandstone, basalt, and other rocks that formed in a deep-water ocean basin known as the Havallah Basin. The end of the Somoma orogeny is marked by unconformity 2 in Fig. 18.23. The Early Triassic−Early Cretaceous rock sequence above the unconformity represents yet another return to miogeoclinal deposition on a passive continental shelf. The Sonoma orogeny must have occurred in the Late Permian-Early Triassic (c. 260−247 Ma) following deposition of the youngest rock in the Golconda allochthon below the unconformity, and prior to deposition of the oldest rock in the miogeocline above the unconformity. Rocks of the Golconda allochthon are exposed in mountain blocks west of and above the Roberts Mountain allochthon including the Toquima, Sonoma, and Tobin Ranges, and along the western flank of Battle Mountain where rocks of the Roberts Mountain allochthon form Antler Peak and Elephant Head. Orogeny is interpreted to have resulted from collision of an offshore volcanic arc with the North American mainland.

Fig. 18.23 shows that the Sevier thrust sheets carry all of the previously described rock units including the Triassic-Early Cretaceous miogeocline, but that the faults are overlain with Eocene rocks, thus creating

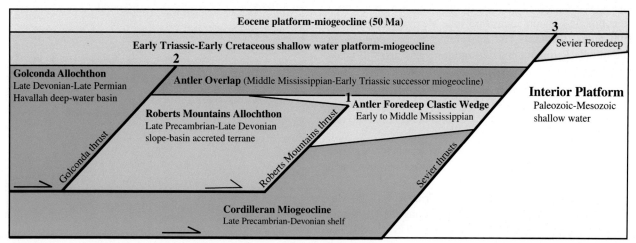

FIGURE 18.23 Schematic cross-section that shows three unconformities numbered 1, 2, and 3 signifying the end of : 1−Antler orogeny (Late Devonian-Early Mississippian), 2−Sonoma orogeny (Late Permian-Early Triassic), and 3−Sevier orogeny (Late Cretaceous-Early Cenozoic).

unconformity 3. These relationships help bracket the age of thrusting between 115 and 52 million years ago. The Sevier thrust sheets represent the final stage in the building of the tectonic wedge. The Antler and Sonoma orogenies are too old to have been directly involved with the building of the tectonic wedge. Instead, there is evidence for thrusting in the Cordillera beginning in the Middle Jurassic 173 million years ago. These thrust faults propagated eastward over the course of more than 100 million years to build the wedge that culminated with the Sevier foreland fold and thrust belt. Some of these early thrust faults are hidden in normal fault-block mountains that include the East, Humbolt, Stillwater, Desatoya, and Shoshone Ranges. As thrusting progressed eastward, erosion of the mountain landscape left a highland terrain, the Nevadaplano, in the western Basin and Range that could have been in existence as early as 110 million years ago. In addition to the thrust faulting that built the tectonic wedge, there is evidence that the accreted terrane belt experienced Early Cretaceous strike-slip faulting between about 140 and 120 million years ago with as much as 140 miles of right-lateral offset. Clearly, the relationships in the Basin and Range indicate a long and complex Cordilleran geologic history.

One additional tectonic episode can be deduced from the geological relationships. Fig. 18.22 shows that the Interior platform and miogeocline both end abruptly in southern California against the belt of accreted terranes. This truncation of rock successions is unusual and is not seen in the Appalachians. The truncation is understood to have resulted from development of a left-lateral transform plate boundary known as the California-Coahuila transform (or Mojave-Sonora megashear) that may have come into existence during the Pennsylvanian (c. 323 Ma) following the Antler orogeny, and persisted into the Late Triassic (c. 220 Ma) following the Sonoma orogeny. Regional relationships suggest that the transform fault displaced the southwestern edge of the North American miogeocline and craton southward into Mexico a distance of 590 miles (950 km). The interpretive paleogeography in the middle Permian circa 270 million years ago following the Antler orogeny is shown in Fig. 18.24. The actual California-Coahuila fault is not well exposed. It trends through the Mojave Desert and Sierra Nevada, but there has been too much subsequent deformation, intrusion, and deposition to really put your finger on it. Instead, the fault is defined as a zone several miles wide where there are marked changes in the strike, composition, age, and degree of deformation in the rocks.

Based on the presence of Late Triassic (c. 220 Ma) and younger subduction-related rocks along the California coast, we can suggest that the entire US Cordilleran continental margin had evolved into a subduction zone boundary by about 220 million years ago that persisted for the next 190 million years until initiation of the San Andreas Fault as discussed in Chapter 20. The change in the southern part of the Cordillera from a passive (miogeoclinal) continental boundary, to a transform boundary, and then to a subduction boundary, must mark times of major plate reorganization. It is perhaps no coincidence that these changes occurred during the Alleghany and Ouachita orogenies, and with subsequent formation of Triassic rift basins in the eastern United States.

Notice in Fig. 18.22 that the interior platform sedimentary rock sequence extends across the entire New Mexico-Arizona Basin and Range. The presence of this relatively thin sedimentary sequence (relative to the much thicker miogeocline) is perhaps one reason why so many metamorphic core complexes are developed in the region. The underlying crystalline shield rocks were already close to the surface prior to core complex formation.

Also present in the Arizona-New Mexico Basin and Range is the Maria fold-and-thrust belt, which is something of an aberration because it extends across the interior

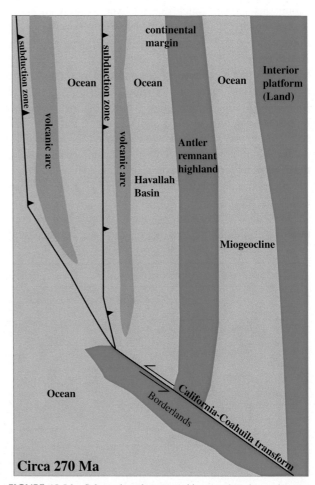

FIGURE 18.24 Schematic paleogeographic map that shows the western Cordilleran coast and the California-Coahuila transform fault circa 270 million years ago. Based partly on Blakey (2011).

platform rock succession in an east-west direction with vergence (thrust direction) dominantly toward the south. The approximate extent of the thrust belt is shown as line 6 in Figs. 18.8, 18.9, and 18.10. The thrust belt likely continues south of El Paso into Mexico, but is severely disrupted by normal faults. Evidence for its existence is seen in the form of folds and faults within mountain blocks that include the Harquahala Mountains. The thrust belt was active between about 90 and 70 million years ago during the early stages of both the Sevier and Laramide orogenies. It may have formed across preexisting weaknesses in the underlying crystalline shield rocks.

In summary, we see that the Basin and Range provides a wealth of information on the geological development of the United States that includes several orogenic events, the formation of a transform fault, and the building of the tectonic wedge that culminated with the Sevier fold and thrust belt. Hidden in the fault block-mountains is evidence for strike-slip faulting and for development of the somewhat unusual Maria fold and thrust belt. We saw in Chapter 17 that the Basin and Range was also the site for extensive volcanism both prior to and during normal faulting.

The $Sr_i = 0.706$ Line

One question that we addressed regarding the Appalachian Mountain belt but could not answer was how far east the North American crystalline shield extends at depth below the orogenic belt. All we can say is that it extends at least as far east as the Coastal Plain. We can ask a similar question with respect to the Cordilleran Mountains, but this time we would like to know how far west shield rocks extend at depth below the orogenic belt. In this case, we can provide an answer based on the geochemistry of intrusive rocks that have pierced the edge of the buried shield on their way to the surface. We can define the western limit of shield rocks as coinciding with a boundary line known as the initial $^{87}Sr/^{86}Sr = 0.706$ line (Sr is the symbol for the element strontium). We will refer to this boundary informally as the 706 line. It is line 4 in Figs. 18.7 and 18.8. Such a line cannot be defined in the Appalachian orogenic belt because intrusions that pierce the edge of the shield do not reach the surface. Instead, they are buried beneath younger sedimentary rock of the Coastal Plain.

The 706 line is based on the chemical properties of magma generated through the partial melting of mantle rock. Magma generated in the mantle will have an initial ^{87}Sr to ^{86}Sr ratio of 0.702 to 0.704. If the magma subsequently rises into ancient North American shield rock, it will be contaminated such that the $^{87}Sr/^{86}Sr$ ratio will increase. The reason for the increase has to do with: (1) the mineral composition of mantle rock versus continental rock, (2) the relatively old age of crystalline shield rock, and (3) the chemical properties of ^{87}Sr and ^{86}Sr.

^{87}Sr is a stable decay product of radioactive ^{87}Rb (rubidium 87). ^{86}Sr is a stable isotope that cannot form by the decay of any radioactive substance. What this means is that if rubidium is present in a rock, the amount of ^{87}Sr will increase over time as ^{87}Rb undergoes radioactive decay, but the amount of ^{86}Sr will remain constant in the rock for all of eternity. Thus, the presence of Rb in a rock will cause the $^{87}Sr/^{86}Sr$ ratio to increase over time.

Rubidium is rarely found in abundance in igneous rocks. However, it does substitute for potassium (K), which means that it is found in small quantities in any K-bearing mineral. The Earth's mantle is composed of minerals such as olivine, pyroxene, spinel, and garnet, none of which contain potassium (and, therefore, no rubidium). In the absence of rubidium, the amount of ^{87}Sr in mantle rock will not increase over time, and because ^{86}Sr also does not increase, the $^{87}Sr/^{86}Sr$ ratio in the mantle remains constant at values between 0.702 and 0.704 irrespective of the age of the mantle rock. A magma that forms via partial melting of mantle rock will have these same values.

The North American shield, on the other hand, contains feldspar and mica, both of which contain potassium, and therefore, contain rubidium. Shield rocks are also very old. This means that the original ^{87}Rb in shield rock has had ample time to decay such that the amount of ^{87}Sr has increased to the point where the $^{87}Sr/^{86}Sr$ ratio can be 0.710 or higher. A magma that forms initially in the mantle and then rises through these old rocks will be contaminated with enough ^{87}Sr to increase the initial $^{87}Sr/^{86}Sr$ ratio to 0.706 or higher. The Cordilleran hinterland is riddled with intrusive rock, and all of it can be analyzed to determine its initial $^{87}Sr/^{86}Sr$ ratio (written as Sr_i). The line across which Sr_i increases above 0.706 thus marks the approximate location where rising magma must have encountered the buried edge of the ancient crystalline shield on its way to the surface. Because the miogeocline overlies the crystalline shield, the 706 line also marks the approximate western limit of the miogeocline. Note that we are interested in the initial $^{87}Sr/^{86}Sr$ ratio of the intrusive rock when it first crystallizes from magma. The $^{87}Sr/^{86}Sr$ ratio will increase as the intrusive rock ages. Luckily there are methods by which the initial ratio can be determined.

An important piece of the story is the age and composition of accreted terranes, and the fact that ^{87}Rb decays to ^{87}Sr very slowly. Accreted terranes in the Cordillera contain substantial amounts of relatively young oceanic volcanic rock derived from partial melting in the mantle. These rocks, partly because of their young age and partly because many are potassium-poor, have $^{87}Sr/^{86}Sr$ ratios less than 0.706. So, even if magma passes through these rocks, or if some of the accreted rock melts, the initial $^{87}Sr/^{86}Sr$ ratio likely would not rise above 0.706. The $^{87}Sr/^{86}Sr$ ratio will rise above 0.706 only if magma encounters rocks of the ancient crystalline shield.

The location of the 706 line (line 4) east of the line of accreted terranes (line 3) in Fig. 18.7 indicates that part of the original North American continental margin was underthrust during Cordilleran orogeny. These lines merge at the Nevada-California boundary before splitting again. In California, the 706 line is disrupted by both young and ancient strike-slip faults. Line 4 in Fig. 18.7 it is shown looping northward along the east side of the Sierra Nevada and then southward along the western side. The loop is possibly due to offset along strike-slip faults active circa 140 to 120 million years ago. Fig. 18.8 shows the 706 line ending against the San Andreas Fault south of Bakersfield. The line reappears on the west side of the fault about 240 miles to the north near San Jose and trends southward along the west side of the San Andreas Fault to Mexico. Part of the southward trend west of the San Andreas Fault is shown in Fig. 18.8. Shield rocks in this area are present between the 706 line (line 4) and the San Andreas Fault (line 5).

Thus, the 706 line provides a first-order approximation to the (now deformed) westward limit of the North American continent prior to Cordilleran accretion. On this basis, it appears that ancient North America extended almost to the Pacific Ocean in southern California but only to central Nevada farther north.

RIO GRANDE RIFT

As seen in Fig. 18.12A, the area that crosses southern Arizona without active faults is abruptly terminated in central New Mexico by a narrow concentration of active normal faults that define the Rio Grande Rift. Active faults associated with the Rio Grande Rift are present within the dark shaded area in Figs. 18.9 and 18.10. The active rift zone is long and linear, stretching for more than 800 miles along the Rio Grande from just south of the Colorado River on the east side of the Gore Range in Colorado, to the vicinity of the Bofecillos Mountains in Texas along the Mexican border. Inactive normal faults extend farther south where they surround the Chisos Mountains in Big Bend National Park. The rift zone is bordered in part by the Basin and Range, Colorado Plateau, and Southern Rocky Mountain physiographic provinces. The northward extension of the rift is unusual because it developed along the crest of the Southern Rockies. Its significance and effect on the Colorado Rockies was discussed in Chapter 14.

The active rift zone is up to 100 miles wide between El Paso and Santa Fe, but narrows to less than 20 miles wide north of Santa Fe and south of El Paso. The rift zone landscape consists of a central trough occupied by the Arkansas and Rio Grande River valleys bordered by a series of mountain ranges and interconnected basins. From north to south, basins include the Arkansas River Valley at Leadville, the San Luis valley, the Jornada Del Muerto, Tularosa, and Hueco Basins, and part of the Salt Basin, all of which are shown in Figs. 18.9 and 18.10. Field analysis suggests that faulting began as early as 30 million years ago, but similar to the Basin and Range, faulting did not become prevalent until 17 to 10 million years ago. This younger phase of faulting is most responsible for present-day landscape. Faulting is generally thought to have migrated northward although areas in the south have remained active. According to the US Geological Survey fault database, the entire rift zone shown with dark shade in Figs. 18.9 and 18.10 has been active within the past 130,000 years except for the southernmost area between the Van Horn and Bofecillos Mountains, where activity has occurred in the past 1.6 million years primarily on the Mexican side of the Rio Grande. Active normal faults are present along the flanks of some of the highest peaks, including the west flanks of the Quitman, Sacramento (Sierra Blanca), Caballos, Manzano/Sandia, Nacimimiento, Sangre de Cristo, and Mosquito Mountains, and along the east flanks of the Franklin, Organ, San Andres, Sawatch, and Gore Mountains. Faulting has helped create a large area of internal drainage that includes parts of the Estancia, Jornada Del Muerto, Tularosa, and Salt Basins (Fig. 2.2).

The area surrounding the Rio Grande south of El Paso consists of low lying mountains with peaks between 4000 and 6000 feet. A few volcanic peaks, such as Chinati and Eagle Peak, rise above 7400 feet. The river drops from 3700 feet at El Paso to 2360 feet at the southern end of the active rift zone near the Bofecillos Mountains along a straight-line distance of 228 miles.

Landscape in southern New Mexico between El Paso and Santa Fe is not terribly different from Basin and Range landscape to the west. River elevation rises from El Paso to 4950 feet near Albuquerque along a straight-line distance of 227 miles. With the exception of Sierra Blanca (11,973 feet), an old eroded volcano, much of the surrounding highland is between 6000 and 9000 feet with a few peaks above 10,000 feet. Wide basins are between 4000 and 4500 feet. The San Andres and Manzano Mountains expose rock of the North American crystalline shield along active faults on their eastern and western sides respectively. An active fault along the east side of the Organ Mountains exposes silicic volcanic and intrusive rocks associated with the Mogollon-Datil volcanic field.

Basins and mountain peaks reach high elevation north of Santa Fe where the rift zone enters the Southern Rocky Mountains. River elevation near Santa Fe, 56 miles north of Albuquerque, is 5400 feet, and the river is incised more than 1000 feet as it crosses the eastern flank of Valles Caldera. Truchas Peak, in the Sangre de Cristo Mountains north of Santa Fe, rises above 13,000 feet. Upon entering the San Luis Valley in southern Colorado, river elevation at Alamosa, 120 miles north of Santa Fe, is 7550 feet, and several peaks in the neighboring Sangre de Cristo Mountains rise above 14,000 feet including Culebra and Blanca Peaks. Whereas the Rio Grande turns westward into the San Juan Mountains, the rift zone continues northward into the topographically high Arkansas River valley where river elevation near Leadville, 125 miles north of Alamosa, is 9700 feet, and peaks on both sides of the valley rise above 14,000 feet, including Mount Elbert, the highest in the Rocky Mountains.

Seismic and gravity studies indicate that crustal thickness is about 20 miles near El Paso, which is somewhat thicker than the 17- to 18-mile-thick crust in southern Arizona. Crustal thickness increases northward in the rift zone to about 25 miles at the New Mexico-Colorado border, and between 31 and 32 miles in the Arkansas River Valley. Crustal thickness on the surrounding Colorado Plateau, Great Plains, and Southern Rockies is between 25 and 31 miles.

There are areas of volcanism surrounding the Rio Grande rift zone that include the San Juan and Mogollon-Datil fields, and the Jemez Lineament (Chapter 17). These volcanic rocks may have heated and weakened the crust, allowing rifting to more easily occur, but a direct link to rifting is debatable. Younger volcanism, including numerous basalt flows from 5 million years old to a few thousand years old, are scattered along the rift zone and are probably more directly associated with rifting.

The rift zone extends south of the Bofecillos Mountains across Big Bend National Park surrounding the Chisos Mountains, but faults are older than 1.6 million years and are not considered active. This area hosts abundant volcanic rock as part of the Trans-Pecos volcanic field. The Rio Grande is a designated wild and scenic river in this area, and has cut several canyons that are more than 1000 feet deep, including the Santa Elena and Boquillas Canyons. Because of its national park status, the following subsections discuss the influence that normal faults have had on landscape in the Big Bend region followed by a short discussion on the origin of the Rio Grande. Additionally, we discuss two major dune fields within the rift zone, White Sands in New Mexico, and the Great Sand Dunes in Colorado.

Monoclines and Normal Faults in the Big Bend Area, Texas

Some of the most widely exposed sedimentary rocks in the Big Bend area are those of the Western interior seaway. Recall from Fig. 8.3 that this was the seaway that covered the Great Plains, Middle-Southern Rockies and Colorado Plateau between about 120 and 75 million years ago. Limestone is the most abundant rock layer with subordinate conglomerate, sandstone, mudstone, and shale. Several thick-bedded limestone rock units, most notably the Del Carmon and Santa Elena Formations, form massive, resistant cliffs. The rocks were variably deformed during Laramide (or Maria) deformation prior to 47 million years ago, which lifted the Big Bend area out of the Cretaceous seaway. The style of deformation is that of monoclines and uplifts, mimicking the deformational style seen on the Colorado Plateau. A major difference is that monoclines in Big Bend area have been modified by normal faults associated with an early period of the Rio Grande rift. We will discuss three of the largest monoclines, the Fresno-Terlingua, the Mesa de Anguila, and the Sierra del Carmen-Santiago Mountains monoclines and their modification. All three originally formed above a reverse fault as shown in Fig. 4.10.

Fig. 18.25 looks just west of north at the Big Bend area. The Rio Grande is shown with a thin yellow line. In addition to Cretaceous limestone and other sedimentary rocks, the area is underlain with volcanic-intrusive rocks, and with young rocks and sediment deposited by streams. Most of the dark-shaded areas, including the Rosillos (r), Christmas (ch), Chisos (c), McKinney (m), Mine Point (MP), Stillwell (st), and Bofecillos (B, Bp) areas, are underlain with volcanic and intrusive rocks that are 38 to 18 million years old and associated with the Trans-Pecos volcanic field (Chapter 17). Mariscal Mountain (mm) is a large anticline underlain with thick Cretaceous limestone.

Normal faults in Big Bend generally do not create large fault-block mountains such as those seen elsewhere in the Basin and Range. The faults, however, are numerous and closely spaced, particularly within the massive Cretaceous limestone where the slow pace of erosion has preserved numerous small but well expressed escarpments and narrow valleys, each marked by a normal fault. The faults were active between 25 and 2 million years ago. The USGS lists just one active fault within the area of Fig. 18.25, the Dugout Wells Fault, active within the past 1.6 million years (shown as short yellow line at Estufa Bolson, eb).

We begin our discussion in the northwestern corner of Fig. 18.25 in the area surrounding the Contrabando Lowland (C) where two of the three monoclines are present. The Fresno-Terlingua monocline separates the Terlingua Uplift (t) from the Contrabando Lowland (C).

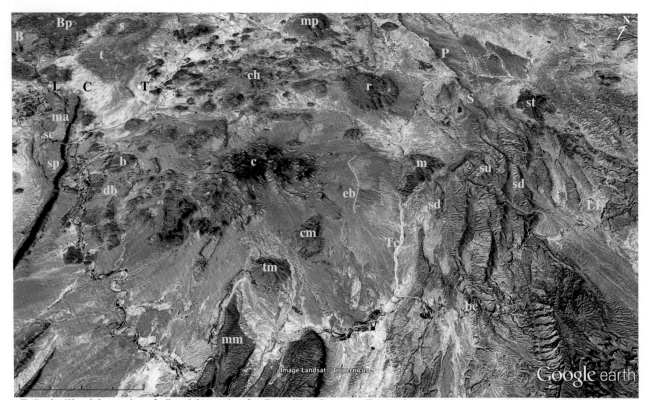

B-Bofecillos Mountains, **b**-Bee Mountain, **bc**-Boquillas Canyon, **Bp**-Bofecillos Plateau, **c**-Chisos Mountains, **C**-Contrabando Lowland, **ch**-Christmas Mountains, **cm**-Chilicotal Mountain, **db**-Delaho Bolson, **eb**-Estufa Bolson, **L**-Lajitas, **LL**-La Linda, **m**-McKinney Hills, **ma**-Mesa de Anguila, **mm**-Mariscal Mountain anticline, **mp**-Mine Point Mesa, **P**-Persimmon Gap, **r**-Rosillos Mtns., **S**-Santiago Mtns., **s**-Solitario, **sc**-Santa Elena Canyon, **sd**-Sierra del Carmen, **sp**-Sierra Ponce, **su**-Sue Peaks, **st**-Stillwell Mountain, **T**-Terlingua, **t**-Terlingua Uplift, **Tc**-Tornillo Creek, **tm**-Talley Mountain. The yellow line at Estufa Bolson is Dugout Wells Fault.

FIGURE 18.25 Google Earth image of Big Bend National Park and vicinity, Texas looking west of north. The thin yellow line is the Rio Grande. The short yellow line at Estufa bolson (eb) is the active Dugout Wells normal fault.

The trace of the Mesa de Anguila monocline is along the prominent 1000- to 1500-foot escarpment visible on the east side of Mesa de Anguila (ma) and extending across the Rio Grande into Mexico as the Sierra Ponce (sp). The Rio Grande cuts through the escarpment to form Santa Elena Canyon (sc).

Fig. 18.26 is a Google Earth image looking northeast that shows the area in more detail. The trace of the Fresno-Terlingua monocline is indicated with a thin yellow line and arrows that point in the direction of dip. This monocline is largely unaltered by normal faults. The trace of the Mesa de Anguila monocline has been disrupted and replaced by the inactive Terlingua normal fault (tf), which extends along the base of the escarpment. Nearly all of the small ridges and escarpments visible on the Terlingua Uplift (t) and Mesa de Anguila (ma) are small, inactive normal faults within massive Cretaceous limestone whose expression is enhanced by water solution and development of karst features.

Fig. 18.27 is a composite of two cross-sections that, together, extend across both the Fresno-Terlingua and Mesa de Anguila monoclines. The location of the composite cross-section is shown in Fig. 18.26 with two thin black lines. Within the Fresno-Terlingua monocline, the Santa Elena Limestone forms a monoclinal slope that separates the high-standing Terlingua Uplift at 3200 to 3700 feet, from the Contrabando Lowland at 2600 to 2800 feet. One can view the monoclinal slope, karst features, and faults by hiking Contrabando canyon (c). The Mesa de Anguila at one time formed the uplifted side of a monoclinal slope that dipped northeastward toward the Contrabando Lowland (C), similar to, but in the opposite direction of the Fresno-Terlingua monocline. The monoclinal slope has since been replaced by a steep escarpment wall of limestone formed by down-dropping along the east side of the Terlingua normal fault. The landscape is seen in Fig. 18.28, which looks northwestward along the escarpment wall. The inactive Terlingua normal fault

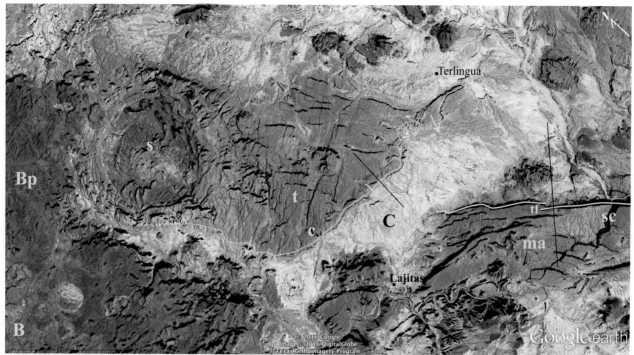

B-Bofecillos Mountains, Bp-Bofecillos Plateau, C-Contrabando Lowland, c-Contrabando Canyon, ma-Mesa de Anguila, s-Solitario, sc-Santa Elena Canyon, t-Terlingua Uplift, tf-Terlingua fault.

Segmented lines for Cross-section shown in Fig. 18.27

FIGURE 18.26 Google Earth image looking northeast at the Terlingua-Mesa de Anguila area, Texas. The Fresno-Terlingua Monocline is shown with a thin yellow line and arrows that point in the direction of dip. The Mesa de Anguila Monocline is largely replaced by the Terlingua normal fault shown with two thick yellow lines.

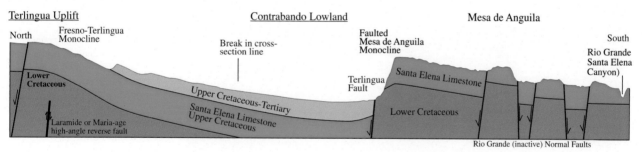

FIGURE 18.27 Composite cross-section of the Fresno-Terlingua monocline and the Mesa de Anguila monocline, which is disrupted by the Terlingua normal fault. Cross-section location is shown with two black lines in Fig. 18.26. Based broadly on DeCamp (1985).

extends along the base of the escarpment. Smaller escarpments on the mesa mark the location of additional inactive normal faults.

The third of the three monoclines, the Sierra del Carman-Santiago monocline, extends along the entire eastern side of Fig. 18.25 through the Sierra del Carman (sd) and Santiago Mountains (S). The Sierra del Carman (sd) forms a series of ridges and escarpments that step down toward the east over a distance of about 12 miles from Sue Peaks (su) at 5854 feet to the Rio Grande at 1677 feet. Here again, all of the escarpments are marked by inactive normal faults primarily within massive Cretaceous Limestone. The Sierra del Carman-Santiago monocline trends along the western side of Sierra del Carman (sd), west of Sue Peaks (su), and northward along the single ridge that forms the Santiago Mountains (S). The monocline dips westward and is cut in some areas by normal faults. We will look at the monocline at Persimmon Gap (P).

Persimmon Gap (P) is a lowland pass through the otherwise continuous Santiago Mountain ridge. It forms the main entrance to Big Bend National Park traversed by route 385. Here, at the entrance to the park, one can see an Ouachita-age thrust fault, a Laramide- (or Maria)-age reverse fault that cuts the monocline, and a normal fault associated with the early stages of the Rio Grande Rift. All of the faults are inactive. Persimmon Gap is thus one of three locations in the Trans-Pecos where aspects of Appalachian-Ouachita orogeny are exposed. The other two are the Marathon Basin (Chapter 15) and the Solitario (Chapter 17).

Fig. 18.29 is a Google Earth image looking across to the southeastern side of Persimmon Gap. The Laramide reverse fault that elsewhere underlies the monocline has breached the surface at this location and is shown on the west (right) side of the hillside with a thick black line and teeth on the up-thrown side. A normal fault is shown on the east side of the image with a thick blue line and tick marks on the down-thrown side. The dark-shaded area outlines two rock units, labeled P1 and P2, that represent Paleozoic rocks involved in the Ouachita fold and thrust belt. Units P1 and P2 are separated by a thrust fault shown in orange that was active in the late Paleozoic during formation of the Ouachita fold and thrust belt. The thrust is cut off on the east (left) side by the normal fault, and on the west side by an unconformity below Cretaceous layers K1. Fig. 18.30 is an east to west cross-section across rock unit P1 that shows the relationships. The monocline is not well represented in this cross-section due to displacement on the reverse fault.

The Rio Grande Bolson Deposits

Evidence suggests that prior to about 2.25 million years ago, the Rio Grande did not exist in Texas. Instead, it flowed into a large lake near the New Mexico-Texas border. The age and location of Rio Grande sediment suggests that the Rio Grande overtopped its terminal lake about 2.25 million years ago and, soon thereafter, flowed all the way to the ocean thereby creating the

FIGURE 18.28 Google Earth image looking northwestward at Mesa de Anguila and Santa Elena Canyon, Big Bend area, Texas. The Rio Grande is shown with a thin yellow line. The inactive Terlingua normal fault extends along the base of the high escarpment. Smaller escarpments on the mesa also mark the location of inactive normal faults.

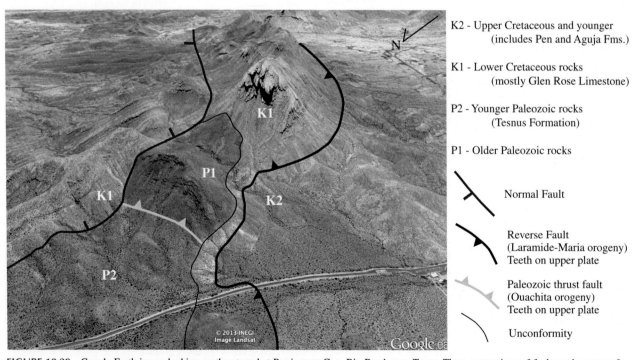

FIGURE 18.29 Google Earth image looking southeastward at Persimmon Gap, Big Bend area, Texas. Three generations of faults and an unconformity are present. The entrance to Big Bend National Park is at lower right.

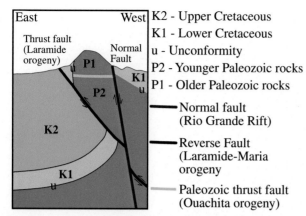

FIGURE 18.30 East to west cross-section across Persimmon Gap, Big Bend area, Texas. A Laramide or Maria reverse fault cuts across the Sierra del Carman-Santiago monocline and exposes thrust faulted Paleozoic rocks. The monocline is also cut by a normal fault associated with the Rio Grande rift. Based on Turner et al., 2011.

present-day river system. Prior to 2.25 million years ago, stream and lake deposits in the Big Bend area accumulated in closed, flat-bottomed, normal fault-bound basins known as bolsons. These were mini Great Basins in the sense that they had no outlet to the sea. As the Rio Grande made its way to the ocean, tributary streams within the bolsons began to integrate (connect) with the Rio Grande, and in doing so they cut downward excavating the bolson deposits. Downcutting and bolson excavation has been intense. It has been suggested that nearly all of the canyons in the Big Bend area, including the Santa Elena and Boquillas Canyons, formed in the past 2.25 million years. Fig. 18.25 locates two bolsons within the Big Bend area, Delaho (db) and Estufa Bolson (eb). A third bolson deposit, the large Presidio Bolson (Pb), extends for more than 40 miles along the Rio Grande north of Presidio (Fig. 18.10). All three are now integrated with the Rio Grande.

Bolson deposits are basin-fill deposits. They are dominantly alluvial (stream-related) and consist of gravel, sand, silt, and clay at various stages of consolidation from unconsolidated to completely cemented. Bolson deposition began with normal faulting, and continued after normal faults became inactive, creating pediments that partly bury the mountains in their own debris in a manner not unlike that shown in Fig. 18.16C. A basalt lava flow interlayered with sediment near the base of Delaho Bolson dates some of the oldest deposits at 23.3 million years ago. Maximum thickness of bolson deposits is between 1500 and 3200 feet, and they contain a variety of mammal, tortoise, lizard, and toad fossils.

Following integration with the Rio Grande, rivers began to cut downward into the bolson deposits creating river valleys. Uneroded remnants of the original bolson surface would be left as small, flat-topped hills above the incised river valleys. These remnants could be considered erosional leftovers similar to flat-topped erosional mountains. The Delaho Bolson was severely dissected leaving only isolated remnants that today are difficult to recognize in Google Earth images. The Presidio and Estufa bolsons were only partly dissected such that preserved bolson surfaces stand as high, flat-topped hills 150 to 250 feet above the valley floor.

During the past 2.25 million years, and especially during the past 800,000 years when glacial advances and retreats were most active, we would expect periods of stream incision to be punctuated with periods of deposition. Streams would cut downward into bolson sediments during periods of incision, and then deposit sediment within the incised stream channel during periods of deposition. If the stream later cuts downward into the newly deposited sediment, it will leave remnants (erosional leftovers) of the old stream surface as high ground in the form of a terrace. These remnants would be topographically higher than the active stream valley, but lower than the original bolson surface. If this process is repeated, it should be possible to see a hierarchy of deposition levels. In other words, it should be possible to see several levels of stream deposition each at a distinct elevation. Within this hierarchy, the depositional surface at highest elevation would be oldest, and the one at lowest elevation or at active stream level would be youngest.

We can illustrate an incision-deposition hierarchy at Estufa Bolson (eb) located in Fig. 18.25. Fig. 18.31 is a Google Earth image looking north-northwest at the area surrounding Estufa Bolson. The thin black line is Big Bend National Park Route 12. The intersection with Main Park Road at Panther Junction (pj) is shown at the top of the image. Also visible are the dark intrusive rocks that form McKinney Hills (Mh) and Chilicotal Mountain (Cm). The many escarpments seen both in the McKinney Hills and the Sierra del Carmen (sd) are normal faults (labeled f) that have chopped the landscape.

Four depositional levels are visible in Fig. 18.31, labeled A1 to A4 from oldest to youngest. The topographically highest and chronologically oldest is A1. This level probably represents the original bolson surface that existed prior to integration with the Rio Grande and prior to stream downcutting. A2 represents sediment deposited following initial integration and stream downcutting and thus, is presumed to be less than 2.25 million years old. At relatively lower elevation we find what is presumably younger stream sediment (A3) deposited following a period of stream incision that cut into A2 sediment. The youngest sediment (A4), at the lowest elevation, represents active deposition in Tornillo Creek (T). In detail, the situation is more complicated because there are several hill levels in addition to the ones visible in this Google Earth image, and there is an active fault, the Dugout Wells Fault

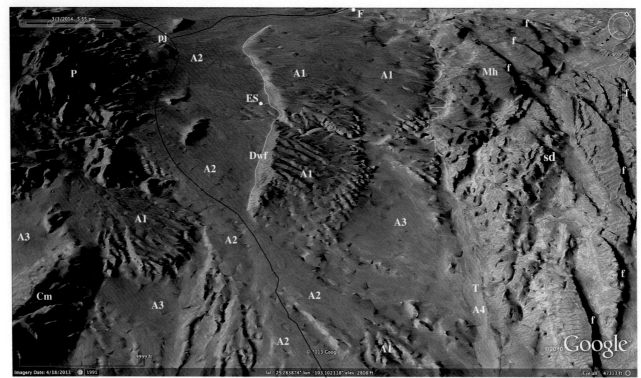

A4 - Youngest Alluvium	Dwf - Dugout Wells Fault	I - Intrusive rocks	pj - Panther Junction
A2, A3 - Younger Alluvium	ES - Estufa Spring	Mh - McKinney Hills intrusive rocks	sd - Sierra del Carmen
A1 - Oldest Alluvium	F - Fossil Bone Exhibit	P - Pine Canyon Caldera	T - Tornillo Creek
Cm - Chilicotal Mountain intrusive rocks	f - normal fault		

FIGURE 18.31 Google Earth image looking north-northwest at the area surrounding Estufa Springs, Big Bend National Park, Texas. The thin yellow line is the active Dugout Wells fault.

(Dwf), which has raised an escarpment at Estufa Springs (ES) thereby elevating and tilting unit A1 and pushing Tornillo Creek against the McKinney Hills. Large areas of internal drainage remain along the Rio Grande in both the United States and Mexico as seen in Fig. 2.2.

White Sands National Monument

Two rather unique dune fields lie within the Rio Grande rift zone, Great Sand Dunes National Park boasts the highest dunes in the United States, and White Sands National Monument is composed of gypsum sand rather than quartz-feldspar sand. Although sand dunes are common throughout the world, most are dominated by the minerals quartz and feldspar. Few are composed of gypsum. Gypsum is a soft, brittle calcium sulfate mineral that will dissolve slowly in fresh water and will chemically precipitate as brine water undergoes evaporation. In order for gypsum dune fields to form, there must be a ready source of gypsum and a mechanism by which gypsum is converted to sand. Such conditions exist in west Texas and eastern New Mexico where the Tularosa Basin, the Salt Basin, and Estancia Valley all boast gypsum dune fields, the only ones in the United States. The largest gypsum dune field in the world is White Sands National Monument in the Tularosa Basin where the dunes are 99% pure gypsum (Fig. 18.9). The Guadalupe dune field in the Salt Basin near El Capitan, and the Estancia dune field southeast of Albuquerque are less pure with both gypsum-rich and quartz-rich dunes.

Gypsum dunes in the Texas-New Mexico area form via a four-step process; dissolution, precipitation, physical weathering, and wind transport. Permian age layers of gypsum (299–260 Ma) are abundant in the Sacramento, San Andres, and Guadalupe mountains surrounding the dune fields. These rocks do not provide the sand for the dunes, but they do provide source material for the calcium sulfate. Rainwater in the mountains partially dissolves the gypsum layers and transports the calcium and sulfate ions downstream to playa (ephemeral) lakes in the Tularosa, Salt, and Estancia basins, all three of which are closed basins with no river outlet to the ocean. The ions become concentrated as the lakes evaporate, forcing second-generation gypsum to precipitate at the bottom of the lake. Once the lake dries up, physical weathering fractures the gypsum into small pieces, which are then transported a short distance by wind and redeposited to form dune fields.

There is no evidence that the process of gypsum dune formation is occurring on a large scale in any of the present-day dune fields. The primary phase of dune formation must have occurred at sometime in the past. So, the question is, when did the dune fields form? Beach ridges and ancient shorelines indicate that large, permanent lakes once existed in all three basins. These lakes can be tied to the last glacial advance circa 18,000 years ago. Beginning 12,000 years ago, these lakes began to dry up, and in doing so, left behind thick layers of second-generation gypsum beds. These are the beds that were broken via physical weathering and carried by wind to form the gypsum fields. An optically stimulated luminescence date from sediments deposited in Lake Otero, the lake that once existed in the Tularosa Basin, suggests that the lake was nearly dry by about 7000 years ago. If we extrapolate to the other two basins, we can suggest that the dune fields could not have formed prior to 12,000 years ago when lakes were present in the valleys, and that much of their formation occurred within the past 7000 years when the lakebeds were dry.

Study of the White Sands dune field suggests that dune formation can be correlated with two specific deflation events associated with the lowering and drying of Lake Otero. Deflation is a term used in geology to describe the removal of sand by wind. The deflation events collectively removed 50 feet (15 m) of gypsum from the bottom of the dry lakebed. It was the removal of these beds that created most of the sand supply for the White Sands dune field. The first event occurred about 7000 years ago, was short-lived, and likely was most important in creating the dunes. The timing for the second event is poorly constrained, but likely occurred less than 4000 years ago.

Our conclusion, therefore, is that all three dune fields could have begun to form as early as 12,000 years ago and were well developed by about 7000 years ago following a major short-lived period of dune formation. Given the abundance of gypsum in the surrounding mountains, and the fact that glacial advances have come and gone periodically for the past 2.4 million years, it is reasonable to suggest that gypsum dune fields were periodically present in this area during dry interglacial periods prior to 12,000 years ago.

Gypsum dune fields have two characteristics that set them apart from quartz-rich dune fields. The first is that gypsum will dissolve in water, and will precipitate from water. As a result, ion-rich water moving through the sand will precipitate gypsum cement, which binds and stabilizes the dunes. Secondly, unlike quartz-rich sand, gypsum sand does not get hot in the blazing sun. At White Sands, you can walk barefoot through the dunes without a problem. You can even rent a sled and slide down the dunes. While in the area, you could also view the spectacular Corrizozo lava flow, which is labeled and highlighted in Fig. 18.9, and discussed in Chapter 17.

Great Sand Dunes National Park

One of the most impressive and highest dune fields in the United States is Great Sand Dunes National Park at the eastern edge of the San Luis Valley along the western margin of the Sangre de Cristo Mountains. The dune field is labeled dunes in Fig. 18.9. As you can appreciate from the photograph shown in Fig. 18.32, the dunes reach a height of 750 feet, making them the tallest in North America.

The great height is due to a combination of opposing wind directions coupled with the location of the dunes at a recess where the Sangre de Cristo Mountains turn from a southeasterly trend to a southwesterly trend. The recess corresponds with Medano Pass (MP), a low point in the mountain range. Several 14,000-foot peaks are present north and south of Medano Pass including Crestone Peak and Blanca Peak. Fig. 18.33 is a Google Earth image looking slightly east of north that shows the relationships. A strong and constant southwest wind blows sand in a northeasterly direction across the San Luis Valley toward the mountain front where it becomes trapped in the recess (shown with a large arrow in Fig. 18.33). The dunes are then piled back on top of each other by an opposing, but less frequent east wind that comes out of the mountains across Medano Pass (the smaller arrow). These opposing winds create the super-high dunes.

Although pinned against the crystalline and sedimentary Sangre de Cristo Mountains, the dune field contains large quantities of volcanic minerals that could only have been derived from the San Juan Mountains 44 miles (70 km) to the west across the San Luis Valley. There are several hypotheses as to how sand was transported across the valley. An early hypothesis suggested that sand was carried by the Rio Grande (RGR) and then abandoned when the river channel migrated to a different location. A more recent hypothesis, based on detailed field mapping, suggests that much of the sand was transported during times of glacial meltwater flooding. The primary transport route was across the Rio Grande fan to a low area known as the Sump located in Fig. 18.33. Glacial meltwater streams from the Sangre de Cristo Mountains also periodically reached the Sump. Sand accumulated principally in the lower (southern) part of the Sump, which is the deepest part of San Luis Valley. This area is considered to be the principal location from which the sand was derived.

Once in the Sump area, sand must dry out before it can be transported by wind. In most areas, and

FIGURE 18.32 Photograph looking north-northwestward toward Great Sand Dunes National Park and the Sangre de Cristo Mountains. Cleveland Peak (13,420 feet) forms the skyline.

FIGURE 18.33 Google Earth image looking slightly east of north across the northern San Luis Valley, Colorado. Arrows indicate wind direction. Thin lines along the western margin of the Sangre de Cristo Mountains are faults active in the past 15,000 years. MP-Medano Pass, RGR-Rio Grande River, S-Great Sand Dunes National Park.

particularly in mountain areas, we know that days are warmer than nights. The volume of glacial meltwater would correspondingly decrease at night, and this would allow some of the sand to dry. The constant southwesterly wind would then move the sand toward the Sangre de Cristo Mountains. Today, there are no permanent streams that reach the sump area, but sand is periodically deposited during especially wet conditions when streams are able to reach the area. This sand continues to be transported toward the Great Sand Dunes during times of drought.

The question then becomes, when did the Great Sand Dune field first develop. Field analysis indicates that an ancient body of water, Lake Alamosa, occupied the Sump area and the eastern half of the Rio Grande Fan from about 3.5 million years ago to 419,000 years ago. The presence of this lake in the Sump area obviously would have prevented too much sand from drying out. It seems logical to conclude, therefore, that the dunes are younger than 419,000 years.

In order to determine when the dunes first appeared, we can look for features that have been modified due to the existence of dunes. Although it is not obvious in Fig. 18.33, small creeks flowing out of the Sangre de Cristo Mountains toward the center of the dune field have been deflected to the margins of the dune field. We can conclude that the dunes must have been in existence in order to cause this deflection. Based on the degree to which some of the deflected stream deposits are weathered, it is estimated that the oldest deposits are at least 130,000 years old. Thus, the dune field is understood to have formed initially between 419,000 and 130,000 years ago.

ROCKY MOUNTAIN BASIN AND RANGE

The Rocky Mountain Basin and Range is located primarily north of the Snake River Plain where variably oriented mountains are separated by wide valleys. A landscape map of the southern part of the area is shown in Fig. 18.34 with the area of active normal faults shown with dark shade. Notice that the Teton Mountains are included. Also shown on this figure are areas of ancient crystalline shield rock, metamorphosed Belt Supergroup rocks, five metamorphic core complexes labeled with yellow highlight, a few additional areas of active normal faults shown with thick green lines, and the northern extension of lines 1, 2, 3 and 4 from Fig. 18.7. Not shown are wide areas of granitic rock associated with the Idaho and Boulder batholiths. Figs. 15.4 and 15.5 show the entire extent of the Rocky Mountain Basin and Range along with other geological features.

The southern part of the Rocky Mountain Basin and Range is part of the Yellowstone crescent of high terrain. In this area, the combination of active normal faults and thermal buoyancy associated with the active Yellowstone hot spot has produced some of the highest elevations in the Northern Rockies. The Teton Mountains rise above 13,000 feet. The Lost River and Lemhi ranges have peaks above 12,000 feet. The Madison and Beaverhead Ranges have peaks above 11,000 feet. This area contains a great variety of rock types, and a complex structural history that includes thrust faults, intrusion, and two primary episodes of normal faulting. The landscape likely had elements of linear, north-trending ridges and valleys circa 115 to 52 million years ago during its involvement in the Sevier fold and thrust belt, however, any mountain linearity at this time would have been disrupted due to the widespread presence of crystalline shield rocks south of Three Forks, and the presence of Idaho and Boulder Batholith granitic rocks surrounding Butte, Montana.

Shortly following the end of thrust faulting, the landscape underwent reincarnation via normal faulting and intense explosive volcanism associated with the Challis and Absaroka volcanic fields. The earliest known Basin and Range-type normal faulting occurred between 56 and 39 million years ago with development of the Pioneer, Anaconda, Bitterroot, and Clearwater crystalline core complexes shown in Fig. 18.34, as well as the three core complexes in northern Washington mentioned in Chapter 16. Normal faulting continued intermittently until at least 30 million years ago. Intrusion continued in some areas until at least 34 million years ago.

The present-day landscape is younger than 30 million years. Following a period of little or no deformation, a new generation of high-angle normal faults began activity beginning sometime after 17 million years ago. These faults have remained active to the present day, and in some cases they reactivate older thrust and normal faults. The primary change to the landscape as a result of recent normal faulting has been a likely increase in mountain elevation and the widening of valleys between the mountains. Normal faults accentuated the somewhat random pattern of mountains and valleys surrounding Butte, thus creating the present-day landscape.

Active faults in the Rocky Mountain Basin and Range are responsible for some of the strongest earthquakes in the region. The strongest recorded earthquake in Idaho was the 1983 magnitude 6.9 Borah Peak earthquake, which created a fault scarp more than 7 feet high on the southwest flank of the Lost River Range visible in Google Earth. Geodetic measurements indicate that the basin dropped an average of four feet and that the mountain uplifted nearly one foot as a result of the earthquake. The largest recorded earthquake in Montana was the 1959 magnitude 7.3 Hebgen Lake earthquake, which created a landslide near the southern end of the Madison Range that blocked the Madison River and created Earthquake (Quake) Lake.

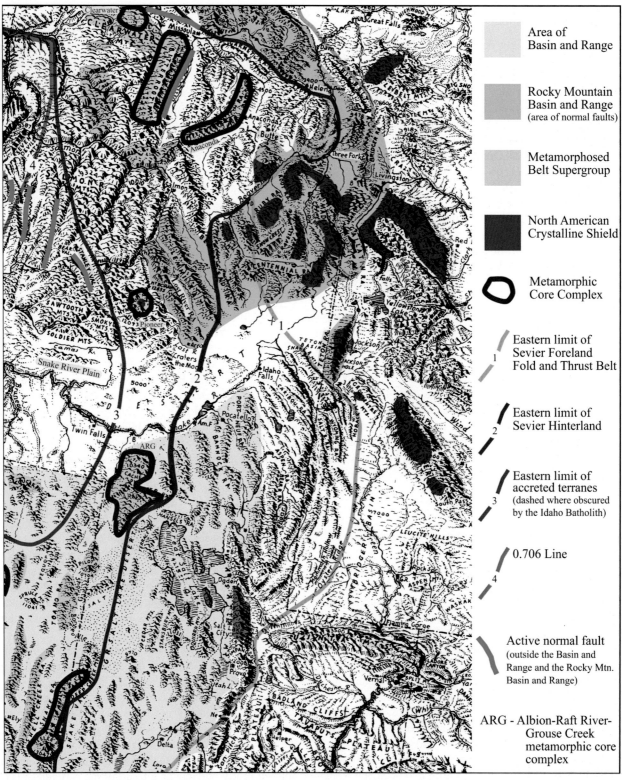

FIGURE 18.34 Landscape map of the southern part of the Rocky Mountain Basin and Range.

THE TETON MOUNTAINS

The Teton Range is an actively uplifting, tilted normal fault block mountain approximately 40 miles long and 10 to 15 miles wide located just south of Yellowstone National Park (Fig. 18.34). The high peaks are composed of ancient (>2500 Ma) North American crystalline shield. Cambrian-Ordovician and Mississippian sedimentary rocks (mostly limestone and dolostone) overlie shield rocks across the Great Unconformity along the northern, southern, and western flanks. As shown in both cross-section (Fig. 18.35) and Google Earth (Fig. 18.36), the mountain block tilts gently westward such that the highest peaks are concentrated near the eastern edge. The steep eastern escarpment is one of the most magnificent sites in the world. From the photograph in Fig. 18.37, one could imagine that a wall of mountain has risen from the flatlands. There is nearly 7000 feet of relief between Jenny Lake (6783 feet), at the base of the escarpment, and Grand Teton (13,775 feet) across a horizontal distance of less than 3.5 miles. Major peaks, in addition to Grand Teton include South Teton (12,519 feet), Middle Teton (12,809 feet), Mount Owen (12,933 feet), Mount Teewinot (12,330 feet), Mount Moran (12,610 feet), and Thor Peak (12,028 feet). The range is sharply terminated to the south by landscape associated with the Idaho-Wyoming fold and thrust belt, and more gradually to the north where crystalline and sedimentary rocks plunge below young volcanic rocks of Yellowstone National Park.

Although located within the Middle Rocky Mountain physiographic province, the Teton Range owes much of its elevation and relief to normal faulting and to its position within the Yellowstone crescent of high terrain. It is considered to be part of the Rocky Mountain Basin and Range even though it is separated from this region by the Yellowstone volcanic field. The Teton fault is an active normal fault that extends along the base of the steep eastern escarpment. The fault has listric geometry such that a low area (Jackson Hole) has formed at the base of the mountain producing Jackson, Leigh, and Jenny Lakes, all partly dammed by glacial moraines (Fig. 18.35).

Initiation of normal faulting and total cumulative vertical displacement associated with the Teton fault are not known with certainty. It is generally agreed that displacement is on the order of 29,500 feet based on the vertical separation of crystalline shield rock on either side of the fault; however, an unknown amount of this

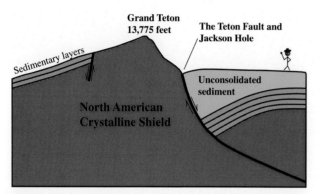

FIGURE 18.35 Cross-section of the Teton Range. The Teton fault is listric and forms a rollover anticline and depression (Jackson Hole) at the base of the escarpment.

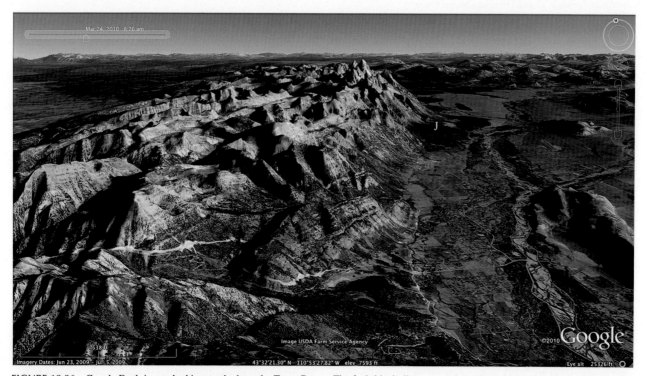

FIGURE 18.36 Google Earth image looking north along the Teton Range. The fault block tilts westward creating a high escarpment along the eastern flank. The highest peak is Grand Teton, J-Jenny Lake.

FIGURE 18.37 Photograph looking westward at the Cathedral Group and the steep eastern front of the Teton Range. The peaks, from left to right are Mount Teewinot, Grand Teton, and Mount Owen.

displacement may have occurred more than 30 million years ago during the early period of Rocky Mountain Basin and Range normal fault activity.

There are two competing theories. Some geologists suggest that normal faulting and initial uplift of the present-day Teton Range began between 13 and 5 million years ago with total cumulative displacement between 19,680 and 29,520 feet (1.8 to 7.1 inches per 100 years). Other geologists suggest two separate periods of normal faulting. The first was active 40 million years ago and the second between 3 and 2 million years ago. Total cumulative displacement on the youngest phase of activity is estimated to be between 8200 and 11,480 feet (3.3 to 6.9 inches per 100 years). In either case, displacement occurs in spurts of 6 feet or more during a single earthquake. Exposure of the fault surface, and offset glacial and landslide deposits along the entire length of the fault, suggest that there has been about 100 feet of displacement during the past 15,000 years (8.0 inches per 100 years). Geological evidence suggests that 8 to 10 earthquakes have occurred during that time span which is equivalent to one every 1500 to 2000 years with an average displacement of 10 to 12 feet per event.

There have been no earthquakes along the northern part of the fault for at least the past 1500 years, and no earthquakes in the southern part for more than 5000 years. Overall, it appears that earthquakes were more common before 8000 years ago. An interesting hypothesis suggests that isostatic adjustment following the melting of glaciers may have accelerated fault displacement. A large ice cap, more than a half-mile thick, covered nearby Yellowstone National Park 22,000 years ago. At the same time, alpine glaciers were present in the valleys of the Teton Range. It is suggested that initial removal of the weight of the glaciers, beginning about 16,000 years ago and ending 14,000 years ago, caused isostatic uplift of the crust. Rather than broad isostatic uplift of the entire region, the hypothesis suggests that movement was concentrated along the Teton normal fault, resulting in an acceleration of earthquakes during the time the crust was isostatically rebounding.

THE WASATCH MOUNTAINS

The Wasatch Mountains extend for about 230 miles from central Utah to southern Idaho directly east of several cities that include Nephi, Provo, Salt Lake City, Ogden, Brigham

City, and Logan (Fig. 18.34). The mountains are a source of water and recreation for more than 80% of Utah's population. Similar to the Teton Range, the Wasatch Range is an asymmetric, tilted, normal fault block mountain with a steep western face that locally towers more than 7000 feet above the populated valleys to the west. In contrast to the Teton Range, the rocks are dominantly sedimentary, but areas of young intrusive rock and crystalline shield rock are also present. The highest peaks, including Mt. Nebo (11,933 feet) and Mt. Timpanogos (11,752feet), occur along the western face. The base of the western face is host to the Wasatch Fault, one of the longest and most active normal faults in the country with 16 major earthquakes in the past 5600 years (one earthquake every 350 years). Paleoseismic field studies suggest that accumulated vertical displacement has averaged between 3.9 and 6.7 in/100 years over the 5600-year period. At the maximum rate of 6.7 in/100 years, total displacement over the 5600-year period is just over 31 feet (9.5 m).

The east side of the mountain is rugged with deeply incised glacial valleys that belie the gentle eastward tilt of the range. Fig. 18.38 shows that the northward trend of the Wasatch Range in the vicinity of Salt Lake City (SLC) is perpendicular to the trend of the adjacent Uinta Mountains anticline. Flexure of the anticline has produced an especially high, broad, and rugged region within the adjacent Wasatch Range where some of the best skiing in the world can be found. The anticlinal flexure, however, does not cross the Wasatch fault.

The Wasatch fault zone is one of the most intensely studied normal fault systems in the world. The fault dips steeply westward at the surface at an angle between 65 and 80 degrees, but the dip angle likely decreases with depth. Several investigations have determined rates of fault displacement, and rates of exhumation and denudation, on time scales from several thousand to several million years, particularly along the central part of the fault trace. Like nearly every major fault in the world, the Wasatch fault zone is segmented in the sense that the surface trace of the fault is not continuous. In some areas, there are gaps between fault segments where no surface trace of a fault can be found. In other areas, two segments overlap, but are separated horizontally by up to several miles. The segmented nature of the fault trace is visible in Fig. 18.38. Segments along a particular fault are often grouped and named based on their earthquake recurrence interval over short time scales of thousands of years, with some segments more active than others. The Wasatch system is composed of 10 segments, each named after a city or mountain in the vicinity. Four of the central segments, from Weber to Nephi, were active between 1250 and 600 years ago, but have no historic earthquakes. The three southern segments have been active during the past 15,000 years. The three northern segments have not been active in the past 15,000 years.

Apatite (U-Th)/He ages obtained from rocks collected at the base of the mountain from the footwall block (the upthrown block of the fault) along five of the central fault segments (Brigham City to Nephi) were used to determine the exhumation history during the past few million years. Fig. 18.39 is a cross-section of the range that shows the location of the rock samples. The average apatite (U-Th)/He age along all five fault segments is 5.3 million years with one anomalous 1.6-million-year date from the southern end of the Salt Lake City segment. These dates, because they are on the up-thrown block close to the fault surface, are interpreted as dating tectonic exhumation of rocks. If we apply depth estimates to the apatite data, then

GSL-Great Salt Lake, P-Provo, SLC-Salt Lake City, UL-Utah Lake; d-area where isotopic studies were conducted.

FIGURE 18.38 Google Earth image looking east at the western front of the Wasatch Range, Utah. The Uinta Mountains form the snow-covered anticlinal mountain at top-center oriented perpendicular to the north-south orientation of the Wasatch Mountains. The vague thin lines are active fault traces from USGS web site: http//earthquake.usgs.gov/regional/qfaults/.

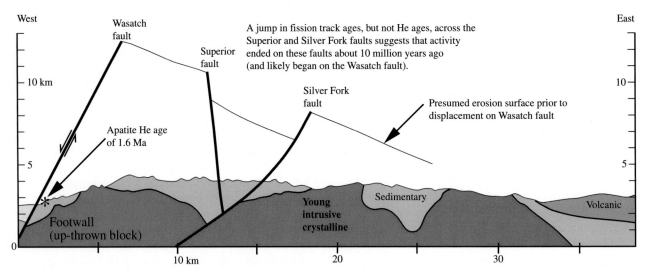

FIGURE 18.39 Cross-section of the Wasatch Range. Based on Perry and Bruhn (1987).

the best estimate on the rate of exhumation is <0.79 to 1.57 inches/100 years over the 5.3-million-year period with a maximum rate of about 4.0 inches/100 years for the rapidly exhumed Salt Lake City segment where the sample with the 1.6 million year old date was obtained. The Salt Lake City segment forms the highest and most rugged part of the range, and it coincides with the location of the Uinta Mountains anticline. It makes sense, therefore, that this is also the most rapidly exhuming part of the range. Because of its anomalous character, this part of the range has been investigated in great detail. Apatite fission track, apatite (U-Th)/He, and zircon fission track ages have been determined along a west-to-east traverse across the range at this location.

All three dating methods show an increase in age from the Wasatch fault eastward across the mountain. Apatite fission track ages increase from 3.4 to 39.6 million years, Zircon fission track ages increase from 9.3 to 37 million years, and apatite He ages increase from 1.6 to 23 million years. As shown in Fig. 18.39, this relationship is consistent with the observation that the fault block is tilted eastward, resulting in greater and more rapid exhumation (and higher elevation) along the western front of the range. The increase in ages from west to east is not smooth. There is a jump in age data across the Silver Fork-Superior fault zone, which lies within of the mountain block east of the Wasatch fault. The age jump across the fault disappears in data that is less than 10 million years old. This relationship suggests that the Silver Fork-Superior fault zone was active prior to 10 million years ago, that it is no longer active, and that the locus of fault activity has jumped eastward to the Wasatch fault beginning between 10 and 12 million years ago. In other words, the Wasatch fault began activity 10 to 12 million years ago, implying that present-day uplift of the Wasatch Mountains did not begin until 10 or 12 million years ago.

A fluid inclusion study of hydrothermally altered rocks was also conducted at the base of the Salt Lake City segment. This type of study is a direct measure of the amount of pressure the rock was under when the fluid inclusion was trapped. The study determined that the base of the western face of the mountain had been exhumed from a depth of 36,080 feet (11 km). This equates to an exhumation rate over the 10- to 12-million-year period of 4.33 to 3.61 inches per 100 years. These rates would be lower if some of the exhumation occurred prior to 12 million years ago. Nevertheless, the rate agrees rather nicely with the exhumation estimate of 4 inches per 100 years deduced from the 1.6 Ma apatite (U-Th)/He age obtained in the same area. Fig. 18.39 shows the amount of rock estimated to have been eroded from the southern Salt Lake City segment. Note the decrease in both total uplift and erosion toward the east.

The Wasatch fault offers a natural laboratory to study the geometry and uplift/exhumation history along a well-exposed, active, normal fault. We are reasonably sure that the Wasatch fault began activity between 10 and 12 million years ago and we can guess that most of the 7000-plus feet of relief between mountain top and valley floor was generated during that time. These figures indicate development of relief at a rate of 0.70 to 0.84 inches per 100 years over the 10- to 12-million-year period. We know that the Wasatch fault is active, but we do not know if the mountain range has reached steady state elevation. Therefore, we do not know if the mountain will continue to gain relief or if it has reached its maximum height.

TRIASSIC LOWLANDS OF THE APPALACHIAN MOUNTAINS

Nestled within crystalline rock of the Piedmont Plateau and New England Highlands are narrow, isolated, lowland

FIGURE 18.40 Landscape Map of the eastern United States showing the location of the largest of the Triassic Lowland valleys between Connecticut and North Carolina.

valleys underlain with clastic sedimentary rocks, basaltic lava flows, and shallow gabbroic intrusions (dikes and sills known as diabase). An old term used for both basalt and diabase is trap rock. The rocks are exposed as far northward as Nova Scotia and southward to the South Carolina border. Fig. 18.40 locates the largest areas in the United States, which include the Hartford Basin in central Connecticut and Massachusetts, and the large Newark-Gettysburg-Culpepper Basin from New York City to Virginia located just east of and parallel with the Great Valley. Less distinctive valleys in Virginia and North Carolina appear to cross older Appalachian structural trends at a small angle. Similar rocks are found in well holes as far south as Florida. Collectively, these areas are known as the Triassic Lowlands or the Triassic Rift Basins.

The rocks were deposited during the Late Triassic-Early Jurassic between 237 and 174 million years ago. They are gently to moderately tilted across normal faults that developed at about the same time as deposition. The faults have not been active since about 180 million years ago, and consequently topography is controlled mostly by erosion and rock hardness. The clastic sedimentary rocks in these valleys consist of reddish sandstone, mudstone, conglomerate, and shale derived from erosion of ancient highland areas. These rocks are not nearly as resistant as the surrounding crystalline rock and are primarily responsible for the lowland topography.

Fig. 18.41 is a Google Earth image looking eastward across the Hartford Basin from Long Island Sound to Massachusetts. The sandstone and shale that forms much of the Triassic Lowlands crops out along road cuts where the brick red color stands out in bold contrast to the light-colored crystalline rocks in the surrounding highlands. Anybody driving from highlands to lowlands would easily notice the difference. The distinctive change in rock type and rock hardness creates sharp boundaries with the crystalline highlands that rise abruptly 500 feet or more on both sides of the Hartford Basin. Both boundaries are marked with normal faults.

Trap rock (basalt and diabase) within the Hartford Basin is resistant enough to form cuestas, hills and cliffs that stand in high relief against the much weaker red sandstone and shale. Fig. 18.42 is a schematic cross-section across the Hartford Basin that shows the relationships. The rocks are tilted eastward along normal faults such that the more resistant trap rock forms west-facing cuestas. Fig. 18.41 shows a long nearly continuous ridge that extends for more than 60 miles from north of New Haven, Connecticut (NH) to well north of Springfield, Massachusetts (S). This is a tilted ridge of basalt known as Metacomet Ridge. It is shown schematically in Fig. 18.42. The northern end of the ridge, where it turns eastward and crosses the Connecticut River, forms the Holyoke Range, which although only 1106 feet in elevation at Mount Norwottuck, towers as much as 1000 feet above surrounding lowlands.

The Metacomet Ridge south of Hartford is broken and offset along normal faults resulting in a series of small north-south ridge segments that trend toward the eastern side of the lowland basin. This area is shown in more detail in Fig. 18.43, which looks just west of north. Note that many of the ridges form westward-facing cuestas. Their presence across the Hartford Basin may have helped push the Connecticut River out of the lowland at Middletown (M) and into crystalline highlands. Notice that the top surfaces of the basalt ridges host numerous elongate lakes. These lakes occupy areas of weak sedimentary rock, excavated during the glaciation, that alternate with the basalt.

The Hartford Basin is also host to about 2000 dinosaur tracks, each from 10 to 16 inches in length and spaced 3.5 to 4.5 feet apart. The tracks are about 200 million years

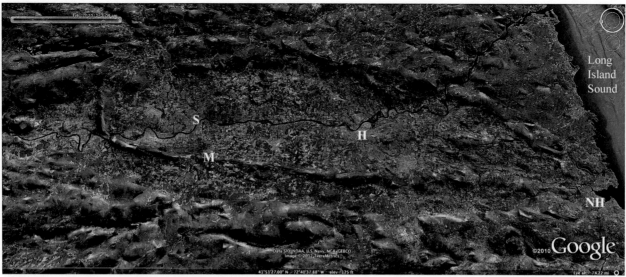

H-Hartford, M-Metacomet Ridge, NH-New Haven, S-Springfield

FIGURE 18.41 Google Earth image looking east across the Triassic Lowlands of Connecticut and Massachusetts. The crystalline margins of the lowland basin are evident. A tilted layer of resistant basalt extends for more than 60 miles through the valley to form Metacomet Ridge (M). Several state attractions including Sleeping Giant and Hanging Hills in Connecticut, and Mount Tom in Massachusetts, are present along the ridge. The large lake at upper left is Quabbin Reservoir. The smaller lake at lower center is Barkhamsted Reservoir. Both reservoirs are situated within crystalline rock.

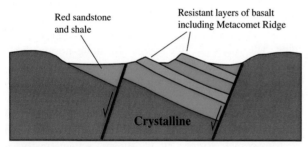

FIGURE 18.42 Schematic cross-section of the Hartford Basin (Triassic Lowlands). Based on Hunt (1974).

old, they are named *Eubrontes*, and although no fossil remains are present at the site, they are thought to have been left by the carnivorous dinosaur *Dilophosaurus*. The tracks are located at Dinosaur State Park in Rocky Hill. It is one of the largest dinosaur track sites in North America.

The Newark Basin is located in the northeastern part of New Jersey approximately between Reading, Pennsylvania and Stony Point, New York. Similar with the Hartford Basin, and visible in Fig. 16.21, hard crystalline rock of the Reading Prong (rp) forms an escarpment up to 500 feet high that marks the northwestern side of the Newark Basin (TR). The escarpment becomes less prominent south of the line of glaciation (toward Morristown, Ms), suggesting it was carved and shaped by glaciers that preferentially excavated soft rocks along the margin of the rift valley.

Normal faulting in the Newark Basin has tilted trap rocks and interlayered sedimentary rocks westward, opposite that of the Hartford Basin. The trap rock forms long, linear, resistant ridges with high cuestas that face (steep-side) east. The most prominent of these are the Watchung Mountains and Palisades Sill. Both are highlighted in yellow in Fig. 18.40 and are visible in Fig. 16.21. The Watchung Mountains consist of three curved ridges of basalt. Preakness Mountain, at the north end of the range rises to 885 feet so that, on a clear day, one can look eastward and see the Manhattan skyline 20 miles away. Basalt flows in the Watchung Mountains have a total thickness of approximately 1850 feet.

The Palisades Sill forms a prominent cuesta for 45 miles along the western shore of the Hudson River from Staten Island almost to Stony Point. A sill is an igneous rock (a diabase in this case) that intrudes parallel with overlying and underlying rock layers (slightly older sandstone and shale in this case). The Palisades Sill is considered to be the finest example of a diabase sill in the United States. The sill is almost 1000 feet thick and is between 192 and 186 million years old. It forms a cliff that in some areas towers more than 600 feet above the Hudson River. The cliff is clearly visible along the west side of the Hudson River in Fig. 18.44, which looks northward at Manhattan between the Hudson and East Rivers. East River is actually a glacier-carved saltwater estuary that connects Long Island Sound with Upper New York Bay. Rocks of the Palisades Sill anchor the western end of the George Washington Bridge leading into Manhattan and are easily accessible at Palisades Interstate Park north of New York City. The Hudson River flows

1-Beseck Lake, 2-Broad Brook Reservoir, 3-Pistapaug Pond, 4-Shuttle Meadow Reservoir 5-Silver Lake, 6-William J Ulbrich Reservoir, C-Cheshire, Cr-Connecticut River, D-Durham, m-Metacomet Ridge. M-Middletown, Nb-New Britain, W-Wallingford. The two yellow lines mark the edges of the Hartford Basin.

FIGURE 18.43 Google Earth image looking just west of north at ridges of basaltic flows and sills in the Hartford Basin, Connecticut.

FIGURE 18.44 Google Earth image looking northward at Manhattan between the Hudson and East Rivers. The Palisades Sill forms the high escarpment along the west side of the Hudson River.

along the contact that separates trap rock, sandstone, and shale of the Palisades Sill from the older Appalachian crystalline rocks that form Manhattan Island and the Western Connecticut Highlands. The river likely took advantage of weak sandstone along the contact.

Lowland valleys south of the Newark Basin are subtle due to the absence of glaciers to carve the valleys and due to the presence of thick soil that has buried bedrock, thus diminishing bedrock influence on landscape. These valleys are less than a few hundred feet below surrounding crystalline highlands and are without distinctive borders. Basalt and diabase are common in the Gettysburg and Culpeper Basins, but they do not form strong landscape features. Trap rock is largely absent in the lowland valleys of southern Virginia and North Carolina, thus contributing to the absence of distinctive landforms.

Why are these lowland valleys of relatively young rock present in the Appalachian Mountain belt? In Chapter 5, we noted that the Appalachian Mountains formed some 265 million years ago as a result of continental collision between North America, Europe, Africa, and South America, and that this collision produced the supercontinent Pangea. It was the rifting of Pangea, and the separation of North America from Africa and Europe beginning about 237 million years ago that eventually produced the Atlantic Ocean. The Triassic Lowlands are a remnant of this ancient rifting event. The Lowlands represent narrow down-dropped basins (grabens and half-grabens) flanked with normal faults much like the Rio Grande Rift. The Triassic Lowland rifts developed too far inland to go to completion. Instead, the eventual split between North America and Africa occurred farther eastward along much larger rift basins that are now buried beneath younger rock of the Coastal Plain and continental shelf. Considering all of the recent tectonic events that have affected the Cordillera, it is perhaps surprising that these lowland valleys represent the final major tectonic landscape-forming event to have affected the Appalachian Mountain belt. And this particular tectonic event affected only a small part of the mountain chain.

QUESTIONS

1. Draw a simple cross-section that shows horst-graben and half-graben geologic structures.
2. Why is average elevation in the Basin and Range so high?
3. How long ago was the earliest known Basan and Range normal faulting and where did it occur?
4. Locate and describe an area in Figs. 18.7, 18.8, or 18.9 where there is a sharp boundary between the Basin and Range province and an adjacent physiographic province. Also locate an area where the boundary is diffuse and poorly defined. Speculate as to why the boundary is sharp in one area and diffuse in other areas.
5. Different mountains are at different stages of erosion and self-burial depending on the recency of activity along range-front faults. In addition to the recency of fault activity, what other factors control the rate of self-burial?
6. In Google Earth, fly to Mount Moses, Nevada. Comment on the stage of erosion and self-burial of the range front.
7. In Google Earth, fly to Mt. Tobin, Nevada. The 1915 Pleasant Valley fault scarp is located along the range front 8 miles southeast of the peak. Compare the morphology of the eastern, faulted, side of Pleasant Valley with its western side. Is there evidence for an active fault along the western side of the valley?
8. In Google Earth, fly to the Toiyabe Range, Nevada. The Big Smoky Valley lies just south of the range. Compare the morphology of the eastern side of the Big Smokey Valley with the western side in terms of recency of faulting.
9. Speculate on the parallelism of the Central Nevada rift and the Walker Lane belt evident in Fig. 18.7. Are the two in any way related?
10. Describe the landscape expression of the Rio Grande Rift in Fig. 18.9. How is the landscape distinctive from the surrounding area?
11. In Google Earth, fly to Dixie Valley, Nevada. This area is just southwest of the Tobin Range. Compare the morphology of the mountain fronts on both sides of the valley. On which side would you expect to find more recent faulting and why? Zoom down to that mountain front and look for the 1954 Dixie Valley Fault scarp. Describe the morphology of this scarp.
12. In Google Earth, fly to Borah Peak, Idaho. What mountain range is this? Examine the morphology of the mountain front directly west of the peak. This is the site of the 1983 Borah Peak earthquake. Zoom down, look for the Lost River Fault scarp, and describe its morphology. Go to the USGS website given in Fig. 18.12, to more accurately locate this scarp.
13. Fly to Hebgen Lake, Montana, site of 1959 Hebgen Lake earthquake. It is very difficult to locate this fault scarp without help from the USGS website given in Fig. 18.12. The earthquake scarp is located above the road that runs along the lake. Also look for a highly visible scar just below the summit of Kirkwood Ridge. Explain the location of the scar so high on the mountain.
14. Fly to Earthquake Lake, Montana. Can you locate the landslide and its source that resulted from the 1959 Hebgen Lake earthquake?
15. Explain why the Rocky Mountain erosion surface discussed in Chapter 14, is, or is not, a pediment surface.
16. What is the Nevadaplano? What happened to it?
17. Describe the fault history of the Teton Range.
18. What evidence suggests that activity on the Wasatch fault did not begin until about 10 or 12 million years ago.
19. What is the Northern Nevada rift zone and why does its history suggest that present-day Basin and Range landscape developed less than 17 Ma.

20. Why isn't Basin and Range landscape present everywhere in areas with active normal faults?
21. What is the origin of high elevation in the southern part of the Rocky Mountain Basin and Range?
22. Generally, what type of evidence is used to determine that widespread Basin and Range normal faulting did not begin until about 17 million years ago?
23. Explain why Death Valley is at such low topographic elevation.
24. Explain why the $^{87}Sr/^{86}Sr$ ratio does not rise above 0.706 when it encounters accreted terrane rocks.
25. What is the significance of the $Sr_i = 0.706$ line?
26. Why does the coincidence of line 4, and line 3, in the vicinity of Mono Lake in Fig. 18.7, suggest a vertical or near vertical contact between the North American crystalline shield and accreted terranes? Why does noncorrespondence suggest a more gently dipping contact?
27. Regarding the 706 line, what does a Sr_i ratio of 0.712 indicate? What does a ratio of 0.702 indicate? Include information about the origin of the rock, its mineralogy, and the presence or absence of crystalline shield.
28. What happens to the 706 line in California? Why?
29. What factors may have caused or contributed to the Antler orogeny?
30. What caused the Sonoma orogeny?
31. What kind of plate boundary was the California-Coahuila fault? When did the California-Coahuila fault exist?
32. How do geologists know about the existence of the California-Coahuila fault? How is it located?
33. Locate the following features in Fig. 18.6. The Canadian border, Rocky Mountain trench, Glacier National Park, Montana Disturbed Belt, Lewis and Clark Line, Front Range, Sierra Nevada, and other features.
34. Using Google Earth, compare the elevation of the Wyoming Basin with the Colorado Plateau.
35. Where is the Rio Grande Rift? Describe its history.
36. Where is the Rocky Mountain Basin and Range? Describe its history.
37. What is the origin of the Triassic Rift Valleys?
38. What is the Metacomet Ridge? Where is it located? How did it form?
39. From a tectonic plate movement point of view, explain why it is not surprising that development of the Triassic Lowlands was the last major tectonic event to affect the Appalachian Mountains.

Chapter 19

Cascadia Volcanic Arc System

The North Cascades, the Central-Southern Cascades, the Washington-Oregon Coast Range and Valleys, the Olympic Mountains, and the Klamath Mountains together constitute the Cascadia volcanic arc system where the Juan de Fuca plate is actively subducting beneath the North American plate. As implied in Fig. 11.8, the entire area is part of the accreted terrane rock succession. The structure is complex. Thrust faults are dominant, but folds along with normal and strike-slip faults are also present, some of which are active. In Chapter 19, we discuss landscape and geology that is directly attributable to the subduction of the Juan de Fuca plate beneath the North American plate. The subduction zone is part of a plate boundary configuration that also involves the Pacific plate and the San Andreas transform fault system as discussed in Chapter 20. Together, these two chapters describe the modern-day orogenic mountain systems group of structural provinces (Table 11.1).

THE JUAN DE FUCA PLATE

The Juan de Fuca plate is separated into three semi-independent segments, two of which are shown in Fig. 19.1, the Juan de Fuca and Gorda segments. The third, the Explorer segment, is off the Canadian coast. The Cascadia trench, which marks the subduction zone of all three segments, is less than 100 miles from the coastline. Although it is the largest of the three segments, the Juan de Fuca segment is less than 275 miles wide measured from spreading ridge to subduction zone. The three segments subduct beneath North America at different rates. The Explorer segment may no longer be subducting. The average rate of convergence between the Juan de Fuca segment and North America over the past 5 million years is between 9.8 and 13.8 feet per 100 years and the direction is about 49 degree east of north, which implies oblique subduction. The much smaller Gorda segment is subducting below the Klamath Mountains at a similar angle, but possibly at a slower rate between about 6.5 and 9.8 feet/100 years, particularly in its southern part. The Gorda segment (and the Juan de Fuca plate) terminates southward at the Mendocino triple junction near Cape Mendocino, where the Mendocino transform fault intersects the Cascadia trench and San Andreas Fault (Fig. 19.1). The triple junction is not stable. It has been migrating northward for the past 29 million years at the expense of the Juan de Fuca plate, which has been getting progressively smaller. The present-day rate of northward migration is 16.4 feet per 100 years. In Chapter 20, we will discuss how the Juan de Fuca plate originated. For now, we can say that the Juan de Fuca plate is a shrinking remnant of the Farallon plate, which was introduced in Chapters 5 and 17 as the paleo-Pacific plate that underthrust the Cordillera to create the Laramide orogeny and then sank to form the ignimbrite flare-up (Fig. 5.13).

Although geologically a trench, the Cascadia Trench does not have the topographic expression of a deep linear valley for two reasons. The first has to do with the age of the subducting plate. The oldest rocks on the Juan de Fuca and Gorda segments are less than 10 million years old. Tectonic plates cool, thicken, and become more dense as they age. The young age implies that the subducting lithosphere is warm and thin, and therefore isostatically buoyant. Seismic modeling suggests that the plate enters the subduction zone at a shallow initial angle of 10 to 15 degrees, which in turn, creates a shallow trench. The second reason has to do with the relatively slow rate of subduction in association with the Willapa, Columbia, Umpqua, Rogue, and other rivers that contribute copious amounts of sediment to the coastline. River sediment fills the shallow, slowly developing trench.

It was once thought that the absence of a trench coupled with the rarity of destructive historic earthquakes indicated that the Juan de Fuca plate was too warm to generate a great earthquake as it slipped below North America. Instead of snapping and fracturing, it was thought that the subducting plate was deforming like warm wax. This assumption has since been proven false. Key evidence for large destructive earthquakes was the discovery of rapidly drowned marshes and tree stands. Analysis of disrupted sediment suggests that as many as 12 powerful subduction-related earthquakes have occurred

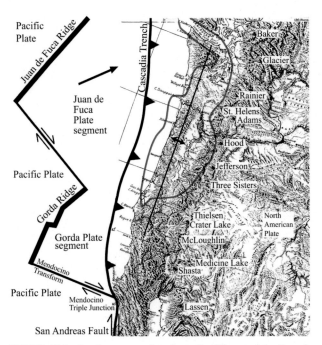

FIGURE 19.1 Landscape map that shows the US part of the Juan de Fuca plate. The Juan de Fuca and Gorda ridges mark the divergent plate boundary (the spreading ridge) with the Pacific plate. The Cascadia trench marks the subduction zone with the North American plate. The arrow shows the direction of convergence. Major Cascade volcanoes are labeled. The heavy line with double arrow along the Coast Range follows the crest of an anticlinal flexure. Area of Siletz River terrane is outlined in blue. *From Wells et al., 2014.*

in the past 7700 years, about one every 642 years. The last great earthquake occurred more than 300 years ago on January 26, 1700. With an estimated magnitude of 8.7 to 9.2, this quake was on par with the famous New Madrid earthquakes. Smaller earthquakes, some large enough to be felt, occur on average every 5 years or so. Some of the more recent earthquakes include a magnitude 6.7 quake in 1949 and again in 1965, a magnitude 5.6 quake in 1996, and a magnitude 6.8 earthquake in 2001. These earthquakes were large enough to destroy buildings and cause injuries, yet they pale with respect to the 1700 earthquake, which released about 1000 times more energy.

The volcanic arc is a direct consequence of subduction of the Juan de Fuca plate. It is not surprising, therefore, that the northern terminus of the volcanic arc at Mt. Meager in British Columbia, and the southern terminus at Mount Lassen in northern California, correspond with the northern and southern termini of the Juan de Fuca plate. It is important to understand that direct melting of the subducting Juan de Fuca plate is not the cause of most of the volcanism. Instead, it is the partial melting of the solid upper mantle directly above the subducting plate. The location where partial melting occurs is shown in Fig. 19.2, which is a cross-section across the subduction zone in central Oregon. Although shallow at the trench, the angle of the subducting plate steepens to between 60 and 80 degrees within about 200 miles inland where it reaches a depth of 62 miles (100 km). Beyond 62 miles

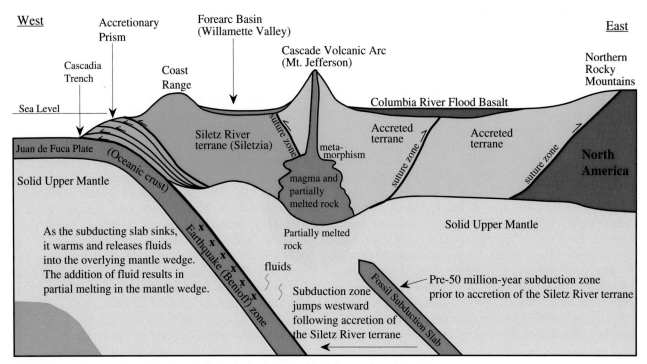

FIGURE 19.2 Cross-section of the Cascadia Volcanic Arc System at Mt. Jefferson, Central Oregon.

depth, the subducting plate becomes hot enough to undergo metamorphic reactions. These reactions release fluids (H2O and CO2) into the solid upper mantle above the subducting plate that disturb equilibrium conditions, causing mantle rock to melt.

The calamity caused by a major earthquake or large volcanic eruption is obvious. Less obvious is the potential damage resulting from landslides and avalanches caused by small eruptions. The problem with landslides and avalanches high on the mountaintop is that they could result in down-valley mudflows capable of destroying towns. Mt. Rainier is probably the most dangerous volcano in this respect because of the relatively large population in its lowland drainages. Landslides and avalanches off the slopes of Mt. Rainier have produced at least seven large mudflows in the past 5600 years, including the Osceola Mudflow 5600 years ago and the Electron Mudflow only 560 years ago. The Osceola Mudflow was by far the largest, extending all the way to the Seattle suburbs.

The overall topography shown in a Raisz landform map in Fig. 19.3, and in a shaded-relief elevation map in Fig. 19.4, is one of coastal mountains, an inland valley, and a tall, active volcanic, mountain range. This is a classic subduction-related tectonic landscape in which accretion has caused uplift along the coast, subduction has created a volcanic highland in the Cascade Mountains, and the intervening Puget Sound/Willamette Valley forms a basin known as the forearc basin. The relationship between topography and tectonics is well displayed in Fig. 19.2.

THE PACIFIC COASTLINE

The morphology of the Pacific coastline is nearly opposite that of the Atlantic coastline. Subsidence relative to sea level along the Atlantic coast has created an embayed shoreline of low relief with an abundance of beaches, barrier islands, lagoons, and marshes, shaped more by deposition than erosion. The Pacific coast is straight and rocky with sea stacks and narrow beaches that end abruptly against cliffs that rise directly into the Olympic Mountains, the Oregon Coast Range, the Klamath Mountains, and farther south, the California Coast Ranges, Transverse Ranges, and Peninsular Ranges. Four of the tide gauge measurements shown in Table 19.1 record a relative drop in sea level, and several others record only a small rise. Such measurements imply shoreline uplift at a higher rate than rising sea level. The Pacific coastline is emerging from the ocean, and as such, it is shaped more by erosion than deposition. Given the presence of active faults along the coastline (Fig. 18.12) we can reasonably assume that uplift is the result of tectonic stress. Constant uplift will produce a straight rocky coastline that migrates oceanward, leaving only a narrow continental shelf as shown in Fig. 1.8.

Fig. 19.5 shows the typical morphology of the Pacific coastline, consisting of a wave-cut platform at sea level, a sea cliff, and one or more uplifted marine terraces. This type of shoreline is created by waves that roll across the platform and undercut the sea cliff causing it to retreat. Once cut, the platform may collect sand through erosion to form a narrow beach that is often inundated during high tide. The base of the cliff is known as the shoreline angle or wave-cut angle. Its elevation approximately corresponds with mean sea level. Fig. 19.6 shows an example of a wave-cut platform that developed across tilted bedrock at Sunset Bay, Oregon. Sea stacks situated on the platform represent the erosional remnants of resistant high ground not yet fully removed by wave action. If shoreline uplift occurs, perhaps generated by a series of earthquakes, a wave-cut platform will be elevated above sea level and left high and dry to become a marine terrace. A younger wave-cut platform would then develop along the shoreline via undercutting of the newly emergent highland. Repetition of this process can create a series of marine terraces that stair-step up a relatively straight coastline as depicted in Fig. 19.5. In addition to shoreline uplift, a marine terrace can also form if a wave-cut platform develops during a high sea level stand, and is subsequently left high and dry following a drop in sea level. According to Fig. 8.6, the most recent high sea level stand occurred about 125,000 years ago with secondary brief stands at about 107,000 and 84,000 years ago.

Nearly the entire Pacific coastline is replete with marine terraces that formed either during high sea level stands, or following uplift associated with folds and faults. Fig. 19.7 looks northward at Cape Blanco along the Oregon coast. The image shows a well developed wave-cut platform and narrow beach at sea level, and a marine terrace more than 200 feet above sea level. Marine terraces, such as the one at Cape Blanco, can be used to determine surface uplift rates if the age of the marine terrace and the height of paleo-sea level are known. The age of the terrace can be estimated using fossils within sediment deposited on the terrace surface. If coral fossils are present, U-series dates could also be used. A third possibility is employing cosmogenic radionuclide dating methods. The height of sea level at the time the terrace was cut can be estimated using worldwide sea level curves. If a marine terrace is no longer flat, or the wave-cut angle is no longer at a constant elevation, then the amount of discordance can be used to gauge the amount of deformation that has occurred since the terrace initially developed.

The North Spit tidal gauge near Eureka shows a rise in sea level of 19.0 inches/100 years, by far the largest rise on the US west coast. The tide gauge is located south of the Klamath Mountains in the California Coast

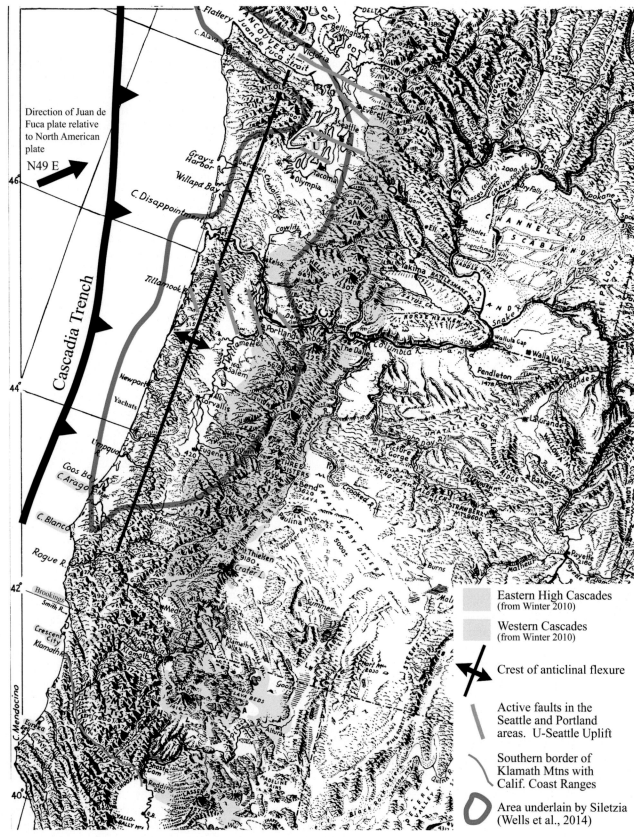

FIGURE 19.3 Landscape map of the Cascadia Volcanic Arc System.

Ranges surrounded by faults associated with the San Andreas Fault. The high rate of subsidence is likely a local phenomena associated with displacement on these faults, perhaps in the form of a normal fault pull-apart basin as explained in Chapter 20.

THE OREGON COAST RANGE

The Coast Range forms a rugged, dissected, low-lying mountain range that extends for more than 200 miles along the Oregon coast from the vicinity of Cape Blanco northward to the Columbia River at the Washington-Oregon border. Fig. 19.8 is an enlarged shaded-relief map that shows the mountain range and place locations mentioned in the text. The highest elevation in the Oregon Coast Range proper is Mary's Peak (M, 4097 feet) located west of Corvallis within 20 miles of the coast. Alternatively, if we extend the Coast Range farther south to the Rogue River, then Mount Bolivar (Bv, 4323 feet), east of Port Orford, is the highest peak. There are, in addition, many peaks above 3000 feet. Much of the moisture that moves inland from the Pacific Ocean is captured particularly in the central and northern Coast Range creating precipitation that averages between 100 and 200 inches per year. As a result, the entire Coast Range is heavily forested and logging is a major industry. Elevation south of the Coast Range rises to more than 7000 feet at Mount Ashland in the Siskiyou Mountains (part of the Klamath Mountains) before reaching the California border. Northward, the Coast Range loses elevation to form a subdued, dissected landscape in southern Washington known as the Willapa Hills, where two peaks rise above 3000 feet and many rise above 2000 feet (Fig. 19.3). The landscape flattens north of Willapa Bay where elevation rarely exceeds 1000 feet before rising to form the Olympic Mountains.

It is estimated based on sediment ages that the Oregon Coast Range first began to emerge from below sea level between 10 and 5 million years ago. The straight coastline, tide gauge measurements, leveling surveys, and marine terraces are all consistent with recent and ongoing uplift. Cape Blanco shows some of the highest uplift rates in the Coast Range with five preserved marine terraces stepping to an elevation above 900 feet. Estimated uplift rates over the past 125,000 years are on the order of 3.4 to 4.5 inches/100 years based on marine terraces. Tide gauge measurements since 1977 at Port Orford (near Cape Blanco) are 0.4 inches/100 years implying that uplift has kept pace with sea level rise (Table 19.1). Leveling surveys from 1941 to 1988 show a rapid uplift rate within the past 100 years on the order of 20 inches/100 years. All five mapped terraces at Cape Blanco are deformed across an actively forming anticline.

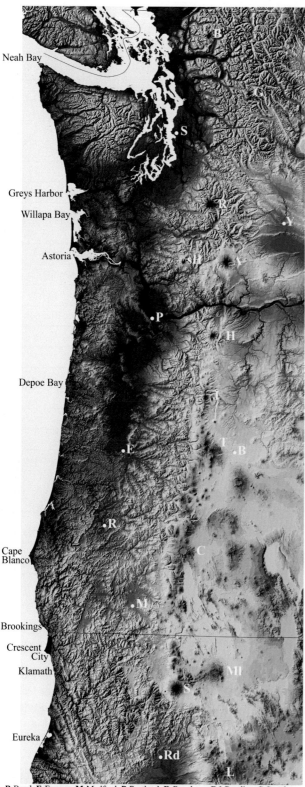

B-Bend, E-Eugene, M-Medford, P-Portland, R-Roseburg, Rd-Reading, S-Seattle, Y-Yakima. A-Mount Adams, B-Mount Baker, C-Crater Lake, G-Glacier Peak, H-Mount Hood, J-Mount Jefferson, L-Lassen Peak, Ml-Medicine Lake, R-Mount Rainier, S-Mount Shasta, sH-Mount St. Helens, T-Three Sisters.

FIGURE 19.4 USGS National Atlas digital shaded relief elevation map of the Cascadia Subduction complex. Nationalmap.gov/small_scale/atlasftp.html.

TABLE 19.1 Relative Sea Level Change Based on Tide Gauge Measurements

Tide Gauge	mm/year	in/100 years	Measurement Interval	
Washington				
Seattle	2.05	8.1	1899–2017	Puget Sound
Port Townsend	1.94	7.6	1973–2017	Puget Sound
Port Angeles	0.37	1.5	1975–2017	Straight of Juan de Fuca
Neah Bay	−1.7	−6.7	1934–2017	Cape Flattery
Toke Point	0.45	1.8	1973–2017	
Oregon				
Hammond	−1.22	−4.8	1983–2014	Near Astoria
Astoria	−0.16	−0.63	1925–2017	
Garibaldi	2.6	10.2	1970–2017	Near Cape Meares
South Beach	1.73	6.8	1967–2017	Near Newport
Charleston	1.12	4.4	1970–2017	Near Cape Argo
Point Orford	0.1	0..4	1977–2017	Near Cape Blanco
California				
Crescent City	−0.78	−3.1	1933–2017	
North Spit	4.81	19.0	1977–2017	Eureka
Arena Cove	0.69	2.7	1978–2017	Point Arena
San Francisco	1.96	7.7	1897–2017	
Monterey	1.48	5.8	1973–2017	
Port San Luis	0.89	3.5	1945–2017	Near San Luis Obispo
Santa Barbara	1.07	4.2	1973–2017	
Santa Monica	1.53	6.0	1962–2017	
Los Angeles	0.98	3.9	1923–2017	
Newport Beach	2.22	8.7	1955–1993	
San Diego	2.17	8.5	1906–2017	

Based on NOAA-Tides&Currents.
http://tidesandcurrents.noaa.gov/sltrends/sltrends.html.

Based on overall lower elevation and an embayed coastline, we can suggest that long-term uplift rates in the Willapa Hills and along the Washington coast south of the Olympic Mountains are less than in the Coast Range. Inland bays shown in Fig. 19.4, such as Grey's Harbor, Willapa Bay, and the wide Columbia River estuary at Astoria create a coastline more suggestive of subsidence and drowning than uplift. However, marine terraces and active faults are present, and tide gauge measurements show minimal sea level rise of 1.8 inches/100 years at Toke Point (Grey's Harbor) and sea level drops of −4.8 inches/100 years and −0.63 inches/100 years at the Hammond and Astoria tide gauges, both of which are in the Columbia River estuary near Astoria. Given the worldwide rise in sea level, these numbers suggest significant active uplift over the past 50 to 100 years.

A unique aspect of the Oregon coast is the presence of sand dunes north of Coos Bay that extend for at least 50 miles. As seen in Fig. 19.9, sand deposited along the shoreline from nearby rivers becomes trapped in a recess between Cape Arago and Heceta Head where it creates dunes that move onshore with the prevailing wind. The shifting dunes have blocked small streams from reaching the coast, creating a chain of freshwater coastal lakes that include Lake Siltcoos and Lake Tahkenitch. Smaller areas of coastal dunes are present farther south near Bandon and farther north on both sides of the Columbia River estuary.

Cascadia Volcanic Arc System **Chapter | 19** 479

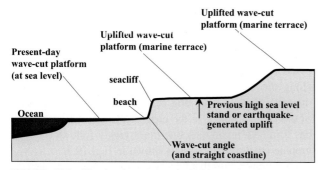

FIGURE 19.5 Sketch of a wave-cut platform and marine terraces. Wave action undercuts a newly uplifted land area to create a wave-cut platform at sea level that ends against a retreating seacliff. The wave-cut platform becomes increasingly wider as the seacliff retreats. The wave-cut platform becomes a marine terrace if uplift or sea level drop elevates the platform above sea level.

FIGURE 19.7 Google Earth image looking north at a wave-cut platform and marine terrace, Cape Blanco, Oregon.

FIGURE 19.6 Photograph of a wave-cut platform across tilted rocks that are part of the circa 38-million-year-old Coaledo Formation. The location is Sunset Bay, Oregon.

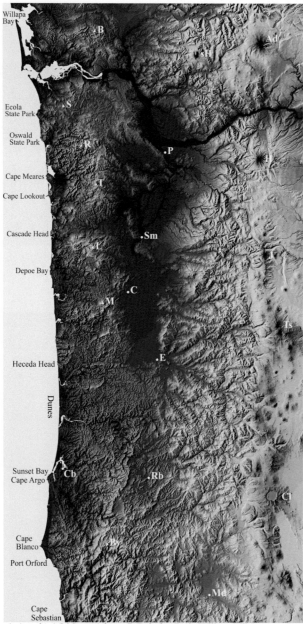

FIGURE 19.8 USGS National Atlas digital shaded relief elevation map of the Oregon Coast Range. See Fig. 19.4 for reference.

A-Astoria, **Ad**-Mount Adams, **B**-Boistfort Peak (Willapa Hills), **Bv**-Mount Bolivar, **C**-Corvallis, **Cb**-Coos Bay, **Cl**-Crater Lake, **E**-Eugene, **H**-Mount Hood, **J**-Mount Jefferson, **L**-Laurel Mtn., **M**-Marys Peak, **Md**-Medford, **P**-Portland, **R**-Rogers Peak, **Rb**-Roseburg, **S**-Saddle Mtn., **S**-Salem, **sH**-Mount St. Helens, **T**-Trask Mtn., **Ts**-Three Sisters.

Cause of Uplift Along the Oregon Coast

The accretion of material to the underside of the North American continent is known as underplating and is one of at least three forces that drive coastal uplift. The mechanism of underplating is outlined schematically in Fig. 5.7. Note in this figure that wedging of material in the subduction zone results in rotation and landward tilting of overlying rock. This, in turn, causes relative uplift near the coastline within the accretionary prism and possible subsidence inland. The accretionary prism is not exposed in the Coast Range, but active landward tilting of coastal rocks is evident from leveling surveys that record greater uplift along the coast than in Willamette Valley. It is possible therefore, that underplating is responsible for at least some of the coastal uplift. In the case of the Coast Range, uplift has formed a broad anticline that roughly follows the topographic crest of the range as shown in Fig. 19.3.

A second factor that causes both uplift and subsidence in the Coast Range is the subduction of seamounts carried to the subduction zone on the Juan de Fuca plate. The presence of a seamount (an extinct, submerged volcanic area) creates a bulge within an otherwise relatively thin down-going plate. These particular seamounts are small and not buoyant enough to accrete. Instead they are subducted with the down-going plate. As the seamount bulge subducts below the Coast Range, it causes uplift. An interesting aspect to this is that, as the seamount subducts farther into the mantle, the Coast Range will subside with its passage. The process brings to mind a snake swallowing a large object such that it creates a bulge as it passes through the snake.

A third factor that may partly be responsible for present-day uplift rates is the degree of coupling between the overriding North American plate and the downgoing Juan de Fuca plate. If the coupling is strong, then we can say that the subduction zone fault is locked to the point where there is little if any displacement between the plates on either side of the fault. The locked plates instead flex upward as stress builds, resulting in rapid uplift rates at the surface. Stress will continually build on a locked fault until the fault finally breaks, at which point the rocks will rebound to their original shape causing rapid subsidence and a large earthquake. The idea is shown schematically in Fig. 19.10. Recall from our earlier discussion that one line of evidence for great prehistoric earthquakes is shoreline drowning of marshes and low-lying areas associated with a sudden drop in elevation. Elastic rebound of a flexed plate during an earthquake could explain the rapid subsidence. Notice also that, in areas where flexural uplift on a locked fault is occurring, we have an example in which measured present-day uplift rates will be more rapid than rates measured over a much longer period where, on average, some of the accumulated uplift is diminished by subsidence following an earthquake. The process of flexural uplift points out the perils of extrapolating present-day rates back in time without corroborating evidence, and at the same time, shows where stress is accumulating for a potentially large earthquake.

If the plates are not locked, then there will be some slippage between the plates that occurs without causing a significant earthquake. This slippage, known as creep, will relieve some of the stress that otherwise would build

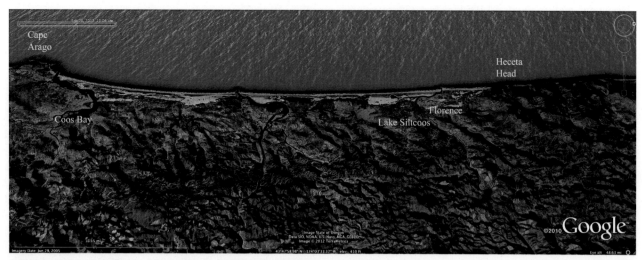

FIGURE 19.9 Google Earth image looking west across Oregon Dunes National Recreation area. The dunes block inland streams, creating Lake Siltcoos (labeled) and Lake Tahkenitch, located south of Lake Siltcoos. The cliffs at Heceta Head are composed of Yachats Basalt and some Tyee Formation. Cape Argo consists of sedimentary rock.

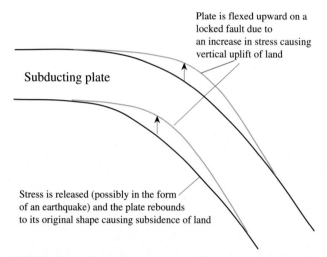

FIGURE 19.10 Sketch that shows upward flexure of a subducting slab due to an accumulation of stress on a locked fault. The subducting slab snaps back to its original shape during earthquake-induced stress release, which causes land subsidence.

up on the fault, and as a result, flexural uplift prior to an earthquake will be less, the resulting earthquake will be lower in magnitude, and subsidence following the earthquake will be less. These are areas where a large earthquake is less likely to happen and where measured present-day uplift rates better approximate long-term rates.

Using tidal gauge data, we can determine to a first approximation where creep verses flexural uplift is occurring. Notice in Table 19.1 that the highest rates of sea level rise (the lowest rates of uplift) over the past 45 years or so are in central Oregon (the Garibaldi, South Beach, and Charleston gauges located between Cape Meares and Cape Argo). Such rates have been confirmed through more rigorous treatment of the data including leveling surveys and global positioning (GPS) measurement. The data suggest that subduction in this area is undergoing a greater amount of inter-seismic creep relative to areas in the north and south where uplift rates are higher. This area, therefore, may not experience as large an earthquake as areas to the north and south.

Geology of the Oregon Coast Range

Some of the rocks that form the Coast Range northward to the Olympic Mountains are part of a single piece of accreted material known by several names that include the Coast Range terrane, the Peripheral rocks, the Crescent terrane, and the Siletz River terrane. For simplicity, we will refer to it as the Siletz River terrane or as Siletzia. The rock that forms Siletzia is oceanic basalt with interbeds of deep marine sedimentary rock. The basaltic rocks were extruded between 56 and 49 million years ago as part of a large oceanic plateau on what was then the Farallon plate (this plate would later become the Juan de Fuca plate; see Chapter 20). The youngest rocks were extruded above sea level as oceanic islands. The entire terrane was accreted to North America between 51 and 49 million years ago as the last major accretion event in the United States. Following accretion, the subduction zone jumped westward to its present location as depicted in Fig. 19.2, thus changing the shape of the coastline. The inferred extent of the Siletz River terrane is shown in Fig. 19.3.

It is not known how much of Siletzia was accreted and how much was subducted. What we do know is that the volume of unsubducted Siletzia is enormous, amounting to 8 to 12 times the volume of the Columbia River basalt. Siletzia is 54,000 feet (16.5 km) thick in the Olympic

Mountains and as much as 105,000 feet (32 km) thick in Oregon. The basalt that forms Siletzia is the oldest rock in the Coast Range. It is well exposed in the area between Corvallis and Mary's Peak (M, Fig. 19.8), in Garden Valley surrounding Roseburg (R), and on the highest peak in the Willapa Hills, Boistfort Peak (B, 3140 feet), among other areas. It is not exposed along the Oregon coast.

Siletzia is overlain with shallow marine and river sandstone of the Tyee Formation. These rocks were deposited between about 49 and 46.5 million years ago. They are less deformed than Siletzia indicating they were deposited following accretion. They indicate that accretion was complete by 49 to 50 million years ago. The Tyee Formation forms a large part of the central Coast Range between the latitudes of Corvallis (C) and Roseburg (Rb). Magmatism continued in the now accreted Siletzia in the form of gabbroic intrusions and basaltic (and some silicic) lava flows. The most prominent of these are the Tillamook, Yachats, Cascade Head, and Greys River volcanic rocks, extruded largely between 42 and 40 million years ago. These rocks are well exposed along the coast at Heceda Head and Cascade Head (Fig. 19.8). They also form and surround Rogers Peak (R, 3710 feet) west of Portland, which is the highest peak in the northern Coast Range. Gabbro intrusions are dispersed across the Coast Range primarily north of the latitude of Salem (Sm). These rocks form many of the higher peaks including Laurel Mountain (L), Trask Mountain (T), and Mary's Peak (M), as well as part of the coastal headlands at Oswald State Park and within and north of Ecola State Park. The favored interpretation for both the 56- to 49-million-year-old Siletz River volcanic rocks and the younger Tillamook and associated rocks, is that they formed above a long-lived Yellowstone hot spot at a time when the hot spot was well to the west of where it is today.

In addition to Siletz River and Tillamook-associated rocks, a few of the headlands along the northern Oregon coast are underlain with much younger circa 16- to 15-million-year-old Columbia River flood basalt. These lavas flowed westward more than 200 miles via the Columbia River channel where they froze and fractured upon entering the ocean. On the coast, they form Cape Lookout, part of Cape Meares, and headlands just south of Depoe Bay (Fig. 19.8). Inland they are abundant along the west side of the Columbia River north of Portland, including Saddle Mountain (S, 3287 feet).

Other than the rocks mentioned above, much of the Oregon coast north of Cape Blanco consists of unconsolidated beach and dune sand along with terrace deposits and young, poorly consolidated, nearly flat-lying sandstone. However, older sedimentary rocks, in the range of 40 million years, are exposed in a few areas including Cascade Head and Sunset Bay. Part of this older sequence is tilted and forms the wave-cut platform shown in Fig. 19.6. These rocks are unconformably overlain with younger nearly flat-lying rocks that are not visible in the photo.

INLAND VALLEYS AND THE FOREARC BASIN

Inland from the Oregon Coast Range and coastal Washington is a series of valleys that step to higher elevation from north to south. In the north, the Puget Sound is at sea level. Southward, we enter the more than 100-mile long Willamette Valley, which rises to more than 400 feet at its southern end near Eugene (E, Fig. 19.8). From a tectonic standpoint, these areas form the forearc basin to the volcanic arc complex as depicted in Fig. 19.2. Low-lying areas surrounding Puget Sound are largely covered in glacial drift. Lake and river deposits cover the Willamette Valley. Farther south there are two small valleys, Garden Valley near Roseburg (Rb) at 400 to 600 feet, and Rogue Valley surrounding Medford (Md) at 1300 to 1400 feet. Rocks in Garden Valley form a fold and thrust belt, active during accretion of Siletzia, in which older Mesozoic rocks were thrust northwestward above Siletzia. The Rogue Valley is underlain with sedimentary and volcanic rocks that post-date accretion.

Rogue Valley contains an excellent example of volcanic reincarnation and subsequent erosional lowering. Fig. 19.11 is a Google Earth image that looks northward at Upper Table Rock, located north of Medford.

FIGURE 19.11 Google Earth image looking northward at horseshoe-shaped Upper Table Rock, located just north of Medford, Oregon.

FIGURE 19.12 Google Earth image looking southeastward across the Boring Lava field at Portland, Oregon. Rocky Butte, Kelly Butte, and Mt. Tabor are all within Portland city limits.

The Rogue River is seen at the bottom of the image. Lava of andesite composition flowed across this area approximately 7 million years ago filling not only the ancestral Rogue River channel, but also covering much of the surrounding river valley. The Rogue River was completely disrupted. Erosion in the past 7 million years has removed most of the lava and stripped away the surrounding weak sedimentary layers. Upper Table Rock is a lava-capped remnant of the lava flow that forms a mesa 800 feet above surrounding land areas. The horseshoe shape of the mesa preserves the shape of a meander bend that was once part of the Rogue River. In less than 7 million years, what was once a river valley is now a distinct highland.

Oblique convergence of the Juan de Fuca and North American plates has generated north-south-oriented compressional stresses that result in 17.3 inches/100 years (4.4 mm/year) of permanent shortening in the Puget Sound area. Shortening is manifested by the development of active northwest- and west-trending reverse, thrust and strike-slip faults that separate structural basins from uplifts. Most important is the 43.5-mile long Seattle Fault zone, which consists of several strands that extend through the southern part of the city. The main strand is a south-dipping thrust fault. This fault, and several others in the Seattle and Portland areas, are shown with heavy orange lines in Fig. 19.3. The Seattle fault separates the Seattle Basin to the north, from the Seattle Uplift (U) to the south. The fault zone is dangerous. Fieldwork indicates that it was the site of a strong earthquake between A.D. 900 and 930 that generated a tsunami, landslides, and caused about 21 feet of vertical uplift. Another active, potentially dangerous fault is the north-dipping Tacoma fault, which separates the southern margin of the Seattle Uplift from the Tacoma Basin. It is shown just north of Tacoma in Fig. 19.3. Also shown in Fig. 19.3 are active, potentially dangerous faults in the Portland region including the East Bank and Portland Hills faults. Both are thrust (or reverse) faults with a component of strike-slip displacement. Both appear to have produced earthquakes in the past 15,000 years.

In addition to earthquake-generating faults, the area in and surrounding Portland is also host to a series of young volcanic cinder cones and vents known as the Boring Lava Field. At least 74 vents have been named. The largest concentration lies east and south of Portland, but they also extend west of the Willamette River. Three volcanic cones lie within Portland city limits, including the 1.2-million-year-old Rocky Butte. Nearly all of the rock is basaltic in composition. Fig. 19.12 shows all three cones within Portland as well as a part of the volcanic field southeast of Portland. Argon dates indicate the rocks were extruded in three phases 2.7 to 2.2 million years ago, 1.7 to 0.5 million years ago, and 350,000 to 50,000 years ago. These young volcanoes are somewhat unusual considering their location in the forearc basin. They are considered to be the westernmost magmatism associated with subduction of the Juan de Fuca plate. Extension in this part of the forearc may have allowed magma to reach the surface.

THE CENTRAL-SOUTHERN CASCADE MOUNTAINS

The Cascade composite (strato) volcanoes are beasts. They grow quickly into a behemoth that dominates the skyline, and then are gone, either blown up or eroded in less than half a million years. Table 19.2 lists 21 volcanic cones in the United States and 3 in Canada. The major stratocones are located in Figs. 19.3 and 19.4. Fig. 19.13 shows several volcanoes towering as much as 1 mile

TABLE 19.2 South to North List of Major Cascade Stratovolcanoes

	Active in Past 2000 Years	Active in Past 200 Years	Elevation (Feet)	
California				
Lassen Peak	xxx	xxx	10,457	
Mount Shasta	xxx	xxx	14,168	
Medicine Lake	xxx		7,795	
Oregon				
Mt. McLoughlin			9,499	
Crater Lake			6,176	
Mt. Thielsen			9,187	Extinct
Mt. Bachelor			9,069	
Broken Top			9,179	Extinct
Three Sisters				
South Sister	xxx		10,363	
Middle Sister			10,052	
North Sister			10,090	
Mt. Washington			7,798	Extinct
Three-Fingered Jack			7,845	Extinct
Mount Jefferson			10,502	
Mount Hood	xxx	xxx	11,243	
Washington				
Mount Adams	xxx		12,281	
Mount St. Helens	xxx	xxx	8,337	9,677 before eruption
Gilbert Peak (Goat Rocks)			8,188	Extinct
Mount Rainier	xxx	xxx	14,411	
Glacier Peak	xxx	xxx	10,525	North Cascades
Mount Baker	xxx	xxx	10,786	North Cascades
Canada				
Mt. Garibaldi			8,776	
Mt. Cayley			7,825	
Mt. Meager			8,694	

above the surrounding landscape. A photograph of Mount Jefferson is shown in Fig. 19.14. During their existence, a stratovolcano can add a 1000 feet of elevation in less than 1000 years and then lose it all in an instant through eruption, or more slowly through erosion. Mount St. Helens, the most active of the stratovolcanoes, grew from about 6000 feet to its pre-1980 eruption height of 9677 feet in less than 4000 years. It then lost 1340 feet in only a few minutes during its famous catastrophic May 18, 1980 eruption. The highest peak, Mount Rainier, may have been higher prior to an eruption 5600 years ago that also produced the Osceola Mudflow. This volcano, 75,000 years ago, may have reached as high as 16,000 feet prior to eruptions that lowered its summit. Crater Lake occupies the location of ancient Mount Mazama, which reached a maximum elevation of 12,000 feet prior to a massive eruption some 6870 years ago when the volcano collapsed into a 6 mile-wide caldera that filled with water

FIGURE 19.13 Google Earth image looking northward along the High Cascades. Major peaks are Broken Top, at foreground right, the Three Sisters (South, Middle, North) at foreground center, Mount Jefferson, Mount Hood, and right to left in far back are Mount Adams, Mount Rainier, and Mount St. Helens. Olympic Mountains form the white speck farthest left. Other volcanic peaks are: b-Black Crater, Bn-Belknap Crater, h-The Husband, t-Three-Fingered Jack, and w-Mount Washington.

FIGURE 19.14 Photograph of Mt. Jefferson, central Oregon.

to form Crater Lake, the deepest freshwater lake in the United States (1996 feet deep). Of the 24 major volcanoes listed in Table 19.2, 10 have erupted in the past 2000 years and 7 in the past 250 years.

Cascade volcanism began soon after the accretion of Siletzia, perhaps as early as 48 million years ago, but volcanism has migrated over time and volcanoes have come and gone. Most of the present-day stratovolcanoes are no more than 500,000 years old including Mount Baker, Glacier Peak, Mount Rainier, and Mount Hood. North Sister is about 400,000 years old and has seen only minor activity in the past 50,000 years. Mount St. Helens is 275,000 years old. Mount Jefferson was built mostly in the past 100,000 years, but hasn't erupted in the past 10,000 years. Middle and South Sister may be no more than 50,000 years old. Mount McLoughlin may not have seen volcanic activity for the past 20,000 years. Mount Mazama was about 420,000 years old prior to its explosive demise to form Crater Lake. Some volcanoes, including Mount Thielsen, Mount Washington, Three-Fingered Jack, Gilbert Peak (Goat Rocks), and Broken Top, have not been active for more than 100,000 years and are considered extinct. These extinct volcanoes have since undergone strong erosion.

The straovolcanoes, themselves are andesite-dacite in composition, but basalt is common especially as a substratum. Rhyolite is less common. Andesite-basalt compositions typically flow. Dacite-rhyolite compositions, on the other hand, can be highly explosive and dangerous. Rather than flow freely, these compositions form lava domes that grow high on the flanks of the mountain. As the dome grows, it either begins to flow down the mountainside, or it collapses into a pyroclastic flow that becomes a mudflow downstream (known as a lahar). Mount Hood has alternated between andesite lava flows and dacite lava domes. Mount St. Helens has erupted a wide range of magma including basalt, andesite, and dacite. The most remote of the volcanoes, Glacier Peak, has erupted considerable dacite domes and pyroclastic flows. Mount Rainer, Mount Hood, and Mount Jefferson are a combination of andesite, dacite, and pyroclastic flows. Mount Lassen compositions are dominantly andesite-rhyolite, but basaltic magma associated with Basin and Range extension has also erupted. North Sister is mostly basalt-andesite whereas Middle Sister is mostly andesite-dacite, and South Sister is dacite-rhyolite.

The stratocones are not solitary volcanoes. In most cases, they are part of a large volcanic field where basaltic compositions are common. Mount Adams is an andesitic stratovolcano in a field with at least 120 basaltic volcanoes, scoria cones, and small shield volcanoes active from 520,000 to 1000 years ago. There are at least 466 volcanic centers associated with the Three Sisters volcanic field, all of which have erupted in the past 1 million years. Within these fields, the stratocones are sites of multiple eruptions whereas many of the surrounding smaller vents, especially the basaltic vents, produce only a single eruption.

There is evidence of past volcanoes on the same site, or in the same vicinity, as some of the present-day volcanoes. Mount Lassen is part of a large volcanic field that has been active nearly continuously for the past 3.5 million years. Lassen Peak itself may have begun formation as early as 825,000 years ago, and has erupted as recently as 1917. The area surrounding Mount Baker has been an active volcanic site for at least 3.7 million years and has undergone nearly continuous volcanism for the past 1.3 million years. The area surrounding Mount Rainier has seen intrusion and volcanism for at least the past 36 million years. Mount Hood and Mount Rainier are both built on the remnants of an earlier volcano that was in existence 1 to 2 million years ago. Mount Shasta, the most voluminous of the stratocones, began forming on the remnants of an older volcano that was destroyed about 300,000 years ago during a massive eruption and collapse that created, on its northeast side, one of the largest landslides known to exist (now mostly covered with younger volcanic rock). A 12,330-foot sub-cone on the west flank of Mount Shasta, known as Shastina, was built mostly between 9700 and 9400 years ago.

A single volcano can undergo a series of closely spaced volcanic spurts that can last more than 100,000 years, and then lie mostly dormant for an equally long time, only to awaken again from its long slumber. During periods of dormancy, or of limited volcanism, the volcano can erode at extremely high rates (feet per 100 years) creating unconformities between volcanic episodes. North Sister began as a shield volcano 400,000 years ago and underwent three additional eruptive stages. The first occurred between 182,000 and 99,000 years ago, the second at 80,000 years, and the third from 70,000 to 55,000 years ago, each separated by an unconformity. Roughly 90% of its volume was built prior to 99,000 years ago. Mount Hood has undergone periods of frequent eruptions lasting decades to centuries separated by quiet periods lasting hundreds to thousands of years.

High rates of erosion result from a variety of factors that include the abundance of fractured and weathered rock, rock that has been weakened and altered by hot fluids, steep slopes, high precipitation, and the presence of glaciers. Magma movement at depth, seismicity, ground movement, and minor eruptions (especially pyroclastic eruptions) have the effect of destabilizing slopes and melting ice, causing landslides that can remove huge sections of a mountain in a matter of seconds. A large earthquake-generated landslide on the north side of Mount St. Helens occurred seconds before the 1980 volcanic eruption. The sudden release of pressure by the landslide caused gases in magma below the surface to violently expand and explode as though somebody had just unlocked a pressure cooker. Field studies suggest that a similar debris avalanche and explosion may have occurred on Mount St. Helens 2800 years ago. The Osceola Mudflow was also initiated by a landslide and volcanic eruption.

In the following section, we will discover that the substratum to many of the stratovolcanoes consists of a thick pile of basalt. The Medicine Lake volcano, located in California northeast of Mount Shasta and shown in Fig. 19.3 as the Modoc lava beds, contains this substratum, but is unique in that it is without a centralized high-elevation silicic stratocone. It is a shield volcano composed dominantly of basalt, but includes everything from andesite to dacite to rhyolite. In some respects, it is similar to the Newberry Caldera (Chapter 17). Its size is enormous. It is 22 miles east to west and 30 miles north to south with a basaltic pile more than 3000 feet thick. The central part of the volcano is a large depression (a caldera) 4 to 8 miles wide that contains Medicine Lake and several large obsidian flows. The flanks of the volcano are littered with cinder cones, fissure vents, and associated lava flows. The caldera formed initially due to subsidence following extrusion of basaltic magma. The center of the caldera continues to subside at a rate of 34 inches/100 years (8.6 mm/year). Volcanism began about 475,000 years ago with rhyolite flows, and then continued with mostly basalt, andesite, some dacite, and rare rhyolite. An interesting feature is Glass Mountain on the eastern flank of the Medicine Lake volcanic field where one can find a mountain of rhyolitic and dacitic obsidian and pumice in the form of several lobes that erupted about 950 years ago. Fig. 19.15 looks west across Glass Mountain at Mount Shasta in the distance. The Medicine Lake volcano is located between the Cascade volcanic arc and the normal fault structures of the Basin and Range. It likely has been influenced by both tectonic regimes as well as by tectonics associated with the Walker Lane Belt discussed in Chapter 20. The presence of andesite and other silicic rocks, and the geochemistry of the magma, suggest that it is part of the Cascade volcanic arc.

Geology of the Central-Southern Cascade Mountains

The North Cascade Mountains and the Central-Southern Cascade Mountains both contain high volcanoes. However, the rock type, elevation, relief, and structure of the surrounding landscape in the two areas are very different. Physiographic differences are clearly evident in Fig. 19.4, which shows the transition just north of Mt. Rainier. The North Cascades are a rugged, glaciated, high relief, snow-capped mountainous terrain with jagged peaks of crystalline rock that reach 9500 feet in elevation. The Central-Southern Cascade landscape is a heavily dissected, forested highland of volcanic/sedimentary rock that, with the exception of the volcanic cones themselves, is mostly less than 6500 feet in elevation, similar to the Southern Appalachian Blue Ridge Mountains. Differences between these two areas are in part related to the recency of uplift and the level of erosion. Accreted terranes, including Siletzia, are not exposed in the Central-Southern Cascades, and instead are buried beneath relatively young volcanic and sedimentary rocks. The wide exposure of crystalline accreted terranes and the mountainous topography in the North Cascades, on the other hand, implies recent and more rapid uplift/erosion in order to exhume these once deeply buried rocks.

As Cascade volcanoes formed and died over the past 48 million years, the locus of volcanic activity in Oregon has narrowed and migrated eastward over time. This migration has created two distinctive belts of Cascade volcanic rock, the Eastern or High Cascades, where all of the active volcanoes are located and where most of the surface rock is less than 5 million years old; and the inactive lower elevation Western Cascades where most of the surface rock is more than 5 million years old. The physiographic distinction between these two areas is evident in Fig. 19.4. The inactive Western Cascades represent the uplifted locus of eroded, extinct, and blown-apart volcanoes. The landscape tilts gently westward such that elevation gradually increases from less than 2000 feet at the edge of the Willamette Valley to about 5800 feet at the base of the active High Cascade stratovolcanoes. Although not especially high, the Western Cascades are rugged due to deep stream dissection. It is the gentle tilt of the mountain range, and not the high volcanic cones, that help to create one of the most pronounced rain

FIGURE 19.15 Google Earth image looking west at Glass Mountain (Medicine Lake volcanic field) with Mount Shasta in the skyline.

shadows in the country. Annual precipitation averages 80 to 140 inches on the western slope in the Washington Cascades and 60 to 120 inches in the Oregon Cascades north of Crater Lake, but is typically less than 35 inches on the eastern slope.

The tilting of the Western Cascades and development of a rain shadow apparently occurred within the past 10 million years. Studies of fossil plants along the east side of the Cascade Range and on the Columbia Plateau suggest that plants growing in these areas had changed from summer-wet to summer-dry species by about 8 million years ago, thus dating the initiation of the rain shadow. Another method of studying the formation of a rain shadow is to look at progressive changes in oxygen isotope ratios. Studies of soil samples from the Columbia Plateau suggest a gradual decrease in precipitation between 16 and 4 million years ago. Presumably, this is the time span over which the Cascade rain shadow developed. A third line of evidence has to do with the Columbia River basalts, which extend almost to the crest of the Western Cascades and are tilted westward. The basalt flows are mostly between 16.8 and 15.6 million years old, and their presence so far up the flank of the Central-Southern Cascade Range suggests that the mountain could not have existed at this time (that is, the basalts could not have flowed uphill). Recall from Chapter 16 that the North Cascades may have already been a highland with significant relief by 16 million years ago and that it underwent an additional period of uplift between 12 and 4 million years ago. This later period of uplift apparently also affected the Southern-Central Cascade Range but perhaps not to the extent and magnitude that it affected the North Cascades.

The Columbia River has probably been in existence since the accretion of Siletzia 50 million years ago, which implies that the river has continually crossed the Cascade Mountains since their inception. Today, the river crosses the crest of the Cascades just north of Mount Hood between The Dalles and Portland where, as seen in Fig. 19.3, it cuts the spectacular 75-mile-long Columbia River gorge. Fig. 17.4 shows that, within the gorge, the river flows close to the northern margin of the Columbia River flood basalt. This relationship is consistent with the idea that basaltic lava has repeatedly filled the river channel, thereby pushing the channel progressively northward to its present location at the edge of the flows. These flows are more than 14.5 million years old; therefore, it is likely that the Columbia River has been in its present-day channel for at least that long. We can speculate that the present-day gorge was cut within the past 10 million years during uplift and tilting of the Western Cascades.

Fig. 19.16 looks westward along the western part of the gorge near Stevenson, Washington. Note the change in topography on either side of the river. Cliffs on the southern (left) side form along the edges of lava flows associated with the Columbia River basalt. Tributary rivers cascade down the cliffs creating more than 70 waterfalls including the 620-foot tall Multnomah Falls. The more subdued landscape on the northern side is underlain with older (Eocene-Oligocene) Cascade volcanic and volcaniclastic rocks topped in a few areas with Columbia

FIGURE 19.16 Google Earth image looking westward along the Columbia River Gorge toward Portland. The Bonneville landslide is labeled.

River basalt. Similar to the Grand Mesa (Chapter 17), weak clay-rich sedimentary rocks below the basalt provide ready surfaces for ground failure and landslides. The largest, youngest, and most obvious is the Bonneville landslide (also known as the Table Mountain slide) labeled in the center of the image. The source of the slide is visible as a scar on the mountainside to the right of the word landslide. The slide itself consists of irregular, hummocky, slumped ground with many lakes. The slide spilled out across the river, temporally damming the river, and eventually creating the lobe of land seen to the left of the word landslide where the river becomes narrow. Dating of drowned trees suggests the slide occurred between 400 and 250 years ago. Several additional slides are present in this area.

In conclusion, we can say that the Cascade Mountains form a relatively young tectonic landscape less than 10 million years old with conspicuous volcanic cones less than 500,000 years old. One thing that is different about this area relative to the North Cascades and many other areas of active tectonics is that the landscape is largely constructional (built by the addition of volcanic/sedimentary rock) rather than erosional (the exhumation of deeply buried rock).

Clockwise Block Rotation

The Cascade Mountains have been affected by several tectonic phenomena that include subduction of the Juan de Fuca plate, the sinking of older, previously subducted tectonic plates, the accretion of Siletzia, the Yellowstone hot spot, and normal faulting associated with extension in the Basin and Range. Part of the response to these phenomena has been clockwise block rotation. As everybody knows, iron is magnetic and can become oriented in a magnetic field. This simple relationship is used in geology to determine whether iron-bearing rocks, particularly volcanic rocks, have been rotated since the time they were deposited. As volcanic rock cools, the iron in the rock becomes oriented parallel with the Earth's magnetic field. Once the rock is cold, the magnetic orientation is locked in place. If a rock is rotated after the magnetic orientation is locked in place, a geologist can determine the amount of rotation. Paleomagnetic evidence indicates that the entire Pacific Northwest, including the Cascade Mountains and the Coast Range, has been rotating clockwise for at least the past 50 million years. Over that time, the rate of rotation in the Oregon Coast Range has averaged 1.19°/my with the rate decreasing to the north, south, and east. The rotation is likely due to a combination of oblique subduction of the Juan de Fuca plate and extension in the Basin and Range.

Rotation is an active, ongoing process. Global positioning (GPS) measurements since the year 2000 have

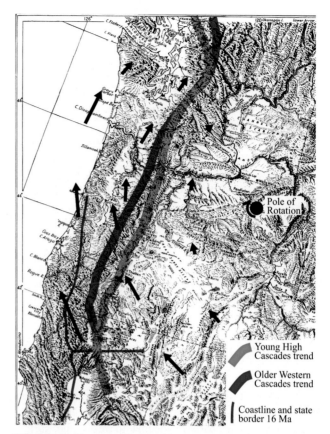

FIGURE 19.17 Map showing present-day rotation of the Pacific Northwest based on Wells and McCaffrey (2013). The heavy arrows show the relative rate of displacement (proportional to the length of the arrow) and the direction of rotation about a pole centered near La Grande, Oregon.

recorded a constant rate of clockwise rotation similar to the long-term average of 1.19°/my implying, in this case, that present-day rates are indicative of long-term rates. Recall from Fig. 5.3 that plates rotate about an Euler pole such that the rate of rotation varies with respect to the pole. Blocks in the Pacific Northwest also rotate about a pole, and this is part of the reason the rate varies across the region. Variable rates are also due to how stress and deformation is distributed across the region and how subblocks interact. The heavy arrows in Fig. 19.17 show the relative rate of displacement (proportional to the length of the arrow) and the direction of rotation about a pole centered near La Grande, Oregon. Displacement is westward and northward in Oregon, resulting in extension in that part of the Cascade volcanic arc. Displacement is eastward and northward in Washington resulting in compression. Rotation-driven compression is likely responsible for development of the Yakama Fold belt (Chapter 17).

Cascade block rotation could potentially explain why volcanism in the Oregon Cascades has migrated eastward. The thin blue line in Fig. 19.17 shows the inferred location of the Oregon coastline and the boundary with

California 16 million years ago. Rotation has since moved these boundaries northward and westward. The thick shaded blue line is the linear trend of the older Western Cascades (corresponding with ancient volcanism) based on exposure of associated intrusive rock. The thick shaded green line is the linear trend of present-day High Cascades volcanism, which corresponds with the crest of the active High Cascade volcanoes. These trends are consistent with clockwise rotation. In Oregon, if we assume northwestward land movement above a fairly stationary subduction zone magma source, then the area of ancient volcanism (the Western Cascades) would now be west of the area of present-day volcanism, just as we see in Fig. 19.17. Notice in Fig. 19.17 that the Western Cascade trend crosses to the east of the active High Cascades trend in northern Washington. The crossing of trend lines can also be explained with respect to rotation. Due to the location of the pole of rotation, the Western Cascade trend in Washington will move eastward rather than westward relative to the active arc.

Normal Faults Along the Crest of the High Cascades

The Oregon Cascade stratovolcanoes have been affected by normal faults that have thinned the deep crust providing conduits for basaltic magma to reach the surface. Active faults along the eastern side of the Cascade Mountains are visible in Fig. 18.12A as far north as the Columbia River. In central Oregon, east-west extension is occurring at a rate of about 3.9 inches per 100 years, and is currently propagating northward, which allows a significant amount of basaltic magma to reach the surface as far north as Mt. Adams in southern Washington. A beautifully exposed basaltic flow, only about 1500 years old, is shown in Fig. 19.18. The flow is part of a basaltic field between Three Sisters and Mt. Jefferson that includes Black Crater and Belknap Crater, both of which are small shield volcanoes (Fig. 19.13).

As a result of normal faults, many of the stratovolcanoes, particularly in Oregon and California, sit in large, basins (grabens) that are, on average, tens of miles wide, 2000 feet deep, and filled with piles of basaltic volcanic rock. It is these basaltic rocks that form the massive substratum to the younger, smaller, explosive, stratovolcaones. The basaltic rocks take the shape of a shield volcano similar to the Hawaiian volcanoes except, in the case of the Cascades, the shield volcanoes are capped with a centralized high-elevation silicic volcanic cone at all but one location (Medicine Lake). The basins formed as a result of clockwise rotation of the Coast Range, incursion of Basin and Range extension into the area, and isostatic sinking due to the weight of the accumulating basaltic pile. The greatest concentration of volcanic vents occurs in the vicinity of Three Sisters. This is also the location where the Brothers fault zone (Chapter 17) intersects the extensional basins of the High Cascades, thus creating many avenues by which magma generated at depth can reach the surface. Based on the age of the basaltic rocks, we can estimate that normal faulting beneath the High Cascade range, and the filling of basins with basaltic lava, began less than 10 million years ago.

FIGURE 19.18 Photograph of the edge of a 1500-year-old basalt flow on McKenzie Pass, Oregon.

The North Cascades have not yet been affected by an active phase of extensional normal faulting and, as a result, volcanic material, particularly basaltic volcanic material, is far less common. In Fig. 19.3, the two volcanoes in the North Cascades (Mt. Baker and Glacier Peak) are separated somewhat from the main volcanic chain to the south. In contrast to the large volcanic fields that surround stratovolcanoes of the Central-Southern Cascades, volcanism in the North Cascades is restricted primarily to these two volcanic cones, which stand alone as sentinels above the jagged crystalline landscape.

THE OLYMPIC MOUNTAINS

The two highest areas along the northern Pacific coast are the Klamath Mountains and the Olympic Mountains (Fig. 19.4). Both are young, active mountain ranges that have undergone recent uplift. The Olympic Mountains form part of the Olympic Peninsula in the remote northwest corner of Washington. A landscape map of the area with simplified geology is shown in Fig. 19.19. Ocean waters surround the peninsula on three sides. Lowland, less than 1000 feet in elevation, forms the south side. The mountain region is circular or somewhat horseshoe-shaped, about 65 miles wide, with rivers radiating from all sides. It rises abruptly to 7973 feet at Mt. Olympus within as little as 25 miles from the coast. The mountain region is isolated enough to have evolved 15 animals and 8 plants that occur nowhere else on Earth. There are many peaks above 6000 feet. Peaks above or approaching 7000 feet are present in the Mount Olympus-Mount Carrie region west of the Elwha River, and in the Mount Deception-Mount Anderson region east of the river. Lowland areas are heavily forested but the forest thins at about 5000 feet and disappears at high elevation. Mountain peaks are steep, rugged, and with the wet climate and loose rock, difficult to summit. Most of the region is within Olympic National Park and surrounding

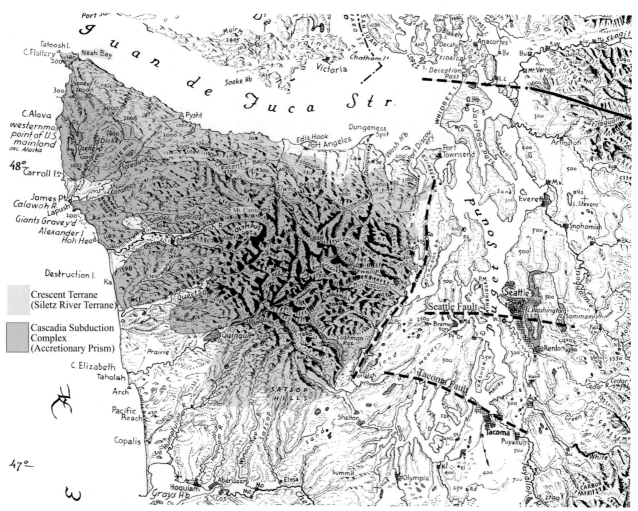

FIGURE 19.19 Simplified geologic map of the Olympic Peninsula showing the location of the Siletz River terrane (Crescent terrane).

designated wilderness. Roads surround the region but none cross the range or even extend very deep into the range. With the many surrounding peaks, Mount Olympus is not visible from surrounding towns. One must climb to get a good view.

Known for its temperate rain forests, the western side of the range boasts the wettest climate in the contiguous United States. The entire western side receives at least 100 inches of precipitation per year. Mt. Olympus receives more than 200 inches, and although not especially high, the Olympic Mountains are far enough north to be home to more than 200 glaciers. The mountains create a rain shadow effect, especially across the northeastern part of the range where precipitation decreases rapidly to less than 30 inches per year. Although considered to be a wet city, Seattle receives only about 38 inches of precipitation per year. The problem, of course, is that rain comes in the form of light drizzle that can last for days on end during the cold winter months.

Geology of the Olympic Mountains

Fig. 19.19 shows two primary rock units, the Siletz River terrane, known here as the Crescent terrane, and the Cascadia Subduction Complex, which represents an exposed part of the active subduction-related accretionay prism. The Siletz River terrane forms a crescent or horseshoe shape along the northern, eastern, and southern margins of the range. The Cascadia Subduction Complex, in the center of the range, and including most of the high peaks, comprises all of the rocks above the subducting Juan de Fuca plate and below the Siletz River terrane. The thickness of the subduction complex is impressive. Geophysical studies suggest the rocks reach a maximum thickness of more than 100,000 feet, and that nearly all of this thickness consists of sedimentary rocks scraped from the top of the subducting Juan de Fuca plate. Virtually none of the underlying oceanic basaltic crust has been accreted. Fig. 19.20 compares a cross-section of the Olympic Mountains with that of the Coast Range just south of the Olympic Mountains. The cross-sections are uncolored above the line of erosion. Notice that the angle of subduction below the Olympic Mountains is relatively gentle when compared to the Coast Range. Evidence for a gentle subduction angle includes the location of subduction-related earthquakes (the Benioff zone), which are shallow below the Olympic Mountains relative to areas in the north and south, and the location of subduction-related volcanoes. Glacier Peak in Fig. 19.3 is further inland than either Mt. Baker to the north or Mt. Rainier to the south. A subducting slab must reach a depth of 62 miles or greater to become hot enough to initiate melting and volcanism. The location of Glacier Peak farther inland, coupled with the presence of a relatively shallow Benioff zone, are both consistent with a gentle angle of subduction below the Olympic Mountains. The angle of subduction, coupled with greater uplift-erosion, results in widespread exposure of the Cascadia Subduction Complex. These rocks are not exposed in the Oregon Coast Range due to insufficient uplift-erosion across a more steeply dipping subduction zone.

A Case for Topographic Steady-State

It is clear from their size and elevation, and from exposure of the Cascadia Subduction Complex, that the Olympic Mountains have undergone a greater amount of uplift and erosion than the Coast Range. An interesting aspect to this conclusion is that there is evidence that the Olympic Mountains have been in topographic steady state for the past 14 million years. Recall from Chapter 10 that a steady-state mountain is one where the rate of uplift is approximately equal to the rate of erosion such that, in this case, even though the mountain has undergone uplift over the past 14 million years, it has remained approximately at its present elevation.

One theory for the origin of the Olympic Mountains has to do with ongoing clockwise rotation of the Pacific Northwest. By about 18 million years ago, the rotation had caught the Olympic area in a vice grip between the rotating Coast Range and nonrotating Vancouver Island. The result is compression in the area of the Olympic Mountains as shown in Fig. 19.21. Compression in the Olympic area may be the primary reason the Juan de Fuca plate has arched upward to create a more gentle subduction angle as shown in Fig. 19.20B.

We noted earlier that rotation appears to have been steady over the past 14 million years. If we assume that rotational compression is ultimately responsible for uplift, then we might expect rates of uplift in the Olympic Mountains to also be steady. As noted in Chapter 9, it is difficult to directly measure ancient uplift rates. Instead, we use measured rates of erosional exhumation as a proxy for uplift. Given a predictable climate, we can make the assumption that any change in the rate of uplift would likely cause a change in the rate of erosional exhumation. The exhumation history of the Olympic Mountains has been intensely studied using a number of different methods that include field data, apatite and zircon fission track ages, apatite (U-Th)/He ages, and river incision rates averaged over time scales that range from 14 million years to 100,000 years. A significant conclusion is that exhumation rates are roughly the same regardless of whether rates are averaged over the entire 14-million-year period or over a short 100,000-year period. Steady exhumation rates averaged over different time scales suggests that uplift has also been steady and that uplift and erosion are balanced such that the mountain has not grown taller or decreased in

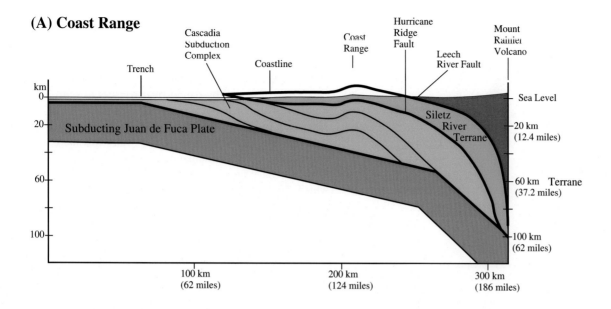

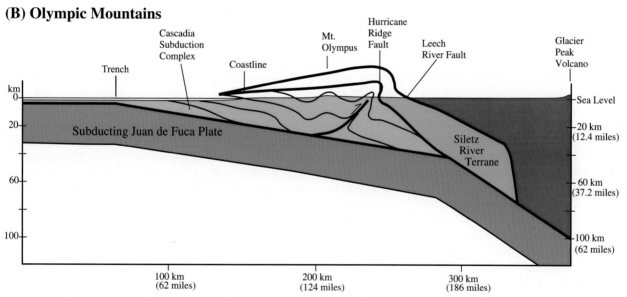

FIGURE 19.20 Comparative cross-sections of (A) the Coast Range and (B) the Olympic Mountains. Areas above the level of erosion are uncolored. The depth to the top of the Juan de Fuca plate is 25 miles below the Coast Range and 18 miles below the Olympic Mountains based on the location of earthquakes. The Cascadia Subduction Complex is exposed in the Olympic Mountains, but not in the Coast Range. The distance from the Coast Range to Mt. Rainier is 40 miles shorter than the distance from the Olympic Mountains to Glacier Peak. *Based on Brandon and Calderwood (1990).*

elevation during the 14-million-year period. The results suggest that the Olympic Mountains have been in topographic steady-state for the past 14 million years.

Let us now look at the consequences of steady-state in more detail. If we assume that exhumation is tied to uplift, and that their rates are about equal, then the rate of both uplift and exhumation should be highest near the central topographic high, and should decrease toward the coastline. In other words, measured exhumation rates should mimic present-day topography. Modeling of fission track and (U-Th)/He ages, coupled with depositional ages, supports this contention. Calculated erosional exhumation rates over the past 14 million years based on fission track and (U-Th)/He ages mimic present-day topography in the sense that the highest rates are in the central, high part of the peninsula, and the lowest rates are close to the shoreline. Additionally, the presence of 5- to 12-million-year-old shallow-marine sedimentary deposits along the shoreline indicates that these coastal areas were submerged 5 to 12 million years ago, and that they have not been elevated significantly above sea level during the past 5 million years.

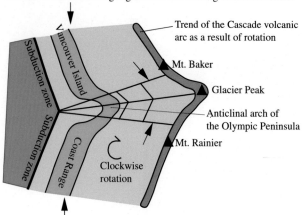

FIGURE 19.21 Sketch that shows how clockwise rotation of the Coast Range against a relatively stable Vancouver Island has caused compression and development of an anticlinal arch in the Juan de Fuca plate below the Olympic Mountains. The arch creates a shallow subduction angle such that Glacier Peak is farther inland than other volcanoes. *Based on Brandon and Calderwood (1990).*

Zircon and apatite fission-track modeling, along with geological field data, suggest that initial emergence of the Olympic Peninsula from below sea level occurred between 18 and 17 million years ago, and that the mountain had reached steady state by 14 million years ago. The time frame implies an uplift rate in the central mountain area of 2.4 to 3.2 inches/100 years over the initial 3 to 4 million-year period in order for the mountain to reach its present elevation by 14 million years ago. Zircon and apatite fission track data from the central elevated part of the mountain suggest an average exhumation rate of 3.2 inches/100 years between 14 and 7 million years ago, and a rate of 2.6 inches/100 years from 7 million years ago to the present. Apatite (U-Th)/He ages averaged over the past 2 to 3 million years record similar values. The data suggest that the mountain gained elevation until 14 million years ago when the rate of erosion became equal to the rate of uplift and the mountain reached steady-state. The data also imply that rates of uplift and erosion have remained steady and approximately equal at 2.4 to 3.2 inches/100 years over the entire 14 million year interval. The data further indicate that these exhumation rates decrease sharply toward the shoreline to rates that are considerably less than 1 inch per 100 years, consistent with the presence of young marine rocks along the shoreline. Integration of all exhumation rates across the entire peninsula (shoreline and mountain area) suggests that erosion (exhumation) has been occurring at an overall rate of about 1 inch per 100 years over the past 7 million years.

Now, let us compare these ancient long-term rates of erosional exhumation with more recent and present-day rates of erosion to see if they also support the idea of a steady state mountain. Ideally, present-day erosion rates should match long-term exhumation rates. Incision rates on the Clearwater River, which drains the western side of the Olympic Mountains, have been estimated over a 140,000-year period by dating river terraces as outlined in Fig. 9.9. The results suggest incision rates of 3.5 inches/100 years in the central mountain area, decreasing to less than 0.39/per 100 years at the coast—rates that are consistent with long-term exhumation rates. Present-day denudation rates have been calculated for the Hoh River system, which drains the west side of the mountain just north of the Clearwater River, and for the Elwha River system, which drains the north side of the mountain. In both cases, the estimates were calculated by dividing the amount of sediment leaving the drainage basin by the area of the drainage basin as outlined in Fig. 9.8. The calculated denudation rates are 1.26 inches/100 years for the Hoh River, and 0.71 inches/100 years for the Elwha River drainage basins. These rates are consistent with an overall exhumation rate of 1 inch per 100 years calculated for the entire peninsula over the past 7 million years. These data suggest that exhumation rates have been relatively constant and steady over the entire history of the Olympic mountains regardless of the time interval measured.

Although we cannot easily measure ancient uplift rates, we can measure present-day uplift rates. In this case, measured present-day rates are inconsistent with long-term rates. Leveling data measured over the past 60 years suggest that the entire Olympic Peninsula is currently uplifting at a rate greater than 3.15 inches/100 years with rates between 7.87 and 9.45 inches/100 years in the vicinity of Mt. Olympus. Maximum uplift rates based on leveling data are on the order of 12.6 inches/100 years. Tide gauge measurements also indicate strong uplift. The highest tide gauge measurements are at Neah Bay (Cape Flattery) in the northwest corner of the Olympic Peninsula where the rate of sea level drop was −6.7 inches/100 years from 1934 to 2017. Given an average worldwide sea level rise of about 8.0 inches/100 years over that time span, we can suggest that the land surface is rising at a rate approaching 15 inches/100 years. These rates are in conflict with long-term exhumation rates, which instead imply negligible exhumation along the coast. One possible explanation for the rather high present-day uplift rates in the Olympic Mountains is that the rates reflect stress buildup on a locked subduction zone fault as suggested in Fig. 19.10, implying that the rates are transient and not indicative of long-term rates.

In contrast to high uplift rates on the Olympic Peninsula, leveling data around the Puget Sound suggest that land areas are rising at rates less than 3.15 inches/100 years along the western side of the Sound, and that land areas are subsiding by an equal amount on the eastern side.

Tide gauge measurements for both Seattle (west side of Puget Sound) and Port Townsend (northern entrance to Puget Sound) are similar at 8.1 and 7.6 inches/100 years respectively. These rates are similar to the rate of sea level rise implying that absolute land elevation change is near zero.

From this discussion we can conclude that increased compression in the Olympic Peninsula has created less room to accommodate the total volume of accreted rock above an arched and more gently dipping subducting plate. This situation has resulted in earlier and more rapid uplift of the Olympic Mountains relative to surrounding areas. Field geology and exhumation data imply that the first significant land area to emerge from below sea level along the Washington-Oregon coast was the central part of the Olympic Peninsula beginning about 18 million years ago. Uplift progressed slowly northward and southward as the Olympic Mountains grew higher. It is estimated, based on sediment ages, that coastal areas of the Olympic Peninsula first emerged from below sea level about 12 million years ago and that the Oregon Coast Range first emerged between 10 and 5 million years ago. If we assume an average exhumation rate between 3.2 and 2.6 inches/100 years, the total thickness of rock removed from the top of the Olympic Peninsula since 18 million years ago is about 7.5 miles. This is considerably more than the estimate for the Coast Range, which is about 1.9 miles. The Coast Range has not yet reached topographic steady state. However, if tectonic and climatic conditions do not change appreciably in the next several million years, it is possible that the Coast Range will follow a similar but slower uplift history as that followed by the Olympic Peninsula, and that the Coast Range will eventually reach steady state. The slower uplift rate, however, implies that steady-state topography will likely be lower in elevation than the Olympic Mountains.

THE KLAMATH MOUNTAINS

The Klamath Mountains form a 150 by 100 mile welt along the Oregon-California coast that separates the Oregon Coast Range and Willamette Valley from the California Coast Ranges and Central Valley. The mountains extend approximately from Cape Blanco, Oregon to the proximity of Klamath and Redding, California (Figs. 19.3 and 19.4). Included are several individual mountain areas such as the Siskiyou, Marble, and Trinity Mountains, and the Trinity Alps.

The Klamath Mountains form part of the Cordilleran rain shadow. The western part of the range receives well over 100 inches of precipitation per year. The eastern part is dry, receiving as little as 20 inches per year. The landscape is rugged and heavily forested, especially in the west. As is visible in Fig. 19.4, the highest peaks are in the southeast part of the range. There is one peak, Mount

FIGURE 19.22 Google Earth image looking north at the sparsely forested Castle Crags granitic intrusion. Mount Shasta is at upper right.

Eddy (9025 feet), in the Trinity Mountains, that exceeds 9000 feet, and more than a dozen that exceed 8000 feet, mostly in the Trinity Alps. Glaciers at one time occupied Mount Eddy, high peaks in the Trinity Alps, and perhaps a few other high peaks, but none exist today.

Although the Klamath Mountains are heavily forested, there is one area nestled in the Trinity Mountains 18 miles south of Mt. Shasta, known as Castle Crags, where lightly forested granitic rock is surrounded by an ocean of forest-covered mountains. The Crags rise to more than 7000 feet in elevation with fractures along which erosion has carved vertical bare-rock spires. The scenery, shown in Fig. 19.22 and visible while driving Interstate 5, is unlike anywhere in the Klamaths. This unique area is likely the result of a combination of relatively hard rock, homogenous rock composition, and the development of vertical fractures that have created slopes too steep for plants to easily take hold.

Uplift History of the Klamath Mountains

The Klamath Mountains boast both a complex landscape history and a complex geological history. The northwestern part of the mountain region, in the vicinity of Crescent City (Fig. 19.4), appears to be a young, uplifted, and heavily dissected plateau surface known as the Klamath peneplain. On the basis of fieldwork, the Klamath peneplain is understood to have existed as a flat erosion surface at or near sea level until as late as 5 million years ago. There has been 6000 feet of uplift since that time, with most of the uplift occurring in the past 2 million years at an average rate between 1.4 and 3.6 inches/100 years. Present-day uplift rates in some areas of the Klamaths are more than 1 foot per 100 years. Rapid

FIGURE 19.23 Google Earth image looking SSE across the southern Klamath Mountains, California. T1 and T2 are wave-cut terraces. T1 is at the elevation of Crescent City, 20 to 50 feet above sea level. T2 is 200 to 500 feet above sea level. The higher terrace extends to the ocean at extreme right to form an escarpment that overlooks the ocean. The town of Klamath is on the Klamath River shown at upper right. Note the flat-topped mountains in the vicinity of the Smith River. The Trinity Alps form the highest part of the Klamaths.

uplift and high rainfall have combined to produce a region of dissected erosional mountains similar to those on the Appalachian Plateau. The Klamath Mountains, however, are younger, topographically higher, and composed of a variety of structurally complex rock types rather than the simple nearly flat-lying sedimentary rocks found on the Appalachian Plateau. Fig. 19.23 is a Google Earth image looking southeastward across the range at Crescent City. Incised river valleys and flat-topped mountains, typical of a dissected plateau surface, characterize the area. The swift-flowing Klamath, Rogue, Trinity, and Smith Rivers have cut canyons 1500 to 2500 feet deep.

Wave-cut terraces are present along the entire Klamath coastline. Two terraces (T1 and T2) are obvious at Crescent City in Fig. 19.23. As many as seven terraces are reported in the Brookings, Oregon area just north of Crescent City (Fig. 19.4) that range in elevation from about 100 feet above sea level to 1050 feet (321 m). Tide gauge measurements indicate a drop in sea level of −3.1 inches/100 years at Crescent City and a slight rise at Point Orford (near Cape Blanco) of 0.4 inches/100 years, implying significant uplift along coastal land areas (Table 19.1). Uplift rates over the past 80,000 to 125,000 years, based on marine terraces, have been estimated primarily for the southern Oregon coast southward to Crescent City. Rates vary from about 1.0 inch/100 years near Crescent City to a maximum between 3.4 and 4.5 inches/100 years at Cape Blanco. The relatively low uplift rate near Crescent City is at odds with tide gauge data. One explanation is that the area is experiencing active deformation such that some areas are rising and others, at the same time, are subsiding.

Deformation and uplift along the coastal Klamath Mountains could be the result of a variety of tectonic processes. Possibilities include shallow subduction, active underplating in the trench area, the passage of a seamount bulge, or a locked, flexing subduction plate. Another potential factor driving uplift may be tectonic activity and compression associated with the Walker Lane Belt discussed in Chapter 20.

Landscape evolution in the high elevation southeastern part of the Klamaths, north of Redding, appears to be older and markedly different than the northwestern mountain region. Apatite fission track and (U-Th)/He ages suggest a period of exhumation between about 126 and 65 million years ago and again between about 50 and 15 million years ago. Field data suggest that both periods, particularly the second period, could have occurred wholly or in part due to displacement and tectonic exhumation along a low-angle normal fault. These data imply the probable existence of topography as far back as 126 million years ago. The elevation of any such topography, however, is unknown.

Geology of the Klamath Mountains

The Klamath Mountains record more than 400 million years of nearly continuous subduction and tectonic accretion. Consequently, they are one of the best places to see remnants of ancient accreted volcanic arc complexes and ocean basin rocks. The structure consists of a series of east-dipping, thrust-faulted terranes composed of Ordovician to Cretaceous oceanic rock that was accreted to the continental margin from east to west in

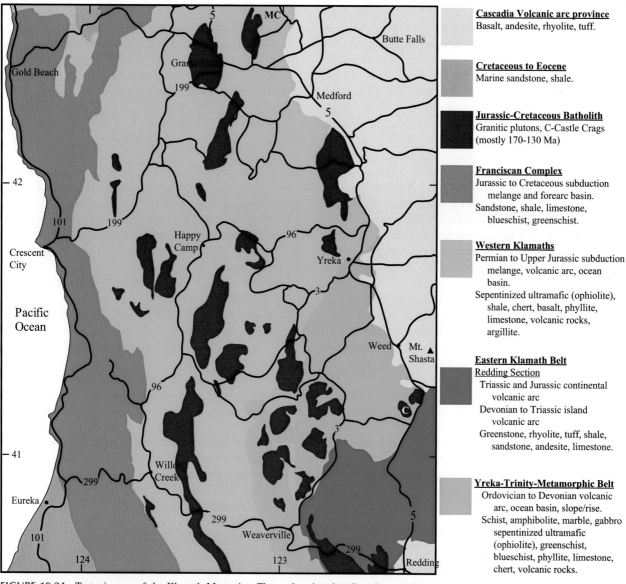

FIGURE 19.24 Tectonic map of the Klamath Mountains. The explanation describes the tectonic interpretation followed by common rock types. *Based on Irwin (1994), and Snoke and Barnes (2006).*

an almost continuous fashion beginning in the Devonian and continuing to the present-day at the Cascadia trench. The oldest rocks are located in the east, at the top of the east-dipping thrust pile. Two primary types of terranes are exposed. The first represents the remains of island volcanic arc systems that were constructed primarily offshore and then transported, deformed, and accreted to the continental margin. The second represents the remains of the surrounding ocean basin floor that formed either at the continental margin or along the flanks of island volcanic arc systems. This type of terrane includes deep-sea sediment (shale, chert) as well as slabs of oceanic lithospheric mantle (ultramafic rock) known as ophiolite. Some of the rocks were metamorphosed in the subduction zone and are now blueschists or greenschists. Rocks similar to those in the Klamaths are present at the northern end of the Sierra Nevada, and in the Blue Mountain region of Oregon.

A geologic map of the Klamath Mountains is presented in Fig. 19.24. The map shows the distribution of rock units, roads, and towns. The explanation lists the tectonic setting and rock types in each map unit. Four major accretionary thrust belts are shown, the Yreka-Trinity metamorphic belt, the Eastern Klamath belt, the Western Klamaths, and the Franciscan Complex. All four are separated by faults, and all except the Franciscan Complex are intruded by subduction-related granitic plutons, mostly between 170 and 130 million years old that includes Castle Crags (C), and which correlate with intrusions in the Sierra Nevada. The Yreka-Trinity metamorphic belt is an area of strongly metamorphosed and deformed ocean basin and volcanic arc rocks composed of schist, amphibolite, marble, gabbro, and ultramafic rock, including the Trinity ophiolite. The Eastern Klamath Belt contains an almost continuous record of volcanic arc deposition from Devonian to Jurassic. Preserved in this sequence is a Devonian-to-Triassic island volcanic arc complex and a Triassic-to-Jurassic continental volcanic arc complex. Rocks include greenstone, andesite-rhyolite, tuff, and sandstone. The Western Klamaths and the Franciscan complex record subduction and accretion during the Jurassic and Cretaceous. The Western Klamaths contain a variety of island arc and ocean basin terranes that include the Josephine ophiolite (ultramafic rock) and the Rattlesnake Creek subduction mélange (shale, chert, basalt). The Franciscan Complex contains accretionary prism and forearc basin deposits (sandstone, shale, limestone) along with blocks of blueschist and greenschist.

QUESTIONS

1. Google "mudflow potential on Mt. Rainier, Washington" or go to: http://vulcan.wr.usgs.gov/Volcanoes/Rainier/Lahars/Historical/description_osceola.html and describe recent activity and the potential effects of a mudflow emanating from Mt. Rainier.
2. Google "Cascades Volcano Observatory" or go to http://vulcan.wr.usgs.gov/ and describe recent activity on any of the following volcanoes: Mt. Lassen, M. Shasta, Mount Hood, Mount St. Helens, Mount Rainier, Glacier Peak, Mt. Baker. What do all of these volcanoes have in common?
3. Has the Juan de Fuca plate been increasing or decreasing in size over the past 29 million years, or has it remained the same size? Explain your answer.
4. Describe the Siletz River volcanic complex. Speculate in its origin.
5. Speculate on the origin of the Juan de Fuca Straight, north of the Olympic Peninsula.
6. Should the town of Corvallis be worried about a volcanic eruption from Mt. Jefferson?
7. What evidence is there, or what evidence would you look for, to determine if the Klamath Mountains were a lowland 5 million years ago? What evidence is there for recent, rapid, uplift?
8. Research the Boring lava field; its age, origin, distribution, and composition.
9. Why are there no volcanoes in the Klamath Mountains?
10. Speculate as to why Oregon has a coastal mountain range and why the coastal area south of the Olympic Mountains is lowland.
11. What is the origin of the Columbia River Gorge?
12. How does one distinguish between an active volcanic cone and one that has been dormant for a long time?
13. Name the highest volcano in Washington, Oregon, and California.
14. Why is there no trench to mark the site of subduction of the Juan de Fuca plate.
15. Why is basalt so abundant in the Cascade Range?
16. Research the 1980 Mount St. Helens volcanic eruption and write a one or two page report.
17. Research the eruption that produced Crater Lake and its aftermath, and write a one or two page report.
18. Research the Seattle or Tacoma fault (Fig. 19.3) and write a one or two page report. Are they active? What is the potential for a major earthquake?
19. Explain why large destructive earthquakes are less likely between Cape Meares and Cape Argo, Oregon, than areas to the north and south.
20. Describe the uplift history of the Olympic Mountains.
21. What is the evidence for the timing of uplift of the Cascade Range?
22. How might the substrate of Glacier Peak and Mt. Baker (in the North Cascades) differ from the substrate of volcanoes to the south?

23. Why are there so many volcanic landforms in the vicinity of Three Sisters?
24. If volcanism has occurred at the site of the Cascade Mountains for the past 36 million years. Why are none of the volcanoes more than 1 million years old?
25. Why was the Siletz River terrane accreted and not subducted?
26. How could one prove that volcanism has occurred in the Cascade Range for the past 36 million years?
27. How does the deep crust of the North Cascades differ from the crust of the Central-Southern Cascades?
28. Provide an explanation as to why short term measured uplift rates may not be indicative of long-term rates.
29. In Google Earth, fly to Allegany, Oregon and then move 10 miles to the east. What is the origin of the checkerboard landscape pattern?
30. In Google Earth, fly to Toke Point, Washington. Zoom out and look for any evidence for a marine terrace in the area.
31. In Google Earth, fly to Home Valley, Washington. What landscape evidence is there for landslides?
32. If it is difficult or impossible to measure ancient uplift rates, how does one determine if a mountain has reached steady state over a period of time.
33. Look closely at Fig. 19.6. Is there any evidence for slope erosion and retreat along the hillside? Speculate as to why there is no sand.
34. Can you name which volcano is seen as a speck on the skyline in the far distant center of Fig. 19.22?
35. Can you name the small volcano in the valley between Castle Crags and Mount Shasta in Fig. 19.22?

Chapter 20

California Strike-Slip System

The California Borderland, the Sierra Nevada, and the western margin of the Basin and Range together constitute the California strike-slip system. This structural province is fundamentally different from other structural provinces in that it is the only one that straddles a plate tectonic boundary. The Pacific plate is sliding northward along the San Andreas Fault relative to the North American plate. In addition, there is a subplate known as the Sierran plate that moves independent of both the North American and Pacific plates. The total rate of displacement is about 16.4 feet (5.0 m) per 100 years. What this means is 16.4 feet of displacement must be accounted for along faults in California every 100 years. Displacement can occur in lurches during earthquakes, or slowly and constantly without earthquakes. The San Andreas Fault system, including all active faults in the Coast Ranges, accounts for about 11.2 feet (3.4 m) of displacement. Faults east of the Sierra Nevada associated with the Sierran plate account for about 3.6 feet (1.1 m). The remaining 1.6 feet (0.5 m) probably occurs on faults off the California coast. Displacement of this magnitude creates one of the most geologically diverse and active regions in North America. A landscape map with highlighted locations and the trace of the San Andreas Fault is shown in Fig. 20.1.

The variety of landscape is evident in the shaded-relief elevation image shown in Fig. 20.2. The province is a grouping of seven landscape areas, each with unique characteristics and structures, but all part of the same active tectonic system. From north to south along the coast, these landscape areas are the Northern Coast Ranges, Central Coast Ranges, Transverse Ranges, and the Peninsular Ranges. Inland from west to east are the Central Valley (Cv), Sierra Nevada (Sn), and Walker Lane Belt (Wl). The Northern and Central California Coast Ranges form a linear landscape associated with the San Andreas Fault. The Transverse Ranges form a series of nearly east-west mountains between Point Sal (Ps) and Los Angeles (La) that merge with the Peninsular Ranges south of Los Angeles. These areas are separated with blue lines in Fig. 20.1. The Central Valley of California (Cv) is a large synclinal valley, one of the largest in the world. The Sierra Nevada (Sn) is a huge, nearly intact asymmetric normal fault block mountain that tilts westward and northward from its high point at Mount Whitney (W) where it overlooks Owens Valley (Ov) more than 10,000 feet below. East of the Sierra Nevada, the Walker Lane Belt (Wl) includes the White Mountains (Wm), the Inyo Mountains, Owens Valley (Ov), Lake Tahoe (t), and Death Valley (Dv). The Walker Lane Belt is distinguished from the Basin and Range by a change in orientation of mountain blocks from a north-northeast trend in central Nevada to a north-northwest trend in the Walker Lane. The change is visible in Figs. 20.1 and 20.2, particularly in the vicinity of Pyramid (p) and Walker Lakes (w). The Walker Lane belt extends southward into the Mojave Desert as the Eastern California shear zone where numerous low-lying mountains dot the landscape. Here, the change in orientation is less obvious, but is visible in Fig. 20.1 along a line trending from the area east of Death Valley through a wide region of dunes and salt flats that includes Soda Lake and Devils Playground. Notice in Fig. 18.12A that active faults in the Mojave Desert associated with the Eastern California shear zone end abruptly eastward across this line.

The rocks are mostly sedimentary and crystalline (plutonic and metamorphic), but volcanic rocks are important within and east of the Sierra Nevada. All of the rock successions outlined in Fig. 11.8 are represented with the exception of the Atlantic miogeocline. The dominant structural form is strike-slip faults, but there are examples of every type of structure mentioned in Chapter 4. Recall from Chapter 4 that shear stress produces strike-slip faults; compressional stress produces folds, reverse, and thrust faults; and tensional stress produces normal faults. Active stress across much of the California landscape is a combination of shear stress and compression, or a combination of shear stress and extension. Geologists use the term transpression for areas where shear is combined with compression. Such a combination produces strike-slip, reverse, and thrust faults along with folds. Conversely, the term transtension is used for areas where shear is

Geology and Landscape Evolution. DOI: https://doi.org/10.1016/B978-0-12-811191-8.00020-8
© 2018 Elsevier Inc. All rights reserved.

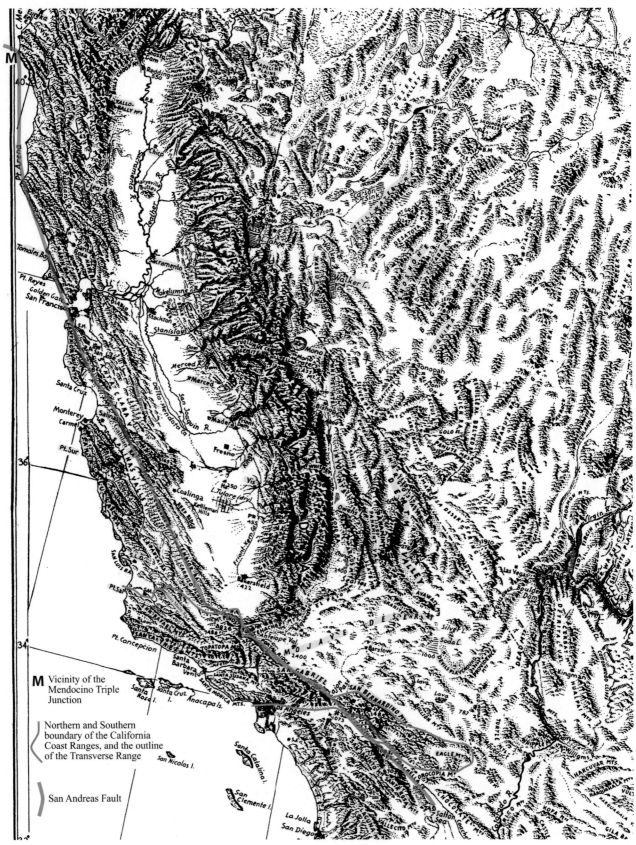

FIGURE 20.1 Landscape map of the California strike-slip system with names highlighted.

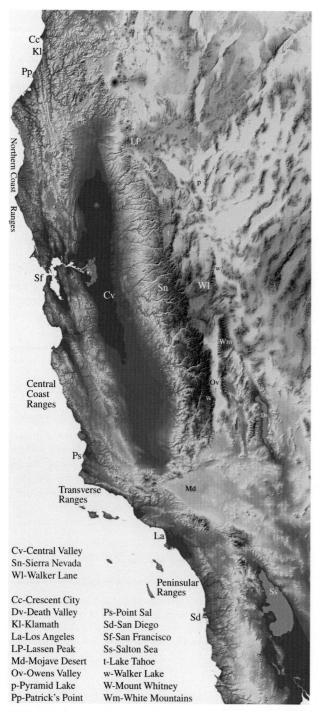

FIGURE 20.2 Shaded relief elevation map of the California strike-slip system. National Atlas of the United States of America, USGS, Nationalmap.gov/small_scale/atlasftp.html.

mountains and basins, and reverse/thrust faults produce asymmetric mountains. These young actively forming structures and associated landforms overprint a succession of relict structural and landscape features that include a volcanic arc terrain more than 80 million years old that was physiographically similar to the Cascadia arc system, a high plateau (the Nevadaplano) that may have existed from at least 80 to 17 million years ago, and Basin and Range landscape that developed between 17 and 12 million years ago. Relict landscape features are preserved to various degrees in different parts of the province, thus complicating the overall landscape picture. The abrupt appearance of active faults in the Mojave Desert as seen in Fig. 18.12A is an example of an actively forming tectonic regime that has superimposed itself on an older, pre-existing Basin and Range landscape.

In this chapter, we begin with a discussion of how strike-slip faults shape landscape. We then discuss individual areas beginning with the San Andreas Fault, followed by discussions of the Coast, Transverse, and Peninsular Ranges, the Sierra Nevada, and the Walker Lane Belt. We end with examples of active faulting in Death Valley.

LANDSCAPE ASSOCIATED WITH STRIKE-SLIP FAULTS

To understand some of the landscape surrounding the San Andreas Fault, we need to look more closely at how stress is distributed across faults and how strike-slip faults, in particular, are capable of generating compressional and tensional stresses. We noted in Figs. 4.2I and 4.2J that there are two types of strike-slip faults: right-lateral, in which one side of the fault is displaced to the right relative to the other side; and left-lateral, in which displacement across the fault is to the left. The San Andreas is both a transform fault (a plate boundary) in which the east side (the North American plate) moves southward relative to the northward-moving Pacific plate, and a right-lateral strike-slip fault that displaces rocks to the right. Chapter 5 discusses transform faults. Here we discuss the impact of strike-slip faults on the California landscape.

Before we embark on how strike-slip faults affect landscape, we need to review the various types of faults, and introduce a type of fault known as an oblique-slip fault that may occur in areas of transpression and transtension. A fault is a fracture along which rock on one side has undergone displacement (slip) parallel to the fracture surface. Slip can occur in any direction along the fracture surface. Reverse/thrust and normal faults displace rock vertically up and down the dip of the fault plane respectively. Ideally, there is no horizontal (lateral) displacement parallel to the strike of the fault. The maximum

combined with extension to produce strike-slip and normal faults.

Much of the landscape created by active transpression and transtension is young, less than 12 million years old. As such, landscape in many areas is a direct reflection of the active structure. Strike-slip faults produce linear mountains and valleys, normal faults produce block-

compressional and tensional stress directions respectively are parallel to the direction of slip as suggested in Figs. 20.3A and 20.3B. Strike-slip faults displace rocks horizontally (laterally) in a direction parallel with the strike of the fault. Ideally, there is no vertical (up-down) displacement parallel to the dip of the fault. Under these ideal conditions the maximum compressional stress direction is oriented 30° to the direction of displacement as shown in Fig. 20.3C. An oblique-slip fault is one in which displacement is oblique to the strike and the dip of the fault such that there is both vertical (thrust- or normal-slip) and strike-slip displacement. In practice, these faults can be named based on the dominant slip direction with or without the modifier, oblique. As an example, an oblique thrust fault is one in which thrust-slip is dominant, but where strike-slip is also significant (perhaps greater than 20% of total displacement).

Strike-slip faults are often segmented in the sense that the fault is broken along its length into smaller fault segments that act independent of each other such that not all of the segments are active during an earthquake. In other words, a strike-slip fault may end such that displacement is transferred to another segment or to a different fault that is present either to the left or to the right of the original one. Fig. 20.3D shows two primary right-lateral strike-slip fault segments (heavy black lines) in which displacement is transferred from one segment to the other. The location where the fault jumps from one segment to the other is known as a step. If you were to walk to the end of one fault segment, and then look for the other segment, you would look either to your left or to your right. In the case of Fig. 20.3D, you would look to your left; therefore, we would say that the fault steps to the left or that it is left-stepping. The area between the two fault segments is a transfer zone because this is where deformation is transferred from one fault to the next.

The geometry of the transfer zone shown in Fig. 20.3D results in transpression, which manifests itself in the form of thrust, strike-slip, and oblique-slip faults, and folds. In this example, oblique-slip thrust faults are oriented in an *en echelon* (parallel, in step) pattern approximately 45° to the primary strike-slip faults. Rocks are displaced upward in the direction of maximum compression (indicated by small arrows) and sheared right-laterally (indicated with half-arrows). These structures create mountains and valleys oriented parallel with the strike of faults and folds.

The geometry of the transfer zone shown in Fig. 20.3E results in transtension, which manifests itself in the form of normal, strike-slip, and oblique-slip faults. In this example, oblique-slip normal faults (yellow lines), are oriented in an *en echelon* pattern approximately 45° to the primary strike-slip faults. Oblique slip is indicated with arrows and half-arrows. The rocks are dropped vertically

Stress Direction and Ideal Fault Orientation

(A) Reverse/thrust fault. Slip is up the dip of the fault plane in the direction of maximum compressive stress.
(B) Normal fault. Slip is down the dip of the fault plane in the direction of tensional stress.
(C) Strike-slip fault. Slip is horizontal (lateral) parallel with the strike of the fault oriented 30° to the direction of maximum compressive stress.

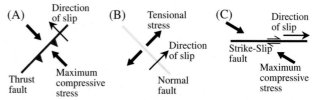

Transpression Between Right-Lateral Strike-Slip Faults

(D) A right-lateral strike-slip fault steps to the left across a transfer zone. In this case, the area between the two strike-slip faults (the transfer zone) is compressed creating oblique thrust faults oriented in an *en echelon* pattern. Landscape in the transfer zone consists of mountain ridges and valleys oriented parallel with the structures.

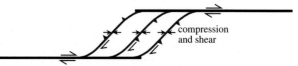

Transtension Between Right-Lateral Strike-Slip Faults

(E) A right-lateral strike-slip fault steps to the right across a transfer zone. In this case, the area between the two strike-slip faults (the transfer zone) is pulled apart creating oblique normal faults in an *en echelon* pattern. Landscape consists of pull-apart basins, sag ponds, and small tilted fault-block mountains oriented parallel with the normal faults.

FIGURE 20.3 Relationship between stress direction and ideal fault orientation. (A) Map view of a thrust fault. (B) Map view of a normal fault. (C) Map view of a strike-slip fault. (D) Map view of a left step along a right-lateral strike-slip fault. (E) Map view of a right step along a right-lateral strike-slip fault.

and sheared right-laterally at the same time. These structures create pull-apart basins (down-dropped basins in the shape of the transfer zone), sag ponds (down-dropped basins filled with water), and tilted, asymmetric normal fault block mountains oriented parallel with the strike of the faults.

Transpression and transtension are not limited to the location of a transfer zone, but can occur anywhere along a strike-slip fault where the strike of a fault is not parallel with the direction of displacement. Any bend in a strike-slip fault could potentially create areas of transpression and transtension. We will discover that transpressional structures are developed in the Coast Range, Transverse Range, and Peninsular Ranges along nearly the entire length of San Andreas Fault, and that transtensional structures are developed in the Walker Lane Belt east of the Sierra Nevada and north of the Garlock Fault. The Eastern California shear zone, in the Mojave Desert, forms a southward continuation of the Walker Lane Belt where both right- and left-lateral strike-slip faults are active. This region contains areas of transtension and transpression. Let us now look at additional characteristics associated with strike-slip faults.

In some instances, strike-slip faults permanently terminate rather than continue across a transfer zone. Fig. 20.4 shows geometric possibilities associated with the termination of a strike-slip fault. Geometry dictates that there must either be extension in the form of normal faults, or compression in the form of thrust faults, at both ends of the strike-slip fault. Fig. 20.4 shows both of these possibilities. Note that if the lower part of either strike-slip fault were to bend in the opposite direction, then normal faults would develop at one end of the strike-slip fault, and thrust faults would develop at the other end.

Additionally, because strike-slip faults form at an angle to the maximum compressive stress (Fig. 20.3C), it is possible for thrust faults and folds to form at an angle to the strike-slip fault (perpendicular to the maximum stress) as shown with thin red lines in Fig. 20.4. Notice that these structures are oriented roughly perpendicular to the maximum compressive stress direction. These structures form an *en echelon* pattern adjacent to the strike-slip fault regardless if the fault is oriented for pure strike-slip displacement or not. Compressional anticlinal ridges that form in this manner are known as pressure ridges. These folds may be dragged (rotated) toward parallelism with the fault with continued displacement as suggested schematically in the figure.

It is not uncommon for any of the faults mentioned above to dip at a high angle (>60°) at the surface. Strike-slip faults, in particular, are near vertical (84—90°) at the surface. These high-angle faults have two important characteristics. The first is that they can undergo reactivation with any change in the direction of maximum stress. A strike-slip fault, for example, could reactivate as a high-angle reverse (or oblique) fault if the maximum stress direction changes toward perpendicular to the strike of the fault. The second has to do with the vertical dip of strike-slip faults. In Fig. 4.11 we saw that a vertical plane cuts a straight-line path through landscape creating a lineament. Strike-slip faults, therefore, are often associated with strong lineaments. Because they displace rocks horizontally, they also tend to slice and shuffle landforms without producing any consistent asymmetry. Given these characteristics, landscape associated with strike-slip faults is one of narrow, linear fault valleys that are not particularly deep, and semi-parallel, symmetrical, linear mountain ridges that are not particularly high. Additional landscape features in areas of transpression and transtension include *en echelon* ridge and valley patterns, *en echelon* normal fault mountain blocks, pull-apart basins, and sag ponds.

THE SAN ANDREAS FAULT SYSTEM

The San Andreas Fault is one of the most intensely studied faults on the planet. It is a continental transform that separates the Pacific plate from the North American plate along its entire length. The trace of the fault through California is shown in Fig. 20.1. Its northern point begins at the Mendocino triple junction (M) near Cape Mendocino. This is the location where the North American, Pacific, and Juan de Fuca plates meet at a point, or more accurately, meet within a small area that includes the Pacific Ocean and the immediate coastline. Here, the San Andreas Fault merges with the Cascadia trench (also known as the Cascadia megathrust), which extends northward out to sea, and with the Mendocino transform fault and fracture zone, which extends westward out to sea (Fig. 19.1). From the Mendocino triple junction, the San Andreas Fault extends southward, touching land at Shelter Cove, but otherwise extends just off the coastline before reaching land at Point Arena. From Point Arena, it forms a straight lineament through Tomales Bay into western San Francisco and then through the Santa Cruz Mountains to an area east of Monterey Bay where the fault makes a slight bend to the east. It

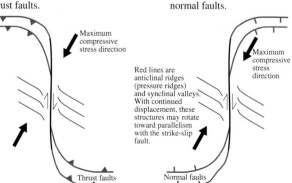

FIGURE 20.4 Map view of right-lateral strike-slip faults that terminate within a zone of (A) thrust faults and (B) normal faults. Also shown with thin red lines are en echelon anticlinal ridges (pressure ridges) and synclinal valleys that form perpendicular to the inferred maximum stress direction (indicated with heavy black arrows). Part of these folds have been dragged toward parallelism with the fault during continued displacement.

then maintains a straight line diagonally through the Coast Ranges to the southern end of the Central Valley between Bakersfield and Santa Barbara. From here the San Andreas Fault makes a second larger and more abrupt "Big Bend" to the east across part of the Transverse Ranges through Frazer Park just east of Mt. Pinos. It then turns southward along the northeastern side of the San Gabriel Mountains. East of San Bernardino, the San Andreas splits into two primary branches for a distance of about 60 miles within and along the southwestern edge of the San Bernardino Mountains before rejoining into a single fault just north of the Salton Sea. After about 672 miles (1082 km), the San Andreas Fault ends along the eastern shoreline of the Salton Sea virtually in the town of Bombay Beach (Fig. 20.1). It is an active fault, however, sections of the fault north of Point Arena and south of the San Gabriel Mountains have not been active historically.

As with any large fault, the surface trace of the San Andreas is not continuous, but instead is segmented into individual strands that can act independent of one another. Segments are separated by as little as a few feet or as much as a few miles. One or several segments may break during an earthquake.

The dip of the San Andreas Fault is vertical or nearly vertical along its entire length at least to the depth of its deepest earthquake, which is less than 12 miles. A zone of weakened, crushed rock that, in some areas can be up to a mile wide, marks the fault zone at the Earth's surface. This zone is easily eroded, and in combination with the vertical dip of the fault, produces a strong topographic lineament that is easily seen in Fig. 20.2 and in areas such as Tomales Bay (Fig. 4.12).

Fig. 20.5 and Table 20.1 identify a few of the many faults in the strike-slip system, all of which have been active in the past 1.6 million years. The distribution of several major fault zones is partly controlled by the two bends in the San Andreas. The Calaveras (10), Greenville (11), Hayward (9), Rogers Creek (7), Bartlett Springs (6), Maacama (2), Eaton Roughs (3), and Grogan (4) faults all extend northwestward from the vicinity of the small bend in the Santa Cruz Mountains near Monterey Bay. The Hayward Fault (9) extends just east of Oakland and was the site of a strong earthquake in 1868. Together, the Hayward and San Andreas faults frame San Francisco Bay.

The Big Bend forms a major transpressional zone responsible for uplift of the Transverse Ranges. Accordingly, the fault system here is complex. The Garlock Fault (20), a major left-lateral strike-slip fault that separates the Sierra Nevada from the Mojave Desert, intersects the San Andreas at a high angle. Several other faults, including the Big Pine (15), Santa Ynez (16), and Mission Ridge—San Cayetano—Del Valle (17) faults also intersect the San Andreas at a high angle. These faults, along with the San Gabriel (19), Santa Cruz Island—Santa Rosa Island—Malibu Coast—Raymond (18), and Pinto Mountain faults (28), are part of the Transverse Ranges.

There are numerous other faults in the Big Bend region that are semiparallel with the San Andreas. The most important is the San Gregorio-Hosgri Fault (8, 13), which skirts the coastline from San Francisco to Point Sal. This fault, together with the Reliz—Rinconada—East Huasna Fault (14), frames the Santa Lucia Range south of Monterey. Additional faults south of the Transverse Ranges include the San Clemente (21), Santa Catalina Ridge (22), Palos Verdes (23), and Elsinore (25) faults. The San Jacinto Fault (26) splays off the San Andreas at San Bernardino, and the two faults frame the Jacinto and Santa Rosa Mountains (part of the Peninsular Ranges).

Although the San Andreas Fault proper ends along the eastern side of the Salton Sea, the greater San Andreas transform system continues southward. Southwest of where the San Andreas Fault ends, another strike-slip fault, the Imperial Valley Fault (27), appears in the Imperial Valley south of the Salton Sea. This fault represents the southward continuation of the San Andreas Fault system. Farther south the Imperial Valley Fault is part of a series of strike-slip (transform) fault segments that extend through the Gulf of California where they connect small, divergent spreading ridges within which new oceanic lithosphere has formed (Fig. 5.12). The San Andreas transform system ends near the southern end of the Gulf of California at the Rivera triple junction where the Pacific, North American, and Cocos tectonic plates meet. With continued displacement, all land areas west of the San Andreas Fault, including Baja California and the city of Los Angeles, will be displaced northward along the California coast. As Baja California is displaced northward, the Gulf of California will, over time, widen to become part of the open Pacific Ocean.

Displacement Along the San Andreas Fault

To understand earthquakes, one must first understand how rocks react to stresses imposed on them. Stress is force per unit area. If you take a small dry stick between your hands and bend it, you are applying stress to that stick. Strain is the distortion (the bending) that results from stress. If you release the stick so that it is no longer under stress, the stick will instantly (elastically) snap back to its original shape. If you apply enough stress to the stick, it will break into two pieces. Once broken, the stick is no longer under stress, and all of the strain (the bending prior to breaking) is recovered. The two broken pieces are not bent and could theoretically fit back together. The type of strain that is present when the stick is under stress, but absent when stress is removed, is known as elastic (or recoverable)

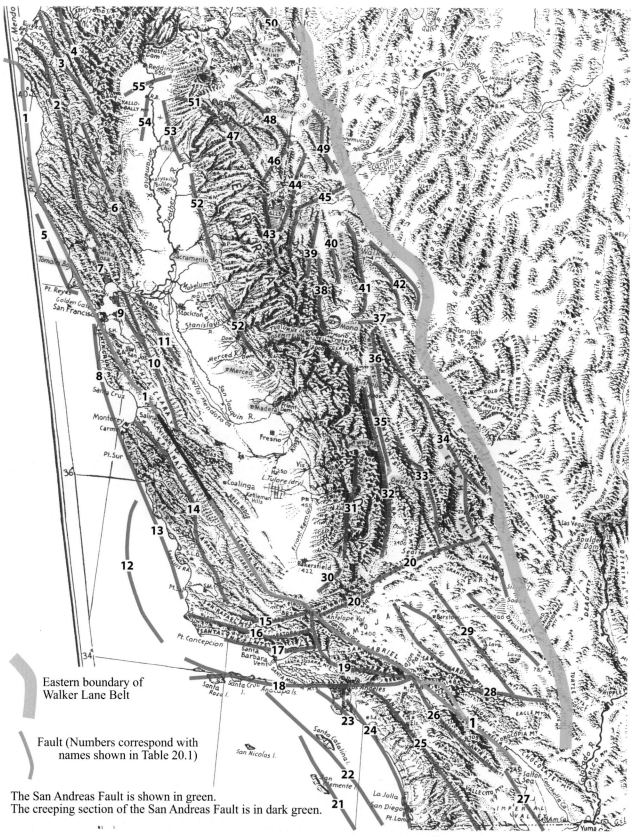

Eastern boundary of Walker Lane Belt

Fault (Numbers correspond with names shown in Table 20.1)

The San Andreas Fault is shown in green.
The creeping section of the San Andreas Fault is in dark green.

FIGURE 20.5 Landscape map that shows major active faults. Fault numbers correspond with Table 20.1.

TABLE 20.1 Major Faults Shown in Fig. 20.5

1. San Andreas
2. Maacama
3. Eaton Roughs
4. Grogan
5. Point Reyes
6. Bartlett Springs
7. Rogers Creek
8. San Gregorio
9. Hayward
10. Calaveras
11. Greenville
12. Santa Lucia Bank
13. Hosgri
14. Reliz–Rinconada–East Huasna
15. Big Pine
16. Santa Ynez
17. Mission Ridge–San Cayetano–Del Valle
18. Santa Cruz Island–Santa Rosa Island–Malibu Coast–Raymond
19. San Gabriel
20. Garlock
21. San Clemente
22. Santa Catalina
23. Palos Verdes
24. Rose Canyon
25. Elsinore
26. San Jacinto
27. Imperial Valley
28. Pinto Mountain
29. Eastern California system (several faults)
30. Breckenridge-White Wolf Fault
31. Kern Canyon
32. Southern Sierra Nevada Frontal
33. Saline Valley-Hunter Mountains-Panamint Valley
34. Fish Lake Valley-Death Valley-Black Mountains
35. Owens Valley
36. White Mountains
37. Coaldale (Mina Deflection)
38. Smith Valley-Mono Lake
39. Pine Mountains-Antelope Valley
40. Pine Grove Flat-Pine Grove Hills
41. Wassuk Range
42. area with numerous faults
43. Tahoe Frontal-West Lake Tahoe
44. Mt. Rose-East Lake Tahoe
45. Genoa-Carson Lineament
46. Dog Valley-Peterson Mountain-Last Chance
47. Mohawk Valley
48. Honey Lake
49. Pyramid Lake
50. Likely
51. Sierra Nevada-Cascade Range Boundary zone
52. Foothills Fault System
53. Chico Monocline
54. Corning
55. Battle Creek

strain. A larger stick will require a greater amount of stress to bend or break, and will snap back with greater force due to the greater amount of stored elastic energy.

Rocks behave the same way. Earthquakes form along a fault surface when rocks on either side of the fault are so tightly stuck together that they bend (and, therefore, accumulate elastic strain) prior to breaking. We alluded to this process in Fig. 19.10 while discussing locked faults associated with the Cascadia Trench. Elastic strain is recovered when the rock breaks and snaps back to its original shape. The difference between a rock and a stick is that rocks are buried deep within the Earth, and when they snap back upon breaking, they snap into adjacent rocks. The force of the snap applied to adjacent rocks creates seismic waves that, in turn, create earthquakes. A stronger rock will require a greater amount of stress to break, and will snap back with greater force. The result is a larger earthquake. Thus, the size of an earthquake is roughly a function of the amount of stress a rock can withstand prior to breaking. Creep occurs where rock is weak or lubricated so that every time stress is applied, the rock breaks or slips along the fault surface with little or no bending, and therefore with little or no accumulated elastic strain. Fault displacement, in this instance, occurs without an earthquake.

Rocks along the San Andreas Fault are stuck tightly together in the vicinity of San Francisco, site of the devastating 1906 earthquake as well as large earthquakes in 1838 and 1989. Rocks are also stuck tightly together near Los Angeles, site of a large 1994 earthquake. However, not all of the fault trace is subject to dangerous earthquakes. Rocks across part of the fault zone move via creep such that a significantly large earthquake never occurs. The main part of the San Andreas Fault that moves via creep is located between Salinas and Coalinga. It is shown in Fig. 20.5 with a slightly darker line. Parkfield is located just south of the area of creep and has been the site of large earthquakes in 1857, 1881, 1901, 1922, 1934, 1966, and 2004. Creep has been reported along the Hayward Fault (9) although apparently not enough to completely relieve the possibility of another large earthquake.

History of the San Andreas Fault

To avoid confusion, keep in mind that the San Andreas Fault extends from the Mendocino triple junction to the Salton Sea and that the San Andreas transform fault system extends from the Mendocino triple junction to the Rivera triple junction. The entire transform system, including the San Andreas Fault, is no more than 29 million years old.

For more than 200 million years prior to the formation of the San Andreas Fault, the entire California coast was a subduction boundary much like the present-day Oregon-

Washington coast. This subduction boundary separated the North American plate from a plate that no longer exists, the Farallon plate, which at that time extended most of the way down the North American coastline. We were previously introduced to the Farallon plate in Chapter 5, and again in Chapter 17, as the plate that underplated the Cordillera to cause the Laramide orogeny, and then sank to cause the ignimbrite flare-up. During subduction, the Farallon plate was moving obliquely eastward toward the North American plate, which was moving westward over the Farallon plate. The Pacific plate was in existence at this time, separated from the Farallon plate by a divergent plate boundary. The relationships are shown in Fig. 20.6A.

The San Andreas Fault began to form between 29 and 27 million years ago when the divergent plate boundary that separated the Pacific from the Farallon plate was itself subducted. It was at this time that the Pacific plate first came into direct contact with the North American plate, and because the Pacific plate was moving north-northwestward at high velocity relative to the slow-moving North American plate, the contact evolved into a transform (strike-slip) boundary as shown in Fig. 20.6B. The transform boundary split the remaining Farallon plate into a northern and southern half. Today, these remnants form the Juan de Fuca and Cocos plates. The initial transform boundary some 27 million years ago was at a latitude near Los Angeles and was probably located offshore at the now extinct subduction zone trench. The transform has grown in length in both a northward and southward direction since 27 million years ago as more and more of the Farallon plate was obliquely subducted. Such growth implies that the Mendocino and Rivera triple junctions have continuously migrated northward and southward respectively, and that the remnants of the Farallon plate (the Juan de Fuca and Cocos plates) have continually become smaller as implied in Figs. 20.6B and 20.6C. Both plates will continue to shrink and will eventually disappear.

The San Andreas Fault is not oriented for perfect strike-slip displacement and the misfit has been unstable enough for the fault to jump eastward several times to compensate. An eastward jump of the San Andreas implies that part of

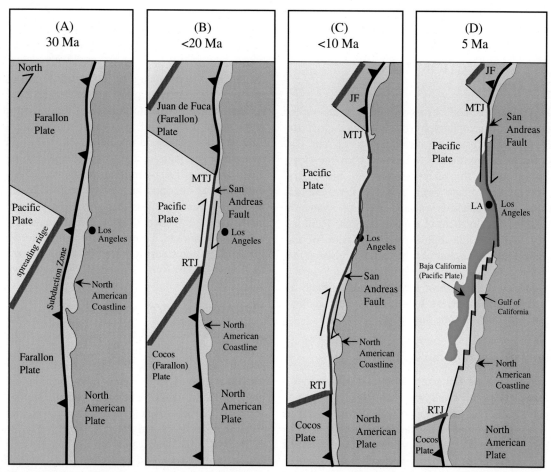

FIGURE 20.6 Sketch maps that show the sequence of development of the San Andreas Fault. MTJ-Mendocino triple junction, RTJ-Rivera triple junction. *Based on Atwater (1970) and Atwater and Stock (1998).*

what was the North American plate has been transferred to the Pacific plate. Several faults currently on the Pacific plate, such as the Santa Lucia Bank (12, Fig. 20.5), San Gregorio-Hosgri (8, 13), the Reliz—Rinconada—East Huasna (14), the San Clemente (21), and the Elsinore (25) faults, in the past may have marked the Pacific-North American transform plate boundary.

The most recent eastward jump occurred about 5 million years ago near Los Angeles. This jump created the Big Bend and the present-day geometry of the San Andreas Fault as depicted in Fig. 20.6D. The jump transferred Baja California from the North American plate to the Pacific plate resulting in initial opening of the Gulf of California. Since 5 million years ago, Baja California has moved northward about 155 miles away from mainland Mexico at a rate of about 16.4 feet per 100 years, thus enlarging the Gulf of California.

The northward and southward migration of triple junctions implies that the San Andreas Fault system is oldest and has its greatest displacement near its center, and that it gets progressively younger with less displacement toward the triple junctions. Total displacement is not well constrained. Offset features across the fault suggest minimum displacement in central California on the order of 200 miles, but it could be much more.

A RELICT SUBDUCTION ZONE LANDSCAPE

Fig. 20.7A is a schematic cross-section that shows the relationship between topography and tectonics circa 100 million years ago prior to development of the San Andreas Fault. In this figure, we can see that the Coast Ranges were host to the accretionary prism, the Central Valley formed the forearc basin, and the Sierra Nevada formed the now vanished volcanic arc. As shown in Fig. 20.7B, the development of the San Andreas Fault has effectively fossilized the topographic expression of this ancient subduction zone, but not the actual landscape that existed during subduction. The present-day landscape is younger than even the birth of the San Andreas Fault. In the following subsections, we look briefly at the geology of this relict subduction zone landscape.

The Ancient Accretionary Prism

The Coast Ranges are characterized by a variety of rock types that includes the Franciscan Complex, a sequence of Jurassic-Cretaceous accretionary prism rocks scraped off the down-going Farallon subduction slab and added to the North American plate during subduction as described in Fig. 5.7. These rocks are partly buried beneath younger rock south of San Francisco, but make up large areas of the Coast Ranges north of San Francisco. They are present in the Marin Range just north of San Francisco and on both sides of the Golden Gate Bridge. They are beautifully exposed in sea cliffs and coastal headlands north of Eureka such as Patrick's Point State Park (Pp, Fig. 20.2). Exposures south of San Francisco are most notable in the Diablo Range, which extends east of San Jose southward to Coalinga, and along the coastal Santa Lucia Range from Point Sur southward to San Luis Obispo (S.L.O. in Fig. 20.1). These are the same rocks as those present along the coastline of the Klamath Mountains in Fig. 19.24.

The rocks form a typical mélange assemblage composed of deformed, fine-grained, oceanic sedimentary rock with embedded lenses of basalt, metamorphosed basalt (greenschist, blueschist), serpentinite, and other parts of the Farallon oceanic lithosphere. Some of the embedded lenses are the size of a house or even larger. The lenses tend to be randomly distributed throughout the Franciscan Complex and are harder than the rocks in which they are embedded. As such, they form sea stacks along the shoreline and resistant knobs in the mountains.

Unlike a normal sedimentary rock unit, there is little in the way of consistent layering in the Franciscan Complex. Given the absence of layering, one might expect a random distribution of mountains and valleys. Instead, the mountains show a fairly regular linear landscape, and on this basis alone, we could suggest that the mountains we see today are unrelated to the subduction process that occurred prior to development of the San Andreas Fault. A clue to the age of the present-day landscape is the presence of very young marine (oceanic) sedimentary layers that lie unconformably above Franciscan rocks in many parts of the Coast Ranges. Some of these rocks are less than 4 million years old. Their presence indicates that the land area now occupied by these rocks was below sea level less than 4 million years ago. Thus, landscape in at least part of the Coast Ranges is less than 4 million years old. Later in this chapter, we will discuss the origin of the Coast Ranges with particular reference to the Central Coast Ranges south of San Francisco.

The Ancient Forearc Basin

The Central Valley of California, as shown in Fig. 20.7A, represents a subduction-related forearc basin landscape similar in origin to the Puget Sound-Willamette valley. The term Central Valley refers to its location and topography. The term Great Valley refers to its geological form, which is that of a syncline (the Great Synclinal Valley). At 400 miles long and 50 miles wide, it is one of the major synclinal depressions of the world. It is a relict landscape that was fossilized during formation of the San Andreas Fault and preserved virtually unchanged since the days of subduction. The Central Valley has undergone

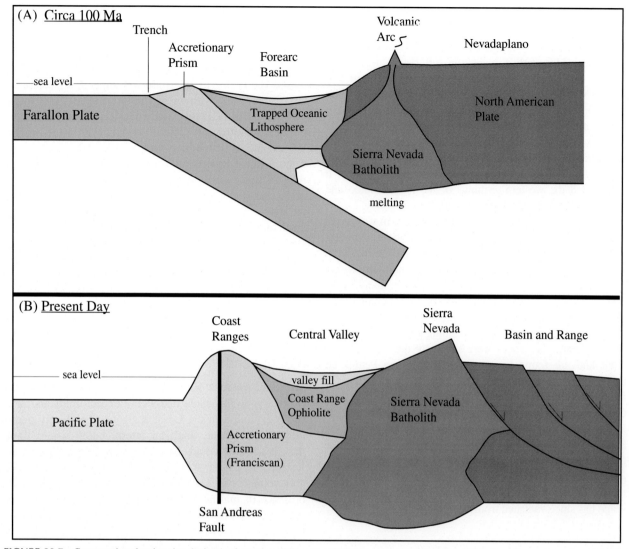

FIGURE 20.7 Cross-section sketches that depict the formation of the relict California subduction landscape. The upper figure shows the subduction landscape that existed circa 100 million years ago. Note the inferred existence of the Nevadaplano. The lower figure shows the present-day topographic expression of the relic subduction landscape. Oceanic lithosphere (Coast Range ophiolite) is trapped between the Coast Ranges and the Sierra Nevada.

nearly continuous deposition over the past 155 million years, accumulating as much as 20,000 feet of rock and sediment. The abundance of marine (oceanic) deposits indicates that for most of its existence, the Central Valley was below sea level with emergent areas primarily along its eastern side that were built up from the deposition of sediment derived from the Sierra Nevada. The present-day apron of sediment shed from the Sierra Nevada can be seen in Fig. 20.2. It has been only during the past 4 or 5 million years that the entire Central Valley has emerged permanently from below sea level. Today it is a flat, highly fertile, heavily irrigated landscape, mostly between 50 and 300 feet in elevation, drained by the Sacramento River in the north and the San Joaquin River in the south. These two rivers join and drain into San Francisco Bay.

The southern third of the Central Valley, south of Fresno, is a closed basin with internal drainage (Fig. 2.2).

The flat-lying sediment that covers the Central Valley is, for the most part very young (<2 Ma), and derived from surrounding highlands. Older rocks are hidden from view in the central part of the valley, but are turned up on edge and exposed in the Coast Ranges along the western margin of the Central Valley. Here, the entire sedimentary sequence can be examined. Surprisingly, oceanic lithosphere (the Coast Range ophiolite) can be found at the base of the sedimentary sequence south of Reading in the Yolla Bolly Mountain area, and in the Diablo Range (Fig. 20.1). Apparently, part of the Farallon oceanic lithosphere became trapped in the forearc basin during the subduction process as depicted in Fig. 20.7B.

The Ancient Volcanic Arc

The Sierra Nevada consists primarily of a large batholith of granitic rock that intruded between 220 and 80 million years ago (Ma) with a magmatic lull between 140 and 120 Ma and particularly strong magmatism between 100 and 80 Ma. The batholith consists of more than 200 individual intrusions (plutons) emplaced at depths between 5 and 20 miles. These rocks represent the feeder batholith to the now vanished overlying, subduction-related volcanic arc. The area likely was a volcanic highland 80 million years ago in close proximity to an ocean (Fig. 20.7A). It represented the western terminus of the Nevadaplano and looked nothing like the Sierra Nevada we see today.

The Sierra Nevada is not the only location in California where 220- to 80-million-year-old subduction-related batholithic rocks exist. These same rocks are widely exposed in the Basin and Range to the immediate east, including within the White, Inyo, and Argus Mountains, and in many of the small peaks that dot the Mojave Desert. They are also present in the San Bernardino and Eagle Mountain area east of Palm Springs (P.S., Fig. 20.1) including part of Joshua Tree National Park. The rocks were sliced by the San Andreas Fault and are found west of the fault in the Peninsular Ranges such as in the Santa Ana, Laguna, Santa Rosa, and Jacinto Mountains where the granite forms the Southern California batholith. Northward, granitic rocks are present in the San Gabriel Mountains north of Los Angeles, in the Santa Lucia Mountains south of Monterey, and in the Gabilan Range east of Salinas where, along with metamorphic rocks of variable age, they form part of a transported terrane known as the Salinian block. The subduction landscape has since vanished from all of these areas.

THE CALIFORNIA COAST RANGES

The California Coast Ranges extend along the coast approximately from the town of Klamath (Kl. Fig. 20.2) near Redwood National Park, to Point Sal (Ps) where the Coast Range turns east-west to form the Transverse Ranges. The mountains are broadly divided into the Northern California Coast Ranges north of San Francisco, and the Central California Coast Ranges south of San Francisco. The Coast Range boundaries vary somewhat with different maps. The northern and southern borders, as defined in this book, are shown in Fig. 20.1 with a thin blue line. The northern extreme of the range extends beyond Fig. 20.1. At the southern end, the San Rafael and Sierra Madre Mountains, and Mount Pinos, are sometimes considered part of the Coast Ranges but are shown here as part of the Transverse Ranges. Defined as such, the Northern Coast Ranges boast one peak above 8000 feet and many above 6000 feet, most of which are in the Yolla Bolly Mountain region. All peaks in the Central Coast Ranges are less than 6000 feet.

A glance at Figs. 20.2 and 20.5 shows that the Coast Ranges have linear trends. Mountain regions alternate with narrow, linear valleys, most of which contain active faults. The northern ranges are wet, rugged, dissected, and heavily forested with precipitation reaching 100 inches per year north of Point Arena, but decreasing to less than 60 inches southward. Outcrop is poor within the heart of the mountain range due to extensive vegetation. The Central Coast Ranges are drier, less densely forested, less rugged, and better exposed with less than 25 inches of precipitation per year. The dominant structural feature is the active San Andreas Fault, which cuts from south to north diagonally across the Central Coast Ranges at a slight angle to the coastline, thereby intersecting the coast at San Francisco. The ranges and valleys are approximately parallel with the fault, and therefore also intersect the coast at a slight angle such that the coastline north of Point Arena consists of alternating rocky mountain headlands and sandy valley inlets.

Age of Landscape

The San Andreas Fault forms an active tectonic plate boundary, and as such, one would expect uplift and landscape evolution in the Coast Ranges to be closely tied to fault displacement. Such an interpretation would imply that landscape could be as old as 29 million years. Instead, the evidence suggests that landscape is considerably younger. Key evidence that landscape is young is the presence of marine (oceanic) rock as young as 4 million years old. These rocks were deposited at or below sea level implying that present-day landscape must be younger than 4 million years even though there is evidence that the region was emergent numerous times prior to 4 million years ago.

Circumstantial evidence for recent uplift includes the nature of the rocks themselves. In addition to Franciscan Complex and granitic rocks, much of the Central Coast Ranges consists of young, poorly consolidated, nonresistant sedimentary rocks. The granitic rocks are more resistant, but even these have been weakened by intense fracture. The combination of weak rock, steep slopes, strong wave action along the coast, and infrequent storms, results in rapid erosion that includes numerous landslides. It is unusual, to say the least, for such weak rock formations to form mountains. The only explanation is that the mountains are young and that they were uplifted at a higher rate than the rather rapid pace of erosion.

Evidence for recent uplift can also be seen along the shoreline where the mountain landscape drops directly into the ocean. Some of the more outstanding examples are the Santa Cruz Mountains south of San Francisco, the

Santa Lucia Mountains south of Monterey, and the San Luis Range west of San Luis Obispo. The Santa Cruz Mountains are especially well known for its flight of five marine terraces that step down to the ocean. The lowest terrace is about 100 feet and the highest 750 feet above sea level. Fig. 20.8 looks southeastward at no fewer than three well-developed marine terraces beveled into young (less than 23 million years old) mudstone and fine-grained sandstone. Cosmogenic ^{10}Be and ^{26}Al dating indicates that the lowest terrace is about 65,000 years old, and the highest is about 226,000 years old. When the age and elevation of each terrace is correlated with sea level changes over the past 250,000 years, a steady uplift rate of 4.3 inches/100 years emerges. This rate is slightly higher than geodetic studies that suggest present-day uplift rates on the order of 3.2 inches/100 years, but one must remember that uplift, although steady in the long-term, is episodic in the sense that at least some of the uplift occurs during earthquakes. The closest tide gauge is across Monterey Bay at Monterey and it indicates a relative sea level rise of 5.8 inches/100 years, a value lower than average worldwide sea level rise and therefore consistent with geodetic studies that indicate active coastal uplift.

Some insight to the landscape history of the Coast Ranges may be gained by an apatite (U-Th)/He study of the Santa Lucia Mountains located south of Monterey. Recall that this dating system has a low closure temperature, which, in this case, allows for the determination of cooling history over the past 10 million years. The data indicate slow cooling between about 10 and 6.1 million years ago, followed by rapid cooling from 6.1 to 2.3 million years ago, followed by even more rapid cooling beginning sometime after 2.3 million years ago. Geological evidence suggests that the cooling history can be correlated directly with erosional exhumation. The data suggest that following a period of little or no exhumation from 10 to 6.1 million years ago, the rocks were exhumed at a rather steady rate of 1.14 to 1.61 inches per 100 years between 6.1 and 2.3 million years ago and that this rate increased to 3.54 inches/100 years or higher sometime after 2.3 million years ago. The data do not constrain exactly when (after 2.3 million years ago) the exhumation rate increased. If we assume that increased rates of erosional exhumation correlate with increased rates of uplift, and we also keep in mind that much of the landscape remained below sea level until at least 4 million years ago, we can suggest that uplift had begun as early as 6.1 million years ago, but that erosion was largely able to keep pace such that surface uplift was minimal and most land areas were not elevated above sea level until 4 million years ago. Rapid exhumation beginning sometime after 2.3 million years ago presumably marks the beginning of rapid surface uplift and development of the present-day landscape. It has been argued based on field data that rapid surface uplift did not begin until as late as 400,000 years ago. Thus, the Santa Lucia Mountains we see today are very likely less than 2.3 million years old and possibly as young as 400,000 years old. Active uplift along the coast suggests the mountains are still in the process of forming.

FIGURE 20.8 Google Earth image looking north-northwest at wave-cut terraces in the Santa Cruz Mountains just north of Santa Cruz, California.

Mountain Alignment Relative to the San Andreas Fault

A pressure ridge is an anticlinal ridge that forms at an angle to a strike-slip fault roughly perpendicular to the maximum stress direction as outlined in Fig. 20.4. The Coast Ranges, because many are approximately parallel with the San Andreas and other major faults, are not at an orientation expected for the development of pressure ridges. There are, however, several exceptions that are visible in Fig. 20.5. One location is just north of Point Arena where the San Andreas Fault trends offshore parallel with the shoreline, but where the mountains intersect the shoreline at an angle. Another location evident in Fig. 20.5 is just east of Point Arena between Clear Lake and the town of Santa Rosa where ridges are oriented at an angle to the Maacama Fault (2). A Google Earth image of this area is shown in Fig. 20.9. If you look closely at Fig. 20.5 you can find additional examples.

The example shown in Fig. 20.9 not withstanding, how do we explain the near parallelism of most mountain trends with the San Andreas Fault? First, we have to realize that the San Andreas Fault was probably close to but never perfectly aligned with the displacement direction of the Pacific plate relative to the North American plate.

There are at least two explanations for the lack of perfect alignment. The first are the existence of pre-existing fractures and faults. Rather than cut a new path through solid rock, faults are likely to form along pre-existing weak zones if they are oriented at a low to moderate angle from perfect alignment. For example, the San Andreas probably formed initially along a zone of weakness created by the pre-existing subduction zone. The second explanation has to do with the direction of relative motion between the Pacific and North American plates, which has changed several times during the history of the San Andreas. Changes in direction of relative plate motion without a corresponding change in the orientation of the plate boundary can result in misalignment. Any form of misalignment will result in either transpression or transtension. In the case of the Coast Ranges, transpression appears to have been the dominant result.

Let us look at the consequences of perfect alignment verses misalignment. Fig. 20.10A assumes that the Pacific plate is approaching a stationary North American plate at a convergence direction of 30 degrees east of north and that the plate boundary is oriented due north. Under these conditions, the plate boundary (a hypothetical San Andreas Fault) will experience pure strike-slip displacement with no additional component of compression or extension. In this

FIGURE 20.9 Google Earth image looking northward along the Northern Coast Ranges toward Santa Rosa, California. The San Andreas, Rogers Creek, and Maacama faults are drawn and labeled. An inferred approximate stress direction for the Rogers Creek Fault is shown with arrows. Mountain ridges between Santa Rosa and Clear Lake are oriented at an angle to the fault lines consistent with pressure ridges formed via compression perpendicular to the stress direction. Note also that ridges in the lower-left and upper-left part of the image are oblique to the San Andreas Fault. These may also have formed via compressional stress. The San Andreas Fault at lower left extends through Tomales Bay. Point Reyes is located just off the image west of the San Andreas Fault.

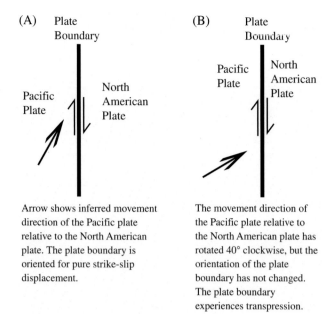

FIGURE 20.10 Sketch maps that show (A) ideal strike-slip faulting along a plate boundary and (B) transpression along a plate boundary caused by a change in the direction of relative plate convergence without a change in the orientation of the plate boundary.

scenario, the fault is in perfect alignment with relative plate motion. A pressure ridge or thrust fault (and resulting mountain) could form perpendicular to the stress direction at an angle of 60 degrees to the fault. With displacement on the fault, it is possible that the mountain will rotate toward parallelism with the fault as shown in Fig. 20.4, but likely would not reach perfect alignment.

Let us now look at the case of misalignment between convergence direction and the fault. Fig. 20.10B shows the Pacific plate approaching a stationary North American plate at a convergence direction 60 degrees east of north, but with the plate boundary still oriented due north. Under these conditions, a pressure ridge or thrust fault (and resulting mountain) will develop perpendicular to the compression direction at an angle that more closely aligns with the plate boundary. Subsequent rotation can now more easily align the resulting mountain with the fault.

Studies suggest that about 8 million years ago the direction of relative plate motion between the Pacific and North American plates rotated clockwise in a manner similar to, but not as extreme, as shown in Fig. 20.10B. One possibility is that compression resulting from this change in plate motion is ultimately responsible for uplift and eventual permanent emergence of the Coast Ranges, as well as for the near-parallelism of mountain trends with the San Andreas Fault.

Misalignment of plate boundaries has additional consequences such as the development of new strike-slip faults and the repositioning of existing transform faults. For example, the formation of the Calaveras (10) and other faults at the small bend near Monterey Bay may have originated as an attempt by the San Andreas to become better aligned with plate motion. In addition to the possible formation of these faults, this bend produces compression and uplift (transpression) in the Santa Cruz Mountains. As shown in Fig. 20.11, the Santa Cruz Mountains appear to be highest in the vicinity of the bend.

We noted earlier that the San Andreas transform has jumped eastward several times throughout its history. These jumps were likely in response to changes in plate motion. The most recent jump 5 million years ago created the Big Bend on the San Andreas north of Los Angeles. This jump may have been in response to the change in relative plate motion that occurred 8 million years ago. The Big Bend itself is unstable. It has created a large zone of compression with numerous folds and faults that has resulted in uplift of the Transverse Ranges. In a later section, we discuss the Walker Lane Belt and the possibility that a major jump of the San Andreas to the east side of the Sierra Nevada is currently in progress. Presumably, this jump will create a more stable plate boundary configuration.

Deformation History Prior to Surface Uplift

Field evidence in the form of young marine rocks indicate that there were few if any areas permanently above sea level prior to 4 million years ago. However, there is evidence in the form of unconformities to indicate that rocks in the Coast Ranges were deformed prior to 4 million years ago, in some areas multiple times. A primary piece of evidence is the observation that fold and fault trends within individual mountain ranges are oriented at an angle to the trend of the mountain itself. In other words, the morphology of a particular mountain range seems to have no direct correlation with internal structures present within the range. We have seen this before. It is the same type of relationship seen in mountain blocks of the Basin and Range where the internal structure of the foreland fold-and-thrust belt has no correlation with the landscape of the range, which instead, is controlled by much younger range-front normal faults.

Another line of evidence in favor of deformation prior to 4 million years ago is the fact that all internal structures are truncated (cut) by a relatively flat erosion surface. Such a surface would be expected if erosion were able to keep pace with deformation-induced uplift and maintain landscape at or just below sea level. Today, this gentle low-relief surface is preserved along summit areas of some of the ranges. An example from the southern Diablo Range is shown in Fig. 20.12. The preservation of this easily eroded mountaintop surface is consistent with

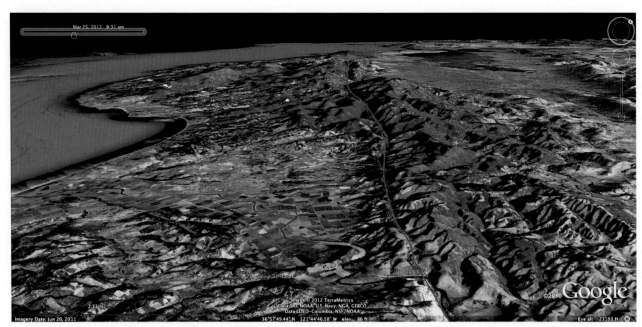

FIGURE 20.11 Google Earth image looking north-northwest along the Santa Cruz Mountains south of San Francisco. The thick line represents the San Andreas Fault, which cuts obliquely through the mountains.

FIGURE 20.12 Google Earth image looking north-northwest at the relatively flat summit surface of the Diablo Range near Coyote Lake (c) southeast of San Jose, California.

the conclusion that present-day uplift is recent. Otherwise, the surface would have been deeply dissected.

Mechanism and Cause of Surface Uplift

We need to ask one final question. How did uplift occur? Insight to this question is illustrated in a cross-section shown in in Fig. 20.13 that extends from the Pacific coastline across the northern Santa Cruz Mountains to the San Francisco Bay south of San Francisco. Individual mountain blocks in the Coast Ranges are bound not by normal faults as in the Basin and Range but by range-front reverse/thrust, and strike-slip faults. In some areas, a pre-existing strike-slip fault has splayed into a flower-like structure of reverse and thrust faults, along which the mountains were uplifted. In Fig. 20.13, the San Gregorio Fault is shown as a single strike-slip fault at depth that has splayed upward into a flower structure composed of

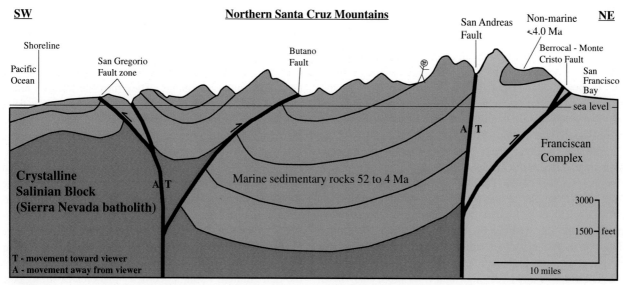

FIGURE 20.13 Southwest-to-northeast interpretive cross-section across the Santa Cruz Mountains, California. *Based on Page et al., 1998.*

multiple reverse/thrust faults, including the Butano thrust, along which the mountain was uplifted. The figure also shows the Berrocal-Monte Cristo reverse fault splaying from the San Andreas Fault to form a range-front reverse fault on the east side of the Santa Cruz Mountains. The mode of uplift is compression (transpression).

Given the presence of flower structures, we can suggest a possible uplift history for the Coast Ranges based on the previously outlined findings from the Santa Lucia Mountains. The dominant form of deformation prior to 8 million years ago was strike-slip faulting with little in the way of uplift. A change in relative plate motion 8 million years ago resulted in transpression, reverse/thrust faulting, and uplift beginning about 6.1 million years ago, but with no significant gain in surface elevation due to simultaneous erosion. Surface uplift and development of present-day landscape began sometime after 2.3 million years ago coincident with a significant increase in the calculated rate of exhumation. Some of the reverse/thrust faults have now reverted back to strike-slip displacement, which suggests that present-day compression and the rate of uplift have slowed.

THE TRANSVERSE RANGES AND THE SALTON SEA

The Transverse Ranges are a set of nearly east-west ranges developed at an angle transverse to both the Coast Ranges to the north and to the Peninsular Ranges in the south. The northern and southern boundaries of the Transverse Ranges as used in this book are shown between the thin blue lines in Fig. 20.1. Included are the Channel Islands (Santa Rosa and Santa Cruz Islands), the San Rafael Mountains, the Sierra Madre Range, and Mount Pinos.

The Transverse Ranges are a young, actively forming mountain system that boasts some of the highest peaks close to the US coastline outside of Alaska. There are two peaks above 11,000 feet, San Gorgonio Mountain and Jepson Peak. Both are less than 70 miles from the coast and both reside in the San Bernardino Mountains. There are, in addition, many peaks above 10,000 feet in the San Bernardino Mountains and one in the San Gabriel Mountains. The high elevation is due in part to the presence of resistant granitic rock and of older rocks of the North American crystalline shield. These rocks form a belt of mostly high terrain from the 8835-foot Mount Pinos in the north, to the Eagle Mountain area at Joshua Tree National Park in the south (Fig. 20.1). Other areas in the Transverse Ranges consist of marine sedimentary rocks mostly less than 34 million years old. These rocks are soft, yet they form two peaks above 7000 feet and many above 6000 feet including peaks north of Santa Barbara in the San Rafael, Topatopa, and Pine Mountains. Volcanic rocks between 25 and 8 million years old form part of the Santa Monica Mountains. Franciscan Complex rocks are present in the San Rafael Mountains. The region overall is dry with lowlands receiving less than 20 inches of precipitation per year. Mountain flanks are forested and relatively wet, receiving as much as 50 inches precipitation per year. None of the mountains were glaciated except for the north side of San Gorgonio Mountain, and none are exceptionally rugged.

Whereas most of the mountains in the Coast Ranges have formed along straight segments of the San Andreas Fault, the Transverse Ranges developed in the transpressional Big Bend region of the fault. As shown in Fig. 20.3D, a step or bend to the left along a right-lateral strike-slip fault results in a zone of unusually strong compression where the two sides of the fault move directly toward each other. Compression in the Big Bend area produces east-west-trending anticlinal and thrust fault mountains that extend offshore as the Channel Islands. Given this compressional origin for the Transverse Ranges, the mountains must be younger than the age of the Big Bend, which formed about 5 million years ago with the most recent the eastward jump of the San Andreas Fault.

The San Andreas Fault is active, therefore compression in the Big Bend area is also active, which implies that at least part of the Transverse Ranges are still growing. One example of an actively growing landform is Wheeler Ridge, a faulted, anticlinal ridge located just north of the San Andreas Fault at the border between the Transverse Ranges and Central Valley just west of the junction between Interstate 5 and Golden State Highway 99. The ridge is 2141 feet high and rises more than 1400 feet above surrounding flatlands. A detailed study of this ridge showed that a single identifiable erosion surface is present at progressively higher elevation with a greater degree of tilt in the higher central part of the mountain than at its eastern end. This observation suggests that the eastern part of the ridge is younger than the central part, and therefore has not yet been elevated or tilted as much as the central part. Such a conclusion implies not only that the ridge is growing taller via uplift; it is also growing longer at its eastern end.

Rates of uplift are provided by radiocarbon and uranium-series dating of charcoal and carbonate in soil horizons. These dates indicate that Wheeler Ridge has been uplifting at a minimum rate of 12 inches per 100 years for at least the past 1000 years. The dates also indicate that the anticlinal ridge is becoming longer at a rate of about 9.8 feet per 100 years. In other words, during a 100-year period, almost 10 feet of flat ground at the east end of the ridge will be uplifted to form a small anticlinal hill, and the mountain as a whole will gain one foot in elevation. Given these rates of growth, it is estimated that the Wheeler Ridge anticline began to develop about 400,000 years ago. Because Wheeler Ridge is located at the northern margin of the Transverse Ranges, its youthful development suggests that deformation may be spreading northward into the southern edge of the Central Valley. Perhaps we are witnessing the very beginning of a long process that will result in the reincarnation of the Central Valley.

Wheeler Ridge is not the only documented actively growing structure in the Transverse Ranges. Another fold, the Ventura Avenue anticline, located near Ventura, is one of the fastest-growing structures known. Dating techniques and soil correlation suggest that this structure grew at an average rate of about 6.5 feet per 100 years between 200,000 and 100,000 years ago. Today it is growing at a rate of nearly 20 inches per 100 years.

It should not be surprising that actively growing mountains close to an ocean produce wave-cut terraces. As many as 14 terraces have been recognized on Santa Catalina Island with the highest as much as 1700 feet above sea level. Terraces at this elevation cannot possibly be due to sea-level change. They most definitely indicate recent uplift. Fig. 20.14 looks northward at a high, tilted, and warped terrace overlooking a beach on the flank of the Santa Monica Mountains at Pacific Palisades west of Los Angeles where uplift rates of 1.18 inches per 100 years for the past 320,000 years have been reported. The Santa Monica tide gauge, close to the Pacific Palisades shows an average rise of 6.0 inches/100 years suggesting a small rise in land elevation relative to average sea level rise, which is consistent with reported uplift rates (Table 19.1). The Santa Ynez Mountains at Santa Barbara form coastal mountains oriented almost perpendicular to the San Andreas Fault, a geometry that creates strong compressional folding and reverse/thrust faulting. A wave-cut terrace, similar in appearance to the Pacific Palisades is present on the flank of the mountain range overlooking a narrow beach. Also similar to the Pacific Palisades, active folding has warped the terrace creating uplift rates west of Santa Barbara on the order of 7.9 inches/100 years with rates decreasing eastward. The Santa Barbara tide gauge shows a sea level rise of 4.42 inches/100 years, a rate consistent with active uplift when compared to worldwide sea level rise.

The opposite effect, transtension, can occur where a right-lateral strike-slip fault steps or bends to the right. This type of structure is shown in Fig. 20.3E. A right step occurs at the southern termination of the San Andreas Fault between the San Andreas on the east side of the Salton Sea and the Imperial Valley Fault (27, Fig. 20.5) on the west side. The result is normal faulting and development of a sag pond in the form of the Salton Sea. As California's largest lake, the Salton Sea covers an area of about 381 square miles, but has a maximum depth of only 51 feet. Like the Great Salt Lake, the Salton Sea has no outlet to an ocean. Water is lost through evaporation such that salinity in the lake is about 25% greater than ocean water. The lake is currently at an elevation 227 feet below sea level.

Rotation of the Transverse Block

The reason the Transverse block has undergone such rapid uplift is that it has been rotated into an orientation consistent with compression. Rotation involved several blocks in addition to the Transverse block, and created at least two pull-apart basins. The present-day distribution of these tectonic features west of the San Andreas Fault is shown in Fig. 20.15. Significant among them are four blocks, the Santa Lucia Bank (*SL*), the Outer Borderland

FIGURE 20.14 Google Earth image looking northward along a wave-cut terrace that overlooks a beach at Pacific Palisades west of Los Angeles, California.

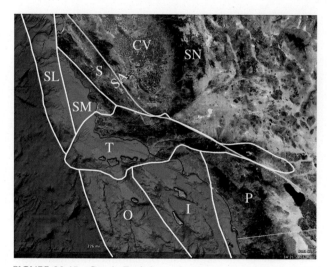

FIGURE 20.15 Google Earth image looking north at tectonic blocks and pull-apart basins along the southern California coastline. CV = Central Valley, I = Inner Borderland Basin, P = Peninsular Ranges and Valleys, O = Outer Borderland block, S = Salinian block, SA = San Andreas Fault, SL = Santa Lucia Bank block, SM = Santa Maria Basin, SN = Sierra Nevada, T = Transverse block.

(O), the Transverse (T), and the Salinian (S), and two pull-apart basins, the Santa Maria (SM) and the Inner Borderland (I), which includes the Los Angeles Basin. Note that some of these features are currently offshore. Note also that the Transverse block is oriented approximately perpendicular to the trend of the other blocks and close to the orientation for maximum compression.

One method by which to understand the origin of the Transverse block is paleomagnetism. When rocks in southern California were analyzed for paleomagmatism, it was discovered that rocks in the Transverse block older than 16 million years had been rotated 80 to 110 degrees clockwise, and that progressively younger rocks in the block were rotated progressively lesser amounts. These relationships imply that clockwise rotation began soon after 16 million years ago and has continued, possibly to the present day. It thus appears that the Transverse Ranges were first rotated into the orientation we see today and then, following formation of the Big Bend about 5 million years ago, underwent rapid uplift.

Fig. 20.16 is a schematic set of sketches that show how and why this rotation occurred, as well as the origin of the two pull-apart basins. Twenty million years ago, the San Andreas Fault likely was positioned offshore and west of all four blocks. Sometime around 16 million years ago the fault jumped eastward, transferring all four of the tectonic blocks onto the Pacific plate. The Santa Lucia Bank, Salinian, and the Outer Borderland blocks were all displaced northward. The northern (now eastern) part of the Transverse block apparently snagged against the continent while the southern (now western) side swung seaward. Much of this rotation likely occurred while the Transverse block was below sea level. The geometry of the rotating block resulted in the opening of the Santa Maria and Inner Borderland basins. Additional room was created by plate kinematics in which the Pacific plate was not only sliding past the North American plate, but also diverging slightly.

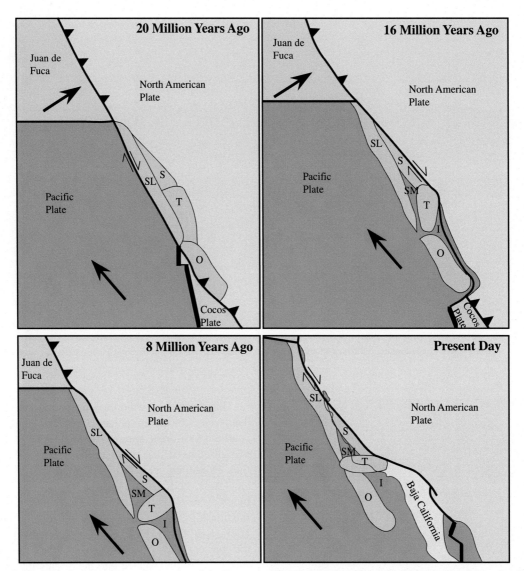

FIGURE 20.16 Sketches that show progressive rotation of the Transverse block and development of the Santa Maria and Inner Borderland basins. *Based on Atwater (1998)*.

The eastward jump 5 million years ago created the Big Bend along the San Andreas Fault and strong compression in the Transverse Ranges leading to uplift.

PENINSULAR RANGES

The US Peninsular Ranges are located along the southern California coast south of Los Angeles and west of the Transverse Ranges (Fig. 20.1). Only the San Jacinto Mountains have peaks above 10,000 feet. Peaks in other ranges are less than 6600 feet. None of the peaks have been glaciated. Many of the smaller mountaintops are relatively flat with roads that reach the summit. The high San Jacinto Mountains form a steep escarpment on their eastern and northern sides that overlooks Palm Springs and Snow Creek more than 10,000 feet below. Fig. 20.17 is a view looking east-southeastward at the steep escarpment and at the high San Bernardino Mountains (SB, Transverse Ranges), which overlook Yuciapa (Y). Although not particularly wet, the mountains form part of a rain shadow to the Mojave Desert.

Most of the ranges, including the Santa Ana, Santa Rosa, San Jacinto, and Laguna Mountains, consist of granitic rocks that form part of a separate southward extension of the Sierra Nevada batholith known as the Southern California batholith. Older volcanic and metamorphic rocks are present within the granitic rocks in many areas. Sedimentary rocks are also present, especially near the coast. The ranges tilt westward and are sliced by active strike-slip faults that include the Rose Canyon (24), Elsinore (25), and San Jacinto (26; Fig. 20.5).

EC-Eastern California Shear Zone, PS-Palm Springs, SA-San Andreas Fault, SB-San Bernardino Mountains SGP-San Gorgonio Pass fault zone, SJ-San Jacinto Mountains, SJF-San Jacinto fault zone, SS-Salton Sea, STA-Santa Anna fault zone, Y-Yucaipa. The thin, barely visible lines are faults active in the past 1.6 million years.

FIGURE 20.17 Google Earth image looking southeastward at part of the Peninsular Ranges. Thin lines are faults from USGS Website: https://earthquake.usgs.gov/hazards/qfaults/.

The landscape history of the Peninsular Ranges is not well understood and probably varies across the region. The general increase in elevation inland from the coastline, coupled with the character of sedimentary rocks and soil horizons, suggests that inland areas were uplifted earlier than coastal areas and that they were above sea level (but not necessarily mountainous) by 40 million years ago. The timing for significant mountain uplift is poorly constrained but likely occurred prior to 3.8 million years ago. This is in contrast to marine sedimentary rocks along the coastline that are between 5 and 2 million years old. These rocks indicate that coastal areas did not emerge from below sea level until after about 2 million years ago. Wave-cut terraces are common along the coastline, including Oceanside north of San Diego and the Palos Verde Hills south of Los Angeles, implying recent uplift or sea level lowering. Tide gauge measurements at San Diego and Newport Beach show a relative sea level rise between 8.45 and 8.7 inches/100 years, similar to worldwide sea level rise, implying a steady coastline (Table 19.1).

SIERRA NEVADA

The Sierra Nevada is roughly 400 miles north to south and 40 to 60 miles east to west measured from Dear Creek (route 32) located about 18 miles south of Lassen Peak in northern California, to the Garlock fault at the northern edge of the Mojave Desert in southern California. It is an asymmetric, westward- and northward-tilted fault-block mountain, the largest in the United States, composed of massive batholithic granitic rock with metamorphic rocks and young volcanic-sedimentary rocks increasingly more abundant north of Mono Lake. The mountain range and surrounding area is shown in Figs. 20.1 and 20.2.

The Sierra Nevada offers an exceptional opportunity to decipher landscape development for two reasons. The first is that the mountain range, to a first approximation, is an internally coherent fault block, meaning it has deformed (tilted) as a semi-rigid mass. Evidence for a semi-rigid block is seen in Fig. 20.5 (see also Fig. 18.12A), where active faults surround the mountain on all sides except the southwestern side south of Merced, and where active faults are largely absent within the mountain itself. The eastern and northern flanks of the mountain range are bordered by a series of faults collectively referred to as the Sierra Nevada Frontal Fault system. In Fig. 20.5, this fault system includes the Southern Sierra Nevada Frontal fault (32), the Smith Valley-Mono Lake fault (38), the Tahoe Frontal-West Lake Tahoe fault (43), the Mohawk Valley fault (47), and the Sierra Nevada-Cascade Range Boundary zone (51). This fault system is primarily responsible for the westward and northward tilt of the range. The Garlock fault (20) borders the south flank of the mountain, and the Foothills fault system (52) and Chico monocline (53) occupy the western flank north of Merced. The Chico Monocline likely overlies a buried (blind) active fault. A fault system is present on the southwest side south of Merced, but it is not active and therefore is not shown in Fig. 20.5. It is known as the Western Sierra Fault System and was active 45 to 40 million years ago. The absence of an active fault south of Merced suggests that this part of the Central Valley is connected to and deforming with the Sierra Nevada. The major exceptions to a semi-rigid block are at the southern and northern ends of the range where the Kern Canyon fault (31) and faults associated with the Sierra Nevada-Cascade Range Boundary zone (51) cut into the Sierra Nevada. The presence of a semi-rigid block allows us to predict how landscape will change along its length. For example, given a westward and northward tilt, we can predict that uplift has been greatest in the southeastern part of the range and that it decreases westward and northward.

The second exceptional aspect is the presence of a well-developed relict erosion surface along the western slope of the range underlain with granitic and metamorphic rock and overlain with young, easily dated volcanic rocks, including the 16 to 3 million-year-old Mehrten Formation. Similar to our analysis of the Southern Rocky Mountains, volcanic rocks can be used to determine the timing of river incision into the western slope of the Sierra Nevada, which in turn, can be used to help decipher recent uplift history. We will explore the incision history later in this chapter.

The westward and northward tilt of the mountain block produces a predicable morphology consisting of a gently tilted western slope that leads from the crest of the range to the Central Valley, and a steep eastern escarpment that leads from the crest of the range down to the Basin and Range. Fig. 20.18 is a shaded relief elevation map of the Sierra Nevada and surrounding area that shows both the eastern escarpment and the gentle western slope. The thin white line shown along the eastern side of the range represents both the range crest and a major drainage divide that separates rivers that flow west toward the Pacific Ocean from rivers that drop down the steep escarpment to the Basin and Range only to be trapped in the Great Basin. The drainage divide diverges from the range crest north of Mount Lola (L) and is shown as a thin yellow line.

Elevation along the range crest increases from north to south as predicted from the tilting of a semi-rigid mass. There are only a few peaks above 8000 feet north of Reno and few above 11,000 feet between Reno and Mono Lake (M). Peaks south of Mono Lake reach a maximum elevation that exceeds 14,000 feet in the Mount Whitney area (W), but peak elevation farther south decreases to about 8000 feet at the southern end of the range.

Rivers and glaciers on the gentle western slope of the mountain have created canyons in the high southern part of the range that are up to 5000 feet deep. Major river canyons are evident in Fig. 20.18 and are named in Fig. 20.1. They include Yosemite Valley along the Merced River, the San Joaquin River Gorge, and Kings Canyon. The deep canyons in the southern part of the range, paired with 13,000 and 14,000-foot peaks, giant sequoia trees, and glacial features, have produced a spectacular wonderland recognized with three national parks, Yosemite, Kings Canyon, and Sequoia.

The steep escarpment on the eastern side of the range reaches its maximum height in the Mount Whitney area where it is truly one of the most impressive sites in the world. There is almost 11,000 feet of elevation change from Mount Whitney (W, Fig. 20.18), the highest peak in the contiguous United States at 14,505 feet, to Owens Lake (ol) at the western edge of the Basin and Range at 3560 feet. This part of the escarpment is prominently displayed in Figs. 17.34 and 18.11. There are 11 universally recognized peaks above 14,000 feet in California (excluding sub-peaks) and 9 of them lie within 43 miles of each other along the range crest in the Mount Whitney area. Mt. Sill (S, Fig. 20.18) is farthest north. Mount Langley (La) is farthest south. Collectively, these peaks form the eastern boundary of Kings Canyon and Sequoia National Parks. All of the peaks are accessible (with some effort) from the Pacific Crest National Scenic trail, which extends the length of the Sierra Nevada through the entire US Cordillera from Mexico to Canada. Glacial erosion at high elevation has produced jagged mountain peaks and glacial basins (cirques) floored with bedrock lakes (tarns) as is evident in a photograph of Mt. Whitney shown as Fig. 20.19 that looks westward at the steep eastern escarpment.

At Mono Lake you can turn westward and drive over the range crest escarpment on route 120 (Tioga Pass Road) into the spectacular glacial and granitic landscape of Yosemite National Park. Southward for 140 miles to Sherman Pass Road at the southern end of the range, the wall of high-mountain peaks and the absence of a through-going river prevent any road from crossing the escarpment. If you wish to travel by road to Sequoia and Kings Canyon National Parks, you must approach from the Central Valley.

Glaciers in the southernmost part of the range have excavated a fault/fracture zone producing a straight, 70-mile long gorge up to 6000 feet deep that cuts southward (rather than westward) through the heart of the highest peaks in the Sierras dividing the western slope lengthwise. Today, the Kern River and the Kern Canyon fault (31), an active, east-side-down normal fault that extends for 85 miles from the Mount Whitney area to an area south of Isabella Lake, occupy this canyon (Figs. 20.5 and 20.18). Similar with many faults across the country, the Kern Canyon fault has partly reactivated an older fault zone that was active as a right-lateral strike-slip fault between 95 and 80 million years ago. The fault was also active between about 20 and 10 million years ago, this time as a normal fault. Recent activity on the fault may have begun less than 3.5 million years ago. The Kern Canyon fault (31) is not shown as an active fault in Fig. 18.12.

Fig. 20.20 is a Google Earth image that looks northward from the southern part of the range at Kern Canyon. The gently tilted western slope of the Sierra Nevada, and the steep eastern escarpment, are visible. This image also

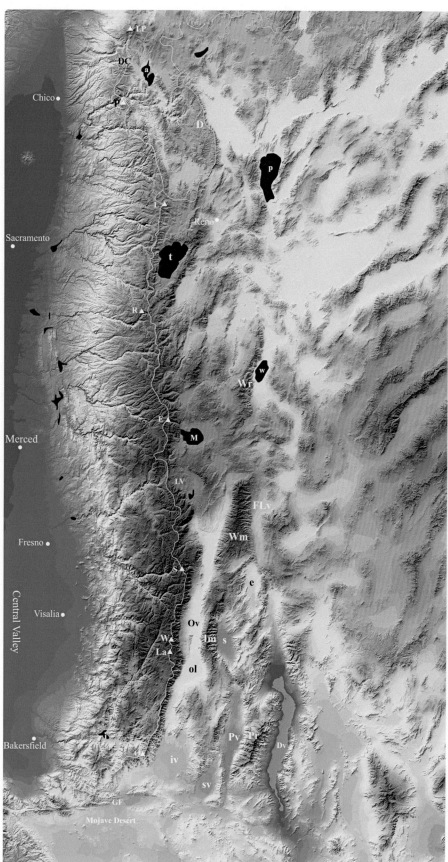

FIGURE 20.18 Shaded-relief elevation map of the Sierra Nevada and surrounding region. Enlarged from Fig. 20.2.

FIGURE 20.19 A photograph looking westward at Mt. Whitney and the great eastern escarpment of the Sierra Nevada.

FIGURE 20.20 Google Earth image looking northward along the southern crest of the Sierra Nevada. The westward tilt of the range is visible. The major northerly trending scar through the middle of the range is Kern Canyon (k). Part of the relict erosion surface (r) is visible above the canyon wall. Kern Canyon separates 13,000-foot peaks of the Great Western Divide on the west side, from 14,000-foot peaks in the Mount Whitney area to the east.

shows part of a relict erosion surface (r) to be discussed later. Visalia (Vis, Fig. 20.1), in the Central Valley, is at an elevation of 325 feet. If you were to travel east from Visalia toward Mount Whitney, you would quickly climb to well over 13,000 feet before you reach Kern Canyon. You would then drop to about 7700 feet in Kern Canyon and then climb again, this time to well over 14,000 feet at Mount Whitney. The mountain peaks west of Kern Canyon are known as the Great Western Divide. It is an imposing sub-range with more than a dozen peaks over 13,000 feet including Mount Kaweah, the highest at 13,807 feet. Uplift of this sub-range is likely the result of recent and active displacement along the Kern Canyon and associated faults.

Global positioning system measurements coupled with other space-based radar techniques indicate that the Sierra Nevada range is undergoing broad uplift along its entire length. The notable aspect to this uplift is that it does not appear to be localized entirely along range-front faults. One possibility is that uplift is occurring in the form of a giant, rising dome. The implication is that the dome structure has only recently begun to form and that it is superimposed on a pre-existing tilted, semi-rigid Sierra Nevada mountain block. The maximum measured rate of uplift is 4 to 8 inches per 100 years. Such rates are at least 10 times greater than rates of erosion implying active elevation gain.

The Sierra Nevada Frontal Fault System

The Sierra Nevada Frontal fault system is a collection of fault segments located along the base of the great eastern escarpment. It is dominantly a normal fault system along which the Sierra Nevada has experienced uplift in the form of westward, and northward tilting. Studies in the Walker Lane belt to the east of the Sierra Nevada suggest that the frontal fault system may have initiated 12 million years ago between Mono Lake and Lake Tahoe, and that other parts of the system may be less than 6 million years old or as young as 3.5 million years old. Fig. 20.21 is a Google Earth image looking southward along the range crest, which can be followed from Mount Pleasant (P) to Mount Lola (L) to Excelsior Mountain (E) to Mount Whitney (W). Some of the larger fault trends in the Sierra Nevada Frontal fault system are shown. Although shown as continuous, each of the faults actually consists of smaller fault segments. The fault zone will be described from south to north using Fig. 20.21. Fault numbers in parentheses correspond to Fig. 20.5.

The primary fault zone in the high southern part of the range, and the one most responsible for the great escarpment and the western tilt of the range, is the Southern Sierra Nevada Frontal Fault (SSNF, 32). This fault extends along the base of the great escarpment from the Garlock Fault (20) to the vicinity of Mono Lake (M) where range-front normal faults become less continuous and the escarpment becomes less imposing (<6200 feet of relief) and not as straight.

From the vicinity of Mono Lake northward to Mount Lola (L), faults that represent the Sierra Nevada Frontal fault system intersect the range-front escarpment at an angle in an *en echelon* (parallel, in-step) pattern. In this area, the Tahoe Frontal-West Lake Tahoe fault zone (TF, 43) is the only major fault oriented parallel with the range-front. The escarpment along the western shore of Lake Tahoe associated with this fault is about 3000 feet high. The primary en echelon faults are numbered 1 through 5. These faults have produced several fault-parallel mountain ridges that flare-out across the Nevada border at an angle to the Sierra Nevada range crest. One example is the Carson Range (C) along the east side of Lake Tahoe. These ridges are visible in Fig. 20.18. Detailed fieldwork has demonstrated that some of these faults were active prior to, during, and following deposition of 11.5- to 9.0-million-year-old volcanic rocks, a relationship that suggests the Sierra Nevada Frontal Fault system in this area had initiated by about 12 million years ago. A few of the faults show strike-slip displacement, a characteristic feature of the Walker Lane Belt discussed later in the chapter. Notice in Fig. 20.5 that the en echelon fault pattern extends southward to include the Wassuk Range (41) and White Mountains (36) faults.

Mountainous topography north of Mount Lola (L) steps eastward to the Diamond Mountains (D) to form two parallel escarpments, one along the eastern front of the Diamond Mountains bordered by the Honey Lake fault (HLF, 48), the other along the eastern front of the Sierra Nevada bordered by the Mohawk Valley Fault (MVF, 47). Peaks in the Diamond Mountains are less than 8000 feet but the escarpment overlooking Honey Lake is 3800 feet high. Mt. Lola is at an elevation of 9152 feet but only one peak farther north along the Sierra Nevada is more than 8000 feet. The escarpment above Mohawk Valley is about 3000 feet high. Fieldwork and the dating of displaced volcanic rocks suggest that both faults initiated less than 6 million years ago.

Mount Lola is also the location where the Sierra Nevada drainage divide (yellow line in Fig. 20.18) diverges from the range crest (white line) and extends to the crest of the Diamond Mountains (D). The Feather River flows westward from the drainage divide in the Diamond Mountains across the crest of the Sierra Nevada to the Central Valley (Fig. 20.1). It is the only river system that crosses the crest of the Sierra Nevada and because of this, some have located the eastern boundary of the Sierra Nevada on the east side of the Diamond Mountains so that the Feather River drainage is contained entirely within the Sierra Nevada. In this book, the boundary is located at the base of the eastern escarpment along

FIGURE 20.21 Google Earth image looking southward at the northern part of the Sierra Nevada. The thin orange line marks the northern limit of crystalline granitic and metamorphic rock in the Sierra Nevada. White lines are faults. The range crest is marked by Mount Pleasant (P), Mount Lola (L), Excelsior Mountain (E), and Mount Whitney (W).

the Mohawk Valley Fault (MVF, 47) such that the Diamond Mountains and Lake Tahoe are both part of the Basin and Range.

The Mohawk Valley fault (MVF, 47) leads northward into a wide zone of faults, collectively referred to in Fig. 20.21 as the Sierra Nevada-Cascade Range Boundary Zone (SCBZ, 51). These faults extend across the northern margin of the Sierra Nevada and possibly across the Foothills fault system (FFS, 52). They mark the northern margin of the Walker Lane Belt. The faults appear to be quite young. Field evidence suggests at least some of them initiated less than 600,000 years ago. They create a series of escarpments that step down toward Lassen Peak. The North Fork Feather River (f) cuts a path directly across the escarpments creating a canyon occupied by State Route 70, the Feather River Highway.

Crystalline rocks continue for a short distance north of the North Fork Feather River (f), but elevation drops below 5000 feet and the Sierra Nevada effectively ends at Deer Creek (DC). Farther north, elevation rises toward Lassen Peak. The thin orange line in Fig. 20.21 marks the northern limit of crystalline rock in the Sierra Nevada. All of the rocks north of this line are volcanic and are part of the Cascade Volcanic arc. Note the subtle change in topography across the line.

Sierra Nevada Uplift History

The Sierra Nevada is, in a sense, a lynchpin between the older subduction-related geology of California and the younger normal fault/strike-slip geology of the Basin and Range and Walker Lane Belt. Needless to say, the landscape history is complex and not fully understood. A more complete interpretation has emerged in the past few years based both on detailed fieldwork and on sophisticated methods of obtaining elevation, paleoclimate, and age data, as well as methods that can reveal relative rates and recency of uplift. Such methods include (U-Th)/He dating, oxygen and hydrogen isotope analysis, fossils, and river incision rate analysis. In a nutshell, the data suggest that the Sierra Nevada has been a topographic highland for more than 80 million years, however, it has not always been a mountain range with a steep eastern escarpment physiographically distinct from the Basin and Range.

In order to understand landscape history in more detail, we need to begin at a time 80 million years ago when subduction-related volcanism ended in the Sierra Nevada even though subduction off the west coast of California continued uninterrupted until initiation of the San Andreas Fault 29 to 27 million years ago. We know that volcanism ended in the Sierra Nevada because there is a gap in the age of volcanic rocks beginning roughly 80 million years ago and ending 16 million years ago when volcanism returned. At this point you should be asking yourself, if subduction off the west coast continued uninterrupted, why did volcanism end in the Sierra Nevada and where did it go? We will address these questions shortly. First, let us trace the landscape history since 80 million years ago.

Recall from Chapter 18 that as far back as 80 million years ago the Sierra Nevada formed the western margin of the Nevadaplano. The great eastern escarpment that today separates the Sierra Nevada from the Basin and Range was not yet in existence as evidenced by ancient west-flowing stream channels that crossed this boundary on their way to the paleo-Pacific ocean (Fig. 18.21). Apatite and zircon (U-Th)/He ages from west of Lake Tahoe and south of Yosemite Valley are consistent with relatively high exhumation rates on the order of 0.78 to 3.14 inches/100 years between 90 and 60 million years ago. Exhumation rates apparently decreased steadily during this time interval followed by much slower rates of 0.08 to 0.20 inches per 100 years from at least 60 million years ago almost to present day.

Apparent rapid exhumation from 90 to as late as 60 million years ago could be due to uplift and concomitant erosion, or it could be due to rapid erosion of a highland composed of weak volcanic rock without strong uplift. In either case, slow exhumation rates from 60 million years ago almost to present-day suggest limited uplift, which supports the idea that the high elevation Nevadaplano was already in existence prior to 60 million years ago. The geological evidence, therefore, suggests that the Sierra Nevada, 80 million years ago, was a westward-sloping, rolling upland probably with embedded volcanoes and a surface covered in volcanic material similar to what is shown in Fig. 18.21. The landscape at that time may have resembled the present-day volcanic landscape of the Central-Southern Cascade Mountains, the major differences being the extension of the upland eastward as the Nevadaplano, and the presence of an ocean to the immediate west at the present-day location of the Central Valley. The appearance of granitic debris in 65-million-year-old sediment that surrounds the Sierra Nevada indicates that, by then, the volcanic carapace had eroded and crystalline (granitic) rock was exposed. Wide exposure of crystalline rock (as opposed to volcanic rock) by 60 million years ago would have slowed erosion rates as implied by the exhumation data.

There is little geological evidence for widespread tectonic activity in the Sierra Nevada from 80 million years ago to as late as 16 million years ago when volcanism returned. The region remained a highland (the Nevadaplano) of uncertain elevation during that time span undergoing a lengthy period of erosion. The erosion surface that developed during this long time period is well preserved along the entire lower western slope of the range as well as in the area surrounding Kern River canyon. This ancient erosion surface was recognized as early as 1904 as a relict surface of abnormally low relief relative to surrounding modern river valleys. A small remnant of this erosion surface (r) is visible in Fig. 22.20 above Kern Canyon.

The erosion surface, although of low relief, certainly was not flat. Imbedded in the surface are ancient, now abandoned river channels partly filled with Eocene (primarily 50 to 40 Ma) gold-bearing river gravels that were exploited during the gold rush. Also present within these ancient, now abandoned river channels are Oligocene (primarily 31 to 23 Ma) volcanic ignimbrite deposits derived from Central Nevada and used to show that rivers once flowed westward from central Nevada across the Sierra Nevada to the Central Valley. These ancient river deposits are best preserved north of the Madera-Bishop area.

The abandoned river channels can be used to estimate paleoelevation and paleorelief as discussed later, but first we need to address our earlier question as to why volcanism ended in the Sierra Nevada 80 million years ago, where it went, and why it returned beginning 16 million years ago. Recall that 18 to 16 million years ago began a time of profound change in the Cordillera that included widespread normal faulting in the Basin and Range and volcanism on the Columbia Plateau and Snake River Plain associated with the Yellowstone hot spot. Although volcanism ended in the Sierra Nevada 80 million years ago, it did not end across the Cordillera. Instead, it shifted eastward to create part of the Idaho Batholith, and by 56 million years ago, the Sanpoil, Challis, and Absaroka volcanic fields coincident with the beginning of the ignimbrite flare-up (Chapter 17). The lithospheric plate that was subducting beneath California 80 million years ago was the Farallon plate. Recall from Chapter 5 that a sinking plate must reach a depth of 62 miles for melting to begin. The eastward shift in volcanism is understood to be the result of the shallowing of the Farallon subducting slab. Instead of sinking into the mantle within a few 100 miles of the trench, the Farallon plate appears to have pushed itself horizontally beneath the North American lithosphere, a process known as underplating as shown schematically in Fig. 20.22, and alluded to in Fig. 5.13. Because of the shallow angle of subduction, the Farallon slab could not generate magma until it was well to the east of the Sierra Nevada. As noted in Chapters 5 and 17, underplating of the Farallon slab may be responsible for the Laramide orogeny, and the sinking of the slab may be responsible for the ignimbrite flare-up.

A critical piece of evidence regarding the ignimbrite flare-up is the progression in the age of andesite and rhyolite volcanism across the Cordillera. Fig. 20.23 shows a set of lines that delineate the successive location of volcanic fronts between 50 and 20 million years ago. The figure shows that volcanism was present in eastern Washington and Idaho 50 million years ago, and that by 20 million years ago volcanism had spread progressively southward across Nevada. The figure also shows the migration of volcanism northwestward from the New Mexico-Mexico border region 43 million years ago to southern Nevada 21 million years ago. One way to explain these features is to suggest that the eastern part of Farallon slab began to sink into the mantle causing the plate to roll back on itself as shown in Fig. 20.22. The sinking plate left a void below the North American lithosphere that was filled with hot, rising mantle rock, which melted the overlying North American crust, causing progressively younger silicic volcanism toward the west as rollback progressed westward.

Rollback reached the Sierra Nevada by 16 million years ago and the sinking Farallon slab was again generating volcanism in and in close proximity to the Sierra Nevada. The San Andreas Fault was in existence by this time but only in southern California. Sixteen million years ago the Mendocino triple junction, which separates the San Andreas Fault from the Juan de Fuca plate (the northern relict of the Farallon plate), was located in the vicinity of San Luis Obispo, well to the south of its present location, implying that most of the Sierra Nevada was above the Juan de Fuca plate subduction zone. 40Ar/39Ar ages indicate that volcanism lasted from about 16 to 3 million years ago ending first in the southern Sierra Nevada and then in the northern Sierra Nevada as the Mendocino triple junction migrated northward toward its present location. These young volcanic rocks are referred to as the Mehrten Formation. However, rather than forming large volcanic cones such as modern day Mt. Hood and Mt. Rainier, field evidence indicates that volcanism was in the form of numerous small andesitic and rhyolitic domes. In some areas, these small domes collapsed into blocky lava flows that entered ancient river channels. Amazingly, these are some of the same river channels previously filled with gold-bearing river gravels and then with volcanic ignimbrite deposits derived from central Nevada.

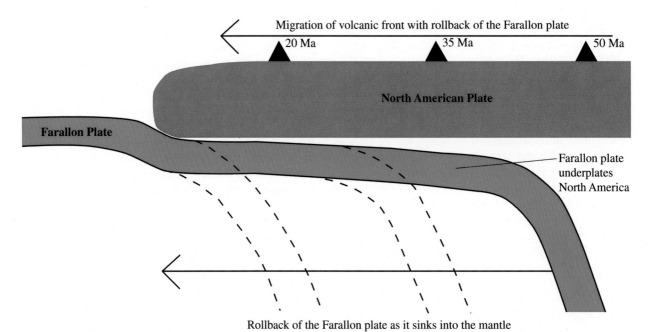

FIGURE 20.22 Cross-section sketch that shows the Farallon plate underplating North American plate. Subsequent rollback of the Farallon plate causes migration of the volcanic front.

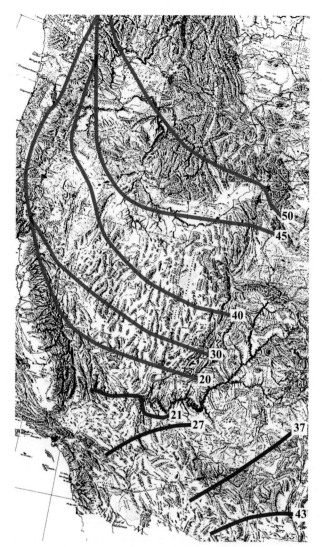

FIGURE 20.23 Landscape map that shows the southward and westward migration of successive volcanic fronts toward California. Numbers (in millions of years) indicate how long ago volcanism began at each location. Based on Dickinson (2006) and Humphreys (1995). Volcanic fronts are superimposed on present-day landscape and not on the narrow pre-Basin and Range landscape in existence at the time of volcanism.

Paleoelevation and paleorelief refer to the elevation and relief of an ancient landscape at some specified time in the past. We cannot easily determine paleoelevation, but clues to paleorelief are provided by detailed study of some of these gold-bearing ancient river channels. First, let's think about modern-day river dynamics. As a modern river incises its channel, it will remove any sediment that may have accumulated at the bottom of the channel such that the river will flow directly over bedrock. Thus, if we find a modern-day river flowing directly over bedrock, we can safely assume that the stream channel is actively attempting to cut downward into bedrock. If the stream is actively incising its channel by cutting into bedrock, the stream is also generating relief. Conversely, if we find a layer of sediment on the channel bottom between the water and bedrock, we can assume that incision occurred prior to deposition of that sediment. Given these relationships, we can estimate paleorelief at the time deposition first begins in an ancient river channel as equal to the elevation of a high point of crystalline rock directly above the river channel minus the elevation of bedrock at the bottom of the river channel. Deposition in ancient river channels of the Sierra Nevada began 50 to 40 million years ago with gold-bearing gravel. Paleorelief 50 to 40 million years ago (just prior to deposition of ancient river gravel), therefore, can be estimated as shown in Fig. 20.24. The analysis assumes there are no major folds or faults between the ancient stream channel and the high point of crystalline rock. It is a minimum estimate because it does not include erosional lowering of the high point since 50 to 40 million years ago.

When river incision analysis is applied to the Sierra Nevada, it is found that paleorelief 50 to 40 million years ago was less than 1300 feet (400 m) north of Sacramento, increasing to 3200 feet (1000 m) southward to the vicinity of Merced. Ancient stream channels are absent south of Merced, but other methods used to estimate paelorelief such as low-temperature isotopic dating, suggest that potentially more than 3200 feet of paleorelief may have been present. Paleoelief is related to paleoelevation in the sense that elevation must be at least as high as relief assuming all land is above sea level. Greater relief favors higher elevation. Given our estimates of paleorelief, we can reasonably estimate that elevation 50 to 40 million years ago was more than 3000 feet north of Sacramento, more than 6000 feet southward to the vicinity of Merced, and possibly more than 10,000 feet farther south. These estimates suggest high elevation 50 to 40 million years ago, particularly in the southern part of the Sierra Nevada. The higher elevation in the south may have prevented rivers from ever crossing the southern Sierra Nevada at any time in the past 80 million years. Alternatively, it could be that rivers did cross the southern Sierra Nevada, but greater uplift and subsequent erosion have removed evidence of their existence.

Ancient stream channels in the Sierra Nevada are now filled with sediment and are no longer part of the modern-day landscape. Presumably, the river channels were abandoned between 16 and 3 million years ago during and following deposition of volcanic rocks of the Mehrten Formation. The great eastern escarpment of the Sierra Nevada is a significant part of the present-day landscape. This escarpment rose concomitant with collapse of the Nevadaplano and development of the Sierra Nevada Frontal fault system. Field analysis in the Basin and Range suggests that collapse of the Nevadaplano began about 17 million years ago in central Nevada and progressed westward into the Walker Lane Belt by 12 million

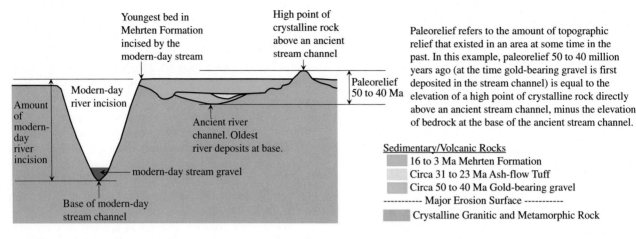

FIGURE 20.24 Schematic diagram that shows how topographic paleorelief and river incision are calculated. The ancient river channel contains deposits of gold-bearing gravel, ash-flow tuff, and Mehrten Formation. Based broadly on Wakabayashi, 2013.

years ago. We noted earlier that the Sierra Nevada Frontal Fault System began activity between 12 and 6 million years ago and that it was fully active by 3.5 million years ago. By 6 million years ago, rivers south of Lake Tahoe that had once crossed the Sierra Nevada were cut off, signaling the presence of an eastern escarpment. Thus, based on the timing for faulting and stream channel abandonment, we can say that the modern-day Sierra Nevada landscape is no older than 12 million years, and potentially younger than 6 million years. Let us now look at evidence for recent river incision and changes in elevation and relief.

We noted earlier that global positioning and space-based radar techniques indicate that the entire Sierra Nevada is undergoing uplift at a maximum rate of 4 to 8 inches per 100 years. We can use the shape of modern-day river valleys on the western slope of the Sierra Nevada to determine where the highest rates of uplift are concentrated. In general, a steep river valley that is also narrow and with high relief suggests more rapid or sustained downcutting (incision) relative to one of similar size without these characteristics. The cause of downcutting could be more rapid or sustained uplift, but it could also be due to climate or rock type variations. In the case of the Sierra Nevada, we can to a certain extent discount climate and rock type variation because climate at a particular elevation is fairly uniform, and because the rocks are uniformly crystalline. We can discount the effects of glaciation by looking only at valleys below the elevation of glaciation. When this type of analysis is applied, we find steep river valleys on the western slope south of Madera and from the vicinity of Sacramento-Lake Tahoe northward, implying that these are areas where recent uplift has been most rapid.

There is additional evidence for recent uplift in these areas. The area north of Sacramento-Lake Tahoe is replete with small escarpments and young active faults associated with the Sierra Nevada-Cascade Range Boundary Zone (SCBZ, 51). The area south of Madera includes the topographically highest part of the range with the active Southern Sierra Nevada Frontal fault (32) along its eastern flank. Beryllium-10 cosmogenic radionuclide surface-exposure dating along this frontal fault suggests that vertical (up and down) displacement over the past several 100,000 years has occurred at a rate of 0.79 to 1.18 inches/100 years.

We can apply river incision analysis to modern-day rivers to estimate when the most recent phase of uplift began. It should be obvious that river incision must be younger than any rock or sediment into which incision has occurred, and older than any sediment deposited at the bottom of the river channel. In Fig. 20.24, modern-day river incision must have occurred after deposition of the youngest bed in the Mehrten Formation that is incised by the river, but prior to deposition of gravel presently at the bottom of the river channel. If the age of deposition of both the gravel and the Mehrten Formation can be determined, then the timing and rate of incision of the modern-day channel can be estimated in a manner similar to Fig. 9.9. Unfortunately, relationships regarding the timing of incision in some areas are not as clear-cut or as precise as we would like. The data that we do have suggest that incision and formation of our modern-day stream channels began earlier in the south than in the north. If we assume incision occurs as a result of uplift, then renewed uplift began in the south and migrated northward consistent with northward tilting of a semi-rigid block. In the northern Sierra Nevada, incision began after 3.0 million years ago. In the area between Lake Tahoe and Mono Lake, incision appears to have begun between 3.6 and 4 million years ago. Incision just south of Mono Lake is estimated to have begun between 6 and 10 million years ago. Farther south, incision could have initiated as early as 20 million years ago.

In the southern Sierra Nevada, south of Mono Lake, there is stream incision evidence for a second, younger period of uplift that began within the past 3.5 million years (coincident with uplift in the northern Sierra Nevada). Since 3.5 million years ago, rivers in the southern Sierra Nevada have incised as much as 1300 feet into the western slope at a rate of about 0.45 inches per 100 years. This presumed active pulse of uplift continues today with faulting along the Southern Sierra Nevada fault system (32) and the along the Kern Canyon fault (31). Total estimated uplift of the Sierra Nevada over the past 20 million years is 3000 to 6500 feet, with greater uplift in the south. This amount of uplift can be added to any pre-existing elevation to bring the Sierra Nevada to its current elevation.

Putting it all together, the Sierra Nevada existed as a volcanic highland at the western edge of a major plateau known as the Nevadaplano beginning prior to 80 million years ago. The granitic rocks that today make up much of the Sierra Nevada were exhumed from below the volcanic carapace by 60 million years ago. Elevation was high across the Basin and Range and Sierra Nevada throughout the existence of the Nevadaplano until at least 17 million years ago when the Nevadaplano in central Nevada began to collapse. The elevation of the Nevadaplano prior to collapse was higher in central Nevada than in the Sierra Nevada in order for streams to flow westward across the Sierra Nevada to the ocean. Although largely tectonically quiet from 80 million to 16 million years ago, river erosion between 50 and 40 million years ago, partly in response to displacement on the now inactive Western Sierra fault system, created more than 3200 feet of relief in the southern Sierra Nevada, and at least 1300 feet of relief in the north. Elevation at that time may have ranged from 6000 feet in the north to more than 10,000 feet in the south. The modern landscape began to develop possibly as early as 20 million years ago when a pulse of river incision, interpreted as uplift, began in the south and progressed northward reaching the northern Sierra Nevada by 3 million years ago. Some of the faults in the northernmost part of the range appear to have initiated less than 600,000 years ago. The Sierra Nevada Frontal fault system was active beginning about 12 million years ago, resulting in development of the great eastern escarpment by about 6 million years ago. A second enhanced uplift pulse has affected the southern Sierra Nevada during the past 3.5 million years. This pulse is active and has contributed to northward tilting and an increase in elevation above 14,000 feet thus putting the final touches on the present-day landscape. Currently, the Sierra Nevada is undergoing broad uplift along its entire length at a maximum rate of 4 to 8 inches per 100 years.

THE WALKER LANE BELT

If you look closely at any physiographic map of eastern California-Nevada, such as Figs. 20.1 and 20.18, you will notice that the trend of mountain ranges in the Basin and Range directly east of the Sierra Nevada is different from mountain ranges in central Nevada. For example, the Stillwater, Shoshone, and Toquima ranges in central Nevada all trend north to northeast, as do surrounding ranges. This trend changes to a variable but dominantly northwest direction along a line that is most clearly seen in the vicinity of Pyramid and Walker Lakes. The break in mountain trends is shown with a thick green line in Fig. 20.5. This line represents the approximate eastern limit of a transtensional zone of active dominantly right-lateral strike-slip faults and normal faults known as the Eastern California—Walker Lane Belt. The belt is not defined by a single through-going fault, or by the thick green line itself, but rather by an array of individual faults that cover the entire area from the Sierra Nevada eastward to the thick green line.

The system of faults associated with the Walker Lane Belt diverges from the San Andreas Fault in the vicinity of the Salton Sea as part of the Eastern California shear zone and extends northward along the east side of the Sierra Nevada to the vicinity of the Madeline Plains where the fault system appears to die out. The complex interaction of strike-slip and normal faulting in this area has produced some of the most recognizable landforms in the region. Death Valley, Mono Lake, and Lake Tahoe are all essentially pull-apart basins rimmed with normal faults that have formed in a manner similar to the Salton Sea (Fig. 20.3C and D). Total displacement on all faults within the system is on the order of 30 to 60 miles in southern California, 35 to 50 miles in the Lake Tahoe region, and essentially zero in the north, where the fault system dies out.

The time for the initiation of faulting can be deduced from field evidence. There is an abundance of sedimentary and volcanic rock as well as unconsolidated and semi-consolidated sediment that spans the age range from 35 million years ago to the present. Each has a slightly different structural history depending on its age and location. In simple terms, if rock or sediment is offset by a fault, faulting must be younger than the age of the rock or sediment. If rock or sediment cuts across a fault without offset, faulting must have occurred before deposition of the rock or sediment. A second line of evidence can be found in the many fault-generated (pull-apart) basins. Sediment and volcanic rocks deposited in a basin must be younger than the initiation of faulting that created the basin. Some examples of cross-cutting relationships are shown in Fig. 20.25. When these relationships are applied to the many faults in the Eastern California–Walker Lane Belt, they suggest that faulting has been active since about 12 million years ago along much of the belt except for the very northern part where faulting may have begun only in the past 3.5 million years. Beginning 11 million years ago, faults began to restrict rivers from crossing the Sierra Nevada thus signaling initial development of the great eastern escarpment. Rivers south of Lake Tahoe were permanently cut off by 6 million years ago.

Satellite measurements indicate that the entire area, from the Walker Lane Belt westward to the San Andreas Fault, is moving as a single block separate from the North American plate. In other words, this area, which includes the Sierra Nevada, the Central Valley, the Mojave Desert, part of the Coast Ranges, and the eastern part of the Transverse Ranges, is a new tectonic plate that is in the process of forming. It is a microplate known as the Sierran microplate. Geodetic and global positioning system measurements indicate that the Sierran microplate is moving northwestward at a rate of 3.6 feet per 100 years and rotating counterclockwise relative to the North American plate. Such movement creates shear and tensional stresses (transtension) in the area east of the Sierra Nevada, which manifest themselves in the form of strike-slip and normal faults, thus creating the Walker Lane Belt. Although the western boundary of the microplate is distinct and sharply defined by the San Andreas Fault, the

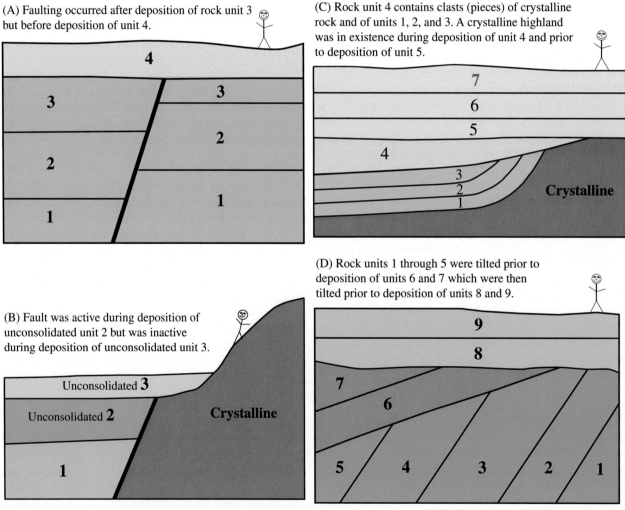

FIGURE 20.25 Examples of cross-cutting relationships and how they are used to determine age of deformation and mountain building. Numbers in all diagrams denote sedimentary rock units.

eastern boundary is not fully detached from the North American plate, and is represented by the entire array of strike-slip and normal faults that form the Walker Lane belt. We humans have come along at just the right time to see the Sierran miocroplate in the process of developing.

There are at least two possible end scenarios regarding the Walker Lane. The first is for the Sierran microplate to be ripped from the North American plate to become a separate plate independent from both the Pacific and North American plates. A second and perhaps more likely scenario is for the San Andreas Fault to jump eastward to the Walker Lane, thereby transferring the Sierran microplate to the Pacific plate. The western, eastern, and southern boundaries of the Sierran microplate are fairly well defined by the San Andreas, the Eastern California, and the Walker Lane fault zones. The northern boundary, however, is poorly defined. Recall that the Mendocino triple junction is migrating northward. If the Walker Lane continues to grow lengthwise in a northwest direction, it could meet up with the triple junction somewhere along the Oregon coast. When this occurs, the eastward jump will be complete. Under this scenario, the Walker Lane Belt would become the new San Andreas transform plate boundary, the Sierran microplate would transfer to the Pacific plate, and the San Andreas Fault would become inactive or would become an active interplate fault. Such a jump would constitute the largest eastward jump in the history of the San Andreas Fault. But, that possibility is for the future. In the present, the Sierran plate is rotating counterclockwise producing compression at its northern end that may in part be responsible for active uplift in the Klamath Mountains and at the northern edge of the Central Valley.

The Walker Lane belt is one of the most tectonically active regions in the United States. Nearly every mountain range and every valley is bordered on one or both sides by a fault that has been active in the past 1.6 million years. It is also an arid region within the Sierra Nevada rain shadow where precipitation, especially in its southern part, rarely exceeds 10 inches per year. The result is landscape where hundreds of well-preserved active fault scarps and lineaments are visible. There is perhaps no better place in the United States to study structure-controlled landscape. We can thus summarize the Walker Lane Belt as a transtensional array of active, intersecting right-lateral strike-slip and normal faults located east of the Sierra Nevada that has produced surface uplift, subsidence, and lateral offset resulting in mountains that trend north-northwest rather than the more typical north-northeast trending mountains of the interior Basin and Range.

Our discussion of the Walker Lane belt will be limited primarily to the region from Mono Lake southward to the Garlock fault, an area referred to in the literature as the Inyo-Mono section or the southern Walker Lane. Most geologists can agree on a general history of the area, but in detail there are several slightly different interpretations. Instead of reciting every possible interpretation, we will look at the landscape, rely on exhumation ages where available, and draw our own conclusions consistent with the geology and consistent with interpretations put forth by geologists who have done the field work and who have studied these problems in detail. We will begin with an overview of the landscape history in the Walker Lane belt, and then look in more detail at individual mountain areas in the Inyo-Mono section.

A Tale of Three Landscapes

There is evidence that the Inyo-Mono section has evolved through three landscape styles within the past 17 million years. This, to me, is quite amazing because it shows how quickly a new phase of tectonic activity can reincarnate an older landscape. The oldest landscape is the Nevadaplano, which was likely in existence prior to 80 million years ago. The Nevadaplano was replaced in at least some areas of the Walker Lane belt with a now relict extensional landscape composed of normal fault-block mountains and basins (without strike-slip faults) that developed primarily between 17 and 12 million years ago. The normal faults associated with this landscape are responsible for initial collapse and destruction of the Nevadaplano. The present-day transtensional landscape began to develop 12 million years ago and remains active today.

The elevation map shown as Fig. 20.18 provides circumstantial evidence to suggest that collapse of the Nevadaplano is farther advanced in the southern part of the Walker Lane belt than in the central and northern parts. The evidence is based on the elevation of intermontane valleys (valleys between the mountains). Valleys in the White Mountains area (Wm) and areas north of Mono Lake (M) are above 3700 feet, and in some areas above 5000 feet. Lake Tahoe (t) and Mono Lake (M) are both above 6200 feet. Walker Lake (w) is above 3900 feet, and Pyramid Lake (p) is above 3700 feet. South of the White Mountains, Owens Lake (ol) is at 3560 feet, but valleys to the east are generally less than 3000 feet. Saline Valley (s) drops to less than 1100 feet, and Death Valley (dv) to less than 100 feet. Indian Wells Valley (Iv) lies below 2300 feet and Searles Valley (sv) below 2000 feet. Valley elevation remains less than 3000 feet across the southern part of the Basin and Range. If we presume that the entire area was part of a high Nevadaplano, and that elevation at that time was reasonably uniform, we can suggest that the southern part of the Walker lane collapsed via extension and normal faulting to a greater extent than the northern part. There are two possibilities. Collapse began earlier in the south, or the rate of collapse was more rapid in the south.

Evidence for early collapse in the south can be found in the Death Valley area and in areas farther southeast where extreme extension began prior to 13 million years ago with development of metamorphic core complexes as outlined in Chapter 18. This early extension can be contrasted with the Wassuk Range (Wr) in the central Walker Lane where field evidence suggests initial development of Basin and Range landscape could have begun as late as 10 million years ago. It is possible, therefore, that the area surrounding the Wassuk Range was not active during the 17 to 12 million-year-old phase of extension that occurred elsewhere in the Basin and Range.

The youngest and presently active phase of landscape development is dominated by transtensional stresses. There is evidence from the north-central Walker Lane that transtension first appeared circa 12 million years ago. However, in the Inyo-Mono section, transtensional landscape began to form less than 6 million years ago and is so young that it has not yet completely reincarnated the earlier normal fault-dominated Basin and Range-style extensional landscape. As a result, todays Inyo-Mono section is a hybrid between an actively forming, very young, structure-controlled, transtensional landscape, and a relict extensional normal fault-dominated Basin and Range-style landscape. These landscapes are not entirely different as both involve normal faults. Because of this similarity, the younger tectonic regime has in some cases enhanced, rather than reincarnated, the older landscape.

The Inyo-Mono Section

Fig. 20.26 uses a color scheme to show major active faults in the Basin and Range and Sierra Nevada. All of the faults shown on this map have been active during the past 1.6 million years and many have been active in the past 130,000 years. The primary faults are numbered 1 through 15 and are listed in Table 20.2 along with other abbreviations. Numbers to the right of the primary faults correspond with numbered faults in Fig. 20.5. The area contained in Fig. 20.26 is approximately 153 miles east to west and 182 miles north to south.

The Inyo-Mono section extends from the Garlock fault (15) in the south to the Coaldale fault (1) in the north. These are the only two left-lateral strike-slip faults shown in Fig. 20.26. The Garlock fault has been active over the past 11 million years with 40 miles (64 km) of displacement that separates mountainous topography of the Inyo-Mono section in the north from the rather subdued topography of the Mojave Desert. The fault is important because left-lateral displacement creates additional room to accommodate extension at the southern end of the Inyo-Mono section. The Coaldale fault (1) is part of a zone of left-lateral strike-slip faults known as the Excelsior-Coaldale Section or Mina Deflection. An interesting oddity is that rocks north of the Coaldale fault appear to be displaced to the right (east) across the Mina Deflection, not to the left as expected along left-lateral faults. The relationships suggest that the Mina Deflection is an ancient (pre-Nevadaplano) right-lateral fault zone that had previously displaced rocks and older structures to the right. Thus, we can explain the apparent contradiction of right-lateral offset along a left-lateral fault by suggesting that the Coaldale Fault has only recently reactivated as a left-lateral fault and that limited left-lateral displacement has preserved some of the geology and landscape characteristics that developed during earlier, more extensive, right-lateral displacement.

Strike-slip faults within the Inyo-Mono section are dominantly right-lateral. There are two major northwest-trending fault zones, the Owens Valley-White Mountains fault zone (3, 4) and the Death Valley-Fish Lake Valley fault zone (11). Together, they form a triangular-shaped area, wide in the south, that converges northward toward the White Mountains. The steep dip of these faults has produced two semi-continuous linear valleys, both of which are clearly evident in Fig. 20.26. Between these two faults are numerous small normal fault-block mountains and valleys oriented in an en echelon orientation consistent with a pull-apart origin within a transtensional regime. These include faults bordering Deep Springs Valley (Dsv), the Last Chance Range (Lcr), the Cottonwood Mountains (Cm), and the northern Panamint Range (Pr). Additionally, several of the strike-slip faults bend into transtensional orientation to create normal faults. The Death Valley strike-slip fault (11) appears to bend into the Black Mountains normal fault (12) in a manner similar with Fig. 20.3E. Similarly the Hunter Mountain (7) and the White Mountains (3) strike slip faults also appear to bend to become the Saline Valley (6) and Queen Valley (2) normal faults respectively. Some of the normal faults are not perfectly oriented, possibly due to the presence of pre-existing fractures, faults, or layering in the rock, all of which can influence and deflect the orientation of subsequent faults. Many of the faults likely include a component of oblique-slip. The addition of these influences, coupled with the presence of relict basin and range-type landforms, creates landscape that is less predictable than otherwise might be expected.

Even at the scale of Fig. 20.26, it is reasonably clear that many of the small normal fault-block mountains are asymmetric with sharp, steep range fronts. They appear to be recently uplifted, such as those depicted in Fig. 18.16A. Such landscape is in contrast to landscape east of the Walker Lane Belt in Fig. 20.26 where we find isolated, low-lying ranges with irregular mountain fronts

FIGURE 20.26 Shaded relief elevation image oriented north showing the Inyo-Mono section of the Walker Lane belt. Refer to Table 20.2 for abbreviations and symbols. Numbers in parentheses in Table 20.2 correspond with active faults shown in Table 20.1 and Fig. 20.5.

TABLE 20.2 List of Abbreviations for Fig. 20.26

a - Location of Fig. 20.32	M - Mono Lake
Ad - Amargosa Desert	ML - Mud Lake
Ar - Argus Range	Mnr - Montezuma Range
Av - Adobe Valley	Mr - Monitor Range
Bm - Black Mountains	NDv - Northern Death Valley
Cf - Cactus Flat	NFLv - North Fish Lake V.
Ch - Candelaria Hills	Om - Owls Head Mountains
Cm - Cottonwood Mtns.	ol - Owens Lake
Cr - Coso Range	OV - Owens Valley
Cr - Cactus Range	Pm - Pahute Mesa
Cv - Clayton Valley	Pmm - Palmetto Mountains
Dv - Death Valley	Pnv - Pilot Knob Valley
Dsv - Deep Springs Valley	Pr - Panamint Range
Ev - Eureka Valley	Pv - Panamint Valley
FLv - Fish Lake Valley	RV - Ralston Valley
Fm - Funeral Mountains	S - Mount Sill
Gf - Gold Flat	SCV - Stone Cabin Valley
Gm - Gold Mountain	Sf - Sarcobatus Flat
Gvm - Grapevine Mountains	Slr - Saline Ranges
Grm - Granite Mountains	Slv - Saline Valley
Im - Inyo Mountains	Sm - Stonewall Mountain
Iv - Indian Wells Valley	SPr - Silver Peak Range
Kc - Kern Canyon	Sr - Slate Range
La - Mount Langley	Sv - Searles Valley
Lcr - Last Chance Range	Swf - Stonewall Flat
Lm - Lone Mountain	VT - Volcanic Tableland
LV - Long Valley Caldera	W - Mount Whitney
Mcr - Monte Cristo Range	Wm - White Mountains

<u>Towns</u>
B-Bishop
Bt-Beatty
BP-Big Pine
I-Independence
LP-Lone Pine
P-Porterville
O-Orosi
R-Ridgecrest
S-Stovepipe Wells
T-Tonopah
W-Wasco

<u>Active Faults</u> (color-coded)
(normal faults written in black ink)
1-Coaldale (37)
2-Queen Valley
3-White Mountains (36)
4-Owens Valley (35)
5-Southern Sierra Nevada Frontal (32)
6-Saline Valley (33)
7-Hunter Mountain (33)
8-Panamint Valley (33)
9-Southern Panamint Valley (33)
10-Ash Hill
11-Death Valley-Fish Lake Valley (34)
12-Black Mountains (34)
13-Southern Death Valley (34)
14-Stateline
15-Garlock (20)

Eastern boundary of the Walker Lane Belt and the Sierran microplate.

<u>Color-Coded Faults</u>

Normal fault Right-lateral strike-slip fault Left-lateral strike-slip fault

that border wide valleys. In spite of the presence of active faults, landscape in this area more closely resembles that of Fig. 18.16C where inactive mountain blocks are buried in their own debris. Perhaps this area is just now awakening from a long period of inactivity.

Volcanic rocks cover 30% to 40% of the area east of the Sierra Nevada in Fig. 20.26. These rocks are important because they are variably offset by faults (or not offset) and because they are easily dated. Such characteristics help immensely in deciphering the tectonic and landscape history of the area. These rocks can be used as a baseline to determine how much extension is occurring in the region. We noted in Chapter 17 that the Long Valley Caldera is surrounded by an abundance of volcanic ejecta in the form of the Bishop Tuff. This rock is brittle and seems to record every inch of extension in the Walker Lane Belt. On the Volcanic Tableland (VT) alone there are at least 226 fault scarps. Most are oriented 10 to 20 degrees west of north and show displacement on the order of a few feet to several hundred feet. Many are readily visible in Google Earth images. The Bishop Tuff, which underlies the Volcanic Tableland, is approximately 758,900 years old. The cumulative amount of extension on all faults across the Tableland is at least 950 feet (290 m). These numbers equate to an average of 1.5 inches of extension on the Volcanic Tableland every 100 years. Let us now look first at the White Mountains-Inyo Mountains region followed by the Death Valley-Panamint valley region.

White Mountains

The White Mountains (Wm, Fig. 20.26) are unique for several reasons. It is the only mountain range in the contiguous US outside the Sierra Nevada, Cascade volcanoes, and Colorado Rockies with a summit area above 14,000 feet (White Mountain Peak, 14,252 feet). In addition, there are seven peaks above 13,000 feet. It is one of the driest high mountain ranges in the United States if not the world, with only about 12 inches of precipitation per year, mostly in winter. Although the dry barren mountain summit is easily reached by foot or mountain bike, it towers more than 9000 feet above valleys to the east and west. The northeastern side of the mountain is glacially scalloped, especially between White Mountain Peak and Boundary Peak (13,146 feet), the highest point in Nevada.

As an interesting aside, the White Mountains are host to the Ancient Bristlecone Pine Forest where one can admire the oldest trees on Earth. Bristlecone pines grow throughout the arid west, and regularly reach ages of 1000 to 2000 years. In the White Mountains, they approach 5000 years old because they grow in a harsh, dry climate at elevations between 10,100 and 11,200 feet on a dolomite soil that lacks important nutrients. Virtually nothing else can grow under these conditions, so the trees have no competition. Bristlecone pines are particularly resilient trees that seemingly refuse to die. If a tree blows over, a branch will burrow into the ground to create a root system. If part of a tree dies, another part survives to the point where the trees resemble living driftwood. An example is shown in Fig. 20.27.

The White Mountains form part of the miogeoclinal sedimentary rock sequence located just south and east of the boundary with accreted terranes (line 3, Fig. 18.7). Miogeoclinal rocks are exposed in the southwestern part of the range underlain with sedimentary rock of the Precambrian sedimentary/volcanic rock succession. The northern part of the range and most of the high peaks consist of Mesozoic granitic rocks associated with the Sierra Nevada batholith. These rocks intrude the miogeoclinal rocks resulting in areas of metamorphism. White Mountain peak itself consists of Mesozoic volcanic rocks, in part metamorphosed. A small area of crystalline shield rock is exposed north of White Mountain Peak surrounding Indian Peak. All of the rocks are older than 66 million years, and are unconformably overlain with young volcanic/sedimentary rocks between 12.3 and 4.0 million years old.

Fig. 20.28 is a Google Earth image of the White Mountains looking south-southeast. The range is less than 60 miles north to south and no more than 25 miles east to west. From this figure it appears that the White Mountains form an asymmetric, east-tilted fault-block mountain flanked on nearly all sides by active, range-front faults. The west side of the mountain boasts the active White Mountains right-lateral strike-slip fault (3), which forms a steep mountain-front rising from 5000 feet in Hammil and Chalfant valleys, to more than 14,000 feet in about 6 miles. At its northern termination, the White Mountains fault, and the mountain itself, bend to the east. The change in orientation results in a change from strike-slip displacement to normal fault displacement, and the fault is renamed the Queen Valley fault (2). Southward, the White Mountains fault (3) steps to the right where right-lateral displacement continues along the Owens Valley Fault (4), which was made famous following a strong 1872 earthquake centered near Lone Pine in which land areas were displaced right-laterally 20.0 ± 6.0 feet and vertically 3.3 ± 1.6 feet. The fault scarp remains visible in Owens Valley.

On the east side of the mountain, the right-lateral Death Valley-Fish Lake Valley strike-slip fault (11) marks an sharp boundary between Fish Lake Valley (FLv) and a less steep eastern White Mountain front that rises from 5000 feet to more than 14,000 feet in 10 miles. The Death Valley-Fish Lake Valley Fault does not extend

FIGURE 20.27 Bristlecone Pine tree in the Patriarch Grove, White Mountains, California.

Fault colors and numbers follow Table 20.2; **1**-Coaldale, **2**-Queen Valley, **3**-White Mountains, **4**-Owens Valley, **5**-Southern Sierra Nevada Frontal, **10**-Death Valley-Fish Lake Valley. Ball and bar on down-thrown side of normal faults. **BP**-Big Pine, .
Dsv -Deep Springs Valley and Deep Springs normal fault; **Epf**-Emigrant Peak normal fault; **Gmf**-Gold Mountain normal fault; **Lf**-Lida fault.
X and X' are the approximate locations of fission track and (U-Th)/He samples (from Stockli et al., 2003).

FIGURE 20.28 Google Earth image looking south-southeast at the White Mountains, California.

the length of the White Mountains. Instead it dies out into a series of small faults along the northeast flank.

Timing of Uplift

Given our understanding of how normal faults create asymmetric mountains, we should have expected to see a normal fault along the base of the steep western front of the White Mountains. Instead we see a strike-slip fault (3). One explanation is that the asymmetric morphology is a relict of the earlier normal fault-controlled landscape and has not yet undergone complete reincarnation under the newer transtensional regime. In other words, it is possible that the White Mountains fault (3) was, at one time, a normal fault that was later reactivated as a strike-slip fault. There is field evidence to support this contention. When rocks are displaced along a fault, they will bend and scrape in a direction consistent with rock displacement. Scratches (known as slickenlines) oriented close to horizontal indicate strike-slip displacement. Scratches oriented down the dip of a fault plane indicate dip-slip displacement (normal- or reverse-slip). Both orientations are present on rocks in the White Mountains fault zone, and they suggest that normal fault displacement occurred prior to strike-slip displacement. Additionally, geophysical evidence indicates that Hammil and Chalfant valleys both tilt eastward toward the White Mountains and contain up to 7000 feet (2 km) of sediment. An asymmetric shape to both the mountain and valleys is a characteristic feature of normal faults (e.g., Fig. 18.4). The correlation of physiography with normal fault processes as outlined in Chapter 18 suggests that the White Mountains formed when the White Mountains fault was active as a normal fault, and that its asymmetric shape has not changed appreciably since that time irrespective of the younger strike-slip regime.

Field relationships and apatite cooling ages can be used to place limits as to when the White Mountains fault was active as a normal fault, and when it was later reactivated as a strike-slip fault. Earlier we stated that the White Mountains consist of older (pre-66 Ma) sedimentary, granitic, and metamorphic rocks unconformably overlain with young (12.3 to 4.0 Ma) volcanic/sedimentary rocks. Field observation suggests that the erosion surface (the unconformity) between these two rock units was close to horizontal, and that there was little in the way of relief during deposition of the 12.3 million-year-old volcanic rocks. In other words, there is nothing to suggest that the mountain range was in existence at this time. The White Mountains had not yet formed.

Because the 12.3 million-year-old volcanic rocks were roughly horizontal when deposited, the present-day tilt of these rocks provides a means to estimate the eastward tilt of the mountain since 12.3 million years ago. Field data suggest 15 degrees of eastward tilt in the central part of the range increasing to 25 degrees in the northern part. In order to produce this much tilt, there must be approximately 5 miles (8 km) of normal fault displacement along the White Mountains fault since 12.3 million years ago. Conversely, the younger, 4.0-million-year-old volcanic rocks were deposited on a sloping surface as indicated by the filling of paleochannels and the presence of associated conglomerate deposits. The relationships suggest that the White Mountains were in existence at the time these rocks were deposited. Thus, we can argue that the White Mountains formed between about 12 and 4 million years ago. The actual rate of uplift, and the time at which the White Mountains reached an elevation similar to today, is not constrained. If we assume 5 miles of normal fault displacement over 8 million years (from 12 to 4 million years), then the average displacement rate is nearly 4 inches per 100 years. If we assume that the mountains grew quickly and maintained their elevation, then the initial rate of uplift could have been significantly greater than 4 inches per 100 years.

We can apply apatite fission track and (U-Th)/He ages in a manner similar to our discussion of the Wasatch fault in Chapter 18 to determine more precisely when normal faulting and uplift occurred, and when strike-slip faulting began. Fig. 18.4 shows how rock x, in the lower plate of a normal fault, can be exhumed to the surface. This is an example of tectonic exhumation where displacement on the fault (not erosion) causes the rock to reach the surface. For our analysis, we will assume that a period of rapid cooling, as indicated by apatite data, implies a period of rapid tectonic exhumation, and that this exhumation corresponds with a period of active fault displacement and surface uplift.

Recall from Chapter 9 that fission tracks are completely erased in apatite at temperatures above about 110°C, partially retained between 110°C and 50°C, and fully retained below 50°C. Recall also that helium is expelled from apatite at temperatures above 70°C and is completely retained within the mineral at temperatures below 30°C. Samples taken from the base of the White Mountains (location X in Fig. 20.28) and analyzed for both fission track and (U-Th)/He ages, produce nearly identical ages of 12 million years. In other words, both dating methods closed at about the same time. One possible scenario by which this could happen is if the rocks cooled very rapidly from greater than 110°C to less than 30°C. Such rapid cooling implies rapid tectonic exhumation, presumably in association with rapid mountain uplift along the White Mountains fault. Depending on the inferred geothermal gradient, and assuming an average temperature of 10°C at the Earth's surface, the data suggest that the rocks were quickly exhumed at about 12 million years ago from a depth between 4.2 and 2.5 miles (6.7 to 4.0 km) to a depth between 0.9 and 0.5 miles (0.8 to 1.4 km). If rapid tectonic exhumation implies rapid uplift, then the White Mountains were uplifted rather quickly around 12 million years ago.

Initiation of Strike-Slip Faulting

Our next step is to gain insight to the present day strike-slip tectonic regime including information regarding how long ago the White Mountains normal fault was reactivated as a strike-slip fault, when the Fish Lake Valley fault (11) came into existence, and how this new tectonic regime altered the landscape. An important observation is that the active White Mountains strike-slip fault becomes an active normal fault (the Queen Valley Fault) at the northern end of the range. Queen Valley appears to be a pull-apart basin associated with displacement on both faults. If we can date the time the Queen Valley normal fault became active, we also date the time the White Mountains fault was reactivated as a strike-slip fault.

Here again, we employ apatite (U-Th)/He ages derived from rocks obtained from the lower plate of the Queen Valley Fault at the base of the White Mountains (location X' in Fig. 20.28). In this case, the ages are mostly in the range of 3 to 5 million years. If we assume that tectonic exhumation along an active Queen Valley normal fault caused these rocks to cool below the apatite (U-Th)/He closure temperature, then the data suggest that the Queen Valley Fault was active 3 to 5 million years ago, and by extension, strike-slip reactivation of the White Mountains fault was in progress by 3 to 5 million years ago.

Southward, the White Mountains fault (3) steps to the right to become the active Owens Valley Fault (4). Both are right-lateral strike-slip faults and both have generated large earthquakes including the massive 1872 earthquake along the Owens Valley fault and a 1986 event along the White Mountains fault. A significant aspect to the Owens Valley Fault is that, in spite of the 1872 earthquake, there is little in the way of total offset along the length of the fault. From this observation, we can suggest that the Owens Valley Fault has not been in existence for a very long time. The right-stepping relationship with the White Mountains fault suggests that it has been active for no more than 3 to 5 million years.

Geological evidence, including the age of deformed volcanic rocks, suggests that the Death Valley-Fish Lake Valley strike-slip fault (11) on the east side of the White Mountains began activity less than 6 million years ago. Given this information, we can assume that two pull-apart basins, north Fish Lake Valley (NFLv) and Deep Springs Valley (Dsv), also began activity less than 6 million years ago.

The development of the Death Valley-Fish Lake Valley Fault along the east side of the White Mountains, and the reactivation of the White Mountains fault along the west side, has resulted in some interesting physical changes to the White Mountains. Faults associated with the Deep Springs Valley pull-apart basin (Dsv) appear to have down-dropped the southern end of the White Mountains, thus creating a basin where part of a mountain once stood. The opposite appears to be true at the northern end of the White Mountains where activity on the Queen Valley normal fault (2) appears to have resulted in renewed uplift and lengthening of the White Mountains in a north-northeastward direction adjacent to the fault.

In summary, we can suggest that the White Mountain area was part of a high elevation low relief Nevadaplano prior to 12 million years ago. Normal fault activity on the White Mountains fault began circa 12 million years ago resulting in sinking and east-tilting of the Hammil and Chalfant valleys and simultaneous uplift and asymmetric east-tilting of the White Mountains. This Basin and Range style extensional deformation was overprinted by active dominantly transtensional deformation (strike-slip and normal faulting) beginning between 6 and 3 million years ago. Transtension has altered the shape of the White Mountains at two locations. Formation of the Deep Springs Valley normal fault has lowered elevation along the southern end of the White Mountains, and formation of the Queen Valley normal fault has lengthened the mountain range and increased its elevation at its northern end. An interesting aspect to the White Mountains fault is field evidence indicating that part of the fault zone was active as a shear zone more than 66 million years. The present-day fault zone partially followed and reactivated this ancient fault zone.

Inyo Mountains

The Inyo Mountains are a long narrow range, about 70 miles north to south and only 12 miles wide. There are two peaks above 11,000 feet. The rocks are part of the North American rock succession and are similar to what is found in the White Mountains. As seen in Fig. 18.7 the line of accreted terranes lies in Owens Valley, just west of the Inyo Mountains. Fig. 20.29 is a Google Earth image of the Inyo Mountains (Im) looking south. The image shows a curved and slightly asymmetric mountain form with a relatively steep western flank. The curvature of the mountain can be followed northward (to the bottom of the figure) to Waucoba Mountain (W), the highest peak in the range. The eastern (left) flank of the range terminates abruptly at Saline Valley (Slv), and the western flank at Owens Valley (OV). The steep Sierra Nevada escarpment is visible along the west (right) side of the image along with Kern Canyon (Kc). Three pull-apart basins marked by normal faults are visible in the east (upper left) of the figure between the Saline (Slr) and Panamint Ranges (Pr).

Here, we have a situation similar to but different than the White Mountains. Based its asymmetry we would expect to find an active or recently active normal fault along the steeper western range-front. Surprisingly we see no such thing. Instead, we find an active normal fault along the eastern (opposite) side of the mountain in the form of the Saline Valley fault (6). This fault has no obvious correspondence with the present-day asymmetry or curvature of the Inyo Mountains. Given little alternative, we are forced to suggest that the asymmetry and curvature of the mountain are relict morphologies. One possibility is that the asymmetry developed circa 12 million years ago at the same time as normal faulting and initial uplift along the White Mountains fault. In other words, we can suggest that a buried, inactive segment of the old White Mountains normal fault extends southward to the west side of the Inyo Range. In support of this interpretation, a geophysical survey suggests that an inactive fault is indeed buried along the western side of the range.

But this explanation has problems. The first is the curvature of the Inyo Range, which does not follow the expected trend of the White Mountains fault. A second problem are zircon and apatite (U-Th)/He cooling ages, which show no sign of exhumation on the west side of the Inyo Mountains at 12 million years ago. The ages, instead, suggest two separate and far older periods of exhumation, one at about 66 million years ago and again about 54 million years ago. The data also imply that there was almost no exhumation between 54 and 15.6 million years ago.

Given these ages, an alternative explanation is that the buried fault along the west (right) side of the Inyo Range was active, not at 12 million years ago, but at 66 and possibly again at 54 million years ago resulting in initial uplift and eastward tilting of the range. The timing of uplift corresponds with the existence of the Nevadaplano. The Inyo Range may have been a location where the ancient plateau was broken, presumably by faults, into an east-tilted asymmetric fault-block mountain. Given the long period of inactivity between 54 and 15.6 million years ago, we could suggest that the Inyo Range was slowly eroded to a low relief mountain range by 15.6 Ma. If this scenario is true, then initial uplift of the Inyo Mountains is considerably older than the White Mountains.

In contrast to the west side of the mountain, apatite (U-Th)/He cooling ages from the east (left) side of the Inyo Mountains tells a story that provides clues to development of the present-day mountain morphology. The apatite ages suggest rapid cooling at 15.6 million years ago and again at 2.8 million years ago separated by a period of slow cooling. If we correlate cooling with exhumation, then exhumation at 15.6 million years ago occurred at rates equal to or greater than 7.5 inches per 100 years. Such rates likely correspond with tectonic exhumation (rather than erosional exhumation) in the form of strong uplift, westward tilting of the mountain range, and downdropping of the valley along an active Saline Valley normal fault (6). Notice that a western tilt at 15.6 million years ago is in the opposite direction of the inferred eastward tilt that occurred 66 and 54 million

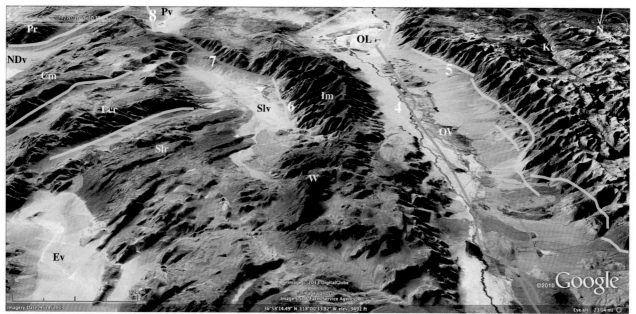

Fault colors and numbers follow Table 20.2 and are as follows. **4**-Owens Valley, **5**-Southern Sierra Nevada Frontal, **6**-Saline Valley, **7**-Hunter Mountain, **8**-Panamint Valley. Geographic features are as follows. **Cm**-Cottonwood Mountains, **Ev**-Eureka Valley, **Im**-Inyo Mountains, **K**-Keynot Peak, **Kc**-Kern Canyon, **Lcr**-Last Chance Range, **NDv**-Northern Death Valley, **OL**-Owens Lake, **OV**-Owens Valley, **Pr**-Panamint Range, **Pv**-Panamint Valley, **Slr**-Saline Range, **Slv**-Saline Valley, **W**-Waucoba Mountain.

FIGURE 20.29 Google Earth image looking southward at the Inyo Mountains (Im). Three normal fault pull-apart basins are present between the Saline (Slr) and Panamint (Pr).

years ago. The effect of this more recent western tilt would be to rotate the mountain block westward thus potentially over-steepening an already steep western flank. The 15.6 Ma period of extension along the east side of the Inyo Mountains likely correlates with Basin and Range extension.

The Saline Valley is deep and tilted to the west. Geophysical measurements suggest that the valley contains as much as 15,000 feet of unconsolidated sediment fill. The location of a playa (a dry lake bed shown as a white area in Fig. 20.29) at the western margin of the valley suggests westward tilting. Given the thickness of unconsolidated sediment, it is likely that the Saline Valley was in existence following the 15.6 million-year event.

The Saline Valley fault (6) is an active normal fault as indicated by offset alluvium. We can suggest based on apatite cooling data that the fault was reactivated in a transtensional regime beginning 2.8 million years ago following a period of inactivity. The geometry of the Saline Valley normal fault relative to the Hunter Mountain right-lateral strike-slip fault (7) suggests the two were active at the same time in a manner similar to that shown in Fig. 20.4B. Renewed displacement on the Saline Valley Fault at 2.8 million years ago would result in renewed uplift and westward tilting of the Inyo Mountains and further widening and tilting of Saline Valley, resulting in development of the present-day landscape. Thus, both the White Mountains and Saline Valley faults were active as normal faults during Basin and Range extension between 15.6 to 12 million years ago, and both were reactivated during the transtensional regime. However, in contrast to the White Mountains fault, which was reactivated as a strike-slip fault, the Saline Valley fault was reactivated as a normal fault that enhanced the landscape first developed during the extensional Basin and Range phase.

In summary, we find that the uplift history of the Inyo Mountains is similar in general to the White Mountains, but different in detail. Inyo Mountain uplift was active 66 and 54 million years ago along a fault on the west side of the range that tilted the mountain block eastward creating a steep, asymmetric western front and possibly also the curvature of the range. Part of this relief was reduced during a long period of inactivity between 54 and 15.6 million years ago, but some relief and the original asymmetry of the range apparently survived. Development of the Saline Valley fault (6) along the east side of the range 15.6 million years ago resulted in significant renewed uplift and westward tilting of the range, which caused oversteepening of the pre-existing western flank. The Saline Valley likely opened as a west-tilted half-graben similar with that shown in Fig. 18.3. The Saline Valley half-graben was subsequently widened as a pull-apart basin beginning 2.8 million years ago associated with strike-slip displacement on the Hunter Mountain fault (7) resulting in renewed uplift of the Inyo Mountains to its present elevation. The activation of the Saline Valley fault at 15.6 million years ago, and the reactivation of this fault along with the Hunter Mountain strike-slip fault at 2.8 million years ago, mimics the change seen in the White Mountains from Basin and Range normal faulting to transtensional strike-slip/normal faulting.

Death Valley-Panamint Valley Region

In this section, we discuss the origin of landscape in the southeast part of Fig. 20.26 including the Black Mountains (Bm), southern Death Valley (Dv), the Panamint Range (Pr), and northern Panamint Valley (Pv). Fig. 20.30 is a Google Earth image of the Death Valley-Panamint Valley region, looking north. Death Valley is on the right side of the image between the massive Panamint Range (Pr) and the smaller Black Mountains (Bm). The Argus Range (Ar) lies to the west. The Black Mountains, together with the Funeral (Fm) and Grapevine (Gvm) Mountains, form part of the Amargosa Range. The Panamint Range is easily the highest in the area with Telescope Peak above 11,000 feet and seven additional peaks above 9000 feet. From the vantage point of Telescope Peak, one can spot the lowest point in North America, Badwater Basin (b, −282 feet) only 16 miles away. There are no peaks above 9000 feet in any of the surrounding mountains.

The region contains a wide variety of rocks. Paleozoic miogeoclinal rocks and the underlying Precambrian sedimentary/volcanic rock succession are especially widespread. Cenozoic granitic rocks are prominent in the Black Mountains along with gneisses of the ancient crystalline shield. Volcanic rocks from 22 to 4 million years old are present in all mountain areas. Mesozoic granitic rocks, similar with those in the Sierra Nevada, form most of the Argus Range along with young intrusive and sedimentary rocks. In this section, we discuss evidence that the area evolved from a metamorphic core complex with low-angle normal faults to an active transtensional regime with high-angle normal faults and strike-sip faults.

Transtensional Regime of Death Valley and the Black Mountains

The most recent deformational phase to affect Death Valley and the Black Mountains is transtension similar in style to Fig. 20.3E. The Black Mountains normal fault (12) is present along the east side of Death Valley linking the Death Valley-Fish Lake Valley strike-slip fault (11) with the Southern Death Valley strike-slip fault (13). Death Valley appears to be a pull-apart basin associated with the Black Mountains normal fault. The Black Mountains form a small, eastward-tilted, asymmetric normal fault-block mountain with a steep western face and a

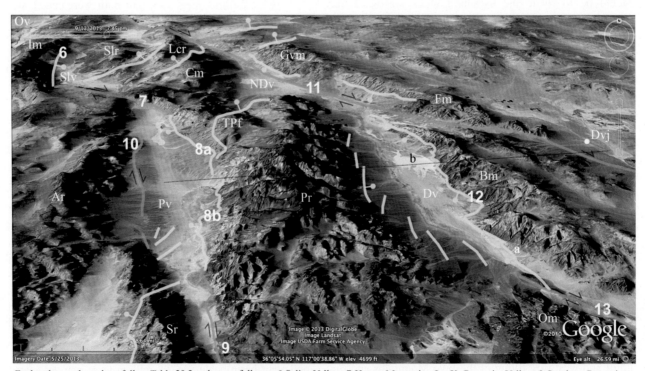

Fault colors and numbers follow Table 20.2 and are as follows. **6**-Saline Valley, **7**-Hunter Mountain, **8a, 8b**-Panamint Valley, **9**-Southern Panamint Valley, **10**-Ash Hill, **11**-Death Valley-Fish Lake Valley, **12**-Black Mountain, **13**-Southern Death Valley. Also **TPf**-Towne Pass fault. Geographic features follow Table 20.2. The thin black line is the line of cross-section shown in Figure 20.32.

FIGURE 20.30 Google Earth image looking north at the Panamint Range-Death Valley area.

gentle eastern slope. The morphology is similar to that shown in the lower panel of Fig. 18.4. Death Valley, like the Black Mountains, shows topographic evidence for an eastward tilt. Note the size and extent of alluvial fans at the base of the Panamint Range relative to alluvial fans along the base of the Black Mountains. The larger, wider Panamint fans suggest that sediment is carried across the valley to a low point close to the Black Mountains. The Amargosa River flows along the east side of the valley, which is also the location of the low point, Badwater Basin (b). One would have to drill through more than 14,000 feet of sediment on the east side of the valley in order to reach bedrock. All of these observations are consistent with an east-tilted valley that is sinking along a high-angle normal fault.

Notice in Fig. 20.30 that there is an unnamed discontinuous fault zone along the west side of Death Valley that cuts alluvial fans near the base of the Panamint Range. This normal fault system is down on the east side, thus contributing to the sinking of the valley. Based on these observations, we can conclude that an active transtensional regime is responsible for at least the present-day pull-apart, tilted morphology of Death Valley and the Black Mountains. The presence of young, uplifted sedimentary/volcanic deposits on the eastern flank of the Black Mountains lends support to the idea that the transtensional regime began as late as 2 to 3 million years ago.

Turtlebacks of the Black Mountains

As seen in Fig. 18.7, a metamorphic core complex forms part of the Black Mountains. Evidence with respect to the formation and age of the core complex includes geologic mapping, the radiometric dating of volcanic ash layers within unconsolidated sediment, and the presence of rather amazing geologic structures known as turtlebacks. Turtlebacks are so named because they are relatively smooth (slightly dissected), curvilinear anticlinal surfaces that resemble the shape of a turtle's shell except, in the case of the Black Mountains, they are tilted rather steeply downward to the northwest. What makes these turtleback surfaces so amazing is that they represent the exposed surface of a normal fault detachment surface within a metamorphic core complex. There are three well-known turtlebacks in the Black Mountains, and possibly additional examples elsewhere in the region. From north to south they are the Badwater, Copper Canyon, and Mormon Point turtlebacks. The Badwater Turtleback is exposed continuously for about 3 square miles and is shown in Fig. 20.31. The other turtlebacks are of similar size. The surface of a turtleback is a relatively smooth brittle fault surface that separates chaotically faulted young sedimentary-volcanic rocks above from strongly deformed, thrust-faulted and much older crystalline rocks below. The thrust faults were active between 61 and 55 million years ago at about the same time deformation was occurring in the Inyo Range. They are not associated with formation of the metamorphic core complex or with development of the turtleback fault surface.

In most areas, the overlying young sedimentary/volcanic rocks are absent such that the brittle normal fault turtleback surface sits directly on crystalline rocks. The sedimentary-volcanic rock sequence, where it is present, shows no signs of metamorphism. The crystalline rocks, on the other hand, reached temperatures approaching or exceeding 500°C at depths of 9 miles or more. The fabric of the underlying crystalline rocks is mylonitic, meaning the rocks show evidence of strong shearing deformation while at elevated temperature in the form of a wide ductile fault zone. Radiometric dating of mylonitic rocks, and of undeformed and weakly deformed granitic rocks that intrude the mylonite, suggest the rocks underwent exhumation and cooling during mylonitization (during faulting) mostly between 16 and 10 million years ago. This is the time the metamorphic core complex was active. Apparently, the turtleback surface evolved from a high-temperature ductile shear zone within the underlying crystalline rock, to a narrow brittle fault zone that formed at low tempterature as the rocks were exhumed toward the surface. Notice that the ductile shear zone was active at about the same time as Basin and Range style normal faulting in the White and Inyo Mountains.

We gain insight to the origin of turtlebacks by reviewing the progression of a crystalline core complex as depicted in Fig. 18.14. In panels A and B of this figure, a mylonitic shear zone forms along the deeply buried, subhorizontal detachment normal fault that separates crystalline from sedimentary rocks. As the fault rises to surface, the rocks cool, and the ductile shear zone transitions to a brittle fault. With continued extension, unmetamorphosed sedimentary rocks rotate and slide down onto the detachment surface. Eventually, all of the sedimentary rocks are removed leaving the fault surface and underlying crystalline rocks exposed at the surface as shown in Fig. 18.14C. The turtlebacks represent the exposed surface of the ductile/brittle normal fault detachment surface at the top of the crystalline core as depicted in Fig. 18.14C. The near absence of erosion of the fault surface is likely a combination of its strong resistance to erosion, the slow rate of erosion in the desert environment, and the possibility that it has only recently been exposed. Any soft sedimentary-volcanic rocks remaining above the fault could have easily been eroded leaving the fault surface and the hard underlying, mylonitic crystalline rocks exposed.

Given the presence of both a turtleback detachment surface and a high-angle normal fault (the Black Mountains Fault, 12) at the base of the Black Mountains, a working hypothesis, based on a compilation of

FIGURE 20.31 Google Earth image looking eastward at Badwater turtleback (outlined in yellow) in Death Valley, California. Note the gently dissected, convex shape of the turtleback surface. The Badwater parking area is labeled.

stratigraphic, sedimentologic, and tectonic considerations, suggests a two-stage development for the landscape. The first stage was likely underway by about 16 million years ago and involves formation of a metamorphic core complex along a deeply buried, sub-horizontal detachment fault that would later become a turtleback. Because the fault was deeply buried, there may not have been very much topographic relief at the surface directly above the fault. It is unlikely that the Black Mountains or Death Valley were in existence at this time. Instead, the entire region may have been one of low relief and unknown elevation associated with the Nevadaplano. The second stage, beginning as late as 2 or 3 million years ago, is the initiation of the high-angle Black Mountains range-front normal fault within an active transtensional regime. To continue our analysis, we will need to consider field relationships and geology in the adjacent Panamint Range (Pr) and Panamint Valley (Pv).

Panamint Mountains

Fig. 20.30 shows that the Panamint Range is nearly symmetrical. It shows none of the form indicative of an asymmetric normal fault-block mountain. Additionally, it forms a large, high mountain range significantly higher than any peak in the Black Mountains. Rocks consist mostly of the Precambrian sedimentary-volcanic succession metamorphosed to low grade with older rocks of the crystalline shield and a few young granitic intrusions. Volcanic rocks from 22 to 4 million years old form most of the southern part of the range. We will explore the possibility that the origin of the Panamint Range has less to do with transtensional range-front normal faults and more to do with the older turtleback detachment fault. In Fig. 18.14C, the detachment surface rises to form a metamorphic core complex. This same process may be responsible for the Panamint Range. In other words, the Panamint Range may form the center a metamorphic core complex. The difference between the Panamint Range and the turtlebacks is that erosion has completely removed the fault surface leaving only the underlying metamorphic core to from the top of the mountain. The absence of basin deposits within the Panamint Range younger than 8 million years old suggests that the mountain was topographically elevated by this time and undergoing erosion. The rise of the Panamint Mountains by 8 million years ago would have created Panamint Valley to the west and Death Valley to the east. Death Valley would have been higher in elevation than today. The Black Mountains (if they existed at all) and the Panamint Mountains were likely lower in elevation.

As seen in Figs. 20.26 and 20.30, the southern part of Panamint Valley adjacent to the Slate Range (Sr) is narrow. Here, the Southern Panamint Valley fault (9) shows dominantly right-lateral strike-slip displacement. Northward, the Panamint Valley widens considerably between the Panamint (Pr) and Argus (Ar) Ranges. Panamint Valley, along much of this distance, is roughly symmetrical in shape. However, at its northern end the valley becomes asymmetric and is tilted to the east as indicated by the location of the playa (dry lake bed) along the east side of the valley. The Ash Hill fault (10), along the west side of the valley shows dominantly right-lateral strike-slip displacement. The Panamint Valley fault along the east side of the valley is divided in Fig. 20.30 into a northern (8a), central (8b) and southern (9) section, all of which show both normal and right-lateral strike-slip (oblique) displacement. Normal displacement along the northern section (8a) in a direction roughly parallel with the Hunter Mountain strike-slip fault (7) implies that northern Panamint Valley is a pull-apart basin.

There are, however, two problems when considering the northern Panamint Valley to be a pull-apart. The first is the thickness of unconsolidated sediment fill in the valley, which amounts to less than 1000 feet, much less than what is seen in Death Valley and entirely inconsistent with the idea that northern Panamint Valley is a tilted, sinking, pull-apart basin. The second problem has to do with the Panamint Valley fault itself, which does not appear to have enough total normal fault displacement to create a wide pull-apart basin.

The thin black line in Fig. 20.30 is a cross-section line that extends from Panamint Valley eastward across the Black Mountains. The corresponding cross-section is shown in Fig. 20.32. The symmetry of the Panamint Range, and the asymmetry of the Black Mountains, are readily apparent. The cross-section shows both the older detachment fault and the high-angle normal faults that offset the detachment surface. Let us follow the detachment surface from east to west. The detachment is buried below young sedimentary-volcanic rock layers along the east side of the Black Mountains, but is exposed along the west side of the range in the form of Badwater turtleback, which is then offset by the younger Black Mountains normal fault. The detachment is buried beneath Death Valley but resurfaces along the eastern base of the Panamint Range where it underlies Early Paleozoic rocks that form part of the upper plate of the detachment. These rocks, and the detachment surface, are eroded such that the detachment extends in the air over the crest of the Range. The lower plate of the metamorphic core complex is exposed across the range below the eroded detachment surface as shown in the cross-section. On the west side of the mountain range, the still eroded detachment surface is offset downward along the high-angle Panamint Valley fault (8a, 8b) and continues westward below Panamint Valley as a buried nearly horizontal boundary.

The favored interpretation to explain the symmetrical shape of the Panamint Range and the shallow depth of Panamint Valley, is that the detachment fault, as a whole, became inactive by about 8 million years ago, but that a section of the detachment below Panamint Valley was reactivated during activation of the Hunter Mountain strike-slip fault beginning 2.8 million years ago. In other words, the Hunter Mountain strike-slip fault activated a remnant part of the detachment fault rather than initially creating a high-angle fault. The detachment below Panamint Valley may still be active, but the presence of the high-angle Panamint Valley fault (8a, 8b) suggests that displacement is currently shifting to the high-angle fault. Thus, the detachment surface as a whole became inactive, but low-angle faulting has continued along the west side of the Panamint Range. A low-angle detachment fault close to the Earth's surface would not cause strong asymmetry or subsidence. Instead, it would produce a shallow basin such as what we see in Panamint Valley. Such an interpretation implies that the steeply dipping Panamint Valley normal faults are recent features consistent with their observed limited displacement.

To summarize, a low relief high elevation area (Nevadaplano) was likely in existence circa 16 million

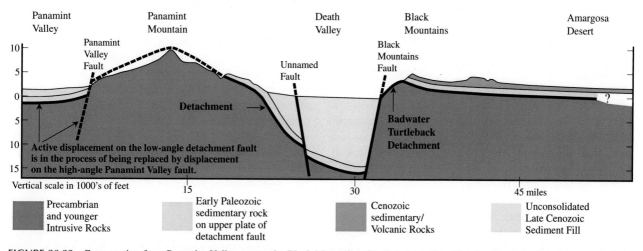

FIGURE 20.32 Cross-section from Panamint Valley across the Black Mountains, CA that shows the older detachment surface and younger high-angle faults. The detachment is eroded above the Panamint Range. The detachment surface is inactive except possibly below Panamint Valley at the extreme left side of the cross-section. Activity on this part of the detachment is in the process of being replaced by high-angle faulting. The extent of the detachment to the east is unknown. Based on Norton (2011). Line of cross-section is shown in Fig. 20.30.

years ago when low-angle detachment faulting initiated. By 8 million years ago, the Panamint Range began to uplift as a crystalline core complex, thus exposing the detachment surface to erosion and creating an early version of Panamint and Death valleys. High-angle normal faults initiated between 2 and 3 million years ago and resulted in uplift of the Black Mountains and the widening and sinking of Death Valley. One could speculate that the gentle slope on the east side of the Black Mountains, as seen in Fig. 20.32, is the original ground surface prior to initiation of the high-angle Black Mountains normal fault. High-angle faulting has recently progressed to the northern Panamint Range where a section of the detachment fault may still be active. Active offset along the detachment fault, however, is currently in the process of being replaced by active offset along the steeply dipping Panamint Valley normal faults (8a, 8b). As offset shifts to high-angle faulting, we would expect Panamint Valley to slowly evolve into an asymmetric, east-tilted valley, and for the Panamint Mountains to become wider and increase in elevation.

EXAMPLE OF ACTIVE FAULTING IN DEATH VALLEY

We end our discussion with an example of landscape in Death Valley that formed via active faults. Fig. 20.33 is a Google Earth image looking north-northeast at the active Black Mountains normal fault in southern Death Valley. The area of this image is identified with the letter a in the southeast corner of Figs. 20.26 and 20.30. An obvious feature indicative of an active fault is the tall, straight, range front along the left side of the image. Also, present are several partly eroded faceted spurs along the mountain front, the largest of which is labeled F. The origin of a faceted spur was discussed in Chapter 4. We can infer from the size and the well-preserved nature of the faceted spurs that the rate of fault displacement outpaces the rate of erosion.

On the right side of Fig. 20.33, we notice that the active fault has stepped to the right, toward the basin, to form a semi-continuous normal fault scarp (a lineament) that displaces elevated older alluvium (unit 3) on one side, from younger alluvium that is not elevated (unit 4). A step (or jump) toward the basin is a characteristic feature of normal faults. The effect is to widen the adjacent mountain range and potentially to increase its elevation. As a result of this jump, deposition of unit 4 has pushed the Amargosa River away from the mountain front. The cause of the basin-ward jump is not immediately clear. It is possible that the active normal fault stepped into a preexisting fracture zone. We note that there is a change in rock type along the mountain front (from unit 2 to unit 1) that coincides with the location of the step. Whether or not a fracture or a change in rock type influenced the location of the jump cannot be known without further detailed analysis.

Along the active fault segment between units 3 and 4, we see several faceted spurs developed in unit 3, each at various states of erosional decay. A well-preserved spur is marked F'. These particular spurs are developed in unconsolidated alluvium (unit 3), not hard bedrock. The well-preserved nature of these relatively delicate landforms indicates recent fault activity. If we were to visit this area, we would notice that stream gullies associated with the more recent alluvial deposits (unit 4) are offset in a few places by the fault. We can infer, therefore, that displacement along the active fault is concurrent with deposition of unit 4, and that the most recent fault displacement must have occurred after at least some of unit 4 was deposited. The US Geological Survey classifies this fault as active in the past 15,000 years. By the same reasoning, the now inactive fault must have been active during deposition of unit 3, and likely became inactive when unit 3 was uplifted along the active fault and incised by streams. If the time of deposition or incision of unit 3 can be determined, then the age of activity on the now inactive fault could be inferred. Finally, notice in the lower right corner of Fig. 20.33 that the active fault segment steps to the left. This left step, in turn, has created a step in the location of uplifted alluvial sediment (unit 3).

There are several additional fault characteristics present in this image, not all of which indicate active faulting, but all are consistent with the presence of active or recently active faults. Note the inactive fault on the right side of Fig. 20.33 between units 1 and 3. According to the US Geological Survey, there is no evidence to suggest that this fault has been active in the past 1.6 million years. We can, however, find several characteristics that point to recent activity including a steep mountain front, partly eroded faceted spurs, and a wineglass canyon (WC). A wineglass canyon is so named because it has the overall shape of a wine glass. It has a wide cup, a narrow stem, and a wide base. In Fig. 20.33, the stem is located where the stream cuts a narrow notch through the base of the inactive fault scarp. The notch widens with elevation in the mountain to form the cup (the WC label is in the cup). Widening occurs because rock higher in the mountain has been exposed to erosion for a longer period of time and, therefore; more of the rock has been removed. If we ignore the uplifted alluvium (unit 3), we could imagine that alluvial fan deposits (unit 4) would spread out at the base of the mountain as they do on the left side of the image. These deposits represent the wide base of the wine glass. Wineglass canyons are common features in desert climates along active and recently active mountain fronts. There are several wineglass canyons on the left side

Unit 1-Sandstone, shale, dolostone, limestone; Unit 2-Granitic Rocks; Unit 3-Older alluvium (loosely consolidated sandstone and conglomerate) Unit 4-Young alluvium (sand and gravel); Unit 5-Young lake and playa deposits (silt and evaporites). Contact between units 2 and 1. F and F'-Faceted spur; WC-Wineglass canyon.

FIGURE 20.33 Google Earth image looking north-northeast at active faults along the west flank of the Black Mountains in the southern part of Death Valley, California. The location of this figure is shown with an "a" in Figs. 20.26 and 20.30.

of Fig. 20.33 but they are not obvious from the camera angle. The fact that active fault-zone features are so well preserved along an inactive fault suggests that the jump toward the basin was a fairly recent phenomenon.

QUESTIONS

1. Draw the trace of the San Andreas Fault onto Fig. 20.2 or a copy of the figure. Where is the trace of the San Andreas most obvious and where is the trace least obvious?
2. What is transpression and transtension?
3. Refer to Fig. 20.5. What types of structures (extensional or compressional) should be expected between the Maacama (2) and Rogers Creek (7) faults and between the Eaton Roughs (3) and Bartlett Springs (6) faults? All four are right lateral strike-slip faults. What type of structures (extensional or compressional) should be expected between the two segments of the left-lateral Garlock fault (20)?
4. Why are strike-slip fault valleys typically linear and straight?
5. What rock, climate, and geographic conditions make coastal California particularly susceptible to landslides?
6. In Google Earth, fly to Rolling Hills, California. Fly around the area and describe evidence for past landslides. Estimate the age of the landslides and state criteria for estimating age. Do any of the surrounding areas appear to be susceptible to future landslides?
7. What is the Big Bend along the San Andreas Fault? When did it form and what is its significance to landscape development in the surrounding area?
8. The San Andreas Fault makes a small bend in the vicinity of the Santa Cruz Mountains. Explain why the fault trends diagonally through the mountain rather than through an adjacent valley.
9. What is fault creep? How is fault creep different from other types of fault displacement?
10. Describe both rock evidence and landscape evidence that indicates a subduction zone was present along the California coast prior to development of the San Andreas fault.
11. What is the Franciscan Complex?

12. Given a surface uplift rate of 0.79 inches/100 years for 4 million years, calculate the average height of a mountain in both feet and meters.
13. The Salton Sea is located between the Imperial Valley and San Andreas faults. Explain its origin. What geologic name is applied to this type of basin?
14. The Salton Sea is below sea level and very close to the coast. Explain why it has not been drowned by the ocean.
15. What is the Walker Lane Belt? Where is it located? When did it first begin to develop? What might be its eventual fate? What is its relationship with the Klamath Mountains?
16. Describe evidence that suggests the Coast Ranges are young.
17. A relict landscape is present along the western slope of the Sierra Nevada. What does the landscape look like? When did it form?
18. Take note of the very straight lineament in Fig. 20.18 that extends northward from Chico along the boundary that separates the Sierra Nevada from the Central Valley. What is the origin of this lineament?
19. What evidence is there to suggest that the northern and southern margins of the Central Valley are undergoing reincarnation?
20. What is the Salinian block? Where is it located?
21. How long ago did the great eastern escarpment of the Sierra Nevada begin to develop and what is the evidence for the timing of its development?
22. Why does the presence of conglomerate suggest a sloping surface at the time the conglomerate was deposited?
23. In Fig. 20.8, draw lines along the wave-cut angle at the base of as many terraces as you can. In Google Earth fly to Santa Cruz. How far northward can you follow some of these terraces? How far above sea level are these terraces?
24. Given the orientations of the Death Valley-Fish Lake Valley Fault (11, Fig. 20.30) and the Southern Death Valley fault (13), is slip on the Black Mountains fault thrust, normal, strike-slip, or oblique? Explain.
25. If we find river water flowing directly over bedrock, explain why we can suggest that the river is actively downcutting (incising) its channel.
26. If we find an incised river channel underlain with a layer of sediment, what can we say regarding the age of river incision?

Chapter 21

The Grand Canyon

The Grand Canyon is without question one of the most recognized and spectacular landforms in North America. A book on landscape would hardly seem complete without some discussion of this, the quintessential icon of the United States. A proper theory on the cutting of the Canyon must take into account all available field data beginning with the legendary 3-month 1869 expedition through the Grand Canyon led by John Wesley Powell, a one-armed professor of geology at Illinois Wesleyan University. Powell lost his right forearm as a result of wounds suffered during the 1862 Battle of Shiloh. In spite of having only one arm, Powell was an extraordinary explorer. He led additional expeditions on the Green and Colorado Rivers in 1871 and 1872 producing photographs, a map, and scientific publications. Powell would ride the lead boat perched in a secured armchair so that he could see upcoming rapids and signal the other boats. His boat, named the Emma Dean after his wife, is shown in Fig. 21.1. Beginning in 1881 Powell became the second director of the US Geological Survey, a position he held for 13 years. He also served as director of the Smithsonian Bureau of Ethnology beginning in 1880 until his death in 1902.

The Grand Canyon has been extensively explored and studied since the time of Powell, and although we have learned a great deal, its age, origin, and evolution have been difficult to unravel. Prior to the 21st century, fieldwork by hundreds of researchers led to the widely accepted conclusion that the Canyon formed largely in the past 6 million years. However, beginning in the 21st century, apatite fission track and (U-Th)/He geochronology has, in a manner of speaking, opened a can of worms, because some of the data imply that the Canyon could be as old as 70 million years. In this chapter, we characterize the physiographic Grand Canyon and discuss what is known about its history. We then look briefly at arguments for a young canyon less than 6 million years old, and for an old Canyon 50 to 70 million years old. We end the chapter with updated interpretations on canyon formation that attempt to rectify field data with geochronologic and other data.

THE PHYSIOGRAPHIC CANYON

Fig. 21.2 is a Google Earth image of the Grand Canyon with major physiographic features labeled. The Grand Canyon sits at the southwestern edge of the Colorado Plateau in northwestern Arizona. It begins at Lee's Ferry (LF) just downstream from where the Colorado River crosses the Echo Cliffs (EC). It ends 277 river miles downstream at Grand Wash Cliffs (G) where the Colorado River drops onto the Basin and Range and across the wide-open space surrounding Lake Mead. The Colorado River in the Canyon area consists of five relatively straight segments separated by four sharp bends. The first and the last bend form half-loops at Upper Granite Gorge (UG) and Lower Granite Gorge (LG).

Along its entire length, the rim of the Grand Canyon consists of nearly flat-lying sedimentary rocks of the interior platform rock succession. Although bench and slope landscape is not ideally developed, cliffs along the river in several areas retreat such that they form two benches, an inner bench whose inner rim rises sharply from river bottom, and an outer bench whose outer rim overlooks the entire Canyon. This type of topography is illustrated in Fig. 21.3, which is a Google Earth image looking eastward, up river, along the central part of the Canyon near the confluence with Cataract Creek (CC, Fig. 21.2). The Colorado River is at the center of the image, but is not visible. Canyon walls rise nearly vertically from the Colorado River to the inner bench. Topography is somewhat flat across the bench except for deep incision along tributary rivers. Near vertical cliffs separate the inner bench from the outer bench marked by the Kanab Plateau (KP) on the north (left) side and the Coconino Plateau (CP) on the south side. The distance between the Kanab and Coconino Plateaus near the front of this image is 3.5 to 5.0 miles. About one mile separates the inner rim on the Kanab side from the inner rim on the Coconino side.

Lee's Ferry is the staging area for raft rides through the Canyon and is one of only two locations where canyon walls step back far enough to allow a road to reach river level. Fig. 21.4 looks northward up the Colorado

FIGURE 21.1 Photograph of the boat (the Emma Dean) and the armchair used by John Wesley Powell to explore the Grand Canyon. Photo by E. O. Beaman, August 20, 1872, in Marble Canyon (cropped). Photo courtesy of the US Geological Survey Denver Library photographic collection. Downloaded at https://library.usgs.gov/photo/#/item/51dc856fe4b097e4d383948f.

River toward Lee's Ferry (LF). Here, the Colorado River has crossed from a higher plateau, marked by the Echo Cliffs (EC), to a lower plateau marked by the Marble Platform (MP). Lee's Ferry sits at the northern edge of the Marble Platform (MP), at river level, at an elevation of 3125 feet. From Lee's Ferry, the Colorado River flows in a southward direction into a narrow gorge known as Marble Canyon (MC). The Marble Platform (MP) slopes northward in a direction opposite that of the Colorado River such that the Colorado River flows progressively deeper into the Marble Platform, thereby creating a very narrow, progressively deeper canyon. In this area, the Marble Platform (MP) forms the inner bench, and the Echo Cliffs (EC) and Paria Plateau (PP) are set back to form the outer bench. Near Lee's Ferry, the Echo Cliffs rise to between 4400 and 4850 feet, and the Paria Plateau (PP) rises to between 6200 and 6400 feet. Two river miles south of Lee's Ferry the inner rim of Marble Platform is already 200 feet above river level and the walls of Marble Canyon are less than a quarter-mile apart.

Marble Canyon remains narrow until the location marked with a small yellow arrow in Fig. 21.2. Here, the Colorado River begins to cross the Kaibab Arch and the western wall of the canyon opens to more than 9 miles wide. The Kaibab Arch (also known as the Kaibab Uplift) is a broad, asymmetric anticline, typical of the Colorado Plateau and similar in form to Fig. 13.41A, but more complex with several nested monoclines and faults. The crest of the arch follows the East Kaibab monocline, shown with a faded white line and opposing arrows in Fig. 21.2. The abrupt widening of the canyon is probably due to the presence of the inactive Butte fault in the crest of the monocline, which provides an easy path for erosion.

The Little Colorado River (Lc) joins the Colorado River along the eastern flank of the Kaibab Arch at a river elevation of 2724 feet. The eastern rim of Marble Platform (MP) directly above the river confluence is more than 6000 feet in elevation, creating a 3300-foot deep canyon measured from river to rim. The Marble Platform (MP) continues south of the Little Colorado River (Lc) where it forms the Coconino Plateau (CP), and west of the East Kaibab monocline where it steps up in elevation to form the Kaibab Plateau (KA). All of these areas are underlain with Permian Kaibab Limestone.

South of the confluence with the Little Colorado River, the Colorado River enters the Upper (Eastern) Granite Gorge (UG) between the Coconino and Kaibab Plateaus where it takes its first major bend, turning sharply to the northwest to create a half-loop across the Kaibab Arch. This is the deepest and most visited part of the Canyon. River elevation at its southernmost point in the gorge is 2550 feet. Sedimentary rocks dip 1 to 2

B-Bright Angel Canyon, CC-Cataract Creek, CP-Coconino Plateau, EC-Echo Cliffs, G-Grand Wash Cliffs, Gv-Grand Canyon Village, H-Hurricane Fault, HP-Hualapai Plateau, KA-Kaibab Arch and Kaibab Plateau, KC-Kanab Creek, KP-Kanab Plateau, Lc-Little Colorado River, LF-Lees Ferry, LG-Lower Granite Gorge, LM-Lake Mead, M-Milkweed-Hindu paleocanyon, Mc-Marble Canyon, MP-Marble Platform, P-Peach Springs Canyon and Paleocanyon, PD-Painted Desert, PP-Paria Plateau, S-Shivwits Plateau and Volcanic Field, SGW-Southern Grand Wash Cliffs, T-Toroweap Fault, U-Uinkaret Plateau and Volcanic field, UG-Upper Granite Gorge, VC-Vermilion Cliffs.

▪ Location of Hindu Fanglomerate   ★ Source rock of Hindu Fanglomerate   ⌒ Paleocanyon

FIGURE 21.2 A Google Earth image of the Grand Canyon oriented north. White lines represent normal faults. White line with opposing arrows follows the crest of the Kaibab Arch.

degrees southward across the gorge such that the Kaibab Plateau (KA) on the north side of the river is higher than the Coconino Plateau (CP) on the south side, and the highest anywhere along the rim. These plateaus now form the outer bench. The North Rim is between 8000 and 8400 feet in elevation reaching 9204 feet at Kaibab Plateau High Point 26 miles from the southern point of the river and less than 15 miles from its closest point to the river. The South Rim is between 6800 and 7400 feet. Relief from river to rim is on the order of 5850 feet (8400 − 2550 feet). The 8 to 10 miles that separate the North and South (outer) rims is a wide area of erosion, especially on the north side of the river, where landscape more closely resembles erosional mountains. There is a poorly developed inner bench at 3600 to 3800 feet that overlooks the inner gorge where elevation drops precipitously more than 1300 feet in little more than a quarter mile. Poorly understood problems on the origin of the Canyon in this area are why the Colorado River flows southward against the slope of the Marble Platform, and what caused the Canyon to turn sharply across the Kaibab Arch rather than continue in a southward direction along its eastern flank.

From its southern point in the Upper Granite Gorge, the Colorado River flows northwestward along the west limb of the Kaibab Arch, past Grand Canyon Village (Gv), where a second sharp bend turns the river toward the west-southwest to flow past Kanab Creek (KC), Cataract Creek (CC), and two active normal faults, the Toroweap (T) and Hurricane (H) faults, shown in Fig. 21.2 with faded white lines. An inner bench begins to develop in the vicinity of the second bend and is well developed along the straight section near Cataract Creek (CC) as seen in Fig. 21.3. Both of the normal faults show active displacement down-dropped to the west. The Colorado River is at an elevation of 1675 feet near where the river crosses the Toroweap fault. The inner bench on both sides of the river in this area varies from about 4000

CP-Coconino Plateau, **IB**-Inner Bench, **IR**-Inner Rim, **KP**-Kanab Plateau

FIGURE 21.3 Google Earth image looking eastward up the Colorado River in the central part of the Canyon near the confluence with Cataract Creek (CC). The image shows an inner bench (IB) and an outer bench represented by the Kanab and Coconino Plateaus.

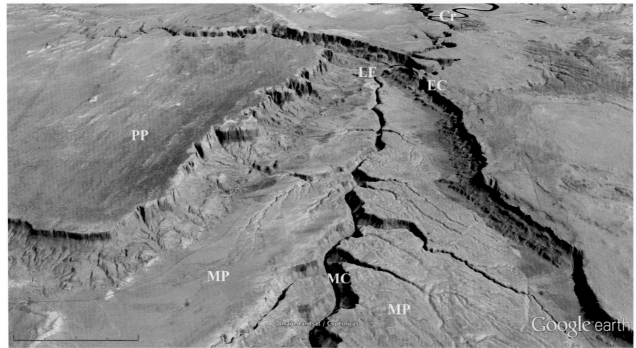

Cr-Colorado River, **EC**-Echo Cliffs, **LF**-Lee's Ferry, **MC**-Marble Canyon, **MP**-Marble Platform, **PP**-Paria Plateau

FIGURE 21.4 Google Earth image looking northward up the Colorado River toward Lee's Ferry (LF). The Marble Platform (MP) forms the inner bench and the Echo Cliffs (EC) and Paria Plateau (PP) form the outer bench.

to 4600 feet, stepping up to between 5700 and 6200 feet on the outer bench Kanab (KP) and Coconino (CP) Plateaus, and to more than 7000 feet on the Unikaret Plateau (U) located on the north rim between the Toroweap and Hurricane faults. Relief to the inner bench is 2925 feet (4600 − 1675 feet) stepping up to 4525 feet (6200 − 1675 feet) on the Kanab−Coconino Plateaus. The Unikaret volcanic field covers most of the Unikaret Plateau (U). It is the location shown in Fig. 17.29 where lava flows, frozen on the north Canyon wall, are seen cascading down to the river. The Colorado River crosses the Hurricane normal fault downstream from the cascading lava flows. The Hurricane fault is down-dropped on the west side, which lowers the inner rim on that side. River elevation is 1595 feet where the river crosses the Hurricane fault. The inner rim west of the fault is between 3500 and 4200 feet in elevation whereas the inner rim east of the fault is between 4700 and 4900 feet.

The river makes a third bend downstream from the Hurricane fault, turning sharply southward to flow along the east flank of Shivwits Plateau (S) toward the Lower (Western) Granite Gorge (LG). The inner rim is now entirely west of the Hurricane fault and at variable elevation from 4500 to 3500 feet. The river flows south at a slight angle to the trend of the Hurricane fault such that the two intersect less than 6 miles east of the southernmost point in the Lower Granite Gorge (LG). The river then flows parallel with the fault for a short distance prior to turning westward. The fault continues a southward path along the Peach Springs paleocanyon (P) as seen in Fig. 21.2.

The river makes the last of its sharp bends at the southernmost point in the Lower Granite Gorge (LG) creating a second half-loop around Shivwits Plateau (S). River elevation is 1295 feet at the southernmost point. The inner bench on the north rim is between 4400 and 4850 feet before stepping up to about 6080 feet on the Shivwits Plateau (S), which forms the outer bench. The Shivwits Plateau (S) is partly covered in volcanic rock, none of which is preserved near the Colorado River. On the south side, the Hualapai Plateau (HP) forms an inner bench at 4400 to 4850 feet in elevation. The Hualapai Plateau continues southward gradually gaining elevation without a defined outer bench. Relief from river to inner rim on both sides of the river at the southernmost point in the Lower Granite Gorge (LG) is on the order of 3555 feet (4850 − 1295 feet) rising to 4785 feet (6080 − 1295 feet) at the Shivwits Plateau. The straight-line distance between southernmost points in the Lower and Upper Granite Gorge is 85.25 miles. Once beyond the Lower Granite Gorge (LG), the river follows a relatively straight path northwestward to the Grand Wash Cliffs (GW) generally maintaining an inner bench at about 4700 feet. River elevation as it leaves the Grand Wash Cliffs is 1195 feet. Elevation on both sides of the river in the vicinity of the Grand Wash Cliffs is about 4700 feet creating 3505 feet of relief.

Although not the deepest canyon in the world, the Grand Canyon is impressively long and wide. Along most of its length, the Colorado River flows through about 5000 feet of nearly flat-lying Cambrian to Permian sedimentary rocks of the interior platform rock succession. The river at two locations cuts below the interior platform into circa 1.84- to 1.66-billion-year-old metamorphic and plutonic rocks of the North American crystalline shield. The first location is just west of the southernmost river point in the Upper Granite Gorge extending northwestward along river level for about 38 miles almost to Kanab Creek. The second location is at the southernmost river point in the Lower Granite Gorge extending westward at river level for 24 miles and eastward for about 7 miles. Roughly 900 to 1000 vertical feet of shield rock is exposed at both locations. The contact between the shield and interior platform is the original Great Unconformity of Powell, shown in Fig. 11.12.

In a few areas east of Kanab Creek (KC), a tilted, faulted, circa 1254- to 735-million-year-old sedimentary rock sequence intervenes between the shield and the interior platform rock succession. These rocks are part of the Grand Canyon Supergroup and are best seen at river level and along canyon walls along a 12-mile stretch from just south of the confluence with the Little Colorado River (Lc) to the southern most part of the Upper Granite Gorge (UG) where the rocks come in contact with underlying rocks of the crystalline shield. These rocks are part of the Precambrian sedimentary-volcanic rock succession. The complete stratigraphy of the Grand Canyon is shown in Fig. 11.13. This figure also shows four prominent cliff-forming formations of the interior platform, the Tapeats Sandstone, Red Wall Limestone, Coconino Sandstone, and Kaibab Limestone.

Deformation in the Grand Canyon area is characteristic of the Colorado Plateau as a whole. Shield rocks developed a prominent north- to northeast-oriented, steeply dipping metamorphic fabric (foliation) during mountain building circa 1.84 to 1.66 billion years ago. All subsequent deformation was brittle, some of it along the same trend as the metamorphic foliation. Normal faults, and a few reverse faults, were active during and following deposition of the Grand Canyon Supergroup. Many of these faults, including the Toroweap and Hurricane faults, were reactivated as reverse faults and monoclines during Laramide orogeny beginning circa 75 million years ago, and again as normal faults associated with Basin and Range extension beginning less than 10 million years ago. The major structures that cross the Colorado River are the aforementioned Laramide-age Kaibab Arch, and the Laramide and Basin and Range-age Toroweap and Hurricane faults. Additional faults and monoclines are also present.

ACTIVE FAULTS AND INCISION RATES

Fig. 21.5 is a shaded relief elevation map of the Grand Canyon. Active and recently active faults create escarpments, river valleys, and other lineaments that are visible in this image. Four major faults that cross the Colorado River are identified with large yellow arrows. They are the Bright Angel (B), Eminence (E), Hurricane (H), and Toroweap (T) faults. Other active faults with topographic expression are identified with a small yellow arrows. The Bright Angel fault (B) forms a strong northeast-trending lineament that extends from the outskirts of Grand Canyon Village (Gv) across the Colorado River and along Bright Angel Canyon. The fault crosses Bright Angel Trail, the main trail into the Canyon, several times. The Eminence fault (E) crosses Marble Canyon (MC) roughly along the same trend as the Bright Angel fault. The Bright Angel fault is seismically active, and both faults show strong topographic expression, however, neither shows evidence of surface displacement in the past 1.6 million years. The Bright Angel fault is oriented parallel with steeply dipping foliation in metamorphic shield rocks allowing relatively easy downcutting and creation of the very straight Bright Angel Canyon visible in Fig. 11.12. The Bright Angel fault was active as a reverse fault during deposition of the Grand Canyon Supergroup, reactivated as a reverse fault during Laramide orogeny, and most recently reactivated as an east-side down normal fault. The history of the Eminence fault is less certain, but it too was active most recently as a normal fault.

The Toroweap (T) and Hurricane (H) faults are active normal faults with strong topographic expression. Both are reactivated faults previously active during the Late Precambrian and again during Laramide orogeny. The Toroweap fault follows a tributary river south of the

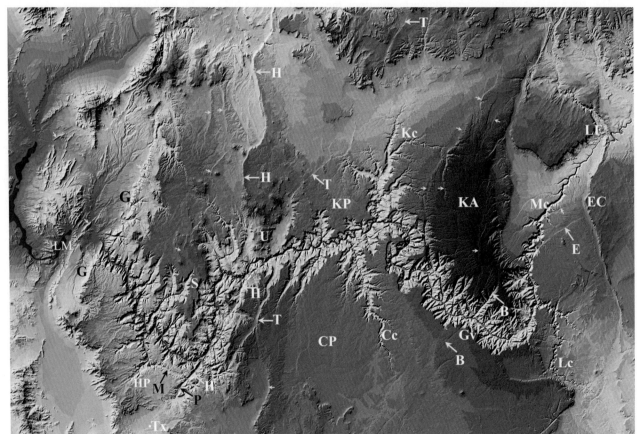

B-Bright Angel Fault, **CC**-Cataract Creek, **CP**-Coconino Plateau, **E**-Eminence Fault, **EC**-Echo Cliffs, **Gv**-Grand Canyon Village, **G**-Grand Wash Cliffs, **H**-Hurricane Fault, **HP**-Hualapai Plateau, **KA**-Kaibab Arch and Kaibab Plateau, **KC**-Kanab Creek, **KP**-Kanab Plateau, **Lc**-Little Colorado River, **LF**-Lees Ferry, **LG**-Lower Granite Gorge, **LM**-Lake Mead, **M**-Milkweed-Hindu paleocanyon, **Mc**-Marble Canyon, **P**-Peach Springs Canyon and Paleocanyon, **S**-Shivwits Plateau and Volcanic Field, **T**-Toroweap Fault, **U**-Uinkaret Plateau and Volcanic field, **UG**-Upper Granite Gorge. ⇐ Yellow arrows locate active faults.

FIGURE 21.5 USGS National Atlas digital shaded relief elevation map of the Grand Canyon. Blue arrows locate paleocanyons. Yellow arrows locate active faults. Nationalmap.gov/small_scale/atlasftp.html.

Colorado River and then extends northward along the east side of the Uinkaret Volcanic field (U) for more than 100 miles. The base of Vulcan's Thone, a cinder cone at the edge of the volcanic field overlooking the North Rim, is cut by the Toroweap fault. The Toroweap Fault (T) has been active over the past 2 to 3 million years. It shows 197 feet (60 m) of displacement in the past 600,000 years where it crosses the Grand Canyon (equal to 0.39 in/100 years). The Hurricane fault (H) follows the excavated Peach Springs paleocanyon (P) south of the Colorado River and then extends northward for more than 100 miles crossing the western side of the Uinkaret Volcanic field (Fig. 17.29) to form the Hurricane Cliffs at the western edge of the Colorado Plateau (Fig. 13.38). The Hurricane Fault (H) has been active over the past 3 to 4 million years as a west-side-down normal fault. It shows 98 to 115 feet (30 to 35 m) of offset in the past 200,000 to 320,000 years (equal to 0.37 to 0.69 in/100 years).

If we assume that there was no canyon in existence prior to 6 million years ago, then the average rate to incise a canyon to a depth of 4000 to 6000 feet over a 6-million-year period is between 0.8 and 1.2 inches per 100 years. This rate is rather fast for an extended period of time, but it is possible. Is there evidence for such rapid incision? Basalt is an easily datable rock, and therefore a handy tool when available. Basalt flows emanating from the Uinkaret volcanic field spilled into the Grand Canyon as shown in Fig. 17.29. These basalts have been used to determine Colorado River incision rates. The Uinkaret field is between 3.7 million and 1000 years old, but basalt that flowed into the Canyon is less than 723,000 years old. Incision rates east of the Toroweap fault were calculated to be between 0.59 and 0.69 inches per 100 years (150 and 175 m/my) over the past 500,000 years with a maximum rate of 0.98 in/100 years (250 m/my). At these rates, between 2950 and 3450 feet of canyon could be cut over a 6-million-year period. A total of 4900 feet could be cut using the maximum rate. Incision rates west of the Hurricane Fault are between 0.20 and 0.32 inches per 100 years (50 and 80 m/my) over the past 720,000 years. At this rate, 1000 to 1575 feet of canyon could be cut in 6 million years. Rates are less due to the dampening effect on the down-dropped western side of the Hurricane fault. Even at the maximum calculated incision rate, according to these incision rates, the Canyon, in its entirety, could not have been cut in 6 million years. It is possible that these rates don't mean very much because incision rates could have been different in the past due to changes in river gradient and water volume. For example, the dampening effect on the west side of the Hurricane fault would not have been present during the first half of the 6-million-year interval.

HUALAPAI PLATEAU

The Hualapai Plateau (HP) is important in the history of the Grand Canyon because it contains ancient canyons (paleocanyons) filled or partially filled with sediment that are no longer part of the Grand Canyon river system. These paleocanyons preserve a relict landscape that dates to the time the canyons were formed and filled with sediment. The two largest are the Milkweed-Hindu (M) and Peach Springs (P) paleocanyons. Both are shown in Fig. 21.2 with a green line. The Peach Springs paleocanyon follows the Hurricane Fault (H) northward to where both the paleocanyon and the fault intersect the Colorado River. This part of the Peach Springs paleocanyon has been excavated to form a deep scar, the Peach Springs Canyon. Diamond Creek Road follows Peach Springs Canyon to the Colorado River. It is the only road to reach river level within the national park (other than at Lee's Ferry). The Milkweed-Hindu (M) paleocanyon was a large tributary to the Peach Springs paleoriver. This paleocanyon remains mostly filled with sediment.

The existence of these paleocanyons requires that river incision and the generation of relief must have occurred in the vicinity of the Grand Canyon prior to the filling of the paleocanyons with sediment. The age of incision, therefore, can be estimated from the age of the sediment. Based on regional considerations and the age of sediment-fill, we can suggest that paleocanyons developed during the Laramide orogeny sometime between 80 and 65 million years ago, and were filled or partially filled with sediment between 50 and 18 million years ago. Clearly these are ancient canyons. And the fact that the Peach Springs paleocanyon intersects the modern-day Colorado River implies that at least part of the Grand Canyon was in existence by 65 million years ago. It also implies that the area was elevated and had substantial relief.

All workers agree that initial uplift of the Grand Canyon region from sea level began during Laramide orogeny when the Colorado Plateau rose out of the Western Interior Seaway (Fig. 8.3). But the amount of elevation gain and the amount of relief is open to debate. Based on measurements of temperature-dependent carbon and oxygen isotopes in carbonate rocks, some researchers argue that present-day elevation was gained between 80 and 60 million years ago. However, these data have enough built-in uncertainty to allow for up to 1500 feet of elevation gain within the past 6 million years. One way to estimate relief at the time of paleocanyon-cutting is to measure the depth of the paleocanyons and the thickness of the sediment fill. Additionally, if we assume that the lowest elevation was above sea level at the time of paleocanyon cutting, then elevation can be no less than relief. The Hualapai Plateau surface at one location close to the

Peach Springs paleocanyon lies at 4700 feet, and the excavated paleocanyon at an elevation of 2200 feet. The difference in elevation suggests minimum paleorelief of 2500 feet. Detailed measurements suggest that paleorelief could have been in excess of 3900 feet, which is within 800 feet of present-day relief between the Shivwits Plateau and the Lower Granite Gorge. This relief must have been generated between 80 and 65 million years ago.

RIVER MORPHOLOGY

The orientation and depositional character of the Peach Springs and Milkweed-Hindu paleocanyons indicate they were incised by eastward and northward-flowing rivers. That is, by rivers that flowed in a direction opposite the modern Colorado River. Given the age of deposition in the paleocanyons, an eastward and northward-flowing river system must have been in existence by 65 million years ago. Unfortunately, the time during which rivers reversed direction is poorly constrained and has been estimated to be between 50 and 6 million years ago.

The morphology of the Canyon visible in Fig. 21.5 provides additional evidence for east and north-flowing rivers. Tributary streams typically enter a larger river in the same direction as flow. An example is the dendritic tributary pattern of the Mississippi River shown in Fig. 2.2. This type of pattern is in contrast to the pattern created by three tributary canyons visible in Fig. 21.5 located east of the Peach Springs paleocanyon (P), including one that contains the Toroweap fault. A second location with this type of river morphology is Marble Canyon, where several tributaries enter in a direction opposite the flow of the Colorado River.

An east and north-flowing river system is consistent with the inferred drainage direction of rivers circa 50 million years ago on the Colorado Plateau as depicted in Fig. 14.28 when many of the rivers emptied into lakes rather than an ocean. A possible relict of the paleoriver system is the north-flowing Little Colorado River, which turns sharply westward to enter the south-flowing Colorado River. Perhaps the Little Colorado River, 50 million years ago, continued northward through Marble Canyon. Notice on the left side of Fig. 14.28 that north-flowing paleorivers are inferred to be present at a location close to the modern-day Colorado River. Given the location and flow direction of these paleorivers, we can conclude that the modern Colorado River, as we know it today, did not exist 50 million years ago.

Evidence is solid that a northeast-flowing river did at one time exist in the Peach Springs paleocanyon. However, evidence is less certain that a north-flowing river at one time existed in Marble Canyon. An alternative possibility is that tributary streams on the Marble Platform have always flowed northward due to the regional slope of the land, but have always emptied into a south-flowing primary river. Marble Platform on both sides of Marble Canyon slopes consistently northward from an elevation above 6000 feet near the Little Colorado River to an elevation of 3125 feet at Lee's Ferry. Thus, the natural flow direction should be downhill to the north. Additionally, many of the tributaries, and Marble Canyon itself, follow fractures in the rock that could have helped channel the tributaries into a northward direction. Note, for example, that Marble Canyon is oriented roughly parallel with the Eminence fault zone. Such a conclusion, however, would imply that Marble Canyon, and its tributaries did not exist until a south- and west-flowing Grand Canyon river system was established. Additionally, if the tributaries are truly artifacts of the regional slope of the land, and there was never a northward-flowing river in Marble Canyon, then it begs the question as to how or why the Colorado River managed to cut progressively deeper into the Marble Platform in a southward direction against the regional slope of the land.

THE MODERN COLORADO RIVER

The Colorado River of today flows from the Colorado Rockies through the Grand Canyon into Lake Mead and then southward all the way to the Gulf of California. Defined as such, the river is only about 5.36 million years old. How do we know this? Here again we rely on geologic principles. The Colorado River flows out of the Rocky Mountains carrying rock fragments that are found only in the Rocky Mountains. These rock fragments, when found in river sediment, are like a fingerprint that indicates how long ago the river flowed across a certain area. Rock fragments indicative of the Rocky Mountains are present near Grand Junction, Colorado in river gravels that underlie the circa 10-million-year-old Grand Mesa basalt. Their presence implies that the Colorado River was in existence near Grand Junction prior to 10 million years ago. These same rock fragments, however, are absent in the Lake Mead area within loosely consolidated rock known as the Muddy Creek Formation, dated between 13 and 6 million years old. At one location, known as Sandy Point, river gravels that contain Rocky Mountain rock fragments are present directly above the Hualapai Limestone, which forms the uppermost rock unit in the Muddy Creek Formation, and which is dated at 5.97 million years old. These same river gravels do not appear at the Gulf of California until about 5.36 million years ago, which from Chapter 20, had recently opened via rifting.

What these relationships indicate is that there was an ancestral upper Colorado River that flowed across Colorado more than 10 million years ago, but this same river did not flow through the Grand Canyon until some time between 5.97 and 5.36 million years ago. There must have been a drainage divide in existence prior to 5.97 million years ago that prevented the upper Colorado River from flowing southward into the Canyon region. A likely location for this drainage divide is the Echo Cliffs near Lee's Ferry. It is here that the topographic expression of the Colorado River changes. North of the Echo Cliffs, the Colorado River and all its tributaries form an intricate network of deeply incised, closely spaced meandering canyons for which Utah is famous. Meanders are present within the Grand Canyon area south of the Echo Cliffs, but they do not characterize the river. The different river pattern on either side of the Echo Cliffs (EC) is visible in Fig. 21.6, which looks northwestward along the Colorado River. These differences suggest the two river segments have had different histories.

Given that the Colorado River did not reach the Pacific Ocean until 5.36 million years ago, it must have emptied into a succession of lakes during most of its existence. The location of these lakes is uncertain. Fig. 14.28 shows a precursor Colorado River circa 50 million years ago flowing a short distance southwestward from the present-day Gore Range in Colorado to where it emptied into a basin near Grand Mesa. During the long interval from 50 to 6 million years ago, before the river flowed through the Grand Canyon, it probably lengthened itself and emptied into lakes farther west. By 24 million years ago, the ignimbrite flare-up had blanketed the region in volcanic ash up to several hundred feet thick. The area north of the Echo Cliffs likely developed into one of lazy, meandering rivers as they flowed over the volcanic debris.

Recall from Chapter 14 that the Middle-Southern Rocky Mountains and Colorado Plateau underwent a period of broad uplift that began 10 to 6 million years ago (and may still be ongoing). The favored interpretation is that by 6 million years ago, broad uplift had triggered intense river incision resulting in the exhumation of the Rocky Mountains, the destruction of the depositional basins (lakes) shown in Fig. 14.28, and the removal of the drainage divide at Echo Cliffs. The breaching of the drainage divide caused rivers in Utah to reverse direction and begin flowing southward across the Echo Cliffs toward the Grand Canyon. The Colorado River, which previously flowed into closed basins north of the Echo Cliffs, became integrated with the new drainage system and now flowed through the breached divide into the Grand Canyon.

Once the divide at Echo Cliffs was breached, a southward-flowing, through-going Colorado River system would have lowered base level in the area north of the Echo Cliffs. Lowered base level, coupled with land uplift, would be the impetus for deep incision into what was previously a lazy meandering river system, thus creating the spectacular meandering canyon country north of the Echo Cliffs. The Grand Canyon area was not part of this lazy

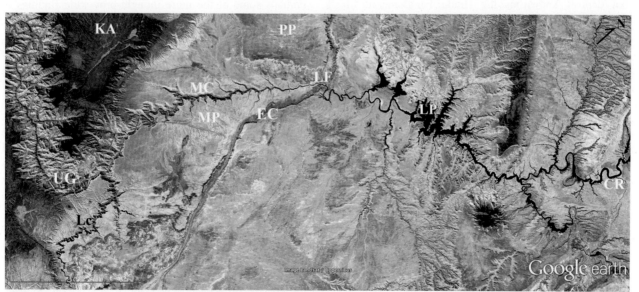

CR-Colorado River, **EC**-Echo Cliffs, **KA**-Kaibab Arch, **LF**-Lee's Ferry, **LP**-Lake Powell, **MC**-Marble Canyon, **MP**-Marble Platform, **PP**-Paria Plateau, **UC**-Upper Granite Gorge

FIGURE 21.6 Google Earth image looking northwest across the Echo Cliffs (EC). Note the change in river morphology from one of incised meanders north (right) of the Cliffs, to straighter channels in the Grand Canyon south of the Cliffs.

meandering river system, and therefore does not show the same density of incised meanders. Following passage of the Colorado River through the Canyon sometime between 5.97 and 5.36 million years ago, the river likely terminated in a lake. Over time, the river lengthened itself by overtopping the lake, or a succession of lakes, as it made its way to the newly opened Gulf of California. Still unexplained is how the Colorado River managed to incise Marble Canyon in a direction opposite the slope of the surrounding land surface.

ARGUMENT FOR A 6 MILLION-YEAR-OLD CANYON

The prevailing view for many of the most seasoned geologists of the Grand Canyon is that the Canyon formed mostly or entirely within the past 6 million years via incision by the Colorado River. The logic and reasoning, in its simplest form, is (1) the floor of the Grand Canyon is currently occupied by the Colorado River and (2) the Colorado River did not flow through the Grand Canyon until some time after 6 million years ago. Thus, if the Colorado River carved the Grand Canyon, the Canyon must be no older than the river that created it. The idea, again in its simplest form, is that the drainage divide at Echo cliffs (Lee's Ferry) was breached beginning 6 million years ago, and the added water provided by a through-going Colorado River was enough to cut the Canyon.

There is, of course, a great deal more evidence for a young Canyon. One prime piece of evidence is an alluvial fan deposit known as the Hindu Fanglomerate. It is located with a yellow box in Fig. 21.2 at the rim of the Milkweed-Hindu paleocanyon. A fanglomerate is a conglomerate deposited in an alluvial fan. The Hindu Fanglomerate appears to have been derived from a tributary stream that entered the Milkweed-Hindu (MH) paleocanyon from the north. Stratigraphic relationships indicate that the rock is less than 66 million years but older than 40 million years. What makes this rock special is that some of the rock fragments appear to have been derived from the Shivwits Plateau (S) located with a yellow star in Fig. 21.2 on the opposite side of the Colorado River. Given this relationship, if the Lower Granite Gorge were in existence during deposition of the fanglomerate, the stream that carried the fanglomerate would have been diverted into the gorge and could not have reached the Milkweed-Hindu paleocanyon. The Hindu Fanglomerate does not prove that the Canyon is less than 6 million years old, but it does imply that the Canyon is younger than the fanglomerate. The implication is that the Lower Granite Gorge could not have existed 70 million years ago at its present location in its present form because the fanglomerate is younger than 70 million years.

Part of the problem in deciphering the history of the Grand Canyon is the absence of sedimentary and volcanic deposits between 80 and 2 million years old. The only location where these deposits occur is on the Hualapai Plateau (HP). We noted previously that paleochannels on the plateau were filled with sediment between 50 and 18 million years ago. Although ages are not tightly constrained in all areas, the overall distribution of sediment across the Hualapai Plateau suggests near continuous deposition from about 50 million years ago to as late as 6 million years ago. Logic dictates that it would be unlikely for deep river incision to occur during periods of widespread river deposition. If we agree with this logic, then the Lower Granite Gorge, and the paleocanyons on the Hualapai Plateau (HP), most have been incised before 50 million years ago or after 6 million years ago. The Peach Springs and Milkweed-Hindu paleocanyons clearly were incised prior to 50 million years ago in response to Laramide orogeny. The Hindu Fanglomerate, on the other hand, could have been deposited on the Hualapai Plateau (HP) prior to 50 million years ago, but only if the Lower Granite Gorge was not yet in existence. Given that the Lower Granite Gorge could not have formed between 50 and 6 million years ago during widespread river deposition, the favored interpretation is for the Lower Granite Gorge to have formed less than 6 million years ago.

ARGUMENT FOR A 70 MILLION-YEAR-OLD CANYON

Perhaps some of the more thought-provoking findings in recent years have been the results of apatite fission track and (U-Th)/He analyses from the Grand Canyon area. Recall from Chapter 9 that both techniques record the time that a buried rock first reaches to within a few miles of the Earth's surface. Apatite fission tracks are preserved at temperatures below about 110°C (equivalent to less than 2.7 miles depth) and helium in apatite closes at about 30°C (equivalent to less than 4000 feet depth). Here, we look at data used for the original 70-million-year-old Canyon hypothesis. Data were obtained from crystalline shield rocks at river level, and from sedimentary rocks on the plateau surface as much as 5000 feet above river level.

Let us look first at the deeper Upper Granite Gorge. Apatite fission track dates from crystalline shield rocks exposed at river level are between 80 and 70 million years old. Apatite fission track dates from sedimentary rocks on the plateau surface 5000 feet above the canyon floor are as much as 50 million years older. An older age for rocks on the plateau would be expected because these rocks were not initially buried as deeply as the crystalline rocks. Fig. 21.7A is a cross-section that shows crystalline rock A

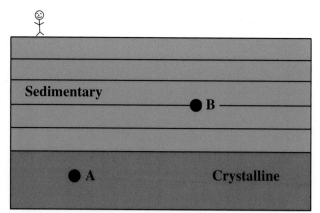

(A) Rock A within the crystalline shield is initially buried deeper than sedimentary rock B. If layers are stripped from both locations at the same rate, then rock B would cool earlier than rock A.

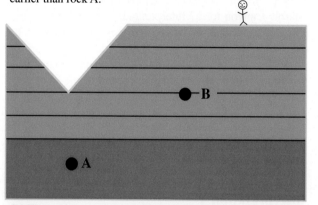

(B) The existence of a canyon allows rock A to be at the same depth below the surface as rock B. Both rocks will cool at the same time consistent with apatite data.

FIGURE 21.7 Cross-sections that show relative location and depth of crystalline and sedimentary rock.

at a greater depth than sedimentary rock B. If we assume that overlying layers are stripped (eroded) at the same rate from above both rocks, then sedimentary rock B should cool earlier than crystalline rock A, which is what we see. But this is not what we see when we look at the apatite (U-Th)/He data. Instead, we find that both rocks record similar ages of about 20 million years, implying the two rocks were at the same depth below the Earth's surface. A simple explanation is that the Upper Granite Gorge was already in existence 20 million years ago directly above crystalline rock A as shown in Fig. 21.7B.

There is actually more to the story regarding the Upper Granite Gorge. Recall again from Chapter 9 that there is a temperature range (known as the partial annealing zone) across which fission tracks are shortened but not destroyed and where some (but not all) of the trapped helium is able to escape from the crystal structure.

Sophisticated modeling of data through these temperatures suggests that crystalline rocks at river level and sedimentary rocks on the plateau were already at about the same depth as early as 70 million years ago. The modeling therefore suggests that an ancestral Upper Granite Gorge, nearly the size of the present-day gorge, was in existence 70 million years ago at the same location as the present-day gorge. Recall that the Colorado River did not flow through the canyon until 6 million years ago. If this interpretation is correct, then an ancient Upper Granite Gorge was cut by a river other than the Colorado.

Now let us look at the smaller but equally impressive Lower Granite Gorge. Data from this area are just as surprising. Both apatite fission track and the apatite (U-Th)/He dates from river level are between 90 and 70 million years old and one apatite (U-Th)/He date from the plateau is 68 million years. These data suggest that river-level crystalline rocks cooled rather quickly from above 110°C to below 30°C prior to about 70 million years ago and that, by about 68 million years ago, they were at about the same depth below the surface as sedimentary rocks on the plateau rim. Here again, the simplest interpretation is that the Lower Granite Gorge was cut at the location where it is today during a period of rapid cooling between 90 and 70 million years ago (Fig. 21.7B). The apatite data therefore suggest that nearly the entire Grand Canyon was cut prior to about 70 million years ago by a river other than the Colorado.

GEOLOGIC HISTORY

Any theory on the origin of the Canyon must be consistent with the erosional and depositional history that was painstakingly gathered over the course of more than 100 years. Different theories arise because not enough data have been obtained, critical field relationships are missing, two pieces of field evidence contradict each other, or the apatite data contradicts field data. From a field perspective, with the exception of the Hualapai Plateau, the absence of sedimentary and volcanic rock in the age range of 80 to 2 million years leaves a hole in the dataset that cannot be filled. From an apatite geochronology perspective, the volume of apatite data so far collected is sparse relative to the size of the Grand Canyon, and the data are complex enough that different assumptions and methods can be used to model the data, which can result in different conclusions. Such problems allow for several valid theories to be constructed. It is then up to the geologist to evaluate each theory, identify problems or weaknesses, and obtain new data that can prove or disprove part of a theory. We will discuss three of several possible interpretations, each based on the most recently gathered field, apatite, and chemical data available at the time the interpretation was made. Each interpretation is consistent with the general history of the Grand Canyon described next.

Among those who have worked in the Grand Canyon, there is general consensus regarding the following depositional and erosional history. Beginning 80 million years ago, before the Grand Canyon came into existence, the nearly flat-lying sedimentary interior platform rock section above the crystalline shield was on the order of 12,000 feet thick, more than twice the present-day thickness of 5000 feet, and nearly as thick as the sedimentary section that underlies the High Plateaus in Fig. 13.38. By 70 million years ago, uplift had migrated into the Grand Canyon region coincident with Laramide orogeny and the rise of the Middle-Southern Rocky Mountains. Uplift produced mild deformation that included development of the Kaibab Arch and the Hurricane and Toroweap faults. Other areas of the Canyon remained relatively undeformed. Uplift also triggered bench and slope erosional stripping of layers and initial incision of the Milkweed-Hindu and Peach Springs paleocanyons. By 50 million years ago, the Hualapai Plateau was stripped of its thick section of nearly flat-lying sedimentary rock all the way down to its present-day land surface. The Milkweed-Hindu and Peach Springs paleocanyons were active and undergoing deposition by this time. The area of the Upper Granite Gorge was also stripped of sedimentary rock, but not all the way down to its present-day land surface. There remained a thicker section of sedimentary rock at this location than what is present today.

The time period between 50 and 25 million years ago was largely one of deposition that includes ash from the ignimbrite flare-up. Apparently there were no major or lasting changes to landscape in the Canyon region during this time interval. A period of broad uplift began sometime after 25 million years ago in the Upper Granite Gorge, and by 15 million years ago bench and slope erosion had stripped the remaining sedimentary rock from the Kaibab and Coconino Plateau areas down to the present-day land surface. This uplift apparently did not affect the Lower Granite Gorge. Beginning around 17 million years ago active normal faults began lowering the Basin and Range region to the west, including the Lake Mead area. Some time after 10 million years ago there began another period of broad uplift, and between 5.97 and 5.36 million years ago the modern Colorado River had established itself as the through-going river in the Canyon region.

REVISED ARGUMENTS

We have seen from our previous discussion that the Colorado River did not occupy the Grand Canyon prior to 5.97 million years ago, that apatite fission track and (U-Th)/He data suggest the existence of a canyon that is far older than 6 million years, and that the Peach Springs paleocanyon was already in existence no later than about 65 million years ago. Given the fact that the Peach Springs paleocanyon intersects the modern-day Colorado River, we can also suggest that at least a small part of the Grand Canyon was already in existence by 65 million years ago.

Three interpretations are presented in this section, each of which is consistent with the geologic history outlined above. Each interpretation is based on publications referenced in associated figure captions, but keep in mind that it is possible that the interpretations presented here differ slightly from the ones presented in those publications. For each interpretation, we will address the following key questions. How do the Hualapai paleochannels fit into the overall history? How did the Colorado River manage to cut across the Kaibab Arch? And how did Marble Canyon form? Interpretation 1 maintains that both the Upper and Lower Granite Gorges were established at their present location during Laramide orogeny prior to 55 million years ago, and that subsequent erosion maintained, modified, and deepened the canyons. Interpretations 2 and 3 argue that the modern Grand Canyon formed in the past 6 million years by incorporating old canyon segments that formed prior to 6 million years ago with young canyon segments that formed less than 6 million years ago.

Interpretation 1

Beginning around 80 million years ago, the Milkweed-Hindu and Peach Springs paleocanyons acted as tributaries to a now vanished, ancient river known as the California River. The California River flowed eastward from southern California through the Lower and Upper Granite Gorges, and then northward through Marble Canyon and the Echo Cliffs. This river, in other words, took the same route through the canyon as the modern-day Colorado River except that it flowed in the opposite direction. The California River is superimposed across present-day landscape in Fig. 21.8A. By 70 million years ago, the California River had incised the Lower Granite Gorge to within 70% to 80% of its present-day depth, and bench and slope erosional stripping had removed sedimentary layers down to the present-day Hualapai Plateau land surface. If we use a present-day depth of 3555 feet from Canyon bottom to the rim of the Hualapai Plateau, we can suggest that between 2480 and 2850 feet of the Canyon was cut prior to 70 million years ago. In the Upper Granite Gorge, the California River cut across the rising Kaibab Arch, and by 55 million years ago (and possibly as early as 70 million years ago) had cut a canyon about 3300 feet deep (1 km), or about 57% of the total depth of the gorge using the depth (5850 feet)

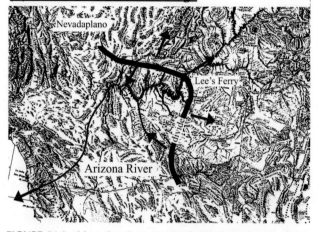

FIGURE 21.8 Maps that show (A) the California River and (B) the Arizona River, superimposed on present-day landscape. *Based on Flowers et al. (2008) and Wernicke (2011).*

provided in our earlier discussion. Thus, both the Lower and Upper Granite Gorges were in existence by 55 million years ago and possibly as early as 70 million years ago. By that time, the Lower Granite Gorge was already incised down to shield rocks and the surrounding present-day Hualapai Plateau surface was already exposed. The Upper Granite Gorge was still entirely within the interior platform rock succession, which was thicker 55 million years ago than it is today.

The rise of the Rocky Mountains by 40 million years ago caused the Colorado Plateau to tilt gently westward. At the same time, the drainage divide in southern California shown in Fig. 21.8A was breached such that there was a clear path from the Colorado Plateau to the Pacific Ocean. The combination of uplift and a breached drainage divide caused rivers in the Grand Canyon to reverse direction and began flowing toward the Pacific Ocean. The California River was destroyed. Given the preexistence of a well-developed east-northeast-flowing California river network, one would expect ponding and choking of river channels and much deposition as water first stagnated and then began to reverse direction. Such a scenario may have contributed to deposition of sediment that filled paleochannels on the Hualapai Plateau.

Coincident with stream reversal was development of a rather small river system originating this time on the Colorado Plateau south of the Echo Cliffs, which was now acting as a drainage divide. This river, known as the Arizona River, is shown circa 40 million years ago in Fig. 21.8B superimposed on present-day topography. The Arizona River flowed westward toward the Pacific Ocean in a direction opposite that of the California River, but probably utilized many of the pre-existing river channels including the Grand Canyon. The Arizona River at this time did not have the cutting power to further incise the Grand Canyon to any substantial degree.

By about 20 million years ago, a pulse of broad uplift in the Upper Granite Gorge provided the Arizona River with greater cutting power. Bench and slope erosional stripping removed sedimentary layers down to the present-day plateau surface, and at the same time, a stronger Arizona River incised into rocks of the crystalline shield. Rather than substantially add to the depth of the Upper Granite Gorge, the Arizona River simply maintained its depth as the surrounding plateau surface was stripped of its top layers down its its present-day surface.

By about 17 million years ago, normal faulting in the Basin and Range west of the Grand Canyon had begun to dismember and destroy the Arizona River leaving a small river system with internal drainage and very little cutting power. The canyon and surrounding plateau underwent little if any change until about 6 million years ago when the Colorado River breached the drainage divide at Echo Cliffs and spilled into the Canyon utilizing the remains of the California and Arizona River systems, including Marble Canyon. Most of the remaining incision (20%–30% in the Lower Granite Gorge, and 43 percent in the Upper Granite Gorge) occurred at this time. A summery of this interpretation is shown in Fig. 21.9A superimposed on a Raisz landform map of the Grand Canyon.

Interpretation 2

The second interpretation, shown in Fig. 21.9B, suggests that the Canyon was built in three stages, 70 to 50 million years ago, 25 to 15 million years ago, and less than 6 million years ago. During the early (Laramide) stage, the Milkweed-Hindu and Peach Springs paleorivers flowed northward along the eastern side of the present-day Lower Granite Gorge exiting the Canyon near the Toroweap fault and continuing northward. In doing so, it created a canyon 1700–1800 feet deep, or about half its present-

The Lower Granite Gorge was incised about 2480-2850 feet (70-80% of total) and the Upper Granite Gorge was incised about 3300 feet (57% of total) prior to 55 Ma (possibly prior to 70 Ma) by the east-flowing California River (dashed where inferred). The Peach Springs River was a tributary to the California River. The west-flowing Arizona River established itself prior to 40 Ma. It maintained the Upper Granite Gorge during bench and slope erosional stripping 20 Ma, but did not significantly deepen it. The remaining incision took place less than 6 Ma by the Colorado River.

The eastern side of Lower Granite Gorge was incised to half its depth (1700-1800 feet) by 50 Ma via the north-flowing Peach Springs River. The remaining incision took place less than 6 Ma by the Colorado River.

The western side of Upper Granite Gorge was incised roughly 3300 feet (57% of total) between 25 and 15 Ma via the Little Colorado River. The remaining incision took place less than 6 Ma by the Colorado River.

Marble Canyon and the western side of Lower Granite Gorge were both incised in the past 6 Ma by Colorado River.

Part of the Canyon was incised via the north-flowing Peach Springs River prior to 50 Ma. The Little Colorado River flowed northward through Marble Canyon prior to 50 Ma.

Headward erosion between 17 and 8 Ma along the western side of both the Lower and Upper Granite Gorge established a west-flowing river system from the west side of Kaibab Plateau to Lake Mead. The Little Colorado River continued northward through Marble Canyon.

Between 8 and 7 Ma, the Little Colorado River was diverted through a cave system (Karst) below the Kaibab Arch to connect with the west-flowing river system.

Northward headward erosion up Marble Canyon resulted in drainage reversal and establishment of the modern-day Colorado River system by 6 Ma. Much of the canyon deepening occurred after 6 Ma.

FIGURE 21.9 Summary maps for three interpretations on the origin of the Grand Canyon. (A) Interpretation 1 from Flowers and Farley (2012). (B) Interpretation 2 based on Karlstrom et al., 2014. (C) Interpretation 3 based on Hill and Polyak (2014).

day depth measured from the rim of the Hualapai Plateau. Other parts of the Grand Canyon had not yet formed. Little change occurred until 25 to 15 million years ago when the Little Colorado River flowed through the Upper Granite Gorge with enough power to cross the Kaibab Arch, exiting near Kanab Creek and incising the Upper Granite Gorge to a depth estimated at 3300 feet (1 km ± 0.5 km), or about 57% of its present-day depth. Marble Canyon and the western side of Lower Granite Gorge developed more or less in their entirety beginning 6 million years ago, after the Colorado River was integrated through the Grand Canyon. This interpretation does not directly address how the Colorado River flowed across the Marble Platform opposite the surface slope, or how the Little Colorado River crossed the Kaibab Arch. Both could have occurred in any number of ways, including those described in Interpretation 3.

Interpretation 3

The third interpretation, shown in Fig. 21.9C, is based on an analysis of karst features primarily in the Mauv and Redwall Limestone. The amount of pre-6-million-year incision is not constrained in this interpretation; however, significant incision is thought to have occurred less than 6 million years ago following integration of the Colorado River through the Canyon.

Prior to 50 million years ago, the Peach Springs paleoriver flowed northward along the eastern side of the present-day Lower Granite Gorge, and then east along the Canyon floor all the way to the east side of the Kaibab Arch before turning north. This river system may have persisted until around 17 million years ago, although it would have been filled or partially filled with gravel. The Kaibab Arch was acting as a drainage divide 50 million years ago that separated the Peach Springs paleoriver from the Little Colorado River, which was already in its modern-day channel except that it continued northward through Marble Canyon, draining into a lake located north of the Echo Cliffs. Other parts of the Grand Canyon had not yet formed.

Normal faulting in the Lake Mead area beginning around 17 million years ago created a drainage divide in which a river drained westward from the newly uplifted Grand Wash Cliffs. This river progressively lengthened in a southeastward direction via headward erosion along western side of the Lower Granite Gorge to where it eventually intersected the Peach Springs drainage system. Once this connection was complete, the Peach Springs drainage system reversed direction, thus creating a western Grand Canyon river system that flowed westward from the west side of the Kaibab Arch to Lake Mead and beyond. This river system was not as deep as the present-day Canyon and did not cross the Kaibab Arch. The Little

Colorado River was still flowing north through Marble Canyon. The Upper Granite Gorge had not yet formed.

By about 8 million years ago, the west-flowing river system had cut eastward and headward into the Kaibab Arch along the west side of the Upper Granite Gorge. At the same time, a cave system developed below the Kaibab Arch. Between 7 and 6 million years ago, the Little Colorado River was captured in sinkholes and redirected into the cave system below the Kaibab Arch such that the Little Colorado no longer flowed through Marble Canyon. The redirected river water would soon collapse the cave system and connect with the existing river system on the west side of the Kaibab Arch creating a through-going Little Colorado River system. The Colorado River had not yet crossed the Echo Cliffs.

Following capture of the Little Colorado River, headward erosion progressed northward from the Little Colorado River up the now abandoned Marble Canyon to where it crossed the Echo Cliffs soon after 6 million years ago. As headward erosion progressed north of the Echo Cliffs, it captured the Colorado River, thereby creating the modern-day river system through the Grand Canyon. In this interpretation, most of Marble Canyon formed after 7 million years ago, and an uncertain percentage of the Grand Canyon was carved by the Colorado River following integration sometime after 6 million years ago.

FINALE

There you have it. Geology is a detective story where the hard truth is difficult to reveal and seldom known with absolute certainty. It is a science where much of the evidence is lost or out of reach, where one has to combine logic with geologic fundamentals, knowledge, and experience without preconceived ideas that may cloud interpretation. It is a science where one must mesh field data with laboratory data and figure out why these two avenues sometimes seemingly contradict. We chip away and ponder, chip away and ponder. The day is long and there is much work to be done. Geology at its best.

QUESTIONS

1. How long would the Toroweap fault have to remain active in order to accumulate 3000 feet of displacement? If such displacement did occur, how would it affect the look of the Grand Canyon?
2. What is the total vertical drop of the Colorado River from Lee's Ferry to the Grand Wash Cliffs?
3. Rivers north of Lee's Ferry on the Colorado Plateau are characterized by incised meanders. What are incised meanders? Suggest a process by which they could form.
4. State very succinctly, what is the best argument in your opinion in favor of a Grand Canyon mostly less than 6 million years old.
5. Describe in one paragraph, the incision history of the Upper Granite Gorge or Lower Granite Gorge based on apatite data.
6. Research John Wesley Powell and the early exploration history of the Grand Canyon.
7. Why is the incision rate of the Colorado River higher east of the Toroweap fault than west of the fault?
8. Describe the history of the California River. The Arizona River. The Colorado River. The Little Colorado River.
9. Use Fig. 21.5 as a guide to locate the Toroweap fault in Google Earth. Describe the topographic expression of this fault from south to north. Do the same for the Hurricane, Bright Angel, and Eminence faults. Describe evidence for a graben along the Eminence fault.
10. Fly to 36° 07′ 02.30″ N 111° 51′ 27.60″ W or to 36° 15′ 2.27″ N 111°53′24.12″W and fly around the area looking at the rocks visible along the canyon walls. Are these rocks part of the interior platform, Precambrian sedimentary-volcanic, or shield rock succession? Explain what you see and how you came to your conclusion.
11. Interpretation 2 does not offer an explicit explanation as to how or why the Colorado River cut downward against the surface slope of the Marble Platform to cut Marble Canyon or why the Little Colorado River crossed the Kaibab Arch. Speculate on the origin of both of these.
12. Compare all three interpretations and describe areas where they are different and where they are the same.
13. Describe an origin for the Grand Canyon that combines Interpretations 2 and 3. Some speculation is required.
14. There are several interpretations as to how the Grand Canyon crossed the Kaibab Arch in addition to those described in this chapter. What are the interpretations described here. Research some additional interpretations?
15. What evidence suggests that the westernmost part of the Canyon is either more than 50 million years old or less than 6 million years old?
16. Research the Bidahochi Formation and Lake Bidahochi. What is the age of this lake and how does it figure in the origin and history of the Grand Canyon?

Appendix

The appendix contains enlarged and uncolored versions of maps found in the text with text figure numbers given in parentheses.

FIGURE A.1 *Color shaded relief map* of the Conterminous United States. Hundred-meter resolution color-sliced elevation image with relief shading added to accentuate terrain features, in an Albers Equal-Area Conic projection. U.S. Geological Survey, The National Atlas of the United States of America, Nationalmap.gov/small_scale/atlasftp.html. Dark blue signifies lowest elevation. Deep brown signifies highest elevation.

FIGURE A.2 Raisz landform *base map* (Fig. 1.3).

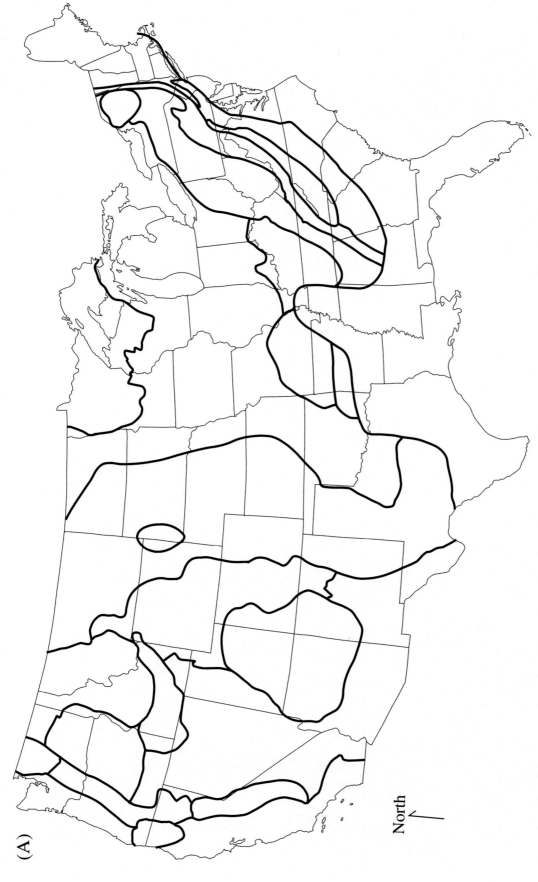

FIGURE A.3 (A) US state map with *physiographic province* boundaries (Fig. 1.6). (B) Raisz landform map with *physiographic province* boundaries (Fig. 1.6).

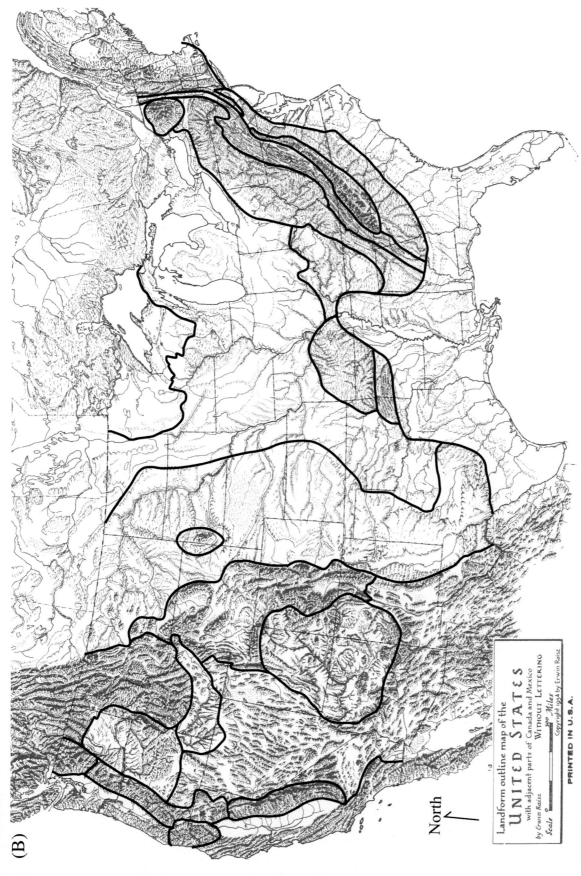

FIGURE A.3 (Continued)

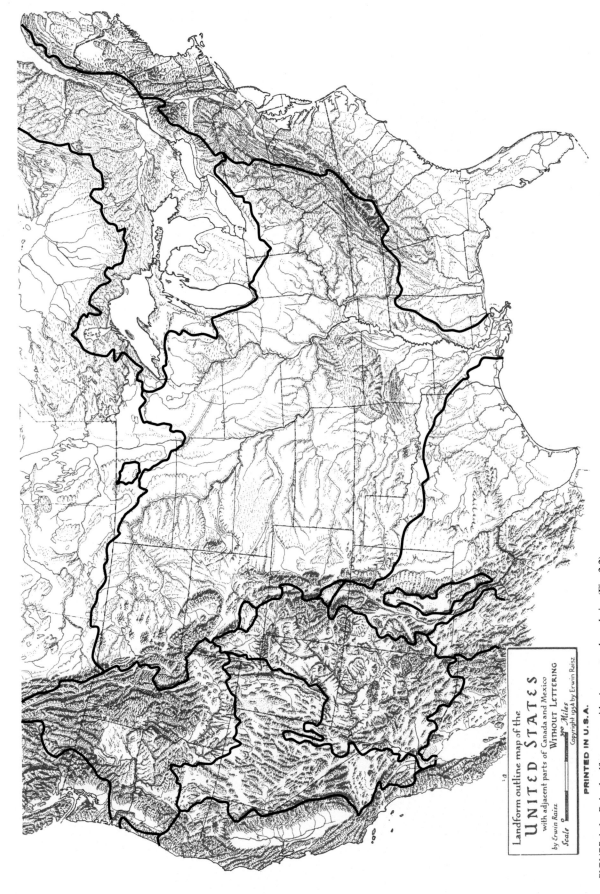

FIGURE A.4 Raisz landform map with *river system* boundaries (Fig. 2.2).

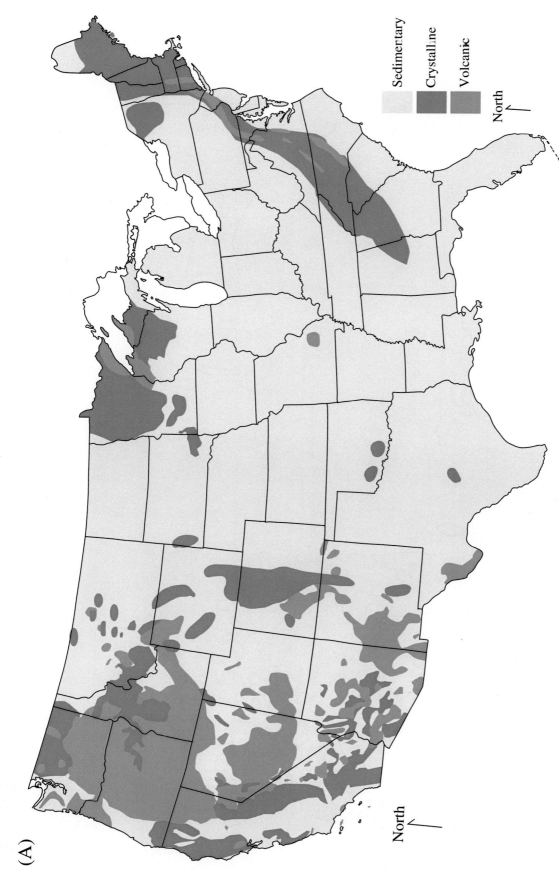

FIGURE A.5 (A) Colored US state map showing *rock type* (Fig. 3.16). (B) Raisz landform map with *rock type* boundaries. A few crystalline and volcanic rocks are colored (Fig. 3.16).

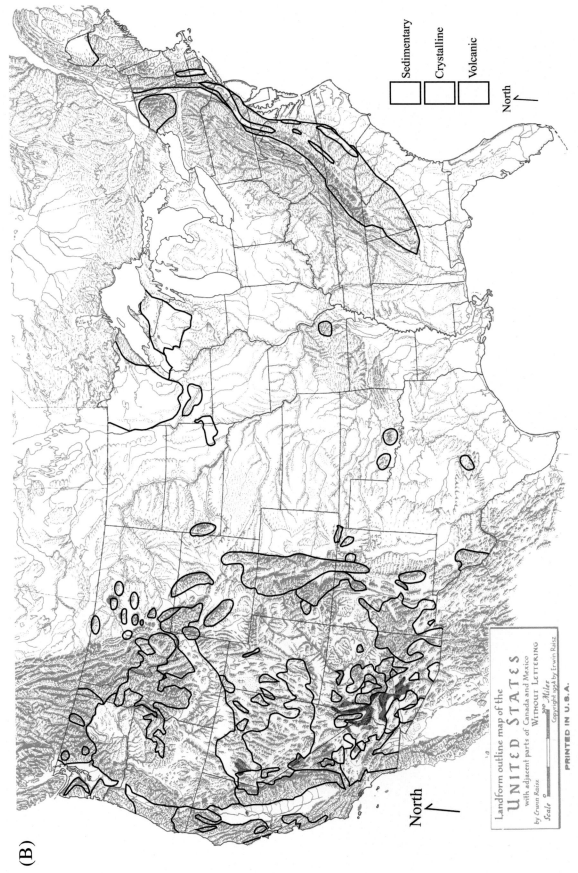

FIGURE A.5 (Continued).

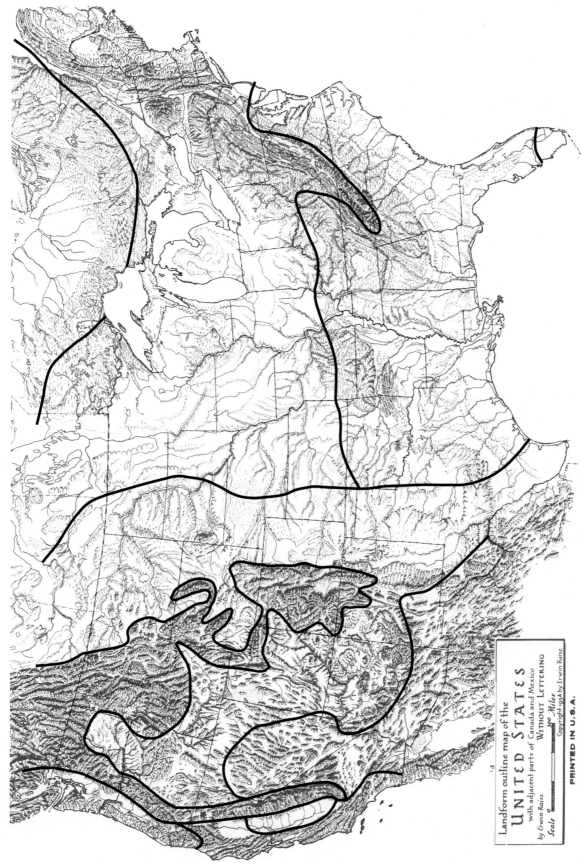

FIGURE A.6 Raisz landform map with *climate zone* boundaries (Fig. 6.1).

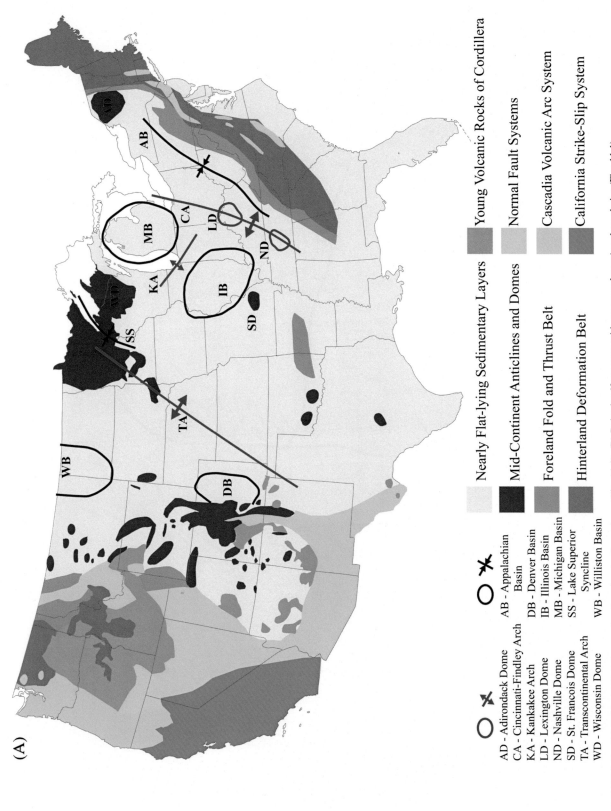

FIGURE A.7 (A) Colored US state map showing *structural provinces* (Fig. 11.1). (B) Raisz landform map with *structural province boundaries* (Fig. 11.1).

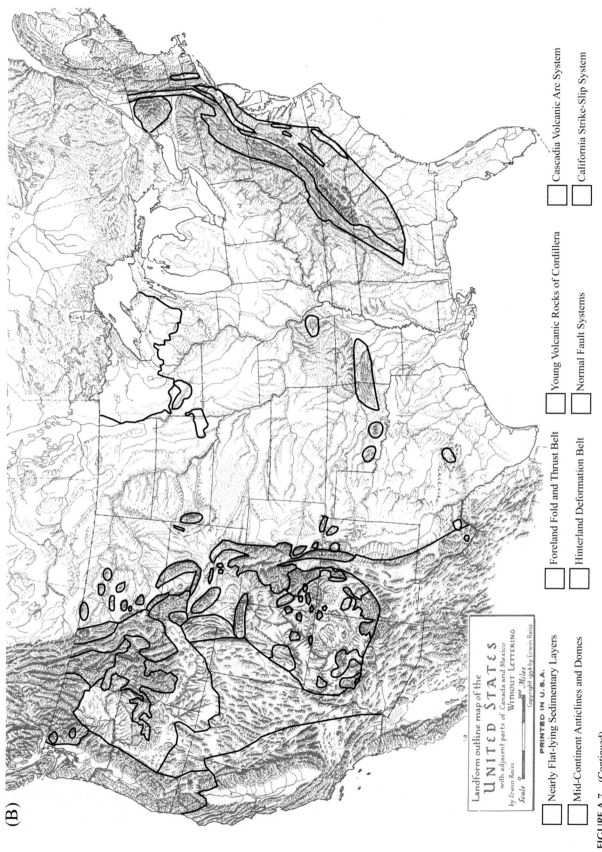

FIGURE A.7 (Continued)

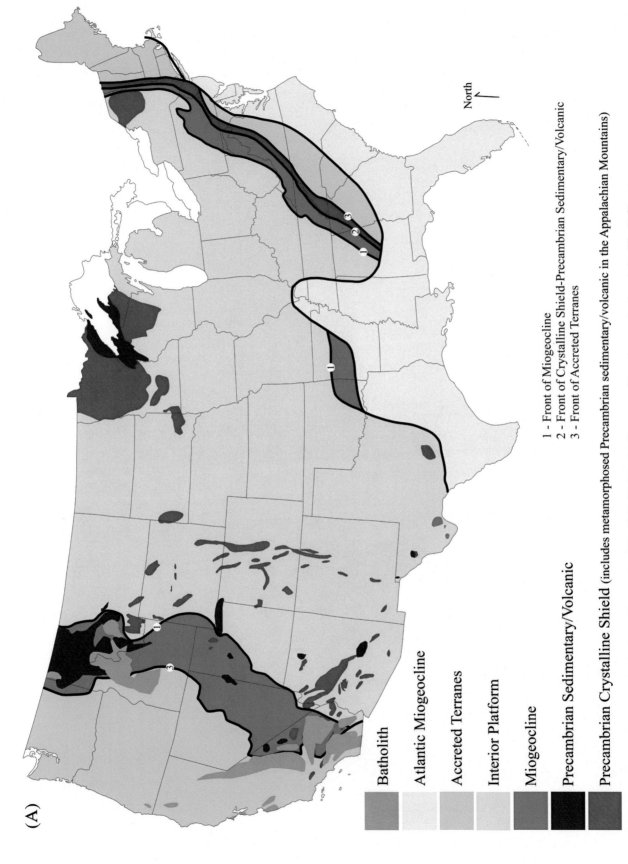

FIGURE A.8 (A) Colored US state map showing *rock successions* (Fig. 11.8). (B) Raisz landform map with *rock succession* boundaries (Fig. 11.8).

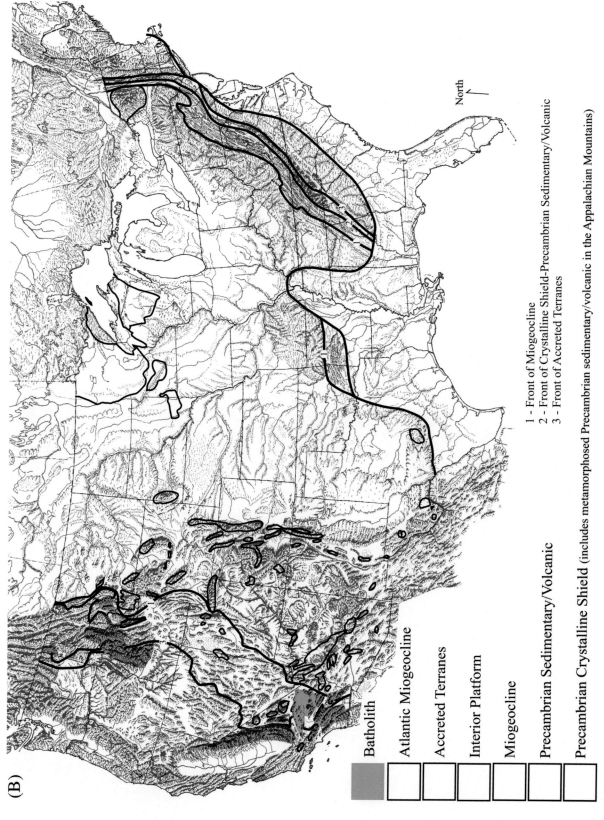

FIGURE A.8 (Continued)

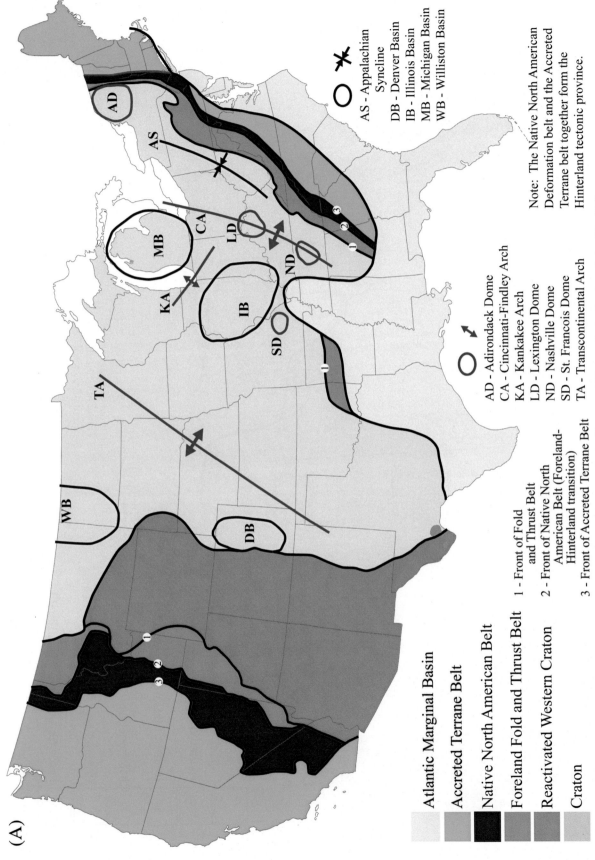

FIGURE A.9 (A) Colored US state map showing *tectonic provinces* (Fig. 11.9). (B) Raisz landform map with *tectonic province* boundaries (Fig. 11.9).

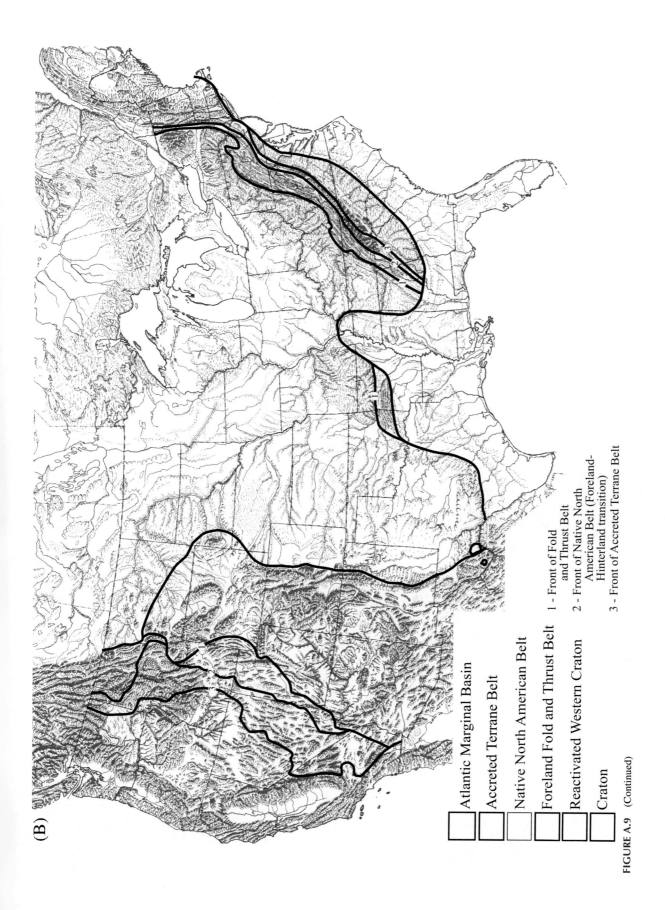

FIGURE A.9 (Continued)

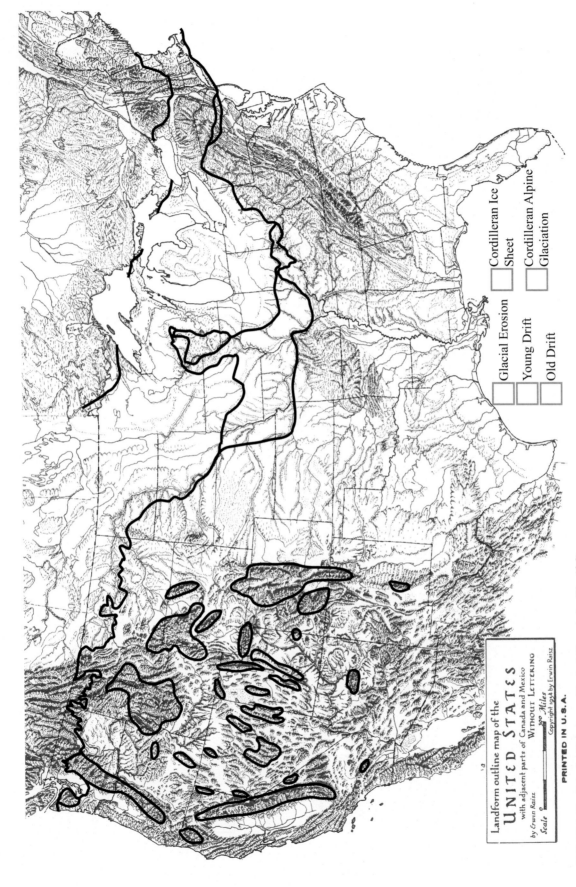

FIGURE A.10 Raisz landform map with *glacial* boundaries (Fig. 12.9).

FIGURE A.11 Raisz landform map *volcanic rock boundaries* (Fig. 17.1).

References

CHAPTERS 1–11: KEYS TO UNDERSTANDING LANDSCAPE EVOLUTION

Antonelis, K., Johnson, D. J., Miller, M. M., & Palmer, R. (1999). GPS determination of current Pacific-North American plate motion. *Geology, 27*, 299–302.

Atwater, T. (2016). Educational multimedia visualization center. <http://emvc.geol.ucsb.edu/1_DownloadPage/Download_Page.html>.

Bally, A. W. (1989). *Geology of North America: An overview* Vol. A. Boulder, CL: GSA DNAG, 619 pages.

Barber, D. C., Dyke, A., Hillaire-Marcel, C., Jennings, A. E., Andrews, J. T., Kerwin, M. W., et al. (1999). Forcing of the cold event of 8,200 years ago by catastrophic drainage of Laurentide lakes. *Nature, 400*, 344–348.

Bennett, R. A., Davis, J. L., & Wernicke, B. P. (1999). Present-day pattern of Cordilleran deformation in the Western United States. *Geology, 27*, 371–374.

Binnie, S. A., & Summerfield, M. A. (2013). 7.6 Rates of Denudation. *Treatise on geomorphology, 7*, 66–72, Chap. 6.

Bloom, A. L. (1998). *Geomorphology: A systematic Analysis of Late Cenozoic Landforms* (3rd ed.). Upper Saddle River, New Jersey: Prentice Hall.

Blum, M.D., & Törnqvist, T.E. (2000) Fluvial response to climate and sea-level change: A review and a look forward. *Sedimentology, 47*, 2–48, <https://doi.org/10.1046/j.1365-3091.2000.00008.x>.

Burbank, D.W., & Anderson, R.S. (2001). *Tectonic geomorphology.* Malden, MA: Blackwell Science.

Burbank, D. W., & Beck, R. A. (1991). Rapid, long-term rates of denudation. *Geology, 19*(12), 1169–1172.

Burbank, D. W., & Pinter, N. (1999). Landscape evolution; the interactions of tectonics and surface processes. *Basin Research, 11*(1), 1–6.

Burchfiel, B. C., Lipman, P. W., & Zoback, M. L. (1992). *Cordilleran Orogen: Conterminous U.S.* Vol. G3. Boulder, CL: GSA DNAG, 724 pages.

Cazenave, A., Dieng, H.-B., Meyssignac, B., von Schuckmann, K., Decharme, B., & Berthier, E. (2014). The rate of sea-level rise. *Nature Climate Change, 4*, 358–361.

Centre for Ice and Climate, Niels Bohr Institute, University of Copenhagen, <http://www.iceandclimate.nbi.ku.dk/research/past_atmos>.

Chernicoff, S., & Whitney, D. (2007). *Geology: An introduction to physical geology* (4th ed.). Upper Saddle River, New Jersey: Pearson Prentice Hall.

Church, J. A., Woodworth, P. L., Aarup, T., & Wilson, S. (Eds.), (2010). *Understanding sea-level rise and variability.* Blackwell Publishing Ltd.

Church, J. A., White, N. J., Aarup, T., Wilson, W. S., Woodworth, P. L., Domingues, C. M., ... Lambeck, K. (2008). Understanding global sea levels: Past, present and future. *Sustainability Science, 3*(1), 9–22.

Clark, P. U., Marshall, S. J., Clarke, G. K. C., Hostetler, S. W., Licciardi, J. M., & Teller, J. T. (2001). Freshwater forcing of abrupt climate change during the last glaciation. *Science, 293*(5528), 283–287.

Colman, S. M., & Mixon, R. B. (1988). The record of major quaternary sea-level changes in a large coastal plain estuary, Chesapeake Bay, Eastern United States. *Palaeogeography, Palaeoclimatology, Palaeoecology, 68*(2–4), 99–116.

Csank, A. Z., Tripati, A. K., Patterson, W. P., Eagle, R. A., Rybczynski, N., Ballantyne, A. P., & Eiler, J. M. (2011). Estimates of Arctic land surface temperatures during the early Pliocene from two novel proxies. *Earth and Planetary Science Letters, 304*(3–4), 291–299.

Curren, H. A., Justus, P. S., Young, D. M., & Garver, J. B. (1984). *Atlas of landforms* (3rd ed.). United States Military Academy, John Wiley & Sons, Inc., 165 p.

England, P., & Molnar, P. (1990). Surface uplift, uplift of rocks, and exhumation of rocks. *Geology, 18*, 1173–1177.

England, P., & Molnar, P. (1991a). Surface uplift, uplift of rocks, and exhumation of rocks: Reply. *Geology, 19*, 1051–1052.

England, P., & Molnar, P. (1991b). Surface uplift, uplift of rocks, and exhumation of rocks: Reply. *Geology, 19*, 1053–1054.

Etheridge, D. M., Steele, L. P., Langenfelds, R. L., Francey, R. J., Barnola, J.-M., & Morgan, V. I. (1996). Natural and anthropogenic changes in atmospheric CO_2 over the last 1000 years from air in Antarctic ice and firn. *Journal of Geophysical Research, 101*, 4115–4128.

Fedorov, A. V., Brierley, C. M., Lawrence, K. T., Liu, Z., Dekens, P. S., & Ravelo, A. C. (2013). Patterns and mechanisms of early Pliocene warmth. *Nature, 496*, 43–49. Available from https://doi.org/10.1038/nature12003.

Geologic maps of US states. US Geological Survey, <http://mrdata.usgs.gov/geology/state/>.

Godt, J.W. (1997). Digital compilation of landslide overview map of the conterminous United States. In: D.H., Radbruch-Hall, R.B. Colton, W.E. Davies, I. Lucchitta, B.A. Skipp, & D.J. Varnes (Eds.), 1982. US Geological Survey Open File Report pp. 97–289. <http://landslides.usgs.gov/hazards/nationalmap/>.

Gornitz, V., Lebedeff, S., & Hansen, J. (1982). Global sea level trend in the past century. *Science, 215*(4540), 1611–1614.

Gornitz, V., & Seeber, L. (1990). Vertical crustal movements along the East Coast, North America, from historic and late Holocene sea level data. *Tectonophysics, 178*(2–4), 127–150.

Graf, W.L. (Ed.). (1987). Geomorphic systems of North America, Vol. 2. Boulder, CL: GSA DNAG, 643 pages.

Grotzinger, J., & Jordan, T. (2010). *Understanding earth.* New York: W. H. Freeman, 654 pages.

Hamblin, W. K., & Christensen, E. H. (2001). *Earth's dynamic systems.* Upper Saddle River, NJ: Prentice-Hall, 735 pages.

Hardwick, S. W., Shelley, F. M., & Holtgrieve, D. G. (2008). *The Geography of North America*. Upper Saddle River, New Jersey: Pearson Prentice Hall.

Harris, A.G., Tuttle, E., & Tuttle, S.D. (1995). *Geology of national parks* (5th ed.). Dubuque, IA: Kendell Hunt.

Hasterok, D., & Chapmann, D. S. (2007). Continental thermal isostasy; 2, Application to North America. *Journal of Geophysical Research*, *112*(B6).

Hatcher, R.D. Jr., Thomas, W.A., & Viele, G.W. (Eds.). (1989). The Appalachian - Ouachita Orogen in The United States, Vol. F2. Boulder, CL: GSA DNAG, 767 pages.

Hatfield, C. B. (1991). Surface uplift, uplift of rocks, and exhumation of rocks: Comment. *Geology*, *19*, 1051.

Haug, G. H., Tiedemann, R., & Keigwin, L. D. (2004). How the Isthmus of Panama put ice in the Arctic. *Oceanus*, *42*(2). 7 p., <http://vishnu.whoi.edu/services/communications/oceanusmag.050826/v42n2/haug.html>.

Henry, J.A., & Mossa, A. (1995). *Natural landscapes of the United States*. Dubuque, IA: Kendall Hunt.

Hillaire-Marcel, C., & Fairbridge, R.W. (1978) Isostasy and eustasy of Hudson Bay. *Geology*, *6*, 117–122.

Hodges, K. V. (2004). *Geochronology and thermochronology in orogenic systems Treatise on geochemistry* (pp. 263–292). Oxford: Elsevier.

Hunt, C. B. (1974). *Natural regions of the United States and Canada*. San Francisco, CA: W.H. Freeman.

Hyndman, R. D., & Currie, C. A. (2011). Why is the North America Cordillera high? Hot backarcs, thermal isostasy, and mountain belts. *Geology*, *39*(8), 783–786.

Idaho Geological Survey, Digital Geology of Idaho. <http://geology.isu.edu/Digital_Geology_Idaho/>.

IPCC. (2013). *Climate change 2013: The physical science basis. Contribution of Working Group I to the Fifth Assessment Report of the Intergovernmental Panel on Climate Change*. In Stocker, T.F., Qin, D., Plattner, G.-K., Tignor, M., Allen, S.K., Boschung, J., Nauels, A., Xia, Y., Bex, V., & Midgley, P.M. (Eds.). Cambridge University Press, Cambridge and New York, NY, 1535 pp., <http://www.ipcc.ch/report/ar5/wg1/>.

Kammerer, J.C., (1990), Largest rivers in the United States. US Geological Survey. <http://pubs.usgs.gov/of/1987/ofr87-242/>.

Keller, E. A., & Pinter, N. (2002). *Active tectonics: Earthquakes, uplift, and landscape* (2nd ed.). Upper Saddle River, New Jersey: Prentice Hall.

Kemp, A. C., Horton, B. P., Culver, S. J., Corbett, D. R., van de Plassche, O., Gehrels, W. R., ... Parnell, A. C. (2009). Timing and magnitude of recent accelerated sea-level rise (North Carolina, United States). *Geology [Boulder]*, *37*(11), 1035–1038.

King, P. B. (1977). *The evolution of North America*. Princeton, NJ: Princeton University Press.

Kirchner, J. W., Finkel, R. C., Riebe, C. S., Granger, D. E., Clayton, J. L., King, J. G., & Megahan, W. F. (2001). Mountain erosion over 10 yr, 10 k.y., and 10 m.y. time scales. *Geology*, *29*(7), 591–594.

Kiver, E.P., & Harris, 1999. *Geology of U.S. Parklands*. New York, NY: John Wiley and Sons.

Kominz, M. A., Browning, J. V., Miller, K. G., Sugarman, P. J., Mizintseva, S., & Scotese, C. R. (2008). Late cretaceous to Miocene sea-level estimates from New Jersey and Delaware coastal plain coreholes: An error analysis. *Basin Research*, *20*, 211–226.

Lambeck, K., & Chappell, J. (2001). Sea level change through the last glacial cycle. *Science*, *292*, 679–686.

Lambeck, K., Esat, T. M., & Potter, E.-K. (2002). Links between climate and sea levels for the past three million years. *Nature*, *419*(6903), 199–206.

Lambeck, K., Yokoyama, Y., & Purcell, A. (2002). Into and out of the last glacial maximum: Sea-level change during oxygen isotope Stages 3 and 2. *Quaternary Science Reviews*, *21*, 343–360.

Learn About The 50 States. <http://www.netstate.com/states/index.html>.

Leech, M. L., Howell, D. G., & Egger, A. E. (2004). A guided inquiry approach to learning the geology of the U.S. *Journal of Geoscience Education*, *52*(4), 368–373.

Leighty, R.D., 2001. Automated IFSAR Terrain Analysis System. <http://www.agc.army.mil/publications/ifsar/lafinal08_01/cover/cover_frame.htm>.

Lillie, R.J. (2005). *Parks and plates*. New York, NY: Norton.

Lisiecki, L. E., & Raymo, M. E. (2005). A Pliocene-Pleistocene stack of 57 globally disturbed benthic delta (super 18) O records. *Paleoceanography*, *20*(1).

Lobeck, A. K. (1956). *Things maps don't tell us*. Chicago, IL: The Uninversity of Chicago Press, 159 pages.

Marcott, S. A., Shakun, J. D., Clark, P. U., & Mix, A. C. (2014). A reconstruction of regional and global temperature for the past 11, 300 years. *Science*, *339*, 1198–1201.

Marshak, S. (2009). Earth: Portrait of a planet (3rd ed.). New York, NY: W.W. Norton and Company.

McKnight, T. L. (2004). *Regional geography of the United States and Canada* (4th ed.). Upper Saddle River, New Jersey: Pearson Prentice Hall.

McKnight, T. L., & Hess, D. (2002). *Physical geography: A landscape appreciation* (7th ed.). Upper Saddle River, New Jersey: Prentice Hall.

Miller, G. H., Geirsdottir, A., Zhong, Y., Larson, D., Otto-Bliesner, B. L., Holland, M. M., ... Thordarson, T. (2012). Abrupt onset of the Little Ice Age triggered by volcanism and sustained by sea-ice/ocean feedbacks. *Geophysical Research Letters*, *39*. L02708. <https://doi.org/10.1029/2011GL050168>.

Miller, J.A. (2000). Ground water atlas of the United States. USGS Hydrologic Atlas 730. <http://water.usgs.gov/ogw/aquifer/atlas.html>.

Miller, K. G., Kominz, M. A., Browning, J. V., Wright, J. D., Mountain, G. S., Katz, M. E., ... Pekar, S. F. (2005). The Phanerozoic record of global sea-level change. *Science*, *310*(5752), 1293–1298.

Molnar, P., & England, P. C. (1990). Late Cenozoic uplift of mountain ranges and global climate change; chicken or egg? *Nature [London]*, *346*(6279), 29–34.

Montgomery, D. R. (1994). Valley incision and the uplift of mountain peaks. *Journal of Geophysical Research, B, Solid Earth and Planets*, *99*, 13,913–13,921.

Moores, E.M., & Twiss, R.J. (1995). *Tectonics*. New York: W.H. Freeman, 415 pages.

Morrison, R.B. (Ed.). (1991). *Quaternary Nonglacial Geology: Conterminous U.S., Vol. K2*. Boulder, CL: GSA DNAG. 672 pages.

Muhs, D. R., Wehmiller, J. F., Simmons, K. R., & York, L. L. (2004). Quaternary sea-level history of the United States. *Developments in Quaternary Science*, *1*, 147–183.

Mulch, A., Teyssier, C., Cosca, M. A., Vanderhaeghe, O., & Vennemann, T. W. (2004). Reconstructing paleoelevation in eroded orogens. *Geology [Boulder]*, *32*(6), 525–528.

NASA. *Global mean CO₂ mixing ratios (ppm): Observations 1850-2011*. <http://data.giss.nasa.gov/modelforce/ghgases/Fig1A.ext.txt>.

NASA. *Ocean and earth system*. (2014). <http://science.nasa.gov/earth-science/oceanography/ocean-earth-system/>.

National Academy of the Sciences. *Ocean acidification*. <http://nas-sites.org/oceanacidification/>.

National Atlas of the United States: *Where we are*. (2012). <https://nationalmap.gov>.

NOAA Current Sea Level Trends. <http://tidesandcurrents.noaa.gov/sltrends/>.

NOAA Physical Sciences Division. *United States interactive climate pages*. (2016). <http://www.esrl.noaa.gov/psd/data/usclimate/>.

Parsons, T., George, A., Thompson, G. A., Norman, H., & Sleep, N. H. (1994). Mantle plume influence on the Neogene uplift and extension of the U.S. western Cordillera? *Geology*, 22, 83–86.

Peakbagger. <http://www.peakbagger.com/>.

Philpotts, A. R. (1990). *Principles of igeaous and metamorphic petrology*. Englewood Cliffs, NJ: Prentice Hall, 498 pages.

Pinter, N., & Keller, E. A. (1991). Surface uplift, uplift of rocks, and exhumation of rocks: Comment. *Geology*, 19, 1053.

Plummer, C. C., McGeary, D., & Carlson, D. H. (2005). *Physical geology* (10th ed.). New York, NY: McGraw Hill.

Radbruch-Hall, D.H., Colton, R.B., Davies, W.E., Lucchitta, I., Skipp, B.A., & Varnes, D.J. (1982). Landslide overview map of the conterminous United States Geological Survey Professional Paper 1183. Washington: U.S. Geological Survey.

Raisz, E.J. (1957). Landforms of the United States. <http://www.raisz-maps.com/>.

RealClimate.org. *Climate science from climate scientists*. <http://www.realclimate.org/index.php/archives/2007/05/start-here/>.

Reed, J.C., & Bush, C.A. (2007). *About the geologic map in the National Atlas of the United States of America*. US Geological Survey Circular 1300. 52 pages.

Reed, J.C. Jr., Wheeler, J.O., & Tucholke, B.E. (2005) *Geologic map of North America*. Geological Society of America, scale 1:5,000,000.

Reiners, P. W., & Brandon, M. T. (2006). Using thermochronology to understand orogenic erosion. *Annual Review of Earth and Planetary Sciences*, 34, 419–466.

Reiners, P. W., & Shuster, D. L. (2009). Thermochronology and landscape evolution. *Physics Today*, 62(9), 31–36.

Ring, U., Brandon, M. T., Lister, G. S., & Willett, S. D. (Eds.), (1999). *Exhumation processes; normal faulting, ductile flow and erosion* (154). *Geological Society Special Publications*.

Rittenour, T. M., Blum, M. D., & Goble, R. J. (2007). Fluvial evolution of the lower Mississippi River valley during the last 100 k.y. glacial cycle: Response to glaciation and sea-level change. *GSA Bulletin*, 119(5/6), 586–608.

Rohling, E. J. (2010). Continuous 520,000-year sea-level record in 250-year time steps, on an independent radiometrically calibrated chronology. *Geochimica Et Cosmochimica Acta*, 74(12, Suppl. 1), A878.

Rohling, E. J., Foster, G. L., Grant, K. M., Marino, G., Roberts, A. P., Tamisiea, M. E., & Williams, F. (2014). Sea-level and deep-sea-temperature variability over the past 5.3 million years. *Nature [London]*, 508(7497), 477–482.

Rohling, E. J., Grant, K., Bolshaw, M., Roberts, A. P., Siddall, M., Hemleben, C. H., & Kucera, M. (2009). Antarctic temperature and global sea level closely coupled over the past five glacial cycles. *Nature Geoscience*, 2, 500–504.

Rohling, E. J., Grant, K., Hemleben, C., Siddall, M., Hoogakker, B. A. A., Bolshaw, M., & Kucera, M. (2008). High rates of sea-level rise during the last interglacial period. *Nature Geoscience*, 1(1), 38–42.

Roy, M., Clark, P. U., Raisbeck, G. M., & Yiou, F. (2004). Geochemical constraints on the regolith hypothesis for the middle Pleistocene transition. *Earth and Planetary Science Letters*, 227(3–4), 281–296.

Royer, D. L. (2006). CO_2-forced climate thresholds during the Phanerozoic. *Geochimica Et Cosmochimica Acta*, 70(23), 5665–5675.

Saucier, R.T. (1974). *Quaternary geology of the lower Mississippi valley*. Arkansas Archeological Survey Research Series 6. 26 p.

Saucier, R.T. (1994a). *Geomorphology and Quaternary geologic history of the Lower Mississippi Valley*. Vicksburg, Mississippi: U.S. Army Corps of Engineers Waterways Experiment Station. 364 p.

Saucier, R. T. (1994b). Evidence of late-glacial runoff in the Lower Mississippi Valley. *Quaternary Science Reviews*, 13, 973–981.

Schulz, M. (2002). On the 1470-year pacing of Dansgaard-Oeschger warm events. *Paleoceanography*, 17(2), 4-1–4-9.

Selley, R.C., Cocks, R.M., & Plimer, I.R. (2004). Encyclopedia of Geology, Elsevier, Amsterdam.

Shimer, J. A. (1972). *Field guide to landforms in the United States*. New York, NY: Macmillian.

Siddall, M., Chappell, J., & Potter, E. K. (2007). Eustatic sea level during past interglacials. *Developments in Quaternary Science*, 7, 75–92.

Siddall, M., Rohling, E. J., Almogi-Labin, A., Hemleben, C., Meischner, D., Schmelzer, I., & Smeed, D. A. (2003). Sea-level fluctuations during the last glacial cycle. *Nature [London]*, 423(6942), 853–858.

Siegenthaler, U., Stocker, T. F., Monnin, E., Luethi, D., Schwander, J., Stauffer, B., ... Jouzel, J. (2005). Stable carbon cycle-climate relationship during the late Pleistocene. *Science*, 310(5752), 1313–1317.

Skeptical Science. *Getting skeptical about global warming skepticism*. <www.skepticalscience.com>.

Sloss, L.L. (Ed.). (1988). *Sedimentary cover – North American Craton: U.S.*, Vol. D2, Boulder, CL: GSA DNAG. 506 pages.

Snoke, A. W. (2005). North America; Southern Cordillera. In R. C. Selley, L. R. M. Cocks, & I. R. Plimer (Eds.), *Encyclopedia of Geology* (pp. 48–61). United Kingdom: Elsevier Academic Press.

State Geologic Maps. <http://www.uwgb.edu/dutchs/StateGeolMaps.htm, http://geology.about.com/od/stategeologicmaps/Geologic_Maps_of_the_US_States.htm, http://mrdata.usgs.gov/geology/state/>.

State Geological Survey Websites. <http://www.stategeologists.org/>.

Sterner, R. (2011). Color landform atlas of the United States. ray.sterner@jhuapl.edu. <http://fermi.jhuapl.edu/states/states.html>.

Summerfield, M. A. (Ed.), (2000). *Geomorphology and global tectonics*. Chichester: United Kingdom: John Wiley & Sons.

SummitPost: Climbing, hiking, Mountaineering. <http://www.summitpost.org/>.

Tarduno, J., Bunge, H.-P., Sleep, N., & Hansen, U. (2009). The bent Hawaiian-Emperor Hotspot track; inheriting the mantle wind. *Science*, 324(5923), 50–53. Available from https://doi.org/10.1126/science.1161256.

Tarduno, J. A., Duncan, R. A., Scholl, D. W., Cottrell, R. D., Steinberger, B., Thordarson, T., ... Carvallo, C. (2003). The Emperor Seamounts; southward motion of the Hawaiian Hotspot plume in Earth's mantle. *Science*, 301(5636), 1064–1069. Available from https://doi.org/10.1126/science.1086442.

Teller, J. T., & Yang, Z. (2015). Mapping and measuring Lake Agassiz strandlines in North Dakota and Manitoba using LiDAR DEM data: Comparing techniques, revising correlations, and interpreting anomalous isostatic rebound gradients. *GSA Bulletin*, *127*, 608–620. Available from https://doi.org/10.1130/B31070.1.

Thelin, G., & Pike, R. J. (1991). *Landforms of the conterminous United States-a digital shaded-relief portrayal*. U.S. Geological Survey Miscellaneous Investigations Series Map I-2206. <https://pubs.usgs.gov/imap/i2206/>.

Thornbury, W. D. (1965). *Regional geomorphology of the United States*. New York, NY: John Wiley and Sons.

Tripati, A. K., Roberts, C. D., & Eagle, R. A. (2009). Coupling of CO_2 and ice sheet stability over major climate transitions of the last 20 million years. *Science*, *326*(5958), 1394–1397.

Twiss, R. J., & Moores, E. M. (2007). *Structural geology*. New York: W.H. Freeman, 736 pages.

United States Geological Survey. (2011). A tapestry of time and terrane. <http://tapestry.usgs.gov/Default.html>.

United States Geological Survey. A tapestry of time and terrain: The union of two maps – geology and topography. <http://ulpeis.anl.gov/documents/dpeis/references/pdfs/USGS_2003.pdf>.

United States Geological Survey. (2004). Geology of the National Parks. <http://geomaps.wr.usgs.gov/parks/>.

United States Geological Survey. (2012). National geologic map database. <http://ngmdb.usgs.gov/ngmdb/ngm_SMsearch.html>.

U.S. Geological Survey. (2015). Geology of National parks, 3D and photographic tours. <http://3dparks.wr.usgs.gov>.

USGS National Geologic Map Database. <http://ngmdb.usgs.gov/ngmdb/ngm_SMsearch.html>.

Vogt, P.R., & W.-Y. Jung. (2007). Origin of the Bermuda volcanoes and the Bermuda Rise; history, observations, models, and puzzles. Special Paper - *Geological Society of America 430*, 553–591.

Water Encyclopedia. <http://www.waterencyclopedia.com/>.

Weary, D.J., & Doctor, D.H. (2014). *Karst in the United States: A digital map compilation and database*. US Geological Survey Open-File Report 2014–1156, 23 p. <http://pubs.usgs.gov/of/2014/1156/>.

Whipple, K. X. (2009). The influence of climate on the tectonic evolution of mountain belts. *Nature Geoscience*, *2*, 97–104.

Whitmeyer, S. J., & Karlstrom, K. E. (2007). Tectonic model for the Proterozoic growth of North America. *Geosphere*, *3*(4), 220–259.

Whittaker, A. C. (2012). How do landscapes record tectonics and climate? *Lithosphere*, *4*, 160–164.

Willett, S. D., & Brandon, M. T. (2002). On steady states in mountain belts. *Geology [Boulder]*, *30*(2), 175–178.

Willett, S. D., Slingerland, R., & Hovius, N. (2001). Uplift, shortening, and steady state topography in active mountain belts. *American Journal of Science*, *301*(4–5), 455–485.

Winter, J. D. (2010). *Principles of igneous and metamorphic petrology* (2nd ed.). Pearson: Prentice Hall.

Wolff, E. W., Chappellaz, J., Blunier, T., Rasmussen, S. O., & Svensson, A. (2010). Millennial-scale variability during the last glacial; the ice core record. *Quaternary Science Reviews*, *29*(21–22), 2828–2838.

Woodroffe, C. D. (2008). Reef-island topography and the vulnerability of atolls to sea-level rise. *Global and Planetary Change*, *62*(1–2), 77–96.

Woodroffe, C. D., & Webster, J. M. (2014). Coral reefs and sea-level change. *Marine Geology*, *352*, 248–267.

Yeats, R. (2012). *Active faults of the world*. New York: Cambridge University Press, 634 p.

Yu, S., Berglund, B., Sandgren, P., et al. (2007). Evidence for a rapid sea-level rise 7600 yr ago. *Geology*, *35*(10), 891–894.

Zachos, J., Pagani, M., Sloan, L., Thomas, E., & Billups, K. (2001). Trends, rhythms, and aberrations in global climate 65 Ma to present. *Science*, *292*(5517), 686–693.

CHAPTERS 12–14, AND 21: THE INTERIOR UNITED STATES AND COASTAL PLAIN

Adams, P. N., Opdyke, N. D., & Jaeger, J. M. (2010). Isostatic uplift driven by karstification and sea-level oscillation; modeling landscape evolution in north Florida. *Geology [Boulder]*, *38*(6), 531–534.

Addison, W. D., Brumpton, G. R., Vallini, D. A., McNaughton, N. J., Davis, D. W., Kisson, S. A., ... Hammond, A. L. (2005). Discovery of distal ejecta from the 1850 Ma Sudbury impact event. *Geology*, *33*(3), 193–196.

Anthony, D. M., & Granger, D. E. (2007). A new chronology for the age of Appalachian erosional surfaces determined by cosmogenic nuclides in cave sediments. *Earth Surface Processes and Landforms*, *32*, 874–887.

Balbas, A. M., Barth, A. M., Clark, P. U., Clark, J., Caffee, M., O'Connor, J., ... Bjornstad, B. (2017). (Super 10) Be dating of late Pleistocene megafloods and Cordilleran ice sheet retreat in the northwestern United States. *Geology*, *45*(7), 583–586.

Balco, G., & Rovey, C. W., II (2010). Absolute chronology for major Pleistocene advances of the Laurentide Ice Sheet. *Geology*, *38*, 795–798.

Bennington, J. B. (2003). New observations on the glacial geomorphology of Long Island from a digital elevation model (DEM). *Long Island Geologists Conference*. Stony Brook, New York, p. 1–12. <http://pbisotopes.ess.sunysb.edu/lig/>.

Beus, S. S., & Morales, M. (2003). *Grand canyon geology*. New York: Oxford University Press, 432 pages.

Bickford, M. E., Wooden, J. L., & Bauer, R. L. (2006). SHRIMP study of zircons from Early Archean rocks in the Minnesota River Valley: Implications for the tectonic history of the Superior Province. *GSA Bulletin*, *118*(1/2), 94–108.

Bird, P. (1998). Kinematic history of the Laramide Orogeny in latitudes 35 degrees – 49 degrees N, Western United States. *Tectonics*, *17*(5), 780–801.

Blair, R. (1996). Geology of the western San Juan Mountains and a tour of the San Juan skyway, southwestern Colorado. In Thompson, R. A., Hudson, M.R., & Pillmore, C.L. (Eds.), *Geologic Excursions to the Rocky Mountains and Beyond, Field Trip Guidebook for the 1996 Annual Meeting*. GSA Sp-44, 8 p.

Bluemle, J.P., 2005. Glacial rebound, warped beaches and the thickness of the glaciers in North Dakota. <https://www.dmr.nd.gov/ndgs/ndnotes/Rebound/Glacial%20Rebound.htm>.

Booth, D. B., Troost, K. G., Clague, J. J., & Waitt, R. B. (2004). The Cordilleran ice sheet. *Developments in Quaternary Science*, *1*, 17–43.

Borns, H. W., Jr., Doner, L. A., Dorion, C. C., Jacobson, G. L., Jr., Kaplan, M. R., Kreutz, K. J., ... Weddle, T. K. (2004). The deglaciation of Maine, U.S.A. In J. Ehlers, & P. L. Gibbard (Eds.), *Quaternary glaciations - extent and chronology, Part II: North America* (pp. 89–109). Amsterdam: Elsevier.

Braile, L. W., William, I., Hinze, J., Keller, G. R., Lidiak, E. G., & Sexton, J. L. (1986). Tectonic development of the New Madrid rift complex, Mississippi embayment, North America. *Tectonophysics*, *131*, 1–21.

Brown, S. (2005). *Ozark Plateau, midcontinental North America; insights into thermochronology using apatite fission track analysis.* Master's Thesis. University of Illinois at Urbana-Champaign, Urbana, IL, 137 pages.

Cannon, W. F., Schulz, K. J., Wright, J. W., Jr., & Kring, D. A. (2010). The Sudbury impact layer in the Paleoproterozoic iron ranges of northern Michigan, USA. *GSA Bulletin*, *122*(1/2), 50–75.

Cather, S. M., Chapin, C. E., & Kelley, S. A. (2012). Diachronous episodes of Cenozoic erosion in southwestern North America and their relationship to surface uplift, paleoclimate, paleodrainage, and paleoaltimetry. *Geosphere*, *8*(6), 1177–1206.

Cerveny, P. F., & Steidtmann, J. R. (1993). Fission track thermochronology of the Wind River Range, Wyoming; evidence for timing and magnitude of Laramide exhumation. *Tectonics*, *12*(1), 77–92.

Chamberlain, C. P., Mix, H. T., Mulch, A., Hren, M. T., Kent-Corson, M. L., Davis, S. J., ... Graham, S. A. (2012). The Cenozoic climatic and topographic evolution of the western North American Cordillera. *American Journal of Science*, *312*, 213–262. Available from https://doi.org/10.2475/02.2012.05.

Chapin, C. E. (2012). Origin of the Colorado mineral belt. *Geosphere*, *8*(1), 28–43.

Chapin, C.E., & Kelley, S.A. (1997). *The Rocky Mountain erosion surface in the Front Range of Colorado*. Denver, CO: Rocky Mountain Association of Geologists. pp. 101–114.

Chapin, C. E., Kelley, S. A., & Cather, S. M. (2014). The Rocky Mountain Front, southwestern USA. *Geosphere*, *10*(5), 1043–1060.

Chronic, H. (1980). *Roadside geology of Colorado*. Missoula, Montana: Mountain Press Publishing Company.

Colorado 14ers. <http://www.14ers.com/photos/photos_14ers1.php>.

Colton, R.B., Lemke, R.W., Lindvall, R.V. (1961). *Glacial map of Montana East of the Rocky Mountains*. USGS miscellaneous geologic investigations Map I-327.

Crandell, D.R., Mullineaux, D.R., & Waldron, H.H. 1965. Age and origin of the Puget Sound Trough in western Washington. *U.S. Geological Survey Professional Paper* B132-b136.

Crossey, L. J., Karlstrom, K. E., Dorsey, R., Pearce, J., Wan, E., Beard, L. S., ... Pecha, M. E. (2015). Importance of groundwater in propagating downward integration of the 6-5 Ma Colorado River system; geochemistry of springs, travertines, and lacustrine carbonates of the Grand Canyon region over the past 12 Ma. *Geosphere [Boulder, CO]*, *11*(3), 660–682.

Crow, R., Karlstrom, K. E., McIntosh, W., Peters, L., & Dunbar, N. (2008). History of Quaternary volcanism and lava dams in western Grand Canyon based on lidar analysis, (super 40) Ar/(super 39) Ar dating, and field studies; implications for flow stratigraphy, timing of volcanic events, and lava dams. *Geosphere*, *4*(1), 183–206.

Crowley, P. D., Reiners, P. W., Reuter, J. M., & Kaye, G. D. (2002). Laramide exhumation of the Bighorn Mountains, Wyoming; an apatite (U-Th)/He thermochronology study. *Geology [Boulder]*, *30*(1), 27–30.

Csontos, R., & Van Arsdale, R. (2008). New Madrid seismic zone fault geometry. *Geosphere*, *4*(5), 802–813.

Cupples, W., & Van Arsdale, R. (2014). The Preglacial "Pliocene" Mississippi river. *Journal of Geology*, *122*, 1–15.

Darling, A., & Whipple, K. (2015). Geomorphic constraints on the age of the western Grand Canyon. *Geosphere [Boulder, CO]*, *11*(4), 958–976.

Davis, G.H. (1999). Structural Geology of the Colorado Plateau Region of Southern Utah: With special emphasis on deformation bands. *Geological Society of America Special Paper* 342, 157 p.

Davis, G.H., & Bump, A.P. (2009). Structural geologic evolution of the Colorado Plateau. In Kay, S.M., Ramos, V.A., & Dickinson, W.R., (Eds.), *Backbone of the Americas: Shallow subduction, plateau uplift, and ridge and terrane collision: Geological Society of America Memoir* vol. 204, p. 99–124.

Davis, S. J., Mulch, A., Carroll, A. R., Horton, T. W., & Chamberlain, C. P. (2009). Paleogene landscape evolution of the central North American Cordillera; developing topography and hydrology in the Laramide foreland. *Geological Society of America Bulletin*, *121*(1–2), 100–116.

Dethier, D. P., Ouimet, W., Bierman, P. R., Rood, D. H., & Balco, G. (2014). Basins and bedrock; spatial variation in (super 10) Be erosion rates and increasing relief in the Southern Rocky Mountains, USA. *Geology*, *42*(2), 167–170.

Dettman, D. L., & Lohmann, K. C. (2000). Oxygen isotope evidence for high-altitude snow in the Laramide Rocky Mountains of North America during the Late Cretaceous and Paleogene. *Geology*, *28*(3), 243–246.

Dickinson, W. R. (2013). Rejection of the lake spillover model for initial incision of the Grand Canyon, and discussion of alternatives. *Geosphere*, *9*(1), 1–20.

Donahue, M. S., Karlstrom, K. E., Aslan, A., Darling, A., Granger, D., Wan, E., ... Kirby, E. (2013). Incision history of the Black Canyon of Gunnison, Colorado, over the past approximately 1 Ma inferred from dating of fluvial gravel deposits. *Geosphere*, *9*(4), 815–826.

Dorsey, R. J., Fluette, A., McDougall, K., Housen, B. A., Janecke, S. U., Axen, G. J., & Shirvell, C. R. (2007). Chronology of Miocene-Pliocene deposits at Split Mountain Gorge, Southern California; a record of regional tectonics and Colorado River evolution. *Geology*, *35*(1), 57–60.

Duller, R. A., Whittaker, A. C., Swinehart, J. B., Armitage, J. J., Sinclair, H. D., Bair, A., & Allen, P. A. (2012). Abrupt landscape change post–6 Ma on the central Great Plains, USA. *Geology*, *40*(10), 871–874.

Dumitru, T. A., Duddy, I. R., & Green, P. F. (1994). Mesozoic-Cenozoic burial, uplift, and erosion history of the west-central Colorado Plateau. *Geology*, *22*, 499–502.

Eaton, G. P. (2008). Epeirogeny in the Southern Rocky Mountains region; evidence and origin. *Geosphere*, *4*(5), 764–784.

Elston, D. P., & Young, R. A. (1991). Cretaceous-Eocene (Laramide) landscape development and Oligocene-Pliocene drainage reorganization of transition zone and Colorado Plateau, Arizona. *Journal of Geophysical Research*, *96*, 12,389–12,406.

Epis, R. C., & Chapin, C. E. (1975). Geomorphic and tectonic implications of the post-Laramide, late Eocene erosion surface in the southern Rocky Mountains. *Memoir - Geological Society of America*, *144*, 45–74.

Epis, R. C., Scott, G. R., Taylor, R. B., & Chapin, C. E. (1976). Cenozoic volcanic, tectonic, and geomorphic features of central Colorado. In R. C. Epis, & R. J. Weimer (Eds.), *Studies in Colorado Field Geology: Colorado* (pp. 301–322). School of Mines Professional Contributions, no. 8.

Evanoff, E. (1990). Early Oligocene paleovalleys in southern and central Wyoming; evidence of high local relief on the late Eocene unconformity. *Geology, 18*(5), 443–446.

Fan, M., DeCelles, P. G., Gehrels, G. E., Dettman, D. L., & Peyton, S. L. (2011). Sedimentology, detrital zircon geochronology, and stable isotope geochemistry of the lower Eocene strata in the Wind River Basin, central Wyoming. *Geological Society of America Bulletin, 123*, 979–996. Available from https://doi.org/10.1130/B30235.1.

Fan, M., & Dettman, D. L. (2009). Late Paleocene high Laramide ranges in northeast Wyoming: Oxygen isotope study of ancient river water. *Earth and Planetary Science Letters, 286*, 110–121. Available from https://doi.org/10.1016/j.epsl.2009.06.024.

Fan, M., Heller, P., Allen, S. D., & Hough, B. G. (2014). Middle Cenozoic uplift and concomitant drying in the Central Rocky Mountains and adjacent Great Plains. *Geology, 42*(6), 547–550.

Fan, M., Quade, J., Dettman, D., & DeCelles, P. G. (2011). Widespread basement erosion during late Paleocene-early Eocene in the Laramide Rocky Mountains inferred from 87Sr/86Sr ratios of freshwater bivalve fossils. *Geological Society of America Bulletin, 123*, 2069–2082.

Flowers, R. M., & Farley, K. A. (2012). Apatite (super 4) He/(super 3) He and (U-Th)/He evidence for an ancient Grand Canyon. *Science, 338*(6114), 1616–1619.

Flowers, R. M., Wernicke, B. P., & Farley, K. A. (2008). Unroofing, incision, and uplift history of the southwestern Colorado Plateau from apatite (U-Th)/He thermochronometry. *Geological Society of America Bulletin, 120*(5–6), 571–587.

Fountain, A.G., Hoffmann, M., Jackson, K., Basagic, H., Nylen, T. & Percy, D. (2006). Digital outlines and topography of the glaciers of the American West. *Open File Report 2006-1340*. US Geological Survey, 23 p.

Gardner, T. W. (1989). Neotectonism along the Atlantic passive continental margin; a review. *Geomorphology, 2*(1–3), 71–97.

Gillespie, A. R., Porter, S. C., & Atwater, B. F. (Eds.), (2004). *The Quaternary period in the United States, Developments in Quaternary Science* (Vol. 1, pp. 351–380). Elsevier.

Gornitz, V., & Seeber, L. (1990). Vertical crustal movements along the East Coast, North America, from historic and late Holocene sea level data. *Tectonophysics, 178*(2–4), 127–150.

Granger, D. E., Fabel, D., & Anthony, D. M. (2001). Pliocene–Pleistocene incision of the Green River, Kentucky, determined from radioactive decay of cosmogenic 26Al and 10Be in Mammoth Cave sediments. *Geological Society of America Bulletin, 113*, 825–836.

Gutentag, E.D., Heimes, F.J., Krothe, N.C., Luckey, R.R., & Weeks, J.B. (1984). Geohydrology of the high plains aquifer in parts of Colorado, Kansas, Nebraska, New Mexico, Oklahoma, South Dakota, Texas, and Wyoming. *U.S. Geological Survey Professional Paper 1400-B*, 63 pp.

Hack, J.T. (1966). Interpretation of Cumberland Escarpment and Highland Rim, south-central Tennessee and northeast Alabama. *U.S. Geological Survey Professional Paper* C1-c16.

Halfen, A.F., & Johnson, W.C. (2013). A review of Great Plains dune field chronologies, Aeolian Research. <https://doi.org/10.1016/j.aeolia.2013.03.001>.

Halley, R. B., Vacher, H. L., & Shinn, E. A. (2004). Geology and hydrogeology of the Florida Keys. In H. L. Vacher, & T. Quinn (Eds.), *Geology and hydrology of Carbonate Islands. Developments in Sedimentology* (54), pp. 217–248. Elsevier Science B.V.

Hansen, M.C. (1987). The Teays River: Ohio Division of Geological Survey. Ohio Geology, Summer, p. 1–6.

Hanson, R. E., Puckett, R. E., Jr., Keller, G. R., Brueseke, M. E., Bulen, C. L., Mertzman, S. A., ... McCleery, D. A. (2013). Intraplate magmatism related to opening of the southern Iapetus Ocean: Cambrian Wichita igneous province in the Southern Oklahoma rift zone. *Lithos, 174*, 57–70.

Heller, P. L., McMillan, M. E., & Humphrey, N. F. (2011). Climate-induced formation of a closed basin; Great Divide Basin, Wyoming. *Geological Society of America Bulletin, 123*(1–2), 150–157.

Hill, C. A., Eberz, N., & Buecher, R. H. (2008). A karst connection model for Grand Canyon, Arizona, USA. *Geomorphology, 95*(3–4), 316–334.

Hill, C. A., & Polyak, V. J. (2014). Karst piracy; a mechanism for integrating the Colorado River across the Kaibab uplift, Grand Canyon, Arizona, USA. *Geosphere, 10*(4), 627–640.

Hill, C. A., & Ranney, W. D. (2008). A proposed Laramide proto-Grand Canyon. *Geomorphology, 102*(3–4), 482–495.

Hine, A.C. (2009). Geology of Florida, Brooks/Cole, Cengage Learning. <http://www.cengage.com/custom/enrichment_modules.bak/data/1426628390_Florida-LowRes_watermarked.pdf>.

Holm, D., Schneider, D., & Coath, C. D. (1998). Age and deformation of Early Proterozoic quartzites in the southern Lake Superior region: Implications for extent of foreland deformation during final assembly of Laurentia. *Geology, 26*, 907–910.

Holm, R. F. (2001). Cenozoic paleogeography of the central Mogollon Rim-southern Colorado Plateau region, Arizona, revealed by Tertiary gravel deposits, Oligocene to Pleistocene lava flows, and incised streams. *Geological Society of America Bulletin, 113*(11), 1467–1485.

Huntington, K., Wernicke, B., & Eiler, J. (2007). Paleoaltimetry from 'clumped' (super 13) C- (super 18) O bonds in carbonates, Colorado Plateau. *Geochimica et Cosmochimica Acta, 71*(15S), A426.

Huntington, K. W., Wernicke, B. P., & Eiler, J. M. (2010). The influence of climate change and uplift on Colorado Plateau paleotemperatures from carbonate 'clumped isotope' thermometry. *Tectonics, 29*, TC3005. Available from https://doi.org/10.1029/2009TC002449.

Huntoon, P. W. (1982). The Meander Anticline, Canyonlands, Utah; an unloading structure resulting from horizontal gliding on salt. *Geological Society of America Bulletin, 93*(10), 941–950.

Huntoon, P.W. (2000). Upheaval Dome, Canyonlands, Utah: Strain indicators that reveal on impact origin. *Utah Geological Association Publication, 28*, 619–628.

Huntoon, P. W., & Sears, J. W. (1975). Bright Angel and Eminence faults, eastern Grand Canyon, Arizona. *Geological Society of America Bulletin, 86*(4), 465–472.

Ingersoll, R. V., Grove, M., Jacobson, C. E., Kimbrough, D. L., & Hoyt, J. F. (2013). Detrital zircons indicate no drainage link between Southern California rivers and the Colorado Plateau from Mid-Cretaceous through Pliocene. *Geology, 41*(3), 311–314.

Isachsen, Y. W. (1975). Possible evidence for contemporary doming of the Adirondack Mountains, New York, and suggested implications for regional tectonics and seismicity. *Tectonophysics, 29*, 169–181.

Jirsa, M.A. (2011). Bedrock geology of the Western Gunflint Trail Area, Northeastern Minnesota. Minnesota Geological Survey, Miscellaneous Map Series Map M-191.

Jirsa, M.A., Boerboom, T.J., Chandler, V.W., Mossler, J.H., Runkel, A. C., & Setterholm, D.R. (2011). Geologic map of Minnesota bedrock geology. *State Map Series S-21, Bedrock Geology, Scale 1:500,000*. St. Paul, MN: Minnesota Geological Survey.

Jirsa, M.A., & Morey, G.B. (1987). Jay Cooke State Park and Grandview areas: Evidence for a major Early Proterozoic-Middle Proterozoic unconformity in Minnesota. Geological Society of America Centennial Field Guide — North-Central Section, number 14–15, p. 67–72.

Jirsa, M.A., & Weiblen, P. (2010). Minnesota's Evidence of an Ancient Meteorite Impact. Minnesota Geological Survey. <http://www.mngs.umn.edu/meteoriteimpact.pdf>.

Karlstrom, K. E., Beard, L. S., House, K., Young, R. A., Aslan, A., Billingsley, G., & Pederson, J. (2012). Introduction; CRevolution 2; Origin and evolution of the Colorado River system II. *Geosphere, 8*(6), 1170–1176.

Karlstrom, K. E., Coblentz, D., Dueker, K., Ouimet, W., Kirby, E., van Wijk, J., … CREST Working Group. (2012). Mantle-driven dynamic uplift of the Rocky Mountains and Colorado Plateau and its surface response: Toward a unified hypothesis. *Lithosphere, 4*(1), 3–22.

Karlstrom, K. E., Crossey, L. J., Embid, E., Crow, R., Heizler, M., Hereford, R., … Kelley, S. (2017). Cenozoic incision history of the Little Colorado River; its role in carving Grand Canyon and onset of rapid incision in the past ca. 2 Ma in the Colorado River system. *Geosphere [Boulder, CO], 13*(1), 49–81.

Karlstrom, K. E., Crow, R., Crossey, L. J., Coblentz, D., & van Wijk, J. W. (2008). Model for tectonically driven incision of the younger than 6 Ma Grand Canyon. *Geology [Boulder], 36*(11), 835–838.

Karlstrom, K. E., Crow, R. S., Peters, L., McIntosh, W., Raucci, J., Crossey, L. J., … Dunbar, N. W. (2007). 40Ar/3Ar and field studies of Quaternary basalts in Grand Canyon and model for carving Grand Canyon; quantifying the interaction of river incision and normal faulting across the western edge of the Colorado Plateau. *Geological Society of America Bulletin, 119*(11–12), 1283–1312.

Karlstrom, K., Darling, A., Crow, R., Lazear, G., Aslan, A., Granger, D., … Whipple, K. (2013). Colorado River chronostratigraphy at Lee's Ferry, Arizona, and the Colorado Plateau bull's-eye of incision; discussion. *Geology [Boulder], 41*(12), @e303.

Karlstrom, K. E., Lee, J. P., Kelley, S. A., Crow, R. S., Crossey, L. J., Young, R. A., & Shuster, D. L. (2014). Formation of the Grand Canyon 5 to 6 million years ago through integration of older palaeocanyons. *Nature Geoscience, 7*(3), 239–244.

Kelley, S. A., & Chapin, C. E. (1995). Apatite fission-track thermochronology of Southern Rocky Mountain-Rio Grande Rift-western High Plains provinces. *Guidebook - New Mexico Geological Society, 46*, 87–96.

Kelley, V. C. (1955). Monoclines of the Colorado Plateau. *Geological Society Of America Bulliten, 66*, 789–804.

Kent-Corson, M. L., Sherman, L. S., Mulch, A., & Chamberlain, C. P. (2006). Cenozoic topographic and climatic response to changing tectonic boundary conditions in western North America. *Earth and Planetary Science Letters, 252*(3–4), 453–466.

Kimbrough, D. L., Grove, M., Gehrels, G. E., Dorsey, R. J., Howard, K. A., Lovera, O., … Pearthree, P. A. (2015). Detrital zircon U-Pb provenance of the Colorado River; a 5 m.y. record of incision into cover strata overlying the Colorado Plateau and adjacent regions. *Geosphere [Boulder, CO], 11*(6), 1719–1748.

Krimmel, R. (2002). Glaciers of the Western United States, U.S. Geological Survey Professional Paper 2002, pp. J329–J381. In Williams, R.S., Jr., & Ferrigno, J.G. (Eds.), *Satellite image atlas of glaciers of the world*: North America.

LaBerge, G.L., & Klasner, J.S. (2001) Geology and tectonic significance of Early Proterozoic rocks in the Monico area, northern Wisconsin. US Geological Survey Geologic Investigations Series Map I-2739.

Lane, E., 1986. Karst in Florida: Florida Geological Survey, Tallahassee, Florida, Special Publication no. 29, 100 p.

Lazear, G., Karlstrom, K. E., Aslan, A., & Kelley, S. A. (2013). Denudation and flexural isostatic response of the Colorado Plateau and Southern Rocky Mountains region since 10 Ma. *Geosphere, 9*(4), 792–814. Available from https://doi.org/10.1130/GES00836.1.

Lee, J. P., Stockli, D. F., Kelley, S. A., Pederson, J. L., Karlstrom, K. E., & Ehlers, T. A. (2013). New thermochronometric constraints on the Tertiary landscape evolution of the central and eastern Grand Canyon, Arizona. *Geosphere, 9*(2), 216–228, GeoRef, EBSCOhost (accessed July 1, 2017).

Leonard, E. M. (2002). Geomorphic and tectonic forcing of late Cenozoic warping of the Colorado piedmont. *Geology [Boulder], 30*(7), 595–598.

Leonard, E. M., Hubbard, W. S., Kelley, S. A., Evanoff, E., Siddoway, C. S., Oviatt, C. G., … Timmons, M. (2002). High plains to Rio Grande Rift: Late Cenozoic evolution of Central Colorado. *Geological Society of America Field Guides, 3*, 59–93.

Levander, A., Schmandt, B., Miller, M. S., Liu, K., Karlstrom, K. E., Crow, R. S., … Humphreys, E. D. (2011). Continuing Colorado Plateau uplift by delamination; style convective lithospheric downwelling. *Nature [London], 472*(7344), 461–465.

Lillegraven, J. A., & Ostresh, L. M., Jr. (1988). Evolution of Wyoming's early Cenozoic topography and drainage patterns. *National Geographic Research, 4*(3), 303–327.

Lindsey, D.A. (2010). The geologic story of Colorado's Sangre de Cristo Range: U.S. Geological Survey Circular 1349, 14 p.

Lindsey, D. A., Andriessen, P. A. M., & Wardlaw, B. R. (1986). Heating, cooling, and uplift during Tertiary time, northern Sangre de Cristo Range, Colorado. *Geological Society of America Bulletin, 97*, 1133–1143.

Lipinski, M.M. (2002). Neotectonic uplift of the Beartooth Plateau assessed using the (U-Th)/He dating method. GSA Abstracts and Programs, 34, p. 45.

Lisenbee, A.L. (1985). *Tectonic map of the Black Hills uplift, Montana, Wyoming, and South Dakota*. The Geological Survey of Wyoming, Map Series 13, 1:250,000.

Lisenbee, A.L. (1988). Tectonic history of the Black Hills uplift. In Diedrich, R.P., Dyka, M.A., & Miller, W.R. (Eds.). *Eastern Powder River Basin-Black Hills. Wyoming Geological Association, 39th Annual field conference guidebook, Vol. 39*, pp. 45–52.

Lisenbee, A. L., & DeWitt, E. (1993). Laramide evolution of the Black Hills Uplift. *Memoir - Geological Survey of Wyoming, 5*, 374–412.

Liu, L., & Gurnis, M. (2010). Dynamic subsidence and uplift of the Colorado Plateau. *Geology [Boulder], 38*(7), 663–666.

Marchetti, D. W., Hynek, S. A., & Cerling, T. E. (2012). Gravel-capped benches above northern tributaries of the Escalante River, south-central Utah. *Geosphere, 8*(4), 835–853.

Marple, R. T., & Talwani, P. (1993). Evidence of possible tectonic upwarping along the South Carolina coastal plain from an examination of river morphology and elevation data. *Geology, 21*, 651–654.

McBride, E. F. (1988). *Geology of the marathon uplift, West Texas Centennial field guide* (pp. 411–416). Boulder, CO: Geological Society of America.

McMillan, M. E., Angevine, C. L., & Heller, P. L. (2002). Postdepositional tilt of the Miocene-Pliocene Ogallala Group on the western Great Plains; evidence of late Cenozoic uplift of the Rocky Mountains. *Geology [Boulder], 30*(1), 63−66.

McMillan, M. E., Heller, P. L., & Wing, S. L. (2006). History and causes of post-Laramide relief in the Rocky Mountain orogenic plateau. *Geological Society of America Bulletin, 118*(3−4), 393−405.

Mears, B., Jr. (1993). Geomorphic history of Wyoming and high-level erosion surfaces. *Memoir - Geological Survey of Wyoming, 5*, 609−626.

Medaris, L. G., Jr., Singer, B. S., Dott, R. H., Jr., Naymark, A., Johnson, C. M., & Schott, R. C. (2003). Late Paleoproterozoic climate, tectonics, and metamorphism in the southern Lake Superior region and proto-North America: Evidence from Baraboo interval quartzites. *Journal of Geology, 111*, 243−257.

Mederos, S., Tikoff, B., & Bankey, V. (2005). Geometry, timing, and continuity of the Rock Springs Uplift, Wyoming, and Douglas Creek Arch, Colorado; implications for uplift mechanisms in the Rocky Mountain foreland, U.S.A. *Rocky Mountain Geology, 40*(2), 167−191.

Mickelson, D. M., & Colgan, P. M. (2004). The southern Laurentide ice sheet. *Developments in Quaternary Science, 1*, 1−16.

Milici, R. C. (1963). Low-angle overthrust faulting, as illustrated by the Cumberland Plateau-Sequatchie Valley fault system. *American Journal of Science, 261*(9), 815−825.

Milici, R. C. (1979). Structural geology along the Eastern Cumberland Escarpment, Tennessee; a second look. *Southeastern Geology, 21*(1), 3−16.

Miller, D. S., & Lakatos, S. (1983). Uplift rate of Adirondack anorthosite measured by fission-track analysis of apatite. *Geology [Boulder], 11*(5), 284−286.

Miller, J.D. (Ed.). (2010). Field guide to the geology of precambrian iron formations in the Western Lake Superior Region, Minnesota and Michigan. Precambrian Research Center Guidebook, 92 p.

Mitra, S. (2005). Case studies; Fault-bend folding; Sequatchie Anticline, Tennessee, U.S.A.; a small displacement fault-bend fold. *AAPG Studies in Geology, 53*, 66−68.

Mix, H. T., Mulch, A., Kent-Corson, M. L., & Chamberlain, C. P. (2011). Cenozoic migration of topography in the North American Cordillera. *Geology, 39*, 87−90.

Moucha, R., Forte, A. M., Rowley, D. B., Mitrovica, J. X., Simmons, N. A., & Grand, S. P. (2008). Mantle convection and the recent evolution of the Colorado Plateau and the Rio Grande Rift valley. *Geology [Boulder], 36*(6), 439−442.

Mullineaux, D. R., Crandell, D. R., & Waldron, H. H. (1957). Multiple glaciation in the Puget Sound Basin, Washington. *Geological Society of America Bulletin, 68*(12, Part 2), 1772.

Nereson, A., Stroud, J., Karlstrom, K., Heizler, M., & McIntosh, W. (2013). Dynamic topography of the western Great Plains; geomorphic and (super 40) Ar/ (super 39) Ar evidence for mantle-driven uplift associated with the Jemez Lineament of NE New Mexico and SE Colorado. *Geosphere, 9*(3), 521−545.

Norris, R. D., Jones, L. S., Corfield, R. M., & Cartlidge, J. E. (1996). Skiing in the Eocene Uinta Mountains? Isotopic evidence in the Green River Formation for snow melt and large mountains. *Geology [Boulder], 24*(5), 403−406.

Omar, G. I., Lutz, T. M., & Giegengack, R. (1994). Apatite fission-track evidence for Laramide and post-Laramide uplift and anomalous thermal regime at the Beartooth overthrust, Montana-Wyoming. *Geological Society of America Bulletin, 106*, 74−85.

Ostrom, M.E. (1987). Precambrian/Paleozoic unconformity at Chippewa Falls, Wisconsin. Society of America Centennial Field Guide − North-Central Section, number 42, p. 177−178.

Paces, J. B., & Miller, J. D. (1993). Precise U-Pb ages of Duluth complex and related mafic intrusions, Northeastern Minnesota' geochronological insights to Physical, Petrogenetic, Paleomagnetic, and Tectonomagmatic processes associated with the 1.1 Ga midcontinent rift system. *Journal of Geophysical Research, 98*(B8), 13,997−14,013.

Pederson, J., Karlstrom, K., Sharp, W., & McIntosh, W. (2002). Differential incision of the Grand Canyon related to Quaternary faulting; constraints from U-series and Ar/Ar dating. *Geology [Boulder], 30*(8), 739−742.

Pederson, J. L. (2008). The mystery of the pre-Grand Canyon Colorado River; results from the Muddy Creek Formation. *GSA Today, 18*(3), 4−10.

Pederson, J. L., Mackley, R. D., & Eddleman, J. L. (2002). Colorado Plateau uplift and erosion evaluated using GIS. *GSA Today, 12*(8), 4−10.

Pelletier, J. D. (2010). Numerical modeling of the late Cenozoic geomorphic evolution of Grand Canyon, Arizona. *Geological Society of America Bulletin, 122*(3−4), 595−608.

Poag, C.W. (1998). The Chesapeake Bay bolide impact: A new view of Coastal Plain Evolution, USGS Fact Sheet 049-98. <http://woodshole.er.usgs.gov/epubs/bolide/>.

Polyak, V. J., Hill, C., & Asmerom, Y. (2008). Age and evolution of the Grand Canyon revealed by U-Pb dating of water table-type speleothems. *Science, 319*(5868), 1377−1380.

Potter, D. B., Jr., & McGill, G. E. (1978). Valley anticlines of the Needles District, Canyonlands National Park, Utah. *Geological Society of America Bulletin, 89*(6), 952−960.

Ranney, W. (2014). A pre-21st century history of ideas on the origin of the Grand Canyon. *Geosphere, 10*(2), 233−242.

Raynolds, R. G. (2002). Upper Cretaceous and Tertiary stratigraphy of the Denver Basin, Colorado. *Rocky Mountain Geology, 37*(2), 111−134.

Redden, J.A. & Lisenbee, A.L. (1990). Geologic setting, Black Hills, South Dakota. In Paterson, C.J., & Lisenbee, A.L. (Eds.). *Metallogeny of gold in the Black Hills, South Dakota; guidebook prepared for Society of Economic Geologists field conference.* Society of Economic Geologists Guidebook Series, Vol. 7, pp. 1−9.

Richmond, G. M., & Fullerton, D. S. (1986). Summation of Quaternary glaciations in the United States of America. *Quaternary Science Reviews, 5*, 183−196.

Roden-Tice, M. K., Tice, S. J., & Schofield, I. S. (2000). Evidence for differential unroofing in the Adirondack Mountains, New York State, determined by apatite fission-track thermochronology. *The Journal of Geology, 108*(2), 155−169.

Rosenberg, R., Kirby, E., Aslan, A., Karlstrom, K., Heizler, M., & Ouimet, W. (2014). Late Miocene erosion and evolution of topography along the western slope of the Colorado Rockies. *Geosphere, 10*(4), 641−663.

Roskowski, J. A., Patchett, P. J., Spencer, J. E., Pearthree, P. A., Dettman, D. L., Faulds, J. E., & Reynolds, A. C. (2010). A late Miocene-early Pliocene chain of lakes fed by the Colorado River; evidence from Sr, C, and O isotopes of the Bouse Formation and related units between Grand Canyon and the Gulf of California. *Geological Society of America Bulletin, 122*(9−10), 1625−1636.

Roy, M., Jordan, T. H., & Pederson, J. (2009). Colorado Plateau magmatism and uplift by warming of heterogeneous lithosphere. *Nature*, *159*, 978–982. Available from https://doi.org/10.1038/nature08052.

Sahagian, D., Proussevitch, A., & Carlson, W. (2002). Timing of Colorado Plateau uplift: Initial constraints from vesicular basalt-derived paleoelevations. *Geology*, *30*(9), 807–810.

Schaffer, J. P., Small, E. E., & Anderson, R. S. (1998). Pleistocene relief production in Laramide mountain ranges, Western United States; discussion and reply. *Geology [Boulder]*, *26*(12), 1150–1152.

Schimitz, M. D., Bowring, S. A., Southwick, D. L., Boerboom, T. J., & Wirth, K. R. (2006). High-precision U-Pb geochronology in the Minnesota River Valley subprovince and its bearing on the Neoarchean to Paleoproterozoic evolution of the southern Superior Province. *GSA Bulletin*, *118*(1/2), 82–93.

Schulz, W.H., Highland, L.M., Ellis, W.L., Gori, P.L., Baum, R.L., Coe, J.A., & Savage, W.Z. (2007). The Slumgullion Landslide, Hinsdale County, Colorado. *1st North American Landslide Conference Vail*. Colorado, June 9–10, 2007, 26 p.

Semken, S. C., & McIntosh, W. L. (1997). (super 40) Ar/(super 39) Ar age determinations for the Carrizo Mountains laccolith, NavajoNation, Arizona. *Guidebook - New Mexico Geological Society*, *48*, 75–80.

Sharp, R. P., Allen, C. R., & Meier, M. F. (1959). Pleistocene glaciers on southern California mountains. *American Journal of Science*, *257*(2), 81–94.

Sims, P.K. (1992). Geologic map of Precambrian rocks, southern Lake Superior region, Wisconsin and northern Michigan. US Geological Survey Miscellaneous Investigations Series Map I-2185.

Sims, P. K., Card, K. D., Morey, G. B., & Peterman, Z. E. (1990). The Great Lakes tectonic zone—A major crustal structure in central North America. *Geological Society of America Bulletin, Part I*, *91*, 690–698.

Sims, P.K., & Carter, L.M.H. (1996). Archean and Proterozoic geology of the Lake Superior region, U.S.A., 1993. US Geological Survey Professional Paper: 1556, 115 p.

Sjostrom, D.J., Hren, M.T., Horton, T.W., Waldbauer, J.R., & Chamberlain, C.P. (2006). Stable isotope evidence for a pre-Miocene elevation gradient in the Great Plains–Rocky Mountains region, USA. In Willett, S.D., et al. (Eds.), *Tectonics, climate, and landscape evolution: Geological Society of America Special Paper (Vol. 398)*. Geological Society of America, pp. 309–319.

Small, E. E., & Anderson, R. S. (1998). Pleistocene relief production in Laramide mountain ranges, western United States. *Geology [Boulder]*, *26*(2), 123–126.

Smithson, S. B., Brewer, J., Kaufman, S., Oliver, J., & Hurich, C. (1978). Nature of the Wind River thrust, Wyoming, from COCORP deep-reflection data and from gravity data. *Geology [Boulder]*, *6*(11), 648–652.

Soreghan, G. S., Sweet, D. E., Thomson, S. N., Kaplan, S. A., Marra, K. R., Balco, G., & Eccles, T. M. (2015). Geology of Unaweep Canyon and its role in the drainage evolution of the northern Colorado Plateau. *Geosphere*, *11*(2), 320–341.

Southwick, D.L. (1987). Geologic highlights of an Archean greenstone belt, western Vermilion district, northeastern Minnesota. Society of America Centennial Field Guide – North-Central Section, number 11, pp. 53–58.

Southwick, D. L., Morey, G. B., & Mossler, J. H. (1986). Fluvial origin of the Lower Proterozoic Sioux Quartzite, southwestern Minnesota. *Geological Society of America Bulletin*, *97*, 1432–1441.

Steidtmann, J. R., & Middleton, L. T. (1991). Fault chronology and uplift history of the southern Wind River Range, Wyoming; implications for Laramide and post-Laramide deformation in the Rocky Mountain foreland. *Geological Society of American Bulletin*, *103*, 472–485.

Steidtmann, J. R., Middleton, L. T., & Shuster, M. W. (1989). Post-Laramide (Oligocene) uplift of the Wind River Range, Wyoming. *Geology*, *17*, 38041.

Stein, C. A., Kley, J., Stein, S., Hindle, D., & Keller, G. R. (2015). North America's Midcontinent Rift: When rift met LIP. *Geosphere*, *11*(5), 1607–1616. Available from https://doi.org/10.1130/GES01183.1.

Sugden, D. E. (1978). Glacial erosion by the Laurentide ice sheet. *Journal of Glacial Geology*, *20*(83), 367–391.

Taylor, J. P., & Fitzgerald, P. G. (2011). Low-temperature thermal history and landscape development of the eastern Adirondack Mountains, New York; constraints from apatite fission-track thermochronology and apatite (U-Th)/He dating. *Geological Society of America Bulletin*, *123*(3–4), 412–426.

Tewksbury, B. J. (1985). Revised interpretation of the age of allochthonous rocks of the Uncompahgre Formation, Needle Mountains, Colorado. *GSA Bulletin*, *96*, 224–232.

Thompson, W.B. (2015). Surficial Geology Handbook for Southern Maine, Bulletin 44. Maine Geological Survey.

Thompson, W.B., & Borns, H.W., Jr. (1985). Surficial geologic map of Maine: Maine Geological Survey, 1:500,000-scale map.

Thorson, R. M. (1989). Glacio-isostatic response of the Puget Sound area, Washington. *Geological Society Of America Bulletin*, *101*(9), 1163–1174.

Timmons, J.M., & Karlstrom, K.E. (2012) Grand Canyon Geology: Two Billion Years of Earth's History. *Geological Society of America Special Paper 489*. Boulder, CL: Geological Society of America.

Trimble, G.E. (1999). The geologic story of the Great Plains. Geological Survey Bulletin 1493, Washington. <http://library.ndsu.edu/exhibits/text/greatplains/text.html>.

Upchurch, S.B. (2007). An introduction to the Cody Escarpment, North-Central Florida. Prepared for the Suwannee River Water Management District, Live Oak, Florida. <http://publicfiles.dep.state.fl.us/FGS/FGS_Publications/FGS%20Library%20Documents/Cody%20Scarp%20Upchurch.pdf>.

Van Arsdale, R. (1997). Hazard in the Heartland: The New Madrid Seismic Zone. *Geotimes*, *42* no. 5, 16–19.

Van Arsdale, R. B., & Cox, R. T. (2007). The Mississippi's curious origins. *Scientific American*, *296*(1), 76, GeoRef, EBSCO*host* (accessed September 2, 2012).

van Wijk, J. W., Baldridge, W. S., van Hunen, J., Goes, S., Aster, R., Coblentz, D. D., ... Ni, J. (2010). Small-scale convection at the edge of the Colorado Plateau; implications for topography, magmatism, and evolution of Proterozoic lithosphere. *Geology [Boulder]*, *38*(7), 611–614.

Ver Steeg, K. (1946). The Teays River. *Ohio Journal of Science*, *46*, 297–307.

Vogt, P. R., & Jung, W.-Y. (2007). Origin of the Bermuda volcanoes and the Bermuda Rise; history, observations, models, and puzzles. *Special Paper - Geological Society of America*, *430*, 553–591.

Wallach, J. L., & Rheault, M. (2010). Uplift of the Tug Hill Plateau in northern New York State. *Canadian Journal of Earth Sciences*, *47*, 1055–1077.

Wernicke, B. P. (2011). The California River and its role in carving Grand Canyon. *Geological Society of America Bulletin, 123*(7−8), 1288−1316.

Winker, C. D., & Howard, J. D. (1977). Correlation of tectonically deformed shorelines on the southern Atlantic Coastal Plain. *Geology [Boulder], 5*(2), 123−127.

Winkler, J. E., Kelley, S. A., & Bergman, S. C. (1999). Cenozoic denudation of the Wichita Mountains, Oklahoma, and southern Mid-continent; apatite fission-track thermochronology constraints. *Tectonophysics, 305*(1−3), 339−353.

Wisconsin Geological and Natural History Survey. (1981) (Revised 2005). Bedrock Geology of Wisconsin, Madison, Wisconsin.

Wolkowinsky, A. J., & Granger, D. E. (2004). Early Pleistocene incision of the San Juan River, Utah, dated with ^{26}Al and ^{10}Be. *Geology, 32*(9), 749−752.

Young, R. A. (2008). Pre-Colorado river drainage in western Grand Canyon; potential influence on Miocene stratigraphy in Grand Wash Trough. *Special Paper - Geological Society of America, 439*, 319−333.

Young, R.A. (2011). Brief Cenozoic Geologic History of the Peach Springs Quadrangle and the Hualapai Plateau, Mohave County, Arizona (Hualapai Indian Reservation). Arizona Geological Survey Contributed Report Series, CR-11-O, 28 p.

Young, R. A., & Crow, R. (2014). Paleogene Grand Canyon incompatible with Tertiary paleogeography and stratigraphy. *Geosphere, 10*(4), 664−679.

Young, R.A., & Spamer, E.E. (Eds.). (2000). *Colorado River: Origin and Evolution*.Grand Canyon, AZ: Grand Canyon Association. 280 pages.

Zaprowski, B. J., Evenson, E. B., Pazzaglia, F. J., & Epstein, J. B. (2001). Knickzone propagation in the Black Hills and northern High Plains: A different perspective on the late Cenozoic exhumation of the Laramide Rocky Mountains. *Geology, 29*, 547−550.

CHAPTERS 15 AND 16: APPALACHIAN-OUACHITA AND CORDILLERAN MOUNTAIN SYSTEMS

Armstrong, R. L. (1968). Sevier orogenic belt in Nevada and Utah. *Geological Society of America Bulletin, 79*(4), 429−458.

Aslan, A., Hood, W. C., Karlstrom, K. E., Kirby, E., Granger, D. E., Kelly, S., ... Asmerom, Y. (2014). Abandonment of Unaweep Canyon (1.4 - 0.8 Ma), western Colorado; effects of stream capture and anomalously rapid Pleistocene river incision. *Geosphere, 10*, 428−446.

Bennington, J. B., & Merguerian, C. (2007). *Geology of New York and New Jersey*. Brooks/Cole: Cengage Learning.

Brandon, M.T., Cowan, D.S., & Vance, J.A. (1988). The Late Cretaceous San Juan thrust system, San Juan Islands, Washington. *Geological Society of America Special Paper 221*, 81 p.

Brown, E.H., & Dragovich, J.D. (2003). Tectonic elements and evolution of northwest Washington. Washington Division of Geology and Earth Resources, Geologic Map GM-52.

Brown, E.H., Housen, B.A., & Schermer, E.R. (2007). Tectonic evolution of the San Juan Islands thrust system, Washington. In Stelling, P., & Tucker, D.S. (Eds.), Floods, Faults, and Fire: Geological Field Trips in Washington State and Southwest British Columbia. *Geological Society of America Field Guide (Vol. 9)*, pp. 143−177.

Bryant, B., & Reed, J.C., Jr. (1970). Geology of the Grandfather Mountain window and vicinity, North Carolina and Tennessee.*U. S. Geological Survey Professional Paper* 615, 190 pages.

Butler, J. R., & Hatcher, R. D., Jr. (1989). *Day 8; Grandfather Mountain Window and vicinity Field trips for the 28th international geological congress* (pp. 73−77). Washington, DC: American Geophysical Union.

Cheney, E. S. (1993). Guide to the geology of northeastern Washington. *Guidebook - Northwest Geological Society, 8*, 35 p.

Clark, G.M., Ciolkosz, E.J., Kite, J.S., & Lietzke, D.A. (1989). Central and Southern Appalachian Geomorphology−Tennessee, Virginia, and West Virginia. *28th International Geological Congress Field Trip Guidebook T150*, AGU, 105 p.

Coogan, J. C., & DeCelles, P. G. (2007). Regional structure and kinematic history of the Sevier fold-and-thrust belt, central Utah; reply. *Geological Society of America Bulletin, 119*(3−4), 508−512.

Cook, F. A., Albaugh, D. S., Brown, L. D., Kaufman, S., Oliver, J. E., & Hatcher, R. D., Jr. (1979). Thin-skinned tectonics in the crystalline southern Appalachians; COCORP seismic-reflection profiling of the Blue Ridge and Piedmont. *Geology [Boulder], 7*(12), 563−567.

Corrie, S. L., & Kohn, M. J. (2007). Resolving the timing of orogenisis in the Western Blue Ridge, southern Appalachians, via in situ ID-TIMS monazite geochronology. *Geology, 35*, 627−630.

Corrigan, J., Cervany, P. F., Donelick, R., & Bergman, S. C. (1998). Postorogenic denudation along the late Paleozoic Ouachita trend, south central United States of America: Magnitude and timing constraints from apatite fission track data. *Tectonics, 17*, 587−603.

Crawford, M.M., & Hunsberger, H. (2011). Geology of Cumberland Gap National Historical Park. *Kentucky Geological Survey, Map and Chart 199, Series XII*.

Currie, B. S. (2002). Structural configuration of the Early Cretaceous Cordilleran foreland-basin system and Sevier thrust belt, Utah and Colorado. *Journal of Geology, 110*(6), 697−718.

DeCelles, P. G. (2004). Late Jurassic to Eocene evolution of the Cordilleran thrust belt and foreland basin system, western U.S.A. *American Journal of Science, 304*(2), 105−168.

DeCelles, P. G., & Coogan, J. C. (2006). Regional structure and kinematic history of the Sevier fold-and-thrust belt, central Utah. *Geological Society of America Bulletin, 118*(7−8), 841−864.

Drake, A.A., Hall, L.M., & Nelson, A.E. (1988). Basement and basement-cover relation map of the Appalachian orogen in the United States, Scale:1:1,000,000, USGS Product Number 27787, ISBN:978-0-607-78788-7.

Foose, R. M., Wise, D. U., & Garbarini, G. S. (1961). Structural geology of the Beartooth Mountains, Montana and Wyoming. *Geological Society of American Bulletin, 72*(8), 1143−1172.

Foti, T. L., & Bukenhofer, G. A. (1998). A description of the sections and subsections of the Interior Highlands of Arkansas and Oklahoma. *Proceedings of The Arkansas Academy of Science, 52*, 53−62.

Fuentes, F., DeCelles, P. G., & Constenius, K. N. (2012). Regional structure and kinematic history of the Cordilleran fold-thrust belt in northwesternMontana, USA. *Geosphere, 8*(5), 1104−1128.

Gallen, S. F., Wegmann, K. W., & Bohnenstiehl, D. R. (2013). Miocene rejuvenation of topographic relief in the Southern Appalachians. *GSA Today, 23*(2), 4−10.

Gilman, R. A., Chapman, C. A., Lowell, T. V., & Borns, H. W., Jr. (1988). *The geology of Mount Desert Island; a visitor's guide to the geology of Acadia National Park*. Augusta, ME: Maine Geological Survey, Department of Conservation.

Giorgis, S., Tikoff, B., & McClelland, W. (2005). Missing Idaho arc; transpressional modification of the (super 87) Sr/(super 86) Sr transition on the western edge of the Idaho Batholith. *Geology [Boulder], 33*(6), 469−472.

Goldberg, S.A. & Butler, R. (1989). Geological cross-section through part of the Southern Appalachian Orogen Inner Piedmont to Valley and Ridge (North Carolina, Tennessee and Virginia). *Field Trip Guidebook T365, July 20−26, 1989*.Washington, D.C.: American Geophysical Union, 62 pages.

Gordon, S. M., Whitney, D. L., Miller, R. B., McLean, N., & Seaton, N. C. A. (2010). Metamorphism and deformation at different structural levels in a strike-slip fault zone, Ross Lake Fault, north Cascades, USA. *Journal of Metamorphic Geology, 28*(2), 117−136.

Hack, J.T. (1973). Drainage adjustment in the Appalachians. 51−69. In Morisawa, M. (Ed.), *Fluvial Geomorphology, State Univversity. NewYork*. Binghamton, New York: Binghamton Publication in Geomorphology, pp. 51−69.

Harris, L.D., & Milici, R.C. (1977). Characteristics of thin-skinned style of deformation in the southern Appalachians, and potential hydrocarbon traps. *U. S. Geological Survey Professional Paper*.

Harris, J.D., & Mixon, R.B. (1970). Geologic Map of the Howard Quarter quadrangle northeastern Tennessee. U.S. Geological Survey Geologic Quadrangle Maps of the United States, Map GQ-842.

Hatcher, R.D. (1987). Tectonics of the Southern and Central Appalachian Internides. In Hatcher, R.D., & Bream, B.R. (Eds.), *North Carolina Geological Survey, Carolina Geological Society annual field trip guidebook*, pp. 1−18.

Hatcher, R.D., Jr. (1989). Day 5; Structure of the western Valley and Ridge and eastern Cumberland Plateau. *Tennessee Field trips for the 28th international geological congress* (pp. 40−48). American Geophysical Union: Washington, DC, 1989.

Hatcher, R.D., Jr. (2001). Rheological partitioning during multiple reactivation of the Palaeozoic Brevard fault zone, Southern Appalachians, USA. *Geological Society Special Publications 186*, 257−271.

Hatcher, R. D., Jr. (2005). North America; Southern and Central Appalachians. In R. C. Selley, L. R. M. Cocks, & I. R. Plimer (Eds.), *Encyclopedia of Geology* (pp. 72−81). Oxford, United Kingdom: United Kingdom: Elsevier Academic Press.

Hatcher, R. D., Jr., Bream, B. R., & Merschat, A. J. (2007). Tectonic map of the Southern and Central Appalachians; a tale of three orogens and a complete Wilson cycle. *Geological Society of America Memoir, 200*, 595−632.

Hatcher, R. D., Jr., Lemiszki, P. J., & Whisner, J. B. (2007). Character of rigid boundaries and internal deformation of the Southern Appalachian foreland fold-thrust belt. *Special Paper - Geological Society of America, 433*, 243−276.

Hatcher, R.D., Jr., Merschat, A.J., & Raymond, L.A. (2006). Geotraverse: Geology of northeastern Tennessee and the Grandfather Mountain region. In Labotka, T.C., & Hatcher, R.D., Jr. (Eds.), *Geological Society of America 2006 Southeastern Section Meeting Field Trip Guidebook*, pp. 129−184.

Hatcher, R.D., Jr., Merschat, A.J., & Thigpen, J. R. (2005). Blue Ridge Primer. In Hatcher, R.D., Jr., & Merschat, A.J. (Eds.), Blue Ridge Geology Geotraverse East of the Great Smoky Mountains National Park, Western North Carolina: Durham. *North Carolina Geological Survey, Carolina Geological Society Annual field trip guide book*, pp. 1−24.

Hibbard J., van Staal C., Miller B. (2007). Links between Carolinia, Avalonia, and Ganderia in the Appalachian peri-Gondwanan Realm. In Sears, J., Harms, T., & Evenchick, C., (Eds.), *From Whence the Mountains? Inquiries into the Evolution of Orogenic Systems: A Volume in Honor of Raymond A. Price, GSA Special Paper, 433*, 2007, pp. 291−311.

Hibbard, J., & Waldron, J. W. F. (2009). Truncation and translation of Appalachian promontories: mid-Paleozoic strike slip tectonics and basin initiation. *Geology, 37*, 487−490.

Hibbard, J.P., van Staal, C.R., & Rankin, D.W. (2006). Lithotectonic map of the Appalachian orogen. USGS Map 2096A, Scale 1:1,500,000 (two parts).

Hill, J. S., & Stewart, K. G. (2012). Correlation of major topographic lineaments in the North Carolina Blue Ridge with regional fracture zones. *Abstracts With Programs - Geological Society of America, 44*(7), 483.

Hnat, J. S., & van der Pluijm, B. A. (2014). Fault gouge dating in the Southern Appalachians, USA. *Geological Society of America Bulletin, 126*(5−6), 639−651.

Karabinos, P., Samson, S. D., Hepburn, J. C., & Stoll, H. M. (1998). Taconian Orogeny in the New England Appalachians; collision between Laurentia and the Shelburne Falls Arc. *Geology [Boulder], 26*(3), 215−218.

King, P.B., & Ferguson, H.W. (1960). Geology of Northeasternmost Tennessee. *U. S. Geological Survey Special Paper 311*, 136 pages.

King, P.B., Neuman, R.B., & Hadley, J.B. (1968). Geology of the Great Smoky Mountains National Park, Tennessee and North Carolina. *U. S. Geological Survey Professional Paper*.

Kruckenberg, S. C., Whitney, D. L., Teyssier, C., Fanning, C. M., & Dunlap, W. J. (2008). Paleocene-Eocene migmatite crystallization, extension, and exhumation in the hinterland of the northern Cordillera; Okanogan Dome, Washington, USA. *Geological Society of America Bulletin, 120*(7−8), 912−929.

Massey, M.A., & Moecher, D.P. (2005). Deformation and metamorphic history of the Western Blue Ridge-Eastern Blue Ridge terrane boundary, southern Appalachian orogen. *Tectonics 24*, TC5010, doi: 10.1029/2004TC001643.

Mazza, S. E., Gazel, E., Johnson, E. A., Kunk, M. J., McAleer, R. J., Spotila, J. A., ... Coleman, D. S. (2014). Volcanoes of the passive margin; the youngest magmatic event in eastern North America. *Geology, 42*(6), 483−486.

McHone, J. G., & Butler, J. R. (1984). Mesozoic igneous provinces of New England and the opening of the North Atlantic Ocean. *Geological Society of America Bulletin, 95*(7), 757−765.

McKeon, R. E., Zeitler, P. K., Pazzaglia, F. J., Idleman, B. D., & Enkelmann, E. (2014). Decay of an old orogen: Inferences about Appalachian landscape evolution from low-temperature thermochronology. *GSA Bulletin, 126*(1/2), 31−46. Available from https://doi.org/10.1130/B30808.1.

Milam, K.A., Evenick, J.C., & Deane, B. (2005). Field guide to the Middlesboro and Flynn Creek impact structures. *Impact Field Studies Group, 69th Annual Meteoritical Society Meeting*, Gatlinburg, TN, 46 p.

Milam, K.A., & Kuehn, K.W. (2003). Field guide to the Middleboro impact structure and beyond. *Annual Field Conference of the Kentucky Society of Professional Geologists* 2003, pp. 30−44.

Miller, B. V., Fetter, A. H., & Stewart, K. G. (2006). Plutonism in three orogenic pulses, eastern Blue Ridge Province, Southern Appalachians. *Geological Society of America Bulletin, 118*(1−2), 171−184.

Miller, E. L., & Gans, P. B. (1989). Cretaceous crustal structure and metamorphism in the hinterland of the Sevier thrust belt, western U. S. Cordillera. *Geology [Boulder], 17*(1), 59−62.

Miller, R. L. (1973). Where and why of Pine Mountain and other major fault planes, Virginia, Kentucky, and Tennessee. *American Journal of Science, 273-A*, 353−371.

Miller, R. B. (1989). The Mesozoic Rimrock Lake Inlier, southern Washington Cascades; implications for the basement to the Columbia Embayment. *Geological Society of America Bulletin, 101* (10), 1289−1305.

Mitchell, S. G., & Montgomery, D. R. (2006). Polygenetic topography of the Cascade Range, Washington State, USA. *American Journal of Science, 306*(9), 736−768.

Mitra, S. (1986). Duplex structures and imbricate thrust systems; geometry, structural position, and hydrocarbon potential. *AAPG Bulletin, 70*(9), 1087−1112.

Mitra, S. (1988). Three-dimensional geometry and kinematic evolution of the Pine Mountain thrust system, Southern Appalachians. *Geological Society of America Bulletin, 100*(1), 72−95.

Mixon, R.B. & Harris, J.D. (1971). Geologic map of the Swan Island quadrangle northeastern Tennessee. U. S. Geological Survey Geologic Quadrangle Maps of the United States, Map GQ-878.

Moecher, D. P., Samson, S. D., & Miller, C. F. (2004). Precise time and conditions of peak granulite facies metamorphism in the southern Appalachian orogen, USA, with implications for zircon behavior during crustal melting events. *Journal of Geology, 112*, 289−304.

Morisawa, M. (1989). Rivers and valleys of Pennsylvania, revisited. *Geomorphology, 2*(1−3), 1−22.

Mudge, M. R. (1970). Origin of the disturbed belt in northwestern Montana. *Geological Society of America Bulletin, 81*, 377−392.

Mudge, M. R. (1982). A resume of the structural geology of the northern disturbed belt, Montana. In R. B. Powers (Ed.), *Geologic studies of the Cordilleran thrust belt* (pp. 91−122). Denver, Colorado: Rocky Mountain Association of Geologists.

Mudge, M.R., & Earhart, R.L. (1980). The Lewis thrust fault and related structures in the Disturbed Belt, Northwestern Montana. *USGS Professional Paper 1174*, 18 p.

Naeser, N.D., Naeser, C.W., Kunk, M.J., Morgan, B.A. III, Schultz, A.P., Southworth, C.S., & Weems, R.E. (2001). Paleozoic through Cenozoic uplift, erosion, stream capture, and deposition history in the Valley and Ridge, Blue Ridge, Piedmont, and Coastal Plain provinces of Tennessee, North Carolina, Virginia, Maryland, and District of Columbia. *Geological Society of America Abstracts with Programs*.

Naeser, N.D., Naeser, C.W., Southworth, C.S., Morgan, B.A., III, & Schultz, A.P. (2004). Paleozoic to recent tectonic and denudation history of rocks in the Blue Ridge Province, central and southern Appalachians—Evidence from fission-track thermochronology. *Geological Society of America Abstracts with Programs, Vol. 36*, No. 2, P. 114.

Nielsen, K. C. (2005). North America; Ouachitas. In R. C. Selley, L. R. M. Cocks, & I. R. Plimer (Eds.), *Encyclopedia of Geology* (pp. 61−71). Oxford: United Kingdom: Elsevier Academic Press, 2005.

Orr, E. L., & Orr, W. N. (2000). *Geology of the Pacific Northwest*. Long Grove, Illinois: Waveland Press, 337 pages.

Paterson, S.R., Miller, R.B., Alsleben, H., Whitney, D.L., Valley, P.M., & Hurlow, H. (2004). Driving mechanisms for >40 km of exhumation during contraction and extension in a continental arc, Cascades core, Washington. *Tectonics 23* (3).

Paterson, S.R., Miller, R.B., Alsleben, H., Whitney, D.L., Valley, P.M., & Vance, J.A. (2000). 30−40 km of Eocene exhumation during orogen-parallel extension in the Cascades core, Washington. *Abstracts with Programs - Geological Society of America 32* (6), p. 61.

Pazzaglia, F. J., & Brandon, M. T. (1996). Macrogeomorphic evolution of the post-Triassic Appalachian Mountains determined by deconvolution of the offshore based sedimentary record. *Basin Research, 8* (3), 255−278.

Pazzaglia, F. J., & Gardner, T. W. (2000). *Late Cenozoic landscape evolution of the US Atlantic passive margin; insights into a North American Great Escarpment* (pp. 283−302). Chichester: United Kingdom: John Wiley & Sons.

Portenga, E. W., Bierman, P. R., Rizzo, D. M., & Rood, D. H. (2013). Low rates of bedrock outcrop erosion in the Central Appalachian Mountains inferred from in situ ^{10}Be. *GSA Bulletin, 125*(1/2), 201−215.

Prince, P. S., Spotila, J. A., & Henika, W. S. (2010). New physical evidence of the role of stream capture in active retreat of the Blue Ridge Escarpment, Southern Appalachians. *Geomorphology, 123* (3−4), 305−319.

Reiners, P. W., Ehlers, T. A., Garver, J. I., Mitchell, S. G., Montgomery, D. R., Vance, J. A., & Nicolescu, S. (2002). Late Miocene exhumation and uplift of the Washington Cascade Range. *Geology [Boulder], 30*(9), 767−770.

Rhodes, B. P., & Cheney, E. S. (1981). Low-angle faulting and the origin of Kettle Dome, a metamorphic core complex in northeastern Washington. *Geology [Boulder], 9*(8), 366−369.

Rich, J. L. (1933). Physiography and structure at Cumberland Gap. *Geological Society of America Bulletin, 44*(6), 1219−1236.

Rich, J. L. (1934). Mechanics of low-angle overthrust faulting as illustrated by Cumberland thrust block, Virginia, Kentucky, and Tennessee. *Bulletin of the American Association of Petroleum Geologists, 18*(12), 1584−1596.

Rogers, J. (1953). Geologic Map of East Tennessee with explanatory text. *State of Tennessee Department of Environment and Conservation Division of Geology, Bulletin 58, Part II*, 168 pages.

Simon-Labric, T., Brocard, G. Y., Teyssier, C., van der Beek, P. A., Reiners, P. W., Shuster, D. L., ... Whitney, D. L. (2014). Low-temperature thermochronologic signature of range-divide migration and breaching in the north Cascades. *Lithosphere, 6*(6), 473−482.

Southworth, S., Drake, A. A., Jr., Brezinski, D. K., Wintsch, R. P., Kunk, M. J., Aleinikoff, J. N., ... Naeser, N. D. (2006). Central Appalachian Piedmont and Blue Ridge tectonic transect, Potomac River corridor. *GSA Field Guide, 8*, 135−167.

Southworth, S., Schultz, A., Aleinikoff, J.N., & Merschat, A. (2012). Geologic map of the Great Smoky Mountains National Park region, Tennessee and North Carolina. U. S. Geological Survey Scientific Investigations Map 2997, 54 pages.

Southworth, S., Schultz, A., & Denenny, D. (2005a). Generalized geologic map of bedrock lithologies and surficial deposits in the Great Smoky Mountains National Park region, Tennessee and North Carolina. *U. S. Geological Survey Open-File Report 2004-1410*, 109 pages.

Southworth, S., Schultz, A., & Denenny, Danielle. (2005b). Geologic map of the Great Smoky Mountains National Park region, Tennessee and North Carolina. *U. S. Geological Survey Open-File Report 2005-1225*, 117 pages.

Spear, F. S., Kohn, M. J., Cheney, J. T., & Florence, F. (2002). Metamorphic, thermal, and tectonic evolution of central New England. *Journal of Petrology, 43*(11), 2097–2120.

Spotila, J. A., Bank, G. C., Reiners, P. W., Naeser, C. W., Naeser, N. D., & Henika, B. S. (2004). Origin of the Blue Ridge Escarpment along the passive margin of eastern North America. *Basin Research, 16*(1), 41–63.

Stanley, R. S., & Ratcliffe, N. M. (1985). Tectonic synthesis of the Taconian Orogeny in western New England. *Geological Society of America Bulletin, 96*(10), 1227–1250.

Stearns, R.G. (1954). The Cumberland Plateau overthrust and geology of the Crab Orchard Mountains area, Tennessee.

Stewart, K. G. (2007). Active fault zones in the Blue Ridge of western North Carolina. *Abstracts With Programs - Geological Society of America, 39*(2), 90.

Stewart, K.G., Adams, M.G., & Trupe, C.H. (1997). Paleozoic structure, metamorphism, and tectonics of the Blue Ridge of western North Carolina. *Carolina Geological Society 1997 Field Trip Guidebook*, 101 pages.

Tabor, R., & Haugerud, R. (1999). Geology of the North Cascades: A Mountain Mosaic. Seattle, WA: The Mountaineers, 144 pages.

Taylor, W. J., Bartley, J. M., Martin, M. W., Geismann, J. W., Walker, J. D., Armstrong, P. A., & Fryxell, J. E. (2000). Relations between hinterland and foreland shortening: Sevier orogeny, central North American Cordillera. *Tectonics, 19*, 1124–1143.

Thigpen, J.R. & Hatcher, R.D. (2009). Geologic map of the western Blue Ridge and portions of the eastern Blue Ridge and Valley and Ridge provinces in southeast Tennessee, southwest North Carolina, and northern Georgia. *Geological Society of America Map and Chart Series MCH097*.

Trupe, C. H., Stewart, K. G., Adams, M. G., Waters, C. L., Miller, B. V., & Hewitt, L. K. (2003). The Burnsville fault: Evidence for the timing and kinematics of the southern Appalachian Acadian dextral transform tectonics. *GSA Bulletin, 115*, 1365–1376.

van Staal, C. R. (2005). North America, Northern Appalachians. In R. C. Selley, L. R. M. Cocks, & I. R. Plimer (Eds.), *Encyclopedia of Geology* (pp. 81–92). United Kingdom: Elsevier Academic Press.

Whitney, D. L., & McGroder, M. F. (1989). Cretaceous crust section through the proposed Insular-Intermontane suture, North Cascades, Washington. *Geology [Boulder], 17*(6), 555–558.

Williams, H. (1978). Tectonic lithofacies map of the Appalachian orogen. Canadian contribution No 5 to the International Geological Correlation Program, Project No 27, the Appalachian-Caladonides orogen, scale 1:1,000,000.

Wilson, C. W., Jr., & Stearns, R. G. (1958). Structure of the Cumberland Plateau, Tennessee. *Geological Society of America Bulletin, 69*, 1283–1296.

Wilson, R.L., & Wojtal, S.F. (1986). *Cumberland Plateau decollement zone at Dunlap, Tennessee Centennial field guide* (pp. 143–148). Boulder, CO: Geological Society of America,.

Wiltschko, D.V. (1989). *Day 8; Review of the tectonics of portions of the Plateau and Valley and Ridge, Southern Appalachians of Virginia and Tennessee Field trips for the 28th international geological congress* (pp. 65–73).: Washington, DC: American Geophysical Union, 1989.

Wiltschko, D. V., Medwedeff, D. A., & Millson, H. E. (1985). Distribution and mechanisms of strain within rocks on the northwest ramp of Pine Mountain block, Southern Appalachian foreland; a field test of theory. *Geological Society of America Bulletin, 96*(4), 426–435.

CHAPTERS 17 AND 18: POSTOROGENIC TECTONIC REINCARNATION

Aldrich, M. J., Jr., & Dethier, D. P. (1990). Stratigraphic and tectonic evolution of the northern Espanola Basin, Rio Grande Rift, New Mexico. *Geological Society of America Bulletin, 102*(12), 1695–1705.

Alpha, T.R., & Vallier, T.L. (1994). Physiography of the Seven Devils Mountains and adjacent Hells Canyon of the Snake River, Idaho and Oregon. *U. S. Geological Survey Professional Paper*, pp. 91–100.

Applegate, J. D. R., Walker, J. D., & Hodges, K. V. (1992). Late Cretaceous extensional unroofing in the Funeral Mountains metamorphic core complex, California. *Geology, 20*, 519–522.

Armstrong, P. A., Ehlers, T. A., Chapman, D. S., Farley, K. A., & Kamp, P. J. J. (2003). Exhumation of the central Wasatch Mountains, Utah; 1, Patterns and timing of exhumation deduced from low-temperature thermochronology data. *Journal of Geophysical Research, 108*(B3).

Armstrong, P. A., Taylor, A. R., & Ehlers, T. A. (2004). Is the Wasatch Fault footwall (Utah, United States) segmented over million-year time scales? *Geology [Boulder], 32*(5), 385–388.

Armstrong, R. L. (1982). Cordilleran metamorphic core complexes; from Arizona to southern Canada. *Annual Review of Earth and Planetary Sciences, 10*, 129–154.

Armstrong, R. L., Taubeneck, W. H., & Hales, P. O. (1977). Rb-Sr and K-Ar geochronometry of Mesozoic granitic rocks and their Sr isotopic composition, Oregon, Washington, and Idaho. *Geological Society of America Bulletin, 88*(3), 397–411.

Baldridge, W. S., Keller, G. R., Haak, V., Wendlandt, E. D., Jiracek, G. R., & Olsen, K. H. (1995). The Rio Grande Rift. *Developments in Geotectonics, 25*, 233–275.

Best, M. G., Barr, D. L., Christiansen, E. H., Gromme, S., Deino, A. L., & Tingey, D. G. (2009). The Great Basin Altiplano during the middle Cenozoic ignimbrite flareup; insights from volcanic rocks. *International Geology Review, 51*(7–8), 589–633.

Best, M. G., Christiansen, E. H., Deino, A. L., Gromme, S., Hart, G. L., & Tingey, D. G. (2013). The 36-18 Ma Indian Peak-Caliente ignimbrite field and calderas, southeastern Great Basin, USA; multicyclic super-eruptions. *Geosphere, 9*, 864–950.

Best, M. G., Christiansen, E. H., de Silva, S. L., & Lipman, P. W. (2016). Slab-rollback ignimbrite flareups in the southern Great Basin and other Cenozoic American arcs; a distinct style of arc volcanism. *Geosphere [Boulder, CO], 12*(4), 1097–1135.

Best, M. G., Christiansen, E. H., & Gromme, S. (2013). Introduction; The 36-18 Ma southern Great Basin, USA, ignimbrite province and flareup; swarms of subduction-related supervolcanoes. *Geosphere, 9* (2), 260–274, GeoRef, EBSCOhost (accessed December 4, 2016).

Best, M. G., Gromme, S., Deino, A. L., Christiansen, E. H., Hart, G. L., & Tingey, D. G. (2013). The 36-18 Ma central Nevada ignimbrite field and calderas, Great Basin, USA; multicyclic super-eruptions. *Geosphere, 9*, 1562–1636.

Bove, D.J., Hon, K., Budding, K.E., Slack, J.F., Snee, L.W., & Yeoman, R.A. (2001). Geochronology and geology of Late Oligocene through Miocene volcanism and mineralization in the western San Juan Mountains, Colorado. *U.S. Geological Survey Professional Paper 1642*, 30 p.

Brueseke, M. E., Heizler, M. T., Hart, W. K., & Mertzman, S. A. (2007). Distribution and geochronology of Oregon Plateau (U.S.A.) flood basalt volcanism; The Steens Basalt revisited. *Journal of Volcanology and Geothermal Research, 161*(3), 187–214.

Camp, V. E., Ross, M. E., & Hanson, W. E. (2003). Genesis of flood basalts and Basin and Range volcanic rocks from Steens Mountain to the Malheur River gorge, Oregon. *Geological Society of America Bulletin*, *115*(1), 105–128.

Cassel, E. J., Breecker, D. O., Henry, C. D., Larson, T. E., & Stockli, D. F. (2014). Profile of a paleo-orogen; high topography across the present-day Basin and Range from 40 to 23 Ma. *Geology [Boulder]*, *42*(11), 1007–1010, *GeoRef*, EBSCOhost (accessed December 4, 2016).

Cassel, E. J., Calvert, A. T., & Graham, S. A. (2009). Age, geochemical composition, and distribution of Oligocene ignimbrites in the northern Sierra Nevada, California: Implications forlandscape morphology, elevation, and drainagedivide geography of the Nevadaplano. *International Geology Review*, *51*, 723–742.

Cassel, E. J., & Graham, S. A. (2011). Paleovalley morphology and fluvial system evolution of Eocene-Oligocene sediments ('auriferous gravels'), northern Sierra Nevada, California; implications for climate, tectonics, and topography. *Geological Society of America Bulletin*, *123*(9–10), 1699–1719, @1sheet. *GeoRef*, EBSCOhost (accessed December 4, 2016).

Cassel, E. J., Graham, S. A., & Chamberlain, C. P. (2009). Cenozoic tectonic and topographic evolution of the northern Sierra Nevada, California, through stable isotope paleoaltimetry in volcanic glass. *Geology [Boulder]*, *37*(6), 547–550, *GeoRef*, EBSCOhost (accessed December 4, 2016).

Cassel, E. J., Graham, S. A., Chamberlain, C. P., & Henry, C. D. (2012). Early Cenozoic topography, morphology, and tectonics of the northern Sierra Nevada and western Basin and Range. *Geosphere*, *8*, 229–249.

Cather, S. M., Connell, S. D., Chamberlin, R. M., McIntosh, W. C., Jones, G. E., Potochnik, A. R., ... Johnson, P. S. (2008). The Chuska Erg; paleogeomorphic and paleoclimatic implications of an Oligocene sand sea on the Colorado Plateau. *Geological Society of America Bulletin*, *120*(1–2), 13–33.

Chamberlain, C. P., & Poage, M. A. (2000). Reconstructing the paleotopography of mountain belts from the isotopic composition of authigenic minerals. *Geology [Boulder]*, *28*(2), 115–118.

Christiansen, R.L. (2001). The Quaternary and Pliocene Yellowstone Plateau volcanic field of Wyoming, Idaho, and Montana. *U.S. Geological Survey Professional Paper, 2001*, pp. G1-G145.

Cole, R.D. (2010). Eruptive history of the Grand Mesa Basalt Field, Western Colorado. *Geological Society of America Abstracts with Programs*, *42*(5), p. 76.

Colgan, J. P. (2013). Reappraisal of the relationship between the northern Nevada rift and Miocene extension in the northern Basin and Range Province. *Geology*, *41*(2), 211–214. Available from https://doi.org/10.1130/G33512.1.

Colgan, J. P., & Henry, C. D. (2009). Rapid middle Miocene collapse of the Mesozoic orogenic plateau in north-central Nevada. *International Geology Review*, *51*(9–11), 920–961.

Coney, P. J. (1980). Cordilleran metamorphic core complexes; an overview. *Memoir - Geological Society of America*, *153*, 7–31.

Coney, P. J., & Harms, T. A. (1984). Cordilleran metamorphic core complexes; Cenozoic extensional relics of Mesozoic compression. *Geology [Boulder]*, *12*(9), 550–554.

Crafford, A. E. J. (2008). Paleozoic tectonic domains of Nevada; an interpretive discussion to accompany the geologic map of Nevada. *Geosphere*, *4*(1), 260–291.

Crafford, A. E.J., & Harris, A.G. (2007). Geologic map of Nevada; with a section on a digital conodont database of Nevada. *U. S. Geological Survey Data Series*.

Crumpler, L. (2015) The volcanoes of New Mexico. New Mexico Museum of Natural History and Science. <http://nmnaturalhistory.org/volcanoes-of-new-mexico.html>.

Dickinson, W. R. (1997). Tectonic implications of Cenozoic volcanism in coastal California. *Geological Society of America Bulletin*, *109*, 936–954.

Dickinson, W. R. (2004). Evolution of the North American Cordillera. *Annual Review of Earth and Planetary Sciences*, *32*, 13–45, GeoRef, EBSCOhost (accessed October 30, 2011).

Dickinson, W. R. (2006). Geotectonic evolution of the Great Basin. *Geosphere*, *2*(7), 353–368.

Dickinson, W. R. (2013). Phanerozoic palinspastic reconstructions of Great Basin geotectonics (Nevada-Utah, USA). *Geosphere*, *9*(5), 1–13.

Dickinson, W. R., & Lawton, T. F. (2001). Carboniferous to Cretaceous assembly and fragmentation of Mexico. *Geological Society of America Bulletin*, *113*(9), 1142–1160.

Donnelly-Nolan, J.M., Stovall, W.K., Ramsey, D.W., Ewert, J.W., & Jensen, R.A., (2011), Newberry Volcano—central Oregon's sleeping giant. *U.S. Geological Survey Fact Sheet 2011–3145*, 6 p., Available at <http://pubs.usgs.gov/fs/2011/3145/>.

Drewes, H. (2008). Table Mountain shoshonite porphyry lava flows and their vents, Golden, Colorado. *U.S. Geological Survey Scientific Investigations Report 2006–5242*, 28 p.

Eaton, G. P. (1982). The Basin and Range Province; origin and tectonic significance. *Annual Review of Earth and Planetary Sciences*, *10*, 409–440.

Ehlers, T.A. (2001). Geothermics of exhumation and erosion in the Wasatch Mountains, Utah.

Epis, R.C., Scott, G.R., Taylor, R.B., & Chapin, C.E. (1976). Cenozoic volcanic, tectonic, and geomorphic features of central Colorado. In Epis, R.C., & Weimer, R.J. (Eds.), *Studies in Colorado Field Geology: Colorado, School of Mines Professional Contributions, no. 8*, pp. 301–322.

Ernst, W. G. (2010). Young convergent-margin orogens, climate, and crustal thickness; a Late Cretaceous-Paleogene Nevadaplano in the American Southwest? *Lithosphere*, *2*(2), 67–75.

Geist, D., & Richards, M. (1993). Origin of the Columbia Plateau and Snake River plain: Deflection of the Yellowstone plume. *Geology*, *21*, 789–792.

Gilbert, H. (2012). Crustal structure and signatures of recent tectonism as influenced by ancient terranes in the Western United States. *Geosphere*, *8*(1), 141–157.

Giorgis, S., Tikoff, B., & McClelland, W. (2005). Missing Idaho arc; transpressional modification of the (super 87) Sr/(super 86) Sr transition on the western edge of the Idaho Batholith. *Geology [Boulder]*, *33*(6), 469–472.

Global Volcanism Program. *National Museum of Natural History, Smithsonian Institution*. (2015). <http://volcano.si.edu>.

Goff, F. (2002). Geothermal potential of Valles Caldera, New Mexico. *Quarterly Bulletin - Oregon Institute Of Technology. Geo-Heat Center*, *23*(4), 7–12.

Hack, J. T. (1942). Sedimentation and Volcanism in the Hopi Buttes, Arizona. *Geological Society Of America Bulletin*, *53*, 335–372.

Halfen, A. F., & Johnson, W. C. (2013). A review of Great Plains dune field chronologies. *Aeolian Research*. Available from https://doi.org/10.1016/j.aeolia.2013.03.001.

Hampel, A., Hetzel, R., & Densmore, A. L. (2007). Postglacial slip-rate increase on the Teton normal fault, northern Basin and Range Province, caused by melting of the Yellowstone ice cap and deglaciation of the Teton Range? *Geology [Boulder]*, *35*(12), 1107−1110.

Henry, C. D. (2008). Ash-flow tuffs and paleovalleys in northeastern Nevada; implications for Eocene paleogeography and extension in the Sevier hinterland, northern Great Basin. *Geosphere*, *4*(1), 1−35.

Henry, C. D., Hinz, N. H., Faulds, J. E., Colgan, J. P., John, D. A., Brooks, E. R., ... Castor, S. B. (2012). Eocene-early Miocene paleotopography of the Sierra Nevada-Great Basin-Nevadaplano based on widespread ash-flow tuffs and paleovalleys. *Geosphere*, *8*(1), 1−27.

Henry, C. D., & John, D. A. (2013). Magmatism, ash-flow tuffs, and calderas of the ignimbrite flareup in the western Nevada volcanic field, Great Basin, USA. *Geosphere*, *9*(4), 951−1008.

Henry, C. D., Kunk, M. J., & McIntosh, W. C. (1994). (super 40) Ar/(super 39) Ar chronology and volcanology of silicic volcanism in the Davis Mountains, Trans-Pecos Texas; with Suppl. Data 9442. *Geological Society of America Bulletin*, *106*(11), 1359−1376.

Hiza, M. M. (1998). The Geologic History of the Absaroka volcanic province. *Yellowstone Science*, *6*(2), 2−7.

Hodges, K. V., McKenna, L. W., & Harding, M. B. (1990). Chapter 19: Structural unroofing of the central Panamint Mountains, Death Valley region, southeastern California. *Memoir of the Geological Society of America*, *176*(1), 377−390.

Hodges, K. V., & Walker, J. D. (1990). Petrologic constraints on the unroofing history of the Funeral Mountains metamorphic core complex, California. *Journal of Geophysical Research*, *95*, 8437−8445.

Hodges, K. V., & Walker, J. D. (1992). Extension in the Cretaceous Sevier orogen, North American Cordillera; with Suppl. Data 92-13. *Geological Society of America Bulletin*, *104*(5), 560−569, GeoRef, EBSCOhost (accessed December 4, 2016).

Hooper P.R., Camp V.E., Reidel S.P., Ross M.E. (2007). The origin of the Columbia River flood basalt province: Plume versus nonplume models. In Foulger G.R., & Jurdy D.M. (Eds.), *Plumes, Plates, and Planetary Processes: Geological Society of America Special Paper 430*, pp. 635−668.

Horton, T. W., & Chamberlain, C. P. (2006). Stable isotopic evidence for Neogene surface downdrop in the central Basin and Range Province. *Geological Society of America Bulletin*, *118*(3−4), 475−490.

Horton, T. W., Sjostrom, D. K., Abruzzese, M. J., Poage, M. A., Waldbauer, J. R., Hren, M., ... Chamberlain, C. P. (2004). Spatial and temporal variation of Cenozoic surface elevation in the Great Basin and Sierra Nevada. *American Journal of Science*, *304*(10), 862−888.

Hubbard, M.S., Oviatt, C.G., Kelley, S., Perkins, M.E., Hodges, K.V., & Robbins, R., 2001, Oligocene-Miocene basin formation and modification in the northern Rio Grande rift; Constraints from 40Ar/39Ar, fission track, and tephrochronology. *GSA Abstracts and Programs*. <https://gsa.confex.com/gsa/2001AM/finalprogram/abstract_22956.htm>.

Hudson, M.R., & Grauch, V.J.S., (2013). Introduction. In Hudson, M.R., & Grauch, V.J.S. (Eds.), *New Perspectives on Rio Grande Rift Basins: From Tectonics to Groundwater: Geological Society of America Special Paper 494*.

John D.A., Wallace, A.R., Ponce, D.A., Fleck, R.B., & Conrad, J.E. (2000). New perspectives on the geology and origin of the northern Nevada Rift. In Cluer, J.K., Price, J.G., Struhsacker, E.M., Hardyman, R.F., & Morris, C.L. (Eds.), *Geology and Ore Deposits 2000: The Great Basin and Beyond: Geological Society of Nevada Symposium Proceedings*, May 15−18, 2000, pp. 127−154.

Jordan, B.T. (2006). The Oregon High Lava Plains: Proof against a plume origin for Yellowstone? <http://www.mantleplumes.org/WebpagePDFs/HighLavaPlains.pdf>.

Jordan, B. T., Grunder, A. L., & Duncan, R. A. (2004). Geochronology of age-progressive volcanism of the Oregon High Lava Plains; implications for the plume interpretation of Yellowstone. *Journal of Geophysical Research*, *109*(B10202). Available from https://doi.org/10.1029/2003JB002776.

Keller, G. R., & Baldridge, W. S. (1999). The Rio Grande Rift; a geological and geophysical overview. *Rocky Mountain Geology*, *34*(1), 121−130.

Kistler, R.W. (1990). Two different lithosphere types in the NevadaSierra. *California. Geological Society of America Memoir 174*, 271−281.

Kluth, C. F. (2007). Inversion and reinversion of the northern Rio Grande Rift, San Luis Basin, southern Colorado. *GSA Abstracts and Programs*, *39*(6), 366.

Koivula, J.I. (1981). San Carlos peridot. *Gems and Gemology*, Gemological Institute of America, Winter, *17*(4), 205−214.

Kudo, A. M. (1974). Outline of the igneous geology of the Jemez Mountains Volcanic Field. *Guidebook - New Mexico Geological Society*, *25*, 287−289.

Kuntz, M.A., Skipp, Betty, Champion, D.E., Gans, P.B., Van Sistine, D. P., and Snyders, S.R., 2007, Geologic map of the craters of the Moon 30' x 60' quadrangle, Idaho: U.S. Geological Survey Scientific Investigations Map 2969, 64-p. pamphlet, 1 plate, scale 1:100,000.

Landman, R. L., & Flowers, R. M. (2013). (U-Th)/He thermochronologic constraints on the evolution of the northern Rio Grande Rift, Gore Range, Colorado, and implications for rift propagation models. *Geosphere*, *9*, 170−187.

Link, P. K., & Janecke, S. U. (1999). Geology of East-Central Idaho: Geologic Roadlogs for the Big and Little Lost River, Lemhi, and Salmon River Valleys. In S. S. Hughes, & G. D. Thackray (Eds.), *Guidebook to the Geology of Eastern Idaho* (pp. 295−334). Pocatello: Idaho Museum of Natural History.

Lipman, P. W. (2007). Incremental assembly and prolonged consolidation of Cordilleran magma chambers; evidence from the Southern Rocky Mountain volcanic field. *Geosphere*, *3*(1), 42−70.

Lipman, P. W., & McIntosh, W. C. (2008). Eruptive and noneruptive calderas, northeastern San Juan Mountains, Colorado: Where did the ignimbrites come from. *Geological Society of America Bulletin*, *120*, 771−795.

Lipman, P. W., Steven, T. A., & Mehnert, H. H. (1970). Volcanic history of the San Juan mountains, Colorado, as indicated by potassium-argon dating. *Geological Society of America Bulletin*, *81*(8), 2329−2351.

Lister, G. S., & Davis, G. A. (1989). The origin of metamorphic core complexes and detachment faults formed during Tertiary continental extension in the northern Colorado River region, U.S.A. *Journal of Structural Geology*, *11*, 65−94.

Machette, M. N., Thompson, R. A., Marchetti, D. W., & Smith, R. S. U. (2013). Evolution of ancient Lake Alamosa and integration of the Rio Grande during the Pliocene and Pleistocene. *Special Paper - Geological Society of America*, *494*, 1−20.

Madole, R. F., Romig, J. H., Aleinikoff, J. N., VanSistine, D. P., & Yacob, E. Y. (2008). On the origin and age of the Great Sand Dunes, Colorado. *Geomorphology*, *99*, 99−119.

Magnani, M. B., Levander, A., Miller, K. C., Eshete, T., & Karlstrom, K. E. (2005). Seismic investigation of the Yavapai-Mazatzal transition zone and the Jemez Lineament in northeastern New Mexico. *Geophysical Monograph*, *154*, 227−238.

McIntosh, W. C., & Chapin, C. E. (2004). Geochronology of the central Colorado volcanic field. *New Mexico Bureau of Geology & Mineral Resources, Bulletin*, *160*, 205−237.

McIntosh, W. C., Chapin, C. E., Ratte, J. C., & Sutter, J. F. (1992). Time-stratigraphic framework for the Eocene-Oligocene Mogollon-Datil volcanic field, Southwest New Mexico. *Geological Society of America Bulletin*, *104*(7), 851−871.

McQuarrie, N., & Wernicke, B. P. (2005). An animated tectonic reconstruction of southwestern North America since 36 Ma. *Geosphere*, *1*(3), 147−172.

Miller, B. A., & Crider, J. G. (2014). Fold form and fault geometry at Umtanum Ridge, Yakima fold belt, Washington. *Geological Society of America Abstracts and Programs*, *46*(6), 607.

Miller, C. F., & Barton, M. D. (1990). Phanerozoic plutonism in the Cordilleran Interior, U.S.A. *Special Paper - Geological Society of America*, *241*, 213−231.

Miller, E. L., & Gans, P. B. (1989). Cretaceous crustal structure and metamorphism in the hinterland of the Sevier thrust belt, western U. S. Cordillera. *Geology [Boulder]*, *17*(1), 59−62.

Molnar, P. (2010). Deuterium and oxygen isotopes, paleoelevations of the Sierra Nevada, and Cenozoic climate. *Geological Society of America Bulletin*, *122*(7−8), 1106−1115.

Morgan, L.A., Pierce, K.L., & Shanks, W.C.P. (2008). Track of the Yellowstone hotspot: Young and ongoing geologic processes from the Snake River Plain to the Yellowstone Plateau and Tetons. In Raynolds, R.G. (Ed.), *Roaming the Rocky Mountains and Environs: Geological Field Trips*. Geological Society of America Field Guide 10.

Moye, F. J., Hackett, W. R., Blakley, J. D., & Snider, L. G. (1988). *Regional geologic setting and volcanic stratigraphy of the Challis volcanic field, central Idaho* (pp. 87−99). Moscow, ID: Idaho Geological Survey.

Mutschler, F.E., Ernst, D.R., Gaskill, D.L., & Billings, P. (1981) Igneous rocks of the Elk Mountains and vicinity, Colorado chemistry and related ore deposits. *New Mexico Geological Society Guidebook, 32nd Field Conference*, Western Slope Colorado, pp. 317−324.

Oldow, J. S. (1984). Evolution of a late Mesoozoic back-arc fold and thrust belt, northwestern Great Basin, USA. *Tectonophysics*, *102*, 245−274.

Ormerod, D. S., Hawkesworth, C. J., Rogers, N. W., Leeman, W. P., & Menzies, M. A. (1988). Tectonic and magmatic transitions in the Western Great Basin, USA. *Nature*, *333*, 349−353.

Parise, M., Coe, J.A., Savage, W.Z., & Varnes, D.J. (2003). The Slumgullion landslide (South-Western Colorado, USA): Investigation and monitoring, http://ww.unina2.it/flows2003/flows2003/articoli/parise.pdf.

Parry, W. T., & Bruhn, R. L. (1987). Fluid inclusion evidence for minimum 11 km vertical offset on the Wasatch Fault, Utah. *Geology [Boulder]*, *15*(1), 67−70.

Parsons, T. (1995). The Basin and Range Province. *Developments in Geotectonics*, *25*, 277−324.

Parsons, T., Thompson, G. A., & Sleep, N. H. (1994). Mantle plume influence on the Neogene uplift and extension of the U.S. western Cordillera? *Geology [Boulder]*, *22*(1), 83−86.

Pierce, K. L., & Morgan, L. A. (1992). The track of the Yellowstone hot spot; volcanism, faulting, and uplift. *Memoir - Geological Society of America*, *179*, 1−53.

Reheis, M. C., Redwine, J., Adams, K., Stine, S., Parker, K., Negrini, R., ... Smoot, J. P. (2003). *Pliocene to Holocene lakes in the western Great Basin; new perspectives on paleoclimate, landscape dynamics, tectonics, and paleodistribution of aquatic species*. Reno, NV: Desert Research Institute.

Rhodes, B. P., & Cheney, E. S. (1981). Low-angle faulting and the origin of Kettle dome, a metamorphic core complex in northeastern Washington. *Geology*, *9*(8), 366−369.

Ruleman, C. A., Miggins, D. P., & Mason, C. C. (2012). Timing and rates of canyon incision and basin integration along the northern Rio Grande, San Luis Valley, New Mexico and Colorado. *GSA Abstracts and Programs*, *44*(6), 79.

Schwartz, J. J., Snoke, A. W., Cordey, F., Johnson, K., Frost, C. D., Barnes, C. G., ... Wooden, J. L. (2011). Late Jurassic magmatism, metamorphism, and deformation in the Blue Mountains Province, northeast Oregon. *Geological Society of America Bulletin*, *123*(9−10), 2083−2111.

Scott, W. E. (2004). Quaternary volcanism in the United States. *Developments in Quaternary Science*, *1*, 351−380.

Semken, S. (2003). Black rocks protruding up: The Navajo volcanic field. *New Mexico Geological Society Guidebook, 54th Field Conference*, Geology of the Zuni Plateau, 2003, pp. 133−138.

Sherrod, B. L., Blakely, R. J., Lasher, J. P., Lamb, A., Mahan, S. A., Foit, F. F., & Barnett, E. A. (2016). Active faulting on the Wallula fault zone within the Olympic-Wallowa Lineament, Washington State, USA. *Geological Society of America Bulletin*, *128*(11−12), 1636−1659.

Smith, R. B., & Siegel, L. J. (2000). *Windows into the Earth: The Geologic Story of the Yellowstone and Grand Teton National Parks*. New York: Oxford University Press, 242 pages.

Snoke, A. W. (2005). North America; Southern Cordillera. In R. C. Selley, L. R. M. Cocks, & I. R. Plimer (Eds.), *Encyclopedia of Geology* (pp. 48−61). Oxford: Elsevier Academic Press.

Sonder, L. J., & Jones, C. H. (1999). Western United States extension; how the west was widened. *Annual Review of Earth and Planetary Sciences*, *27*, 417−462.

Spencer, J.E., S.J. Reynolds, P. Anderson, & Anderson, J.L. (1985). Reconnaissance geology of the crest of the Sierra Estrella, central Arizona. *Open-File Report - University of Arizona*. Bureau of Geology and Mineral Technology, pp. 85−11.

Stock, G. M., Frankel, K. L., Ehlers, T. A., Schaller, M., Briggs, S. M., & Finkel, R. C. (2009). Spatial and temporal variations in denudation of the Wasatch Mountains, Utah, USA. *Lithosphere*, *1*(1), 34−40.

Tizzani, P., Maurizio, B., Giovanni, Z., Simone, A., Paolo, B., & Riccardo, L. (2009). Uplift and magma intrusion at Long Valley Caldera from InSAR and gravity measurements. *Geology [Boulder]*, *37*(1), 63−66.

Thompson, T. B. (1972). Sierra Blanca igneous complex, New Mexico. *Geological Society of America Bulletin*, *83*, 2341−2356.

Tolan, T.L., Beeson, M.H., and Lindsey, K.A., 2002, The Effects of volcanism and tectonism on the Columbia River system, Northwest Geological Society Field Trip Guidebook Series, no. 19, 74 p, Seattle.

U.S. Geological Survey. *Geology of the New York City Region*. (2003). <http://3dparks.wr.usgs.gov/nyc/index.html>.

Vogl, J. J., Min, K., Carmenate, A., Foster, D. A., & Marsellos, A. (2014). Miocene regional hotspot-related uplift, exhumation, and extension north of the Snake River Plain: Evidence from apatite (U-Th)/He thermochronology. *Lithosphere*, *6*(2), 108–123. Available from https://doi.org/10.1130/L308.1.

Volcano Discovery.(2015) <http://www.volcanodiscovery.com/utah.html>.

Volcano Hazards Program. *US geological survey*. (2015). <http://volcanoes.usgs.gov>.

Wernicke, B., & Snow, J. K. (1998). Cenozoic tectonism in the central Basin and Range; motion of the Sierran-Great Valley block. *International Geology Review*, *40*, 403–410.

Wolfe, J. A., Forest, C. E., & Molnar, P. (1998). Paleobotanical evidence of Eocene and Oligocene paleoaltitudes in midlatitude western North America. *Geological Society of America Bulletin*, *110*, 664–678.

Wolfe, J. A., Schorn, H. E., Forest, C. E., & Molnar, P. (1997). Paleobotanical evidence for high altitudes in Nevada during the Miocene. *Science*, *276*, 1672–1675.

Yeend, W.E. (1969). Quaternary geology of the Grand and Battlement Mesas Area, Colorado. *US Geological Survey Professional Paper 617*, 50 p.

Zoback, M. L., McKee, E. H., Blakely, R. J., & Thompson, G. A. (1994). The northern Nevada rift; regional tectono-magmatic relations and middle Miocene stress direction. *Geological Society of America Bulletin*, *106*(3), 371–382.

CHAPTERS 19 AND 20: MODERN-DAY OROGENIC MOUNTAIN SYSTEMS

Aalto, K. R. (2006). The Klamath peneplain; a review of J. S. Diller's classic erosion surface. *Special Paper - Geological Society of America*, *410*, 451–463.

Amos, C. B., Kelson, K. I., Rood, D. H., Simpson, D. T., & Rose, R. S. (2010). Late Quaternary slip rate on the Kern Canyon Fault at Soda Spring, Tulare County, California. *Lithosphere*, *2*(6), 411–417.

Anderson, R. E., Berger, B. R., & Miggins, D. (2012). Timing, magnitude, and style of Miocene deformation, west-central Walker Lane Belt, Nevada. *Lithosphere*, *4*(3), 187–208.

Anderson, R. S., & Menking, K. M. (1994). The Quaternary marine terraces of Santa Cruz, California; evidence for coseismic uplift on two faults. *Geological Society of America Bulletin*, *106*(5), 649–664.

Andrew, J. E., & Walker, J. D. (2009). Reconstructing late Cenozoic deformation in central Panamint Valley, California; evolution of slip partitioning in the Walker Lane. *Geosphere*, *5*(3), 172–198.

Andrew, J. E., Walker, J. D., & Monastero, F. C. (2014). Evolution of the central Garlock fault zone, California; a major sinistral fault embedded in a dextral plate margin. *Geological Society of America Bulletin*, *127*(1–2), 227–249.

Atwater, B. F., & Yamaguchi, D. K. (1991). Sudden, probably coseismic submergence of Holocene trees and grass in coastal Washington State. *Geology [Boulder]*, *19*(7), 706–709.

Atwater, T. (1970). Implications of plate tectonics for the Cenozoic tectonic evolution of western North America. *Geological Society of America Bulletin*, *81*, 3513–3536.

Atwater, T., & Stock, J. (1998). Pacific-North America plate tectonics of the Neogene southwestern United States: An update. *International Geology Review*, *40*, 375–402.

Azor, A., Keller, E. A., & Yeats, R. S. (2002). Geomorphic indicators of active fold growth; South Mountain-Oak Ridge Anticline, Ventura Basin, Southern California. *Geological Society of America Bulletin*, *114*(6), 745–753.

Bacon, S. N., Jayko, A. S., & McGeehin, J. P. (2005). Holocene and latest Pleistocene oblique dextral faulting on the Southern Inyo Mountains Fault, Owens Lake basin, California. *Bulletin of the Seismological Society of America*, *95*(6), 2472–2485.

Bacon, S. N., & Pezzopane, S. K. (2007). A 25,000-year record of earthquakes on the Owens Valley Fault near Lone Pine, California; implications for recurrence intervals, slip rates, and segmentation models. *Geological Society of America Bulletin*, *119*(7–8), 823–847.

Batt, G. E., Brandon, M. T., Farley, K. A., & Roden-Tice, M. (2001). Tectonic synthesis of the Olympic Mountains segment of the Cascadia wedge, using two-dimensional thermal and kinematic modeling of thermochronological ages. *Journal of Geophysical Research*, *106*(B11), 26.

Batt, G. E., Cashman, S. M., Garver, J. I., & Bigelow, J. J. (2010). Thermotectonic evidence for two-stage extension on the Trinity detachment surface, eastern Klamath Mountains, California. *American Journal of Science*, *310*(4), 261–281.

Brandon, M. T. (2004). The Cascadia subduction wedge: the role of accretion, uplift, and erosion. In B. A. van der Pluijmand, & S. Marshak (Eds.), *Earth Structure, An Introduction to Structural Geology and Tectonics* (2nd ed., pp. 566–574). WCB/McGraw Hill Press.

Brandon, M. T., & Calderwood, A. R. (1990). High-pressure metamorphism and uplift of the Olympic subduction complex. *Geology*, *8*, 1252–1255.

Brandon, M. T., Roden-Tice, Mary K., & Garver, J. I. (1998). Late Cenozoic exhumation of the Cascadia accretionary wedge in the Olympic Mountains, northwest Washington State. *GSA Bulletin*, *110*(8), 985–1009.

Brossy, C. C., Kelson, K. I., Amos, C. B., Baldwin, J. N., Kozlowicz, B., Simpson, D., ... Rose, R. (2012). Map of the late Quaternary active Kern Canyon and Breckenridge faults, southern Sierra Nevada, California. *Geosphere*, *8*(3), 581–591.

Burchfiel, B. C., Hodges, K. V., & Royden, L. H. (1987). Geology of Panamint Valley-Saline Valley pull-apart system, California; palinspastic evidence for low-angle geometry of a Neogene range-bounding fault. *Journal of Geophysical Research*, *92*(B10), 10.

Burchfiel, B. C., & Stewart, J. H. (1966). 'pull-apart' origin of the central segment of death valley, California. *Geological Society of America Bulletin*, *77*(4), 439–441.

Busby, C. J., Andrews, G. D. M., Koerner, A. K., Brown, S. R., Melosh, B. L., & Hagan, J. C. (2016). Progressive derangement of ancient (Mesozoic) east-west Nevadaplano paleochannels into modern (Miocene-Holocene) north-northwest trends in the Walker Lane Belt, central Sierra Nevada. *Geosphere [Boulder, CO]*, *12*(1), 135–175.

Busby, C. J., DeOreo, S. B., Skilling, I., Gans, P. B., & Hagan, J. C. (2008). Carson Pass-Kirkwood paleocanyon system; paleogeography of the ancestral Cascades arc and implications for landscape evolution of the Sierra Nevada (California). *Geological Society of America Bulletin*, *120*(3–4), 274–299.

Busby, C. J., Hagan, J. C., Putirka, K., Pluhar, C. J., Gans, P. B., Wagner, D. L., ... Skilling, I. (2008). The ancestral Cascades Arc; Cenozoic evolution of the central Sierra Nevada (California) and the birth of the new plate boundary. *Special Paper - Geological Society of America*, *438*, 331–378.

Busby, C. J., Hagan, J. C., & Renne, P. (2013). Initiation of Sierra Nevada range front-Walker Lane faulting ca. 12 Ma in the ancestral Cascades Arc. *Geosphere*, *9*, 1125–1146.

Busby, C. J., Koerner, A. K., Melosh, B. L., Hagan, J. C., & Andrews, G. D. M. (2013). Sierra crest graben-vent system; a Walker Lane pull apart within the ancestral Cascades Arc. *Geosphere*, *9*(4), 736–780.

Busby, C. J., & Putirka, K. (2009). Miocene evolution of the western edge of the Nevadaplano in the central and northern Sierra Nevada; palaeocanyons, magmatism, and structure. *International Geology Review*, *51*(7–8), 670–701.

Butler, P. R., Troxel, B. W., & Verosub, K. L. (1988). Late Cenozoic history and styles of deformation along the southern Death Valley fault zone, California. *Geological Society of America Bulletin*, *100*(3), 402–410.

Cascades Volcano Observatory. <http://volcanoes.usgs.gov/observatories/cvo/>.

Cashman, P. H., Trexler, J. H., Jr., Widmer, M. C., & Queen, S. J. (2012). Post-2.6 Ma tectonic and topographic evolution of the northeastern Sierra Nevada; the record in the Reno and Verdi Basins. *Geosphere*, *8*(5), 972–990.

Cecil, M. R., Ducea, M. N., Reiners, P. W., & Chase, C. G. (2006). Cenozoic exhumation of the northern Sierra Nevada, California, from (U-Th)/He thermochronology. *Geological Society of America Bulletin*, *118*(11–12), 1481–1488.

Cichanski, M. (2000). Low-angle, range-flank faults in the Panamint, Inyo, and Slate ranges, California; implications for recent tectonics of the Death Valley region. *Geological Society of America Bulletin*, *112*(6), 871–883.

Clark, M. K., Maheo, G., Saleeby, J., & Farley, K. A. (2005). The nonequilibrium landscape of the southern Sierra Nevada, California. *GSA Today*, *15*(9), 4–10.

Coney, P. J., Jones, D. L., & Monger, J. W. H. (1980). Cordilleran suspect terranes. *Nature [London]*, *288*(5789), 329–333.

DeMets, C., Gordon, R. G., Argus, D. F., & Stein, S. (1990). Current plate motions. *Geophysical Journal of the Interior*, *101*, 425–478.

Densmore, A. L., & Anderson, R. S. (1997). Tectonic geomorphology of the Ash Hill Fault, Panamint Valley, California. *Basin Research*, *9*(1), 53–63.

Dickinson, W. R. (2008). Accretionary Mesozoic-Cenozoic expansion of the Cordilleran continental margin in California and adjacent Oregon. *Geosphere*, *4*(2), 329–353.

Donnelly-Nolan, J. M., Grove, T. L., Lanphere, M. A., Champion, D. E., & Ramsey, D. W. (2008). Eruptive history and tectonic setting of Medicine Lake Volcano, a large rear-arc volcano in the southern Cascades. *Journal of Volcanology and Geothermal Research*, *177*(2), 313–328.

Dorsey, R. J., & Roering, J. J. (2006). Quaternary landscape evolution in the San Jacinto fault zone, Peninsular Ranges of Southern California; transient response to strike-slip fault initiation. *Geomorphology*, *73*(1–2), 16–32.

Ducea, M., House, M. A., & Kidder, S. (2003). Late Cenozoic denudation and uplift rates in the Santa Lucia Mountains, California. *Geology*, *31*, 139–142.

Faulds, J.E., & Henry, C.D. (2008). Tectonic influences on the spatial and temporal evolution of the Walker Lane: An incipient transform fault along the evolving Pacific – North American plate boundary. In Spencer, J.E., & Titley, S.R. (Eds.), *Ores and orogenesis: Circum-Pacific tectonics, geologic evolution, and ore deposits*, Arizona Geological Society Digest 22, pp. 437–470.

Faulds, J. E., Henry, C. D., & Hinz, N. H. (2005). Kinematics of the northern Walker Lane; an incipient transform fault along the Pacific-North American Plate boundary. *Geology [Boulder]*, *33*(6), 505–508.

Figueroa, A. M., & Knott, J. R. (2010). Tectonic geomorphology of the southern Sierra Nevada Mountains (California); evidence for uplift and basin formation. *Geomorphology*, *123*(1–2), 34–45.

Fleck, R. J., Hagstrum, J. T., Calvert, A. T., Evarts, R. C., & Conrey, R. M. (2014). (super 40) Ar/(super 39) Ar geochronology, paleomagnetism, and evolution of the Boring volcanic field, Oregon and Washington, USA. *Geosphere*, *10*(6), 1283–1314. Available from https://doi.org/10.1130/GES00985.1.

Garrison, N. J., Busby, C. J., Gans, P. B., Putirka, K., & Wagner, D. L. (2008). A mantle plume beneath California? The mid-Miocene Lovejoy flood basalt, Northern California. *Special Paper - Geological Society of America*, *438*, 551–572.

Guest, B., Niemi, N., & Wernicke, B. (2007). Stateline fault system; a new component of the Miocene-Quaternary Eastern California shear zone. *Geological Society of America Bulletin*, *119*(11–12), 1337–1346.

Gurrola, L. D., Keller, E. A., Chen, J. H., Owen, L. A., & Spencer, J. Q. (2013). Tectonic geomorphology of marine terraces; Santa Barbara fold belt, California. *Geological Society of America Bulletin*, *126*(1–2), 219–233.

Hammond, W. C., Blewitt, G., Li, Z., Plag, H.-P., & Kreemer, C. (2012). Contemporary uplift of the Sierra Nevada, western United States, from GPS and InSAR measurements. *Geology [Boulder]*, *40*(7), 667–670.

Heaton, T. H., & Kanamori, H. (1984). Seismic potential associated with subduction in the Northwestern United States. *Bulletin of the Seismological Society of America*, *74*(3), 933–941.

Henry, C. D. (2009). Uplift of the Sierra Nevada, California. *Geology*, *37*(6), 575–576, GeoRef, EBSCOhost.

Henry, C. D., Hinz, N. H., Faulds, J. E., Colgan, J. P., John, D. A., Brooks, E. R., ... Castor, S. B. (2012). Eocene-early Miocene paleotopography of the Sierra Nevada-Great Basin-Nevadaplano based on widespread ash-flow tuffs and paleovalleys. *Geosphere*, *8*(1), 1–27.

Hildreth, W., Fierstein, J., Champion, D., & Calvert, A. (2014). Mammoth Mountain and its mafic periphery; a late Quaternary volcanic field in eastern California. *Geosphere*, *10*(6), 1315–1365.

Hladky, F. R. (1998). Age, chemistry, and origin of capping lava at Upper Table Rock and Lower Table Rock, Jackson County, Oregon. *Oregon Geology*, *60*(4), 81.

House, M. A., Wernicke, B. P., & Farley, K. A. (2001). Paleogeomorphology of the Sierra Nevada, California, from (U-Th)/He ages in apatite. *American Journal of Science*, *301*(2), 77–102.

Hren, M. T., Pagani, M., Erwin, D. M., & Brandon, M. (2010). Biomarker reconstruction of the early Eocene paleotopography and paleoclimate of the northern Sierra Nevada. *Geology [Boulder]*, *38*(1), 7–10.

Hughes, S. S. (1990). Mafic magmatism and associated tectonism of the central high Cascade Range, Oregon. *Journal of Geophysical Research*, *95*(B12), 19.

Humphreys, E. D. (1995). Post-Laramide removal of the Farallon Slab, Western United States. *Geology [Boulder]*, *23*(11), 987–990.

Irwin, W.P. (1994). Geologic map of the Klamath Mountains, California and Oregon. *Miscellaneous Investigations Series - U. S. Geological Survey* IMAP 2148, Scale 1:500,000.

Jayko, A.S. (2009). Deformation of the late Miocene to Pliocene Inyo Surface, eastern Sierra region, California. In Oldow, J.S., & Cashman, P. (Eds.), *Late Cenozoic structure and evolution of the Great Basin–Sierra Nevada transition: Geological Society of America Special Paper 447*, pp. 313–350.

Jones, C. H., Farmer, G. L., & wakab, J. (2004). Tectonics of Pliocene removal of lithosphere of the Sierra Nevada, California. *Geological Society of America Bulletin, 116*(11–12), 1408–1422.

Keller, E. A., Seaver, D. B., Laduzinsky, D. L., Johnson, D. L., & Ku, T. L. (2000). Tectonic geomorphology of active folding over buried reverse faults; San Emigdio Mountain front, southern San Joaquin Valley, California. *Geological Society of America Bulletin, 112*(1), 86–97.

Kelsey, H. M. (1990). Late Quaternary deformation of marine terraces on the Cascadia subduction zone near Cape Blanco, Oregon. *Tectonics, 9*(5), 983–1014.

Kelsey, H. M., & Bockheim, J. G. (1994). Coastal landscape evolution as a function of eustasy and surface uplift rate, Cascadia margin, southern Oregon; with Suppl. Data 9426. *Geological Society of America Bulletin, 106*(6), 840–854.

Kelsey, H. M., Engebretson, D. C., Mitchell, C. E., & Ticknor, R. L. (1994). Topographic form of the Coast Ranges of the Cascadia margin in relation to coastal uplift rates and plate subduction. *Journal of Geophysical Research, 99*(B6), 12.

LaMaskin, T. A., Vervoort, J. D., Dorsey, R. J., & Wright, J. E. (2011). Early Mesozoic paleogeography and tectonic evolution of the western United States; insights from detrital zircon U-Pb geochronology, Blue Mountains Province, northeastern Oregon. *Geological Society of America Bulletin, 123*(9–10), 1939–1965.

Lamb, A. P., Liberty, L. M., Blakely, R. J., Pratt, T. L., Sherrod, B. L., & van Wijk, Kasper (2012). Western limits of the Seattle fault zone and its interaction with the Olympic Peninsula, Washington. *Geosphere, 8*(4), 915–930.

Le, K., Lee, J., Owen, L. A., & Finkel, R. (2007). Late Quaternary slip rates along the Sierra Nevada frontal fault zone, California; slip partitioning across the western margin of the Eastern California shear zone-Basin and Range Province. *Geological Society of America Bulletin, 119*(1–2), 240–256.

Lechler, A. R., & Galewsky, J. (2012). Refining paleoaltimetry reconstructions of the Sierra Nevada, California, using air parcel trajectories. *Geology [Boulder], 41*(2), 259–262.

Lee, J., Stockli, D. F., Owen, L. A., Finkel, R. C., & Kislitsyn, R. (2009). Exhumation of the Inyo Mountains, California; implications for the timing of extension along the western boundary of the Basin and Range Province and distribution of dextral fault slip rates across the Eastern California shear zone. *Tectonics, 28*(1).

Lee, J., Stockli, D., Schroeder, J., Tincher, C., Bradley, D., Owen, L., ... Garwood, J. (2006). Fault slip transfer in the Eastern California Shear Zone–Walker Lane Belt. *Geological Society of America Penrose Conference Field Trip Guide (Kinematics and Geodynamics of Intraplate Dextral Shear in Eastern California and Western Nevada, Mammoth Lakes, California, 21–26 April 2005)*, 26 p.

Maheo, G., Saleeby, J., Saleeby, Z., & Farley, K. A. (2009). Tectonic control on southern Sierra Nevada topography, California. *Tectonics, 28*(6), @PaperTC6006.

Manley, C. R., Glazner, A. F., & Farmer, G. L. (2000). Timing of volcanism in the Sierra Nevada of California; evidence for Pliocene delamination of the batholithic root? *Geology, 28*(9), 811–814.

Martel, S. J., Stock, G. M., & Ito, G. (2014). Mechanics of relative and absolute displacements across normal faults, and implications for uplift and subsidence along the eastern escarpment of the Sierra Nevada, California. *Geosphere, 10*(2), 243–263.

McInelly, G. W., & Kelsey, H. M. (1990). Late Quaternary tectonic deformation in the Cape Arago-Bandon region of coastal Oregon as deduced from wave-cut platforms. *Journal of Geophysical Research, 95*(B5), 6699–6713.

McPhillips, D., & Brandon, M. T. (2012). Topographic evolution of the Sierra Nevada measured directly by inversion of low-temperature thermochronology. *American Journal of Science, 312*(2), 90–116.

Meigs, A., Brozovic, N., & Johnson, M. L. (1999). Steady, balanced rates of uplift and erosion of the Santa Monica Mountains, California. *Basin Research, 11*(1), 59–73.

Miller, M. B., & Pavlis, T. L. (2005). The Black Mountains turtlebacks; Rosetta stones of Death Valley tectonics. *Earth-Science Reviews, 73*(1–4), 115–138.

Miller, M. G. (1991). High-angle origin of the currently low-angle Badwater turtleback fault, Death Valley, California. *Geology [Boulder], 19*(4), 372–375.

Molnar, P. (2010). Deuterium and oxygen isotopes, paleoelevations of the Sierra Nevada, and Cenozoic climate. *Geological Society of America Bulletin, 122*(7–8), 1106–1115.

Muhs, D. R., Prentice, C. S., & Merritts, D. J. (2003). *Marine terraces, sea level history and Quaternary tectonics of the San Andreas Fault on the coast of California* (pp. 1–18). Reno, NV: Desert Research Institute, 2003.

Muhs, D. R., Rockwell, T. K., & Kennedy, G. L. (1992). Late Quaternary uplift rates of marine terraces on the Pacific coast of North America, southern Oregon to Baja California Sur. *Quaternary International, 15–16*, 121–133.

Mulch, A., Graham, S. A., & Chamberlain, C. P. (2006). Hydrogen isotopes in Eocene river gravels and paleoelevation of the Sierra Nevada. *Science, 313*(5783), 87–89.

Nadin, E. S., & Saleeby, J. B. (2010). Quaternary reactivation of the Kern Canyon fault system, southern Sierra Nevada, California. *Geological Society of America Bulletin, 122*(9–10), 1671–1685.

Nelson, C. H., Bacon, C. R., Robinson, S. W., Adam, D. P., Bradbury, J. P., Barber, J. H., Jr., ... Vagenas, G. (1994). The volcanic, sedimentologic, and paleolimnologic history of the Crater Lake caldera floor, Oregon; evidence for small caldera evolution. *Geological Society of America Bulletin, 106*(5), 684–704.

Norton, I. (2011). Two-stage formation of Death Valley. *Geosphere, 7*(1), 171–182.

Orr, E. L., & Orr, W. N. (2000). *Geology of Oregon*. Dubuque, IA: Kendall/Hunt Publishing, 254 pages.

Page, B. M., Thompson, G. A., & Coleman, R. G. (1998). Late Cenozoic tectonics of the central and southern Coast Ranges of California. *Geological Society of America Bulletin, 110*(7), 846–876.

Parsons, T., Blakely, R.J., Brocher, T.M., Christensen, N.I. Fisher, M.A., Flueh, E., ... Wells, R.E. (2005). Crustal Structure of the Cascadia Fore Arc of Washington. *U.S. Geological Survey Professional Paper 1661-D*, 45 p.

Pazzaglia, F. J., & Brandon, M. T. (2001). A Fluvial record of long-term steady-state uplift and erosion across the Cascadia Forearc high, Western Washington State. *American Journal of Science, 301*, 385–431.

Pazzaglia, F. J., Thackray, G. D., Brandon, M. T., Wegmann, K. W., Gosse, J., McDonald, E., ... Prothero, D. (2003). Tectonic geomorphology and the record of Quaternary plate boundary deformation in the Olympic Mountains. *GSA Field Guide, 4*, 37−67.

Perg, L. A., Anderson, R. S., & Finkel, R. C. (2001). Use of a new (super 10) Be and (super 26) Al inventory method to date marine terraces, Santa Cruz, California, USA. *Geology [Boulder], 29*(10), 879−882.

Peryam, T. C., Dorsey, R. J., & Bindeman, I. N. (2011). Plio-Pleistocene climate change and timing of Peninsular Ranges uplift in Southern California; evidence from paleosols and stable isotopes in the Fish Creek-Vallecito Basin. *Palaeogeography, Palaeoclimatology, Palaeoecology, 305*(1−4), 65−74.

Phillips, F. M., McIntosh, W. C., & Dunbar, N. W. (2011). Chronology of late Cenozoic volcanic eruptions onto relict surfaces in the south-central Sierra Nevada, California. *Geological Society of America Bulletin, 123*(5−6), 890−910.

Pinter, N. (1995). Faulting on the volcanic Tableland, Owens Valley, California. *Journal of Geology, 103*, 73−83.

Piotraschke, R., Cashman, S. M., Furlong, K. P., Kamp, P. J. J., Danisik, M., & Xu, G. (2015). Unroofing the Klamaths; blame it on Siletzia? *Lithosphere, 7*(4), 427−440.

Polenz, M., & Kelsey, H. M. (1999). Development of the late Quaternary marine terraced landscape during on-going tectonic contraction, Crescent City coastal plain, California. *Quaternary Research, 52*(2), 217−228.

Raisz, E. J. (1945). The Olympic-Wallowa lineament. *American Journal of Science, 243-A*, 479−485.

Reilinger, R., & Adams, J. (1982). Geodetic evidence for active landward tilting of the Oregon and Washington coastal ranges. *Geophysical Research Letters, 9*(4), 401−403.

Reiners, P. W. (Ed.), (2012). Paleotopography in the Western U.S. Cordillera. *American Journal of Science, 312*(2), 81−262.

Rockwell, T. K., Keller, E. A., & Dembroff, G. R. (2002). Quaternary rate of folding of the Ventura Avenue Anticline, western Transverse Ranges, southern California. *Annual Field Trip Guidebook. South Coast Geological Society, 30*, 319−327.

Saleeby, J. B., Saleeby, Z., Nadin, E., & Maheo, G. (2009). Step-over in the structure controlling the regional west tilt of the Sierra Nevada Microplate; eastern escarpment system to Kern Canyon system. *International Geology Review, 51*(7−8), 634−669.

Schmalzle, G. M., McCaffrey, R., & Creager, K. C. (2014). Central Cascadia subduction zone creep. *Geochemistry, Geophysics, Geosystems - G [Super 3], 15*(4), 1515−1532.

Schmandt, B., & Humphreys, E. (2011). Seismically imaged relict slab from the 55 Ma Siletzia accretion to the northwest United States. *Geology [Boulder], 39*(2), 175−178.

Schmidt, M. E., Grunder, A. L., & Rowe, M. C. (2008). Segmentation of the Cascade Arc as indicated by Sr and Nd isotopic variation among diverse primitive basalts. *Earth and Planetary Science Letters, 266*(1−2), 166−181.

Shaller, P. J., & Heron, C. W. (2004). Proposed revision of marine terrace extent, geometry and rates of uplift, Pacific Palisades, California. *Environmental & Engineering Geoscience, 10*(3), 253−275.

Sieh, K. E., & Jahns, R. H. (1984). Holocene activity of the San Andreas Fault at Wallace Creek, California. *Geological Society of America Bulletin, 95*(8), 883−896.

Small, E. E., & Anderson, R. S. (1995). Geomorphically driven Late Cenozoic uplift in the Sierra Nevada, California. *Science, 270*, 277−280.

Smith, G. A., Snee, L. W., & Taylor, E. M. (1987). Stratigraphic, sedimentologic, and petrologic record of late Miocene subsidence of the central Oregon High Cascades. *Geology, 15*(5), 389−392.

Snoke, A. W., & Barnes, C. G. (2006). The development of tectonic concepts for the Klamath Mountains province, California and Oregon. *Special Paper - Geological Society of America, 410*, 1−29.

Soreghan, M.J. & Gehrels, G.E. (Eds.). (2000). Paleozoic and Triassic Paleogeography and Tectonics of Western Nevada and Northern California. *Geological Society of America Special Paper 347*, 252 p.

Sousa, F. J., Saleeby, J., Farley, K. A., Unruh, J. R., & Lloyd, M. K. (2016). The southern Sierra Nevada pediment, Central California. *Geosphere [Boulder, CO], 13*(1), 82−101.

Southern California Earthquake Data Center. <http://www.data.scec.org/about/index.html>.

Spotila, J. A., Niemi, N., Brady, R., House, M., Buscher, J., & Oskin, M. (2007). Long-term continental deformation associated with transpressive plate motion; the San Andreas Fault. *Geology [Boulder], 35*(11), 967−970.

Stewart, R. J., & Brandon, M. T. (2004). Detrital-zircon fission-track ages for the "Hoh Formation": Implications for late Cenozoic evolution of the Cascadia subduction wedge. *GSA Bulletin, 116*(1/2), 60−75. Available from https://doi.org/10.1130/B22101.1.

Stock, G. M., Anderson, R. S., & Finkel, R. C. (2005). Rates of erosion and topographic evolution of the Sierra Nevada, California, inferred from cosmogenic (super 26) Al and (super 10) Be concentrations. *Earth Surface Processes and Landforms, 30*(8), 985−1006.

Stockli, D. F., Dumitru, T. A., McWilliams, M. O., & Farley, K. A. (2003). Cenozoic tectonic evolution of the White Mountains, California and Nevada. *Geological Society of America Bulletin, 115*(7), 788−816.

Trehu, A. M., Blakely, R. J., & Williams, M. C. (2011). Subducted seamounts and recent earthquakes beneath the central Cascadia forearc. *Geology [Boulder], 40*(2), 103−106.

Trexler, J., Cashman, P., & Cosca, M. (2012). Constraints on the history and topography of the northeastern Sierra Nevada from a Neogene sedimentary basin in the Reno-Verdi area, western Nevada. *Geosphere, 8*(3), 548−561.

Unruh, J., Humphrey, J., & Barron, A. (2003). Transtensional model for the Sierra Nevada frontal fault system, eastern California. *Geology [Boulder], 31*(4), 327−330.

Unruh, J. R. (1991). The uplift of the Sierra Nevada and implications for late Cenozoic epeirogeny in the western Cordillera. *Geological Society of America Bulletin, 103*(11), 1395−1404.

U.S. Geological Survey. (2013). 100-Meter resolution color shaded relief of the Conterminous United States, National Atlas of the United States. <http://nationalatlas.gov/atlasftp-1m.html>.

Van Buer, N. J., Miller, E. L., & Dumitru, T. A. (2009). Early Tertiary paleogeologic map of the northern Sierra Nevada Batholith and the northwestern Basin and Range. *Geology [Boulder], 37*(4), 371−374.

VanLandingham, S. L., Smith, G. A., Snee, L. W., Taylor, E. M., & Bradbury, J. P. (1987). Stratigraphic, sedimentologic, and petrologic record of late Miocene subsidence of the central Oregon High Cascades; discussion and reply. *Geology [Boulder], 15*(11), 1082−1084.

VanLaningham, S., Meigs, A., & Goldfinger, C. (2006). The effects of rock uplift and rock resistance on river morphology in a subduction zone forearc, Oregon, USA. *Earth Surface Processes and Landforms, 31*(10), 1257−1279.

Wakabayashi, J. (2013). Paleochannels, stream incision, erosion, topographic evolution, and alternative explanations of paleoaltimetry, Sierra Nevada, California. *Geosphere, 9*(2), 191–215.

Wakabayashi, J., & Sawyer, T. L. (2001). Stream incision, tectonics, uplift, and evolution of topography of the Sierra Nevada, California. *Journal of Geology, 109*(5), 539–562, *GeoRef*, EBSCO*host* (accessed December 29, 2013).

Walker, J. D., Kirby, E., & Andrew, J. E. (2005). Strain transfer and partitioning between the Panamint Valley, Searles Valley, and Ash Hill fault zones, California. *Geosphere, 1*(3), 111–118.

Wells, R., Bukry, D., Friedman, R., Pyle, D., Duncan, R., Haeussler, P., & Wooden, J. (2014). Geologic history of Siletzia, a large igneous province in the Oregon and Washington Coast Range; correlation to the geomagnetic polarity time scale and implications for a long-lived Yellowstone Hotspot. *Geosphere, 10*(4), 692–719.

Wells, R. E., & Heller, P. L. (1988). The relative contribution of accretion, shear, and extension to Cenozoic tectonic rotation in the Pacific Northwest. *Geological Society of America Bulletin, 100*(3), 325–338.

Wells, R. E., & McCaffrey, R. (2013). Steady rotation of the Cascade Arc. *Geology [Boulder], 41*(9), 1027–1030.

Wells, R. E., & Simpson, R. W. (2001). Northward migration of the Cascadia forearc in the northwestern U.S. and implications for subduction deformation. *Earth, Planets and Space, 53*(4), 275–283.

Wright, L. A., Otton, J. K., & Troxel, B. W. (1974). *Turtleback surfaces of Death Valley viewed as phenomena of extensional tectonics* (pp. 79–80). Shoshone, CA: Death Valley Publ. Co., 1974.

Wyld, S. J. (2002). Structural evolution of a Mesozoic backarc fold-and-thrust belt in the U.S. Cordillera; new evidence from northern Nevada. *Geological Society of America Bulletin, 114*(11), 1452–1468.

Wyld, S. J., & Wright, J. E. (2001). New evidence for Cretaceous strike-slip faulting in the United States Cordillera and implications for terrane-displacement, deformation patterns, and plutonism. *American Journal of Science, 301*(2), 150–181.

Yeats, R.S., & Grigsby, F.B. (1987). Ventura Avenue Anticline; Amphitheater locality, California Centennial field guide (pp. 219–223). Boulder, CO: Geological Society of America.

Index

Note: Page numbers followed by "*f*" and "*t*" refer to figures and tables, respectively.

A

Ablation. *See* Zone of ablation
Absaroka Mountains, 293
Absaroka volcanic field, 409–411
Absolute hardness or softness of rock, 32
Acadia National Park, 368–369, 368*f*
Acadian Ocean Basin Flysch and Plutons rock unit, 369
Acadian orogeny, 362–363
Accreted terrane, 67, 74, 141, 144, 148, 353, 357, 361, 451–452
Accretion. *See* Tectonic accretion
Accretionary prism, 65–68, 150
    in Coast Range, 480, 510
Accretionary wedge. *See* Accretionary prism
Accumulation. *See* Zone of accumulation
Active continental margins, 65–66, 67*f*, 71–74
Active faults/faulting
    in Death Valley, 546–547, 547*f*
    and incision rates, 554–555
    recognition, 50–52
    USGS National Atlas digital shaded relief elevation map of Grand Canyon, 554*f*
Active tectonic stress, 60
Active tectonics, 489
Adirondack Mountains, 7, 17–18, 33, 142
    ancient North American crystalline shield of, 261–263
    esker system in, 176–177
    high-grade metamorphic shield rocks in, 383
    NNE across west-central, 263*f*
    Raisz landform map of, 262*f*
    Superior Upland in, 298–299
AFT. *See* Apatite fission track
Agassiz, Lake, 20, 175
Alberta Clippers, 84
Aldrich Mountains, 394, 398–399
Aleutian Islands, 74–76
Allamoore Formation, 385–386
Alleghany orogeny, 70–71
Allegheny Front, 253–254
Allegheny Plateau, 253–254
    Catskill Front along Hudson Valley, 255*f*
Allochthon, 321–322, 449
Alluvial. *See* Alluvium
Alluvium, 444
Alpine glaciation, landscape development in areas of, 159–162
Alpine glaciers, 157, 159, 159*f*, 161*f*, 164
Alpine-Fort Davis area, 417–418
Amphibolite, 29, 354, 385
Ancestral Rocky Mountains, 74, 288, 288*f*, 289*f*
Ancient accretionary prism, 510
Ancient canyons, 555
Ancient forearc basin, 510–511
Ancient North American crystalline shield, 227
Ancient Teays River valley, 181
Ancient uplift rates and elevation, measuring, 110
Ancient volcanic arc, 512
Andesite, 30, 389, 391–392
Andesite-rhyolite-tuff family, 30
Animikie Basin, 302
Anorthosite, 262
Antarctic ice core, 105
Antarctic ice sheet, 92, 101
Anthracite Valley, 253–254
Anticlinal mountains, 50, 261, 397
    of Middle Rockies, 289–298
        Beartooth Mountains, 293–295
        Bighorn Mountains, 295–296
        Black Hills, 296–298
        Wind River Range, 290–293
Anticlinal valley, 50
Anticlines, 41–42, 46, 93, 131, 352
    Meander, 93, 93*f*
    Rock Springs, 220–221
    rollover, 431*f*, 432, 432*f*
Anticlinorium, 346, 363–364
Antler orogeny, 449–450
Antler overlap sequence, 449
Apalachee Bay, 202–203, 203*f*
Apatite, 122, 359
Apatite (U-Th)/He ages, 494
    cooling ages, 380–381
Apatite fission track (AFT), 286, 558
    and (U-Th)/He ages, 497
    analysis, 244
    dating techniques, 359, 360*f*
Apatite fission-track, 494
Appalachian and Cordilleran Mountain belts, 69
Appalachian and Ouachita Mountains, 70
Appalachian foreland deformation distribution, 333
Appalachian Mountains, 7–8, 67, 70–71, 84, 110, 125, 337, 342–372
Asheville Basin, 354–355, 354*f*
Balsam Mountains, 353–354
Blue Ridge, 342
Blue Ridge at Roanoke, 345–346
Blue Ridge escarpment, 359–360, 359*f*
Blue Ridge North of Roanoke, 346–347
Blue Ridge South of Roanoke, 347–349
    erosional history, 369–372
        Relict Erosion Surfaces in Southern Blue Ridge, 371–372
Fall Line, 360–361
geologic overview of Blue Ridge, 342–345
Grandfather Mountain Area, 355–358
Great Smoky Mountains, 350–353
level of exhumation across Great Smoky Mountains, 349–350
New England Highlands, 361–369
northwestward at northern Blue Ridge, 346*f*
Piedmont Plateau, 358–359
Rome Formation, 344, 344*f*
triassic lowlands of, 466–470
Appalachian orogeny, 344
Appalachian Plateau, 7–8, 130, 251–258. *See also* Ozark Plateau
    Allegheny Plateau, 253–254
    comparison of Pottsville and Cumberland escarpments, 257–258
    Cumberland Plateau, 254–257
    landscape map of, 252*f*
    shaded-relief elevation map of Appalachian region, 254*f*
Appalachian rock successions, 150
Appalachian Valley and Ridge landscape, 130
Appalachian-Ouachita
    fold and thrust belt, 318–319
    fronts, 383
    Mountain belt, 131
    Mountain system, 70–71
Appalachians Mountain, 91, 397
Aquifer, 200
    Ogallala, 216–218, 218*f*
AR Valley. *See* Arkansas River Valley
Arbuckle Mountains, 267
Arbuckle structural domes, 266–269
Arch, 42–44
Arches National Park, 231–234, 233*f*
Arctic Ocean, 15–16
Argon, 121, 271–272
Argus Range (Ar), 542

Arizona River, 561, 561f
Arizona-New Mexico Basin and Range, 450–451
Arkansas River Valley (AR Valley), 281–282, 281f, 333–334
　AR Valley–Northern Mountains, 334–335
　at Leadville, 452
Arkansas River Valley–Northern Mountains, 333–335, 334f
Aseismic creep, 108–109
Ash-flow tuff, 391
Ashby Gap, 346
Asheville (Av), 347
Asheville Basin (ab), 347, 353–355, 354f
Asian carp, 18
Asthenosphere, 62, 64, 90
Athens Plateau, 129–130, 336
Atlanta lobe, 372–374
Atlantic Marginal Basin, 147
Atlantic miogeocline, 141, 144
Atlantic Ocean, 70–71, 341
Atlantic passive continental margin, 69–71
Atlantic seaboard, 66, 69–70
Atmosphere, CO_2 in, 103–105
Atmospheric circulation, 79
Aulacogen. See Failed rift
Austin Chalk Cuesta, 207
Authigenic minerals, 110
Avalon encroachment, 369

B

Badlands National Park, 213
Bajada, 444
Baker, Mt., 379–380, 491–492
Baker City (Bc), 399
Balcones Escarpment, 207
Balcones fault zone, 207
Balds, 347
Balsam Mountains, 353–354
Baltimore Canyon Trough, 369–370
Bandelier National Monument, 423–424
Bar scale, 4, 4f
Barrens, 247–248
Barrier islands, 190–191
Barring volcanic catastrophe, 125
Barron and Baraboo Quartzite, 303
Basal décollement. See Thrust faults
Basalt(ic), 29–30, 422, 555
　composition, 389
　crust, 62
　magma, 74, 389, 404–405
　volcanic field, 389
　volcanic rocks, 318
　volcanism, 402–403
Basaltic crust. See Oceanic crust
Base level, 111
　changes, 111–112
　　longitudinal profiles that graded river's response, 112f
　　uplift along longitudinal stream profile results, 112f
Basher Kill (BK), 323–324
Basin (b), 414, 546
Basin and Range, 431f, 432–452

cause of basin and range extension, 447
crustal thinning and volcanism, 440–441
expansion into surrounding areas, 438
geology, 447–451
horizontal extension, 440
landscape characteristics, 438–440
metamorphic core complexes, 441, 442f
Nevadaplano, 446–447
normal fault activity verses erosion, 443–446
physiographic limit, 432–437
$Sr_i = 0.706$ line, 451–452
timing of normal faulting, 442–443
vertical displacement, 440
Batholith, 29, 33
　Idaho, 33, 187–188, 341, 372–375, 402
　Salinian, 512, 518–519
　in Sierra Nevada, 512
Baxter Basin, 220–221
Baxter Peak, 366
Baxter State Park, 366
bcm. See Buffalo-Cherokee Mountains
Bear Lake, 318
Bear Lodge Mountain, 273
Beartooth Mountains (B), 293–295, 294f, 375, 405
Bed, 27
Bedrock, 31, 126
　incision rate, 116
　influence on landscape, 32–35
　　landscape in crystalline rocks, 33
　　landscape in sedimentary rocks, 32–33
　　landscape in unconsolidated sediment, 34–35
　　landscape in volcanic rocks, 33–34
　uplift, 107–108, 108f, 119–120
　　rate, 122
　uplift/subsidence, 107–108
Belt Supergroup, 314–315
　at GNP, 314–315
　major rock unit within, 315–316
Bench-and-slope landscape, 185–187, 186f, 187f, 224–226, 247–248
　on Colorado Plateau, 224f
　Grand Staircase and Paunsaugunt Plateau, 225f
　Knobstone escarpment near New Albany, Indiana, 247f
　mesas and buttes in Monument Valley, 226f
　NNE along Dripping Springs Escarpment, 249f
　north to south cross-section from High Plateaus, 226f
　tree-covered knobs of eroded Muldraugh Hills escarpment, 248f
Bench-and-slope topography, 244
Benioff zone, 66
Berkshire Mountains (bk Mountains), 363
Bermuda hot spot, 76, 206–207
BG fault. See Burnsville-Gossen Lead fault
BHB. See Big Horn Basins
Big bend area, monoclines and normal faults in, 453–456
Big Bend region, 506, 517–518

Big Cypress Swamp, 200–202
Big Horn Basins (BHB), 405
Bighorn Mountains, 295–296, 295f, 296f
Bitterroot Range, 375, 376f
Bitterroot River valley, 375
BK. See Basher Kill
Black Belt, 189, 204–205, 205f, 207
Black Hills, 7–8, 296–298, 296f, 297f
Black Mountains, 542–543, 545f
　turtlebacks, 543–544
Black River (BR), 264–265
Blind fault, 46–47
Block faulting, 44–46
Blue Mountains (BM), 392–393, 397–401, 398f
Blue Ridge anticlinorium, 342–344, 346
Blue Ridge Escarpment (BRE), 347, 349, 359–360, 359f
Blue Ridge Mountains, 342, 343f
　Blue Ridge North of Roanoke, 346–347
　Blue Ridge South of Roanoke, 347–349
　geologic overview, 342–345
　physiographic overview, 342
　at Roanoke, 345–346
Blueschist, 29, 497–498, 510
Bluff Hills, 204
BM. See Blue Mountains; Buffalo Mountain
Bofecillos, 453
　Mountains, 416
　Vent, 416
Bolsons, 456–457
Bonaparte, Mt., 376–377
Boone, 347
Borah Peak earthquake, 461
Bordas–Oakville Escarpment, 207
Borderland, 518–519
Boring Lava Field, 483
Boston Mountains, 241, 243
Boundary Ranges, 361–362
Boundary Ranges Highland Trend, 364–365
BR. See Black River
Braided river, 114
BRE. See Blue Ridge Escarpment
Breccia, 27–28, 303, 391
Brevard fault zone, 354
Bright Angel faults, 554
Bristlecone pines, 536–537, 537f
Brittle faults, 47–48
Broad folds, 227
Broken Bow Lake (b Lake), 336
Brothers fault zone, 400–402
Brushy Mountain, 358
Bryce Canyon, 44, 222
Bryce Canyon National Park, 234, 236f
Buffalo Mountain (BM), 357
Buffalo-Cherokee Mountains (bcm), 355
Burial beneath unconsolidated sediment, 132
Buried faults, 71
Burlington escarpment. See Eureka Springs escarpment
Burnsville-Gossen Lead fault (BG fault), 358
Buttes, 224, 389
Buttress unconformity, 264, 265f

C

Cabinet Mountains (Cb), 317
Calamity, 475
Calcium carbonate, 103–104
Calcium carbonate-shelled animals, 103–104
Caldera, 390–391
    Long Valley, 426–427, 426f
    Newberry, 401, 401f
    Pine Canyon, 414–415
    Silicic, 391
    Toledo, 423–424
    Valles, 423–424
    Yellowstone, 405, 406f
Caliche, 216
California
    central valley of, 447
    coast ranges, 512–517
    Current, 81
    landscape, 501–503
    River, 19–20, 560–561, 561f
    706 Line in, 452
California coast ranges, 512–517
    age of landscape, 512–513
    mountain alignment relative to San Andreas Fault, 514–515
    surface uplift
        deformation history prior to, 515–516
        mechanism and cause, 516–517
California strike-slip system, 141, 501, 502f, 503f
    active faulting in Death Valley, 546–547
    California coast ranges, 512–517
    landscape associated with strike-slip faults, 503–505
    Peninsular Ranges, 520–521
    relict subduction zone landscape, 510–512
    San Andreas Fault system, 505–510
    Sierra Nevada, 521–531
    Transverse Ranges and Salton Sea, 517–520
    Walker Lane Belt, 531–546
California-Coahuila transform, 450, 450f
Calloway Peak, 355
Cambrian–Mississippian rock succession, 288–289
Camels Hump, 363–364
Canadian River (CR), 333–334
Canal, 18
Canton (Ct), 347
Canyonlands National Park, 44–46, 47f, 224
Cape Blanco, 477
Cape Fear Arch, 199–200, 208–209
Cape Hatteras, 198–200
Cape Mendocino, 505–506
Cape Romano, 202
Capital Reef National Park, 222, 229
Capitol Lake (CL), 161
Capitol Peak (CP), 160–161
Caprock, 186
Caprock Escarpment, 215–216, 215f, 218, 219f
Caprock layer, 216
    Ogallala, 217
    resistant, 391, 407–409, 421
Capulin (Cp), 424
Carbon cycle, 104

Carbon dioxide (CO_2) in atmosphere, 103–105
Carbonate rocks, 28
Carlsbad Caverns, 218–219
Carolina Mountain Club, 347
Carolina Sand Hills, 208
Carolina Trough, 369–370
Carrizo Mountain Group, 385–386
Carrizo Plain National Monument, 48f
Carrizozo lava flow, 424–425
Carson Range, 525
Cascade block rotation, 489–490
Cascade Mountains, 10, 118, 389, 392–393, 489
    active faults in, 490
    alpine glaciers in, 164
    forming tectonic landscape, 489
    geology of Central-Southern, 487–489
    North, 377–381, 380f, 487
    sparse vegetation east of, 85f
    volcanic highland in, 475
Cascade volcanic arc, 73–74, 393
Cascade volcanism, 486
Cascade volcanoes, 72, 487–488
Cascade-Klamath-Sierra Nevada ranges, 84, 84f
Cascadia Subduction Complex, 492
Cascadia Trench, 66
Cascadia volcanic arc system, 141. See also River systems
    Central-Southern Cascade Mountains, 483–491, 485f
    Inland Valleys and Forearc Basin, 482–483
    Juan de Fuca plate, 473–475, 474f
    Klamath Mountains, 495–498
    landscape map, 476f
    Olympic Mountains, 491–495
    Oregon coast range, 477–482
    Pacific coastline, 475–477
    SGS National Atlas digital shaded relief elevation map, 477f
Castle Crags, 495, 498
Castle Rock Conglomerate, 413
Cataract Creek (CC), 549, 551–553
Cathedral Group, 464f
Catoctin Formation, 344
Catoctin Mountain (CM), 346–347
Catskill Delta, 253
Catskill Front, 324
Catskill molasse, 322, 324f
Catskill Mountains, 7–8, 187–188, 253, 322
Caves, 36, 267
    Carlsbad Caverns, 218–219
    Evaporite caves, 36
    Jewel Cave, 298
    Mammoth Cave, 36, 244–245, 248–251, 249f, 251f
    Wind Cave, 298
CC. See Cataract Creek
Cebolitta Mesa (CM), 422–423
Central Appalachian fold and thrust belt, 325–326
Central High Plains, 218
    Google earth image at Post, Texas, 219f

    headward erosion of the Pecos River, Texas, 219f
Central Idaho, 375–376
Central Longfellow Mountains, 366
Central Lowlands, 7, 237–241
    deposition and erosion across basins and arches in, 238f
    Great Lakes region, landscape map of, 238f
    Niagara escarpment and Niagara Gorge, New York–Canada border, 239f
    Osage Plains section of, 240f
Central Montana alkalic province, 407
Central Nevada-Utah volcanic fields, 413
Central thrust belt, 332–333
Central Valley (Cv), 501, 510–511
Central-Southern Cascade Mountains, 483–491, 485f, 487f. See also Klamath Mountains
    Clockwise Block Rotation, 489–490
    geology of, 487–489
    normal faults along crest of high cascades, 490–491
    South to North list of Cascade Stratovolcanoes, 484t
Chain Lakes Massif, 364–365
Challis volcanic field, 409
Channeled Scablands, 175, 394–397
Chelan, Lake, 380
Chemical weathering, 25–26
Cheraw normal fault, 438
Cherokee Lake, 332–333
Chert, 241–243, 334
Chesapeake Bay, 199
Cheyenne Table, 216
Chief Mountain, 314
Chihuahuan Desert, 414
Chilhowee Group, 344–345, 351, 355
Chilhowee Mountain, 351
Chilicotal Mountain, 457
China Hat, 401
Chinati peak, 452
Chisos Mountains, 414, 415f, 453
Chuska Mountains, 418
Cincinnati Arch, 244
Cinder cone, 389
Circle Cliffs Uplift (Ccu), 229, 230f
Citadel Rock, 273
CL. See Capitol Lake
Clark Fork River, 394–395
Clarks Fork Yellowstone River (CFYr), 293
Claron Formation, 225
Clastic rocks, 27–28
Clearwater River, 494
Climat–driven processes, 92
Climate, 10, 79, 87
    controls on, 81–86
        global wind patterns, 81–82
        latitude, 81
        mountains, 84–86
        proximity to large water bodies, 81
        tilt of Earth's axis of rotation, 82–84
    sculptor, 60–61
    Sierra Nevada and, 521–531
Climate change, 95

Climate system, 59, 125
  controls on climate, 81–86
  present-day climate zones, 79–80
Climate zones, 79, 80t
  Raisz landform map with climate zone boundaries, 573f
Climate-driven processes, 61, 87
  erosion-deposition processes, 61
Climatic effect, 84
Clinch Mountain, 327, 332
Clingmans Dome (Cd), 351
Clockwise Block Rotation, 489–490
Closure temperature, 120–121
CM. See Catoctin Mountain; Cebolitta Mesa
Coal, 28
Coast Range, 419, 481, 504, 512
  accretionary prism in, 480, 510
  California, 512–517
  Oregon, 477–482
  terrane, 481
Coastal Plain, 8–10, 62, 185, 189–209
  ancient shorelines of Coastal Plain, 207–209
  barrier islands, 190–191
  Mississippi embayment, 203–207
  New England, 191–198
  New Jersey to North Carolina, 198–200
  relative sea level change based on tide-gauge measurements, 191t
  South Carolina to Florida, 200–203
  Texas, 207
Coastal Plutonic and Volcanic belt, 369
Coconino Plateau (CP), 549–550
Cocos tectonic plates, 506
Cody Scarp, 202–203
Collision process, 89–90
Colorado Mineral Belt, 269, 271–272, 287
Colorado Piedmont, 218–219
Colorado Plateau, 98, 127, 128f, 222–235, 224f, 228f, 229f, 244–245
  bench and slope landscape, 224–226
  cause of accelerated erosion in, 287
  fractures and impact features, 231–235
    Arches National Park, 231–234, 233f
    Bryce Canyon National Park, 234
    Meteor impact features, 234–235
    Zion National Park, 234
  incised meanders, 224
  landscape history of, 283–286
  Mogollon Rim, 227
  physiographic boundary of, 223f
  uplifts and monoclines, 227–231
Colorado River, 19, 286, 549–553, 557
Colorado Rocky Mountains, 10
Columbia Basin (CB), 392–397, 395f, 400
  Channeled Scablands, 394–397
  Yakama fold belt, 397
Columbia Plateau, 389, 392–402, 392f, 393f, 406–407, 407f, 419
  Blue Mountains, 397–400
  Columbia Basin, 394–397
  Columbia river flood basalt, 393–394
  High Lava Plains, 400–402
  OW lineament, 400
Columbia River, 394, 488
  flood basalt, 393–394, 397–398
  Plateau, 91
  system, 19
Comb Ridge, 230–231
Comby Ridge (Cb), 332–333
Components, 10–11, 11t, 12f
  rock/sediment type, 23
  structural form, 41
Composite volcano, 391–392
Compression
  in Big Bend area, 517–518
  in Olympic Mountains, 494f
  in Olympic Peninsula, 495
  stress and, 41
Compressional mountain system, 69
Compressional stress, 41–42, 44
Conglomerate, 27–28
Connecticut River Valley, 362
Continental crust. See Granitic crust
Continental Divide, 15–16, 19, 221, 280–281, 290
Continental glaciation, 107
  landscape development in areas, 157–159
Continental glaciers. See Glaciers
Continental lithosphere, 62
Continental shelf, 62
Continents elevation, 89
Convergence of tectonic plates, 63, 89
Cool air, 81
Copper Ridge, 332–333
Cordillera, 10, 74, 80, 131, 432, 433f
  accreted terranes in, 67–68, 448, 451–452
  active faults, 50
  active tectonic landscape, 73
  change in southern part, 450
  crystalline rocks and, 36–37
  elevation and elevation changes, 110
  evaporite caves in, 36
  ignimbrite flare-up, 528
  older volcanic rocks, 36–37
  postorogenic reincarnation, 139–140
  rock successions in, 448f
  terrane accretion events, 74
  thrusting in Middle Jurassic 173 million years ago, 449–450
  topographic reincarnation, 131
  volcanic rocks
    Columbia Plateau, 392–402
    landscape map showing volcanic areas, 390f
    magma types and common volcanic landforms, 389–392
    origin of volcanism on Columbia Plateau and High Lava Plains, 406–407
    Snake River Plain, 402–406
  weathering, 80
Cordillera, 376–377
Cordilleran and Appalachian Mountain systems, 67–68
Cordilleran fold and thrust belt, 310–318. See also Ouachita fold and thrust belt; Valley and ridge fold and thrust belt
  Idaho-Wyoming fold and thrust belt, 317–318, 319f
  Northern Rocky Mountains, 314–317
  Rocky Mountain Trench, 317, 318f
Cordilleran Ice Sheet, 157, 163–164, 313, 394–395, 396f
Cordilleran Mountains, 451
  building events, 372
  System, 10
Cordilleran orogenic belt, 140, 143–144
Cordilleran orogeny, 441
Cordilleran region, 10
Cordilleran volcanic areas
  70–20 million years old, 407–419, 408f
    ignimbrite flare-up, 409–418
    Navaho volcanic field and Shiprock, 418–419
    North and South Table Mountain, 407–409
    Northern Great Plains, 407
    Pinnacles, Neenach, and Nine Sisters, 419
  younger than 20 million years, 419–427
    Carrizozo lava flow, 424–425
    Grand Mesa, 421–422
    Hopi Buttes volcanic field, 420–421
    Jemez lineament, 422–424
    Long Valley caldera and Inyo-Mono Craters, 426–427
    Northern Nevada rift zone, 425
    Northwest basin and range and Northern Sierra Nevada, 425–426
    SF volcanic field, 420
    Sutter Buttes, 427
    Uinkaret and Markagunt volcanic fields, 419–420
Cortez Range, 444–445, 445f, 448
Corvallis, 482
Cosmic rays, 251
Cosmogenic ^{10}Be measurements, 286
Cosmogenic radionuclide dating technique, 116–117
Coteau Du Missouri, 172–173, 173f, 175–176, 183, 213f
Cottonwood Mountains (Cm), 534
Coulees, 395
Coves, in Blue Ridge Mountains, 191, 352
CP. See Capitol Peak; Coconino Plateau
CR. See Canadian River
Crab Orchard Mountains, 256
Crag Crest, 422
Cramer, Mt., 374–375
Cranberry Lake, 262
Crater, 391–392
Crater Lake, 3, 391–392, 401–402, 483–486
Craters of the Moon, 403, 404f
Cratons, 144–145, 147
Crazy Mountains, 273, 407
Creede, 411–412
Creep, 480–481
Crescent terrane, 481, 491f, 492
Cretaceous limestone, 453
Crocker Mountain, 367
Cross-cutting relationships, 173–174, 442, 532f
Cross-sections, 4, 4f
Crossnore Plutonic suite, 355

Crow Peak, 273
Crowder's Mountain, 358f
Crowley's Ridge, 203
Crust, 62, 63f, 89, 90f
　asthenosphere and, 89f
　isostasy and, 88f, 89
　mountains and, 88f
Crustal thinning, 440–441
Crystal Mountains, 336
Crystalline basement, 142
Crystalline rock, 25, 28–29, 28f, 32, 160, 448. *See also* Ancient North American crystalline shield
　anticlines of, 261
　erodability of, 265–266
　landscape in, 33
Crystalline shield rocks, 142, 261
Crystalline-cored anticlinal structure, comparison with, 310
Crystalline-cored anticlines and domes, 269–271
　Middle-Southern Rocky Mountains, 270f
　northwest Great Plains, 271f
Crystalline-cored mid-continent anticlines and domes
　Adirondack Mountains, 261–263
　anticlinal mountains of Middle Rockies, 289–298
　Colorado Plateau
　　cause of accelerated erosion in, 287
　　landscape history of, 283–286
　intrusive domal mountains, 271–274
　mountain, 287–289
　Southern Rocky Mountains, 274–289
　　cause of accelerated erosion in, 287
　　landscape history of, 283–286
　St. Francois Mountains, 263–266
　Superior Upland, 298–306
　water gaps in Rocky Mountains, 298
　western margin of crystalline-cored anticlines and domes, 269–271
　Wichita, Arbuckle, and Llano structural domes, 266–269
　　Arbuckle Mountains, 267
　　landscape development, 268–269
　　Llano Uplift, 268
　　Wichita Mountains, 266–267
Cuesta (slope), 49
Cullasaja River, 371, 372f
Cumberland escarpments, comparison of Pottsville escarpments and, 257–258
Cumberland Front, 328f, 329–330
Cumberland Gap, 339–340
Cumberland Mountains, 255
Cumberland Plateau, 253–257
　cross-section from Nashville Basin southeastward to Valley and Ridge, 258f
　fault zones on, 333
　SSW across dissected Cumberland Plateau in southeastern Kentucky, 257f
Cumberland/Stone Mountain, 329–330
Cut bank, outer bank, 113–114

D

Dacite-rhyolite compositions, 486
Dakota Hogback, 276–277, 296f, 297–298
Dampening effect, 555
Dansgaard-Oeschger events, 102
Daytona Beach, 190, 192f
De Gray Lake (dg Lake), 336
De Queen Lake (q Lake), 129–130, 130f, 336
Death Valley (Dv), 4, 438, 440, 501
　active faulting example in, 546–547
　transtensional regime, 542–543
Death Valley-Fish Lake Valley fault zone, 534
Death Valley-Panamint Valley region, 542–546, 542f
　Panamint Mountains, 544–546
　transtensional regime of Death Valley and Black Mountains, 542–543
　turtlebacks of Black Mountains, 543–544
Decker Peak, 374–375, 375f
Décollement, 256, 309
Deep creek range, 438
Deep Springs Valley (Dsv), 534, 539
Deflation, 459
Deformation, 41, 42f, 341 *See also* Crystalline deformation belts
　of Grand Canyon, 224
　history prior to surface uplift, 515–516
　of Llano area, 268, 337
　orogenesis and, 69
　of structural form, 41–46
Deformed rocks of Shawnee Hills, 248–249
Deglaciation, 95
Delaho bolson (db), 456–457
Delamination, 440–441, 442f
Delaware River (DR), 323–324
Delaware Water Gap, 337–338
DEM. *See* Digital elevation model
Dendritic pattern, 17
Dense rock layer, 91
Density, 87–88
Denudation rate, 116–118, 116f
Deposition, 10, 25–26, 79, 92, 107, 110–115, 129
　base level changes, 111–112
　changes in discharge and sediment supply, 113
　elevation and, 119
　by glaciers, 158
　graded rivers and base level, 111
　knickpoint migration, 112–113
　lower Mississippi River Valley, 113–115
　rates, 118–119
　of rock/sediment type, 25–26
　of sedimentary rock, 118–119
Desert climate, 80, 546–547
Detachments, 431–432
Detritus, 27–28
　from erosion, 91
Deuterium, 97
Devil's Postpile, 426
Devils Mountain fault (DF), 378
Devils Tower, 273–274, 274f
Devonian rocks, 251–253
DF. *See* Devils Mountain fault

Diabase, 466–467
Diablo Range, 510, 515–516, 516f
Diamond Creek Road, 555
Diamond Mountains, 425–426
Diatreme, 392
Diffusion process, 121–122
Digital elevation model (DEM), 173–174
Dike, 272–273
Dinosaur Ridge, 276–277
Diorite, 315–316
Dip, 41, 44f, 49, 49f
Dipping layers
　influence on landscape, 48–49, 49f
　　cuestas and hogbacks, 49
　　horizontal to gently dipping rock layers, 49
　　response of dipping layers to erosional lowering, 49, 52f
　　vertical to steep-dipping rock layers, 48–49
　response to erosional lowering, 49, 52f
Dirks Lake (d Lake), 336
Discharge, changes in, 113
Displacement, 46
　along San Andreas Fault, 506–508
Dissolved silica, 26
Divergence, 63
Doldrums, 81–82
Dolostone, 28, 36, 241–243
Domes, 41–42
　lava, 391, 423–424, 486
　Lexington, 244
　Nashville, 244
　Upheaval, 234–235, 237f
Dominguez Mountain, 415–416
Dorset Mountain, 364
Doubly plunging anticlines, 227
Downcutting
　in Grand Canyon, 186–187
　process, 111
　in Sierra Nevada, 530
DR. *See* Delaware River
Drainage divides, 15, 15f, 20–21, 557, 562–563
　at EC, 558
Driftless area, 37f, 164, 172, 178
Dripping Springs-Chester escarpment, 246f, 248
Drumlins, 163
　fields, 175
　and kettles, 160
Dry Falls, 395–397
Ductile faults, 47–48
Dugout Wells Fault (Dwf), 457–458
Duluth gabbro, 304

E

Eagle peak, 452
Early Eocene
　climatic optimum, 99
　peak, 99
Early Ordovician, 266–267
Earth
　landscape, 59

Earth (*Continued*)
  lithosphere, 87
  ocean basins, 95
  orbital parameters on glaciations, 100–101
  rotation, 79
  surface phenomena, 79
  surface
    distribution of elevation on, 89, 89*f*
    elevation, 89
  tilt of Earth's axis of rotation, 82–84
Earthquakes, 66, 263
  Pleasant Valley, 443
  San Francisco, 508
  in Teton Range, 464
East Baylor Mountains-Carrizo Mountain fault (EBC), 384
East Butte, 401
East Fork Black River, 266*f*
Eastern and western segments, 311
Eastern California-Walker Lane belt, 437, 531
Eastern Continental Divide, 16
Eastern Granite Gorge, 550, 561–562
Eastern Mojave Desert, 503
Eastern Seaboard river system, 369–370
Eastern segment of Northern Rocky Mountains, 314–317
Eastern Snake River Plain, 402, 405
EBC. *See* East Baylor Mountains-Carrizo Mountain fault
EC. *See* Echo Cliffs
Eccentricity, Earth's orbital parameters, 100–101
Echo Cliffs (EC), 549–550, 557–558
  North, 557
  uplift, 230–231, 232*f*
Eclogite, 357
Edwards Plateau, 211, 218–219
Elastic strain, 506–508
Electron Mudflow, 475
Elevation, 4
  deposition and, 119
  earthquakes and, 480
  erosion and, 117–118
  rivers and, 111
  subsidence and, 119
  of Suffolk scarp, 208
  of Trail Ridge-Orangeburg-Coates-Broad Rock scarp, 208–209
  uplift and, 122
Elk Ridge, 346
Elkhorn Mountains, 376, 407
Elkhorn Ridge, 376, 399
Ellenville, 323–324
Elwha River system, 494
Eminence faults, 554
Emma Dean (boat), 549, 550*f*
Enchanted Rock, 268, 268*f*
Entrada Formation, 231
Equinox Mountain, 364
Erodability, 31
  of crystalline rock, 265–266
Erosion, 10, 25–26, 69, 79, 92–93, 107–108, 110–115, 120, 129, 443–444
  of anticlines, 57*f*
  of Appalachian Mountains, 369–372
  base level changes, 111–112
  of Basin and Range, 444, 447*f*
  of BRE, 349
  changes in discharge and sediment supply, 113
  controls on erosion rates, 117–118
  of crystalline rock, 32, 32*f*
  dip and, 49
  drainage divides and, 20–21, 21*f*
  elevation and, 117–118
  erosion/deposition rates, 115–116
  exhumation from, 119–120
  by glaciers, 163
  Google Earth image looking NNE along Colorado River, 93*f*
  graded rivers and base level, 111
  knickpoint migration, 112–113
  lower Mississippi River Valley, 113–115
  of Olympic Mountains, 492–493
  of Piedmont Plateau, 358
  in plains, 118
  of plateaus, 118
  by rivers, 111
  of rock/sediment type, 25–26
  of Rocky Mountains, 277–278, 278*f*
  of sedimentary rock, 32*f*
  uplift and, 444
  vegetation and, 117–118
  of volcanoes, 33
Erosion-controlled landscape, 52–56, 56*f*, 57*f*
Erosional decay, 125
Erosional exhumation, 119–120, 119*f*
  calculating rates of erosional exhumation, 120–122
Erosional lowering, 125
  dipping layers response to, 49, 52*f*
Erosional Mountains, 185, 187–188
  edge of Edwards Plateau in area west of Austin, Texas, 189*f*
  Edwards Plateau, Texas, 188*f*
  intermediate stream dissection, 188*f*
  nearly flat-lying sedimentary layers from, 185
  plateaus and, 188
Erosional power, 113
Erratics, 178–179
ES. *See* Estufa Springs
Escalante River (Er), 222–224, 229, 230*f*
Escarpments, 204, 241, 455
  face, 360
Eskers, 176–177, 262
  snake-like esker in Lows Lake and Hitchins Pond, 178*f*
Estufa bolson (eb), 456–457
Estufa Springs (ES), 457–458
Eugene, 482
Euler pole, 63
Eureka Springs escarpment, 241
Eustatic sea level, 103
  change effect, 96–97
Evaporite caves, 36
Excelsior-Coaldale Section, 534
Exfoliation joint, 44

Exhumation, 108, 119–122
  from erosion, 119
  erosional, 119–120
    calculating rates, 120–122
  rates, 120–121
  tectonics and, 122
  from uplift, 119
Exmore breccia, 199
Explorer segment, 473
Extension, 431–432
  in Basin and Range, 440, 440*f*, 442–443
  process, 441
  in San Andreas Fault, 518
Extensional stress, 46*f*
Extensive drilling, 189–190
Extrusive igneous rock. *See* Volcanic rocks

F

Faceted spur, 51–52
Failed rift, 64, 71
Fall Line, 189, 208, 360–361, 361*f*
Fan, shape of deposit, 444
Fanglomerate, 558
Farallon plate, 73, 447, 508–509, 527–528, 528*f*
  shrinking remnant, 473
Faulting/faults, 41, 43*f*, 44–46, 309–310
  blocks, 430–431
  in eastern California-Walker Lane Belt, 437
  fault-block
    mountain, 431
    rotation, 432
  reactivation, 46–47, 48*f*
  of Wichita Mountains, 267
  zones on Cumberland Plateau, 333
Feldspar, 26, 451
Fins, 231
Fish Lake Valley (FLv), 537–538
Fission tracks, 121
Fissures, 389
Flathead Lake, 317
Flathead Range, 314
Flatland, 125, 130*f*, 268–269
Flats, 309
  flat-lying layers, 42–44
  flat-ramp-flat, 46
Flexural rebound, 430–431, 431*f*
Flood
  basalt, 74, 393
  waters, 395–397
Floodplain, 113–114
  deposits, 115
Florida, 98, 202
  South Carolina to, 200–203
Florida Keys, 27, 201–202, 202*f*
Flysch, 324
Fold and thrust belt. *See also* Foreland fold and thrust belt
  Cordilleran, 310–318
  Idaho-Wyoming, 311, 317–318, 319*f*, 462–463
  Marathon Basin, 337
  normal faults and, 447*f*
  Ouachita, 333–336

of Ouachita Mountains, northern front, 333–334
    Sevier, 447f, 448
    Valley and ridge, 319–333
Folds, 41–44, 43f
    belt, 310
    waterpocket, 229
Foliation, 441
Forcing agents, 10–11, 11t, 12f, 59–62
    climate, sculptor, 60–61
    isostasy, equalizer, 61
    sea level, baseline, 61–62
    tekton
        builder, 59–60
        carpenter, 59–60
Forearc basin, 66, 475, 482–483
    ancient, 510–511
Foredeep, 91, 92f, 146
Foreland, 145–146
    basins, 146–147
    structural form foreland thrust faults, 309–310, 310f
    tectonic provinces, 147
    thrust faulting, 310
Foreland clastic wedge, 146
Foreland fold and thrust belt, 327
    Appalachian-Ouachita fold and thrust belt, 318–319
    comparison with crystalline-cored anticlinal structure, 310
    Cordilleran fold and thrust belt, 310–318
    Marathon Basin fold and thrust belt, 337
    Ouachita fold and thrust belt, 333–336
    structural form of foreland thrust faults, 309–310, 310f
    valley and ridge fold and thrust belt, 319–333
    water gaps in Valley and Ridge and Ouachita Mountains, 337–340
Foreland-hinterland transition, 302, 448
Formations, 288, 316, 334, 512, 553
Fort Smith (FS), 335
Fossils, 11, 13
    in Yellowstone National Park, 410, 410f
Fourche Mountains, 335–336
FR. *See* Front Royal
Fracture, 33, 44, 48–49, 231–235
    Arches National Park, 231–234
    Bryce Canyon National Park, 234
    Zion National Park, 234
Franciscan complex, 498, 510
Franklin Mountains, 383–384
Fraser glaciation, 163–164
French Broad River, 354
Front Range, of Rocky Mountains, 277, 278f, 279f
Front Royal (FR), 346
Frost cracking, 25, 26f
FS. *See* Fort Smith
Funeral Mountains, 542

G

Gabbro, 29, 304
    intrusions, 482

Gabilan Mountains, 419
Gallatin Range, 405, 409–411
Ganderia microcontinent, 369
Ganderia passive margin, 369
Gangplank, 216, 279, 279f, 293
Garden of the Gods, 248–249, 250f
Garlock Fault, 506, 521–522, 534
Garnet Range (Gt), 311
Gatlinburg fault, 350–351
Geologic basin. *See* Structural basin
Geologic history of Grand Canyon, 559–560
Geologic Time scale, 13, 13f
Geology, 11–13, 563
    of Central-Southern Cascade Mountains, 487–489
    of Klamath Mountains, 497–498
    of Olympic Mountains, 492, 493f
    of Oregon coast range, 481–482
    of Southern Hudson River Valley, 322–325
Geometry, 49
Georges Bank, 76
Geothermal gradient, 47–48, 120–121
Gila River, 445
Glacial advance, 113–115
Glacial boundaries, Raisz landform map with, 580f
Glacial depositional landforms, 159
Glacial drift, 180, 180f
Glacial erosion
    boundary across North America, 165–171
        Google earth image oriented northward at Canada and US, 172f
        northwest at kettle lakes on Coteau Du Missouri, 173f
    boundary in United States, 164–165
        north across Minnesota and Canada, 171f
        northwest at area of young drift, 171f
Glacial lakes, 174–175, 182f
Glacial landscape
    area south of glacial limit, 178–179
    drumlin fields, 175
    eskers, 176–177
    glacial erosion boundary
        across North America, 165–171
        in United States, 164–165
    glaciation effect on landscape, 157–162
    kame–kettle fields, 175–176
    Lake Agassiz, 175
    loess deposition, 177–178
    Marine incursions, 175
    Missouri River, 181–183
    moraines, 171–174
    pluvial lakes of cordillera, 183
    proglacial lakes, 174–175
    Sand Dune fields, 177
    Teays River, 179–181
    in United States, 162–164, 163f
        glacial zones, 163t
        Raisz landform map, 165f, 166f, 167f, 168f, 169f, 170f
Glacial limit, area south of, 178–179
Glacial moraines, 171–172
Glacial terrain, 4–5
Glacial zones, 163t

Glaciation(s), 61, 79, 95, 103, 313, 367
    Earth's orbital parameters on, 100–101
    effect on landscape, 157–162
        landscape development in areas of alpine glaciation, 159–162
        landscape development in areas of continental glaciation, 157–159
    reincarnation to, 131–132
Glacier National Park (GNP), 314, 316f, 376–377
Glacier Peak, 377–378, 492
Glaciers, 92, 101, 104–105, 118, 157, 159f, 522
    of Adirondack Mountains, 261–263
    alpine, 159, 161f
    climatic system and, 161–162
    crystalline rock and, 130f
    erosion by, 118
    isostasy and, 87
    of Minnesota, 158, 158f
    of Mt. Rainier, 160f
    of Red River Valley, 158, 158f
Glass Mountain (G Mountain), 426–427
Global effect, 102
Global positioning systems (GPS), 481
Global sea level, 95
Global wind patterns, 81–82, 83f
    present-day wind and ocean current patterns, 82f
GMW. *See* Grandfather Mountain window
Gneiss, 29
GNP. *See* Glacier National Park
Golconda allochthon, 449
Gold Mountain (Gm), 534
Gooseberry Falls State Park, 304–305
Gorda segments, 473
GP. *See* Great Plains
GPS. *See* Global positioning systems
Graben(s), 44–46, 47f, 49
    structure, 429–430
Graded rivers, 111
Grand Canyon, 549
    active faults and incision rates, 554–555
    age, 549
    argument
        for 6 million-year-old canyon, 558
        for 70 million-year-old canyon, 558–559
    bench-and-slope landscapes near, 185–186
    geologic history, 559–560, 563
    HP, 555–556
    incision, 555
    meanders, 557
    modern Colorado river, 556–558
    physiographic canyon, 549–553
    revised arguments, 560–563
        interpretations, 560–563
    river morphology, 556
    river system, 555
    rock successions, 153f
Grand Canyon Supergroup, 150–154, 554
Grand Canyon Village (Gv), 551–554
Grand Coulee, 395
Grand Mesa, 421–422, 422f
Grand Staircase, 225–226, 225f, 226f

Grand Wash Cliffs, 549, 553
Grandfather Mountain Area, 355–358, 356f, 357f
Grandfather Mountain Formation, 355
Grandfather Mountain window (GMW), 357
Granitic batholiths, 33
Granitic crust, 62, 89
Granitic intrusive rocks, 369
Granitic rocks, 29, 33f, 45f, 160, 376, 381
Granitoid, 29
Grapevine Mountains (Gvm), 534, 542
Gravitational equilibrium, 87
Grays River Valley, 318
Great Appalachian Valley. See Great Valley
Great Balsam Mountains, 353
  boast peaks, 347
Great Basin, 182f, 367
  Nevada-Utah, 182f
Great Basin River system, 20, 432
Great Divide. See Continental Divide
Great Divide Basin, 15
Great Gulf, 365–366
Great Lakes, 18
  isostatic adjustment of, 109–110
  uplift in, 109
Great Meteor hot spot, 76
Great Plains (GP), 7, 80, 98, 131, 211–219, 337
  Colorado Piedmont, Pecos Valley, Plains Border, and Edwards Plateau, 218–219
  High Plains, 215–218
  Missouri Plateau, 211–214
  northwest Great Plains, 212f
  uplift, 221
Great Rift, 403, 404f
Great Salt Lake, 20, 438
Great Salt Lake Desert, 183
Great Sand Dunes National Park, 459–461
Great Sandy Desert (GSD), 401
Great Smoky, 327
  faults, 347–349
  thrust, 342–344, 352
  fault, 351
Great Smoky Mountains, 350–353, 350f, 351f
  boast peaks, 347
  in broad syncline, 352f
  level of exhumation across, 349–350
Great Smoky Mountains National Park, 345, 350f
Great Unconformity, 137, 150–154, 269, 276–277, 288, 291
  northeastward across Grand Canyon, 153f
  simplified cross-section of Grand Canyon, 153f
  structural province map, 152f
Great Valley, 320–321, 362, 510–511
Great Valley of California, 109
Green Mountain foothills (GMf), 321
Green Mountains, 363, 365f
  anticlinorium, 363–364
  Highland Trend, 363–364
  trends, 361–362
Green River, 224, 251f
Green River Basin, 293

Green River glacial valley (Gr), 290–291
Green River in Canyonlands National Park, 93
Greenbrier thrust, 352
Greenhouse gases, 103–104
Greenland Ice Sheet, 103
Greenschists, 497–498
Greenstone, 29, 302, 344
Greenville (Gv), 354
Greeson Lake (g Lake), 336
Gregory Bold (Gb), 351
Grenville front, 383–386
  Van Horn Area, 384–386, 385f
Grenville orogeny, 142, 351, 383
Grey's Harbor, 478
Groundwater, 389
Growing old type, 125
  reincarnation, 129–131
GSD. See Great Sandy Desert
Gt. See Garnet Range
Guadalupe Mountains, 417–418
Gulf Coast, 81
Gulf Stream, 81
Gvm. See Grapevine Mountains
GW. See Grand Wash Cliffs
Gypsum, 458
  cement, 459
  dune fields, 459

H

Half-grabens, 430–431
Hamlin Peak, 367
Hard rocks, 126
Harpers Ferry, 346
Harquahala Mountains, 450–451
Harrisburg (Hb), 346–347
Hartford Basin, 362
Havallah Basin, 449
Hawaii, 74–76, 91, 91f
Hawaiian basalt, 389
Hawaiian hot spot, 74–76
Hawaiian Island–Emperor Seamount Chain, 74–76, 75f
Hazel Formation, 385–386
He Devil, 399–400
Headwall scarp (HW scarp), 413
Headward erosion, 20–21, 21f
Heat, 62
Hebgen Lake earthquake, 461
Helena Formation, 315–316
Helium, 121
Hells Canyon, 399–400, 399f
Hendersonville, 347, 349
Hicks Dome, 249
High Lava Plains (HP), 392–393, 397–398, 400–402, 406–407
High Peaks and Bigelow Ranges, 366
High Plains, 215–218
  central Great Plains, 214f
  central High Plains, 218
  Llano Estacado, 218
  Nebraska Sand Hill Region and Ogallala Aquifer, 216–218

  Sand Hill region at Hyannis, Nebraska, 217f
  northeast Great Plains, 213f
  southern Great Plains, 215f
High Plateaus, 159, 224–226, 420
Highland Rim-Pennyroyal Plateau, 247–248
Hilton Head, 200, 201f
Himalayan Mountains, 67
Hindu Fanglomerate, 558
Hinge line, 143–144
Hinterland, 146
  deformation belts, 341
    Appalachian Mountains, 342–372
    Grenville front, 383–386
    Northern Rocky Mountains and North cascades, 372–383
    rocks within, 341–342
  tectonic provinces, 146
Hogback Ridge, 188
Hogbacks, 49, 220f, 227–228, 231, 276–277, 297
Hoh River system, 494
Holston Mountains, 327, 355
  faults, 347–349
Holyoke range, 467
Hoodoos, 234
Hoosac Mountains. See Berkshire Mountains
Hopi Buttes, 420–421
  volcanic field, 420–421
Horizontal extension, 440
Horizontal to gently dipping rock layers, 49
Horse Latitudes, 81–82
Horst(s), 431–432, 446
  structure, 44–46, 429–430
Hot spots, 74–76
  in Aleutian Islands, 74–76
  Bermuda, 76
  in Georges Bank, 76
  in Hawaiian Island-Emperor Seamount Chain, 74–76, 75f
  origins of, 75f
  in Snake River Plain, 76
  Thermal Plumes in United States, 74–76
  volcanism, 74
  in Yellowstone National Park, 76
Hot Springs (HS), 336
HP. See High Lava Plains; Hualapai Plateau
HR. See Hudson River
HS. See Hot Springs
Hualapai Limestone, 556
Hualapai Plateau (HP), 553, 555–556, 558
Hudson Bay, 15–16, 92
Hudson Bay River system, 20
Hudson River (HR), 323–324
Hudson Valley, 253
Hueco basins, 452
Humphreys Peak, 420
Hurricane faults (H), 438, 551–555
HW scarp. See Headwall scarp
Hydrogen isotope, 221
  analysis, 446–447

I

Iapetus Ocean, 351

Ice Age, 162, 164
Ice cores, 97–98
Ice sheets. *See* Continental glaciers
Ice wedging. *See* Frost cracking
Idaho and adjacent Montana, Google Earth image of, 311
Idaho Batholith, 33, 341, 372–375, 381
Idaho-Wyoming fold and thrust belt, 311, 317–318, 319f
Idaho-Wyoming overthrust belt (IW overthrust belt), 405
Idaho-Wyoming thrust belt, 311
Idealized orogenic belt, 144–145
Igneous belt, 407
Igneous extrusive surface, 28–29
Igneous intrusive surface, 28–29
Igneous plutonic rocks, 28
Igneous rock. *See* Plutonic rock(s)
Ignimbrite, 391, 407
Ignimbrite Flare-up, 407, 409–418, 528
  Absaroka volcanic field, 409–411
  Central Nevada-Utah volcanic fields, 413
  Mogollon-Datil volcanic field, 414
  San Juan Mountains and surrounding areas, 411–413
  Sanpoil and Challis volcanic fields, 409
  SB, 418
  Trans-Pecos volcanic field, 414–418
Illinois Drift Plains, 175, 178, 179f
Imbricate zones, 316–317
Imperial Valley Fault, 506, 518
Inactive accretionary prism, 67–68
Inactive subduction zone. *See* Inactive accretionary prism
Incision
  process, 111
  rates, 116–117, 117f
    active faults and, 554–555
  of river, 117
  isostasy, 91
Inclination, 41
Inland continental rocks, 190
Inland Valleys, 482–483
Inner bench (IB), 549, 551–553, 552f
Inner Bluegrass, 244
Inner cuesta (IC), 235
Interior Low Plateaus, 7–8, 245f, 246f
  bench and slope landscape, 247–248
  deformed rocks of Shawnee Hills, 248–249
  glacial deposition in, 127
  karst in, 36
  Mammoth Cave, 249–251
  monoclines of, 185
  nearly flat-lying sedimentary layers of, 185, 244–251
  structural overview of, 4–10
Interior Plains, 4–10, 71, 189
Interior platform, 143
Intermontane, 5
Internal drainage, 15, 18
Intervening Keweenawan Rift system, 299–300
Intervening synclinal basins, 290
Intrusion, 461

active tectonics, 69
Intrusive domal mountains, 271–274
  abandoned mining camp in Mosquito Range near Leadville, Colorado, 272f
  cross-section across central Colorado, 275f
  Devils Tower, 274f
  vertical dike with West Spanish Peak, 273f
Intrusive domes, 261
Intrusive rocks, 31, 210f, 249, 271–272
Inyo Mountains, 540–542, 541f
Inyo-Mono Craters, 426–427, 427f, 534–536, 535f, 536t
Iron formations, 302–303
Iron Mountains, 355, 357
Ironton, 412
Isolated Mountain Peaks, 367–368
Isostasy, 87
  continental lithosphere and, 89
  crust and, 89, 89f
  deposition, 92
  elevation of continents and ocean basins, 89
  equalizer, 61
  erosion, 92–93
  forcing agent, 87
  glaciers, 92
  of Great Lakes, 109
  in Hudson Bay, 92
  lithosphere and, 87
  mantle and, 90
  in Mississippi River, 92
  mountain building and preservation, 89–91
  in normal faults, 442f
  oceanic lithosphere and, 89
  rain shadow and, 92–93
  river incision and, 92–93
  tectonic compression *versus*, 87–89
  tectonic loads, 91
  tectonic *vs.* isostatic uplift, 87–89
  thermal isostasy, 91–92
Isostatic adjustment, 10–11, 87–89
Isostatic changes, 61
Isostatic compensation, 87, 90f
Isostatic equilibrium, 87, 88f, 90, 108–109
Isostatic rise amount, 90
Isostatic subsidence, 87–89
Isostatic uplift, 87–89, 108–109
  of crustal root, 90
Isostatic uplift/subsidence, 108–109
Isotopic composition, 110
IW overthrust belt. *See* Idaho-Wyoming overthrust belt

J

Jacinto Mountains, 512, 520
James River (Jr), 346
Jefferson, Mt., 483–486, 485f
Jemez lineament, 422–424, 423f
  Jemez volcanic field and Valles Caldera, 423–424
  Raton-Clayton volcanic field, 424
  volcanic fields in Southern part, 422–423
Jemez volcanic field, 423–424
Jennings Mountain, 335

Jewel Cave, 298
John Day River valley, 398
Johnson Shut-ins rocks (JS rocks), 264–265
Joint sets, 33, 43f, 44, 45f, 46f, 48–49
Joints, 44
  types, 44
JS rocks. *See* Johnson Shut-ins rocks
Juan de Fuca plate, 66, 72, 473–475, 474f
Jurassic Morrison Formation, 276–277

K

K-feldspar, 29, 121
Kaibab Arch, 550, 562
Kaibab Plateau (KA), 225–226, 550
Kaibab Uplift, 225–226, 228–229, 230f
Kame–kettle fields, 175–176
Kames, 159, 176
Kanab Creek (KC), 551–553
Kanab Plateau (KP), 549
Kanawha River, 254
Karst landscape, 36
Katahdin, Mt., 362–363, 366–367, 366f
Katahdin highland, 366–367
Kern Canyon River (Kc River), 527, 540
Kettle River Range, 376–377, 381–382
Kettles, 159
Keweenawan Rift system, 299–300, 304–306
  eastward at southeast-facing cuestas on Keweenaw Peninsula, 306f
  Porcupine Mountains, 305f
Keweenawan rock sequence, 143
Key Biscayne National Park, 201–202
Key Largo Limestone, 201–202
Kibby Mountain, 364–365
Kings Bowl, 403
Kings Mountain, 358, 358f
Kingsport (Kp), 327
Klamath Mountains, 10, 495–498, 496f. *See also* Central-Southern Cascade Mountains
  alpine glaciers of, 159–162
  Cascadia volcanic arc system and, 495–498
  geology, 497–498
  rain shadow of, 84
  tectonic map, 497f
  tectonics of, 497f
  terranes of, 497–498
  thrust faults of, 497–498
  uplift history, 495–497
Klamath peneplain, 495–496
Klippen, 314, 316f
Klippen Mountain. *See* Chief Mountain
Knickpoint migration, 112–113
Knobs, 247
Knobstone Escarpment, 247, 247f
Knobstone-Muldraugh Hills escarpment, 247, 247f
Knoxville, 327

L

La Grande (Lg), 399
Laguna Mountains, 520

Lahar (mudflows), 410, 486
Lake Agassiz, 20, 175
Lake Allison, 394
Lake Bonneville, 183
Lake Champlain (LC), 321
Lake City (L), 412
Lake George (LG), 321
Lake Michigan, 109
Lake Missoula, 394, 396f
Lake of Clouds (LC), 304–305
Lake Pend Oreille, 313
Lake Region, 18
Lake San Cristobal (LSC), 412–413
Lake-effect snow, 81
Landform types, 185, 186f
Landform(s), 4–5, 52, 54. See also Physiographic province
Landscape, 4–5, 11–13, 49–50, 52, 54, 503, 507f
  age, 512–513
  approaches, 126
  bedrock influence on, 32–35
    in crystalline rocks, 33
    in sedimentary rocks, 32–33
    in unconsolidated sediment, 34–35
    in volcanic rocks, 33–34
  characteristics, 438–440
  development, 268–269
    in areas of alpine glaciation, 159–162
    in areas of continental glaciation, 157–159
    sequence in Wichita, Arbuckle, and Llano areas, 269f
  dipping layers influence on, 48–49, 49f
    cuestas and hogbacks, 49
    horizontal to gently dipping rock layers, 49
    response of dipping layers to erosional lowering, 49, 52f
    vertical to steep-dipping rock layers, 48–49
  evolution, 107, 497
    reincarnation, 129–132
    rejuvenation, 127–129
  grows old, 125–126, 126f
  impart change to
    controls on erosion rates, 117–118
    deposition rates, 118–119
    erosion, deposition, and rivers, 110–115
    exhumation, 119–122
    present-day erosion rates, 115–117
    tectonic exhumation, 122
    uplift and subsidence, 107–110
    volcanism, 122
  map, 311f, 474f
    distribution of rock types, 40f
  modification, 79
  in nearly flat-lying layers, 185–188
    bench and slope landscape, 185–187, 186f, 187f
    Erosional Mountains, 187–188
    monoclinal slopes and hogback ridges, 188
  perspective, 96–97

  response, 10–11
  with strike-slip faults, 503–505
  tale of, 533–534
  at topographic steady-state, 126–127
  types, 185
  visible changes to, 12f
Laramide orogeny, 283, 288, 555–556
Laramie Range, 274–275
Last Chance Range (Lcr), 534
Late Cretaceous to Early Eocene Sevier orogeny (115–52 Ma), 74
Late Cretaceous to Middle Eocene Laramide orogeny (75–45 Ma), 74
Late Devonian to Early Mississippian Antler orogeny (360–347 Ma), 74
Late Devonian-Early Mississippian Neoacadian orogeny (385–345 Ma), 70, 354
Late Jurassic Nevadan orogeny (164–145 Ma), 74
Late Ordovician to Middle Silurian Salinic orogeny (446–423 Ma), 70
Late Pennsylvanian-Early Permian (c. 300 Ma), 74
Late Permian to Early Triassic Sonoma orogeny (260–247 Ma), 74
Late Silurian to Middle Devonian Acadian orogeny (421–387 Ma), 70
Latitude, 81
Laurel Mountain, 482
Laurentia, 142, 142f
Laurentian Divide. See Northern Divide
Laurentide Ice Sheet, 162
Lava Butte, 401, 402f
Lava domes, 391
Lava flows, 34
Lava Mountain (LM), 401–402
  basalt flow, 401
Lava plateaus, 389
Layered metamorphic rocks, 33
Leaky transform, 63
Ledge, 238–239
Lee's Ferry (LF), 549–550
Left-lateral strike-slip faults, 46, 506, 534
Lemhi (L), 405, 461
Leucite Hills, 220–221
Lewis and Clark Line (LC), 311, 314
Lewis thrust, 314, 332
Lexington Dome, 244, 246f
Lexington Plain subprovince, 244
LF. See Lee's Ferry; Llano Front
LG. See Lake George
Lime Sink, 36, 39f
Limestone, 27–28, 32, 36, 201–202, 453
  of Arbuckle Mountains, 190–191
  in arid environments, 36
  caves in, 36
  in nearly flat-lying sedimentary layers, 186
  in Ozark Plateau, 241–243
  sinkholes and, 36, 202–203
Lineaments, 48–49
  Jemez, 209, 408f, 422–424
Linear scarp, 51
Linville Falls (Lf), 355
  thrust, 357–358

Linville River, 355
Listric fault, 431–432
Lithosphere, 62, 87
  continental, 62, 74
  isostasy and, 87
  oceanic, 62, 65–66, 89
  response of lithosphere to addition of weight, 88f
  strength, 88–89
Little Colorado River (Lc), 550, 562
Llano area, 268–269
Llano Estacado, 218
Llano Front (LF), 383–386
Llano structural domes, 266–269
Llano Uplift, 266, 268, 337, 383, 384f
Local base level, 111, 111f
Loess, 177–178
Loess deposition, 177–178
Loihi, 74–76
Long Lake Valley, 262, 263f
Long Valley caldera (LV caldera), 426–427, 426f
Longitudinal profile of river, 111, 111f
Longshore drift, 190, 192f
Looking Glass Rock (lg Rock), 353–354
Lopez Island (l), 382
Los Alamos (LA), 423–424
Los Angeles (La), 501
Lost River (Lr), 405
Lovejoy Basalt, 425–426
Low-lying Umatilla Range, 398
Lower Granite Gorge (LG), 549, 553, 558
Lower Mississippi River Valley, 113–115
Lower Reservoir (LR), 264–265
LSC. See Lake San Cristobal
Lynchburg (L), 346

M
Macks Mountain, 347
Macon Ridge, 115
Madison Range, 303, 405, 461
Magazine Mountain, 335
Magma, 31
  basaltic, 74, 403–404
  cools, 28–29, 62–63
  igneous plutonic rocks from, 28
  movement, 427, 486–487
  silicic, 391
  types, 389–392
  volcanism, 122
  Yellowstone National Park, 47–48
Magnesian Cuesta, 238–239
Mahomet River. See Teays River
Maine, 368
Mammoth Cave, 36, 249–251, 251f
Mammoth Mountain (M Mountain), 426–427, 426f
Manassas Gap, 346
Manchester (M), 321
Mangrove, 202
Mangrove swamps, 202
Manhattan Prongs (M Prongs), 361–363, 362f
Mansfield, Mt., 363–364
Mantle, 62, 89–90, 451

Maps, 4, 4f
　comparison of map scales, 4f
　Raisz landform outline map of United States, 5f
MAR. *See* Mid-Atlantic Ridge
Marathon Basin (M Basin), 337
Marathon region, 71
Marble, 29
Marble Canyon (MC), 549–550, 554, 554f, 556
Marble Platform (MP), 549–550, 556
Maria fold and thrust belt, 448
Marine environments, 26
　shallow, 29, 385–386
Marine incursions, 175
Marine terrace, 97
Mariscal Mountain, 453
Markagunt field (M field), 419–420
Markagunt volcanic field, 419–420
Marquette Iron Range (Mq), 302–303, 303f
Marsh hammocks, 202–203, 203f
Mary's Peak (M), 482
Mass wasting processes, 79
Massive volcanic pile, 91
Mauna Kea, 91
Mauv Limestone, 562
Mazama, Mt., 483–486
Mc. *See* Mill Creek; Mountain City
McKenzie Pass, 490, 490f
McKinney Hills, 457
McKinney rocks, 453
MCW. *See* Mountain City window
MDB. *See* Montana Disturbed Belt
Meander, 557
　bends, 114–115
　with Grand Canyon, 557
　incised, 224
　of Mississippi River, 113–114
　Red River, 175
Meander Anticline, 93, 93f
Meandering levels process, 113–114
Mechanisms, 10–11, 11t, 12f, 107
Medano Pass (MP), 459
Medicine Bow Mountains, 274–275, 279f
Medicine Lake volcano, 487, 487f
Mediterranean climate, 80
Meers Fault, 267, 267f
Meguma, 369
Mehrten Formation, 528
Meiji, 74–76
Mélange zones. *See* Suture zones
Mendocino triple junction, 473, 505–506, 528
Mesa, 224
Mesa Prieta, 422–423
Mesabi Range, 303–304
Mesozoic era, 13
Mesozoic volcanic rocks, 537
Metacomet ridge, 467
Metamorphic core complex, 349, 381, 441, 442f, 545
Metamorphic rocks, 28–29, 33, 262. *See also* Volcanic rocks; Sedimentary rock
　of Adirondack Mountains, 262f
　of Montana, 375–376
　types, 29

Metamorphism, 441
　active continental margins, 65
　in Appalachian Mountains, 344
　deformation and, 362
Meteor Crater, 234
Meteor impact features, 234–235
Meteorite impact, 79
Methow Basin, 377–379
Mexico, Rio Grande Rift in, 429, 436f
MH paleocanyon. *See* Milkweed-Hindu paleocanyon
Miami Limestone, 200–201
Michigan, Lake, 18, 37f, 164, 169f
Michigan Basin, 42–44
Microcontinent, 67, 67f, 368–369
Mid-Atlantic Ridge (MAR), 62–66, 66f
Mid-Continent Rift. *See* Keweenawan Rift system
Middle and Southern Rocky Mountains, 98
Middle Mississippian to Middle Permian Alleghany orogeny (335–265 Ma), 70
Middle to Late Ordovician Taconic orogeny (480–446 Ma), 70
Middle-Southern Rocky Mountain, 261, 269, 397
　anticlinal structure of rocks in, 274
　landscape map of, 270f
Middlebury Synclinorium, 321–322
Middlesboro, 332–333
Middlesboro Syncline, 332, 332f
Milankovitch
　climate cycle, 100–101
　parameters, 101
Milk River, 181
Milkweed-Hindu paleocanyon (MH paleocanyon), 555, 558
Mill Creek (Mc), 294
Miller Cove, 352
Mina Deflection, 534
Mine Point area (MP area), 453
Minette, 418
Minneapolis, 81
Minnesota
　glaciers of, 158, 158f
　Superior Upland in, 298–299
Miogeocline, 143–144, 325
　Appalachian Mountains, 449
　Atlantic, 141, 144
　of Cordilleran orogenic belt, 143–144
　deformation of, 144
　rock successions, 141t
　rocks, 143–144
Mission Ridge, 506
Mississippi Delta, 38f, 204–207
Mississippi Embayment, 71, 203–207
　Black Belt region, landscape map of, 205f
　NNE at Bluff Hills, Vicksburg, 204f
　salt domes, 206f
Mississippi River, 71, 114–115, 114f, 127, 556
　delta, 92
　drainage, 346
　pathway, 114–115
　system, 16–17
Mississippian and Pennsylvanian rocks, 326

Missoula, Lake, 311, 394, 396f
Missouri Plateau, 211–214
Missouri River, 181–183
Mitchell, Mt., 347, 354
ML. *See* Mono Lake
Modern Colorado River, 556–558, 557f
Modern-day river system, 563
Modoc lava beds, 487
Mogollon Rim, 227, 227f
Mogollon-Datil (MD), 409, 418
　volcanic field, 414
Mohawk Valley Fault (MVF), 526
Moist coastal climate, 80, 80t
Mojave Desert, 419, 440–441, 446f, 501–503, 532–533
Mojave-Sonora megashear. *See* California-Coahuila transform
Molar-tooth structure, 315–316
Molasse basins, 146, 322
Monadnock, 358
Mono Lake (ML), 426–427, 525, 533–534, 536t
Monoclinal hogback ridges, 185, 188
Monoclinal slopes, 185, 188
Monoclines, 42, 227–231, 550
　bench-and-slope landscapes and, 188
　on Colorado Plateau, 553
　nearly flat-lying sedimentary layers and, 227–231
　Waterpocket monocline, 231f
Montana, 375–376
　triple divide in, 16
Montana Disturbed Belt (MDB), 314, 315f
Montezuma Range (Mnr), 534
Monticello rock (Mc rock), 322
Monument Valley, 222, 226, 226f
Moores Knob, 358
Moraines, 37f, 159–160, 161f, 171–174
　DEM of Long Island, 174f
　northwest at kettle lakes on Coteau Du Missouri, 173f
Morros Mountains, 419
Morton Gneiss, 301–302
Moses Coulee, 395, 396f
Mosquito Range, 272f, 276, 280f
Mount Adams, 486
Mount Daly, 160–161
Mount Hood, 486
Mount Katahdin, 366, 366f
Mount Lassen, 486
Mount Monadnock, 367–368
Mount Oglethorpe, 347
Mount Redington, 367
Mount Shasta, 486
Mount Taylor (MT), 422, 423f
Mount Washington, 365
Mount Whitney, 522–525, 526f
Mountain City (Mc), 357
Mountain City window (MCW), 357
Mountain ranges
　definition of, 5
　erosion of, 70–71
　with volcanoes, 73

Mountain(s), 84–86. *See also* Erosional Mountains
  alignment relative to San Andreas Fault, 514–515, 514f
  alpine glaciers and, 159f
  belts
    crystalline rock in, 120
    definition of, 5
    erosion of, 69
    uplift in, 267
  building and preservation, 89–91
    example of low-density crustal root below mountain, 89f
    isostatic compensation following mountain building event, 90f
  building of, 89–91
  climate and, 59–61
  collapse of, 533–534
  crust and, 92f
  crystalline rock in, 28–29
  definition of, 5
  erosion of, 322
  magma and, 401
  orogenesis and, 69
  root, 91
  structural forms in, 41–46
  system, 5, 69
  tectonic plates and, 89
  tectonic wedge and, 145–146
  in vicinity of La Junta Peak near Telluride, 84f
Mountainous topography, 256
Mountains, compression and, 91, 139–140
MP area. *See* Mine Point area
Muddy Creek Formation, 556
Mudflows, 410, 426–427
Muir, Mt., 34f
Muldrough's Hills Escarpment, 247, 248f
Multnomah Falls, 394
MVF. *See* Mohawk Valley Fault
Mylonite, 47–48

N

Nanaimo Group rocks, 382–383
Nashville Dome, 244, 255–256
National Oceanic and Atmospheric Administration (NOAA), 190
Native North American rocks in Appalachian hinterland, 341
Native rock successions, 141
Navaho volcanic field, 411–413
Nearly flat-lying sedimentary layers
  Appalachian Plateau, 251–258
  Central Lowlands, 237–241
  Coastal Plain, 189–209
  Colorado Plateau, 222–235
  Great Plains, 211–219
  interior low plateaus, 244–251
  landscape in nearly flat-lying layers, 185–188
  Ozark Plateau, 241–244
  sedimentary-cored anticlinal and domal mountains, 235–236

  uplift of Wyoming Basin and northern Great Plains, 221–222
  western margin of, 209–211
  Wyoming Basin, 219–221
Nebo, Mt., 335, 335f, 464–465
Nebraska Sand Hill Region, 216–218
Neenach, 419
Nevada-Utah Basin and Range, 434f
Nevadaplano, 446–447, 533
Neversink River (NR), 323–324
New England, 365f
  hot spot. *See* Great Meteor hot spot
  nor'easters in, 84
New England Coastal Plain, 191–198
  landscape map of New England and adjacent areas, 193f
New England Highlands, 342, 361–369, 361f
  Acadia National Park, 368–369
  Boundary Ranges Highland Trend, 364–365
  Green Mountain Highland Trend, 363–364
  Isolated Mountain Peaks, 367–368
  White Mountain Highland Trend, 365–367
New Jersey to North Carolina, 198–200
  drowned Maine coastline from Portland, 198f
  landscape map of Mid-Atlantic Coastal from New Jersey to Georgia, 194f
  Mississippi Embayment area, landscape map of, 196f
  north to south cross-section from Connecticut to long Island, 199f
  South Carolina–Florida coastal area, landscape map of, 195f
  Texas coast, landscape map of, 197f
New Madrid fault, 175–176
Newberry Caldera, 401, 401f
Newberry volcanic field, 400–401
NFLv. *See* North Fish Lake Valley
Nine Sisters, 419, 419f
Nittany Valley, 408f
NOAA. *See* National Oceanic and Atmospheric Administration
Nonglacial climate processes, 132
Nonmarine environments, 26
Nonresistant rocks, 253, 355
Nor'easters, 84
Normal faults/faulting, 49, 429, 562–563
  basin and range, 432–452
    cause of basin and range extension, 447
    crustal thinning and volcanism, 440–441
    expansion into surrounding areas, 438
    geology, 447–451
    horizontal extension, 440
    landscape characteristics, 438–440
    metamorphic core complexes, 441, 442f
    Nevadaplano, 446–447
    normal fault activity verses erosion, 443–446
    physiographic limit, 432–437
    $Sr_i = 0.706$ line, 451–452
    timing of normal faulting, 442–443
    vertical displacement, 440
  along crest of high cascades, 490–491
  forms, 44–46

  mechanisms by extension, 430f
  Rio Grande Rift, 452–461
  structural character and terminology, 429–432
    detachments, 431–432
    fault-block rotation and rollover anticlines, 432
    horst and graben structure, 429–430
    tilted fault blocks, half-grabens, and flexural rebound, 430–431
  timing of, 442–443
North America, glacial erosion boundary across, 165–171
North American crystalline shield, 142–143
  landscape map, 142f
North American lithosphere, 527–528
North American plate, 62, 73, 501
North Atlantic Ocean, 102
North Cascades, 491
  Highway, 378, 379f
  Mountains, 377–381, 380f, 487
North Fish Lake Valley (NFLv), 534, 539
North Park, 274–275
North Rim, 550, 554–555
North Spit tidal gauge, 475–477
North Table Mountain, 407–409, 409f
North-flowing
  Little Colorado River, 556
  paleoriver, 556
  river system, 556
Northern Appalachian Fold and Thrust Belt, 321–325
  geology of Southern Hudson River Valley, 322–325
Northern Divide, 15–16
Northern Great Plains, 407
  uplift of, 221–222
Northern Minnesota, 158, 158f
Northern Nevada rift zone, 425
Northern Panamint Valley, 544–545
Northern Rocky Mountains, 314–317, 392–393
  eastern segment, 314–317
  and North cascades, 372–383, 378f
    Central Idaho, Montana, and Oregon, 375–376
    Northern Washington, 376–383
    Southern Idaho, 372–375
  western segment, 317
Northern Sierra Nevada, 425–426
Northern Washington, 376–383, 377f
  Kettle River Range, 381–382
  North Cascade Mountains and Okanogan Range, 377–381
  Okanogan Highlands, 381–382
  San Juan Islands, 382–383
  Selkirk Range, 381–382
Northwest basin and range, 425–426
Northwest-trending inactive faults, 386
Norumbega Fault, 369
NR. *See* Neversink River

O

^{16}O isotope, 97

^{18}O isotope, 97
Obliquity, Earth's orbital parameters, 100–101
OC. *See* Outer Cuesta
Ocala Uplift, 200
Ocean basins, 64, 89, 95
Ocean-atmosphere exchange, 104
Oceanic circulation, 79
Oceanic crust, 62, 89
Oceanic lithosphere, 62, 89, 91–92
Oceanic trench, 65–66
Ochoco Mountains, 394, 398–399
Ocoee Supergroup, 345, 351–353
Ogallala
   aquifer, 216–218, 218*f*
   Formation, 215–216, 222
Ohio River, 180
Okanogan Range, 377–381
Okeechobee, Lake, 200
Old drift, 162, 177–178
Olistostromes, 386
Olympic Mountains, 491–495, 493*f*, 495*f*. *See also* Central-Southern Cascade Mountains; Klamath Mountains
   Cascadia volcanic arc system and, 491–495
   case for topographic steady-state, 492–495, 494*f*
   compression, 492
   erosion, 494
   exhumation, 494
   geology, 492, 493*f*
   Juan de Fuca plate and, 492
   subduction, 492
   uplift, 492
Olympic Peninsula, 491–492, 491*f*
Olympic-Wallowa lineament (OW lineament), 400
Olympus, Mt., 491–492, 494
Oolites, 201–202
Ophiolite, 497–498
Orcas Island, 382
Oregon, 375–376
   Alps, 376
   coast range, 477–482
      cause of uplift along Oregon coast, 480–481
      geology, 481–482
      relative sea level change based on tide gauge measurements, 478*t*
   Oregon Cascade stratovolcanoes, 490
   Oregon Dunes National Recreation area, 478, 481*f*
   Oregon-Washington coast, 73
Orogenesis. *See* Orogeny
Orogenic belt, 69
   idealized, 144–145
Orogenic system. *See* Orogenic belt
Orogeny, 144
   in Appalachian Mountains, 141*f*
   in Cordillera, 141*f*
   tectonic style and, 69
Osage Plains, 239–241
Osceola Mudflow, 475, 483–486
Ossipee Mountains, 367–368

Ouachita fold and thrust belt, 333–336. *See also* Cordilleran fold and thrust belt; Valley and ridge fold and thrust belt
   Arkansas River Valley–Northern Mountains, 334–335
   Athens Plateau, 336
   Central Mountains, 336
   Fourche Mountains, 335–336
Ouachita Mountains, 5–7, 143, 236, 335–336
   feature, 334
   Ouachita Mountains oriented north Google Earth image, 334*f*
   Raisz landform map, 333*f*
   water gaps in Valley and Ridge and, 337–340
Ouachita orogeny, 70–71
Ouachitas Mountain, 397
Outcrop, 31
Outer bench, 549
Outer Cuesta (OC), 235
OW lineament. *See* Olympic-Wallowa lineament
Owens Valley (OV), 437*f*, 501, 540
   Owens Valley-White Mountains fault zone, 534
Owyhee Upland, 402, 405
Oxygen isotope, 101–102
   analysis, 98
   ratios, 97, 101
   record over past 1.8 million years, 101
      stable oxygen isotope ratios, 101*f*
   record over past 67 million years, 99–100
Ozark Plateau, 7, 241–244, 264*f*. *See also* Appalachian Plateau
   Boston Mountains, 243
   dolostone, 241–243
   landscape map, 242*f*
   limestone, 241–243
   Salem and Springfield Plateaus, 241–243
   uplift history, 243–244

P

Pacific active continental margin, 71–74
   plate tectonic configuration, 72*f*
   shallow subduction, 73*f*
   tectonic setting of Gulf of California, 72*f*
Pacific coastline, 475–477
Pacific Northwest, 372, 373*f*
Pacific Ocean, 557, 561
Pacific plate, 66, 73, 501, 514–515
Painted Desert, 224
Paleocanyons. *See* ancient canyons
Paleoclimate, 95
Paleozoic rocks, 456
Palisades Sill, 468–469
Panamint Mountains, 544–546
Panamint Range (Pr), 534, 540, 541*f*
Pangea, 70–71, 71*f*
Panther Junction (pj), 457
Paoha Island, 427
Paria Plateau (PP), 549–550
Partial annealing zone, 121, 559
Pasayten fault (PF), 378–379

Passive continental margins, 64–65
   Atlantic, 69–71
Peach Springs Canyon, 555
Peach Springs drainage system, 562–563
Peach Springs paleocanyon, 553–556, 558
Peach Springs paleoriver, 562
Pecos River, 18
Pecos Valley, 218–219
Pediment, 283–284
Pediment surface, 444
Pee Dee River, 200
Peninsular Arch, 200
Peninsular Ranges, 504, 520–521, 521*f*
Pennsylvanian
   Pennsylvanian-Early Permian Alleghany orogeny, 354
   rock, 251–253
Penokean province, 302
Peripheral rocks, 481
Persimmon Gap, 455–456
PF. *See* Pasayten fault; Pigeon Forge
Phanerozoic eon, 69
Phonolite porphyry, 273–274
Physical weathering, 25
Physiographic canyon, 549–553
   Colorado River in central part of canyon, 552*f*
   Colorado River toward LF, 552*f*
   grand canyon oriented north, 551*f*
Physiographic limit, 432–437
Physiographic map, 11–12
Physiographic province, 4–5, 6*f*, 8*t*, 9*f*, 569*f*
   comparison of river systems with, 20–21
   distribution, 151*f*
   rock/sediment type, 36–40
Physiographic region
   Appalachian mountain, 8
   and provinces, 4–10
      Appalachian Mountain System, 8
      Coastal Plain, 8–10
      Cordilleran Mountain System, 10
      Interior Plains and Plateaus, 7–8
Physiographic subdivision, 137
   of United States, 15
Piedmont Plateau (PP), 342, 349, 358–359, 358*f*
   province, 120
Pigeon Forge (PF), 351, 351*f*
Pikes Peak, 279
Pine Canyon (P), 414
   caldera, 414–415
Pine Mountain (pm), 329–330, 401
   thrust, 254–255, 327
Pine Ridge escarpment, 211, 215–216, 296
Pineville, 329–330
Pink Cliffs, 234, 236*f*
Pinnacles, 419
Pinnacles National Park, 419
Plagioclase, 29
Plain(s), 5, 389
Plains Border, 218–219
Plant
   boundaries, 62–63
      interaction of plate tectonic, 64*f*

Plant (*Continued*)
   collision, 62
   photosynthesis, 104
   tectonic
      events, 70
      movement, 99–100
      theory, 62
Plateau, 5, 118, 188
Playa (temporary lake), 15, 443–444
Pleasant Valley earthquake, 443
Plot Balsams, 353, 353*f*
Plott Balsams boast peaks, 347
Pluton, 29
Plutonic rock(s), 28–29, 262
   of Adirondack Mountains, 262
Pluvial lakes of Cordillera, 183
Plymouth Rock, 175
Pocono Mountains, 253–254, 325
Point bar, inner bank, 113–114
Point Sal (Ps), 501
Polar Easterlies, 82, 84
Pontotoc Ridge–Ripley cuesta, 205
Poplar Camp, 347
Porcupine Mountains, 304–305, 305*f*
Port Jarvis (PJ), 322
Potassium (K), 26, 121, 451
   partial dissolution of potassium
      feldspar, 25
   potassium-rich basalt, 407–409
Potomac River (Pr), 346
Pottsville escarpments, comparison of
   Cumberland escarpments and, 257–258
Poughkeepsie (Pk), 322
Powell, John Wesley, 150, 549, 550*f*
Powell Mountain (pwm), 332
PP. *See* Paria Plateau; Piedmont Plateau
Precambrian
   rocks, 384–385
   sedimentary-volcanic rocks, 345
      succession, 143
Precession, Earth's orbital parameters,
   100–101
Present-day climate zones, 79–80, 80*f*, 80*t*
Present-day erosion rates, 115–117
Present-day uplift
   present-day uplift/subsidence rates, 109–110
      areas of active uplift and subsidence, 109*f*
   rates, 480
Presidential Range, 365–366
Presidio bolson (Pb), 456–457
Pressure ridges, 505, 514
Princeton, Mt., 411, 413
Profile, 4
Proglacial lakes, 174–175
Proto-Pangea, 142
Proximity to large water bodies, 81
Ps. *See* Point Sal
Pull-apart basin, 518–519
Pumpkin Patch metamorphic suite, 355–357
Purcell anticlinorium, 317
Purcell Lava, 315–316
Purcell Mountains, 317
Pyroclastic flow, 391
Pyroxene, 29–30, 418, 451

Q
Quartz, 26
Quartzite, 29
Quaternary fault, 439*f*
Queen Charlotte transform fault, 73

R
Racetrack. *See* Red Valley
Radiometric dating, 304
Rain shadow, 84, 84*f*, 85*f*
Rainier, Mt., 19, 159, 160*f*, 391–392, 475, 528
Rainy River, 20
Raisz, Erwin, 5–7, 400
Raisz landform base map, 567*f*
   with climate zone boundaries, 573*f*
   with glacial boundaries, 580*f*
   with physiographic province boundaries,
      569*f*
   with river system boundaries, 570*f*
   with rock successions boundaries, 577*f*
   with rock type boundaries, 572*f*
   with structural province boundaries, 575*f*
   volcanic rock boundaries, 581*f*
Raisz landform map, 383, 384*f*
Raisz landscape map, 414, 414*f*
Rampart Range, 274–275
Range-front fault, 444
Rangeley-Stratton Ranges, 362–363
Rangeley-Stratton region, 366–367
Rate of erosion and deposition, 126
Rates of erosion, 118
Raton-Clayton (R), 422
   volcanic field, 424, 425*f*
Reactivated Western Craton, 147, 236
Reading Prong, 165, 362, 362*f*, 468
Recoverable strain. *See* Elastic strain
Red Bluff Granite, 383–384
Red Desert, 15, 221
Red River Valley, 176*f*
   meanders of, 175
Red Valley, 296*f*, 297
Redwall Limestone, 562
Reef, 231
Reelfoot Rift, 206–207, 248
Reincarnation, 129–132, 130*f*, 132*f*
   to burial beneath unconsolidated sediment,
      132
   to glaciations, 131–132
   growing old, 129–131
   type, 125
   to volcanism and tectonic stress, 131
Rejuvenation, 125, 127–129, 224
Relative sea level, 96–97, 102*f*
Relict subduction zone landscape, 510–512,
   511*f*
   ancient accretionary prism, 510
   ancient forearc basin, 510–511
   ancient volcanic arc, 512
Reservation Divide Mountains (Rd), 311
Resistant Pennsylvanian rocks, 331–332
Resistant rocks, 256–257
Reverse faults, 43*f*, 46, 49
RF. *See* Ross Lake fault

RGR. *See* Rio Grande River
Rhyolite, 389, 391, 486
   of Jemez lineament, 422–424
   of Snake River Plain, 402–406
   of Yellowstone National Park, 404–405
Rich Mountains, 355
Richland Balsam, 353–354
Richland Knobs, 332–333
Rift system, 305–306
Rift zone landscape, 452
Rifting, 64–65, 65*f*, 71
Right-lateral strike-slip faults, 46, 504, 505*f*,
   518, 539
Rim of Grand Canyon, 549
Rio Grande Bolson deposits, 456–458
Rio Grande Rift, 414, 452–461
   in Central Colorado, 281–283
   Great Sand Dunes National Park, 459–461
   monoclines and normal faults in big bend
      area, Texas, 453–456
   Rio Grande Bolson deposits, 456–458
   rocky mountain basin and range, 461, 462*f*
   system, 209
   Teton Mountains, 462–464
   triassic lowlands of Appalachian Mountains,
      466–470
   Wasatch Mountains, 464–466
   White Sands National Monument, 458–459
   zone, 453
Rio Grande River (RGR), 459
   system, 222–224, 432
Rio Grande–West Texas River system, 18
River of No Return, 372–374
River systems, 15. *See also* Cascadia volcanic
   arc system
   Atlantic Seaboard–Gulf Coast river system,
      17–18
   California River system, 19–20
   Colorado River system, 19
   Columbia River system, 19
   comparison with physiographic provinces,
      20–21
   divides, 15–16
   Great Basin River system, 20
   Hudson Bay River system, 20
   Mississippi River system, 16–17
   Raisz landform base map with river system
      boundaries, 570*f*
   Rio Grande–West Texas River system, 18
   separated by drainage divides, 15*f*
   St. Lawrence River system, 18
   of United States, 16*f*, 17*t*
Rivera triple junction (RTJ), 506, 508
Rivers, 110–115, 522
   base level changes, 111–112
   changes in discharge and sediment supply, 113
   elevation, 551–553
   graded rivers and base level, 111
   incision, 117, 224, 264
   knickpoint migration, 112–113
   longitudinal profile, 111, 111*f*
   lower Mississippi River Valley, 113–115
   morphology, 556
   valley, 391

Roan Mountains, 347
Roanoke
  Basin, 342, 345
  Blue Ridge at, 345–346, 345f
  Blue Ridge North of, 346–347
  Blue Ridge South of, 347–349, 348f
Roberts Mountain
  allochthon, 449
  thrust, 150, 448–449
Rock Springs Uplift, 220–221
Rock/sediment type, 10, 26–31, 110–111
  of components, 10, 23
  crystalline rock, 28–29, 28f
  deposition, 25–26
  distribution among 26 physiographic provinces, 36–40
  erosion, 25–26
  influence of bedrock on landscape, 32–35
  Karst landscape, 36
  landscape-forming rock/sediment types, 25t
  rock cycle, 31, 31f
  rock hardness and differential erosion, 31–32
  sedimentary rock, 27–28
  unconsolidated sediment, 30–31
  volcanic rock, 29–30
  weathering, 25–26
Rock(s), 17, 69, 88–89, 309, 316, 429, 498
  cycle, 31, 31f
  deformation style, 41–46
    faults, 44–46
    folds, 41–44
    structural form—landscape-forming structures, 42t
    vertical joint sets, 44, 45f, 46f
  hardness and differential erosion, 31–32
  layers, 309
  resistance, 367
  successions, 137, 141–144, 141t
    accreted terranes, 144
    Atlantic miogeocline, 144
    colored US state map showing, 577f
    conceptual west-to-east cross-section across United States, 141f
    distribution, 148–150, 148f
    distribution and correlation, 151f
    interior platform, 143
    Laurentia and, 15–16, 153f
    miogeocline rocks, 143–144
    North American crystalline shield, 142–143
    Precambrian sedimentary/volcanic rocks succession, 143
    Raisz landform map with rock successions boundaries, 577f
  types, 25
    colored US state map showing, 572f
    Raisz landform map with rock type boundaries, 572f
  unit, 314–315
  uplift, 517
Rocky Mountain(s), 7, 556, 561
  anticlines in, 41–42
  Basin and Range, 314, 429, 438, 461, 462f

erosion surface, 277–279, 280f, 283–284, 287
foreland fold-and-thrust belts of, 139f, 147
frost cracking in, 25, 26f
Great Plains and, 296
region, 209f
reverse faults of, 46
Trench, 317, 318f
uplift of, 221
water gaps in, 298
Rodinia, 142–143
Rogue Valley, 482–483
Rollover anticlines, 432, 432f
Rome Formation, 332–333, 344, 344f
Roseburg (Rb), 482
Rosillos, 453
Ross Lake, 380
Ross Lake fault (RF), 377–379
Rotation process, 489
Rough Creek-Pennyrile fault, 248, 250f
RTJ. See Rivera triple junction
Rubidium, 451
Rubidium 87 (^{87}Rb), 451
Rutland (R), 321

S

Saddle Mountain, 397
Saddleback Mountain, 367
Salem and springfield plateaus, 241–243, 243f
  Table Rock Lake, 242f
Salem Plateau, 241
Salem Upland, 241, 264
Saline Valley (Slv), 540
SalineRange (Slr), 540, 541f
Salinian block, 512, 518–519
Salmon Mountains, 374f
Salmon River Mountains, 372–374, 374f
Salt basins, 452
Salt domes, 205–206, 206f
Salt River Range, 318
Salton Sea, 506, 517–520
San Andreas Fault, 3, 48f, 50f, 72, 72f
  California transpressional system and, 505–510
  Coast Range and, 514f, 515
  compression in, 514–515, 515f
  development, 509f
  displacement along, 506–508
  extension in, 514–515, 515f
  faults, 508t
  history, 508–510
  mountain alignment relative to, 514–515, 514f
  North American plate and, 508–509
  Pacific plate and, 508–509
  sedimentary rock in, 510
  segmentation of, 506
  stress in, 506–508
  strike-slip faults and, 509
  subduction and, 510–512, 511f
  system, 46, 66, 72–74, 448, 501, 505–510, 509f, 516f
San Antonio escarpment, 207
San Bernardino Mountains, 506

San Carlos-Springerville (SS), 422
San Francisco
  earthquake, 506–508
  volcanic field, 420, 421f
San Gabriel Mountains, 512
San Gregorio Fault, 516–517
San Gregorio-Hosgri Fault, 506
San Jacinto Fault, 506
San Juan Islands, 382–383, 382f
San Juan Mountains (SJ Mountains), 409, 411–413, 411f
San Juan River, 230–231
San Luis Valley, 269, 452
San Rafael Swell, 228, 232f
Sand Dune fields, 177
Sand Hills
  Carolina Sand Hills, 208
  of Nebraska, 216–218, 217f
Sandstone, 27–28, 27f, 32, 45f, 186
Sandy Point, 549
Sangre de Cristo Mountains, 274–275, 282–283
Sanpoil volcanic field, 409
Santa Cruz Mountains, 516–517, 516f, 517f
Santa Lucia Bank (SL Bank), 518–519
Santa Lucia Mountains, 513, 513f
Santa Maria (SM), 518–519
Santa Rosa Mountains, 506, 512
Santiago Mountains (S Mountains), 455
Saranac Lake, 262, 263f
Sawtooth Mountains, 304–305, 374–375
Sawtooth Range. See Montana Disturbed Belt (MDB)
SB. See Sierra Blanca
SC. See Slumgullion Creek
Scale, 4, 4f
  comparison of map scales, 4f
Scarps, 207–208
SCBZ. See Sierra Nevada-Cascade Range Boundary Zone
Schist, 29
Scioto River, 180
Scoria cone. See Cinder cone
Scotchman Peaks (Sp), 317
Scratches, 538
Sea level, baseline, 61–62
Sea level change, 64, 95, 96f
  cause, 95
  climate and, 98
  Coastal Plain and, 98
  Florida and, 101–102
  as forcing variable, 95
  history of, 99f
  history of CO_2 in atmosphere, 103–105
  Ice Age and, 100
  influence of earth's orbital parameters on glaciations, 100–101
  measurement of, 95–98
  measuring sea level and sea level changes, 95–98, 98f
  oxygen isotope record over past 1.8 million years, 101
  oxygen isotope record over past 67 million years, 99–100

Sea level change (Continued)
  Pacific Ocean and, 100
  sea level changes over past 100 million years, 98, 99f
  sea level over past 150,000 years, 101–102
  sea level response to temperature history, 103
  temperature history, 102–103
  in Texas, 207
Seasons
  Horse Latitudes and, 84
  Polar Easterlies and, 84
  wind patterns and, 83f
Seattle Uplift, 483
Sediment, 30–31
  core, 101–102
  sediment-dwelling (benthic) foraminifera, 97
  supply, 113, 115
Sedimentary rock, 25, 27–28, 31, 41, 441. See also Nearly flat-lying sedimentary layers; Rock/sediment type; Unconsolidated sediment
  of Appalachian Mountains, 369–372
  climatic system and, 79
  of Coastal Plain, 138–139
  of Colorado Plateau, 559f
  of Cordilleran orogenic belt, 144
  crystalline rock and, 32
  deposition and, 118
  erosion of, 32f
  landscape in, 32–33
  of San Andreas Fault, 510
  sedimentary rock B, 558–559, 559f
  stratification in, 27
Sedimentary Series, 304
Sedimentary-cored anticlinal and domal mountains, 235–236
Seismic modeling, 473
Seismic profiling, 189–190
Selway-Bitterroot Wilderness, 375
Semihumid climate, 117–118
Sequatchie Valley, 253, 256
Seven Devils Mountain, 399, 399f
706 Line, 451
Sevier fold-and-thrust belt, 447–448, 449f, 461
Sevier–Blockhouse flysch, 324
SF. See Straight Creek fault
Shady Dolomite, 344
Shale, 27–28
  in nearly flat-lying sedimentary layers, 186
Shallow ocean environments, 190
Shape of land vs. rock structure, 49–50
Shastina, 486
Shaw Island, 382
Shawangunk Ridge (SR), 324
Shawnee Hills
  deformed rocks, 248–249
  at Equality, 250f
  Shawneetown-Rough Creek fault zone and Eagle Valley syncline, 250f
She Devil, 399
Shear, 62
Shear stress, 46, 501–503
Shenandoah River, stream piracy by, 320

Shield rocks, 553
Shield volcano, 390
Shield volcanoes, 390–391
Shiprock, 411–413, 418f
Shivwits Plateau (S Plateau), 553
Shoreline angle, 475
Shoreline rocks, 190
Shoshonite, 407–409
Shrinking remnant of Farallon plate, 473
Sierra Blanca (SB), 418
Sierra del Carmen (sdc), 455, 457
Sierra Estrella, 445, 445f
Sierra Nevada (Sn), 10, 73, 501, 512, 521–531, 523f, 524f
  alpine glaciers of, 164
  Basin and Range and, 430, 432–452
  batholiths in, 33, 512
  California transpressional system and, 521–531
  climate and, 530
  downcutting in, 530
  escarpment, 437f
  exhumation in, 527
  Frontal Fault system, 521–522, 525–526
  Great Basin and, 20
  ignimbrite flare-up in, 527–528
  rain shadow of, 84, 84f
  strike-slip faults and, 506
  tectonics in, 527
  underplating in, 527–528
  uplift history, 526–531
    Farallon plate underplating North America, 528f
    landscape map that southward and westward migration, 529f
    topographic paleorelief and river incision, 530f
  volcanism in, 527
Sierra Nevada-Cascade Range Boundary Zone (SCBZ), 526, 530
Sierra Quemada (Q), 415
Sierran microplate, 532–533
Sierran plate, 501
Siletz terrane, 481, 492
Siletzia, 482
Silica (SiO_2), 26
Silicic, 391
  calderas, 391
  composition, 389
  rocks, 30, 400
  volcanic rocks, 30, 391
  volcanism, 402–403
Silicic acid (H_4SiO_4). See Dissolved silica
Siltcoos, Lake, 478, 481f
Silurian period, 364–365
Silurian-Early Devonian, 369
Silver Fork-Superior fault zone, 466
Silverton (S), 412
Sinkholes, 32f, 36, 203f
  of Arbuckle Mountains, 267
  limestone and, 32, 202–203
Siyeh Limestone. See Helena Formation
SJ Mountains. See San Juan Mountains
Skyline Arch, 233f

SL Bank. See Santa Lucia Bank
Slick Rock Sandstone, 234
Slickenlines, 538
Slope. See Cuesta
Slumgullion Creek (SC), 413
Slumgullion Slide, 412–413, 412f
Snake River, 394, 399–400, 399f
Snake River Plain, 10, 91, 389, 392f, 393, 393f, 399–400, 402–406, 403f, 404f
  aquifers in, 405
  caldera of, 403f
  Columbia River Basalts and, 403
  flood basalts of, 406–407
  hot spots in, 74–76
  Owyhee Upland, 405
  region, 419
  volcanic rocks of, 402–406, 403f
  Yellowstone Plateau volcanic field, 405–406
Snickers Gap, 346
Snowmass Peak (SP), 160–161
Snowshoe Mountain Tuff, 412
Soil respiration, 104
Solar luminosity, 104–105
Solid-state recrystallization process, 29
Solitario, 416, 416f, 417f
Sonoma orogeny, 449, 449f
South Carolina to Florida, 200–203
  downward at marshlands in Apalachee Bay, 203f
  Google earth image of Florida Keys oriented north, 202f
  NNE at Hilton Head Island, South Carolina, 201f
South Mountain (SM), 346–347
South Park, 274–275, 275f
South Rim, 550
South Table Mountain, 407–409, 409f
Southern Appalachian (Tennessee) fold and thrust belt, 327–333, 329f
  central thrust belt, 332–333
  western thrust belt, 329–332
Southern Blue Ridge, 129
  relict erosion surfaces in, 371–372, 371f
Southern Gore Range, 282
Southern Hudson River Valley, geology of, 322–325
Southern Idaho, 372–375
Southern Rocky Mountains, 274–289
  cause of accelerated erosion in, 287
  front range, 276–279
    Garden of Gods, Colorado Springs, 278f
    High Plains, 280f
    Morrison (M) along Front Range near Denver, 277f
    south of Denver at slope-flat-slope topography, 278f
    south-southwest at Front Range (FR) between Cheyenne and Laramie, 279f
  landscape history of, 283–286
    criteria to interpret landscape evolution, 284f
    evolution of Southern Rockies, 283f

northeast at jagged peaks of Grenadier Range, 276f
Rio Grande Rift in Central Colorado, 281–283
Sawatch Mountains, 279–281, 280f
South Park synclinal valley from Wilkerson Pass, 275f
Southern Sierra Nevada Frontal Fault (SSNF), 525
Southern Virginia (SVVR), 320
Southern-Central Cascade Mountains, 389
Spanish Peaks (Sp), 272–273, 424
Spearfish Formation, 297
Specimen Ridge, 410, 410f
Split Mountain, 50, 129
Springerville field, 422–423
Springfield Plateau, 467
SR. See Shawangunk Ridge; Susquehanna River
SS. See San Carlos-Springerville
SSNF. See Southern Sierra Nevada Frontal Fault
ST. See Streeruwitz thrust
St. Albans City, 321–322
St. Francois Mountains, 241, 263–266, 264f
　buttress unconformity in, 264, 265f
　crystalline rock domes of, 263–266
　East Fork Black River, 266f
　Ozark Plateau, 264f
　volcanic rocks of, 389–392
St. Helens, Mount, 3, 30, 483–486
St. Lawrence River system, 18
Stacking, 309–310
Stair-step geometry, 309–310, 310f
Steady-state, 126–127
　as end-product of growing old, 127
Steens Mountain, 394
Steep relief, 380
Steep-dipping rock layers, vertical to, 48–49
Steppe climate, 80
Stillwater Reservoir, 262, 263f
Stillwell (st), 453
Stone Mountain-Unaka Mountain thrust (SU thrust), 357
Stone Mountains (sm), 355
Stony Man (S), 346
Straight Creek fault (SF), 377–378
Strandline, 97
Straovolcanoes, 486
Strath lines (terrace), 116–117
Stratification, in sedimentary rocks, 27
Stratified drift, 158–159
Stratigraphic sequence, 27
Stratigraphy, 27
Strato volcanoes, 30, 391
Stratocone. See Composite volcano
Stratovolcano. See Composite volcano
Strawberry Mountains, 376
Stream capture. See Stream piracy
Stream channel, 112–113
Stream piracy, 20–21
　drainage divides and, 20, 21f
　by Shenandoah River, 346
Stream reversal, 561

Stream velocity, 113
Streeruwitz thrust (ST), 384–386
Stress, 41
　in San Andreas Fault, 506–508
　types of, 42f
Strike, 41, 42f, 44f
Strike-slip faulting initiation, 539–540
Strike-slip faults, 43f, 46, 48–49, 341
　of California transpressional system, 501, 512, 515f
　of Coast Range, 514
　landscape associating with, 503–505, 504f, 505f
　Mojave Desert and, 521
　San Andreas Fault and, 509
　Sierra Nevada and, 521
Stromatolites, 385–386
Strong rocks, 31
Strontium (Sr), 451
　^{86}Sr, 451
　^{87}Sr, 451
　strontium-initial 0.706 line, 448
Structural basin, 41–42
Structural form, 10, 41, 43f
　brittle and ductile faults, 47–48
　of components, 10–11
　deformation of, 41–46
　fault reactivation, 46–47, 48f
　of foreland thrust faults, 309–310, 310f
　influence of, 44
　influence of dipping layers on landscape, 48–49
　recognition of active faults, 50–52
　structure-controlled and erosion-controlled landscape, 52–56, 56f, 57f
　style of rock deformation, 41–46
　topographic form and, 49–50
　types of, 53f
Structural high, 50
Structural provinces, 137–141, 140t. See also Tectonic provinces
　boundaries of physiographic provinces, 140f
　colored US state map, 572f
　distribution and correlation of, 151f
　landscape map, 138f
　Raisz landform map with structural province boundaries, 575f
　and unconsolidated sediment, 139f
Structure of rock. See Structural form
Structure-controlled landscape, 52–56, 56f, 57f
　versus erosion-controlled landscapes, 52–56
　physiographic provinces with, 137, 138f, 140t
　reincarnation of, 129–132
Style of deformation. See Structural form
SU thrust. See Stone Mountain-Unaka Mountain thrust
Subducting lithosphere, 66
Subduction, 68f
　in Cascade Mountains, 473, 474f
　in Coast Range, 481
　earthquakes and, 481f
　of oceanic lithosphere, 65–66
　in Olympic Mountains, 492

San Andreas Fault and, 510–512, 511f
　of seamounts, 480
　in Sutter Buttes, 427
　of tectonic plates, 62, 72
　of tectonic wedge, 145–146
Subduction complex of Cordillera, 450
Subpolar climate, 79
Subsidence, 10, 91–92, 107–110, 119
　active areas of, 109f
　elevation and, 118–119
　measuring ancient uplift rates and elevation, 110
　present-day uplift/subsidence rates, 109–110
　surface uplift/subsidence and bedrock uplift/subsidence, 107–108
　tectonics and, 59
　in Texas, 207
Sudbury Meteorite impact event, 303
Suffolk scarp, 208, 208f
　elevation of, 208–209
Sugarloaf Mountain, 367
Summit Peak (SP), 304–305
Sun, wind patterns and, 83
Sunset Crater (S Crater), 420
Superior, Lake, 109
Superior province, 301–302
Superior Upland, 7, 298–306
　Barron and Baraboo Quartzite, 303
　geologic overview, 300, 301f
　glacial erosion of, 158
　iron formations, 302–303
　Keweenawan Rift system, 304–306
　in Minnesota, 301–302
　nearly flat-lying sedimentary layers of, 152f
　Penokean province, 302
　province, 120
　Raisz landform map of, 299f
　Sudbury Meteorite impact event, 303
　Superior province, 301–302
　volcanic rocks of, 389–392
　volcanism in, 487
Surface subsidence, 107
Surface uplift, 107–108, 108f
　deformation history prior to, 515–516
　mechanism and cause, 516–517
　rate, 122
Surficial material. See Unconsolidated sediment
Susquehanna River (SR), 325
Sutter Buttes, 427
Suture zones, 67–68
　TSZ, 150
SVVR. See Southern Virginia
Swannanoa lineament, 354–355
Swells, 227–228
Synclines, 41–42, 53f, 304–305
　in Rocky Mountains, 73

T

Table Mountain slide, 488
Table Rock Lake, 241, 242f
Taconic Allochthon, 321–322

Taconic flysch, 324
Taconic molasse, 324−325
Taconic Mountains, 364, 364f
Taconic Overlap sequence, 323−324
Taconic suture zone (TSZ), 353, 358
Taconic-age deformation, 326
Tahkenitch, Lake, 478, 481f
Tahoe, Lake, 20, 501
Tahoe Frontal-West Lake Tahoe fault zone, 525
Tallahassee, 202−203, 203f
Teays River, 179−181
Tectonic accretion, 67−69, 68f
Tectonic exhumation, 119, 122, 122f
Tectonic landscape. *See* Structure-controlled landscape
Tectonic loads, 91
　example of foredeep depositional basin in front, 92f
　isostatic depression of oceanic crust, 91f
Tectonic plates, 59, 62, 95 *See also specific plates*
　interactions with, 64f, 65f
　mountains and, 89−91
　movement, 63−64, 95
　oceanic lithosphere and, 89
　subduction of, 63, 72
　triple junction of, 65f
Tectonic provinces, 137, 144−147, 145t. *See also* Structural provinces
　accreted terrane belt and, 144
　characteristics of idealized orogenic tectonic wedge and correlation, 145f
　colored US state map showing, 579f
　distribution, 148−150, 149f, 151f
　foreland, 147
　foreland fold-and-thrust belt and, 147
　formation of tectonic wedge, 146f
　hinterland, 146
　Raisz landform map with tectonic provinces boundaries, 579f
　reactivated Western Craton and Atlantic Marginal Basin, 147
Tectonic reincarnation, 311
Tectonic stress(es), 60−62, 65, 87, 125, 131
Tectonic System, 10, 59−62, 87, 125, 447, 470
　accretion, 67−69
　active continental margins, 65−66
　Atlantic passive continental margin, 63f, 69−71
　components, forcing agents, mechanisms, and visible changes to landscape, 60f
　forcing agents, 59−62
　movement of tectonic plate, 63−64
　orogeny, 69
　Pacific active continental margin, 71−74
　plate, 62
　plate boundaries, 62−63
　rifting and passive continental margins, 64−65
　thermal plumes and hot spots, 74−76
　unconformities, 69
Tectonic terrane, 68−69
Tectonic uplift (subsidence), 87−89, 108

Tectonic wedge, 145−146
　accretionary prism and, 67−68
　theory, 146
Tectonics, 11−13, 87 *See also specific tectonic processes and structures*
　of Acadia National Park, 368f
　of Colorado Plateau, 49
　exhumation and, 121−122
　as forcing agents, 10, 59
　of Klamath Mountains, 497f
　physiographic provinces and, 36−40
　of Sierra Nevada, 527
Tekton
　builder, 59−60
　carpenter, 59−60
Temperature
　heavy line shows global temperature difference, 103f
　history, 102−103
　sea level response to, 103
　temperature-controlled density, 91−92
Temporary lake (*playa*), 15, 443−444
Ten Thousand Islands, 202
TennesSee River (TVR), 256−257, 320
Terlingua normal fault (tf), 454
Terraces, 475
Terrain, 4−5
Terranes, 68−69
　accreted, 144
　of Coast Range, 481−482
　Crescent, 481−482, 491f
　Siletz, 481
Teton (T), 405
　fault, 463
　Mountains, 461−464
Teton Range, 405, 462−463, 463f
　Cathedral Group of, 464f
　earthquakes in, 464
　escarpments in, 463f
　normal faults of, 462−464, 463f
Texas, 207, 207f
　Coastal Plain of, 207, 207f
　Google Earth image northeast at Big Bend, 455f
　monoclines and normal faults in Big Bend area, 453−456
　sea-level change in, 207
　subsidence in, 207
　Suffolk scarp, 208f
Thermal anomalies, 62
Thermal isostasy, 91−92
Thermal isostatic compensation, 92
Thermal isostatic uplift, 91−92
Thermal plumes, 74−76
　in United States, 74−76
Thermodynamic principles, 120
Thompson Peak, 374−375
Thrust belt, 310
Thrust faults, 42f, 43f, 145, 309, 314, 333, 341, 461. *See also* Fold and thrust belt
　development of, 310f
　form, 46
　of Klamath Mountains, 497−498
Thrust ramps in MDB, 314

Thrust sheet, 448
Thrusting, 449
Tidal gauge data, 481
Tide gauge stations, 96−97
Tides, 132
Tiffany Mountain, 380−381
Tillamook-associated rocks, 482
Tilt of Earth's axis of rotation, 82−84
Timing of uplift, 538−539
Timpanogos, Mt., 464−465
Tobin Range, 443, 443f
Toledo Caldera, 423−424
Tomales Bay, California, 48−49, 50f
Topographic form and structural form, 49−50
Topographic high, 50
Topographic lineament, 350−351
Topographic steady-state
　case for, 492−495, 494f
　as end-product of growing old, 127
　landscape at, 126−127
　type, 125
Topography, 4, 243, 549
Toreva blocks, 421
Tornillo Creek (T Creek), 457−458
Toroweap faults (T faults), 551−555
Trade Winds, 82
Trail Ridge-Orangeburg-Coates-Broad Rock scarp, 208
　elevation of, 208−209
Trans-Pecos volcanic field, 414−418
Transform(s), 62
　fault, 63, 66
　in Juan de Fuca plate, 480
　leaky, 63
　plate boundaries, 66
Transition Zone, 227
Transitional environment, deposition in, 26
Transpression, 504, 515f
Transtension, 501−503
Transtensional regime of Death Valley and Black Mountains, 542−543
Transverse (T), 518−519
　block, 518−520, 520f
　　north at tectonic blocks, 519f
　　progressive rotation of Transverse block and development, 520f
　　rotation, 518−520
Transverse Ranges, 504, 517−520, 532−533
　in California transpressional system, 501−503, 517−520
　northward along wave-cut terrace, 519f
　rotation of Transverse block, 518−520
Transverse strike-slip faults, 333
Trap rock, 467
Trask Mountain (T Mountain), 482
Traveler, 367
Trenches
　Cascadia Trench, 473, 497−498
　oceanic, 65−66
　of Rocky Mountains, 317, 318f
Triangular facet, 51−52
Triassic Lowlands, 131, 429, 466−467
Triassic Rift Basins, 429, 466−467
Triassic Rift Valleys, 71, 131

rocks of, 346
Tributary streams, 556
Trinity-Yreka-metamorphic belt, 498
Triple divides, 15f, 16
Triple junctions, 64
  Mendocino, 505–506, 508, 528
  Rivera, 506
  of tectonic plates, 65f
Tropic of Cancer, 82–83
Tropic of Capricorn, 82–83
TSZ. See Taconic suture zone
Tuff, 391
Tug Hill Plateau, 263
Tugaloo terrane, 357
Tularosa basins, 452
Tumbledown Formation, 385–386
Turtlebacks of Black Mountains, 543–544, 544f

U

UG. See Upper Granite Gorge
Uinkaret field (U field), 419–420
Uinkaret Volcanic field, 419–420, 420f, 554–555
Uinta Mountains, 44f, 118, 236, 269, 298, 465f
Ultimate base level, 111, 111f
Umatilla Range, 398
Umtanum Ridge, 397
Unaka Mountains, 355
Unconformity, 69, 70f, 150
Unconsolidated sediment, 25, 30–31, 36f, 50, 137, 139f, 185
  landscape in, 34–35
  Reincarnation due to Burial Beneath, 132
Underplating, 65–66, 68f, 480
Unikaret Plateau, 551–553
Unikaret volcanic field, 551–553
United States
  glacial erosion boundary in, 164–165
  glacial landscape in, 162–164
  thermal plumes in, 74–76
Units, 4
Unstratified drift, 158–159
Upheaval Dome landform, 234–235, 237f
Uplands subprovince, 244
Uplift, 10, 107–110, 120
  Bedrock, 107–108, 108f, 119–120
  Colorado Plateau, 227–231
  Echo Cliffs, 230–231, 232f
  Isostatic, 87–89, 108–109
  Kaibab, 225–226, 228–229, 230f
  Llano, 268
  measuring ancient uplift rates and elevation, 110
  and monoclines, 227–231
  Ocala, 200
  Oregon, 480–481
  Ozark Plateau, 243–244
  of plateau, 127
  present-day uplift/subsidence rates, 109–110
  Sierra Nevada, 526–531
  surface uplift/subsidence and bedrock uplift/subsidence, 107–108
  Tectonic, 87–89, 108

timing, 538–539
White Mountains, 538–539
Wyoming Basin, 221–222
Upper Eocene–Lower Oligocene rocks, 291
Upper Granite Gorge (UG), 549–553, 558–559
Upper plate, 309
Upper Table Rock, 482–483
Upwarps, 227–228, 236, 289–290
Uranium, 97
Uranium–thorium/helium (U-Th/He), 121, 558–559
  ages, 493
  dating techniques, 286
  geochronology, 549
US Geological Survey, 5–7
  fault database, 452

V

Valles Caldera, 423–424
Valley and Ridge, 130, 309, 344
  water gaps in Valley and Ridge and Ouachita Mountains, 337–340
Valley and ridge fold and thrust belt, 319–333. See also Cordilleran fold and thrust belt; Ouachita fold and thrust belt
  central Appalachian fold and thrust belt, 325–326
  distribution of Appalachian foreland deformation, 333
  fault zones on Cumberland Plateau, 333
  Great Valley, 320–321
  northern Appalachian Fold and Thrust Belt, 321–325
  southern Appalachian (Tennessee) fold and thrust belt, 327–333, 329f
Valley of Vermont, 364
Van Horn area, 383–386, 385f
Vegetation, 80, 84
Vergence, 146
Vermilion Range, 302–303
Vermont Uplands, 364–365
Vertical joint sets, 44, 45f, 46f
Vertical to steep-dipping rock layers, 48–49
Voids, 36
Volcanic arc, 474–475
  ancient, 512
  terrane, 302
Volcanic arc complexes, 302, 497–498. See also Cascadia volcanic arc system
Volcanic breccia, 391–392, 409, 415, 425–426
Volcanic center, 390
Volcanic chain stretches, 74–76
Volcanic fields in Southern part of Jemez lineament, 422–423
Volcanic landforms, 389–392
  basaltic landform types, 390f
Volcanic landscapes, 4–5, 122
Volcanic neck, 392
Volcanic rocks, 25, 29–30, 32, 405, 441, 501–503, 536
  of Cordillera
    Columbia Plateau, 392–402

Cordilleran volcanic areas 70–20 million years old, 407–419
Cordilleran volcanic areas younger than 20 million years, 419–427
landscape map showing volcanic areas, 390f
magma types and common volcanic landforms, 389–392
origin of volcanism on Columbia Plateau and High Lava Plains, 406–407
Snake River Plain, 402–406
landscape in, 33–34
Raisz landform map of volcanic rock boundaries, 581f
Volcanic Tableland (VT), 536
Volcanism, 10, 74, 107, 122, 131, 403–404, 440–441
  hot spots and, 74–76
  origin on Columbia Plateau and High Lava Plains, 406–407
  in Sierra Nevada, 527
  tectonics and, 62
Volcanoes, 62, 377–378, 427
  andesite in, 391
  Cascade, 72, 474f
  Cascade composite, 483–486
  Cascade Mountains, 389
  Cascade stratovolcanoes, 484t
  composite, 30, 391–392, 410
  of Hawaii, 29–30
  hot spot, 74
  shield, 390–391
  straovolcanoes, 397–398
  strato, 30, 407, 410
VT. See Volcanic Tableland

W

Wabash River, 181, 181f
Wadena-Brainerd-Pierz field, 175
Walker Lane Belt (Wl), 501, 531–546
  cross-cutting relationships, 532f
  Death Valley-Panamint Valley region, 542–546
  Inyo Mountains, 540–542
  Inyo-Mono section, 534–536
  tale of three landscapes, 533–534
  White Mountains, 536–540
Wall Mountain Tuff, 413
Walland, 352
Wallen Ridge, 332
Wallowa Mountains, 376, 394, 398–399
War Ridge, 332–333
Warm air, 81
Wasatch fault, 466
Wasatch Mountains, 464–466
Wasatch Range, 464–465, 466f
Washburn, 335
Washington-Oregon coast, 83–84
Wassuk Range, 534
Water cycle, 104
Water gaps
  in Blue Ridge, 363
  in Green Mountain, 363
  in Ouachita Mountains, 337–340

Water gaps (*Continued*)
  in Rocky Mountains, 298
  in Valley and Ridge, 337–340
Water table, 217–218
Waterfalls, 304–305, 361, 391
Waterpocket fold, 229, 230f
Waterrock Knob, 347
Waucoba Mountain, 540
Wave-cut angle, 475
Wave-cut terraces, 496
Waves, 132
WC. *See* Wineglass canyon
Weak sedimentary rocks, 190
Wear Cove window, 352
Weathering, 25–26, 80
West Fork Stillwater River (Wr), 294
West of Twin Falls, 402
West-flowing river system, 563
Western Cascades, 488–490
Western Coal Field-Chester Uplands, 244, 248
Western Grand Canyon river system, 562–563
Western Granite Gorge, 553
Western Interior Seaway, 98
Western margin of crystalline-cored anticlines and domes, 269–271
Western margin of nearly flat-lying sedimentary layers, 209–211
Western segment of Northern Rocky Mountains, 317
Western thrust belt, 329–332
Wet Mountains, 33–34
Wheeler Geologic area, 411–412
Wheeler Peak, 438
Wheeler Ridge, 518

White Mountains, 501, 536–540, 537f, 538f
  Highland Trend, 365–367
  initiation of strike-slip faulting, 539–540
  peak, 438–440
  timing of uplift, 538–539
  trends, 361–362
White Plains (WP), 362
White Rock escarpment, 207
White Sands National Monument, 458–459
Whitefish Range, 317
Whitney, Mt., 4, 10, 435–437, 501, 522
Wichita Mountains, 266–267
Wichita structural domes, 266–269
Wide Columbia River estuary at Astoria (A), 478
Willamette Valley, 482
Willapa Bay, 478
Willapa Hills, 477–478
Wind, 81–82
Wind Cave, 298
Wind gaps, 339–340, 346, 347f
Wind patterns, global, 81–83
  seasons and shift, 83f
Wind River Range (WR), 290–293, 405
  crest of, 291f
  southward over Green River Lakes, Wyoming. Peaks, 292f
  Wind River Range with glacial erosion removed, 292f
Windy Peak, 380–381
Wineglass canyon (WC), 546–547
Wl. *See* Walker Lane Belt
Woodbury Heights cuesta, 199
WP. *See* White Plains

Wyomide Ranges, 317–318
Wyoming Basin, 209, 219–221, 279, 289–290, 405
  uplift of, 221–222
  Wyoming Range, 318
  Wyoming Valley, 253–254

Y

Yakama fold belt, 394, 397, 397f
Yellowstone National Park, 3, 92, 410, 410f
  caldera of, 404–405, 406f
  fossils in, 410, 410f
  hot spot in, 76, 410
  Lake, 406
  Yellowstone Plateau, 402
    volcanic field, 405–406
  Yellowstone River Valley, 293
Yosemite Valley, 522, 527
Young drift, 162–163
  northwest at area of, 171f
Young glacial drift, 163, 179
Young Volcanic Rocks of Cordillera, 76, 183
Yreka-Trinity metamorphic belt, 498

Z

Zion Canyon, 186–187, 222–224, 234, 235f
Zion National Park, 234, 235f
Zircon, 271–272
  fission-track, 122, 494
Zone of ablation, 157–159, 158f
Zone of accumulation, 157–159, 162–163
Zuni-Bandera field (ZB field), 422–423

Printed in the United States
By Bookmasters